MICRONUTRIENTS AND BRAIN HEALTH

OXIDATIVE STRESS AND DISEASE

SERIES EDITORS

LESTER PACKER, PH.D.
ENRIQUE CADENAS, M.D., PH.D.

UNIVERSITY OF SOUTHERN CALIFORNIA SCHOOL OF PHARMACY
LOS ANGELES, CALIFORNIA

1. Oxidative Stress in Cancer, AIDS, and Neurodegenerative Diseases, *edited by Luc Montagnier, René Olivier, and Catherine Pasquier*
2. Understanding the Process of Aging: The Roles of Mitochondria, Free Radicals, and Antioxidants, *edited by Enrique Cadenas and Lester Packer*
3. Redox Regulation of Cell Signaling and Its Clinical Application, *edited by Lester Packer and Junji Yodoi*
4. Antioxidants in Diabetes Management, *edited by Lester Packer, Peter Rösen, Hans J. Tritschler, George L. King, and Angelo Azzi*
5. Free Radicals in Brain Pathophysiology, *edited by Giuseppe Poli, Enrique Cadenas, and Lester Packer*
6. Nutraceuticals in Health and Disease Prevention, *edited by Klaus Krämer, Peter-Paul Hoppe, and Lester Packer*
7. Environmental Stressors in Health and Disease, *edited by Jürgen Fuchs and Lester Packer*
8. Handbook of Antioxidants: Second Edition, Revised and Expanded, *edited by Enrique Cadenas and Lester Packer*
9. Flavonoids in Health and Disease: Second Edition, Revised and Expanded, *edited by Catherine A. Rice-Evans and Lester Packer*
10. Redox–Genome Interactions in Health and Disease, *edited by Jürgen Fuchs, Maurizio Podda, and Lester Packer*
11. Thiamine: Catalytic Mechanisms in Normal and Disease States, *edited by Frank Jordan and Mulchand S. Patel*
12. Phytochemicals in Health and Disease, *edited by Yongping Bao and Roger Fenwick*
13. Carotenoids in Health and Disease, *edited by Norman I. Krinsky, Susan T. Mayne, and Helmut Sies*
14. Herbal and Traditional Medicine: Molecular Aspects of Health, *edited by Lester Packer, Choon Nam Ong, and Barry Halliwell*
15. Nutrients and Cell Signaling, *edited by Janos Zempleni and Krishnamurti Dakshinamurti*
16. Mitochondria in Health and Disease, *edited by Carolyn D. Berdanier*
17. Nutrigenomics, *edited by Gerald Rimbach, Jürgen Fuchs, and Lester Packer*
18. Oxidative Stress, Inflammation, and Health, *edited by Young-Joon Surh and Lester Packer*
19. Nitric Oxide, Cell Signaling, and Gene Expression, *edited by Santiago Lamas and Enrique Cadenas*

MICRONUTRIENTS AND BRAIN HEALTH

EDITED BY
LESTER PACKER
HELMUT SIES
MANFRED EGGERSDORFER
ENRIQUE CADENAS

CRC Press
Taylor & Francis Group
Boca Raton London New York

CRC Press is an imprint of the
Taylor & Francis Group, an **informa** business

CRC Press
Taylor & Francis Group
6000 Broken Sound Parkway NW, Suite 300
Boca Raton, FL 33487-2742

© 2010 by Taylor and Francis Group, LLC
CRC Press is an imprint of Taylor & Francis Group, an Informa business

Library of Congress Cataloging-in-Publication Data

Micronutrients and brain health / [edited by] Lester Packer ... [et al.].
        p. ; cm. -- (Oxidative stress and disease ; 26)
    Includes bibliographical references and index.
    ISBN 978-1-4200-7351-5 (hardcover)
    1. Brain--Metabolism. 2. Trace elements in nutrition. 3. Nutrition. 4. Neuroprotective agents. 5. Oxidative stress. I. Packer, Lester. II. Series: Oxidative stress and disease ; 27.
        [DNLM: 1. Micronutrients--therapeutic use. 2. Neuroprotective Agents--therapeutic use. 3. Brain--physiology. 4. Neurodegenerative Diseases--prevention & control. W1 OX626 v.27 2009 / WL 359 M626 2009]

    QP356.3.M535 2009
    612.8'2--dc22   2009010251

Visit the Taylor & Francis Web site at
http://www.taylorandfrancis.com

and the CRC Press Web site at
http://www.crcpress.com

Contents

Series Preface

OXYGEN BIOLOGY AND MEDICINE

Through evolution, oxygen—itself a free radical—was chosen as the terminal electron acceptor for respiration. The two unpaired electrons of oxygen spin in the same direction; thus, oxygen is a biradical. Other oxygen-derived free radicals such as superoxide anion or hydroxyl radicals formed during metabolism or by ionizing radiation are stronger *oxidants,* that is, endowed with higher chemical reactivities. Oxygen-derived free radicals are generated during metabolism and energy production in the body and are involved in regulation of signal transduction and gene expression, activation of receptors and nuclear transcription factors, oxidative damage to cell components, antimicrobial and cytotoxic actions of immune system cells, as well as in aging and age-related degenerative diseases. Conversely, cells conserve antioxidant mechanisms to counteract the effects of oxidants; these *antioxidants* may remove oxidants either in a highly specific manner (e.g., by superoxide dismutases) or in a less specifi c manner (e.g., through small molecules such as vitamin E, vitamin C, and glutathione). *Oxidative stress* as classically defined is an *imbalance between oxidants and antioxidants.* Overwhelming evidence indicates that oxidative stress can lead to cell and tissue injury. However, the same free radicals that are generated during oxidative stress are produced during normal metabolism and, as a corollary, are involved in both human health and disease.

UNDERSTANDING OXIDATIVE STRESS

In recent years, the research disciplines interested in oxidative stress have grown and enormously increased our knowledge of the importance of the cell redox status and the recognition of oxidative stress as a process with implications for many pathophysiological states. From this multi- and inter-disciplinary interest in oxidative stress emerges a concept that attests to the vast consequences of the complex and dynamic interplay of oxidants and antioxidants in cellular and tissue settings. Consequently, our view of oxidative stress is growing in scope and new future directions. Likewise, the term *reactive oxygen species*—adopted at some stage in order to highlight nonradical oxidants such as H_2O_2 and O_2—now fails to refl ect the rich variety of other reactive species in free radical biology and medicine encompassing nitrogen-, sulfur-, oxygen-, and carbon-centered radicals. With the discovery of nitric oxide, nitrogen-centered radicals gathered momentum and have matured into an area of enormous importance in biology and medicine. Nitric oxide or nitrogen monoxide (NO), a free radical generated in a variety of cell types by nitric oxide synthases (NOSs), is involved in a wide array of physiological and pathophysiological phenomena such as vasodilation, neuronal signaling, and inflammation. Of great importance is the radical–radical reaction of nitric oxide with superoxide anion. This is among the most rapid nonenzymatic reactions in biology (well over the diffusion-controlled limits) and yields the potent nonradical oxidant, peroxynitrite. The involvement of this species in tissue injury through oxidation and nitration reactions is well documented.

Virtually all diseases thus far examined involve free radicals. In most cases, free radicals are secondary to the disease process, but in some instances causality is established by free radicals. Thus, there is a delicate balance between oxidants and antioxidants in health and disease. Their proper balance is essential for ensuring healthy aging.

Both reactive oxygen and nitrogen species are involved in the redox regula-tion of cell functions. Oxidative stress is increasingly viewed as a major upstream component in the signaling cascade involved in inflammatory responses, stimulation of cell adhesion molecules, and chemoattractant production and as an early component in age-related neurodegenerative disorders such as Alzheimer's, Parkinson's, and Huntington's diseases, and amyotrophic lateral sclerosis. Hydrogen peroxide is probably the most important redox signaling molecule that, among others, can activate NFκB, Nrf2, and other universal transcription factors. Increasing steady-state levels of hydrogen peroxide have been linked to a cell's redox status with clear involvement in adaptation, proliferation, differentiation, apoptosis, and necrosis.

The identification of oxidants in regulation of redox cell signaling and gene expression was a significant breakthrough in the field of oxidative stress: the classical definition of oxidative stress as an *imbalance between the production of oxidants and the occurrence of cell antioxidant defenses* proposed by Sies in 1985 now seems to provide a limited concept of oxidative stress, but it emphasizes the signifi cance of cell redox status. Because individual signaling and control events occur through discrete redox pathways rather than through global balances, a new definition of oxidative stress was advanced by Dean P. Jones (*Antioxidants & Redox Signaling* [2006]) as a disruption of redox signaling and control that recognizes the occurrence of compartmentalized cellular redox circuits. Recognition of discrete thiol redox circuits led Jones to provide this new definition of oxidative stress. Measurements of GSH/GSSG, cysteine/cystine, or $thioredoxin_{reduced}$/$thioredoxin_{oxidized}$ provide a quantitative defi nition of oxidative stress. Redox status is thus dependent on the degree to which tissue-specific cell components are in the oxidized state.

In general, the reducing environments inside cells help to prevent oxidative damage. In this reducing environment, disulfi de bonds (S–S) do not spontaneously form because sulfhydryl groups are maintained in the reduced state (SH), thus preventing protein misfolding or aggregation. The reducing environment is maintained by metabolism and by the enzymes involved in maintenance of thiol/disulfi de balance and substances such as glutathione, thioredoxin, vitamins E and C, and enzymes such as superoxide dismutases, catalase, and the seleniumdependent glutathione reductase and glutathione and thioredoxin-dependent hydroperoxidases (periredoxins) that serve to remove reactive oxygen species (hydroperoxides). Also of importance is the existence of many tissue- and cell compartment-specific isoforms of antioxidant enzymes and proteins.

Compelling support for the involvement of free radicals in disease development originates from epidemiological studies showing that enhanced antioxidant status is associated with reduced risk of several diseases. Of great signifi cance is the role that micronutrients play in modulation of redox cell signaling; this establishes a strong linking of diet and health and disease centered on the abilities of micronutrients to regulate redox cell signaling and modify gene expression.

These concepts are anticipated to serve as platforms for the development of tissue-specific therapeutics tailored to discrete, compartmentalized redox circuits. This, in essence, dictates principles of drug development-guided knowledge of mechanisms of oxidative stress. Hence, successful interventions will take advantage of new knowledge of compartmentalized redox control and free radical scavenging.

OXIDATIVE STRESS IN HEALTH AND DISEASE

Oxidative stress is an underlying factor in health and disease. In this series of books, the importance of oxidative stress and diseases associated with organ systems of the body is highlighted by exploring the scientifi c evidence and clinical applications of this knowledge. This series is intended for researchers in the basic biomedical sciences and clinicians. The potential of such knowledge for healthy aging and disease prevention warrants further knowledge about how oxidants and antioxidants modulate cell and tissue function.

Lester Packer
Enrique Cadenas

Preface

Micronutrients and Brain Health addresses cutting-edge areas of research of high significance for public health and translational medicine. The book identifies brain-specific micronutrients that support function as well as the molecular mechanisms underlying their neuroprotectant activity. This topic is among the most interesting and challenging areas of contemporary translational biological and medical research, with implications for preventive and therapeutic approaches in age-related neurodegenerative disorders.

This book explores the molecular mechanisms of brain micronutrients including age-related metabolic pathways, mitochondrial nutrients, neurodegeneration and micronutrients, cell signaling, and neuronal functions. General chapters are included on brain structure, function and metabolism, and other chapters are devoted to specific micronutrients, for example, vitamins, minerals, metals and chelators, bioflavonoids from fruits, green tea polyphenols, resveratol, sirtuins and curcumin, choline, amino acids, and the long-chain omega-3 fatty acid, docosahexaenoic acid. Emphasis is also given to lipoic acid and its role as a therapeutic agent in neuropathies, a nutrient that has further matured into a distinctive position in aging research. Of note, impairment of metabolic pathways is an early event in the pathophysiology of age-related neurodegenerative disorders. Hence, identification of brain-specific mitochondrial nutrients acquires further significance.

This book is highly relevant for the oxidative stress and disease series on several accounts: first, the micronutrients addressed possess antioxidant activity and, in many instances, molecular mechanisms of action are known and/or under current investigation; second, new concepts are advanced in many of the chapters where brain micronutrients are found to be both metabolic and redox regulators, thereby establishing a cross-talk between these major pathways involved in modulation of cell signaling and gene expression.

The editors gratefully appreciate the initial stimulus for this book—a workshop on "Micronutrients and Brain Health," conducted March 10 through 13, 2008, as part of the Meeting of the Oxygen Club of California (OCC) at Santa Barbara, CA, cosponsored by the Linus Pauling Institute. We would like to acknowledge the sponsorship of that workshop from DSM Nutritional Products and by the leading experts who have provided information on state-of-the-art research of their own specialty. The help of Marlies Scholtes in this project is gratefully acknowledged. We hope this volume will contribute to the developments in this rapidly expanding and highly important area of biomedical research.

Lester Packer, Helmut Sies, Manfred Eggersdorfer, and Enrique Cadenas

The Editors

Lester Packer received his Ph.D. in microbiology and biochemistry from Yale University, New Haven, CT, USA, and an Honorary M.D. from the University of Buenos Aires, Argentina. He was professor at the University of California at Berkeley, and senior scientist at Lawrence Berkeley National Laboratory between 1961 and 2000, and is currently adjunct professor in the Department of Pharmacology and Pharmaceutical Sciences at the University of Southern California in Los Angeles. His research interests include the molecular, cellular, and physiological role of oxidants, free radicals, antioxidants, and redox regulation in health and disease.

Professor Packer is the recipient of numerous scientific achievement awards including three honorary doctoral degrees. He was a visiting distinguished professor at the National University of Singapore (NUS) and distinguished professor at the Institute of Nutritional Sciences of the Chinese Academy of Sciences, Shanghai, China. He served as president of the Society of Free Radical Research International, vice president of UNESCO–the United Nations Global Network on Molecular and Cell Biology (MCBN), and was the founder and is presently honorary president of the OCC.

Helmut Sies, M.D., Ph.D. (hon), is an emeritus professor of biochemistry and molecular biology at Heinrich Heine University in Dusseldorf, Germany. He also is an adjunct professor at the University of Southern California and a professor of biochemistry and biology at King Saud University, Riyadh, Saudi Arabia. He served as president of the Society for Free Radical Research International and of the OCC. His research interests include oxidative stress, redox signaling, and micronutrients (notably flavonoids, selenium).

Manfred Eggersdorfer is senior vice president of DSM Nutritional Products and head of Research & Development. DSM Nutritional Products is the world leader in vitamins, carotenoids and nutritional ingredients for Human Nutrition, Animal Nutrition and Personal Care. Prior to this, Eggersdorfer worked for BASF, Ludwigshafen in different positions including head of Research and Development Fine Chemicals. Eggersdorfer studied chemistry at the Technical University Munich and did his post-doc at Stanford University in California. He is a member of the Kuratorium of the Fraunhofer-Gesellschaft and author of numerous publications in the fields of vitamins, innovation in nutritional ingredients, and renewable resources.

Enrique Cadenas, M.D., Ph.D., is professor of pharmacology and pharmaceutical sciences and associate dean of Research Affairs at the University of Southern California School of Pharmacy, professor of biochemistry and molecular biology at the School of Medicine, University of Southern California, and doctor honoris causa (Medicine) at the University of Linköping, Sweden. He is the immediate past president of the International Society for Free Radical Research and vice president of the Oxygen Club of California. His research interests include energy metabolism and free radical biology of neurodegenerative diseases, redox signaling, and nitric oxide biology.

Contributors

Bharat B. Aggarwal
University of Texas M.D. Anderson Cancer
    Center
Houston, Texas

Christopher D. Aluise
Department of Chemistry and Sanders–Brown
    Center on Aging
University of Kentucky
Lexington, Kentucky

Bruce N. Ames
Children's Hospital Oakland Research
    Institute
Oakland, California

Tamar Amit
Eve Topf and USA NPF Centers of Excellence
Technion - Faculty of Medicine
Department of Pharmacology
Haifa, Israel

Roberta Diaz Brinton, Ph.D.
School of Pharmacy
University of Southern California
Los Angeles, California

Ashley I. Bush, M.D., Ph.D.
Oxidation Biology Laboratory
Mental Health Research Institute of Victoria
University of Melbourne
Victoria, Australia
Department of Psychiatry
Massachusetts General Hospital East
Charlestown, Massachusetts

D. Allan Butterfield
Center of Membrane Sciences and Sanders–
    Brown Center on Aging
University of Kentucky
Lexington, Kentucky

Antoni Camins
Unitat de Farmacologia i Farmacognòsia i
    Institut de Biomedicina
Facultat de Farmàcia, Universitat de Barcelona
Nucli Universitari de Pedralbes
Barcelona, Spain
Centro de Investigación de Biomedicina en
    Red de Enfermedades Neurodegenerativas
Instituto de Salud Carlos III
Madrid, Spain

David Carlson
GeroNova Research, Inc.
Reno, Nevada

Gemma Casadesus
Department of Neurosciences
Case Western Reserve University
Cleveland, Ohio

Margarida Castell
Department of Physiology, Faculty of Pharmacy
University of Barcelona
Barcelona, Spain

Yves Christen
Institut Ipsen
Boulogne-Billancourt, France

Rosa Cristofol
Department d'Isquèmia Cerebral I
    Neurodegenració
Institut d'Investigacions Biomèdiques de
    Barcelona
Barcelona, Spain

Alan D Dangour, Ph.D.
Nutrition and Public Health Intervention
    Research Unit
London School of Hygiene and Tropical
    Medicine
London, United Kingdom

Sanjit Dey
University of Texas M.D. Anderson Center
Houston, Texas

Dr. Ralf Dringen
Center for Biomolecular Interactions Bremen
University of Bremen
Bremen, Germany

Jürgen Engel
Zentaris GmbH
Frankfurt am Main, Germany

Gary Fiskum
University of Maryland Medical Center
Baltimore, Maryland

Anita J. Fuglestad
Institute of Child Development
University of Minnesota
Minneapolis, Minnesota

Michael K. Georgieff, M.D.
Center for Neurobehavioral Development
University of Minnesota School of Medicine
Minneapolis, Minnesota

Klaus Hager
Klinik für Medizinische Rehabilitation und
    Geriatrie der Henriettenstiftung
Hannover, Germany

Dr. Bernd Hamprecht
Interfaculty Institute for Biochemistry
Universtity of Tuebingen
Tuebingen, Germany

Kuzhuvelil B. Harikumar
University of Texas M.D. Anderson Cancer
    Center
Houston, Texas

Richard E. Hartman, Ph.D.
Department of Psychology
Loma Linda University
Loma Linda, California

Elizabeth Head
Institute for Brain Aging and Dementia
University of California–Irvine
Irvine, California

Johannes Hirrlinger
Interdisciplinary Centre for Clinical Research
University of Leipzig
Leipzig, Germany

Andrea Lisa Holme, Ph.D.
National University Medical Institutes
Yong Loo Lin School of Medicine
Singapore, Singapore

Harry Ischiropoulos
Stokes Research Institute
Children's Hospital of Philadelphia
Philadelphia, Pennsylvania

James A. Joseph, Ph.D.
USDA, HNRCA at Tufts University
Boston, Massachusetts

Marlene Kenklies
Klinik für Medizinische Rehabilitation und
    Geriatrie der Henriettenstiftung
Hannover, Germany

Savita Khanna
Laboratory of Molecular Medicine
Department of Surgery, Davis Heart and Lung
    Research Institute
Ohio State University Medical Center
Columbus, Ohio

Martina Krautwald
Deparment of Pharmacology, School of
    Medicine
University of Western Sydney
Penrith South, Australia

Stuart A. Lipton
Center for Neuroscience, Aging and Stem Cell
    Research
Burnham Institute for Medical Research
La Jolla, California

Gerardo G. Mackenzie
Department of Medicine
Stony Brook University
Stony Brook, New York

Annette Maczurek
Department of Pharmacology, School
   of Medicine
University of Western Sydney
Penrith South, Australia
Department of Biochemistry and Molecular
   Biology
James Cook University
Townsville, Australia

Pamela Maher
The Salk Institute for Biological Studies
La Jolla, California

Kristen Malkus
The Children's Hospital of Philadelphia, PA
   and University of Pennsylvania
Philadelphia, Pennsylvania

Silvia Mandel
Eve Topf and USA NPF Centers of Excellence
Technion - Faculty of Medicine
Department of Pharmacology
Haifa, Israel

Ralph Martins
Centre of Excellence for Alzheimer's Disease
   Research and Care
Edith Cowan University
Perth, Australia

Ana Rodriguez Mateos
Molecular Nutrition Group, School of Chemistry
Food and Pharmacy, University of Reading
Reading, United Kingdom

Joyce C. McCann
Children's Hospital Oakland Research
   Institute
Oakland, California

Andrew McShea
Research and Development
Theo Chocolate
Seattle, Washington

Dr. Paul E. Milbury, Ph.D., CFII
Gerald Human Nutrition Center on Aging at
   Tufts University
Boston, Massachusetts

Su San Mok, Ph.D.
Oxidation Biology Laboratory
Mental Health Research Institute of Victoria
University of Melbourne
Victoria, Australia

Gerald Münch
Department of Pharmacology, School
   of Medicine
University of Western Sydney
Penrith South, Australia

Radovan Murín, Ph.D.
Interfaculty Institute for Biochemistry
Universtity of Tuebingen
Tuebingen, Germany

Dr. Laura E. Murray-Kolb
Bloomberg School of Public Health
Johns Hopkins University
Baltimore, Maryland

Tomohiro Nakamura
Center for Neuroscience, Aging and Stem Cell
   Research
Burnham Institute for Medical Research
La Jolla, California

Wycliffe O. Opii, Ph.D.
Institute of Brain Aging and Dementia
University of California–Irvine
Irvine, California

Patricia I. Oteiza
Department of Nutrition
University of California, Davis
Davis, California

Mercè Pallàs
Unitat de Farmacologia i Farmacognòsia i
   Institut de Biomedicina
Facultat de Farmàcia, Universitat de Barcelona
Nucli Universitari de Pedralbes
Barcelona, Spain
Centro de Investigación de Biomedicina en
   Red de Enfermedades Neurodegenerativas
Instituto de Salud Carlos III
Madrid, Spain

Carme Pelegri
Centro de Investigación de Biomedicina en
    Red de Enfermedades Neurodegenerativas
Instituto de Salud Carlos III
Madrid, Spain
Departament de Fisiologia, Facultat de
    Farmàcia
Universitat de Barcelona, Nucli Universitari de
    Pedralbes
Barcelona, Spain

George Perry
College of Sciences
University of Texas at San Antonio
San Antonio, Texas
Departments of Pathology
Case Western Reserve University
Cleveland, Ohio

Shazib Pervaiz, M.B.B.S., Ph.D.
Yong Loo Lin School of Medicine
Graduate School for Integrative Sciences and
    Engineering
National University of Singapore
Singapore, Singapore

Sara E. Ramel, M.D.
Fellow, Neonatal-Perinatal Medicine
Division of Neonatology
Department of Pediatrics
University of Minnesota
Minneapolis, Minnesota

Emma Ramiro-Puig
Department of Physiology, Faculty of Pharmacy
University of Barcelona
Barcelona, Spain

Robert E. Rosenthal
University of Maryland Medical Center
Baltimore, Maryland

Sashwati Roy
Laboratory of Molecular Medicine
Department of Surgery, Davis Heart and Lung
    Research Institute
Ohio State University Medical Center
Columbus, Ohio

Coral Sanfeliu
Department d'Isquèmia Cerebral I
    Neurodegenració
Institut d'Investigacions Biomèdiques de
    Barcelona
Barcelona, Spain

Chandan K. Sen
Ohio State University Medical Center
Columbus, Ohio

Matt Sharman
Centre of Excellence for Alzheimer's Disease
    Research and Care
Edith Cowan University
Perth, Australia

Barbara Shukitt-Hale
USDA-ARS
Human Nutrition Research Center on Aging at
    Tufts University
Boston, Massachusetts

Mark A. Smith, Ph.D.
Departments of Pathology
Case Western Reserve University
Cleveland, Ohio

Renã A. Sowell
Department of Chemistry and Sanders-Brown
    Center on Aging
University of Kentucky
Lexington, Kentucky

Jeremy P. E. Spencer
Molecular Nutrition Group, School of Chemistry
Food and Pharmacy
University of Reading
Reading, United Kingdom

Megan Steele
Department of Pharmacology, School
    of Medicine
University of Western Sydney
Penrith South, Australia
Department of Biochemistry and Molecular
    Biology
James Cook University
Townsville, Australia

Elpida Tsika
University of Pennsylvania
Philadelphia, Pennsylvania

Ricardo Uauy
Nutrition and Public Health Intervention
    Research Unit
London School of Hygiene and Tropical Medicine
London, United Kingdom

Katerina Vafeiadou
Molecular Nutrition Group, School of Chemistry
Food and Pharmacy
University of Reading
Reading, United Kingdom

David Vauzour
Molecular Nutrition Group, School of Chemistry
Food and Pharmacy
University of Reading
Reading, United Kingdom

Jordi Vilaplana
Centro de Investigación de Biomedicina en
    Red de Enfermedades Neurodegenerativas
Instituto de Salud Carlos III
Madrid, Spain
Departament de Fisiologia, Facultat de Farmàcia
Universitat de Barcelona, Nucli Universitari de
    Pedralbes
Barcelona, Spain

Orly Weinreb
Eve Topf and USA NPF Centers of
    Excellence
Technion - Faculty of Medicine
Department of Pharmacology
Haifa, Israel

Lauren M. Willis
USDA-ARS
Human Nutrition Research Center on Aging at
    Tufts University
Boston, Massachusetts

Moussa B.H. Youdim
Department of Pharmacology
Technion - Faculty of Medicine
Haifa, Israel

Baolu Zhao
Institute of Biophysics
Academia Sinica, China
Beijing, China

Liqin Zhao, Ph.D.
School of Pharmacy
University of Southern California
Los Angeles, California

1 Neuroprotection after Cardiac Arrest by Avoiding Acute Hyperoxia and by Antioxidant Genomic Postconditioning

Gary Fiskum[1,2] and Robert E. Rosenthal[1–4]

[1]Department of Anesthesiology, [2]Trauma and Anesthesiology Research Center, [3]Department of Emergency Medicine, [4]Division of Hyperbaric Medicine, R. Adams Cowley Shock Trauma Center, University of Maryland, Baltimore, Maryland, USA

CONTENTS

1.1 CARDIAC ARREST-INDUCED BRAIN INJURY AND NEUROPROTECTIVE INTERVENTIONS

Cardiac arrest (CA) accounts for more than 400,000 of the 650,000 deaths caused by cardiac disease in the United States each year. Discharge rates for individuals resuscitated after in-hospital CA are only approximately 20%,[1] and less than 10% of patients survive following out-of-hospital CA.[2] Following interruption of blood flow to the brain, neurologic sequelae are frequent complications following initially successful resuscitation from CA. In one study, for example, severe brain injury was the cause of death in 37% of CA survivors and only 14% of patients were neurologically normal or near-normal at 12 months.[3] Permanent, significant brain damage resulting from ischemia/reperfusion contributes substantially to the mortality and neurologic morbidity of more than 250,000 patients "successfully" resuscitated from CA each year in the United States alone.

It is now well known that injury to the brain initiated during CA continues and is further magnified during the postresuscitative period of days to many weeks. This reperfusion injury is caused by many processes, including excitotoxicity, oxidative stress, metabolic failure, inflammation, and apoptosis, among others. Although many drugs provide neuroprotection in animal models of global cerebral ischemia, prevention of post-CA human brain injury through pharmacologic intervention has not yet been achieved.[4] Considerable excitement was generated in 2002, however, when two large clinical trials demonstrated significant improvement in neurologic outcome and

1

reduced mortality in patients treated with moderate hypothermia following resuscitation from CA.[5,6] This breakthrough helped renew the conviction that lessons learned in the laboratory can be successfully translated into improved long-term resuscitation outcomes following human CA. Neuroprotective interventions can be initiated almost immediately following resuscitation or at least as soon as CA survivors reach an emergency department, typically within 45 min after resuscitation. Because of such rapid transit to a critical care setting, the time-to-target temperature in one successful hypothermia trial was less than 2 h.[5] This early window of therapeutic opportunity provides additional optimism that interventions other than, or in addition to, hypothermia will prove successful.

In order for any neuroprotective strategy (including hypothermia) to make an impact, however, the technique must be safe, easy to implement, and widely accepted. Despite successful hypothermia clinical trials, the emergency medicine/critical care community has not yet generally implemented the use of hypothermia in the clinical arena. Even if hypothermia is widely accepted, one recent editorial suggests that it should be only one part of a strategy designed to attack different components of the ischemia/reperfusion cascade.[7] Additional measures are clearly needed to promote neuroprotection for survivors of CA. We have established over many years of work that we can significantly reduce oxidative stress, metabolic dysfunction, delayed neuronal death, and neurobehavioral impairment by avoiding unnecessary hyperoxia during the first hour following resuscitation.[8–10] Moreover, we have demonstrated that a clinically practical pulse oximetry-guided O_2 ventilation protocol can achieve these goals.[11] Although normoxic resuscitation dramatically reduces oxidative stress and improves cerebral energy metabolism within the first few hours of reperfusion, its inhibition of cellular inflammatory responses (e.g., microglial activation) and delayed oxidative DNA/RNA damage is relatively modest. Further development of safe and effective interventions that reduce secondary oxidative stress caused by inflammation is therefore needed.

1.2 INFLAMMATION AND DELAYED OXIDATIVE STRESS IN GLOBAL CEREBRAL ISCHEMIA

A wealth of literature has documented the important role of inflammation in neurologic injury after stroke and traumatic brain injury.[12,13] More recently, a growing body of research has begun to define the role of inflammation in global cerebral ischemia, including exacerbation of injury by peripheral inflammation,[14] as seen in stroke.[15,16] In contrast to research with stroke models, there has been minimal evaluation of the role of anti-inflammatory agents for neuroprotection following global ischemia, and no study to date has used a clinically relevant model of CA. Much of the work on postischemic brain inflammation has focused on microglial activation,[17,18] although inflammatory gene expression and cytokine production have also been reported.[14,19,20] Ironically, one of the first reports of microglial "proliferation" following cerebral ischemia was an autopsy case report for a man who died 10 days after having CA and who also had aplastic bone marrow; it strongly suggested that native brain microglia become phagocytic rather than peripheral macrophages invading damaged brain parenchyma.[21] The proliferation of both microglia and astrocytes after global cerebral ischemia was subsequently verified in monkeys using immunocytochemistry for bromo-deoxyuracil and costaining with cell-specific markers.[22] Substantial evidence indicates that although normal microglia perform neuroprotective functions [e.g., the expression of brain-derived neurotrophic factor, glial cell line-derived neurotrophic factor, and heme oxygenase-1 (HO-1)],[23,24] it appears that postischemic microglial activation is deleterious, via the production of inflammatory cytokines [e.g., interleukin (IL)-1β and IL-6],[25,26] the C1q subcomponent of complement,[27] and reactive oxygen and nitrogen species (hereafter combined and termed ROS).[28,29] Effects of ischemic preconditioning on activation of microglia are controversial, with reports indicating neuroprotection by both inhibition and stimulation of microglial activation.[30,31]

The strongest evidence supporting a pathologic role for microglial activation comes from experiments performed with both in vivo and *in vitro* cerebral ischemia/reperfusion models demonstrating a close relationship between inhibition of microglial activation and protection against the death of neurons, astrocytes, and endothelial cells by the tetracycline drug minocycline.[29,32,33] There is only one report demonstrating neuroprotection by postischemic administration of minocycline after global cerebral ischemia, however.[32] An alternative approach to inhibition of postischemic brain inflammation and associated oxidative stress is pharmacologic stimulation of the expression of endogenous genes that code for proteins with anti-inflammatory and antioxidant activities.

1.3 NEUROPROTECTION BY ACTIVATION OF ENDOGENOUS ANTIOXIDANT AND ANTI-INFLAMMATORY GENE EXPRESSION

Several laboratories are pursuing the hypothesis that brain inflammation and delayed oxidative stress can be suppressed by interventions that alter endogenous gene expression in ways that boost multiple antioxidant and anti-inflammatory defense mechanisms. Based on this hypothesis, one "genomic postconditioning strategy" is based on pharmacologic activation of the nuclear factor erythroid 2-related factor 2 (Nrf2)-mediated pathway of gene expression. Nrf2 is a transcription factor that belongs to the cap-n-collar family, whose members share a highly conserved basic leucine zipper structure.[34] Nrf2 activation provides cytoprotection by up-regulation of a large battery of genes whose expression is controlled by the antioxidant response element (ARE) (Figure 1.1). These more than 100 genes include NAD(P)H:quinone reductase (NQO1), glutathione *S*-transferase, gamma-glutamylcysteine synthetase, genes encoding for cellular NADPH-regenerating enzymes (glucose 6-phosphate dehydrogenase, 6-phosphogluconate dehydrogenase, and malic enzyme), antioxidants (glutathione peroxidase, glutathione reductase, ferritin, and haptaglobin), and biosynthetic enzymes of the glutathione and glucuronidation conjugation pathways.[35,36]

In response to low levels of ROS or electrophiles, specific cysteine sulfhydryl groups present in Keap1, an Nrf2 binding protein, become oxidized, promoting the release of Nrf2 into the cytosol. Nrf2 then translocates to the nucleus, where it transcriptionally activates ARE-dependent genes after recruiting Maf proteins.[37] The release of Nrf2 from Keap1 also requires phosphorylation of Nrf2 at Ser-40 by such kinases as protein kinase C and phosphatidylinositol 3-kinase[38-40] (Figure 1.1).

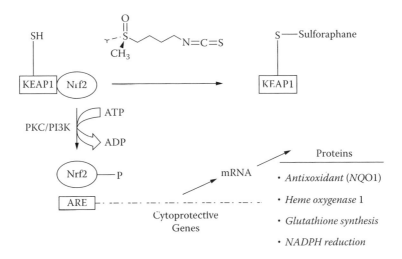

FIGURE 1.1 Activation of Nrf2/ARE pathway of gene expression. The pathway is activated by reaction of SFP with Keap1 and by serine phosphorylation of Nrf2 by protein kinase C or phosphatidylinositol 3-kinase.

Two lines of evidence support the concept that neuroprotection after global cerebral ischemia can be achieved through Nrf2-dependent genomic postconditioning. First, compared with wild-type mice, Nrf2 knockout mice exhibit a larger infarct volume induced by transient focal cerebral ischemia.[41,42] Second, pharmacologic agents that activate Nrf2-mediated gene induction also display neuroprotection in small animal stroke and traumatic brain injury models;[41–46] however, the effects of pharmacologic Nrf2 activation have not been reported for global cerebral ischemia.

Sulforaphane (SFP; 4-methyl-sulfinylbutyl isothiocyanate) is a relatively potent Nrf2-activating agent derived from a glucosinolate compound found in cruciferous vegetables, such as sprouts of broccoli, cabbage, and cauliflower. SFP protects cortical neurons in mixed primary culture from hydrogen peroxide and glutamate toxicity[47] and dopaminergic neurons from cytotoxicity of 6-hydroxydopamine.[48] SFP also provides neuroprotection in rats when given shortly after the onset of reversible occlusion of the middle cerebral artery,[43] as well as following experimental traumatic brain injury.[46,49] Using a model more relevant to global cerebral ischemia, we found that both pre-treatment and posttreatment of rat cortical astrocytes with SFP protect against oxidative DNA damage and delayed death in a transient oxygen/glucose deprivation model.[50] In addition, Talalay et al.[51–55] have repeatedly shown SFP to be effective against carcinogenesis and inflammation. The anti-inflammatory effects are not entirely clear, but they could involve inhibition of inflammatory cytokine expression (e.g., IL-8, monocyte chemoattractant protein-1, and vascular cell adhesion molecule-1),[56,57] inhibition of macrophage infiltration,[58] and stimulated expression of HO-1, which has potent anti-inflammatory activities[59,60] and may be an important mediator of the neuroprotective effects of microglia.[24] HO-1 expression increases after global cerebral ischemia, including that caused by CA.[61] Much of the evidence indicates that this expression is neuroprotective and may be an important mechanism of action for estradiol, which further enhances HO-1 expression after cerebral ischemia.[62] Interestingly, catechol estrogens (e.g., 4-hydroxyestradiol) stimulate the expression of ARE-regulated genes through the Nrf2 pathway.[63] In contrast, 17β-estradiol can inhibit the Nrf2 pathway through binding of estrogen-related receptor β to Nrf2.[64] To our knowledge, there has been no direct comparison between the effects of Nrf2 activators on gene expression or cytoprotection in males and those in females.

Success at achieving neuroprotection with SFP or with other agents and conditions that either directly or indirectly activate the Nrf2 pathway is often dependent on their ability to oxidize Keap1 sulfhydryl groups without exerting toxicity. The timing of postischemic SFP administration must therefore be carefully contemplated, considering the existence of oxidative stress that occurs both during and briefly after complete cerebral ischemia. This first wave of oxidative stress is promoted by many abnormal intracellular conditions, such as elevated $[Ca^{2+}]_i$, known to stimulate the production of both superoxide and nitric oxide.[65,66] Residual lactic acidosis promotes release of iron from ferritin, potentially promoting hydroxyl radical formation from H_2O_2.[67] Tissue concentrations of O_2, the substrate for ROS production, are also abnormally high during the first 30 min of reperfusion, due to vasodilation and associated hyperemia.[68] However, within 30−60 min of reperfusion, ionic homeostasis is reestablished,[69] the tissue redox state normalizes,[70] the production of ROS subsides,[71] and tissue O_2 falls due to conversion of cerebral blood flow from hyperemia to hypoperfusion.[72] We hypothesize that the period between 30 min and 24 h of reperfusion is the window of opportunity for the safest use of SFP or other conditions that can activate the Nrf2/ARE pathway in time to both inhibit and counteract delayed brain inflammation and secondary oxidative stress, which become evident on a cellular level between 6 h and several days later.

1.4   OXIDATIVE STRESS AND MITOCHONDRIAL DYSFUNCTION

A large body of evidence indicates that mitochondrial dysfunction plays a critical role in the pathophysiology of ischemic brain injury.[73–78] Consequences of mitochondrial dysfunction are numerous and include oxidative stress, loss of cellular Ca^{2+} homeostasis, promotion of apoptosis, and metabolic failure.

There are many possible causes of mitochondrial metabolic impairment, most of which involve oxidative modifications to proteins, lipids, or DNA. Identification of the sites at which oxidative stress impairs respiration can guide the development of counteractive interventions with neuroprotective potential. Complex I of the electron transport chain (ETC), which catalyzes the oxidation of NADH and the reduction of ubiquinone, is particularly sensitive to inhibition by both oxidative stress and ischemia/reperfusion and is generally considered to be the rate-limiting component of the ETC.[79–82] Another cause of impaired ETC activity is the release of cytochrome *c* through the outer mitochondrial membrane into the cytosol, an event that is also often followed by caspase-dependent apoptosis.[83] Oxidative stress promotes cytochrome *c* release by several mechanisms, including those promoting translocation of Bax and Bak to the mitochondrial outer membrane.[84,85] These proteins form megapores within the outer membrane when prompted to oligomerize by the binding of BH3 domain-only proteins (e.g., tBid) to these proteins or to antiapoptotic proteins (e.g., Bcl2 or Bclx$_L$) that normally block megapore formation by heterodimerizing with Bax and Bak. In addition to impaired ETC activities, oxidative phosphorylation can also be obstructed by inhibition of other mitochondrial enzymes and membrane transporters (Figure 1.2). Thus, oxidative inactivation of mitochondrial matrix enzymes, such as pyruvate and α-ketoglutarate dehydrogenases and aconitase, is implicated in metabolic failure.[9,86,87] Our work with a clinically relevant CA model suggests that early reperfusion-dependent loss of pyruvate dehydrogenase immunoreactivity and enzyme activity is largely responsible for the tissue lactic acidosis and the inhibition of aerobic cerebral energy metabolism that occurs for many hours and is exacerbated by hyperoxic resuscitation.[9,10,86–88] Oxidative modification and loss of activity of this bridge between glycolysis and the tricarboxylic acid cycle have also been observed in models of traumatic brain injury and neurodegenerative diseases.[89–95]

Much interest is also currently focused on the availability of the metabolic cofactor NAD$^+$, which is necessary for the numerous dehydrogenases present within the mitochondrial matrix.[96] Cellular

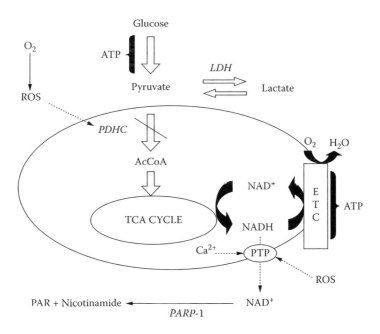

FIGURE 1.2 Oxidative stress and mitochondrial metabolic dysfunction. Two proposed mechanisms are inhibition of pyruvate dehydrogenase complex and loss of NAD(H) through the PTP into the cytosol where NAD$^+$ can be consumed by poly(ADP-ribose) polymerase-1. Inhibition of oxidative phosphorylation can stimulate anaerobic glycolysis via lactate dehydrogenase, resulting in chronic lactic acidosis.

NAD$^+$ can be rapidly catabolized by nuclear and possibly mitochondrial poly(ADP-ribose) polymerase in response to activation of this enzyme by oxidative DNA modifications and extracellular signal-regulated kinase-mediated enzyme phosphorylation.[97–99] NAD$^+$ in its oxidized or reduced form (NADH) can be lost from the mitochondrial matrix following opening of the inner membrane permeability transition pore (PTP), which results in transmembrane equilibration of small ions and molecules of up to approximately 1500 Da.[100,101] The PTP is activated by abnormally high concentrations of Ca^{2+} and by oxidative stress.[102]

Contribution of PTP opening to ischemic brain injury is supported by the neuroprotection observed with PTP inhibitors (e.g., cyclosporins)[103–106] that bind to cyclophilin D, the one well-established protein associated with pore opening. Cyclophilin D knockout mice are resistant to ischemic brain injury[107] and to spinal cord axonal damage associated with experimental autoimmune encephalomyelitis, an animal model characterized by extensive inflammation.[108] Taken together with observations that postischemic inflammation is accompanied by considerable ROS formation via inducible nitric oxide synthase and NADPH oxidase,[109] these findings strongly suggest that the PTP participates in inflammatory neurodegeneration after global cerebral ischemia.

1.5   SUMMARY

The clinically successful use of postischemic mild hypothermia after CA provides optimism that other approaches to neuroprotection for this form of acute brain injury will also be effective. Avoiding unnecessary hyperoxia during the early phase of reperfusion within the first hour of resuscitation after CA is consistently neuroprotective in clinically relevant animal models and should therefore be tested in clinical trials. Despite significant improvement with normoxic resuscitation, additional strategies are necessary to inhibit delayed brain inflammation and associated secondary oxidative stress. One such strategy that has proven successful in animal models of stroke and traumatic brain injury and with *in vitro* ischemia models is genomic postconditioning through pharmacologic activation of the Nrf2/ARE pathway of cytoprotective gene expression. Experiments testing these strategies have also provided insight into the role that oxidative stress plays in postischemic cerebral metabolic dysfunction. The most important molecular metabolic targets of oxidative stress in global cerebral ischemia appear to be the mitochondrial pyruvate dehydrogenase complex and the PTP; however, many additional targets may also contribute to cell death and neurologic outcome.

REFERENCES

1. George, A. L., Jr., Folk, B. P., III, Crecelius, P. L., and Campbell, W. B. Pre-arrest morbidity and other correlates of survival after in-hospital cardiopulmonary arrest. *Am. J. Med.* 87, 28, 1989.
2. Stueven, H. A., Waite, E. M., Troiano, P., and Mateer, J. R. Prehospital cardiac arrest—a critical analysis of factors affecting survival. *Resuscitation* 17, 251, 1989.
3. Brain Resuscitation Clinical Trial I Study Group. Randomized clinical study of thiopental loading in comatose survivors of cardiac arrest. *N. Engl. J. Med.* 314, 397, 1986.
4. Harukuni, I., and Bhardwaj, A. Mechanisms of brain injury after global cerebral ischemia. *Neurol. Clin.* 24, 1, 2006.
5. Bernard, S. A., Gray, T. W., Buist, M. D., Jones, B. M., Silvester, W., Gutteridge, G., and Smith, K. Treatment of comatose survivors of out-of-hospital cardiac arrest with induced hypothermia. *N. Engl. J. Med.* 346, 557, 2002.
6. Hypothermia after Cardiac Arrest Study Group. Mild therapeutic hypothermia to improve the neurologic outcome after cardiac arrest. *N. Engl. J. Med.* 346, 549, 2002.
7. Broccard, A. Therapeutic hypothermia for anoxic brain injury following cardiac arrest: A "cool" transition toward cardiopulmonary cerebral resuscitation. *Crit. Care Med.* 34, 2008, 2006.
8. Liu, Y., Rosenthal, R. E., Haywood, Y., Miljkovic-Lolic, M., Vanderhoek, J. Y., and Fiskum, G. Normoxic ventilation after cardiac arrest reduces oxidation of brain lipids and improves neurological outcome. *Stroke* 29, 1679, 1998.

9. Richards, E. M., Rosenthal, R. E., Kristian, T., and Fiskum, G. Postischemic hyperoxia reduces hippocampal pyruvate dehydrogenase activity. *Free Radic. Biol. Med.* 40, 1960, 2006.

10. Richards, E. M., Fiskum, G., Rosenthal, R. E., Hopkins, I., and McKenna, M. C. Hyperoxic reperfusion after global ischemia decreases hippocampal energy metabolism. *Stroke* 38, 1578, 2007.

11. Balan, I. S., Fiskum, G., Hazelton, J., Cotto-Cumba, C., and Rosenthal, R. E. Oximetry-guided reoxygenation improves neurological outcome after experimental cardiac arrest. *Stroke* 37, 3008, 2006.

12. Stoll, G., Jander, S., and Schroeter, M. Inflammation and glial responses in ischemic brain lesions. *Prog. Neurobiol.* 56, 149, 1998.

13. Danton, G. H., and Dietrich, W. D. Inflammatory mechanisms after ischemia and stroke. *J. Neuropathol. Exp. Neurol.* 62, 127, 2003.

14. Spencer, S. J., Mouihate, A., and Pittman, Q. J. Peripheral inflammation exacerbates damage after global ischemia independently of temperature and acute brain inflammation. *Stroke* 38, 1570, 2007.

15. Offner, H., Subramanian, S., Parker, S. M., Afentoulis, M. E., Vandenbark, A. A., and Hurn, P. D. Experimental stroke induces massive, rapid activation of the peripheral immune system. *J. Cereb. Blood Flow Metab.* 26, 654, 2006.

16. Hurn, P. D., Subramanian, S., Parker, S. M., Afentoulis, M. E., Kaler, L. J., Vandenbark, A. A., and Offner, H. T- and B-cell-deficient mice with experimental stroke have reduced lesion size and inflammation. *J. Cereb. Blood Flow Metab.* 27, 1798, 2007.

17. Sugawara, T., Lewen, A., Noshita, N., Gasche, Y., and Chan, P. H. Effects of global ischemia duration on neuronal, astroglial, oligodendroglial, and microglial reactions in the vulnerable hippocampal CA1 subregion in rats. *J. Neurotrauma* 19, 85, 2002.

18. Lin, B., Ginsberg, M. D., Busto, R., and Dietrich, W. D. Sequential analysis of subacute and chronic neuronal, astrocytic and microglial alterations after transient global ischemia in rats. *Acta Neuropathol.* 95, 511, 1998.

19. Ooboshi, H., Ibayashi, S., Shichita, T., Kumai, Y., Takada, J., Ago, T., Arakawa, S., Sugimori, H., Kamouchi, M., Kitazono, T., and Iida, M. Postischemic gene transfer of interleukin-10 protects against both focal and global brain ischemia. *Circulation* 111, 913, 2005.

20. Hedtjarn, M., Mallard, C., and Hagberg, H. Inflammatory gene profiling in the developing mouse brain after hypoxia-ischemia. *J. Cereb. Blood Flow Metab.* 24, 1333, 2004.

21. Neuwelt, E. A., Garcia, J. H., and Mena, H. Diffuse microglial proliferation after global ischemia in a patient with aplastic bone marrow. *Acta Neuropathol.* 43, 259, 1978.

22. Tonchev, A. B., Yamashima, T., Zhao, L., and Okano, H. Differential proliferative response in the postischemic hippocampus, temporal cortex, and olfactory bulb of young adult macaque monkeys. *Glia* 42, 209, 2003.

23. Imai, F., Suzuki, H., Oda, J., Ninomiya, T., Ono, K., Sano, H., and Sawada, M. Neuroprotective effect of exogenous microglia in global brain ischemia. *J. Cereb. Blood Flow Metab.* 27, 488, 2007.

24. Lee, S., and Suk, K. Heme oxygenase-1 mediates cytoprotective effects of immunostimulation in microglia. *Biochem. Pharmacol.* 74, 723, 2007.

25. Dimayuga, F. O., Wang, C., Clark, J. M., Dimayuga, E. R., Dimayuga, V. M., and Bruce-Keller, A. J. SOD1 overexpression alters ROS production and reduces neurotoxic inflammatory signaling in microglial cells. *J. Neuroimmunol.* 182, 89, 2007.

26. Simi, A., Tsakiri, N., Wang, P., and Rothwell, N. J. Interleukin-1 and inflammatory neurodegeneration. *Biochem. Soc. Trans.* 35, 1122, 2007.

27. Schafer, M. K., Schwaeble, W. J., Post, C., Salvati, P., Calabresi, M., Sim, R. B., Petry, F., Loos, M., and Weihe, E. Complement C1q is dramatically up-regulated in brain microglia in response to transient global cerebral ischemia. *J. Immunol.* 164, 5446, 2000.

28. Chong, Z. Z., Kang, J. Q., and Maiese, K. Essential cellular regulatory elements of oxidative stress in early and late phases of apoptosis in the central nervous system. *Antioxid. Redox Signal.* 6, 277, 2004.

29. Yenari, M. A., Xu, L., Tang, X. N., Qiao, Y., and Giffard, R. G. Microglia potentiate damage to blood-brain barrier constituents: Improvement by minocycline in vivo and *in vitro*. *Stroke* 37, 1087, 2006.

30. Perez-Pinzon, M. A., Vitro, T. M., Dietrich, W. D., and Sick, T. J. The effect of rapid preconditioning on the microglial, astrocytic and neuronal consequences of global cerebral ischemia. *Acta Neuropathol.* 97, 495, 1999.

31. Liu, J., Bartels, M., Lu, A., and Sharp, F. R. Microglia/macrophages proliferate in striatum and neocortex but not in hippocampus after brief global ischemia that produces ischemic tolerance in gerbil brain. *J. Cereb. Blood Flow Metab.* 21, 361, 2001.

32. Yrjanheikki, J., Keinanen, R., Pellikka, M., Hokfelt, T., and Koistinaho, J. Tetracyclines inhibit microglial activation and are neuroprotective in global brain ischemia. *Proc. Natl. Acad. Sci. U S A* 95, 15769, 1998.

33. Lai, A. Y., and Todd, K. G. Hypoxia-activated microglial mediators of neuronal survival are differentially regulated by tetracyclines. *Glia* 53, 809, 2006.
34. Motohashi, H., Shavit, J. A., Igarashi, K., Yamamoto, M., and Engel, J. D. The world according to Maf. *Nucleic Acids Res.* 25, 2953, 1997.
35. Thimmulappa, R. K., Mai, K. H., Srisuma, S., Kensler, T. W., Yamamoto, M., and Biswal, S. Identification of Nrf2-regulated genes induced by the chemopreventive agent sulforaphane by oligonucleotide microarray. *Cancer Res.* 62, 5196, 2002.
36. Lee, J. M., Calkins, M. J., Chan, K., Kan, Y. W., and Johnson, J. A. Identification of the NF-E2-related factor-2-dependent genes conferring protection against oxidative stress in primary cortical astrocytes using oligonucleotide microarray analysis. *J. Biol. Chem.* 278, 12029, 2003.
37. Nguyen, T., Sherratt, P. J., Huang, H. C., Yang, C. S., and Pickett, C. B. Increased protein stability as a mechanism that enhances Nrf2-mediated transcriptional activation of the antioxidant response element. Degradation of Nrf2 by the 26S proteasome. *J. Biol. Chem.* 278, 4536, 2003.
38. Huang, H. C., Nguyen, T., and Pickett, C. B. Phosphorylation of Nrf2 at Ser-40 by protein kinase C regulates antioxidant response element-mediated transcription. *J. Biol. Chem.* 277, 42769, 2002.
39. Bloom, D. A., and Jaiswal, A. K. Phosphorylation of Nrf2 at Ser40 by protein kinase C in response to antioxidants leads to the release of Nrf2 from INrf2, but is not required for Nrf2 stabilization/accumulation in the nucleus and transcriptional activation of antioxidant response element-mediated NAD(P)H:quinone oxidoreductase-1 gene expression. *J. Biol. Chem.* 278, 44675, 2003.
40. Okouchi, M., Okayama, N., Alexander, J. S., and Aw, T. Y. NRF2-dependent glutamate-l-cysteine ligase catalytic subunit expression mediates insulin protection against hyperglycemia-induced brain endothelial cell apoptosis. *Curr. Neurovasc. Res.* 3, 249, 2006.
41. Shih, A. Y., Li, P., and Murphy, T. H. A small-molecule-inducible Nrf2-mediated antioxidant response provides effective prophylaxis against cerebral ischemia in vivo. *J. Neurosci.* 25, 10321, 2005.
42. Shah, Z. A., Li, R. C., Thimmulappa, R. K., Kensler, T. W., Yamamoto, M., Biswal, S., and Dore, S. Role of reactive oxygen species in modulation of Nrf2 following ischemic reperfusion injury. *Neuroscience* 147, 53, 2007.
43. Zhao, J., Kobori, N., Aronowski, J., and Dash, P. K. Sulforaphane reduces infarct volume following focal cerebral ischemia in rodents. *Neurosci. Lett.* 393, 108, 2006.
44. Satoh, T., Okamoto, S. I., Cui, J., Watanabe, Y., Furuta, K., Suzuki, M., Tohyama, K., and Lipton, S. A. Activation of the Keap1/Nrf2 pathway for neuroprotection by electrophilic [correction of *electrophillic*] phase II inducers. *Proc. Natl. Acad. Sci. U S A* 103, 768, 2006.
45. Zhao, X., Sun, G., Zhang, J., Strong, R., Dash, P. K., Kan, Y. W., Grotta, J. C., and Aronowski, J. Transcription factor Nrf2 protects the brain from damage produced by intracerebral hemorrhage. *Stroke* 38, 3280, 2007.
46. Zhao, J., Moore, A. N., Redell, J. B., and Dash, P. K. Enhancing expression of Nrf2-driven genes protects the blood–brain barrier after brain injury. *J. Neurosci.* 27, 10240, 2007.
47. Kraft, A. D., Johnson, D. A., and Johnson, J. A. Nuclear factor E2-related factor 2-dependent antioxidant response element activation by *tert*-butylhydroquinone and sulforaphane occurring preferentially in astrocytes conditions neurons against oxidative insult. *J. Neurosci.* 24, 1101, 2004.
48. Han, J. M., Lee, Y. J., Lee, S. Y., Kim, E. M., Moon, Y., Kim, H. W., and Hwang, O. Protective effect of sulforaphane against dopaminergic cell death. *J. Pharmacol. Exp. Ther.* 321, 249, 2007.
49. Zhao, J., Moore, A. N., Clifton, G. L., and Dash, P. K. Sulforaphane enhances aquaporin-4 expression and decreases cerebral edema following traumatic brain injury. *J. Neurosci. Res.* 82, 499, 2005.
50. Danilov, C. A., Chandrasekaran, K., Racz, J., Soane, L., Zielke, C., and Fiskum, G. Sulforaphane protects astrocytes against oxidative stress and delayed death caused by oxygen and glucose deprivation. *Glia* 57, 645, 2009.
51. Zhang, Y., Talalay, P., Cho, C. G., and Posner, G. H. A major inducer of anticarcinogenic protective enzymes from broccoli: Isolation and elucidation of structure. *Proc. Natl. Acad. Sci. U S A* 89, 2399, 1992.
52. Zhang, Y., Kensler, T. W., Cho, C. G., Posner, G. H., and Talalay, P. Anticarcinogenic activities of sulforaphane and structurally related synthetic norbornyl isothiocyanates. *Proc. Natl. Acad. Sci. U S A* 91, 3147, 1994.
53. Fahey, J. W., Haristoy, X., Dolan, P. M., Kensler, T. W., Scholtus, I., Stephenson, K. K., Talalay, P., and Lozniewski, A. Sulforaphane inhibits extracellular, intracellular, and antibiotic-resistant strains of *Helicobacter pylori* and prevents benzo[a]pyrene-induced stomach tumors. *Proc. Natl. Acad. Sci. U S A* 99, 7610, 2002.

54. Dinkova-Kostova, A. T., Holtzclaw, W. D., Cole, R. N., Itoh, K., Wakabayashi, N., Katoh, Y., Yamamoto, M., and Talalay, P. Direct evidence that sulfhydryl groups of Keap1 are the sensors regulating induction of phase 2 enzymes that protect against carcinogens and oxidants. *Proc. Natl. Acad. Sci. U S A* 99, 11908, 2002.

55. Talalay, P., Fahey, J. W., Healy, Z. R., Wehage, S. L., Benedict, A. L., Min, C., and Dinkova-Kostova, A. T. Sulforaphane mobilizes cellular defenses that protect skin against damage by UV radiation. *Proc. Natl. Acad. Sci. U S A* 104, 17500, 2007.

56. Chen, X. L., Dodd, G., Thomas, S., Zhang, X., Wasserman, M. A., Rovin, B. H., and Kunsch, C. Activation of Nrf2/ARE pathway protects endothelial cells from oxidant injury and inhibits inflammatory gene expression. *Am. J. Physiol. Heart Circ. Physiol.* 290, H1862, 2006.

57. Ritz, S. A., Wan, J., and az-Sanchez, D. Sulforaphane-stimulated phase II enzyme induction inhibits cytokine production by airway epithelial cells stimulated with diesel extract. *Am. J. Physiol. Lung Cell Mol. Physiol.* 292, L33, 2007.

58. Wu, L., Noyan Ashraf, M. H., Facci, M., Wang, R., Paterson, P. G., Ferrie, A., and Juurlink, B. H. Dietary approach to attenuate oxidative stress, hypertension, and inflammation in the cardiovascular system. *Proc. Natl. Acad. Sci. U S A* 101, 7094, 2004.

59. Abraham, N. G., and Kappas, A. Pharmacological and clinical aspects of heme oxygenase. *Pharmacol. Rev.* 60(1), 79, 2008.

60. Alcaraz, M. J., Fernandez, P., and Guillen, M. I. Anti-inflammatory actions of the heme oxygenase-1 pathway. *Curr. Pharm. Des.* 9, 2541, 2003.

61. Geddes, J. W., Pettigrew, L. C., Holtz, M. L., Craddock, S. D., and Maines, M. D. Permanent focal and transient global cerebral ischemia increase glial and neuronal expression of heme oxygenase-1, but not heme oxygenase-2, protein in rat brain. *Neurosci. Lett.* 210, 205, 1996.

62. Lu, A., Ran, R. Q., Clark, J., Reilly, M., Nee, A., and Sharp, F. R. 17-beta-Estradiol induces heat shock proteins in brain arteries and potentiates ischemic heat shock protein induction in glia and neurons. *J. Cereb. Blood Flow Metab.* 22, 183, 2002.

63. Lee, J. M., Anderson, P. C., Padgitt, J. K., Hanson, J. M., Waters, C. M., and Johnson, J. A. Nrf2, not the estrogen receptor, mediates catechol estrogen-induced activation of the antioxidant responsive element. *Biochim. Biophys. Acta* 1629, 92, 2003.

64. Zhou, W., Lo, S. C., Liu, J. H., Hannink, M., and Lubahn, D. B. ERRbeta: A potent inhibitor of Nrf2 transcriptional activity. *Mol. Cell. Endocrinol.* 278, 52, 2007.

65. Lewen, A., Matz, P., and Chan, P. H. Free radical pathways in CNS injury. *J. Neurotrauma* 17, 871, 2000.

66. Scorziello, A., Pellegrini, C., Secondo, A., Sirabella, R., Formisano, L., Sibaud, L., Amoroso, S., Canzoniero, L. M., Annunziato, L., and Di Renzo, G. F. Neuronal NOS activation during oxygen and glucose deprivation triggers cerebellar granule cell death in the later reoxygenation phase. *J. Neurosci. Res.* 76, 812, 2004.

67. Li, P. A., and Siesjo, B. K. Role of hyperglycaemia-related acidosis in ischaemic brain damage. *Acta Physiol. Scand.* 161, 567, 1997.

68. Singh, N. C., Kochanek, P. M., Schiding, J. K., Melick, J. A., and Nemoto, E. M. Uncoupled cerebral blood flow and metabolism after severe global ischemia in rats. *J. Cereb. Blood Flow Metab.* 12, 802, 1992.

69. Erecinska, M., and Silver, I. A. Relationship between ions and energy metabolism: Cerebral calcium movements during ischaemia and subsequent recovery. *Can. J. Physiol. Pharmacol.* 70 [Suppl], S190, 1992.

70. Feng, Z. C., Sick, T. J., and Rosenthal, M. Oxygen sensitivity of mitochondrial redox status and evoked potential recovery early during reperfusion in post-ischemic rat brain. *Resuscitation* 37, 33, 1998.

71. Lei, B., Adachi, N., and Arai, T. The effect of hypothermia on H_2O_2 production during ischemia and reperfusion: A microdialysis study in the gerbil hippocampus. *Neurosci. Lett.* 222, 91, 1997.

72. Tanahashi, N., Fukuuchi, Y., Tomita, M., Kobari, M., Takeda, H., and Yokoyama, M. Pentoxifylline ameliorates postischemic delayed hypoperfusion of the cerebral cortex following cardiac arrest in cats. *J. Neurol. Sci.* 132, 105, 1995.

73. Blomgren, K., and Hagberg, H. Free radicals, mitochondria, and hypoxia-ischemia in the developing brain. *Free Radic. Biol. Med.* 40, 388, 2006.

74. Fiskum, G., Murphy, A. N., and Beal, M. F. Mitochondria in neurodegeneration: Acute ischemia and chronic neurodegenerative diseases. *J. Cereb. Blood Flow Metab.* 19, 351, 1999.

75. Rizzuto, R., Simpson, A. W., Brini, M., and Pozzan, T. Rapid changes of mitochondrial Ca^{2+} revealed by specifically targeted recombinant aequorin [published erratum appears in *Nature* 1992 Dec 24–31, 360(6406):768]. *Nature* 358, 325, 1992.

76. Chang, L. H., Shimizu, H., Abiko, H., Swanson, R. A., Faden, A. I., James, T. L., and Weinstein, P. R. Effect of dichloroacetate on recovery of brain lactate, phosphorus energy metabolites, and glutamate during reperfusion after complete cerebral ischemia in rats. *J. Cereb. Blood Flow Metab.* 12, 1030, 1992.

77. Kuroda, S., Katsura, K. I., Tsuchidate, R., and Siesjo, B. K. Secondary bioenergetic failure after transient focal ischaemia is due to mitochondrial injury. *Acta Physiol. Scand.* 156, 149, 1996.

78. Starkov, A. A., Chinopoulos, C., and Fiskum, G. Mitochondrial calcium and oxidative stress as mediators of ischemic brain injury. *Cell Calcium* 36, 257, 2004.

79. Rosenthal, R. E., Hamud, F., Fiskum, G., Varghese, P. J., and Sharpe, S. Cerebral ischemia and reperfusion: Prevention of brain mitochondrial injury by lidoflazine. *J. Cereb. Blood Flow Metab.* 7, 752, 1987.

80. Sims, N. R. Selective impairment of respiration in mitochondria isolated from brain subregions following transient forebrain ischemia in the rat. *J. Neurochem.* 56, 1836, 1991.

81. Hillered, L., and Ernster, L. Respiratory activity of isolated rat brain mitochondria following *in vitro* exposure to oxygen radicals. *J. Cereb. Blood Flow Metab.* 3, 207, 1983.

82. Rosenthal, R. E. F. G. Brain mitochondrial function in cerebral ischemia and resuscitation. In: *Cerebral Ischemia and Resuscitation* (A. Schurr, Ed.). CRC Press, New York, pp. 289, in press.

83. Polster, B. M., Kinnally, K. W., and Fiskum, G. Bh3 death domain peptide induces cell type-selective mitochondrial outer membrane permeability. *J. Biol. Chem.* 276, 37887, 2001.

84. Castino, R., Bellio, N., Nicotra, G., Follo, C., Trincheri, N. F., and Isidoro, C. Cathepsin D-Bax death pathway in oxidative stressed neuroblastoma cells. *Free Radic. Biol. Med.* 42, 1305, 2007.

85. Perier, C., Tieu, K., Guegan, C., Caspersen, C., Jackson-Lewis, V., Carelli, V., Martinuzzi, A., Hirano, M., Przedborski, S., and Vila, M. Complex I deficiency primes Bax-dependent neuronal apoptosis through mitochondrial oxidative damage. *Proc. Natl. Acad. Sci. U S A* 102, 19126, 2005.

86. Bogaert, Y. E., Rosenthal, R. E., and Fiskum, G. Postischemic inhibition of cerebral cortex pyruvate dehydrogenase. *Free Radic. Biol. Med.* 16, 811, 1994.

87. Vereczki, V., Martin, E., Rosenthal, R. E., Hof, P. R., Hoffman, G. E., and Fiskum, G. Normoxic resuscitation after cardiac arrest protects against hippocampal oxidative stress, metabolic dysfunction, and neuronal death. *J. Cereb. Blood Flow Metab.* 26, 821, 2006.

88. Bogaert, Y. E., Sheu, K. F., Hof, P. R., Brown, A. M., Blass, J. P., Rosenthal, R. E., and Fiskum, G. Neuronal subclass-selective loss of pyruvate dehydrogenase immunoreactivity following canine cardiac arrest and resuscitation. *Exp. Neurol.* 161, 115, 2000.

89. Bartnik, B. L., Hovda, D. A., and Lee, P. W. Glucose metabolism after traumatic brain injury: Estimation of pyruvate carboxylase and pyruvate dehydrogenase flux by mass isotopomer analysis. *J. Neurotrauma* 24, 181, 2007.

90. Robertson, C. L., Saraswati, M., and Fiskum, G. Mitochondrial dysfunction early after traumatic brain injury in immature rats. *J. Neurochem.* 101, 1248, 2007.

91. Opii, W. O., Nukala, V. N., Sultana, R., Pandya, J. D., Day, K. M., Merchant, M. L., Klein, J. B., Sullivan, P. G., and Butterfield, D. A. Proteomic identification of oxidized mitochondrial proteins following experimental traumatic brain injury. *J. Neurotrauma* 24, 772, 2007.

92. Bambrick, L. L., and Fiskum, G. Mitochondrial dysfunction in mouse trisomy 16 brain. *Brain Res.* 1188, 9, 2008.

93. Bunik, V. I., Raddatz, G., Wanders, R. J., and Reiser, G. Brain pyruvate and 2-oxoglutarate dehydrogenase complexes are mitochondrial targets of the CoA ester of the Refsum disease marker phytanic acid. *FEBS Lett.* 580, 3551, 2006.

94. Perluigi, M., Poon, H. F., Maragos, W., Pierce, W. M., Klein, J. B., Calabrese, V., Cini, C., De, M. C., and Butterfield, D. A. Proteomic analysis of protein expression and oxidative modification in r6/2 transgenic mice: A model of Huntington disease. *Mol. Cell. Proteomics* 4, 1849, 2005.

95. Bubber, P., Haroutunian, V., Fisch, G., Blass, J. P., and Gibson, G. E. Mitochondrial abnormalities in Alzheimer brain: Mechanistic implications. *Ann. Neurol.* 57, 695, 2005.

96. Ying, W. NAD$^+$ and NADH in ischemic brain injury. *Front. Biosci.* 13, 1141, 2008.

97. Eliasson, M. J., Sampei, K., Mandir, A. S., Hurn, P. D., Traystman, R. J., Bao, J., Pieper, A., Wang, Z. Q., Dawson, T. M., Snyder, S. H., and Dawson, V. L. Poly(ADP-ribose) polymerase gene disruption renders mice resistant to cerebral ischemia. *Nat. Med.* 3, 1089, 1997.

98. Kauppinen, T. M., and Swanson, R. A. The role of poly(ADP-ribose) polymerase-1 in CNS disease. *Neuroscience* 145, 1267, 2007.

99. Du, L., Zhang, X., Han, Y. Y., Burke, N. A., Kochanek, P. M., Watkins, S. C., Graham, S. H., Carcillo, J. A., Szabo, C., and Clark, R. S. Intra-mitochondrial poly(ADP-ribosylation) contributes to NAD^+ depletion and cell death induced by oxidative stress. *J. Biol. Chem.* 278, 18426, 2003.

100. Crompton, M., Barksby, E., Johnson, N., and Capano, M. Mitochondrial intermembrane junctional complexes and their involvement in cell death. *Biochimie* 84, 143, 2002.

101. Halestrap, A. P., McStay, G. P., and Clarke, S. J. The permeability transition pore complex: Another view. *Biochimie* 84, 153, 2002.

102. Bernardi, P., Scorrano, L., Colonna, R., Petronilli, V., and Di Lisa, F. Mitochondria and cell death. Mechanistic aspects and methodological issues. *Eur. J. Biochem.* 264, 687, 1999.

103. Uchino, H., Minamikawa-Tachino, R., Kristian, T., Perkins, G., Narazaki, M., Siesjo, B. K., and Shibasaki, F. Differential neuroprotection by cyclosporin A and FK506 following ischemia corresponds with differing abilities to inhibit calcineurin and the mitochondrial permeability transition. *Neurobiol. Dis.* 10, 219, 2002.

104. Sullivan, P. G., Keller, J. N., Bussen, W. L., and Scheff, S. W. Cytochrome *c* release and caspase activation after traumatic brain injury. *Brain Res.* 949, 88, 2002.

105. Alessandri, B., Rice, A. C., Levasseur, J., DeFord, M., Hamm, R. J., and Bullock, M. R. Cyclosporin A improves brain tissue oxygen consumption and learning/memory performance after lateral fluid percussion injury in rats. *J. Neurotrauma* 19, 829, 2002.

106. Hansson, M. J., Mattiasson, G., Mansson, R., Karlsson, J., Keep, M. F., Waldmeier, P., Ruegg, U. T., Dumont, J. M., Besseghir, K., and Elmer, E. The nonimmunosuppressive cyclosporin analogs NIM811 and UNIL025 display nanomolar potencies on permeability transition in brain-derived mitochondria. *J. Bioenerg. Biomembr.* 36, 407, 2004.

107. Schinzel, A. C., Takeuchi, O., Huang, Z., Fisher, J. K., Zhou, Z., Rubens, J., Hetz, C., Danial, N. N., Moskowitz, M. A., and Korsmeyer, S. J. Cyclophilin D is a component of mitochondrial permeability transition and mediates neuronal cell death after focal cerebral ischemia. *Proc. Natl. Acad. Sci. U S A* 102, 12005, 2005.

108. Forte, M., Gold, B. G., Marracci, G., Chaudhary, P., Basso, E., Johnsen, D., Yu, X., Fowlkes, J., Rahder, M., Stem, K., Bernardi, P., and Bourdette, D. Cyclophilin D inactivation protects axons in experimental autoimmune encephalomyelitis, an animal model of multiple sclerosis. *Proc. Natl. Acad. Sci. U S A* 104, 7558, 2007.

109. Wang, Q., Tang, X. N., and Yenari, M. A. The inflammatory response in stroke. *J. Neuroimmunol.* 184, 53, 2007.

2 The Neuroprotective Role of Micronutrients in Parkinson's Disease

Kristen Malkus[1], Elpida Tsika[1,4]*, and Harry Ischiropoulos[1,2,3]*

[1]Joseph Stokes Jr. Research Institute, The Children's Hospital of Philadelphia, Philadelphia, Pennsylvania, USA

[2]Department of Pediatrics, The Children's Hospital of Philadelphia, Philadelphia, Pennsylvania, USA

[3]Department of Pharmacology, University of Pennsylvania, Philadelphia, Pennsylvania, USA

[4]Department of Molecular Biology and Genetics, Democritus University of Thrace, Alexandroupolis, Greece

*Authors contributed equally to this chapter

CONTENTS

2.1 INTRODUCTION

Oxidative stress has been implicated in the pathogenesis of a variety of neurodegenerative disorders, including amyotrophic lateral sclerosis, Alzheimer's disease, Huntington's disease, and Parkinson's disease (PD).[1] PD provides an excellent paradigm to explore the role of oxidants in disease development—oxidative biochemistry, protein aggregation, and neural degeneration are all present in the postmortem analysis of patients' brains. PD, the second most common neurodegenerative disorder, is typified by motor impairments that include resting tremor, postural rigidity, instability, and bradykinesia.[2] Selective loss of dopaminergic neurons occurs in the substantia nigra pars compacta, leading to greatly reduced dopamine levels in the striatum, and is accompanied by the presence of cytoplasmic inclusions known as Lewy bodies (LBs) in brain-stem regions and the dopaminergic neurons of the substantia nigra pars compacta.[2] While many hypotheses have been generated to explain the causes of PD, the pathogenic mechanisms remain enigmatic and treatment is consequently focused on alleviating symptoms. This chapter discusses the role of oxidative stress in PD and the potential benefit of micronutrients, focusing on the use of mouse models to explore the efficacy of possible therapies.

2.2   EVIDENCE OF OXIDATIVE STRESS IN PARKINSON'S DISEASE

Initial evidence that oxidants and oxidative modifications of neuronal macromolecules are involved in PD pathogenesis came from the accidental use of 1-methyl-4-phenyl-1,2,3,6-tetrahydropyridine (MPTP), a synthetic analog of heroin. Individuals that had injected MPTP developed a rapid-onset parkinsonian syndrome characterized by many clinical features of PD.[3] As discussed below, one of the actions of 1-methyl-4-phenylpyridinium (MPP+), the active metabolite of MPTP, is to inhibit complex I of the mitochondrial respiratory chain and increase the production of reactive species.

Furthermore, epidemiological studies have revealed an increased risk of developing PD among populations exposed to pesticides and herbicides.[4,5] Examination of postmortem brain tissue from patients with PD has shown increased protein carbonyls, lipid and DNA oxidation, elevated total iron concentration, and reduced glutathione levels,[6–11] supporting the role of oxidative processes in the disease's development and progression. Additionally, studies examining mitochondria isolated from PD patients have shown decreased complex I activity and elevated oxidant production with increased sensitivity to MPP+.[12] Extensive nitration of tyrosine residues of α-synuclein, which has been documented by immunohistochemical staining of PD brains with monoclonal antibodies that specifically recognize nitrated α-synuclein, provides additional evidence for oxidative and nitrative chemistry in PD.[13]

Based on this body of evidence, many studies have focused on uncovering the mechanisms by which oxidation may contribute to the development of disease. Attempts to decipher the precise role of oxidative stress in PD have focused on two principal areas: the genetics of the disease and chemical manipulations that induce symptoms.

2.3   INSIGHTS FROM GENETICS

While the majority of cases of PD are sporadic, rare familial forms of the disease have helped provide insights into the underlying pathogenic molecular mechanisms. Currently, thirteen genetic loci, denoted PARK1 to PARK13, have been associated with PD.[14] Among these, five genes corresponding to six loci are known and have been established as causing familial PD: α-synuclein (PARK1/PARK4), parkin (PARK2), PINK1 (PARK6), DJ-1 (PARK7), and LRRK2 (PARK8).[14] Experiments probing into the function of these proteins and the disease-causing mutations support causative involvement of oxidative processes and impaired mitochondrial function in the pathogenesis of PD.

Among the genes associated with familial PD, recessive mutations occur in two genes that encode mitochondrial-associated proteins, DJ-1 and PINK1.[15,16] DJ-1 is a protein with a putative antioxidant function that has been associated with PD through mutations and deletions that suggest a loss of function. Cysteine residues in DJ-1 that have been found oxidized in humans and in animal model systems have implicated DJ-1 as an antioxidant molecule with peroxiredoxin-like activity.[17,18] In support of this hypothesis, DJ-1-deficient mice display nigrostriatal dopaminergic dysfunction and motor deficits.[19–21] Additionally, flies and mice in which DJ-1 is knocked out or mutated are more sensitive to insults from mitochondrial toxins, such as MPTP, rotenone, and paraquat.[22–25] Conversely, overexpression of DJ-1 in rats exposed to 6-hydroxydopamine or mice exposed to MPTP was protective against dopaminergic neural degeneration.[26,27] Together, these findings suggest that pathogenic mutations leading to a loss of proper function for DJ-1 may reduce neuronal ability to cope with consequences of oxidative stress.

Loss of function is similarly the proposed pathogenic mechanism for mutations in PINK1, a mitochondrial serine/threonine kinase that has been hypothesized to phosphorylate mitochondrial proteins in response to cellular stress and thus protect against mitochondrial dysfunction.[15,28] Lymphoblasts of patients with mutations in PINK1 show increased lipid peroxidation and defects in mitochondrial complex I activity.[29,30] Knockdown of PINK1 or expression of the PINK1 with a pathogenic mutation in neurons resulted in mitochondrial pathology and increased the vulnerability of the cells to rotenone and MPP+.[29,31,32] Furthermore, mice that are deficient in PINK1 demonstrate impaired mitochondrial respiration, which is specific to the dopaminergic circuit and progresses with aging, and have increased sensitivity to oxidative stress when they are challenged with H_2O_2.[33]

α-Synuclein has attracted considerable attention following the combined discovery that this protein is one of the principal components of LBs and that autosomal dominant missense mutations in the α-synuclein gene lead to early-onset PD.[34–37] While the function of this protein relates in part to vesicular regulation and chaperone-like activity,[38–40] there is substantial evidence suggesting an interplay between α-synuclein, oxidative stress, and mitochondria in the pathogenesis of PD.[41–43] α-Synuclein, normally a soluble, relatively unstructured protein, undergoes orderly assembly to make amyloid fibers that eventually form protein inclusions and LBs.[44–46] Experiments *in vitro* and in cell culture suggested that covalent cross-linking of α-synuclein or nitration of tyrosine residues results in acceleration of the aggregation process to amyloid fibers.[47–51]

In cell culture studies, expression of α-synuclein was associated with increased oxidant production and mitochondrial dysfunction.[43] Transgenic mice expressing α-synuclein with the disease-causing A53T mutation exhibited mitochondrial dysregulation.[42] Exposure of A53T-expressing mice to both paraquat and maneb exacerbated α-synuclein pathology and mitochondrial degeneration and increased levels of nitrated proteins.[52] Accordingly, transgenic mice expressing human wild-type or A30P α-synuclein similarly had enhanced neuronal degeneration and mitochondrial dysfunction compared with nontransgenic controls upon exposure to MPTP.[53,54] Recent evidence also indicated that α-synuclein accumulates within the mitochondria, leading to impaired mitochondrial complex I activity and increased production of reactive species.[41] Interestingly, mitochondria isolated from the substantia nigra and striatum of patients with sporadic PD showed significantly greater accumulation of α-synuclein than controls.[41]

Mutations in the E3 ubiquitin ligase parkin are also associated with familial PD. While its role in the ubiquitin proteasome system has helped highlight the potential involvement of protein degradation impairment in disease, additionally, parkin has been implicated in mitochondrial function.[14] Parkin-deficient flies have increased sensitivity to paraquat, and proteomic analysis of these flies indicates differences in proteins regulating energy metabolism and oxidative regulation.[55,56] *Drosophila* models expressing parkin with a pathogenic mutation similarly exhibit mitochondrial dysfunction and transcriptional differences in oxidative response components.[57,58] Knockout of parkin in murine models leads to nigrostriatal degeneration, synaptic dysfunction, dopaminergic behavioral deficits, and aberrant accumulation of parkin substrates, including many proteins involved in mitochondrial function and oxidative stress.[59,60] Conversely, viral overexpression of parkin is able to inhibit dopaminergic neural loss in mice exposed to MPTP.[27]

Genetic animal models of PD genes have been powerful tools in dissecting the molecular pathways involved in the degeneration of dopaminergic neurons. Insights for the contribution of oxidative processes in PD have been gained through deciphering the normal role of the proteins and how the pathogenic mutations lead to neurodegeneration.

2.4 INTOXICATION MODELS OF PARKINSON'S DISEASE

The discovery that MPTP intoxication causes parkinsonian syndrome has led to its use in establishing several PD models in nonhuman primates and mice.[3] MPP+, the glial metabolite of MPTP, is selectively taken up by the dopamine transporter of dopaminergic neurons, where it inhibits mitochondrial complex I, activates nearby microglia, and ultimately induces cell death.[61] MPP+ toxicity can be prevented in vivo by replacement of complex I with the single-subunit NADH dehydrogenase of *Saccharomyces cerevisiae* (NDI1), implicating dysfunctional mitochondrial metabolism in pathogenesis.[62] Microglial activation and NADPH oxidase-derived oxidants may also contribute to the neuron dysfunction after MPTP challenges.[63,64] Mice overexpressing the antioxidant

Cu/Zn superoxide dismutase 1[65] are protected against MPTP toxicity, indicating that oxidative processes contribute to the degeneration of dopaminergic neurons. The involvement of reactive nitrogen species in MPTP-induced neuron injury has been revealed by the use of nitric oxide synthase (NOS)-deficient animals. MPTP toxicity and 3-nitrotyrosine staining are attenuated in either inducible NOS- or neuronal NOS-deficient mice[66,67] or mice treated with neuronal NOS inhibitors.[68,69] Similar results have been obtained from baboons exposed to the toxin,[70] indicating that nitric oxide-derived oxidants are participants in the processes leading to neurodegeneration. Additionally, it has consistently been found that either acute or chronic systemic administration of MPTP to mice diminished the striatal content of dopamine and its metabolites prior to neuron death.[71] A possible explanation for the observed demise of dopamine levels comes from studies showing that MPTP exposure resulted in nitration of tyrosine hydroxylase and loss of its enzymatic activity.[72] This is an additional biochemical event that may contribute to MPTP toxicity.

MPTP studies have also provided further support for the intersection of oxidative stress and α-synuclein in PD. Exposure of nonhuman primates to MPTP has led to the formation of filamentous α-synuclein-positive inclusions.[73,74] Contradictory results had been reported from mouse models studying the interplay between α-synuclein and MPTP, probably due to differences in genetic backgrounds and experimental designs.[75] However, several groups have found that mice that lack α-synuclein are protected against MPTP toxicity[76–78] and, conversely, mice expressing the disease-associated α-synuclein mutations are more vulnerable to MPTP toxicity.[54,79] These findings, as well as the fact that α-synuclein is nitrated after MPTP exposure,[65,80] are an indication that α-synuclein and MPTP may interact in a common pathogenic pathway.[81]

Another prototypical oxidative stress toxin that has been used in animal models for more than 30 years is 6-hydroxydopamine. In the presence of oxygen and transition metals, it oxidizes into *para*-quinone and hydrogen peroxide, while superoxide (O_2^-) and semi-quinone radicals are intermediate species of the reaction.[82] These reactive species and strong electrophiles damage macromolecules and induce neurodegeneration.[83,84]

The herbicide paraquat, a biologically active redox molecule, is another toxin implicated in causing oxidative stress and having deleterious effects on neurons. Paraquat is widely used in mouse models and causes reduced motor activity, death of the dopaminergic neurons of the substantia nigra, and degeneration of the striatal fibers in a dose-dependent manner.[85,86] Overexpression of superoxide dismutase and superoxide dismutase mimetics protects against paraquat neuronal injury, reinforcing the concept of reactive species participation in neurotoxicity.[87,88]

Chronic infusion of rotenone, an insecticide used as a fish poison, has resulted in selective degeneration of the dopaminergic neurons of the nigrostriatal region, motor impairment, and fibrillar inclusions in rats.[89] Rotenone is a well-characterized inhibitor of mitochondrial complex I. Similar to the MPP[+] model, transfection with NDI1 prevented cell death, suggesting that the two chemicals share a common mechanism of toxicity.[90] Interestingly, in addition to α-synuclein accumulation and aggregation, rotenone also induces DJ-1 oxidation that is restricted to dopaminergic regions and can be prevented by pretreatment with α-tocopherol.[91]

2.5 MICRONUTRIENTS AND ANTIOXIDANTS AS POTENTIAL THERAPIES

Mitochondrial defects and oxidative stress have emerged as common pathogenic causes for many diverse conditions and neurodegenerative disorders, including PD. While there are symptomatic therapies for PD, there are no effective treatments that can restore neuronal function or offer neuroprotection. Therefore, the use of micronutrients and antioxidants to improve mitochondrial function and prevent oxidant injury may be beneficial for neurodegenerative diseases. Animal models of PD have been useful in exploring pharmacological interventions, such as the metabolic modifiers creatine, coenzyme Q_{10} (CoQ_{10}), lipoic acid, as well as the antioxidants Ginkgo biloba extract, *N*-acetyl-cysteine, nicotinamide, riboflavin, acetyl-carnitine, and resveratrol.[92]

CoQ_{10}, also known as ubiquinone, serves as an acceptor of electrons from mitochondrial complexes I and II, potentially acts as an antioxidant, and is capable of regenerating α-tocopherol (vitamin E).[93] It is a molecule that has been tested for the treatment of mitochondrial disorders and neurodegenerative diseases in a variety of animal models and in clinical trials with patients having PD and Huntington's disease. CoQ_{10} and CoQ_9 levels were found increased in the nigrostriatal tract of mice 1 wk after acute treatment with MPTP.[94] Additionally, administration of CoQ_{10} attenuated the loss of striatal dopamine and decreased tyrosine hydroxylase immunoreactivity in the striatum of aged mice treated with MPTP.[95–97] CoQ_{10} has also been shown to be neuroprotective against other mitochondrial toxins, such as malonate, 3-nitropropionic acid, and rotenone.[98–100]

Lipoic acid is a coenzyme for pyruvate dehydrogenase and α-ketoglutarate dehydrogenase. It functions as an antioxidant through the chelation of transition metals and the regeneration of endogenous antioxidants, such as ascorbic acid, glutathione, and α-tocopherol. Dihydrolipoic acid, which is the reduced product of lipoic acid, also interacts with CoQ. This interaction was shown to increase the antioxidant capacity of CoQ by reducing ubiquinone to ubiquinol, thus maintaining a normal ratio of reduced and oxidized CoQ following MPTP administration in mice.[101]

While the application of antioxidant micronutrients to cell and animal models of PD has produced encouraging results, extension of these treatments to clinical trials has produced variable findings. One of the most extensive clinical trials of antioxidants to treat PD was the Deprenyl and Tocopherol Antioxidative Therapy of Parkinsonism (DATATOP) study.[102] Eight hundred patients presenting with early stages of PD were randomly assigned to receive both deprenyl and α-tocopherol, deprenyl with an α-tocopherol placebo, α-tocopherol with a deprenyl placebo, or two placebos. The end point of the trial was the onset of parkinsonian disability to the degree that levodopa therapy was needed. Deprenyl, also known as selegiline, is a monoamine oxidase type B inhibitor. α-Tocopherol is a biologically active component of vitamin E.[102] At the end of 14 ± 6 mo of treatment and observation, the researchers found that deprenyl was able to significantly delay the onset of PD symptoms, warranting levodopa therapy by a median time of 9 mo. Conversely, α-tocopherol did not delay the onset or severity of symptoms, and the combination of α-tocopherol and deprenyl did not provide any benefit above that achieved from deprenyl alone.[102]

Rasagiline is another selective monoamine oxidase type B inhibitor that has been successful in drug trials. In a series of large-scale studies, the Parkinson Study Group found that rasagiline slowed the progression of symptoms in individuals with early PD as measured by the Unified Parkinson's Disease Rating Scale (UPDRS) and was more effective when administered immediately compared with after a 6-mo delay, showing that the benefit is not due to an immediate symptomatic effect but rather to an actual influence on disease progression.[103,104] Assdditionally, rasagiline was able to potentiate the beneficial effects of levodopa on disease progression in PD patients with motor fluctuations who were receiving levodopa treatment.[105]

In a study of eighty subjects with early PD, the Parkinson Study Group found that a dose of 1200 mg/d of CoQ_{10} reduced the rate of functional decline as measured by the UPDRS.[106] However, this dose of 1200 mg/d of CoQ_{10} did not change the amount of time until disability requiring treatment with levodopa and lower doses of 300 and 600 mg/d of CoQ_{10} did not significantly improve PD symptoms.[106] In a separate study, mid-stage PD patients who were administered 300 mg/d of CoQ_{10} did not differ from patients administered a placebo in progression of symptoms as measured by the UPDRS.[107]

Recently, pilot trials of the compound creatine have also been undertaken. Creatine, through its conversion to phosphocreatine, is responsible for ATP homeostasis and potentially has antioxidant properties. While these small studies did not have the power to discriminate statistical significance, a definite trend of slower decline as measured by the UPDRS was noted with creatine treatment.[108,109] However, creatine was not able to delay the amount of time needed before progression to dopaminergic replacement therapy.[109] Another small study produced slightly different results, where patients who were administered creatine did not exhibit any significant difference in overall UPDRS scores compared with controls. An improvement in mood did occur, however, as well as smaller dose increases in dopaminergic therapy.[110]

2.6 SUMMARY

Genetic and environmental factors may contribute to the pathogenesis of PD by altering mitochondrial function and accelerating oxidative processes. The use of molecules capable of restoring metabolic homeostasis and alleviating the oxidative burden shows promise as therapy. Additional trials in progress with CoQ_{10} and creatine hold promise for effectively improving disease outcome.

REFERENCES

1. Beal, M. F. Oxidatively modified proteins in aging and disease. *Free Radic. Biol. Med.* **32**, 797, 2002.
2. Lang, A. E., and Lozano, A. M. Parkinson's disease—First of two parts. *N. Engl. J. Med.* **339**, 1044, 1998.
3. Langston, J. W., et al. Chronic parkinsonism in humans due to a product of meperidine-analog synthesis. *Science* **219**, 979, 1983.
4. Elbaz, A., and Tranchant, C. Epidemiologic studies of environmental exposures in Parkinson's disease. *J. Neurol. Sci.* **262**, 37, 2007.
5. Olanow, C. W., and Tatton, W. G. Etiology. and pathogenesis of Parkinson's disease. *Annu. Rev. Neuroscience* **22**, 123, 1999.
6. Alam, Z. I., et al. A generalised increase in protein carbonyls in the brain in Parkinson's but not incidental Lewy body disease. *J. Neurochem.* **69**, 1326, 1997.
7. Good, P. F., et al. Protein nitration in Parkinson's disease. *J. Neuropathol. Exp. Neurol.* **57**, 123, 1998.
8. Dexter, D. T., et al. Basal lipid peroxidation in substantia nigra is increased in Parkinson's disease. *J. Neurochem.* **52**, 381, 1989.
9. Alam, Z. I., et al. Oxidative DNA damage in the parkinsonian brain: An apparent selective increase in 8-hydroxyguanine levels in substantia nigra. *J. Neurochem.* **69**, 1196, 1997.
10. Sofic, E., et al. Increased iron (III) and total iron content in post mortem substantia nigra of parkinsonian brain. *J. Neural Transm.* **74**, 199, 1988.
11. Sian, J., et al. Alterations in glutathione levels in Parkinson's disease and other neurodegenerative disorders affecting basal ganglia. *Ann. Neurol.* **36**, 348, 1994.
12. Swerdlow, R. H., et al. Origin and functional consequences of the complex I defect in Parkinson's disease. *Ann. Neurol.* **40**, 663, 1996.
13. Giasson, B. I., et al. Oxidative damage linked to neurodegeneration by selective alpha-synuclein nitration in synucleinopathy lesions. *Science* **290**, 985, 2000.
14. Thomas, B., and Beal, M. F. Parkinson's disease. *Hum. Mol. Genet.* **16** (Spec No. 2), T183, 2007.
15. Valente, E. M., et al. Hereditary early-onset Parkinson's disease caused by mutations in PINK1. *Science* **304**, 1158, 2004.
16. Bonifati, V., et al. Mutations in the DJ-1 gene associated with autosomal recessive early-onset parkinsonism. *Science* **299**, 256, 2003.
17. Andres-Mateos, E., et al. DJ-1 gene deletion reveals that DJ-1 is an atypical peroxiredoxin-like peroxidase. *Proc. Nat. Acad. Sci.* USA **104**, 19807, 2007.
18. Taira, T., et al. DJ-1 has a role in antioxidative stress to prevent cell death. *EMBO Rep.* **5**, 213, 2004.
19. Goldberg, M. S., et al. Nigrostriatal dopaminergic deficits and hypokinesia caused by inactivation of the familial parkinsonism-linked gene DJ-1. *Neuron* **45**, 489, 2005.
20. Chen, L., et al. Age-dependent motor deficits and dopaminergic dysfunction in DJ-1 null mice. *J. Biol. Chem.* **280**, 21418, 2005.
21. Borrelli, E. Without DJ-1, the D2 receptor doesn't play. *Neuron* **45**, 479, 2005.
22. Park, J., et al. *Drosophila* DJ-1 mutants show oxidative stress-sensitive locomotive dysfunction. *Gene* **361**, 133, 2005.
23. Menzies, F. M., Yenisetti, S. C., and Min, K. T. Roles of *Drosophila* DJ-1 in survival of dopaminergic neurons and oxidative stress. *Curr. Biol.* **15**, 1578, 2005.
24. Meulener, M., et al. *Drosophila* DJ-1 mutants are selectively sensitive to environmental toxins associated with Parkinson's disease. *Curr. Biol.* **15**, 1572, 2005.
25. Kim, R. H., et al. Hypersensitivity of DJ-1-deficient mice to 1-methyl-4-phenyl-1,2,3,6-tetrahydropyridine (MPTP) and oxidative stress. *Proc. Natl. Acad. Sci.* USA **102**, 5215, 2005.
26. Inden, M., et al. PARK7 DJ-1 protects against degeneration of nigral dopaminergic neurons in Parkinson's disease rat model. *Neurobiol. Dis.* **24**, 144, 2006.

27. Paterna, J. C., et al. DJ-1 and parkin modulate dopamine-dependent behavior and inhibit MPTP-induced nigral dopamine neuron loss in mice. *Mol. Ther.* **15**, 698, 2007.

28. Petit, A., et al. Wild-type PINK1 prevents basal and induced neuronal apoptosis, a protective effect abrogated by Parkinson disease-related mutations. *J. Biol. Chem.* **280**, 34025, 2005.

29. Exner, N., et al. Loss-of-function of human PINK1 results in mitochondrial pathology and can be rescued by parkin. *J. Neurosci.* **27**, 12413, 2007.

30. Hoepken, H. H., et al. Mitochondrial dysfunction, peroxidation damage and changes in glutathione metabolism in PARK6. *Neurobiol. Dis.* **25**, 401, 2007.

31. Deng, H., et al. Small interfering RNA targeting the PINK1 induces apoptosis in dopaminergic cells SH-SY5Y. *Biochem. Biophys. Res. Commun.* **337**, 1133, 2005.

32. Tang, B., et al. Association of PINK1 and DJ-1 confers digenic inheritance of early-onset Parkinson's disease. *Hum. Mol. Genet.* **15**, 1816, 2006.

33. Gautier, C. A., Kitada, T., and Shen, J. Loss of PINK1 causes mitochondrial functional defects and increased sensitivity to oxidative stress. *Proc. Nat. Acad. Sci. USA* **105**, 11364, 2008.

34. Spillantini, M. G., et al. Alpha-synuclein in Lewy bodies. *Nature* **388**, 839, 1997.

35. Polymeropoulos, M. H., et al. Mutation in the alpha-synuclein gene identified in families with Parkinson's disease. *Science*, **276**, 2045, 1997.

36. Kruger, R., et al. Ala30Pro mutation in the gene encoding alpha-synuclein in Parkinson's disease. *Nat. Genet.* **18**, 106, 1998.

37. Zarranz, J. J., et al. The new mutation, E46K, of alpha-synuclein causes Parkinson and Lewy body dementia. *Ann. Neurol.* **55**, 164, 2004.

38. Iwai, A., et al. The precursor protein of non-A[beta] component of Alzheimer's disease amyloid is a presynaptic protein of the central nervous system. *Neuron* **14**, 467, 1995.

39. Chandra, S., et al. Alpha-synuclein cooperates with CSPalpha in preventing neurodegeneration. *Cell* **123**, 383, 2005.

40. Maroteaux, L., Campanelli, J. T., and Scheller, R. H. Synuclein: A neuron-specific protein localized to the nucleus and presynaptic nerve terminal. *J. Neurosci.* **8**, 2804, 1988.

41. Devi, L., et al. Mitochondrial import and accumulation of alpha-synuclein impair complex I in human dopaminergic neuronal cultures and Parkinson disease brain. *J. Biol. Chem.* **283**, 9089, 2008.

42. Martin, L. J., et al. Parkinson's disease alpha-synuclein transgenic mice develop neuronal mitochondrial degeneration and cell death. *J. Neurosci.* **26**, 41, 2006.

43. Hsu, L. J., et al. Alpha-synuclein promotes mitochondrial deficit and oxidative stress. *Am. J. Pathol.* **157**, 401, 2000.

44. Goldberg, M. S., and Lansbury Jr., P. T. Is there a cause-and-effect relationship between [alpha]-synuclein fibrillization and Parkinson's disease? *Nat. Cell Biol.*, **2**, E115, 2000.

45. Giasson, B. I., et al. A hydrophobic stretch of 12 amino acid residues in the middle of alpha-synuclein is essential for filament assembly. *J. Biol. Chem.* **276**, 2380, 2001.

46. Norris, E. H., Giasson, B. I., and Lee, V. M. Alpha-synuclein: Normal function and role in neurodegenerative diseases. *Curr. Top. Dev. Biol.* **60**, 17, 2004.

47. Paik, S. R., Shin, H. J., and Lee, J. H. Metal-catalyzed oxidation of alpha-synuclein in the presence of copper(II) and hydrogen peroxide. *Arch. Biochem. Biophys.* **378**, 269, 2000.

48. Paxinou, E., et al. Induction of {alpha}-synuclein aggregation by intracellular nitrative insult. *J. Neurosci.* **21**, 8053, 2001.

49. Souza, J. M., et al. Dityrosine cross-linking promotes formation of stable alpha-synuclein polymers. Implication of nitrative and oxidative stress in the pathogenesis of neurodegenerative synucleinopathies. *J. Biol. Chem.* **275**, 18344, 2000.

50. Norris, E. H., et al. Effects of oxidative and nitrative challenges on alpha-synuclein fibrillogenesis involve distinct mechanisms of protein modifications. *J. Biol. Chem.* **278**, 27230, 2003.

51. Krishnan, S., et al. Oxidative dimer formation is the critical rate-limiting step for Parkinson's disease alpha-synuclein fibrillogenesis. *Biochemistry* **42**, 829, 2003.

52. Norris, E. H., et al. Pesticide exposure exacerbates alpha-synucleinopathy in an A53T transgenic mouse model. *Am. J. Pathol.* **170**, 658, 2007.

53. Song, D. D., et al. Enhanced substantia nigra mitochondrial pathology in human alpha-synuclein transgenic mice after treatment with MPTP. *Exp. Neurol.* **186**, 158, 2004.

54. Nieto, M., et al. Increased sensitivity to MPTP in human [alpha]-synuclein A30P transgenic mice. *Neurobiol. Aging* **27**, 848, 2006.

55. Pesah, Y., et al. *Drosophila* parkin mutants have decreased mass and cell size and increased sensitivity to oxygen radical stress. *Development* **131**, 2183, 2004.

56. Periquet, M., et al. Proteomic analysis of parkin knockout mice: Alterations in energy metabolism, protein handling and synaptic function. *J. Neurochem.* **95**, 1259, 2005.
57. Greene, J. C., et al. Mitochondrial pathology and apoptotic muscle degeneration in *Drosophila* parkin mutants. *Proc. Natl. Acad. Sci. USA* **100**, 4078, 2003.
58. Greene, J. C., et al. Genetic and genomic studies of *Drosophila* parkin mutants implicate oxidative stress and innate immune responses in pathogenesis. *Hum. Mol. Genet.* **14**, 799, 2005.
59. Goldberg, M. S., et al. Parkin-deficient mice exhibit nigrostriatal deficits but not loss of dopaminergic neurons. *J. Biol. Chem.* **278**, 43628, 2003.
60. Palacino, J. J., et al. Mitochondrial dysfunction and oxidative damage in parkin-deficient mice. *J. Biol. Chem.* **279**, 18614, 2004.
61. Przedborski, S., and Ischiropoulos, H. Reactive oxygen and nitrogen species: Weapons of neuronal destruction in models of Parkinson's disease. *Antioxid. Redox Signal.* **7**, 685, 2005.
62. Seo, B. B., et al. In vivo complementation of complex I by the yeast Ndi1 enzyme: Possible application for treatment of Parkinson disease. *J. Biol. Chem.* **281**, 14250, 2006.
63. Wu, D. C., et al. Blockade of microglial activation is neuroprotective in the 1-methyl-4-phenyl-1,2,3,6-tetrahydropyridine mouse model of Parkinson disease. *J. Neurosci.* **22**, 1763, 2002.
64. Wu, D.-C., et al. NADPH oxidase mediates oxidative stress in the 1-methyl-4-phenyl-1,2,3,6-tetrahydropyridine model of Parkinson's disease. *Proc. Nat. Acad. Sci. USA* **100**, 6145, 2003.
65. Przedborski, S., et al. Oxidative post-translational modifications of α-synuclein in the 1-methyl-4-phenyl-1,2,3,6-tetrahydropyridine (MPTP) mouse model of Parkinson's disease. *J. Neurochem.* **76**, 637, 2001.
66. Liberatore, G. T., et al. Inducible nitric oxide synthase stimulates dopaminergic neurodegeneration in the MPTP model of Parkinson disease. *Nat. Med.* **5**, 1403, 1999.
67. Matthews, R. T., et al. MPP+ induced substantia nigra degeneration is attenuated in nNOS knockout mice. *Neurobiol. Dis.* **4**, 114, 1997.
68. Schulz, J. B., et al. Inhibition of neuronal nitric oxide synthase by 7-nitroindazole protects against MPTP-induced neurotoxicity in mice. *J. Neurochem.* **64**, 936, 1995.
69. Przedborski, S., et al. Role of neuronal nitric oxide in 1-methyl-4-phenyl-1,2,3,6-tetrahydropyridine (MPTP)-induced dopaminergic neurotoxicity. *Proc. Natl. Acad. Sci. USA* **93**, 4565, 1996.
70. Hantraye, P., et al. Inhibition of neuronal nitric oxide synthase prevents MPTP-induced parkinsonism in baboons. *Nat. Med.* **2**, 1017, 1996.
71. Shimoji, M., et al. Absence of inclusion body formation in the MPTP mouse model of Parkinson's disease. *Mol. Brain Res.* **134**, 103, 2005.
72. Ara, J., et al. Inactivation of tyrosine hydroxylase by nitration following exposure to peroxynitrite and 1-methyl-4-phenyl-1,2,3,6-tetrahydropyridine (MPTP). *Proc. Nat. Acad. Sci. USA* **95**, 7659, 1998.
73. Forno, L. S., et al. An electron microscopic study of MPTP-induced inclusion bodies in an old monkey. *Brain Res.* **448**, 150, 1988.
74. Kowall, N. W., et al. MPTP induces alpha-synuclein aggregation in the substantia nigra of baboons. *NeuroReport* 11, **211**, 2000.
75. Schluter, O. M., et al. Role of [alpha]-synuclein in 1-methyl-4-phenyl-1,2,3,6-tetrahydropyridine-induced parkinsonism in mice. *Neuroscience* **118**, 985, 2003.
76. Dauer, W., et al. Resistance of α-synuclein null mice to the parkinsonian neurotoxin MPTP. *Proc. Natl. Acad. Sci. USA* **99**, 14524, 2002.
77. Klivenyi, P., et al. Mice lacking alpha-synuclein are resistant to mitochondrial toxins. *Neurobiol. Dis.* **21**, 541, 2006.
78. Drolet, R. E., et al. Mice lacking [alpha]-synuclein have an attenuated loss of striatal dopamine following prolonged chronic MPTP administration. *NeuroToxicology* **25**, 761, 2004.
79. Yu, W., et al. Increased dopaminergic neuron sensitivity to 1-methyl-4-phenyl-1,2,3,6-tetrahydropyridine (MPTP) in transgenic mice expressing mutant A53T α-synuclein. *Neurochem. Res.* **33**, 902, 2008.
80. McCormack, A. L., et al. Pathologic modifications of alpha-synuclein in 1-methyl-4-phenyl-1,2,3,6-tetrahydropyridine (MPTP)-treated squirrel monkeys. *J. Neuropathol. Exp. Neurol.* **67**, 793, 2008.
81. Hodara, R., et al. Functional consequences of alpha-synuclein tyrosine nitration: Diminished binding to lipid vesicles and increased fibril formation. *J. Biol. Chem.* **279**, 47746, 2004.
82. Cohen, G., and Heikkila, R. E. The generation of hydrogen peroxide, superoxide radical, and hydroxyl radical by 6-hydroxydopamine, dialuric acid, and related cytotoxic agents. *J. Biol. Chem.* **249**, 2447, 1974.
83. Dunnett, S. B., and Bjorklund, A. Prospects for new restorative and neuroprotective treatments in Parkinson's disease (cover story). *Nature* **399**, A32, 1999.

84. Jenner, P., and Olanow, C. W. Understanding cell death in Parkinson's disease. *Ann. Neurol.* **44**, 572, 1998.

85. Brooks, A. I., et al. Paraquat elicited neurobehavioral syndrome caused by dopaminergic neuron loss. *Brain Res.* **823**, 1, 1999.

86. McCormack, A. L., et al. Environmental risk factors and Parkinson's disease: Selective degeneration of nigral dopaminergic neurons caused by the herbicide paraquat. *Neurobiol. Dis.* **10**, 119, 2002.

87. Patel, M., et al. Requirement for superoxide in excitotoxic cell death. *Neuron* **16**, 345, 1996.

88. Day, B. J., et al. A metalloporphyrin superoxide dismutase mimetic protects against paraquat-induced endothelial cell injury, *in vitro. J. Pharmacol. Exp. Ther.* **275**, 1227, 1995.

89. Betarbet, R., et al. Chronic systemic pesticide exposure reproduces features of Parkinson's disease. *Nat. Neurosci.* **3**, 1301, 2000.

90. Sherer, T. B., et al. Mechanism of toxicity in rotenone models of Parkinson's disease. *J. Neurosci.* **23**, 10756, 2003.

91. Betarbet, R., et al. Intersecting pathways to neurodegeneration in Parkinson's disease: Effects of the pesticide rotenone on DJ-1, [alpha]-synuclein, and the ubiquitin-proteasome system. *Neurobiol. Dis.* **22**, 2006.

92. Beal, M. F. Bioenergetic approaches for neuroprotection in Parkinson's disease. *Ann. Neurol.* **53**, 539, 2003.

93. Lass, A. and Sohal, R. S. Electron transport-linked ubiquinone-dependent recycling of [alpha]-tocopherol inhibits autooxidation of mitochondrial membranes. *Arch. Biochem. Biophys.* **352**, 229, 1998.

94. Dhanasekaran, M., et al. Effect of dopaminergic neurotoxin MPTP/MPP$^+$ on coenzyme Q content. *Life Sciences* **83**, 92, 2008.

95. Schulz, B. J., et al. Coenzyme Q_{10} and nicotinamide and a free radical spin trap protect against MPTP neurotoxicity. *Exp. Neurol.* **132**, 279, 1995.

96. Beal, M. F., et al. Coenzyme Q_{10} attenuates the 1-methyl-4-phenyl-1,2,3,6-tetrahydropyridine (MPTP) induced loss of striatal dopamine and dopaminergic axons in aged mice. *Brain Res.* **783**, 109, 1998.

97. Cleren, C., et al. Therapeutic effects of coenzyme Q_{10} and reduced CoQ_{10} in the MPTP model of parkinsonism. *J. Neurochem.* **104**, 1613, 2008.

98. Matthews, R. T., et al. Coenzyme Q_{10} administration increases brain mitochondrial concentrations and exerts neuroprotective effects. *Proc. Natl. Acad. Sci. USA* **95**, 8892, 1998.

99. Moon, Y., et al. Mitochondrial membrane depolarization and the selective death of dopaminergic neurons by rotenone: Protective effect of coenzyme Q_{10}. *J. Neurochem.* **93**, 1199, 2005.

100. Beal, M. F., et al. Coenzyme Q_{10} and nicotinamide block striatal lesions produced by the mitochondrial toxin malonate. *Ann. Neurol.* **36**, 882, 1994.

101. Gotz, M., et al. Effect of lipoic acid on redox state of coenzyme Q in mice treated with 1-methyl-4-phenyl-1,2,3,6-tetrahydropyridine and diethyldithiocarbamate. *Eur. J. Pharmacol.* **266**, 291, 1994.

102. The Parkinson Study Group. Effects of tocopherol and deprenyl on the progression of disability in early Parkinson's disease. *N. Engl. J. Med.* **328**, 176, 1993.

103. The Parkinson Study Group. A controlled, randomized, delayed-start study of rasagiline in early Parkinson disease. *Arch. Neurol.* **61**, 561, 2004.

104. The Parkinson Study Group. A controlled trial of rasagiline in early Parkinson disease: The TEMPO study. *Arch. Neurol.* **59**, 1937, 2002.

105. The Parkinson Study Group. A randomized placebo-controlled trial of rasagiline in levodopa-treated patients with Parkinson disease and motor fluctuations: The PRESTO study. *Arch. Neurol.* **62**, 241, 2005.

106. Shults, C. W., et al. Effects of coenzyme Q_{10} in early Parkinson disease: Evidence of slowing of the functional decline. *Arch. Neurol.* **59**, 1541, 2002.

107. Storch, A., et al. Randomized, double-blind, placebo-controlled trial on symptomatic effects of coenzyme Q(10) in Parkinson disease. *Arch. Neurol.* **64**, 938, 2007.

108. The NINDS NET-PD Investigators. A randomized, double-blind, futility clinical trial of creatine and minocycline in early Parkinson disease. *Neurology* **66**, 664, 2006.

109. The NINDS NET-PD Investigators. A pilot clinical trial of creatine and minocycline in early Parkinson disease: 18-month results. *Clin. Neuropharmacol.* **31**, 141, 2008.

110. Bender, A., et al. Creatine supplementation in Parkinson disease: A placebo-controlled randomized pilot trial. *Neurology* **67**, 1262, 2006.

3 Phytoestrogens and Brain Health

Liqin Zhao and Roberta Diaz Brinton

Department of Pharmacology and Pharmaceutical
Sciences, School of Pharmacy, University of Southern
California, Los Angeles, California , USA

CONTENTS

3.1 PHYTOESTROGENS AND HUMAN HEALTH

Phytoestrogens are a diverse group of naturally occurring polyphenolic compounds that structurally resemble mammalian estrogens although they are distributed in plants.[1] Due to their structural similarities, phytoestrogens can bind to mammalian estrogen receptors (ERs); some of them have a binding preference to ERβ, but, overall, they have weaker binding affinities when compared with the female endogenous estrogen 17β-estradiol. In the mammalian system, through their interactions with ERs, phytoestrogens can moderately interfere with the endogenous estrogen-responsive signaling and result in either estrogenic or antiestrogenic bioactivities, depending on the status of the endogenous estrogens and the distribution of two ER subtypes, ERα and ERβ.[2,3]

Three major subclasses of phytoestrogens have been identified and chemically defined as isoflavones, lignans, and coumestans (see Table 3.1).[1] Isoflavones are predominantly enriched in red clover and legumes, such as soybeans; lignans are largely distributed in oilseeds, such as flaxseeds; and coumestans are widely distributed in plant sprouts, such as red clover sprouts and alfalfa sprouts. Mature red clover leaf is the known richest source of total isoflavones, mainly biochanin A and formononetin, along with lesser amounts of genistein and daidzein. Biochanin A and formononetin are *O*-methylated precursor molecules of genistein and daidzein, respectively, and are enzymatically converted to genistein and daidzein once ingested in mammalians.[4] Soy is the richest source of genistein, rich in daidzein, and the known exclusive source of glycitein. Besides red clover and soy, kudzu root is another major botanical source for isoflavones, serving as the richest source of daidzein among all plants.[5] Flaxseed provides the richest dietary source of lignans, with a large amount of secoisolariciresinol and a lesser amount of matairesinol.[6] Once ingested, these plant lignans are enzymatically converted to mammalian lignans—secoisolariciresinol to enterodiol, and matairesinol to enterolactone. Coumestrol is a major coumestan, with a structure quite different from that of isoflavones while highly estrogenic due to the presence of two phenol groups at both ends, which

TABLE 3.1
Three main classes of naturally occurring phytoestrogens

Classification	Main source	Main constituent (s)	Concentration (mg/100 g dry weight)[+]	Structure
Isoflavones	Soybeans	Genistein	26.8–84.1	
		Daidzein	10.5–56.0	*Intestinal flora* Equol[++]
		Glycitein	6.72–20.40	
	Red clover	Biochanin A	417	
		Formononetin	647	
Lignans	Flaxseeds	Matairesinol	1.087	
		Secoisolariciresinol	369.9	
Coumestans	Red clover sprouts	Coumestrol	28.1	

[+]Data from [4–7].
[++]Equol is not present in nature but can be produced exclusively from daidzein metabolism in 30–60% of human adults.

allows it to bind strongly to ERs as does genistein. Only a small trace amount of coumestrol exists in human diets.[6]

Among all known phytoestrogens, soy-based isoflavones, due to their enrichment in human diets, especially in Asian diets, are the most extensively studied. Substantial research had attributed the estrogenic activity of soy products mainly to genistein and daidzein, until glycitein, the third main constitutive component, was characterized.[7] Glycitein accounts for 5–10% of the total isoflavones in soy foods, and the rate may be up to 40% in some soy supplements made from soy germ.[7] These soy-derived isoflavones commonly exist as inactive but water-soluble glucosides (genistin, daidzin, and glycitin) that, inside the mammalian digestive system, are converted to estrogenically active genistein, daidzein, and glycitein, respectively, by intestinal glucosidases prior to absorption. Daidzein can be further metabolized to equol and O-desmethylangiolesin.[8] Because the bioavailability of estrogenically active soy isoflavones is largely dependent upon the metabolic conversion in the gastrointestinal tract, which can be significantly different between individuals, the estrogenic responses derived from the intake of even an equal amount of the same soy product could be varied in different individuals. One typical case known to date is the production of equol, a structural mimic of daidzein. Unlike genistein and daidzein, equol is not of plant origin, yet it can be exclusively produced through the metabolism of daidzein, which is catalyzed by intestinal microbial flora following the intake of pure daidzein or soy products.[8] Interestingly, wide variations in the ability to produce equol from daidzein metabolism exist between animals and humans and across human populations. It has been found that almost all rodents and monkeys can produce equol in large quantities; however, only about 20–35% of human adults have such equol-producing ability.[9] Since it was first identified in human urine,[10] equol has been widely recognized for its highly potent estrogenic activity in some ER-distributed tissues, largely through strong binding to both ERs (with an approximately 10-fold binding preference to ERβ), that are roughly equal to genistein. In contrast, its precursor isoflavone, daidzein, only exhibits weak ER binding affinity and estrogenic activity.[9] Therefore, it is conceivable that the variation in response to soy products could be significant in two distinct subpopulations, "equol producers" and "non-equol producers." It can be speculated that the clinical efficacy of a soy product in humans could be significantly enhanced by the presence of equol in equol producers as compared with non-equol producers."

Historically, research interests in the health effects of soy isoflavones were spurred in large part by a number of early comparative observational studies—for example, studies conducted in Asian and Western populations, where large differences in the daily dietary intake of total isoflavones and other types of phytoestrogens exist. The daily consumption of total phytoestrogens, mainly soy isoflavones, in Asian populations, such as the Japanese and Chinese, is estimated to range from 20 to 80 mg, whereas the daily dietary intake of phytoestrogens in Caucasians in the United States has been estimated to be <1 mg, and among whom lignans account for 80% of the total phytoestrogen content, followed by isoflavones (20%) and a trace amount of coumestans (<0.1%).[11] Although it has been argued that high consumption of soy foods is only one of the many potentially protective lifestyle factors that distinguish Asian and Western women, it has been associated, to a large extent, with the low incidence rates of a number of hormone-dependent conditions in Asian populations. For example, it was found that only 25% and 18% of Japanese and Chinese postmenopausal women, respectively, suffer from hot flashes as compared with 85% and 70% of North American and European women, respectively.[12] In addition, historically, breast cancer rates in the United States have been 4–7 times higher than those in Asia.[13,14] An interesting study conducted in Asian Americans revealed a strong link between a low risk for breast cancer and the intake of tofu in these women.[15] Specifically, a lower breast cancer rate was observed in those Asian American women who were born in Asia and then immigrated to the United States yet still consumed substantial soy foods, as compared with those women born in the United States and had a more Americanized dietary structure, with lower intake of soy isoflavones. Moreover, the same study revealed that the reduction of risk was only seen in perimenopausal women, not postmenopausal women.[15] As an example of how demography and culture might change the dietary structure, in a study in 274 Japanese American women aged 65 years or older living in King County, Washington, the mean dietary intake of soy isoflavones (10.2 mg/day) was about a quarter to a half that of women living in Japan but higher than that consumed by Caucasian women.[16]

The high content of soy isoflavones in Asian diets may have also contributed to the low prevalence of AD in Japan and China compared with that in the United States and Europe. Prevalence rates for all-cause dementias and subtypes AD and vascular dementia (VaD) were reported in 22 studies across continents—13 studies conducted in Asia (9 in Japan[17–23] and 4 in China[24–27]), 4 studies in North America (3 in the United States[28–30] and 1 in Canada[31]), and 5 studies in Europe

(2 in Great Britain,[32,33] 1 in France,[34] 1 in Italy,[35] and 1 in Sweden[36]). In spite of a few exceptional cases, it is apparent that the overall prevalence rates for all-cause dementias in Japan and China are similar to the rates in North America, although they are lower than the rates in Europe. However, the prevalence ratios of AD to VaD in Japan and China are lower than the ratios in both North America and Europe, due to lower prevalence rates for AD and similar rates for VaD in Japan and China compared with North America and Europe. These data suggest that AD is more prevalent in Western than Asian populations (~2.5-fold), while the prevalence rates of VaD are similar. Moreover, AD is more prevalent than VaD in Western populations (~2.2-fold). In comparison, the prevalence of AD and that of VaD are similar in Asian populations, although the AD/VaD ratio has been increasing due to greater longevity and a higher percentage of the elderly, which lead to increased risk for AD.[37] In addition to the age factor, the AD prevalence rates in women have been consistently found to be higher than those in men across both Asian and Western nations.[22,27–29]

Additional evidence for a potential role of dietary intake of soy isoflavones in preventing the onset of AD in Asian populations can be found in two large-scale studies conducted in Asian Americans who had adapted toward Western style diets.[16] The Honolulu-Asia Aging Study in a cohort of 3734 Japanese American men (71–93 years old) living in Hawaii found the overall prevalence rates for all-cause dementias (9.3%) and AD (5.4%)[38] to be similar to the rates reported in Europe (9.1% and 5.1%, respectively),[32–36] plus a prevalence rate for VaD (4.2%)[38] higher than that in Europe (2.5%).[32–36] Similarly, the Kame Project in a cohort of 3045 Japanese Americans (aged 65 years or older) living in King County, Washington, found the overall prevalence rates for all-cause dementias (6.8%), AD (3.9%), and VaD (1.6%)[39] to be similar to the rates reported in North America (5.5%, 3.1%, and 1.3%, respectively).[28–31] In agreement with these studies conducted in a single-cohort population, a retrospective analysis on data collected at the AD Research Centers of California compared the frequency of dementia etiologies between 1992 and 2002 in four ethnic groups.[40] Cases for all the neurodegenerative causes of dementia were identified for 452 Asian and Pacific Islander patients, 472 Black patients, 675 Latino patients, and 2926 White patients. No statistical difference was found among study groups, but there was a trend of less AD in Asian and Pacific Islanders.[40] Taken together, these comparisons suggest that the prevalence of AD in older Japanese immigrants is similar to that among Caucasians in the United States and Europe and higher than that among the matching population living in Japan. This trend correlates well with the daily intake of soy isoflavones among these three groups of populations. The dietary change associated with migration from Japan to the United States may have contributed to the increased risk for AD. However, this dietary factor appears to have less or no effect on the pathogenesis of VaD. Despite population-based epidemiological research that suggests a positive correlation between soy isoflavone intake and a reduction of AD prevalence, an adverse relationship was reported as part of the Honolulu-Asia Aging Study, where higher midlife tofu consumption was associated with cognitive impairment and brain atrophy in late life.[41]

In the last 5 years, research interests in soy isoflavones and other phytoestrogens have greatly expanded, largely due to the revelations of health risks associated with estrogen-containing hormone therapy (HT), particularly those seen in the Women's Health Initiative (WHI) and WHI Memory Study (WHIMS) trials. In view of the promising results from *in vitro* culture and animal studies, it has been proposed that phytoestrogens could play a positive role in those areas that estrogen has demonstrated its effectiveness, particularly in postmenopausal women, such as alleviation of the climacteric symptoms,[12] preservation of bone mineral density,[42,43] reduction of the risks of cardiovascular diseases,[44] and improvement of cognitive function (reviewed below). However, at this time, the mixed results across studies remain problematic.

In addition to the impact of soy isoflavones on a woman's health, recent research revealed a reduction of the risk for prostate cancer associated with soy products in men. A case-controlled study to compare the percentage of equol producers between prostate cancer patients and cancer-free control subjects residing in Japan, Korea, and the United States revealed that the incidence of prostate cancer was significantly higher in non-equol producers, suggesting that the ability to produce equol or equol itself may be closely associated with reduced risk for prostate cancer.[45] The underlying mechanism in support of clinical findings indicated that equol can effectively block the action of the potent male hormone dihydrotestosterone (DHT), which normally stimulates prostate growth.[46] Unlike many conventional DHT blockers that inhibit a certain enzyme that converts testosterone to DHT, equol does not prevent DHT from being made but does prevent it from functioning. It directly binds to DHT, which prevents DHT from binding to the androgen receptor and thereby preventing the prostate from growing.[46] Since DHT is also relevant to male pattern baldness, acne, and excess body hair, it has been proposed that equol may offer some help in treating these androgen-mediated conditions.

In addition to the research and discussions around the health-promoting effects of phytoestrogens, another important area of debate is the relationship between soy intake and breast cancer, especially in women at high risk.[47] Concern has arisen from a number of *in vitro* and animal studies revealing that treatment with soy isoflavones increased the growth of breast cancer cells.[48,49] However, there is no clinical information in support of such a link. In fact, as reviewed earlier, epidemiological studies suggested that high dietary intake of soy isoflavones may be associated with reduced risk for breast cancer in Japanese and Chinese women. It is argued that the differences between nonhumans and human (e.g., metabolism) and differences in experimental design and methods would induce different outcomes. Thus, it is inadequate to extrapolate the conclusions from *in vitro* and animal data to humans.[50] At this time, before further clinical data are available for a definite conclusion, it is recommended that "the impact of isoflavones on breast tissue needs to be evaluated at the cellular level in women at high risk for breast cancer."[51]

3.2 HUMAN STUDIES OF PHYTOESTROGENS ON BRAIN HEALTH

The effects of phytoestrogens on the nervous system and cognitive function have received much less attention than their effects on the peripheral systems, although an increasing body of evidence suggests that, as mammalian estrogen does, phytoestrogens may play a role in the brain.[52,53] Both ER subtypes, ERα and ERβ, are enriched throughout the brains of both rodents and humans, providing the structural basis required for estrogen actions in the brain.[54–56] It has been demonstrated that activation of either ERα or ERβ is able to activate the downstream signaling cascades leading to estrogen promotion of neuroprotection against various neurodegenerative insults.[57,58] ERβ appears to play a more central role in the regulation of brain development and mediation of estrogen promotion of neuronal plasticity and memory function.[59] Results of both *in vitro* cellular and *in vivo* animal studies indicate that phytoestrogens at a physiologically relevant concentration are effective at promoting neuroprotection, neurogenesis, and memory;[60–64] however, the neuroendocrine actions of high-dose phytoestrogens are largely antiestrogenic[52,65] or cytotoxic.[66,67] Clinical research on the impact of phytoestrogens on the brain function has been sparse. Few available observational studies showed that short-term consumption of soybeans had a small positive effect on the cognitive function in women aged between 25 and 40 years.[68,69] Eight randomized and placebo-controlled intervention studies of soy isoflavones on cognition, which were reported in 2001–2007, are identified (Table 3.2) and reviewed below in an attempt to identify a trend for the correlation in particularly postmenopausal women.

Among the eight human intervention studies, five revealed a positive effect of soy isoflavone administration on select cognitive activities. The first study, conducted in London, United Kingdom, examined the effects of supervised intake of high- versus low-soy diets on attention, memory, and frontal lobe function in young healthy adults (average age of 25 years) of both sexes.[70] Results from this 10-week intervention study showed that intake of the high soy diet (100 mg of total isoflavones per day) significantly improved both verbal and nonverbal short-term memory, long-term memory, and mental flexibilities in both sexes. In comparison, a significantly improved performance in two other tests of frontal lobe function, letter fluency, and planning associated with the high-soy diet were observed only in females. There was no effect of diet on tests of attention and in a category-generation task related to semantic memory.[70] Despite the small sample size and other limitations,

TABLE 3.2
Human Intervention Studies of Soy Isoflavones on Cognition (2000–2007)

Design	Location	Population	Intervention	Duration	Outcomes	References
RP	London, UK	27 healthy young adults (15 males and 12 females); mean age = 25 years	High-soy diet (100 mg of total isoflavones/day) versus low-soy diet (0.5 mg of total isoflavones/day); two diets were matched for protein, carbohydrates, and fat composition	10 weeks	In both males and females: ↑ DMTS-CANTAB (short-term nonverbal memory, $P < .05$) ↑ Logical Memory and Recall-Weschler Revised (short-term verbal memory, $P < .05$) ↑ Picture Recall (long-term episodic memory, $P < .05$) → Category Fluency (verbal fluency and semantic memory) → PASAT (sustained attention) ↑ IDED-CANTAB (mental flexibility, $P < .05$) ↑ Mood In females only: ↑ Letter Fluency (frontal function, $P < .01$) ↑ SoC-CANTAB (planning ability, $P < .05$)	File et al. [70]
RDBP, Pa	London, UK	50 healthy PMW; 51–66 years (mean age = 58.0 years); >1 year of menopause (mean = 9.1 years, 66% > 5 years); no HT in the previous 12 months	Soy supplement (60 mg of total isoflavones/day)	6 weeks	↑ DMTS-CANTAB (short-term nonverbal memory, $P < .03$) → Logical Memory and Recall-Weschler Revised (short- and long-term verbal memory) → Picture Recall (long-term episodic memory) → Category Fluency (verbal fluency and semantic memory) → PASAT (sustained attention) ↑ IDED-CANTAB (mental flexibility; simple rule reversals, $P < .05$; compound rule reversals, $P < .03$) ↑ SoC-CANTAB (planning ability, $P < .05$)	File et al. [71]

RDBP, Pa	London, UK	33 healthy PMW; 50–65 years (mean age = 57.8 years); >1 year of menopause (mean = 8.1 years); no HT in the past 12 months	Soy supplement (60 mg of total isoflavones/day)	12 weeks	→ DMTS-CANTAB (short-term nonverbal memory) → Logical Memory and Recall-Weschler Revised (short-term verbal memory) → Picture Recall (long-term episodic memory, $P < .03$) → Category Fluency (verbal fluency and semantic memory) ↑ PASAT (sustained attention, $P < .03$) ↑ IDED-CANTAB (mental flexibility; simple rule reversals, $P < .05$; compound rule reversals, $P < .003$) ↑ SoC-CANTAB (planning ability, $P < .05$)	Duffy et al. [72]
RDBP	San Diego, CA, USA	53 healthy PMW; 55–74 years (mean age = 60.7 years); > 2 years of menopause (mean = 10.9 years); no HT currently (43% past users for 1–6 years, 0–13 years since last use)	Soy supplement (110 mg of total isoflavones/day)	6 months	→ Logical Memory and Recall-Weschler (short and long-term verbal memory) ↑ Category Fluency (verbal fluency and semantic memory, $P = .03$) → Trails A-Halstead-Reitan Trails A (visuomotor tracking and attention) ↑ Trails B-Halstead-Reitan Trails B (visuomotor tracking and attention, $P = .08$ in all women, $P = .007$ in women at ages between 50 and 59 years old)	Kritz-Silverstein et al. [73]
RDBP, CO	Milan and Rome, Italy	76 healthy PMW; mean age = 49.5 years; >1 year of menopause (mean = 5.7 years); no HT use in the past 8 weeks	Soy supplement (60 mg of total isoflavones/day; 40–45% genistein; 40–45% daidzein; 10–20% glycitein)	6 months	↑ Six of 8 measures in Digit Symbol Test (psychomotor performance; pairs recalled correctly, $P < .05$), Digit Span Test (immediate auditory attention and mental flexibility; backward recall of digits, $P = .05$), and Visual Scanning Test (distractibility and visual inattentiveness) ↑ Eight of 9 self-rating mood scales, 5 significant, $p < .05–.001$; 8 of all 8 visual analogue scales, 7 significant, $P < .05–0.001$ ↑ preference to for isoflavone treatment (49 forty-nine participants) versus 9 with preference to for placebo versus and 18 with no preference	Casini et al. [74]

(continued)

TABLE 3.2 (CONTINUED)
Human Intervention Studies of Soy Isoflavones on Cognition (2000–2007)

Design	Location	Population	Intervention	Duration	Outcomes	References
RDBP	Utrecht, The Netherlands	175 healthy PMW; 60–75 years (mean age = 66.6 years); >1 year of menopause (mean = 18.0 years); no HT in the previous 6 months	Soy protein supplement (99 mg of total isoflavones/day; 52 mg of genistein, 41 mg of daidzein, and 6 mg of glycitein)	12 months	→ Sixteen measures in tests of memory: Rey Auditory Verbal Learning Test, Digit Span Test, Doors Test; complex attention: Trail-making Making Test, Digit Symbol Substitution Test; verbal skills: Verbal Fluency Test, Boston Naming Task; global cognition and dementia: MMSE	Kreijkamp-spers et al. [75]
RDBP, Pa	Hong Kong, China	176 (168 complete) healthy PMW; 55–76 years (mean age = 63.5 years); > 5 years of menopause (mean = 13.8 years); no HT in the past 6 months	Soy supplement (80 mg of total isoflavones/day)	6 months	→ Fifteen measures in tests of learning and memory: Hong Kong List Learning Test, Rey-Osterrieth Complex Figure Test, and Wechsler Memory Scale-Revised; executive function: Trail-making Making Test, Verbal Fluency Test; attention and concentration: Digit Span Test; motor control: Finger Tapping Test; language: Boston Naming Test; Visual perception: Rey-Osterrieth Copy Trial; global cognition and dementia: MMSE	Ho et al. [76]
RDBP	Pullman, WA, USA	79 healthy PMW; 48–65 years (mean age = 56.1 years); 1–35 years of menopause (mean = 7.7 years); no HT in the past 6 months	Soy milk (72 mg of total isoflavones/day; 37 mg of genistein, 31 mg of daidzein, and 4 mg of glycitein) or soy supplement (70 mg of total isoflavones/day; 33 mg of genistein, 30 mg of daidzein, and 7 mg of glycitein)	16 weeks	→ Sixteen measures in tests of memory: Digit Ordering Test, Color Matching Test, Benton Visual Retention Test, and Visual Pattern Recognition; selective attention: Stroop Test	Fournier et al. [77]

→ indicates no change in performance; ↑ improvement in performance.
CANTAB, Cambridge Neuropsychological Test Automated Battery; CO, crossover design; DMTS, delayed matching to sample; HT, hormone therapy; IDED, attentional set shifting; MMSE, Mini-Mental State Examination; Pa, parallel design; PASAT, Paced Auditory Serial Addition Test; PMW, postmenopausal women; RDBP, randomized, double-blind, placebo-controlled; RP, randomized, placebo-controlled; SoC, Stockings of Cambridge.

this study clearly indicates that soy isoflavones have the potential to benefit both women and men on cognition.

Following the first study, the same investigators conducted two other studies using the same cognitive test battery but in postmenopausal women aged between 50 and 66 years.[71,72] These two studies revealed a similar beneficial effect of soy isoflavones on some of the measures of cognitive function in older women. In summary, women receiving 60 mg of total isoflavones per day for 6 or 12 weeks had a better overall performance on tests of episodic memory, although there were variations with regard to the specific tasks. Furthermore, in comparison with a less robust effect in memory tests, it was consistently found in both studies that soy isoflavone administration significantly improved performance on both tests of frontal lobe function, mental flexibility, and planning ability.[71,72] In addition, the 12-week study found an additional benefit of soy isoflavones on sustained attention,[72] which was however not observed in the 6-week study,[71] suggesting that longer duration of the treatment may offer greater health benefits.

Concurrently, the SOy and Postmenopausal Health In Aging Study, a separate study, was launched in the United States.[73] In comparison with the studies conducted by File et al.,[71,72] in this study, Kritz-Silverstein et al. examined a higher dose of soy isoflavone administration (110 mg of total isoflavones per day) for a longer duration (6 months) in an average older population of postmenopausal women (age range of 55 to 74 years).[73] Kritz-Silverstein et al. [73] reported that due to a learning effect over time, women in both the isoflavone treatment group and the placebo group showed an improvement in performance on all five cognitive tests with a focus on attention and verbal memory. However, women in the active treatment group tended to show greater improvement for four of the five tests than women in the placebo group, with a significant difference in category fluency (23% improvement in the treatment group versus 3% improvement in the placebo group), a test of verbal memory. Results were similar after adjustment for age and education. Furthermore, the investigators conducted a stratified analysis of both groups on change in test scores in younger (age range of 50 to 59 years) versus older (age range of 60 to 74 years) women after adjustment for education, and they found a significant improvement in Trails B, a test of visuomotor tracking and attention, in the active treatment group compared with the placebo group among the younger women. No such improvement was observed among the older women.[73] These results suggest that soy isoflavones may have a favorable effect on cognitive function, particularly verbal memory, in postmenopausal women. Furthermore, soy isoflavones may have a greater impact in perimenopausal to early menopausal women on select cognitive functions relative to the impact in older postmenopausal women.

In support of those early studies, a recent study conducted in Milan and Rome, Italy, revealed that daily administration of 60 mg of total isoflavones in postmenopausal women with a mean age of 49.5 years and a mean duration of menopause of 5.7 years for 6 months was associated with a small to moderate improvement in performance in three standard cognitive tests.[74] More impressively, women receiving the soy supplement treatment had a significantly better mood than women on placebo treatment, as reflected by the three self-rating mood scales and the eight visual analogue scales, where five of nine and seven of eight scores, respectively, were statistically significant between the two treatment groups. Furthermore, forty-eight participants reported a preference for isoflavone treatment, as compared with eight reporting a preference for placebo and eighteen reporting no preference.[74] These results are particularly significant in that they provide strong support for a possible role of soy isoflavones in alleviating or reversing the psychological disturbances often associated with menopause and improving the overall quality of life in postmenopausal women.

In contrast to the above-mentioned studies, three studies failed to demonstrate a favorable effect of soy isoflavone intake on cognitive function, including the study with the longest study duration[75] among all eight studies published to date. In this study, conducted in Utrecht, The Netherlands, Kreijkamp-Kaspers et al. investigated the effects of a 12-month intervention of a soy protein supplement containing 99 mg of total isoflavones per day (52 mg of genistein, 41 mg of daidzein, and 6 mg of glycitein) on cognitive function, along with measures of bone mineral density and plasma lipids, in postmenopausal women aged between 60 and 75 years.[75] These women were more advanced in age (mean age of 66.0 years) and had been postmenopausal for a mean of 18 years. In addition, approximately 21% and 19% of women were on active cholesterol-lowering and antihypertensive medications, respectively.[75] No significant difference in performance was found in all measures that are sensitive to cognitive aging, including memory, verbal skills, complex attention functions, and risk for dementia, between the soy protein supplement group and the placebo group.[75] Similarly, no significant change was found in measures of bone mineral density and lipid profiles between the two groups.[75]

In agreement with the findings of Kreijkamp-Kaspers et al., a similar study conducted in Hong Kong, China, found no significant effect of daily intake of a total of 80 mg of soy isoflavones in Chinese postmenopausal women aged between 55 and 76 years (mean age of 63.5 years) and with at least 5 years (mean of 13.8 years) of menopause on 13 standardized neuropsychological tests of memory, executive function, attention, motor control, language, visual perception, and a global function assessment, as compared with the placebo group.[76] This is the first and only study conducted in a population for which soy is a staple food throughout life. The mean usual dietary isoflavone intake in this study population was approximately 20 mg/day. The investigators postulated that the insignificant finding of a positive effect of soy supplement in a Chinese population of postmenopausal women may be related to their lifetime habitual intake of soy foods that could have had some overall protective effects on cognitive function, even among the placebo group, and thus could have lessened the between-group differences of the trial results.[76]

In another recent study conducted in the United States, Fournier and colleagues examined the respective effects of soy milk and a soy isoflavone supplement intervention over a 16-week period on several cognitive tasks that measured behavioral responses (accuracy and response time) in 79 healthy postmenopausal women.[77] In comparison with the study population included in the two other studies that failed to show a effect,[75,76] the women recruited in this study were relatively younger at a mean age of 56.1 years (age range of 48 to 65 years) but had a similar wide range of years since the onset of menopause (1–35 years, mean = 8 years). It was found neither soy milk containing 72 mg of total isoflavones per day nor soy isoflavone supplement containing 70 mg of total isoflavones per day yielded any appreciable improvement on performance in all tests of selective attention or working (verbal and visuospatial) memory, visuospatial short-term memory, and visuospatial long-term memory.[77]

In summary, eight intervention studies addressing the relationship between soy isoflavones and cognition in humans are identified. Of these, one study conducted in young adults of both sexes showed a beneficial effect of a high-soy diet on select cognitive activities. Among the seven studies conducted in postmenopausal women, four demonstrated a beneficial effect of soy supplementation primarily on verbal memory and frontal lobe function. Three studies failed to demonstrate such an effect, including a study conducted in Chinese postmenopausal women with a high exposure to soy foods throughout their lives.[76] To conclude, although a subset of studies report promising findings, a definite conclusion cannot be drawn yet due to the inconsistencies across studies. Thus, it is important to address the potential factors that may have contributed to the mixed outcomes and take these factors into the design of future studies.

3.3   STANDARDIZATION OF HUMAN STUDIES ON PHYTOESTROGENS

As exemplified in Table 3.2, the data remain sparse and inconclusive with respect to the effects of phytoestrogens on human health. The overall results derived from basic science research in cell cultures and animals are encouraging; however, either there is a lack of clinical data in support of preclinical findings or the results derived from clinical studies have been inconsistent. Multiple factors could have contributed to the disparities between the *in vitro/in vivo* preclinical and clinical data, or data from various clinical studies. One main impact factor that could have contributed to the discrepant clinical outcomes is the lack of standardization across studies, including but not limited to the composition of the study interventions, the profile of the characteristics of the study populations, and the outcomes measured. Elaborated below is the potential impact from the variations in these key elements on clinical results. It is anticipated that these analyses would emphasize the importance of standardization of these elements in order to reach a clinical consensus in search for a definite answer for phytoestrogen use in relation to human health.

3.3.1 Phytoestrogen Composition

Significant research has related the health effects of soy-based foods and products to two most extensively studied isoflavones, genistein and daidzein. An increasing body of evidence indicates that besides genistein, daidzein, and the recently discovered glycitein, there are a number of additional structurally similar and hormonally active phytoestrogenic molecules present in soy foods and other products, and the types and amounts of these molecules can be significantly different from source to source. First, the complex biosynthetic pathway leading to the production of various phytoestrogenic molecules in soy plants could be readily altered in response to external stimuli, such as environmental factors and growth conditions, adding more complexities in the phytoestrogen profile. For instance, soy plants grown under stressed conditions were found to involve increased biosynthesis of isoflavonoid phytoalexin compounds, glyceollins, which can represent up to 56% of the total isoflavone composition.[78] Increased accumulation of glyceollins can be also present in soy foods prepared from stress-treated soy, such as freezing and fermentation.[78] Second, the contents and present forms (aglycons versus glucosides, solid versus liquid) of the three main constitutive phytoestrogens, genistein, daidzein, and glycitein, naturally present in different soy foods, could be significantly different, leading to marked variations in bioavailability. As revealed in a comparative analysis by Cassidy et al., consumption of tempeh (~50% as aglycons) resulted in higher serum peak levels of both genistein and daidzein compared with textured vegetable protein (predominantly isoflavone glucosides; <15% as aglycons). However, soy milk (<15% as aglycons) was absorbed faster and peak levels of isoflavones were attained earlier than with either tempeh or textured vegetable protein.[79] Additional variability in phytoestrogen composition in different soy supplements could be introduced from the preparation protocols. An extensive analysis of thirty-three commercially available phytoestrogen supplements or extracts revealed that there were not only considerable differences in actual isoflavone content from what was claimed by the manufacturers but also an abundance of peaks of unknown origin and chemical structure found in many of the supplements.[80] Although the bioactivity of these unknown molecules remains undefined, their potential impact on the overall estrogenic activity and effects of a soy product on human health must not be neglected. Based on the facts and clinical data summarized in Table 3.2, one could reasonably conceive that the discrepant outcomes across studies could, in large part, originate from the heterogeneity in the types and relative amounts of both known and unknown but intrinsically hormonally active molecules present in the soy products studied.

A complex formulation comprising a mixture of various phytoestrogenic molecules could induce either a synergistic or an antagonistic effect depending on the composition and interactions among these molecules. One possible antagonism could be derived from the antagonistic interactions among components that have preference to bind to and activate either ERα or ERβ. Research in primary neuron cultures revealed that coadministration of an ERα-selective agonist and an ERβ-selective agonist was less effective than treatment with either single agonist in various neuroprotective measurements, suggesting that although both ERα and ERβ contribute to estrogen promotion of neuronal survival, simultaneous activation of both ERα and ERβ in the same context may diminish the efficacy.[57] Thus, it can be extrapolated that such an antagonism may occur in a complex formulation composed of various known and unknown constituents, such as a soy-derived supplements or extracts, which may offset the overall effect and cause clinical insignificance.

Another possible antagonism present in a complex formulation could be induced by molecules that are predominantly antiestrogenic. For instance, glyceollins that are rich in soy plants grown under stressed conditions and processed, such as fermented soy foods, exerted potent antiestrogenic effects in a number of hormone-responsive systems, whereas genistein and daidzein were found to be estrogenic at similar concentrations.[78] In addition, most phytoestrogens possess a biphasic estrogenic and antiestrogenic activity depending on the dose, target tissue, and/or estrogen status in a mammalian system. For instance, studies have shown that some phytoestrogens tend to exert an estrogenic effect at relatively low doses and develop antiestrogenic[81] or cytotoxic[66,67] effects at higher doses. This dose and tissue-related biphasic effect has significant implications for the design of a clinical study with a specific clinical end point. For example, in a study addressing the relationship between phytoestrogen use and cancer prevention, an overall antiestrogenic activity would be advantageous, whereas in a study that investigates the protective effects of phytoestrogens on the cardiovascular system, bone mass, and cognition, an estrogenic activity would be desired. Moreover, the dose of phytoestrogens studied could significantly impact efficacy and safety. On one hand, a low dose could be insufficient to induce a clinically significant effect. As revealed in the observational Study of Women's Health Across the Nation that included a cohort of midlife Chinese and Japanese women aged between 42 and 52 years, low dietary genistein intake (6.8 and 3.5 mg/day, respectively) could be a factor for the lack of association between genistein intake and measures of cognitive performance in either ethnic group.[82] On the other hand, a high dose during long-term treatment could raise the red flag for safety reasons. This can be illustrated by results from two clinical studies that assessed the effects of phytoestrogens on histological characteristics of the endometrium, a target tissue with a high risk for neoplasia associated with the use of steroidal estrogen-containing HT. One study revealed that a 5-year exposure to 150 mg of soy isoflavones significantly increased the occurrence of endometrial hyperplasia, although the absolute number was relatively small, 6 cases out of 154 participants on isoflavone treatment (3.8%) versus 0% in placebo group.[83] However, at a low dose that was sufficient to alleviate menopausal hot flashes to a large degree (56.4% reduction in the mean number of hot flashes)[84] and to reduce cardiovascular risk markers,[85] no adverse effect in the endometrium was observed in women exposed to a daily intake of 54 mg of pure genistein after 2 years.[85]

In summary, phytoestrogen composition may serve as a major determinant of the overall effects of a soy-based product on human health. Optimal composition and dose should be determined based on the target clinical outcomes investigated. Standardization of such a composition would eliminate the variation related to the phytoestrogens studied and make the data from various studies comparative.

3.3.2 STUDY POPULATION

The characteristics of the study population are another potential factor that may affect study outcomes. Recent reanalyses of the findings from the WHI and WHIMS trials evidently indicate that estrogen-containing HT may benefit younger women who start HT treatment within 10 years of the onset of menopause, while it appears to have an insignificant or even detrimental effect in older women who start the treatment 10 or more years after the onset of menopause. Due to the structural and functional resemblances between mammalian estrogens and phytoestrogens, it can be extrapolated that, as in the WHI and WHIMS trials, age and hormonal history of women may regulate the health effects of phytoestrogens as well. In support of this hypothesis, Tempfer et al. reviewed that in selected studies involving women with early natural postmenopause and mild to moderate vasomotor symptoms, a strong link between phytoestrogen treatment and reduced hot flashes was observed.[86] Health benefits associated with early interventions with soy-based products are to some extent in concordance with the habitual consumption of a substantial amount of soy foods in Asian populations throughout their lives.

The impact of the characteristics of the study population could also be derived from the interindividual differences in equol production in humans. As reviewed earlier, equol binds to both ERα and ERβ, although with a preference for ERβ, at affinities comparable with genistein. In comparison, daidzein, the precursor isoflavone of equol, mainly binds to ERβ at a 10- to 100-fold lower affinity than equol.[87] Furthermore, it was found that equol was most potent in inducing the transcriptional expression of β-galactosidase in yeast cells, especially in those transfected with ERα, among all test isoflavones.[87] In comparison, daidzein only induced a very weak transcription.[87] The substantial differences in the binding and estrogenic activity between daidzein and equol could dramatically influence the overall responses to soy consumption in humans. Therefore, failure to factor these differences into human studies could be another major cause of the disparity in outcomes across studies. The more potent estrogenic activity associated with equol implicates that individuals with equol-producing ability could be more sensitive to phytoestrogen treatment.[9] In support of such a theory, a recent 1-year randomized trial revealed that intake of 75 mg of isoflavone conjugates per day in Japanese postmenopausal women offered a more favorable effect in preserving the bone mineral density in the equol producers, with an annualized change of −0.04%, compared with the nonproducers, with a change of −0.46%. Overall, women in the isoflavone group exhibited a smaller change than the control group (−2.28% and −2.6%, respectively).[88] A similar result was observed in an independent study conducted in an American postmenopausal study.[89] These studies confirm that the presence of equol in the circulating system can enhance the overall efficacy of isoflavones, at least on the bone system.

In addition to the intrinsic metabolic environment, equol production can be affected by other external factors. One factor is the form in which daidzein exists in a soy product. In soy food or a soy product that is made from soy plants grown under a normal condition, daidzein, along with other isoflavones, is present mainly in the inactive sugar-conjugated glycoside form. The glycosides of genistein and daidzein are referred to as genistin and daidzin, respectively. However, in stress-processed soy food or a soy product made from soy germ, daidzein, along with other isoflavones, exists mainly in the unconjugated hormonally active form, aglycon. A study by Setchell et al. revealed that equol appeared in the plasma of half of the women who ingested daidzin, the glucoside conjugate of daidzein, while it was not found in the plasma of the women who consumed daidzein.[80] The other known factor is the dietary matrix, which may impact the composition of the intestinal microflora and thus the amount of the equol produced in equol producers. Two human studies showed that high equol production was associated with a high-fiber diet.[90,91] In addition, since rodents have a much greater capacity to produce equol, it may explain the difference between results derived from animal models that largely showed significant benefits of phytoestrogens, for example, on brain function, and data from human studies that demonstrated a heterogeneous picture.

On the other hand, due to the potent estrogenic activity of equol through activation of ERα, the predominant ER subtype mediating estrogen functions in female reproductive tissues, the enhanced responsiveness to soy isoflavones in equol producers could lead to increased risk for estrogenic stimulation in the breast and endometrium upon long-term use. In a study reported by Unfer et al., 6 (3.8%) of 154 participants receiving soy tablets containing a total of 150 mg of isoflavones daily for 5 years were diagnosed with endometrial hyperplasia, which was statistically significant compared with the placebo group (0%).[83] It would be insightful to investigate whether the increased occurrence of endometrial hyperplasia is linked to the equol-producing status of individuals.

In summary, heterogeneity in the characteristic profile of a study population could obscure the positive response from a favorable subpopulation and bring the overall effects of a soy product down below a clinically detectable range. A stratified investigation justified by defined characteristics, including age, hormone status, equol-producing capacity, and even habitual diet, would likely provide insights into a "window of opportunity" for the use of phytoestrogens to maximally reap their health benefits.

3.4   SUMMARY

Phytoestrogens structurally resemble mammalian estrogens and thus may provide similar functional outcomes. Clinical data on the impact of phytoestrogens on human health are promising while not convincing. Identification of factors that may have caused discrepant outcomes across studies will be crucial for future design of human research. Standardization of the study interventions and populations may be essential in reaching a clinical consensus. Phytoestrogens bind to ERs at modest affinities, which may account for their weak to moderate estrogenic activity and low benefit to human health. Such small effects could be significantly impacted by such external factors as the composition of the study interventions and the heterogeneity of the characteristics of the human subjects. Thus, a high bar for the standardization of the key elements will be required in future human studies in search for a definite answer.

ERβ has been suggested as a novel therapeutic target for the development of a novel estrogen alternative therapy.[59] Selective targeting of ERβ has three major advantages: (1) to reduce antagonistic interactions that may occur in a complex preparation; (2) to minimize adverse effects associated with the activation of ERα in reproductive tissues; and (3) to take advantage of additional beneficial health effects exclusively mediated by ERβ.[59,92] One can envision that (1) a formulation composed of rationally selected ERβ-selective phytoestrogens would offer a greater effect than a complex formulation mixed with both ERα- and ERβ-selective components, (2) a composition composed of phytoestrogens with a rationally defined content that can induce a synergistic rather than an antagonistic effect would likely generate enhanced and comparative results across studies that seek an estrogenic effect, (3) a composition with inclusion of equol would impair the potential influence of the interindividual differences in the capacity to produce equol on the study outcomes and thus elevate the estrogenic response across the study subjects, and (4) inclusion of equol would be beneficial in men to help prevent male hormone-dependent disorders, such as prostate cancer.

In conclusion, phytoestrogens have potential therapeutic promise particularly for brain health in both women and men. Criticism of such promise may lead to loss of the opportunity to take advantage of the potential health benefits of these products, which have been widely consumed in Asian populations for centuries. Standardization of future human studies may be the only path that leads to a clinical consensus.

ACKNOWLEDGMENTS

This work was supported by grants from the Alzheimer Association (L. Zhao) and the Kenneth T. and Eileen L. Norris Foundation and Bensussen Fund (R.D. Brinton).

REFERENCES

1. Kurzer MS, Xu X. Dietary phytoestrogens. *Annu Rev Nutr* 1997;17:353–381.
2. Dixon RA. Phytoestrogens. *Annu Rev Plant Biol* 2004;55:225–261.
3. Zhao L, Brinton RD. Structure-based virtual screening for plant-based ERbeta-selective ligands as potential preventative therapy against age-related neurodegenerative diseases. *J Med Chem* 2005;48:3463–3466.
4. De Rijke E, Zafra-Gomez A, Ariese F, Brinkman UA, Gooije C. Determination of isoflavone glucoside malonates in *Trifolium pratense* L. (red clover) extracts: Quantification and stability studies. *J Chromatogr A* 2001;932:55–64.
5. Mazur W, Adlercreutz H. Naturally occurring oestrogens in food. *Pure Appl Chem* 1998;70:1759–1776.
6. Common phytoestrogens and their sources. http://wwwherbalchemnet/Introductoryhml.
7. Song TT, Hendrich S, Murphy PA. Estrogenic activity of glycitein, a soy isoflavone. *J Agric Food Chem* 1999;47:1607–1610.
8. Setchell KD, Borriello SP, Hulme P, Kirk DN, Axelson M. Nonsteroidal estrogens of dietary origin: Possible roles in hormone-dependent disease. *Am J Clin Nutr* 1984;40:569–578.
9. Setchell KD, Brown NM, Lydeking-Olsen E. The clinical importance of the metabolite equol—A clue to the effectiveness of soy and its isoflavones. *J Nutr* 2002;132:3577–3584.
10. Axelson M, Kirk DN, Farrant RD, Cooley G, Lawson AM, Setchell KD. The identification of the weak oestrogen equol [7-hydroxy-3-(4'-hydroxyphenyl)chroman] in human urine. *Biochem J* 1982;201: 353–357.
11. de Kleijn MJ, van der Schouw YT, Wilson PW, et al. Intake of dietary phytoestrogens is low in postmenopausal women in the United States: The Framingham study (1–4). *J Nutr* 2001;131:1826–3182.
12. Maskarinec S. The effect of phytoestrogens on hot flashes. *Nutr. Bytes* 2003;9:5.
13. Ziegler RG. Phytoestrogens and breast cancer. *Am J Clin Nutr* 2004;79:183–184.
14. Henderson BE, Bernstein L. The international variation in breast cancer rates: An epidemiological assessment. *Breast Cancer Res Treat* 1991;18:S11–S17.

15. Wu AH, Ziegler RG, Horn-Ross PL, et al. Tofu and risk of breast cancer in Asian-Americans. *Cancer Epidemiol Biomarkers Prev* 1996;5:901–906.

16. Rice MM, LaCroix AZ, Lampe JW, et al. Dietary soy isoflavone intake in older Japanese American women. *Public Health Nutr* 2001;4:943–952.

17. Kawano H, Ueda K, Fujishima M. Prevalence of dementia in a Japanese community (Hisayama): Morphological reappraisal of the type of dementia. *Jpn J Med* 1990;29:261–265.

18. Hasegawa K. The clinical issues of age-related dementia. *Tohoku J Exp Med* 1990;161:29–38.

19. Shibayama H, Kasahara Y, Kobayashi H, et al. Prevalence of dementia in a Japanese elderly population. *Acta Psychiatr Scand* 1986;74:144–151.

20. Ichinowatari N, Tatsunuma T, Makiya H. Epidemiological study of old age mental disorders in the two rural areas of Japan. *Jpn J Psychiatry Neurol* 1987;41:629–636.

21. Ueda K, Kawano H, Hasuo Y, Fujishima M. Prevalence and etiology of dementia in a Japanese community. *Stroke* 1992;23:798–803.

22. Kiyohara Y, Yoshitake T, Kato I, et al. Changing patterns in the prevalence of dementia in a Japanese community: The Hisayama study. *Gerontology* 1994;40:29–35.

23. Fukunishi I, Hayabara T, Hosokawa K. Epidemiological surveys of senile dementia in Japan. *Int J Soc Psychiatry* 1991;37:51–56.

24. Liu HC, Chou P, Lin KN, et al. Assessing cognitive abilities and dementia in a predominantly illiterate population of older individuals in Kinmen. *Psychol Med* 1994;24:763–770.

25. Li G, Shen YC, Chen CH, Zhao YW, Li SR, Lu M. An epidemiological survey of age-related dementia in an urban area of Beijing. *Acta Psychiatr Scand* 1989;79:557–563.

26. Zhang MY, Katzman R, Salmon D, et al. The prevalence of dementia and Alzheimer's disease in Shanghai, China: Impact of age, gender, and education. *Ann Neurol* 1990;27:428–437.

27. Wang W, Wu S, Cheng X, et al. Prevalence of Alzheimer's disease and other dementing disorders in an urban community of Beijing, China. *Neuroepidemiology* 2000;19:194–200.

28. Bachman DL, Wolf PA, Linn R, et al. Prevalence of dementia and probable senile dementia of the Alzheimer type in the Framingham Study. *Neurology* 1992;42:115–119.

29. Folstein MF, Bassett SS, Anthony JC, Romanoski AJ, Nestadt GR. Dementia: Case ascertainment in a community survey. *J Gerontol* 1991;46:M132–M138.

30. Evans DA, Funkenstein HH, Albert MS, et al. Prevalence of Alzheimer's disease in a community population of older persons. Higher than previously reported. *JAMA* 1989;262:2551–2556.

31. Group CSoHaAW. Canadian Study of Health and Aging: Study methods and prevalence of dementia. *Can Med Assoc J* 1994;150:899–913.

32. Brayne C, Calloway P. An epidemiological study of dementia in a rural population of elderly women. *Br J Psychiatry* 1989;155:214–219.

33. O'Connor DW, Pollitt PA, Hyde JB, et al. The prevalence of dementia as measured by the Cambridge Mental Disorders of the Elderly Examination. *Acta Psychiatr Scand* 1989;79:190–198.

34. Dartigues JF, Gagnon M, Michel P, et al. The Paquid research program on the epidemiology of dementia. Methods and initial results. *Rev Neurol* 1991;147:225–230.

35. Rocca WA, Bonaiuto S, Lippi A, et al. Prevalence of clinically diagnosed Alzheimer's disease and other dementing disorders: A door-to-door survey in Appignano, Macerata Province, Italy. *Neurology* 1990;40:626–631.

36. Fratiglioni L, Grut M, Forsell Y, et al. Prevalence of Alzheimer's disease and other dementias in an elderly urban population: Relationship with age, sex, and education. *Neurology* 1991;41:1886–1892.

37. Shigeta M. Epidemiology: Rapid increase in Alzheimer's disease prevalence in Japan. *Psychogeriatrics* 2004;4:117–119.

38. White L, Petrovitch H, Ross GW, et al. Prevalence of dementia in older Japanese-American men in Hawaii: The Honolulu-Asia Aging Study. *JAMA* 1996;276:955–960.

39. Graves AB, Larson EB, Edland SD, et al. Prevalence of dementia and its subtypes in the Japanese American population of King County, Washington State. The Kame Project. *Am J Epidemiol* 1996;144:760–771.

40. Hou CE, Yaffe K, Perez-Stable EJ, Miller BL. Frequency of dementia etiologies in four ethnic groups. *Dement Geriatr Cogn Disord* 2006;22:42–47.

41. White L, Petrovitch H, Ross G, et al. Brain aging and midlife tofu consumption. *J Am Coll Nutr* 2000;19:242–255.

42. Reinwald S, Weaver CM. Soy isoflavones and bone health: A double-edged sword? *J Nat Prod* 2006;69:450–459.

43. Weaver CM, Cheong JM. Soy isoflavones and bone health: The relationship is still unclear. *J Nutr* 2005;135:1243–1247.

44. Geller SE, Studee L. Soy and red clover for mid-life and aging. *Climacteric* 2006;9:245–263.

45. Akaza H, Miyanaga N, Takashima N, et al. Comparisons of percent equol producers between prostate cancer patients and controls: Case-controlled studies of isoflavones in Japanese, Korean and American residents. *Jpn J Clin Oncol* 2004;34:86–89.

46. Lund TD, Munson DJ, Haldy ME, Setchell KD, Lephart ED, Handa RJ. Equol is a novel anti-androgen that inhibits prostate growth and hormone feedback. *Biol Reprod* 2004;70:1188–1195.

47. Rice S, Whitehead SA. Phytoestrogens and breast cancer—Promoters or protectors? *Endocr Relat Cancer* 2006;13:995–1015.

48. Ju YH, Allred KF, Allred CD, Helferich WG. Genistein stimulates growth of human breast cancer cells in a novel, postmenopausal animal model, with low plasma estradiol concentrations. *Carcinogenesis* 2006;27:1292–1299.

49. Ju YH, Fultz J, Allred KF, Doerge DR, Helferich WG. Effects of dietary daidzein and its metabolite, equol, at physiological concentrations on the growth of estrogen-dependent human breast cancer (MCF-7) tumors implanted in ovariectomized athymic mice. *Carcinogenesis* 2006;27:856–863.

50. Setchell KD. Assessing risks and benefits of genistein and soy. *Environ Health Perspect* 2006;114:A332–A333.

51. Messina M, McCaskill-Stevens W, Lampe JW. Addressing the soy and breast cancer relationship: Review, commentary, and workshop proceedings. *J Natl Cancer Inst* 2006;98:1275–1284.

52. Patisaul HB. Phytoestrogen action in the adult and developing brain. *J Neuroendocrinol* 2005;17:57–64.

53. Lephart ED, Setchell KD, Lund TD. Phytoestrogens: Hormonal action and brain plasticity. *Brain Res Bull* 2005;65:193–198.

54. Shughrue PJ, Lane MV, Merchenthaler I. Comparative distribution of estrogen receptor-alpha and -beta mRNA in the rat central nervous system. *J Comp Neurol* 1997;388:507–525.

55. Taylor AH, Al-Azzawi F. Immunolocalisation of oestrogen receptor beta in human tissues. *J Mol Endocrinol* 2000;24:145–155.

56. Perez SE, Chen EY, Mufson EJ. Distribution of estrogen receptor alpha and beta immunoreactive profiles in the postnatal rat brain. *Brain Res Dev Brain Res* 2003;145:117–139.

57. Zhao L, Wu TW, Brinton RD. Estrogen receptor subtypes alpha and beta contribute to neuroprotection and increased Bcl-2 expression in primary hippocampal neurons. *Brain Res* 2004;1010:22–34.

58. Zhao L, Brinton RD. Estrogen receptor alpha and beta differentially regulate intracellular Ca^{2+} dynamics leading to ERK phosphorylation and estrogen neuroprotection in hippocampal neurons. *Brain Res* 2007;1172;48–59.

59. Zhao L, Brinton RD. Estrogen receptor beta as a therapeutic target for promotion of neurogenesis and prevention of neurodegeneration. *Drug Dev Res* 2006;66:103–117.

60. Perez-Martin M, Salazar V, Castillo C, et al. Estradiol and soy extract increase the production of new cells in the dentate gyrus of old rats. *Exp Gerontol* 2005;40:450–453.

61. Burguete MC, Torregrosa G, Perez-Asensio FJ, et al. Dietary phytoestrogens improve stroke outcome after transient focal cerebral ischemia in rats. *Eur J Neurosci* 2006;23:703–710.

62. Schreihofer DA, Do KD, Schreihofer AM. High-soy diet decreases infarct size after permanent middle cerebral artery occlusion in female rats. *Am J Physiol Regul Integr Comp Physiol* 2005;289:R103–R108.

63. Zhao L, Chen Q, Brinton RD. Neuroprotective and neurotrophic efficacy of phytoestrogens in cultured hippocampal neurons. *Exp Biol Med* 2002;227:509–519.

64. Lund TD, West TW, Tian LY, et al. Visual spatial memory is enhanced in female rats (but inhibited in males) by dietary soy phytoestrogens. *BMC Neurosci* 2001;2:20.

65. Patisaul HB, Dindo M, Whitten PL, Young LJ. Soy isoflavone supplements antagonize reproductive behavior and estrogen receptor alpha- and beta-dependent gene expression in the brain. *Endocrinology* 2001;142:2946–2952.

66. Linford NJ, Dorsa DM. 17beta-Estradiol and the phytoestrogen genistein attenuate neuronal apoptosis induced by the endoplasmic reticulum calcium-ATPase inhibitor thapsigargin. *Steroids* 2002;67:1029–1040.

67. Linford NJ, Yang Y, Cook DG, Dorsa DM. Neuronal apoptosis resulting from high doses of the isoflavone genistein: Role for calcium and p42/44 mitogen-activated protein kinase. *J Pharmacol Exp Ther* 2001;299:67–75.

68. Ostatnikova D, Celec P, Hodosy J, Hampl R, Zdenek P, Kudela M. Short-term soybean intake and its effect on steroid sex hormones and cognitive abilities. *Fertil Steril* 2007;88;1632–1636.

69. Celec P, Ostatnikova D, Caganova M, et al. Endocrine and cognitive effects of short-time soybean consumption in women. *Gynecol Obstet Invest* 2005;59:62–66.

70. File SE, Jarrett N, Fluck E, Duffy R, Casey K, Wiseman H. Eating soya improves human memory. *Psychopharmacology* 2001;157:430–436.

71. File SE, Hartley DE, Elsabagh S, Duffy R, Wiseman H. Cognitive improvement after 6 weeks of soy supplements in postmenopausal women is limited to frontal lobe function. *Menopause* 2005;12:193–201.

72. Duffy R, Wiseman H, File SE. Improved cognitive function in postmenopausal women after 12 weeks of consumption of a soya extract containing isoflavones. *Pharmacol Biochem Behav* 2003;75:721–729.

73. Kritz-Silverstein D, Von Muhlen D, Barrett-Connor E, Bressel MA. Isoflavones and cognitive function in older women: The SOy and Postmenopausal Health In Aging (SOPHIA) Study. *Menopause* 2003;10:196–202.

74. Casini ML, Marelli G, Papaleo E, Ferrari A, D'Ambrosio F, Unfer V. Psychological assessment of the effects of treatment with phytoestrogens on postmenopausal women: A randomized, double-blind, crossover, placebo-controlled study. *Fertil Steril* 2006;85:972–978.

75. Kreijkamp-Kaspers S, Kok L, Grobbee DE, et al. Effect of soy protein containing isoflavones on cognitive function, bone mineral density, and plasma lipids in postmenopausal women: A randomized controlled trial. *JAMA* 2004;292:65–74.

76. Ho SC, Chan AS, Ho YP, et al. Effects of soy isoflavone supplementation on cognitive function in Chinese postmenopausal women: A double-blind, randomized, controlled trial. *Menopause* 2007;14:489–499.

77. Fournier LR, Ryan Borchers TA, Robison LM, et al. The effects of soy milk and isoflavone supplements on cognitive performance in healthy, postmenopausal women. *J Nutr Health Aging* 2007;11:155–164.

78. Burow ME, Boue SM, Collins-Burow BM, et al. Phytochemical glyceollins, isolated from soy, mediate antihormonal effects through estrogen receptor alpha and beta. *J Clin Endocrinol Metab* 2001;86:1750–1758.

79. Cassidy A, Brown JE, Hawdon A, et al. Factors affecting the bioavailability of soy isoflavones in humans after ingestion of physiologically relevant levels from different soy foods. *J Nutr* 2006;136:45–51.

80. Setchell KD, Brown NM, Desai P, et al. Bioavailability of pure isoflavones in healthy humans and analysis of commercial soy isoflavone supplements. *J Nutr* 2001;131:1362S–1375S.

81. Wang TT, Sathyamoorthy N, Phang JM. Molecular effects of genistein on estrogen receptor mediated pathways. *Carcinogenesis* 1996;17:271–275.

82. Huang MH, Luetters C, Buckwalter GJ, et al. Dietary genistein intake and cognitive performance in a multiethnic cohort of midlife women. *Menopause* 2006;13:621–30.

83. Unfer V, Casini ML, Costabile L, Mignosa M, Gerli S, Di Renzo GC. Endometrial effects of long-term treatment with phytoestrogens: A randomized, double-blind, placebo-controlled study. *Fertil Steril* 2004;82:145–8.

84. D'Anna R, Cannata ML, Atteritano M, et al. Effects of the phytoestrogen genistein on hot flushes, endometrium, and vaginal epithelium in postmenopausal women: A 1-year randomized, double-blind, placebo-controlled study. *Menopause* 2007;14;648–655.

85. Atteritano M, Marini H, Minutoli L, et al. Effects of the phytoestrogen genistein on some predictors of cardiovascular risk in osteopenic, postmenopausal women: A 2-year randomized, double-blind, placebo-controlled study. *J Clin Endocrinol Metab* 2007; 92;3068–3075.

86. Tempfer CB, Bentz EK, Leodolter S, et al. Phytoestrogens in clinical practice: A review of the literature. *Fertil Steril* 2007;87;1243–1249.

87. Morito K, Hirose T, Kinjo J, et al. Interaction of phytoestrogens with estrogen receptors alpha and beta. *Biol Pharm Bull* 2001;24:351–356.

88. Wu J, Oka J, Ezaki J, et al. Possible role of equol status in the effects of isoflavone on bone and fat mass in postmenopausal Japanese women: A double-blind, randomized, controlled trial. *Menopause* 2007;14;866–874.

89. Frankenfeld CL, McTiernan A, Thomas WK, et al. Postmenopausal bone mineral density in relation to soy isoflavone-metabolizing phenotypes. *Maturitas* 2006;53:315–324.

90. Lampe JW, Karr SC, Hutchins AM, Slavin JL. Urinary equol excretion with a soy challenge: Influence of habitual diet. *Proc Soc Exp Biol Med* 1998;217:335–339.

91. Rowland IR, Wiseman H, Sanders TA, Adlercreutz H, Bowey EA. Interindividual variation in metabolism of soy isoflavones and lignans: Influence of habitual diet on equol production by the gut flora. *J Nutr* 2000;36:27–32.

92. McCarty MF. Isoflavones made simple—Genistein's agonist activity for the beta-type estrogen receptor mediates their health benefits. *Med Hypotheses* 2006;66:1093–1114.

4 Food Antioxidants and Alzheimer's Disease

Emma Ramiro-Puig[1], Margarida Castell[2], Andrew McShea[3], George Perry[4,5], Mark A. Smith[5], and Gemma Casadesus[6]

[1]Institut National de la Sante et de la Recherche Medical (Inserm), U793, Paris, France

[2]Department of Physiology, Faculty of Pharmacy, University of Barcelona, Barcelona, Spain

[3]Research and Development, Theo Chocolate, Seattle, Washington, USA

[4]College of Sciences, University of Texas at San Antonio, San Antonio, Texas, USA

[5]Departments of Pathology, Case Western Reserve University, Cleveland, Ohio, USA

[6]Department of Neurosciences, Case Western Reserve University, Cleveland, Ohio, USA

CONTENTS

4.1 ALZHEIMER'S DISEASE

4.1.1 HALLMARKS OF THE DISEASE

Alzheimer's disease (AD) is the most prevalent neurodegenerative disorder, affecting more than 15 million people worldwide, with almost 5 million patients located in the United States alone [1,2]. AD is the primary cause of senile dementia and is characterized by progressive memory loss; impairments in language and visual-spatial skills; episodes of psychosis, aggressiveness, and agitation; and, ultimately, death [1,3]. The severity and the chronicity of this disease lead to institutionalization of patients and thus result in a tremendous cost for families and for society.

The diagnosis of AD is based on medical and familial history and performance on various psychometric cognitive tests, with the Mini-Mental State Examination being the most frequently used. Definitive diagnosis can be only made at autopsy, where the pathological hallmarks are neuronal loss, amyloid-β (Aβ) peptides, and neurofibrillary tangles (NFTs; tau proteins). The exact mechanisms involved in the onset and progression of the disease are still unclear [4].

Proteolytic processing of the Aβ protein precursor (AβPP) to form peptides is implicated in the pathogenesis of AD [5,6]. AβPP can be processed through two pathways: (1) a non-amyloidogenic pathway that involves the cleavage of AβPP to soluble AβPP (sAβPP) by α-secretase enzymes and (2) an amyloidogenic pathway regulated by β- and γ-secretases, which promote the formation of Aβ peptides, constituents of senile plaques in AD brain [1]. Promotion of α-secretase processing leads to both a reduction in Aβ and an increase in sAβPPα with neuroprotective properties [7].

The accumulation of highly phosphorylated tau protein contributes to the formation of intraneuronal NFTs. The development of tau pathology progresses through many phosphorylation and conformational changes [8–10], which are thought to occur very early in the disease alongside increased oxidative stress and damage. Many neurons succumb to neurodegeneration and eventual death. Furthermore, much evidence reflects a strong correlation between NFT number and severity of clinical dementia.

4.1.2 RISK FACTORS

4.1.2.1 Oxidative Stress

Reactive oxygen species (ROSs) are generated in the body as a consequence of cellular metabolism and are involved in many physiological functions, such as vasodilatation and cellular proliferation. Since ROSs are potent oxidant compounds, they must be maintained within a certain range to avoid the oxidation of biomolecules, including DNA, lipids, and proteins, which leads to cell damage and death [11]. Although organisms are endowed with protective mechanisms to prevent oxidative damage, certain situations can shift the oxidant/antioxidant balance toward ROS production, known as oxidative stress. In brain aging, oxidative stress is increased, and this may contribute to the onset and progression of neurodegenerative disorders, such as AD [12–14]. Reactive carbonyls are increased in senile plaques [15,16], NFTs [16,17], and the main component of the latter, tau protein [14,17,18].

4.1.2.2 Sex

The prevalence rate in women is twice that in men, and this skewed sex ratio is specific for AD, and not other dementias. Many lines of evidence suggest that metabolic and hormonal differences among men and women play an important role in the pathogenesis of AD. For instance, estrogen deficiency following menopause may contribute to the etiology of AD in women [19,20] since hormone replacement therapy following menopause [21] decreases the incidence [22] and delays the onset [23] of AD. Other hormones of the hypothalamic-pituitary-gonadal axis may also play a role in AD [24,25].

4.1.2.3 Genetic Polymorphisms

For a small percentage of AD cases, the individual's genetic profile plays a more direct role. Mutations in AβPP, apolipoprotein E, and presenilin, for instance, have all been documented and shown to reflect not only higher prevalence rates of disease but often earlier onset and increased severity as well [26–28]. Many animal models based on these genetic polymorphisms have provided tools with which to study the altered pathways and mechanisms, yet the underlying cause for the majority of cases, or sporadic AD, remains elusive [29].

Presently, only modest palliative treatments are available. The most often recommended is the use of cholinesterase inhibitors, which strive to correct cholinergic deficiency in the central nervous system of AD patients [30–32]. Other therapies commonly used despite the lack of strong scientific proof are various hormone replacement therapies and the use of anti-inflammatory drugs and natural remedies, such as *Ginkgo biloba*. Moreover, based on *in vitro* and in vivo studies in animals, natural antioxidants, including polyphenols and certain vitamin supplements, and secondary antioxidants, such as acetyl-l-carnitine, have been proposed as alternative therapeutic agents for AD. An increasing number of studies suggest the efficacy of primary antioxidants to reduce or block neuronal death occurring in the pathophysiology of this disorder [33].

4.2 FOOD COMPOUNDS AND ALZHEIMER'S DISEASE

Epidemiological evidence linking nutrition to the incidence and risk for AD is rapidly growing (Figure 4.1). Certain nutritional deficiencies observed in patients with AD may suggest supplementation in specific macro- and micronutrients combined to the traditional drugs. These nutrients include omega-3 fatty acids, several B vitamins, and antioxidants such as vitamin E, vitamin C, and carotenoids.

4.2.1 B VITAMINS

Various studies support the potential beneficial effects of B vitamin [vitamin B_1 (thiamine), vitamin B_2 (riboflavin), vitamin B_6 (primarily pyridoxine), vitamin B_{12} (cobalamin), and folate (folic acid, tetrahydrofolate)] supplementation on neurocognitive function and gene expression in AD. Thiamine and riboflavin can be found in a great variety of foods, including whole grain cereals, vegetables, milk products, and liver. Thiamine is critically involved in glucose metabolism, and it is also implicated in oxidative stress (acting as a radical scavenger), protein processing, peroxisomal function, and gene expression [34]. Thiamine-dependent enzyme activities, such as pyruvate dehydrogenase and α-ketoglutarate dehydrogenase, are diminished in AD [34], and the reductions in AD brain are well correlated with the extent of dementia. Although thiamine itself has not been shown to have dramatic benefits in AD patients, the available data are scarce. Further testing on developing more absorbable forms of thiamine or adding thiamine to tested treatments for the abnormality in glucose metabolism in AD may increase their efficacy.

While the data on other compounds, including folate and cobalamin, are often conflicting with regard to whether levels in AD are significantly changed or if these compounds correlate with disease onset and severity, the levels are often reported to be lower in AD. These findings suggest that nutrition monitoring to at least keep the levels to within normal limits may be of worth.

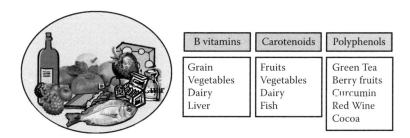

B vitamins	Carotenoids	Polyphenols
Grain	Fruits	Green Tea
Vegetables	Vegetables	Berry fruits
Dairy	Dairy	Curcumin
Liver	Fish	Red Wine
		Cocoa

FIGURE 4.1 **(See color insert following page 234.)** Many foods contain compounds that have been shown to have neuroprotective effects. Maintaining a healthy lifestyle is one way to preserve a healthy brain.

4.2.2 CAROTENOIDS

Levels of many naturally occurring antioxidants have been shown to be decreased in AD patients. These include the class of antioxidants of carotenoids, including β-carotene and lutein, which protect polyunsaturated fatty acids from oxidation. Decreased levels have been shown to correlate with the severity of disease [35]. Maintaining the levels of these compounds may at the least stabilize cognitive function.

4.2.3 POLYPHENOLS

Polyphenolic compounds protect plants against various biotic and abiotic stresses, and there are many pieces of evidence showing that their intake contributes to preventing and ameliorating certain diseases, including neurodegenerative disorders [36,37]. These compounds present potent antioxidant and anti-inflammatory properties (reviewed by Pietta [38]), and they are also capable of modulating cell signaling and enzyme activities (reviewed by Scalbert et al. [39] and Ramiro-Puig et al. [40]). Studies regarding brain health are mainly focused on polyphenolic compounds from green tea (catechins), berry fruits (anthocyanins), curcumin, red wine (resveratrol), and, recently, cocoa (procyanidins).

4.2.3.1 Green Tea Catechins

Epigallocatechin-3-gallate (EGCG) is the main polyphenol found in green tea, and its beneficial properties related to neuroprotection may be linked to its metal chelating, antioxidant, and cell-signaling modulatory properties [41]. These multiple functions make EGCG a novel preventive and therapeutic approach to AD. Importantly, EGCG reduces Aβ and neurite plaque formation through the promotion of the non-amyloidogenic AβPP pathway [42–44]. Thus, EGCG elevates α-secretase activity, specifically, tumor necrosis factor α-converting enzyme and ADAM10 (a desintegrin and metalloprotease 10), with the latter being critical for EGCG-mediated AβPP cleavage to sAPPα [44]. EGCG also activates protein kinase C [42], a well-known pathway that leads to sAPPα release [45].

Additionally, Weinreb et al. have recently shed light on the mechanisms of EGCG underlying its neuroprotective and neurorescue activities. Their transcriptomic study provides information about the effect of EGCG on the expression of several genes involved in neurite outgrowth, cell survival, and iron chelation [46].

Although all these studies demonstrate a beneficial effect of catechins on AD, clinical trials are required to prove their effectiveness and safety in humans.

4.2.3.2 Berry Fruits (Anthocyanins)

Anthocyanins and/or proanthocyanidins are the main polyphenolic compounds found in berry fruits (blueberries, blackberries, cranberries, strawberries, etc.). Joseph and colleagues have provided an extensive knowledge of the neuronal effects of these compounds. Thus, long-term feeding with diets containing strawberry or blueberry extracts (1–2% diet) improves age-related impairments in the cognitive and neuronal functions of rodents [47–49]. Anthocyanins can cross the brain barrier and reach areas associated with cognitive performance [49]. Moreover, blueberry extract supplementation reduces ROS levels in the striata [48] and enhances neurogenesis [50].

With respect to AD, an *in vitro* study showed a protective effect of five berry extracts (blueberry, black currant, boysenberry, strawberry, and cranberry) on the putative toxic effects of Aβ [51]. In AD transgenic mice (AβPP and presenilin-1 mutations), long-term blueberry extract supplementation (2%) prevents deficits on the Y-maze performance test, although it does not affect Aβ deposits [52]. As with green tea and other polyphenolic compounds, the action of blueberry extract goes beyond the radical scavenging capacity. Indeed, blueberry extract is capable of modulating cell-signaling pathways associated with cognitive function—for example, enhancement of the extra-cellular signal-regulated kinase activity and protein kinase Cγ activation [52].

4.2.3.3 Curcumin

Curcumin (diferuloylmethane) is an active polyphenolic compound present in the herb *Curcuma longa* (commonly known as turmeric or curry powder). Traditionally, it has been used as a food spice, a cosmetic, and a natural therapeutic drug. In Asian folk medicine, curcumin is a well-described treatment for various illnesses (e.g., respiratory disorders, abdominal pain, swelling, etc.). Many of these therapeutic effects of curcumin have been confirmed in the last decades by research studies, and they appear to be linked not only to its potent antioxidant and anti-inflammatory properties but also to its ability to bind proteins and modulate the activity of various kinases (reviewed by Goel et al. [53]).

Additionally, some recent studies have pointed out the potential benefits of this spice in neuro-degenerative disorders. In AD transgenic mice, curcumin and its metabolite tetrahydrocurcumin reduce amyloid plaque burden, insoluble Aβ, reactive carbonyls, and inflammatory markers [54–56]. To date, clinical studies on AD are still forthcoming; however, an epidemiological study carried out in elderly Asians suggests better cognitive performance in curry consumers [57]. Moreover, the extreme safety and tolerance of curcumin reported in humans (reviewed by Goel et al. [53]) make it an exciting candidate for AD therapy.

4.2.3.4 Resveratrol

Resveratrol (3,4′,5-trihydroxy-*trans*-stilbene) is a phytoalexin that is found in peanuts, red grape skins, and red wine. Research indicates that resveratrol may act as an antioxidant, promote nitric oxide production, inhibit platelet aggregation, and increase high-density lipoprotein cholesterol, thereby serving as a cardioprotective agent [58,59]. Importantly, resveratrol is an SIRT-1 activator that has been shown to stimulate mitochondrial biogenesis and deliver health benefits in rodents and increase longevity and protects against neurodegenerative and neurotoxic insults in animal models [60]. As such, resveratrol has been shown to be protective against kainate-induced seizures [61] and against brain injury due to ischemia/reperfusion in a gerbil model [62]. Similarly, resveratrol is protective in Parkinson's disease models, such as after 1-methyl-4-phenyl-1,2,3,6-tetrahydropyridine MPTP treatment [63,64], and has been found to protect neurons in Huntington's disease models, such as from 3-nitropropionic acid treatment [65]. Furthermore, with regard to AD, recent studies demonstrate that resveratrol reduced neurodegeneration in the hippocampus and prevented learning impairment in an inducible p25 transgenic mouse model of AD and tauopathies [66] and that consumption of Cabernet Sauvignon attenuated Aβ neuropathology in the Tg2576 mouse model of AD [67,68].

4.2.3.5 Cocoa

Cocoa's beneficial effects on brain health have recently become an object of interest given the potent antioxidant, anti-inflammatory, and iron-chelating properties of its polyphenolic fraction (mainly composed of epicatechin, catechin, and procyanidins) [69–73]. Thus, a serving size of certain cocoa-derived products provides more phenolic antioxidants than beverages and fruits, such as tea and blueberries, traditionally considered to be high in antioxidants [74,75].

Cocoa intake has been shown to improve cerebral blood flow in humans, and this may have a positive impact in aging and cerebrovascular diseases, such as stroke and dementia, in which endothelial function is impaired [76,77]. Regarding AD, it has been found that cocoa extract, epicatechin, and catechin reduce the toxic effects of Aβ *in vitro* through membrane and mitochondrial protective mechanisms [78]. Although no studies on AD animal models have been performed yet, oral administration of cocoa extract (100 mg/kg/day) has been recently shown to attenuate nigrostriatal dopaminergic cell loss in a murine model of Parkinson's disease induced by the infusion of the neurotoxin 6-hydroxidopamine [79].

4.3 SUMMARY

The variety of compounds present in the foods we eat and those that are measurably changed in the course of disease, specifically neurodegenerative disease, are staggering. The lack of normal levels of these compounds has been associated with worsening memory and poor judgment and even correlated with declines in strict, clinically administered cognitive testing. While a single compound has not been deemed the magic bullet for combating mental deterioration, the fact that so many are shown to be changed in afflicted patients provides the basis for continued study. Encouragingly, most of the compounds are readily available from a normal diet; thus, maintaining levels through the consumption of specific foods is a viable option.

ACKNOWLEDGMENTS

Work in the authors' laboratories is supported by the Fundació La Marató (SPN-1554), the National Institutes of Health (1R01 AG032325-01), and the Alzheimer's Association (NIRG-07-59514).

REFERENCES

1. Smith, M.A., Alzheimer disease, *Int. Rev. Neurobiol.* 42, 1, 1998.
2. Hebert, L.E. et al., Alzheimer disease in the US population: Prevalence estimates using the 2000 census, *Arch. Neurol.* 60, 1119, 2003.
3. Robert, P.H. et al., Grouping for behavioral and psychological symptoms in dementia: Clinical and biological aspects. Consensus Paper of the European Alzheimer Disease Consortium, *Eur. Psychiatry* 20, 490, 2005.
4. Castellani, R.J. et al., Neuropathology of Alzheimer disease: Pathognomonic but not pathogenic, *Acta Neuropathol. (Berl.)* 111, 503, 2006.
5. Hardy, J., Alzheimer's disease: The amyloid cascade hypothesis: An update and reappraisal, *J. Alzheimers Dis.* 9, 151, 2006.
6. Masters, C.L., and Beyreuther, K., Pathways to the discovery of the Abeta amyloid of Alzheimer's disease, *J. Alzheimers Dis.* 9, 155, 2006.
7. Obregon, D.F. et al., ADAM10 activation is required for green tea (−)-epigallocatechin-3-gallate-induced alpha-secretase cleavage of amyloid precursor protein, *J. Biol. Chem.* 281, 16419, 2006.
8. Liu, Q. et al., Alzheimer-specific epitopes of tau represent lipid peroxidation-induced conformations, *Free Radic. Biol. Med.* 38, 746, 2005.
9. Takeda, A. et al., In Alzheimer's disease, heme oxygenase is coincident with Alz50, an epitope of tau induced by 4-hydroxy-2-nonenal modification, *J. Neurochem.* 75, 1234, 2000.
10. Mondragon-Rodriguez, S. et al., Cleavage and conformational changes of tau protein follow phosphorylation during Alzheimer's disease, *Int. J. Exp. Pathol.* 89, 81, 2008.
11. Langseth, L., *Oxidants, Antioxidants and Disease Prevention*, International Life Science Institute, Brussels, Belgium, 1996.
12. Marcus, D.L. et al., Increased peroxidation and reduced antioxidant enzyme activity in Alzheimer's disease, *Exp. Neurol.* 150, 40, 1998.
13. Smith, C.D. et al., Excess brain protein oxidation and enzyme dysfunction in normal aging and in Alzheimer disease, *Proc. Natl. Acad. Sci. U. S. A.* 88, 10540, 1991.
14. Zhu, X. et al., Causes of oxidative stress in Alzheimer disease, *Cell. Mol. Life Sci.* 64, 2202, 2007.
15. Vitek, M.P. et al., Advanced glycation end products contribute to amyloidosis in Alzheimer disease, *Proc. Natl. Acad. Sci. U. S. A.* 91, 4766, 1994.
16. Smith, M.A. et al., Advanced Maillard reaction end products are associated with Alzheimer disease pathology, *Proc. Natl. Acad. Sci. U. S. A.* 91, 5710, 1994.
17. Yan, S.D. et al., Glycated tau protein in Alzheimer disease: A mechanism for induction of oxidant stress, *Proc. Natl. Acad. Sci. U. S. A.* 91, 7787, 1994.
18. Ledesma, M.D. et al., Analysis of microtubule-associated protein tau glycation in paired helical filaments, *J. Biol. Chem.* 269, 21614, 1994.
19. Jorm, A.F., Korten, A.E., and Henderson, A.S., The prevalence of dementia: A quantitative integration of the literature, *Acta Psychiatr. Scand.* 76, 465, 1987.

20. McGonigal, G. et al., Epidemiology of Alzheimer's presenile dementia in Scotland, 1974–88, *BMJ (Clinical Research Ed.)* 306, 680, 1993.

21. Kawas, C. et al., A prospective study of estrogen replacement therapy and the risk of developing Alzheimer's disease: The Baltimore Longitudinal Study of Aging, *Neurology* 48, 1517, 1997.

22. Henderson, V.W. et al., Estrogen replacement therapy in older women. Comparisons between Alzheimer's disease cases and nondemented control subjects, *Arch. Neurol.* 51, 896, 1994.

23. Tang, M.X. et al., Effect of oestrogen during menopause on risk and age at onset of Alzheimer's disease, *Lancet* 348, 429, 1996.

24. Genazzani, A.R. et al., The brain as a target organ of gonadal steroids, *Psychoneuroendocrinology* 17, 385, 1992.

25. Casadesus, G. et al., Menopause, estrogen, and gonadotropins in Alzheimer's disease, *Adv. Clin. Chem.* 45, 139, 2008.

26. Goate, A., Segregation of a missense mutation in the amyloid beta-protein precursor gene with familial Alzheimer's disease, *J. Alzheimers Dis.* 9, 341, 2006.

27. Roses, A.D., On the discovery of the genetic association of apolipoprotein E genotypes and common late-onset Alzheimer disease, *J. Alzheimers Dis.* 9, 361, 2006.

28. Rogaeva, E., Kawarai, T., and George-Hyslop, P.S., Genetic complexity of Alzheimer's disease: Successes and challenges, *J. Alzheimers Dis.* 9, 381, 2006.

29. Ashe, K.H., Molecular basis of memory loss in the Tg2576 mouse model of Alzheimer's disease, *J. Alzheimers Dis.* 9, 123, 2006.

30. Summers, W.K., Tacrine and Alzheimer's treatments, *J. Alzheimers Dis.* 9, 439, 2006.

31. Marlatt, M.W. et al., Alzheimer's disease: Cerebrovascular dysfunction, oxidative stress, and advanced clinical therapies, *J. Alzheimers Dis.* 15, 199, 2008.

32. Marlatt, M.W. et al., Therapeutic opportunities in Alzheimer disease: One for all or all for one? *Curr. Med. Chem.* 12, 1137, 2005.

33. Mancuso, C. et al., Natural antioxidants in Alzheimer's disease, *Expert Opin. Investig. Drugs* 16, 1921, 2007.

34. Gibson, G.E., and Blass, J.P., Thiamine-dependent processes and treatment strategies in neurodegeneration, *Antioxid. Redox Signal.* 9, 1605, 2007.

35. Wang, W. et al., Nutritional biomarkers in Alzheimer's disease: The association between carotenoids, n-3 fatty acids, and dementia severity, *J. Alzheimers Dis.* 13, 31, 2008.

36. Commenges, D. et al., Intake of flavonoids and risk of dementia, *Eur. J. Epidemiol.* 16, 357, 2000.

37. Dai, Q. et al., Fruit and vegetable juices and Alzheimer's disease: The Kame project, *Am. J. Med.* 119, 751, 2006.

38. Pietta, P.G., Flavonoids as antioxidants, *J. Nat. Prod.* 63, 1035, 2000.

39. Scalbert, A., Johnson, I.T., and Saltmarsh, M., Polyphenols: Antioxidants and beyond, *Am. J. Clin. Nutr.* 81, 215S, 2005.

40. Ramiro-Puig, E. et al., Cocoa-enriched diet enhances antioxidant enzyme activity and modulates lymphocyte composition in thymus from young rats, *J. Agric. Food Chem.* 55, 6431, 2007.

41. Mandel, S.A. et al., Multifunctional activities of green tea catechins in neuroprotection. Modulation of cell survival genes, iron-dependent oxidative stress and PKC signaling pathway, *Neurosignals* 14, 46, 2005.

42. Levites, Y. et al., Neuroprotection and neurorescue against Abeta toxicity and PKC-dependent release of nonamyloidogenic soluble precursor protein by green tea polyphenol (−)-epigallocatechin-3-gallate, *FASEB J.* 17, 952, 2003.

43. Reznichenko, L. et al., Reduction of iron-regulated amyloid precursor protein and beta-amyloid peptide by (−)-epigallocatechin-3-gallate in cell cultures: Implications for iron chelation in Alzheimer's disease, *J. Neurochem.* 97, 527, 2006.

44. Rezai-Zadeh, K. et al., Green tea epigallocatechin-3-gallate (EGCG) modulates amyloid precursor protein cleavage and reduces cerebral amyloidosis in Alzheimer transgenic mice, *J. Neurosci.* 25, 8807, 2005.

45. Skovronsky, D.M. et al., Protein kinase C-dependent alpha-secretase competes with beta-secretase for cleavage of amyloid-beta precursor protein in the trans-Golgi network, *J. Biol. Chem.* 275, 2568, 2000.

46. Weinreb, O., Amit, T., and Youdim, M.B., A novel approach of proteomics and transcriptomics to study the mechanism of action of the antioxidant-iron chelator green tea polyphenol (−)-epigallocatechin-3-gallate, *Free Radic. Biol. Med.* 43, 546, 2007.

47. Joseph, J.A. et al., Long-term dietary strawberry, spinach, or vitamin E supplementation retards the onset of age-related neuronal signal-transduction and cognitive behavioral deficits, *J. Neurosci.* 18, 8047, 1998.

48. Joseph, J.A. et al., Reversals of age-related declines in neuronal signal transduction, cognitive, and motor behavioral deficits with blueberry, spinach, or strawberry dietary supplementation, *J. Neurosci.* 19, 8114, 1999.

49. Andres-Lacueva, C. et al., Anthocyanins in aged blueberry-fed rats are found centrally and may enhance memory, *Nutr. Neurosci.* 8, 111, 2005.

50. Casadesus, G. et al., Modulation of hippocampal plasticity and cognitive behavior by short-term blueberry supplementation in aged rats, *Nutr. Neurosci.* 7, 309, 2004.

51. Joseph, J.A., Fisher, D.R., and Carey, A.N., Fruit extracts antagonize Abeta- or DA-induced deficits in Ca^{2+} flux in M1-transfected COS-7 cells, *J. Alzheimers Dis.* 6, 403, 2004.

52. Joseph, J.A. et al., Blueberry supplementation enhances signaling and prevents behavioral deficits in an Alzheimer disease model, *Nutr. Neurosci.* 6, 153, 2003.

53. Goel, A., Kunnumakkara, A.B., and Aggarwal, B.B., Curcumin as "Curecumin": From kitchen to clinic, *Biochem. Pharmacol.* 75, 787, 2008.

54. Yang, F. et al., Curcumin inhibits formation of amyloid beta oligomers and fibrils, binds plaques, and reduces amyloid in vivo, *J. Biol. Chem.* 280, 5892, 2005.

55. Garcia-Alloza, M. et al., Curcumin labels amyloid pathology in vivo, disrupts existing plaques, and partially restores distorted neurites in an Alzheimer mouse model, *J. Neurochem.* 102, 1095, 2007.

56. Begum, A.N. et al., Curcumin structure-function, bioavailability, and efficacy in models of neuroinflammation and Alzheimer's disease, *J. Pharmacol. Exp. Ther.* 326, 196, 2008.

57. Ng, T.P. et al., Curry consumption and cognitive function in the elderly, *Am. J. Epidemiol.* 164, 898, 2006.

58. Cherubini, A. et al., Dietary antioxidants as potential pharmacological agents for ischemic stroke, *Curr. Med. Chem.* 15, 1236, 2008.

59. Pallas, M. et al., Modulation of sirtuins: New targets for antiageing, *Recent Pat. CNS Drug Discov.* 3, 61, 2008.

60. Anekonda, T.S., Resveratrol—A boon for treating Alzheimer's disease? *Brain Res. Rev.* 52, 316, 2006.

61. Gupta, Y.K., Briyal, S., and Chaudhary, G., Protective effect of *trans*-resveratrol against kainic acid-induced seizures and oxidative stress in rats, *Pharmacol. Biochem. Behav.* 71, 245, 2002.

62. Wang, Q. et al., Resveratrol protects against global cerebral ischemic injury in gerbils, *Brain Res.* 958, 439, 2002.

63. Wen, Y. et al., Transient cerebral ischemia induces aberrant neuronal cell cycle re-entry and Alzheimer's disease-like tauopathy in female rats, *J. Biol. Chem.* 279, 22684, 2004.

64. Lu, K.T. et al., Neuroprotective effects of resveratrol on MPTP-induced neuron loss mediated by free radical scavenging, *J. Agric. Food Chem.* 56, 6910, 2008.

65. Kumar, P. et al., Effect of resveratrol on 3-nitropropionic acid-induced biochemical and behavioural changes: Possible neuroprotective mechanisms, *Behav. Pharmacol.* 17, 485, 2006.

66. Kim, D. et al., SIRT1 deacetylase protects against neurodegeneration in models for Alzheimer's disease and amyotrophic lateral sclerosis, *EMBO J.* 26, 3169, 2007.

67. Wang, J. et al., Moderate consumption of Cabernet Sauvignon attenuates Abeta neuropathology in a mouse model of Alzheimer's disease, *FASEB J.* 20, 2313, 2006.

68. Ho, L. et al., Heterogeneity in red wine polyphenolic contents differentially influences Alzheimer's disease-type neuropathology and cognitive deterioration, *J. Alzheimers Dis.* 16, 59–72, 2009.

69. Morel, I. et al., Antioxidant and iron-chelating activities of the flavonoids catechin, quercetin and diosmetin on iron-loaded rat hepatocyte cultures, *Biochem. Pharmacol.* 45, 13, 1993.

70. Arteel, G.E., and Sies, H., Protection against peroxynitrite by cocoa polyphenol oligomers, *FEBS Lett.* 462, 167, 1999.

71. Hatano, T. et al., Proanthocyanidin glycosides and related polyphenols from cacao liquor and their antioxidant effects, *Phytochemistry* 59, 749, 2002.

72. Ramiro, E. et al., Flavonoids from *Theobroma cacao* down-regulate inflammatory mediators, *J. Agric. Food Chem.* 53, 8506, 2005.

73. McShea, A. et al., Clinical benefit and preservation of flavonols in dark chocolate manufacturing, *Nutr. Rev.* 66, 630, 2008.

74. Vinson, J.A. et al., Chocolate is a powerful ex vivo and in vivo antioxidant, an antiatherosclerotic agent in an animal model, and a significant contributor to antioxidants in the European and American diets, *J. Agric. Food Chem.* 54, 8071, 2006.

75. Lee, K.W. et al., Cocoa has more phenolic phytochemicals and a higher antioxidant capacity than teas and red wine, *J. Agric. Food Chem.* 51, 7292, 2003.

76. Francis, S.T. et al., The effect of flavanol-rich cocoa on the fMRI response to a cognitive task in healthy young people, *J. Cardiovasc. Pharmacol.* 47 Suppl 2, S215, 2006.
77. Fisher, N.D., Sorond, F.A., and Hollenberg, N.K., Cocoa flavanols and brain perfusion, *J. Cardiovasc. Pharmacol.* 47 Suppl 2, S210, 2006.
78. Heo, H.J., and Lee, C.Y., Epicatechin and catechin in cocoa inhibit amyloid beta protein induced apoptosis, *J. Agric. Food Chem.* 53, 1445, 2005.
79. Datla, K.P. et al., Short-term supplementation with plant extracts rich in flavonoids protect nigrostriatal dopaminergic neurons in a rat model of Parkinson's disease, *J. Am. Coll. Nutr.* 26, 341, 2007.

5 Micronutrient Antioxidants, Cognition, and Neuropathology

A Longitudinal Study in the Canine Model of Human Aging

Wycliffe O. Opii[1] and Elizabeth Head[1, 2]

[1]Institute for Brain Aging and Dementia, University of California–Irvine, Irvine, California, USA

[2]Department of Neurology, University of California–Irvine, Irvine, California, USA

CONTENTS

5.1   INTRODUCTION

Aging is a key risk factor for a number of age-related neurodegenerative disorders, such as Alzheimer's disease and Huntington's disease [1]. The aging brain is under severe oxidative stress, which leads to a progressive accumulation of oxidatively damaged proteins/enzymes, neuronal loss, and cognitive dysfunction. In addition, a reduction in the endogenous antioxidant reserves associated with aging can also contribute to increased oxidative stress. Hence, oxidative stress plays a critical role in the mechanisms leading to the development of neuropathology typically observed in age-related neurodegeneration [2,3]. As a result, there is considerable interest in developing interventions against these disorders. Recently, there has been increasing support for the use of antioxidants in the form of micro- and macronutrients that can slow the progression of pathological aging, maintain cognitive health, or improve overall brain function [4,5]. The use of antioxidants has beneficial effects in various models of age-related neurodegenerative disorders, resulting in maintenance, improvement, or restoration of memory and cognitive function [4,6].

To understand some of the neurobiological mechanisms underlying the effects of antioxidants/ micronutrients on the maintenance and improvement of cognitive function, we used the canine model of human aging. Aging canines develop aspects of neuropathology and cognitive dysfunction similar to those seen in humans, including beta-amyloid (Aβ) deposition, oxidative stress, and learning and memory impairments [5,7,8]. We describe a longitudinal study in the aging canine that investigated the effects of an intervention with a diet fortified with a broad range of micronutrient antioxidants, a behavioral enrichment paradigm, or a combination of both. We observed a significant reduction in the levels of oxidative damage, an increase in the expression and activity of endogenous antioxidants, and a reduction in Aβ deposition in the canine brain [9,10]. These neurobiological outcomes were correlated with improvement and maintenance in cognition as detected by a wide range of tasks in the aging canine. As a result, the current study suggests that the combination of an antioxidant diet that includes micronutrients and behavioral enrichment may improve neuronal function and could be beneficial to the ever-growing aging human population.

5.2   THE CANINE MODEL

5.2.1   NEUROBIOLOGICAL FEATURES

Studies that try to elucidate the mechanisms of human aging and age-related disorders are usually hampered by the lack of a good animal model that can be easily translated to the human population. Recently, the use of the canine model of human aging has gained considerable interest. The aging canine provides one of the best models for human aging for a number of reasons: aged canines develop aspects of neuropathology similar to those observed in aged humans; specifically, canines develop extensive Aβ deposition within neurons and their synaptic fields that appears to give rise to senile plaques. The Aβ deposited in the aged beagle brain is of the same sequence as humans [11,12] and is correlated with decline in cognitive function with age [7,13]. Just as in the human brain, the aging canine brain is also under significant oxidative stress [8]. It has previously been established that Aβ protein plays a central role in the elevation of oxidative stress levels in AD and in animal models of AD, with the key to this link being the amino acid residue methionine 35 [14,15]. As a result, it is possible that the deposition of Aβ in the aging canine brain could play a significant role in molecular pathways involving free radical generation and oxidative stress [15]. In addition, in parallel with oxidative damage to proteins and lipids, the aging canine brain has reduced levels of endogenous antioxidant enzymes, as measured by the activity of glutathione synthase and glutathione levels [8].

5.2.2   COGNITIVE FEATURES

In parallel with neuropathological changes, aged dogs also develop learning and memory deficits. There are several features of cognitive aging in canines that are similar to human aging, including domain-specific vulnerabilities and increased individual variability in function. Simple learning ability and procedural learning remain relatively intact with age in dogs and humans [16]. In contrast, executive function, dependent upon the prefrontal cortex, deteriorates with age, as also seen in humans [16,17]. In a measure of complex learning ability, such as concept learning (oddity task), differential strategy use by young and old animals can be seen [5]. Memory impairments are detected in a subset of aged dogs, and with increasing age, there is increasing variability in the ability of individual animals to remember spatial information [18]. Interestingly, with a more difficult spatial memory task, more subtle age effects can be detected in animals as young as 6 years of age (middle age) [19]. In addition to changes in cognitive function, the other types of behavioral change with age in dogs include reduced locomotor activity and exploration, similar to humans [20,21].

In the canine model, individual variability in cognitive aging can provide useful insights when identifying possible underlying brain mechanisms leading to neuronal dysfunction. For example, the severity of cognitive dysfunction is correlated with the extent of Aβ pathology [7,22]. In particular, these associations are further strengthened if the extent of prefrontal cortex Aβ neuropathology is correlated with measures of prefrontal function [23]. In addition, more extensive accumulation of oxidative damage in the brain is also associated with more severe cognitive impairments [9,24]. In parallel, low brain levels of vitamin E are associated with clinical signs of behavioral dysfunction in pet dogs [25]. Aβ and the accumulation of oxidative damage may lead to neuronal loss and loss of the ability to generate new neurons. Indeed, fewer neurons in the hippocampi of aged dogs are correlated with spatial memory deficits and learning impairments in hippocampus-dependent tasks [26]. However, most of these observations have been correlative in nature and require hypothesis-driven intervention studies to establish a cause-and-effect relationship. Thus, will decreasing oxidative damage or increasing neuron number lead to cognitive improvements in aged dogs?

5.3 LONGITUDINAL INTERVENTION STUDY DESIGN AND RATIONALE

To determine if reducing oxidative damage or if stimulating neurogenesis or growth factors to maintain neuron health may have benefits on cognition and markers of neuropathology, we initiated a longitudinal study of aging in dogs. We selected two intervention protocols, a dietary intervention and a behavioral intervention, and additionally evaluated the effects of combining these two treatments. In the first treatment study, we tested whether a diet enriched in antioxidant micronutrients and mitochondrial cofactors reduces oxidative damage and leads to improved cognition [3,27]. In the second treatment, we provided animals with behavioral enrichment in order to increase growth factors in the brain, maintain neuron number, and potentially increase hippocampal neurogenesis [28–30]. Lastly, we predicted that the combination of these two treatments would lead to additive improvements in cognition and further reductions in neuropathology, beyond the results of either intervention alone.

To test these hypotheses, aged beagles received a long-term treatment (>2.5 years) with a diet rich in antioxidants (AOX), behavioral enrichment (BEH), or a combination of the two treatments (AOX/BEH). Twenty-four aged beagles (mean age of 10.69 years) were placed into one of four treatment groups: (1) control diet/control enrichment (C/C); (2) BEH/control diet (E/C); (3) control enrichment/AOX diet (C/A); and (4) BEH/AOX diet (E/A). The AOX diet contained vitamin E [α-tocopherol (800IU)], vitamin C (10 mg/kg), and fruit and vegetable extracts (1%). In addition, the diet contained two mitochondrial cofactors, dl-lipoic acid (2.7 mg/kg) and l-carnitine (6 mg/kg), thought to improve mitochondrial function [27]. Vitamin E is a lipid-soluble vitamin that protects cell membranes from oxidative damage, while vitamin C is a water-soluble vitamin that helps replenish vitamin E cellular levels [31,32]. L-carnitine is a precursor to acetyl-l-carnitine, which is involved in mitochondrial lipid metabolism and maintaining efficient mitochondrial function [33]. Lipoic acid is an antioxidant nutrient capable of redox recycling of other antioxidants and raising the intracellular levels of glutathione [27,33], the major intracellular water-soluble antioxidant. Fruits and vegetables are sources of carotenoids and flavonoids that appear to have additional antioxidant activity [34,35].

The BEH enrichment consisted of social enrichment (housing with a kennel mate), cognitive enrichment (landmark discrimination, oddity discrimination, size concept learning with testing on a daily basis for 5 days per week), and physical exercise (2–20 min of outdoor walks per week) [36]. The rationale for including behavioral enrichment is based on the evidence that environmental enrichment and physical exercise can lead to improved behavioral function and increased neurogenesis in rodent models [28–30] and that cognitive training and participation in cognitively stimulating activities reduce the risk of developing cognitive impairments and dementia in humans [37]. In addition to the aged beagles, a group of younger beagles (age range of 3.5–6.5 years) was split into two treatment groups and fed either the AOX diet or the control diet for the same period to determine if intervention benefits occur across the life span in both old and young dogs. All young animals were housed under the behavioral enrichment protocol [38].

5.4 COGNITIVE BENEFITS OF ANTIOXIDANT MICRONUTRIENTS AND BEHAVIORAL ENRICHMENT

Treatment with the AOX diet led to cognitive improvements that were rapid, occurring within 2 weeks of beginning the diet. In particular, aged animals showed significant improvements in spatial attention [39]. Subsequent testing of animals with more difficult cognitive tasks, including complex learning task and oddity discrimination, also revealed benefits of the diet [40]. Improved learning ability was maintained over time with the AOX treatment, while untreated animals showed a progressive decline in cognitive function [36]. We observed similar improvements in cognition in animals provided with BEH intervention. Furthermore, the combination treatment resulted in consistently greater benefits than either treatment alone [36,41]. For example, while spatial memory showed a trend toward improvement in singly treated animals, long-term treatment with a combination of the two interventions was needed in order for these differences to be statistically significant [42]. Interestingly, cognitive improvements from dietary intervention were limited to aged animals, as young dogs treated with the AOX diet were not different from dogs that were fed the control diet [38]. The selective improvement in cognitive performance in aged dogs fed an AOX diet suggests that dietary antioxidants repair age-associated oxidative damage.

5.5 EFFECTS OF ANTIOXIDANT MICRONUTRIENTS AND BEHAVIORAL ENRICHMENT ON NEUROBIOLOGICAL OUTCOMES

We next conducted a series of studies to determine the neurobiological mechanisms underlying improved cognition in response to AOX and BEH treatments. Our study in canines suggests that the AOX and BEH treatments may individually have an impact on independent molecular cascades in the brain but that they may also act together to further improve cognition and reduce pathology. We hypothesized that oxidative damage may influence Aβ production and accumulation [14], and thus we measured the extent of Aβ plaque accumulation in several cortical regions in AOX-, BEH-, and combination-treated animals. Aβ deposition in the AOX-treated animals was reduced by 27–84% in the parietal, entorhinal, and occipital cortices, but not in the prefrontal cortex [10]. Aβ-lowering effects were not observed in animals receiving the BEH intervention alone. Because the AOX treatment was initiated at an age at which Aβ pathology should already be present [23], these results suggest that the AOX treatment can reduce new Aβ accumulation but cannot reverse existing pathology.

Interestingly, the cognitive benefits of the BEH treatment may be mediated through a molecular cascade separate from the AOX treatment. While BEH did not lower Aβ plaque load, animals provided with BEH treatment showed less neuronal loss in the hippocampus compared with untreated dogs and compared with those treated with the AOX diet alone [43]. Interestingly, maintenance of neuron number associated with behavioral enrichment was not linked to increased neurogenesis [26], suggesting slowed neuronal loss as a possible mechanism.

Additional neuropathologies were selectively reduced in the combination treatment (BEH + AOX) condition, suggesting that molecular cascades affected by individual treatments may converge to have additional benefits. Age-related disorders are often accompanied by an increase in oxidative stress, a reduction in antioxidant reserves, and loss in energy metabolism, among other features [2,3]. The use of antioxidants and/or related compounds reduces the level of oxidative damage and delays or reduces age-related cognitive decline in animal models and in humans [4,39,44,45]. Previous studies in aged canines show that attenuation of oxidative damage may be critically involved in the maintenance of cognitive function with long-term treatment with antioxidants and a program of behavioral enrichment [36,40,41]. In the current study, levels of toxic brain oxidative stress biomarkers (i.e., 3-nitrotyrosine and protein carbonyls) were reduced with all interventions, but only the combined BEH/AOX treatment resulted in a significant reduction. In addition, while levels of brain lipid peroxidation, as measured by 4-hydroxynonenal, were marginally reduced in all treatments, none was significantly reduced compared with control [9]. This still does support the use of antioxidants in the reduction of lipid peroxidation-mediated oxidative damage in the canine brain.

Using proteomics, we identified key proteins from the canine parietal cortex that showed significantly increased expression or reduced oxidation (as measured by protein carbonyls) following a program of BEH and AOX intervention [9]. Importantly, the proteins identified are involved in energy metabolism, in antioxidant systems, and in the maintenance and stabilization of cell structure. For example, the identified proteins included fructose-bisphosphate aldolase C, creatine kinase, glutamate dehydrogenase, glyceraldehyde-3-phosphate dehydrogenase, and Cu/Zn superoxide dismutase (SOD). In addition, the increased expression of Cu/Zn SOD correlates with improvements in cognition as shown by an improvement in the black/white reversal and spatial learning. Furthermore, there is a significant increase in the activity of antioxidant enzymes glutathione S-transferase and total SOD in the combined BEH/AOX treatment and a significant increase in the expression of HO-1 protein, an important defense system in neurons under oxidative stress [9,46]. Finally, decreased oxidation of some of these key brain proteins correlates with improved cognitive function in the aged canines undergoing these treatments. These results suggest possible mechanisms for the improved memory and cognitive function previously reported [40, 41], including maintenance and stabilization of cell structure, an increase in antioxidant capacity, and efficient energy metabolism.

5.6 SUMMARY

Our study in canines is unique in that two interventions were compared and combined. A diet rich in a broad spectrum of micronutrients (antioxidants and mitochondrial cofactors) either alone or particularly when combined with behavioral enrichment can improve cognition and reduce human-type brain pathology in aged beagles. The antioxidant dose levels used in the current study are consistent with human clinical trials and indicate that dietary supplementation with a broad spectrum of antioxidants may reduce protein oxidative damage and shift Aβ precursor protein processing in favor of nonamyloidogenic pathways (Figure 5.1). In parallel with reduced oxidative damage to these proteins, there was a significant increase in the expression and activity of key proteins related to energy metabolism and antioxidant activity. The increase in protein levels was a good predictor of frontal cortex-dependent learning and hippocampus-dependent spatial memory, suggesting that these proteins are important for the intervention's benefits to cognitive function. Behavioral enrichment improved neuronal function independently of Aβ but may serve to maintain neuronal number or plasticity (Figure 5.1). Thus, the combination of an antioxidant diet with behavioral enrichment may improve neuronal function through separate yet complementary molecular pathways. The results of this study also suggest that even when animals are older, with existing cognitive impairments and brain pathology, modifying environmental factors and diet can have substantial effects on the aging process. The present findings therefore provide a neurobiological basis for improved neuronal function and cognition in canines treated with either an antioxidant-enriched diet and behavioral enrichment or a combination of the two interventions.

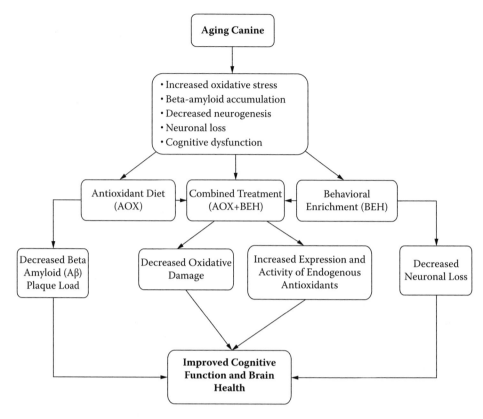

FIGURE 5.1 A summary of the pathways leading to improved cognitive function and brain health in the aging canine brain following intervention with an antioxidant diet and program of behavioral enrichment, or a combination of the two.

ACKNOWLEDGMENTS

Funding for this study was provided by the National Institutes of Health/National Institute on Aging (AG12694) and the U.S. Department of the Army (Contract No. DAMD17-98-1-8622).

REFERENCES

1. Poon, H. F., et al., Free radicals and brain aging, *Clin Geriatr Med 20*, 329, 2004.
2. Hensley, K., et al., Reactive oxygen species as causal agents in the neurotoxicity of the Alzheimer's disease-associated amyloid beta peptide, *Ann N Y Acad Sci, 786*, 120–34, 1996.
3. Markesbery, W. R., Oxidative stress hypothesis in Alzheimer's disease, *Free Radic Biol Med 23*, 134, 1997.
4. Joseph, J. A., et al., Long-term dietary strawberry, spinach, or vitamin E supplementation retards the onset of age-related neuronal signal-transduction and cognitive behavioral deficits, *J Neurosci 18*, 8047, 1998.
5. Cotman, C. W., et al., Brain aging in the canine: A diet enriched in antioxidants reduces cognitive dysfunction, *Neurobiol Aging 23*, 809, 2002.
6. Laurin, D., et al., Physical activity and risk of cognitive impairment and dementia in elderly persons, *Arch Neurol 58*, 498, 2001.
7. Cummings, B. J., et al., Beta-amyloid accumulation correlates with cognitive dysfunction in the aged canine, *Neurobiol Learn Mem 66*, 11, 1996.
8. Head, E., et al., Oxidative damage increases with age in a canine model of human brain aging, *J Neurochem 82*, 375, 2002.

9. Opii, W. O., et al., Proteomic identification of brain proteins in the canine model of human aging following a long-term treatment with antioxidants and a program of behavioral enrichment: Relevance to Alzheimer's disease, *Neurobiol Aging 29*, 51, 2008.

10. Pop, V., et al., Reduced Aβ deposition with long-term antioxidant diet treatment in aged canines. In *Society for Neuroscience, Abstract Viewer/Itinerary Planner*, Vol. 525.4, Washington, DC (2003).

11. Johnstone, E. M., et al., Conservation of the sequence of the Alzheimer's disease amyloid peptide in dog, polar bear and five other mammals by cross-species polymerase chain reaction analysis, *Brain Res Mol Brain Res 10*, 299, 1991.

12. Selkoe, D. J., et al., Conservation of brain amyloid proteins in aged mammals and humans with Alzheimer's disease, *Science 235*, 873, 1987.

13. Head, E., et al., Visual-discrimination learning ability and beta-amyloid accumulation in the dog, *Neurobiol Aging 19*, 415, 1998.

14. Butterfield, D. A., beta-Amyloid-associated free radical oxidative stress and neurotoxicity: Implications for Alzheimer's disease, *Chem Res Toxicol 10*, 495, 1997.

15. Butterfield, D. A., Amyloid beta-peptide (1–42)-induced oxidative stress and neurotoxicity: Implications for neurodegeneration in Alzheimer's disease brain. A review, *Free Radic Res 36*, 1307, 2002.

16. Milgram, N. W., et al., Cognitive functions and aging in the dog: Acquisition of nonspatial visual tasks, *Behav Neurosci 108*, 57, 1994.

17. Tapp, P. D., et al., Size and reversal learning in the beagle dog as a measure of executive function and inhibitory control in aging, *Learn Mem 10*, 64, 2003.

18. Adams, B., et al., Use of a delayed non-matching to position task to model age-dependent cognitive decline in the dog, *Behav Brain Res 108*, 47, 2000.

19. Studzinski, C. M., et al., Visuospatial function in the beagle dog: An early marker of cognitive decline in a model of human aging and dementia, *Neurobiol Learn Mem 86*, 197–204, 2006.

20. Head, E., et al., Open field activity and human interaction as a function of age and breed in dogs, *Physiol Behav 62*, 963, 1997.

21. Siwak, C. T., Tapp, P. D., & Milgram, N. W., Effect of age and level of cognitive function on spontaneous and exploratory behaviors in the beagle dog, *Learn Mem 8*, 317, 2001.

22. Colle, M.-A., et al., Vascular and parenchymal Aβ deposition in the aging dog: Correlation with behavior, *Neurobiol Aging 21*, 695, 2000.

23. Head, E., et al., Region-specific age at onset of beta-amyloid in dogs, *Neurobiol Aging 21*, 89, 2000.

24. Rofina, J. E., et al., Cognitive disturbances in old dogs suffering from the canine counterpart of Alzheimer's disease, *Brain Res 1069*, 216, 2006.

25. Skoumalova, A., et al., The role of free radicals in canine counterpart of senile dementia of the Alzheimer type, *Exp Gerontol 38*, 711, 2003.

26. Siwak-Tapp, C. T., et al., Neurogenesis decreases with age in the canine hippocampus and correlates with cognitive function, *Neurobiol Learn Mem 88*, 249, 2007.

27. Ames, B. N., A role for supplements in optimizing health: The metabolic tune-up, *Arch Biochem Biophys 423*, 227, 2004.

28. Cotman, C. W., & Berchtold, N. C., Exercise: A behavioral intervention to enhance brain health and plasticity, *TINS 25*, 295, 2002.

29. van Praag, H., Kempermann, G., & Gage, F. H., Neural consequences of environmental enrichment, *Nat Rev Neurosci 1*, 191, 2000.

30. van Praag, H., et al., Exercise enhances learning and hippocampal neurogenesis in aged mice, *J. Neurosci 25*, 8680, 2005.

31. Ames, B. N., Shigenaga, M. K., & Hagen, T. M., Oxidants, antioxidants, and the degenerative diseases of aging, *Proc Natl Acad Sci U S A 90*, 7915, 1993.

32. Jewell, D., et al., Effect of increasing dietary antioxidants on concentrations of vitamin E and total alkenals in serum of dogs and cats, *Vet Ther, 1*, 264, 2000.

33. Liu, J., et al., Delaying brain mitochondrial decay and aging with mitochondrial antioxidants and metabolites, *Ann N Y Acad Sci 959*, 133, 2002.

34. Cantuti-Castelvetri, I., Shukitt-Hale, B., and Joseph, J. A., Neurobehavioral aspects of antioxidants in aging, *Int J Dev Neurosci 18*, 367, 2000.

35. Cao, G., et al., Increases in human plasma antioxidant capacity after consumption of controlled diets high in fruit and vegetables, *Am J Clin Nutr 68*, 1081, 1998.

36. Milgram, N. W., et al., Learning ability in aged beagle dogs is preserved by behavioral enrichment and dietary fortification: A two-year longitudinal study, *Neurobiol Aging 26*, 77, 2005.

37. Wilson, R.S., et al., Participation in cognitively stimulating activities and risk of incident Alzheimer disease, *JAMA 287*, 742, 2002.
38. Siwak, C. T., et al., Chronic antioxidant and mitochondrial cofactor administration improves discrimination learning in aged but not young dogs, *Prog Neuropsychopharmacol Biol Psychiatry 29*, 461, 2005.
39. Milgram, N. W., et al., Landmark discrimination learning in the dog: Effects of age, an antioxidant fortified diet, and cognitive strategy, *Neurosci Biobehav Rev 26*, 679, 2002.
40. Cotman, C. W., et al., Brain aging in the canine: A diet enriched in antioxidants reduces cognitive dysfunction, *Neurobiol Aging 23*, 809, 2002.
41. Milgram, N. W., et al., Long-term treatment with antioxidants and a program of behavioral enrichment reduces age-dependent impairment in discrimination and reversal learning in beagle dogs, *Exp Gerontol 39*, 753, 2004.
42. Nippak, P. M., et al., Enhanced spatial ability in aged dogs following dietary and behavioural enrichment, *Neurobiol Learn Mem 87*, 610, 2007.
43. Siwak-Tapp, C. T., et al., Region specific neuron loss in the aged canine hippocampus is reduced by enrichment, *Neurobiol Aging 29*, 521, 2008.
44. Bickford, P. C., et al., Antioxidant-rich diets improve cerebellar physiology and motor learning in aged rats, *Brain Res 866*, 211, 2000.
45. Farr, S. A., et al., The antioxidants alpha-lipoic acid and *N*-acetylcysteine reverse memory impairment and brain oxidative stress in aged SAMP8 mice, *J Neurochem 84*, 1173, 2003.
46. Calabrese, V., et al., Redox regulation of heat shock protein expression in aging and neurodegenerative disorders associated with oxidative stress: A nutritional approach, *Amino Acids 25*, 437, 2003.

6 Excitatory Amino Acids, S-Nitrosylation, and Protein Misfolding in Neurodegenerative Disease

Protection by Memantine and NitroMemantine at NMDA-Gated Channels

Tomohiro Nakamura[1] and Stuart A. Lipton[1,2]

[1]Center for Neuroscience, Aging, and Stem Cell Research, Burnham Institute for Medical Research, La Jolla, California, USA

[2]Department of Neurosciences, University of California at San Diego, La Jolla, California, USA

CONTENTS

6.1 INTRODUCTION

Excessive generation of reactive nitrogen species (RNS) and reactive oxygen species (ROS), which lead to neuronal cell injury and death, is a potential mediator of neurodegenerative disorders, including Parkinson's disease (PD), Alzheimer's disease (AD), amyotrophic lateral sclerosis (ALS), polyglutamine diseases (e.g., Huntington's disease), glaucoma, human immunodeficiency virus-associated dementia, multiple sclerosis, and ischemic brain injury, to name but a few.[1–5] While many intra- and extracellular molecules may participate in neuronal injury, accumulation of nitrosative stress due to excessive generation of nitric oxide (NO) appears to be a potential factor contributing to neuronal cell damage and death.[6,7] A well-established model for NO production entails a central role of the N-methyl-D-Aspartate (NMDA)-type glutamate receptors in the nervous system. Excessive activation of NMDA receptors (NMDARs) drives Ca^{2+} influx, which in turn activates neuronal NO synthase (nNOS) and the generation of ROS.[8,9] Accumulating evidence suggests that NO can mediate both protective and neurotoxic effects by reacting with cysteine residues of target proteins to form S-nitrosothiols (SNOs), a process termed S-nitrosylation because of its effects on the chemical biology of protein function. Importantly, normal mitochondrial respiration may also generate free radicals, principally ROS, and one such molecule, superoxide anion (O_2^-), reacts rapidly with free radical NO to form the very toxic product peroxynitrite ($ONOO^-$).[10,11]

An additional feature of most neurodegenerative diseases is accumulation of misfolded and/ or aggregated proteins.[12–15] These protein aggregates can be cytosolic, nuclear, or extracellular. Importantly, protein aggregation can result from either (1) a rare mutation in the disease-related gene encoding the protein or (2) posttranslational changes to the protein engendered by nitrosative/ oxidative stress, which may well account for the more common sporadic cases of the disease.[16] A key theme of this article, therefore, is the hypothesis that nitrosative or oxidative stress contributes to protein misfolding in the brains of the majority of neurodegenerative patients. In this review, we discuss specific examples showing that S-nitrosylation of (1) ubiquitin E3 ligases, such as parkin, or (2) endoplasmic reticulum (ER) chaperones, such as protein disulfide isomerase (PDI), is critical for the accumulation of misfolded proteins in neurodegenerative diseases such as PD and other conditions.[17–20] We also discuss the neuroprotective mechanism of action of NMDA open-channel blockers such as memantine and NO-related drugs for the treatment of neurodegenerative disorders.[7,21]

6.2 PROTEIN MISFOLDING IN NEURODEGENERATIVE DISEASES

Many neurodegenerative diseases are characterized by the accumulation of misfolded proteins that adversely affect neuronal connectivity and plasticity and trigger cell death signaling pathways.[12,15] For example, degenerating brain contains aberrant accumulations of misfolded, aggregated proteins, such as α-synuclein and synphilin-1 in PD and amyloid-β (Aβ) and tau in AD. The inclusions observed in PD are called Lewy bodies and are mostly found in the cytoplasm. AD brains show intracellular neurofibrillary tangles, which contain tau, and extracellular plaques, which contain Aβ. Other disorders manifesting protein aggregation include Huntington's disease (polyglutamine), ALS, and prion disease.[14] The above-mentioned aggregates may consist of oligomeric complexes of nonnative secondary structures and demonstrate poor solubility in aqueous or detergent solvent.

In general, protein aggregates do not accumulate in unstressed, healthy neurons due in part to the existence of cellular "quality control machineries." For example, molecular chaperones are believed to provide a defense mechanism against the toxicity of misfolded proteins because chaperones can prevent inappropriate interactions within and between polypeptides and can promote refolding of proteins that have been misfolded because of cell stress. In addition to the quality control of proteins provided by molecular chaperones, the ubiquitin-proteasome system (UPS) and autophagy/ lysosomal degradation are involved in the clearance of abnormal or aberrant proteins. When chaperones cannot repair misfolded proteins, they may be tagged via addition of polyubiquitin chains for degradation by the proteasome. In neurodegenerative conditions, intra- or extracellular protein aggregates are thought to accumulate in the brain as a result of a decrease in molecular chaperone or proteasome activities. In fact, several mutations that disturb the activity of molecular chaperones or UPS-associated enzymes can cause neurodegeneration.[15,22,23] Along these lines, postmortem samples from the substantia nigra of PD patients (versus non-PD controls) manifest a significant reduction in proteasome activity.[24] Moreover, overexpression of the molecular chaperone HSP70 can prevent neurodegeneration in vivo in models of PD.[25]

Historically, lesions that contain aggregated proteins were considered to be pathogenic. Recently, several lines of evidence have suggested that aggregates are formed through a complex multi-step process by which misfolded proteins assemble into inclusion bodies; currently, soluble oligomers of these aberrant proteins are thought to be the most toxic forms via interference with normal cell activities, while frank aggregates may be an attempt by the cell to wall off potentially toxic material.[8,26]

6.3   N-METHYL-D-ASPARTATE RECEPTOR-MEDIATED GLUTAMATERGIC SIGNALING PATHWAYS INDUCE CA^{2+} INFLUX

It is well known that the amino acid glutamate is the major excitatory neurotransmitter in the brain. Glutamate is present in high concentrations in the adult central nervous system and is released for milliseconds from nerve terminals in a Ca^{2+}-dependent manner. After glutamate enters synaptic cleft, it diffuses across the cleft to interact with its corresponding receptors on the postsynaptic face of an adjacent neuron. Excitatory neurotransmission is necessary for the normal development and plasticity of synapses and for some forms of learning or memory; however, excessive activation of glutamate receptors is implicated in neuronal damage in many neurological disorders, ranging from acute hypoxic/ischemic brain injury to chronic neurodegenerative diseases. It is currently thought that overstimulation of extrasynaptic NMDARs mediates this neuronal damage, while synaptic activity predominantly activates survival pathways.[27–29] Intense hyperstimulation of excitatory receptors leads to necrotic cell death, but more mild or chronic overstimulation can result in apoptotic or other forms of cell death.[30–32]

There are two large families of glutamate receptors in the nervous system, ionotropic receptors (representing ligand-gated ion channels) and metabotropic receptors (coupled to G-proteins). Ionotropic glutamate receptors are further divided into three broad classes — NMDARs, α-amino-3-hydroxy-5 methyl-4-isoxazole propionic acid receptors, and kainate receptors, which are each named after synthetic ligands that can selectively activate these receptors. The NMDAR has attracted attention for a long time because it has several properties that set it apart from other ionotropic glutamate receptors. One such characteristic, in contrast to most α-amino-3-hydroxy-5 methyl-4-isoxazole propionic acid and kainate receptors, is that NMDAR-coupled channels are highly permeable to Ca^{2+}, thus permitting Ca^{2+} entry after ligand binding if the cell is depolarized in order to relieve block of the receptor-associated ion channel by Mg^{2+}.[33,34] Subsequent binding of Ca^{2+} to various intracellular molecules can lead to many significant consequences. In particular, excessive activation of NMDARs leads to the production of damaging free radicals (e.g., NO and ROS) and other enzymatic processes, contributing to cell death.[6,11,31,32,35,36]

6.4   CA^{2+} INFLUX AND GENERATION OF REACTIVE NITROGEN SPECIES/REACTIVE OXYGEN SPECIES

Excessive activation of glutamate receptors is implicated in neuronal damage in many neurological disorders. John Olney coined the term "excitotoxicity" to describe this phenomenon.[37,38] This form of toxicity is mediated at least in part by excessive activation of NMDA-type receptors,[6,7,39] resulting in excessive Ca^{2+} influx through a receptor's associated ion channel.

Increased levels of neuronal Ca^{2+}, in conjunction with the Ca^{2+}-binding protein CaM, trigger the activation of nNOS and subsequent generation of NO from the amino acid L-arginine[8,40] (Figure 6.1).

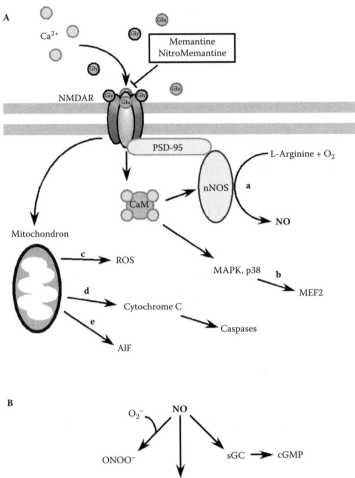

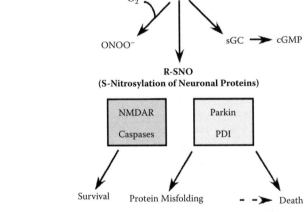

FIGURE 6.1 Activation of the NMDAR by glutamate (Glu) and glycine (Gly) induces Ca^{2+} influx and consequent NO production via activation of nNOS. (A, a) nNOS is part of a protein complex attached to the NR2 subunit of the NMDAR via binding of its PDZ domain to postsynaptic density protein (PSD-95). Many subsequent effects of NO are mediated by chemical, enzymatic, and redox reactions within neurons. NMDAR hyperactivation also triggers (b) activation of the p38 mitogen-activated kinase-MEF2 (transcription factor) pathway, (c) generation of ROS, (d) cytochrome C release from mitochondria and subsequent activation of caspases, and (e) release from mitochondria and activation of apoptosis-inducing factor. (B) NO activates soluble guanylate cyclase to produce cGMP, and cGMP can activate cGMP-dependent protein kinase. Excessive NMDAR activity, leading to the overproduction of NO, can be neurotoxic. For example, S-nitrosylation of parkin and PDI can contribute to neuronal cell damage and death and can also trigger accumulation of misfolded proteins. Neurotoxic effects of NO are also mediated by peroxynitrite ($ONOO^-$), a reaction product of NO and superoxide anion (O_2^-). In contrast, S-nitrosylation can mediate neuroprotective effects, for example, by inhibiting caspase activity and by preventing overactivation of NMDARs.

NO is a gaseous free radical (thus highly diffusible) and a key molecule that plays a vital role in normal signal transduction, but in excess, it can lead to neuronal cell damage and death. The discrepancy of NO effects on neuronal survival can also be caused by the formation of different NO species or intermediates: NO radical (NO·), nitrosonium cation (NO^+), and nitroxyl anion (NO^-, with high-energy singlet and lower-energy triplet forms).[11] Three subtypes of NOS have been identified; two constitutive forms of NOS—nNOS and endothelial NOS—take their names from the cell type in which they were first found. The name of the third subtype—inducible NOS (iNOS)—indicates that expression of the enzyme is induced by acute inflammatory stimuli. All three isoforms are widely distributed in the brain. Each NOS isoform contains an oxidase domain at its amino-terminal end and a reductase domain at its carboxy-terminal end, separated by a Ca^{2+}/CaM-binding site.[8,40–43] Constitutive NOS and iNOS are also further distinguished by CaM binding: nNOS and endothelial NOS bind CaM in a reversible Ca^{2+}-dependent manner. In contrast, iNOS binds CaM so tightly at resting intracellular Ca^{2+} concentrations that its activity does not appear to be affected by transient variations in Ca^{2+} concentration.

Recent studies further pointed out the potential connection between ROS/RNS and mitochondrial dysfunction in neurodegenerative diseases, especially in PD.[5,44] Pesticide and other environmental toxins that inhibit mitochondrial complex I result in oxidative and nitrosative stress and consequent aberrant protein accumulation.[17,18,20,45,46] Administration to animal models of complex I inhibitors, such as MPTP, 6-hydroxydopamine, rotenone, and paraquat, which result in overproduction of ROS/RNS, reproduces many of the features of sporadic PD, such as dopaminergic neuron degeneration, up-regulation and aggregation of α-synuclein, Lewy body-like intraneuronal inclusions, and behavioral impairment.[5,44] In addition, it has recently been proposed that mitochondrial cytochrome oxidase can produce NO in a nitrite (NO_2^-)- and pH-dependent but non-Ca^{2+}-dependent manner.[47]

Increased nitrosative stress and oxidative stress are associated with chaperone and proteasomal dysfunction, resulting in accumulation of misfolded aggregates.[16,48] However, until recently little was known regarding the molecular and pathogenic mechanisms' underlying contribution of NO to the formation of inclusion bodies such as amyloid plaques in AD or Lewy bodies in PD.

6.5 PROTEIN S-NITROSYLATION AND NEURONAL CELL DEATH

Early investigations indicated that the NO group mediates cellular signaling pathways, which regulate broad aspects of brain function, including synaptic plasticity, normal development, and neuronal cell death.[35,49–51] In general, NO exerts physiological and some pathophysiological effects via stimulation of guanylate cyclase to form cyclic guanosine-3′,5′-monophosphate (cGMP) or through S-nitros(yl)ation of regulatory protein thiol groups.[9,11,48,52–54] S-Nitrosylation is the covalent addition of an NO group to a critical cysteine thiol/sulfhydryl (RSH or, more properly, thiolate anion, RS^-) to form an SNO derivative (R-SNO). Such modification modulates the function of a broad spectrum of mammalian, plant, and microbial proteins. In general, a consensus motif of amino acids composed of nucleophilic residues (generally an acid and a base) surrounds a critical cysteine, which increases the cysteine sulfhydryl's susceptibility to S-nitrosylation.[55,56] Our group first identified the physiological relevance of S-nitrosylation by showing that NO and related RNS exert paradoxical effects via redox-based mechanisms—NO is neuroprotective via S-nitrosylation of NMDARs (as well as other subsequently discovered targets, including caspases) and yet can also be neurodestructive by formation of peroxynitrite (or, as later discovered, reaction with additional molecules such as MMP-9 and GAPDH)[11,57–64] (Figure 6.1). Over the past decade, accumulating evidence has suggested that S-nitrosylation can regulate the biological activity of a great variety of proteins, in some ways akin to phosphorylation.[11,17,18,20,56,63–71] Chemically, NO is often a good "leaving group," facilitating further oxidation of critical thiol to disulfide bonds among neighboring (vicinal) cysteine residues or, via reaction with ROS, to sulfenic (–SOH), sulfinic (–SO_2H), or sulfonic (–SO_3H) acid derivatization of the protein.[18,20,63,72] Alternatively, S-nitrosylation may possibly produce a nitroxyl disulfide in which the NO group is shared by close cysteine thiols.[73]

Analyses of mice deficient in either nNOS or iNOS confirmed that NO is an important mediator of cell injury and death after excitotoxic stimulation; NO generated from nNOS or iNOS is detrimental to neuronal survival.[74,75] In addition, inhibition of NOS activity ameliorates the progression of disease pathology in animal models of PD, AD, and ALS, suggesting that excess generation of NO plays a pivotal role in the pathogenesis of several neurodegenerative diseases.[76–79] Although the involvement of NO in neurodegeneration has been widely accepted, the chemical relationship between nitrosative stress and accumulation of misfolded proteins has remained obscure. Recent findings, however, have shed light on molecular events underlying this relationship. Specifically, we recently mounted physiological and chemical evidence that S-nitrosylation modulates the (1) ubiquitin E3 ligase activity of parkin[17–19] and (2) chaperone and isomerase activities of PDI,[20] contributing to protein misfolding and neurotoxicity in models of neurodegenerative disorders.

6.6   PARKIN AND THE UBIQUITIN-PROTEASOME SYSTEM

Recent studies on rare genetic forms of PD have found that mutations in the genes encoding parkin (*PARK2*), PINK1 (*PARK6*), α-synuclein (*PARK1/4*), DJ-1 (*PARK7*), ubiquitin C-terminal hydrolase L1 (UCH-L1) (*PARK5*), leucine-rich repeat kinase-2 (LRRK2) (*PARK8*), or ATP13A2 (*PARK9*) can cause PD pathology.[80–87] The discovery that mutations in these genes predispose patients to very rare familial forms of PD has allowed us to begin to understand the mechanism of protein aggregation and neuronal loss in the more common sporadic form of PD. For instance, the identification of α-synuclein as a familial PD gene led to the recognition that one of the major constituents of Lewy bodies in sporadic PD brains is α-synuclein. In addition, identification of errors in the genes encoding parkin (a ubiquitin E3 ligase) and UCH-L1 in rare familial forms of PD has implicated possible dysfunction of the UPS in the pathogenesis of sporadic PD as well. The UPS represents an important mechanism for proteolysis in mammalian cells. Formation of polyubiquitin chains constitutes the signal for proteasomal attack and degradation. An isopeptide bond covalently attaches the C-terminus of the first ubiquitin in a polyubiquitin chain to a lysine residue in the target protein. The cascade of activating (E1), conjugating (E2), and ubiquitin-ligating (E3) types of enzymes catalyzes the conjugation of the ubiquitin chain to proteins. In addition, individual E3 ubiquitin ligases play a key role in the recognition of specific substrates.[88]

Mutations in the parkin gene can cause autosomal recessive juvenile parkinsonism, accounting for some cases of hereditary PD manifest in young patients with onset beginning anywhere from the teenage years through the 40s.[22,80,89] Parkin is a member of a large family of E3 ubiquitin ligases that are related to one another by the presence of RING finger domains. Parkin contains a total of 35 cysteine residues, the majority of which reside within its RING domains, which coordinate a structurally important zinc atom often involved in catalysis.[90] Parkin has two RING finger domains separated by an "in-between RING" (IBR) domain (Figure 6.2). This motif allows parkin to recruit substrate proteins as well as an E2 enzyme (e.g., UbcH7, UbcH8, or UbcH13). Point mutations, stop mutations, truncations, and deletions in both alleles of the parkin gene will eventually cause dysfunction in its activity and are responsible for many cases of autosomal recessive juvenile parkinsonism and rare adult forms of PD. Parkin mutations usually do not facilitate the formation of Lewy bodies, although there is at least one exception—familial PD patients with the R275W parkin mutant manifest Lewy bodies.[91] Biochemical characterization of parkin mutants shows that not all parkin mutations result in loss of parkin E3 ligase activity; some of the familial-associated parkin mutants (e.g., the R275W mutant) have increased ubiquitination activity compared with the wild type.[92–94] Additionally, parkin can mediate the formation of nonclassical and "nondegradative" lysine 63-linked polyubiquitin chains.[95,96] Likewise, parkin can monoubiquitinate Eps15, HSP70, and possibly itself at multiple sites. This finding may explain how some parkin mutations induce formation of Lewy bodies and why proteins are stabilized within the inclusions.

Several putative target substrates have been identified for parkin E3 ligase activity. One group has reported that mutant parkin failed to bind glycosylated α-synuclein for ubiquitination, leading

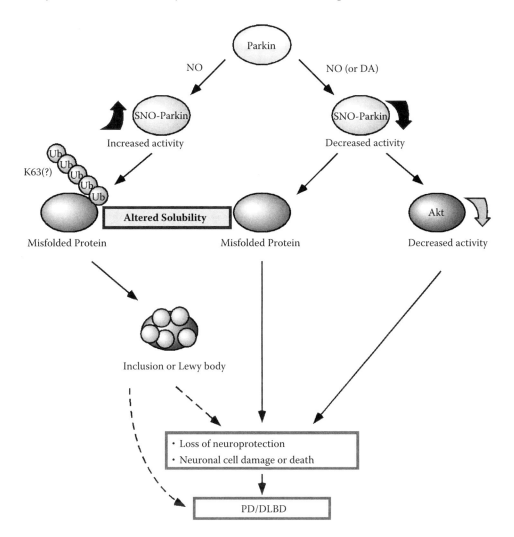

FIGURE 6.2   Possible mechanism of SNO-parkin contributing to the accumulation of aberrant proteins and damage or death of dopaminergic neurons. Nitrosative stress leads to S-nitrosylation of parkin and to an initial dramatic increase that is followed by a decrease in its E3 ubiquitin ligase activity.[17–19] The initial increase in this E3 ubiquitin ligase activity leads to enhanced ubiquitination of parkin substrates (e.g., synphilin-1, Pael-R, and parkin itself). Increased parkin E3 ubiquitin ligase activity may contribute to Lewy body formation and impair parkin function, as also suggested by Sriram et al. [92] The subsequent decrease in parkin activity may allow misfolded proteins to accumulate. Down-regulation of parkin may also result in decreased Akt neuroprotective activity because of enhanced EGFR internalization.[108] DA quinone can also modify cysteine thiols of parkin and reduce its activity.[118] Ub, ubiquitin; DLBD, diffuse Lewy body disease.

to α-synuclein accumulation,[97] but most authorities do not feel that α-synuclein is a direct substrate of parkin. Synphilin-1 (α-synuclein interacting protein), on the other hand, is considered to be a substrate for parkin ubiquitination, and it is included in Lewy body-like inclusions in cultured cells when co-expressed with α-synuclein.[98] Other substrates for parkin include parkin-associated endothelin receptor-like receptor (Pael-R),[99] cell division control related protein,[100] cyclin E,[101] p38 tRNA synthase,[102] synaptotagmin XI,[103] and α/β tubulin heterodimers,[104] as well as possibly parkin itself (auto-ubiquitination). It is generally accepted that accumulation of these substrates can lead to disastrous consequences for the survival of dopaminergic neurons in familial PD and possibly in sporadic PD. Therefore, characterization of potential regulators that affect parkin E3 ligase activity may reveal important molecular mechanisms for the pathogenesis of PD. Heretofore, two cellular components have been shown to regulate the substrate specificity and ubiquitin E3 ligase activity of parkin. The first represents posttranslational modification of parkin through S-nitrosylation (see below for details) or phosphorylation;[105] the second, binding partners of parkin, such as CHIP[106] and BAG5.[107] CHIP enhances the ability of parkin to inhibit cell death through up-regulation of parkin-mediated ubiquitination, while BAG5-mediated inhibition of parkin E3 ligase activity facilitates neuronal cell death. In addition, several groups have recently reported that parkin-mediated monoubiquitination could contribute to neuronal survival via a proteasome-independent pathway.[93,94,108,109] For example, parkin monoubiquitinates the epidermal growth factor receptor (EGFR)-associated protein Eps15, leading to inhibition of EGFR endocytosis.[108] The resulting prolongation of EGFR signaling via the phosphoinositide-3 kinase/Akt (PKB) signaling pathway is postulated to enhance neuronal survival.

Another important molecule that links aberrant UPS activity and PD is the ubiquitin hydrolase UCH-L1, a deubiquitinating enzyme that recycles ubiquitin. Autosomal dominant mutations of UCH-L1 have been identified in two siblings with PD.[84] Interestingly, a recent study suggested that a novel ubiquitin-ubiquitin ligase activity of UCH-L1 might also be important in the pathogenesis of PD.[110] Additional mutations in α-synuclein, DJ-1, PINK1, and LRRK2 may contribute to UPS dysfunction and subsequently lead to PD.

6.7 S-NITROSYLATION AND PARKIN

PD is the second most prevalent neurodegenerative disease and is characterized by the progressive loss of dopamine (DA) neurons in the substantia nigra pars compacta. Appearance of Lewy bodies that contain misfolded and ubiquitinated proteins generally accompanies the loss of dopaminergic neurons in the PD brain. Such ubiquitinated inclusion bodies are the hallmark of many neurodegenerative disorders. Age-associated defects in intracellular proteolysis of misfolded or aberrant proteins might lead to accumulation and, ultimately, deposition of aggregates within neurons or glial cells. Although such aberrant protein accumulation had been observed in patients with genetically encoded mutant proteins, recent evidence from our laboratory suggests that nitrosative stress and oxidative stress are potential causal factors for protein accumulation in the much more common sporadic form of PD. As illustrated below, nitrosative/oxidative stress, commonly found during normal aging, can mimic rare genetic causes of disorders, such as PD, by promoting protein misfolding in the absence of a genetic mutation.[17–19] For example, S-nitrosylation and further oxidation of parkin or UCH-L1 result in dysfunction of these enzymes and thus of the UPS.[17,18,111–114] We and others recently discovered that nitrosative stress triggers S-nitrosylation of parkin (forming SNO-parkin) not only in rodent models of PD but also in the brains of human patients with PD and diffuse Lewy body disease, the related α-synucleinopathy. SNO-parkin initially stimulates ubiquitin E3 ligase activity, resulting in enhanced ubiquitination as observed in Lewy bodies, followed by a decrease in enzyme activity, producing a futile cycle of dysfunctional UPS[18,19,95] (Figure 6.2). We also found that rotenone led to the generation of SNO-parkin and thus dysfunctional ubiquitin E3 ligase activity. Moreover, S-nitrosylation appears to compromise the neuroprotective effect of parkin.[17] These mechanisms involve S-nitrosylation of critical cysteine residues in the first RING domain of parkin.[18] Nitrosative stress and oxidative stress can also alter the solubility of parkin via posttranslational modification of cysteine residues, which may concomitantly compromise its protective function.[115–117] Additionally, it is likely that other ubiquitin E3 ligases with RING-finger thiol motifs are S-nitrosylated in a manner similar to parkin to affect their enzymatic function; hence, S-nitrosylation of E3 ligases may be involved in a number of degenerative conditions.

The neurotransmitter DA may also impair parkin activity and contribute to neuronal demise via the modification of cysteine residues.[118] DA can be oxidized to DA quinone, which can react with and inactivate proteins through covalent modification of cysteine sulfhydryl groups; peroxynitrite has been reported to promote oxidation of DA to form DA quinone.[119] DA quinone can preferentially attack cysteine residues (C268 and C323) in the RING1 and IBR domains of parkin, forming a covalent adduct that abrogates its E3 ubiquitin ligase activity.[116,118] DA quinone also reduces the solubility of parkin, possibly inducing parkin misfolding after disruption of the RING-IBR-RING motif. Therefore, oxidative/nitrosative species may either directly or indirectly contribute to altered parkin activity within the brain, and subsequent loss of parkin-dependent neuroprotection results in increased cell death.

6.8 THE UNFOLDED PROTEIN RESPONSE AND PROTEIN DISULFIDE ISOMERASE

The ER normally participates in protein processing and folding but undergoes a stress response when immature or misfolded proteins accumulate.[120–123] ER stress stimulates two critical intracellular responses (Figure 6.3). The first represents expression of chaperones that prevent protein aggregation via the unfolded protein response (UPR) and is implicated in protein refolding, posttranslational

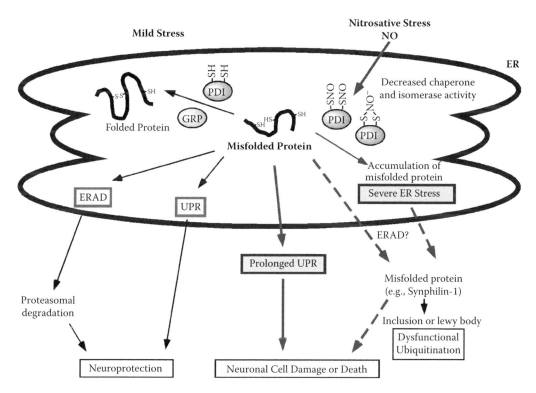

FIGURE 6.3 Possible mechanism of SNO-PDI contributing to the accumulation of aberrant proteins and neuronal cell damage or death. ER stress is triggered when misfolded proteins accumulate within the ER lumen, inducing the UPR. The UPR is usually a transient homeostatic mechanism for cell survival, while a prolonged UPR elicits neuronal cell death. PDI modulates the activity of UPR sensors by mediating proper protein folding in the ER. Proteins that fail to attain their native folded state are eventually retrotranslocated across the ER membrane to be disposed of by cytosolic proteasomes. This process, known as ER associated degradation, is essential in preventing protein accumulation and aggregation in the ER. Under conditions of severe nitrosative stress, S-nitrosylation of neuronal PDI inhibits normal protein folding in the ER, activates ER stress, and induces a prolonged UPR, thus contributing to aberrant protein accumulation and cell damage or death. For simplicity, S-nitrosylation of only one (of two) TRX domain of PDI is shown, resulting in formation of SNO-PDI or possibly nitroxyl-PDI, as described by Uehara et al.[20] and Forrester et al.[143]

assembly of protein complexes, and protein degradation. This response is believed to contribute to adaptation during altered environmental conditions, promoting maintenance of cellular homeostasis. At least three ER transmembrane sensor proteins are involved in the UPR: PKR-like ER kinase, activating transcription factor 6, and inositol-requiring enzyme 1. The activation of all three proximal sensors results in the attenuation of protein synthesis via the eukaryotic initiation factor-2 kinase and increased protein folding capacity of the ER.[124–127] The second ER stress response, termed ER-associated degradation, specifically recognizes terminally misfolded proteins for retrotranslocation across the ER membrane to the cystosol, where they can be degraded by the UPS. Additionally, although severe ER stress can induce apoptosis, the ER withstands relatively mild insults via expression of stress proteins such as glucose-regulated protein and PDI. These proteins behave as molecular chaperones that assist in the maturation, transport, and folding of secretory proteins.

During protein folding in the ER, PDI can introduce disulfide bonds into proteins (oxidation), break disulfide bonds (reduction), and catalyze thiol/disulfide exchange (isomerization), thus facilitating disulfide-bond formation, rearrangement reactions, and structural stability.[128] PDI has four domains that are homologous to thioredoxin (TRX) (termed a, b, b′, and a′). Only two of the four TRX-like domains (a and a′) contain a characteristic redox-active CXXC motif, and these two-thiol/disulfide centers function as independent active sites.[129–132] These active-site cysteines can be found in two redox states: oxidized (disulfide) or reduced (free sulfhydryls or thiols). During oxidation of a target protein, oxidized PDI catalyzes disulfide formation in the substrate protein, resulting in the reduction of PDI. In contrast, the reduced form of the active-site cysteines can initiate isomerization by attacking the disulfide of a substrate protein and forming a transient intermolecular disulfide bond. As a consequence, an intramolecular disulfide rearrangement occurs within the substrate itself, resulting in the generation of reduced PDI. The recently determined structure of yeast PDI revealed that the four TRX-like domains form a twisted "U" shape, with the two active sites facing each other on opposite sides of the "U."[133] Hydrophobic residues line the inside surface of the "U," facilitating interactions between PDI and misfolded proteins. Several mammalian PDI homologues, such as ERp57 and PDIp, also localize to the ER and may manifest similar functions.[134,135] Increased expression of PDIp in neuronal cells under conditions mimicking PD suggests the possible contribution of PDIp to neuronal survival.[134]

In many neurodegenerative disorders and cerebral ischemia, the accumulation of immature and denatured proteins results in ER dysfunction,[134,136–138] but up-regulation of PDI represents an adaptive response promoting protein refolding and may offer neuronal cell protection.[134,135,139,140] In a recent study, we reported that the S-nitrosylation of PDI (to form SNO-PDI) disrupts its neuroprotective role.[20]

6.9 S-NITROSYLATION OF PROTEIN DISULFIDE ISOMERASE MEDIATES PROTEIN MISFOLDING AND NEUROTOXICITY IN CELL MODELS OF PARKINSON'S DISEASE OR ALZHEIMER'S DISEASE

Disturbance of Ca^{2+} homeostasis within the ER plays a critical role in the accumulation of misfolded proteins and ER stress because the function of several ER chaperones requires high concentrations of Ca^{2+}. In addition, it is generally accepted that excessive generation of NO can contribute to activation of the ER stress pathway, at least in some cell types.[141,142] Molecular mechanisms by which NO induces protein misfolding and ER stress, however, have remained enigmatic until recently. The ER normally manifests a relatively positive redox potential in contrast to the highly reducing environment of the cytosol and mitochondria. This redox environment can influence the stability of protein S-nitrosylation and oxidation reactions.[143] S-Nitrosylation can enhance the activity of the ER Ca^{2+} channel-ryanodine receptor,[144] which may provide a clue as to how NO disrupts Ca^{2+} homeostasis in the ER and activates the cell death pathway. Interestingly, we have recently reported that excessive NO can also lead to S-nitrosylation of the active-site thiol groups of PDI, and this reaction inhibits both its isomerase and chaperone activities.[20] Mitochondrial complex I insult by rotenone can also result in S-nitrosylation of PDI in cell culture models. Moreover, we found that PDI is S-nitrosylated in the brains of virtually all cases of sporadic AD and PD examined. Under pathological conditions, it is possible that both cysteine sulfhydryl groups in the TRX-like domains of PDI form SNOs. Unlike formation of a single SNO, which is commonly seen after denitrosylation reactions catalyzed by PDI,[67] dual nitrosylation may be relatively more stable and prevent subsequent disulfide formation on PDI. Therefore, we speculate that these pathological S-nitrosylation reactions on PDI are more easily detected during neurodegenerative conditions. Additionally, it is possible that vicinal (nearby) cysteine thiols reacting with NO can form nitroxyl disulfide,[73] and such reaction may potentially occur in the catalytic side of PDI to inhibit enzymatic activity. In order to determine the consequences of SNO-PDI formation in neurons, we exposed cultured cerebrocortical neurons to neurotoxic concentrations of NMDA, thus inducing excessive Ca^{2+} influx and consequent NO production from nNOS. Under these conditions, we found that PDI was S-nitrosylated in an NOS-dependent manner. SNO-PDI formation led to the accumulation of polyubiquitinated/misfolded proteins and activation of the UPR. Moreover, S-nitrosylation abrogated the inhibitory effect of PDI on aggregation of proteins observed in Lewy body inclusions.[20,98] S-Nitrosylation of PDI also prevented its attenuation of neuronal cell death triggered by ER stress, misfolded proteins, or proteasome inhibition (Figure 6.3). Further evidence suggested that SNO-PDI may in effect transport NO to the extracellular space, where it could conceivably exert additional adverse effects.[67] Additionally, NO can possibly mediate cell death or injury via S-nitrosylation or nitration reactions on other TRX-like proteins, such as TRX itself and glutaredoxin.[66,145,146]

In addition to PDI, S-nitrosylation is likely to affect critical thiol groups on other chaperones, such as HSP90 in the cytoplasm[147] and possibly glucose-regulated protein in the ER. Normally, HSP90 stabilizes misfolded proteins and modulates the activity of cell signaling proteins, including NOS and calcineurin.[15] In AD brains, levels of HSP90 are increased in both the cytosolic and membranous fractions, where HSP90 is thought to maintain tau and Aβ in a soluble conformation, thereby averting their aggregation.[148,149] Martinez-Ruiz et al.[147] recently demonstrated that S-nitrosylation of HSP90 can occur in endothelial cells, and this modification abolishes its ATPase activity, which is required for its function as a molecular chaperone. These studies imply that S-nitrosylation of HSP90 in neurons of AD brains may contribute to the accumulation of tau and Aβ aggregates.

The UPS is apparently impaired in the aging brain. Additionally, inclusion bodies similar to those found in neurodegenerative disorders can appear in brains of normal aged individuals or those with subclinical manifestations of disease.[150] These findings suggest that the activity of the UPS and molecular chaperones may decline in an age-dependent manner.[151] Given that we have not found detectable quantities of SNO-parkin and SNO-PDI in normal aged brain,[17,18,20] we speculate that S-nitrosylation of these and similar proteins may represent a key event that contributes to susceptibility of the aging brain to neurodegenerative conditions.

6.10 POTENTIAL TREATMENT OF EXCESSIVE NMDA-INDUCED CA²⁺ INFLUX AND FREE RADICAL GENERATION

One mechanism that could potentially curtail excessive Ca^{2+} influx and resultant overstimulation of nNOS activity would be inhibition of NMDARs. Until recently, however, drugs in this class blocked virtually all NMDAR activity, including physiological activity, and therefore manifest unacceptable side effects by inhibiting normal functions of the receptor. For this reason, many previous NMDAR antagonists have disappointingly failed in advanced clinical trials conducted for a number of neurodegenerative disorders. In contrast, studies in our laboratory first showed that the adamantane derivative memantine preferentially blocks excessive (pathological) NMDAR activity while relatively sparing normal (physiological) activity. Memantine does this in a surprising fashion because of its low (micromolar) affinity, even though its actions are quite selective for the NMDAR at that concentration (Figure 6.4). "Apparent" affinity of a drug is determined

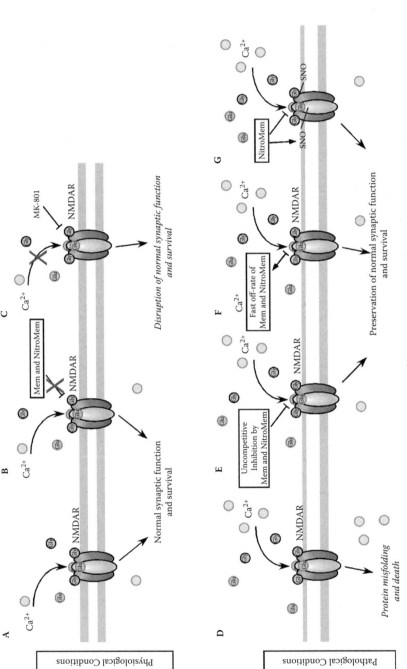

FIGURE 6.4 Memantine/NitroMemantine preferentially blocks excessive/extrasynaptic NMDAR activity. (A) Physiological (synaptic) activation of NMDARs mediates normal function and survival. (B) Memantine (Mem) and NitroMemantine (NitroMem) relatively spare physiological activity of the NMDAR. (C) Usage of classical NMDAR antagonists, such as MK-801, completely blocks receptor activity, including physiological synaptic activity, and thus eventually results in clinical side effects. (D) Excessive activation of the NMDAR, predominantly at extrasynaptic sites, is thought to induce neuronal cell injury and death and is associated with the accumulation of misfolded proteins. (E and F) Mem and NitroMem preferentially block excessive (pathological/extrasynaptic) NMDAR activity while relatively sparing normal (physiological/synaptic) activity. This unique inhibitory action of Mem is achieved by (1) blockade of NMDAR-associated ion channels when they are excessively open in the face of increasing agonist concentrations (uncompetitive inhibition by open-channel block) (E) and (2) relatively low affinity for the target ion channels, predominantly because of a relatively fast off-rate, thus avoiding accumulation in the channels (F). (G) NitroMemantines target NO to the NMDAR by tethering a nitro group to the memantine moiety and thus add to the neuroprotective action of memantine by S-nitrosylating the receptor.

by the ratio of its "on-rate" to its "off-rate" for the target. The on-rate is not only a property of drug diffusion and interaction with the target but also the drug's concentration. In contrast, the off-rate is an intrinsic property of the drug-receptor complex, unaffected by drug concentration. A relatively fast off-rate is a major contributor to memantine's low affinity for the NMDAR. The inhibitory activity of memantine involves blockade of the NMDAR-associated ion channel when it is excessively open (termed open-channel block). The unique and subtle differences of the memantine blocking sites in the channel pore may explain the advantageous properties of memantine action.

Also critical for the clinical tolerability of memantine is its uncompetitive mechanism of action. An uncompetitive antagonist can be distinguished from a noncompetitive antagonist, which acts allosterically at a noncompetitive site (i.e., at a site other than the agonist-binding site). An uncompetitive antagonist is defined as an inhibitor whose action is contingent upon prior activation of the receptor by the agonist. Hence, the same amount of antagonist blocks higher concentrations of agonist relatively better than lower concentrations of agonist. Some open-channel blockers function as pure uncompetitive antagonists, depending on their exact properties of interaction with the ion channel. This uncompetitive mechanism of action coupled with a relatively fast off-rate from the channel yields a drug that preferentially blocks NMDAR-operated channels when they are excessively open while relatively sparing normal neurotransmission. In fact, the relatively fast off-rate is a major contributor to a drug, like memantine's low affinity for the channel pore. While many factors determine a drug's clinical efficacy and tolerability, it appears that the relatively rapid off-rate is a predominant factor in memantine's tolerability in contrast to other NMDA-type receptor antagonists.[7,39] Thus, the critical features of memantine's mode of action are its Uncompetitive mechanism and Fast Off-rate, or what we call a UFO drug—a drug that is present at its site of inhibitory action only when you need it and then quickly disappears.

Interestingly, memantine, which is chemically an adamantine, was first synthesized and patented in 1968 by Eli Lilly and Company, as described in the *Merck Index*. Memantine has been used for many years in Europe to treat PD, spasticity, convulsions, vascular dementia, and later AD.[152,153] Interestingly, the efficacy of adamantine-type drugs in the brain was first discovered by serendipity in a patient taking amantadine for influenza (amantadine is chemically similar to memantine but lacks two side group chains composed of methyl groups). This led scientists to believe that these drugs were dopaminergic or possibly anti-cholinergic, although, as stated above, we later discovered that memantine acts as an open-channel blocker of NMDAR-coupled channel pore; memantine is more potent in this action than amantadine. A large number of studies with *in vitro* and in vivo animal models demonstrated that memantine protects cerebrocortical neurons, cerebellar neurons, and retinal neurons from NMDAR-mediated excitotoxic damage.[154–158] Importantly, in a rat stroke model, memantine, which was given up to 2 hr after the ischemic event, reduced the amount of brain damage by approximately 50%.[155,156] Furthermore, regulatory agencies in both Europe and the United States recently voted its approval as the first treatment for moderate-to-severe AD. It is currently under study for a number of other neurodegenerative disorders, including HIV-associated dementia, Huntington's disease, and ALS, as well as depression.

As promising as the results with memantine are, we are continuing to pursue ways to use additional modulatory sites on the NMDAR to block excitotoxicity even more effectively and safely than memantine alone. New approaches in this regard are explored below.

6.11 FUTURE THERAPEUTICS: NITROMEMANTINES

NitroMemantines are second-generation memantine derivatives that are designed to have enhanced neuroprotective efficacy without sacrificing clinical tolerability. An S-nitrosylation site is located on the extracellular domain of the NMDAR, and S-nitrosylation of this site (i.e., NO reaction with the sulfhydryl group of a critical cysteine residue) down-regulates (but does not completely shut off) receptor activity (Figure 6.4).[7,21] The drug nitroglycerin, which generates NO-related species, can act at this site to limit excessive NMDAR activity. In fact, in rodent models, nitroglycerin can limit ischemic damage,[159] and there is some evidence that patients taking nitroglycerin for other medical reasons may be resistant to glaucomatous visual field loss.[160] Consequently, we carefully characterized the S-nitrosylation sites on the NMDAR in order to determine if we could design a nitroglycerin-like drug that could be more specifically targeted to the receptor. In brief, we found that five cysteine residues on the NMDAR could interact with NO. One of these, located at cysteine residue 399 (Cys399) on the NR2A subunit of the NMDAR, mediates $\geq 90\%$ of the effect of NO under our experimental conditions.[62] From crystal structure models and electrophysiological experiments, we further found that NO binding to the NMDAR at Cys399 may induce a conformational change in the receptor protein that makes glutamate and Zn^{2+} bind more tightly to the receptor. The enhanced binding of glutamate and Zn^{2+} in turn causes the receptor to desensitize and, consequently, the ion channel to close.[71] Electrophysiological studies have demonstrated this inhibitory effect of NO on the NMDAR-associated channel.[11,53,62] Moreover, as the oxygen tension is lowered (a pO_2 of 10–20 Torr is found in normal brain, and even lower levels are found under hypoxic/ischemic conditions), the NMDAR becomes more sensitive to inhibition by S-nitrosylation.[161]

Unfortunately, nitroglycerin itself is not very attractive as a neuroprotective agent. The same cardiovascular vasodilator effect that makes it useful in the treatment of angina could cause dangerously large drops in blood pressure in patients with dementia, stroke, traumatic injury, or glaucoma. However, the open-channel block mechanism of memantine not only leads to a higher degree of channel blockade in the presence of excessive levels of glutamate but also can be used as a homing signal for targeting drugs (e.g., the NO group) to hyperactivated, open NMDA-gated channels. We have therefore been developing combinatorial drugs (NitroMemantines) that theoretically should be able to use memantine to target NO to the nitrosylation sites of the NMDAR in order to avoid the systemic side effects of NO. Two sites of modulation would be analogous to having two volume controls on your television set for fine-tuning the audio signal.

Preliminary studies have shown NitroMemantines to be highly neuroprotective in both *in vitro* and in vivo animal models.[7] In fact, they appear to be more effective than memantine at lower dosage. Moreover, because of the targeting effect of the memantine moiety, NitroMemantines appear to lack the blood pressure-lowering effects typical of nitroglycerin. More research still needs to be performed on NitroMemantine drugs, but by combining two clinically tolerated drugs (memantine and nitroglycerin), we have created a new and improved class of UFO drugs that should be both clinically tolerated and neuroprotective.

6.12   SUMMARY AND CONCLUSIONS

Excessive nitrosative stress and oxidative stress triggered by excessive NMDAR activation and/or mitochondrial dysfunction may result in malfunction of the UPS or molecular chaperones, thus contributing to abnormal protein accumulation and neuronal damage in sporadic forms of neurodegenerative diseases. Our elucidation of an NO-mediated pathway to dysfunction of parkin and PDI by S-nitrosylation provides a mechanistic link between free radical production, abnormal protein accumulation, and neuronal cell injury in neurodegenerative disorders such as PD. Elucidation of this new pathway may lead to the development of additional new therapeutic approaches to prevent aberrant protein misfolding by targeted disruption or prevention of nitrosylation of specific proteins such as parkin and PDI. This article also describes the action of memantine via uncompetitive antagonism of the NMDAR with a fast off-rate. NitroMemantines enhance the neuroprotective efficacy over memantine at a given dose owing to its additional ability to S-nitrosylate the NMDAR. These drugs preferentially inhibit pathologically activated NMDAR while preserving its normal synaptic function; thus, they are clinically tolerated. In this chapter, we propose that the next generation of CNS drugs will interact with their target only during states of pathological activation and not interfere with the target if it is functioning properly. In the future, such perspectives should lead to additional novel, clinically tolerated neuroprotective therapeutics.

ACKNOWLEDGMENTS

This work was supported in part by the Japan Society for the Promotion of Science through a Postdoctoral Fellowship for Research Abroad (to T.N.); the National Institutes of Health through grants P01 HD29587, R01 EY05477, and R01 EY09024; the American Parkinson's Disease Association, San Diego Chapter; and the Ellison Medical Foundation through a Senior Scholars Award in Aging (to S.A.L.).

REFERENCES

1. Lin, M. T., and Beal, M. F., Mitochondrial dysfunction and oxidative stress in neurodegenerative diseases, *Nature* 443, 787, 2006.
2. Barnham, K. J., Masters, C. L., and Bush, A. I., Neurodegenerative diseases and oxidative stress, *Nat Rev Drug Discov* 3, 205, 2004.
3. Muchowski, P. J., Protein misfolding, amyloid formation, and neurodegeneration: A critical role for molecular chaperones? *Neuron* 35, 9, 2002.
4. Emerit, J., Edeas, M., and Bricaire, F., Neurodegenerative diseases and oxidative stress, *Biomed Pharmacother* 58, 39, 2004.
5. Beal, M. F., Experimental models of Parkinson's disease, *Nat Rev Neurosci* 2, 325, 2001.
6. Lipton, S. A., and Rosenberg, P. A., Excitatory amino acids as a final common pathway for neurologic disorders, *N Engl J Med* 330, 613, 1994.
7. Lipton, S. A., Paradigm shift in neuroprotection by NMDA receptor blockade: Memantine and beyond, *Nat Rev Drug Discov* 5, 160, 2006.
8. Bredt, D. S. et al., Cloned and expressed nitric oxide synthase structurally resembles cytochrome P-450 reductase, *Nature* 351, 714, 1991.
9. Garthwaite, J., Charles, S. L., and Chess-Williams, R., Endothelium-derived relaxing factor release on activation of NMDA receptors suggests role as intercellular messenger in the brain, *Nature* 336, 385, 1988.
10. Beckman, J. S. et al., Apparent hydroxyl radical production by peroxynitrite: Implications for endothelial injury from nitric oxide and superoxide, *Proc Natl Acad Sci U S A* 87, 1620, 1990.
11. Lipton, S. A. et al., A redox-based mechanism for the neuroprotective and neurodestructive effects of nitric oxide and related nitroso-compounds, *Nature* 364, 626, 1993.
12. Bence, N. F., Sampat, R. M., and Kopito, R. R., Impairment of the ubiquitin-proteasome system by protein aggregation, *Science* 292, 1552, 2001.
13. Chaudhuri, T. K., and Paul, S., Protein-misfolding diseases and chaperone-based therapeutic approaches, *FEBS J* 273, 1331, 2006.
14. Ciechanover, A., and Brundin, P., The ubiquitin proteasome system in neurodegenerative diseases: Sometimes the chicken, sometimes the egg, *Neuron* 40, 427, 2003.
15. Muchowski, P. J., and Wacker, J. L., Modulation of neurodegeneration by molecular chaperones, *Nat Rev Neurosci* 6, 11, 2005.
16. Zhang, K., and Kaufman, R. J., The unfolded protein response: A stress signaling pathway critical for health and disease, *Neurology* 66, S102, 2006.
17. Chung, K. K. et al., S-nitrosylation of parkin regulates ubiquitination and compromises parkin's protective function, *Science* 304, 1328, 2004.
18. Yao, D. et al., Nitrosative stress linked to sporadic Parkinson's disease: S-nitrosylation of parkin regulates its E3 ubiquitin ligase activity, *Proc Natl Acad Sci U S A* 101, 10810, 2004.
19. Lipton, S. A. et al., Comment on "S-nitrosylation of parkin regulates ubiquitination and compromises parkin's protective function," *Science* 308, 1870; author reply 1870, 2005.
20. Uehara, T. et al., S-nitrosylated protein-disulphide isomerase links protein misfolding to neurodegeneration, *Nature* 441, 513, 2006.
21. Lipton, S. A., Pathologically activated therapeutics for neuroprotection, *Nat Rev Neurosci* 8, 803, 2007.
22. Cookson, M. R., The biochemistry of Parkinson's disease, *Annu Rev Biochem* 74, 29, 2005.
23. Zhao, L. et al., Protein accumulation and neurodegeneration in the woozy mutant mouse is caused by disruption of SIL1, a cochaperone of BiP, *Nat Genet* 37, 974, 2005.
24. McNaught, K. S. et al., Systemic exposure to proteasome inhibitors causes a progressive model of Parkinson's disease, *Ann Neurol* 56, 149, 2004.
25. Auluck, P. K. et al., Chaperone suppression of alpha-synuclein toxicity in a *Drosophila* model for Parkinson's disease, *Science* 295, 865, 2002.

26. Arrasate, M. et al., Inclusion body formation reduces levels of mutant huntingtin and the risk of neuronal death, *Nature* 431, 805, 2004.

27. Hardingham, G. E., Fukunaga, Y., and Bading, H., Extrasynaptic NMDARs oppose synaptic NMDARs by triggering CREB shut-off and cell death pathways, *Nat Neurosci* 5, 405, 2002.

28. Papadia, S. et al., Nuclear Ca^{2+} and the cAMP response element-binding protein family mediate a late phase of activity-dependent neuroprotection, *J Neurosci* 25, 4279, 2005.

29. Papadia, S. et al., Synaptic NMDA receptor activity boosts intrinsic antioxidant defenses, *Nat Neurosci* 11, 476, 2008.

30. Ankarcrona, M. et al., Glutamate-induced neuronal death: A succession of necrosis or apoptosis depending on mitochondrial function, *Neuron* 15, 961, 1995.

31. Bonfoco, E. et al., Apoptosis and necrosis: Two distinct events induced, respectively, by mild and intense insults with *N*-methyl-d-aspartate or nitric oxide/superoxide in cortical cell cultures, *Proc Natl Acad Sci U S A* 92, 7162, 1995.

32. Budd, S. L. et al., Mitochondrial and extramitochondrial apoptotic signaling pathways in cerebrocortical neurons, *Proc Natl Acad Sci U S A* 97, 6161, 2000.

33. Nowak, L. et al., Magnesium gates glutamate-activated channels in mouse central neurones, *Nature* 307, 462, 1984.

34. Mayer, M. L., Westbrook, G. L., and Guthrie, P. B., Voltage-dependent block by Mg^{2+} of NMDA responses in spinal cord neurones, *Nature* 309, 261, 1984.

35. Dawson, V. L. et al., Nitric oxide mediates glutamate neurotoxicity in primary cortical cultures, *Proc Natl Acad Sci U S A* 88, 6368, 1991.

36. Lafon-Cazal, M. et al., NMDA-dependent superoxide production and neurotoxicity, *Nature* 364, 535, 1993.

37. Olney, J. W., Brain lesions, obesity, and other disturbances in mice treated with monosodium glutamate, *Science* 164, 719, 1969.

38. Olney, J. W., Wozniak, D. F., and Farber, N. B., Excitotoxic neurodegeneration in Alzheimer disease. New hypothesis and new therapeutic strategies, *Arch Neurol* 54, 1234, 1997.

39. Chen, H. S., and Lipton, S. A., The chemical biology of clinically tolerated NMDA receptor antagonists, *J Neurochem* 97, 1611, 2006.

40. Abu-Soud, H. M., and Stuehr, D. J., Nitric oxide synthases reveal a role for calmodulin in controlling electron transfer, *Proc Natl Acad Sci U S A* 90, 10769, 1993.

41. Forstermann, U., Boissel, J. P., and Kleinert, H., Expressional control of the "constitutive" isoforms of nitric oxide synthase (NOS I and NOS III), *FASEB J* 12, 773, 1998.

42. Boucher, J. L., Moali, C., and Tenu, J. P., Nitric oxide biosynthesis, nitric oxide synthase inhibitors and arginase competition for l-arginine utilization, *Cell Mol Life Sci* 55, 1015, 1999.

43. Groves, J. T., and Wang, C. C., Nitric oxide synthase: Models and mechanisms, *Curr Opin Chem Biol* 4, 687, 2000.

44. Betarbet, R. et al., Chronic systemic pesticide exposure reproduces features of Parkinson's disease, *Nat Neurosci* 3, 1301, 2000.

45. He, Y. et al., Role of nitric oxide in rotenone-induced nigro-striatal injury, *J Neurochem* 86, 1338, 2003.

46. Abou-Sleiman, P. M., Muqit, M. M., and Wood, N. W., Expanding insights of mitochondrial dysfunction in Parkinson's disease, *Nat Rev Neurosci* 7, 207, 2006.

47. Castello, P. R. et al., Mitochondrial cytochrome oxidase produces nitric oxide under hypoxic conditions: Implications for oxygen sensing and hypoxic signaling in eukaryotes, *Cell Metab* 3, 277, 2006.

48. Isaacs, A. M. et al., Acceleration of amyloid beta-peptide aggregation by physiological concentrations of calcium, *J Biol Chem* 281, 27916, 2006.

49. O'Dell, T. J. et al., Tests of the roles of two diffusible substances in long-term potentiation: Evidence for nitric oxide as a possible early retrograde messenger, *Proc Natl Acad Sci U S A* 88, 11285, 1991.

50. Bredt, D. S., and Snyder, S. H., Nitric oxide: A physiologic messenger molecule, *Annu Rev Biochem* 63, 175, 1994.

51. Schuman, E. M., and Madison, D. V., Locally distributed synaptic potentiation in the hippocampus, *Science* 263, 532, 1994.

52. Stamler, J. S. et al., S-nitrosylation of proteins with nitric oxide: Synthesis and characterization of biologically active compounds, *Proc Natl Acad Sci U S A* 89, 444, 1992.

53. Lei, S. Z. et al., Effect of nitric oxide production on the redox modulatory site of the NMDA receptor-channel complex, *Neuron* 8, 1087, 1992.

54. Kandel, E. R., and O'Dell, T. J., Are adult learning mechanisms also used for development? *Science* 258, 243, 1992.

55. Stamler, J. S. et al., (S)NO signals: Translocation, regulation, and a consensus motif, *Neuron* 18, 691, 1997.
56. Hess, D. T. et al., Protein S-nitrosylation: Purview and parameters, *Nat Rev Mol Cell Biol* 6, 150, 2005.
57. Melino, G. et al., S-nitrosylation regulates apoptosis, *Nature* 388, 432, 1997.
58. Tenneti, L., D'Emilia, D. M., and Lipton, S. A., Suppression of neuronal apoptosis by S-nitrosylation of caspases, *Neurosci Lett* 236, 139, 1997.
59. Dimmeler, S. et al., Suppression of apoptosis by nitric oxide via inhibition of interleukin-1beta-converting enzyme (ICE)-like and cysteine protease protein (CPP)-32-like proteases, *J Exp Med* 185, 601, 1997.
60. Mannick, J. B. et al., Fas-induced caspase denitrosylation, *Science* 284, 651, 1999.
61. Kim, W. K. et al., Attenuation of NMDA receptor activity and neurotoxicity by nitroxyl anion, NO, *Neuron* 24, 461, 1999.
62. Choi, Y. B. et al., Molecular basis of NMDA receptor-coupled ion channel modulation by S-nitrosylation, *Nat Neurosci* 3, 15, 2000.
63. Gu, Z. et al., S-nitrosylation of matrix metalloproteinases: Signaling pathway to neuronal cell death, *Science* 297, 1186, 2002.
64. Hara, M. R. et al., S-nitrosylated GAPDH initiates apoptotic cell death by nuclear translocation following Siah1 binding, *Nat Cell Biol* 7, 665, 2005.
65. Jaffrey, S. R. et al., Protein S-nitrosylation: A physiological signal for neuronal nitric oxide, *Nat Cell Biol* 3, 193, 2001.
66. Haendeler, J. et al., Redox regulatory and anti-apoptotic functions of thioredoxin depend on S-nitrosylation at cysteine 69, *Nat Cell Biol* 4, 743, 2002.
67. Sliskovic, I., Raturi, A., and Mutus, B., Characterization of the S-denitrosation activity of protein disulfide isomerase, *J Biol Chem* 280, 8733, 2005.
68. Stamler, J. S., Singel, D. J., and Loscalzo, J., Biochemistry of nitric oxide and its redox-activated forms, *Science* 258, 1898, 1992.
69. Stamler, J. S., Redox signaling: Nitrosylation and related target interactions of nitric oxide, *Cell* 78, 931, 1994.
70. Stamler, J. S., Lamas, S., and Fang, F. C., Nitrosylation. The prototypic redox-based signaling mechanism, *Cell* 106, 675, 2001.
71. Lipton, S. A. et al., Cysteine regulation of protein function — As exemplified by NMDA-receptor modulation, *Trends Neurosci* 25, 474, 2002.
72. Stamler, J. S., and Hausladen, A., Oxidative modifications in nitrosative stress, *Nat Struct Biol* 5, 247, 1998.
73. Houk, K. N. et al., Nitroxyl disulfides, novel intermediates in transnitrosation reactions, *J Am Chem Soc* 125, 6972, 2003.
74. Huang, Z. et al., Effects of cerebral ischemia in mice deficient in neuronal nitric oxide synthase, *Science* 265, 1883, 1994.
75. Iadecola, C. et al., Delayed reduction of ischemic brain injury and neurological deficits in mice lacking the inducible nitric oxide synthase gene, *J Neurosci* 17, 9157, 1997.
76. Hantraye, P. et al., Inhibition of neuronal nitric oxide synthase prevents MPTP-induced parkinsonism in baboons, *Nat Med* 2, 1017, 1996.
77. Przedborski, S. et al., Role of neuronal nitric oxide in 1-methyl-4-phenyl-1,2,3,6-tetrahydropyridine (MPTP)-induced dopaminergic neurotoxicity, *Proc Natl Acad Sci U S A* 93, 4565, 1996.
78. Liberatore, G. T. et al., Inducible nitric oxide synthase stimulates dopaminergic neurodegeneration in the MPTP model of Parkinson disease, *Nat Med* 5, 1403, 1999.
79. Chabrier, P. E., Demerle-Pallardy, C., and Auguet, M., Nitric oxide synthases: Targets for therapeutic strategies in neurological diseases, *Cell Mol Life Sci* 55, 1029, 1999.
80. Kitada, T. et al., Mutations in the parkin gene cause autosomal recessive juvenile parkinsonism, *Nature* 392, 605, 1998.
81. Valente, E. M. et al., Hereditary early-onset Parkinson's disease caused by mutations in PINK1, *Science* 304, 1158, 2004.
82. Polymeropoulos, M. H. et al., Mutation in the alpha-synuclein gene identified in families with Parkinson's disease, *Science* 276, 2045, 1997.
83. Bonifati, V. et al., Mutations in the DJ-1 gene associated with autosomal recessive early-onset parkinsonism, *Science* 299, 256, 2003.
84. Leroy, E. et al., The ubiquitin pathway in Parkinson's disease, *Nature* 395, 451, 1998.
85. Paisan-Ruiz, C. et al., Cloning of the gene containing mutations that cause PARK8-linked Parkinson's disease, *Neuron* 44, 595, 2004.

86. Zimprich, A. et al., Mutations in LRRK2 cause autosomal-dominant parkinsonism with pleomorphic pathology, *Neuron* 44, 601, 2004.
87. Ramirez, A. et al., Hereditary parkinsonism with dementia is caused by mutations in ATP13A2, encoding a lysosomal type 5 P-type ATPase, *Nat Genet* 38, 1184, 2006.
88. Ross, C. A., and Pickart, C. M., The ubiquitin-proteasome pathway in Parkinson's disease and other neurodegenerative diseases, *Trends Cell Biol* 14, 703, 2004.
89. Shimura, H. et al., Familial Parkinson disease gene product, parkin, is a ubiquitin-protein ligase, *Nat Genet* 25, 302, 2000.
90. Marin, I., and Ferrus, A., Comparative genomics of the RBR family, including the Parkinson's disease-related gene parkin and the genes of the Ariadne subfamily, *Mol Biol Evol* 19, 2039, 2002.
91. Farrer, M. et al., Lewy bodies and parkinsonism in families with parkin mutations, *Ann Neurol* 50, 293, 2001.
92. Sriram, S. R. et al., Familial-associated mutations differentially disrupt the solubility, localization, binding and ubiquitination properties of parkin, *Hum Mol Genet* 14, 2571, 2005.
93. Hampe, C. et al., Biochemical analysis of Parkinson's disease-causing variants of parkin, an E3 ubiquitin-protein ligase with monoubiquitylation capacity, *Hum Mol Genet* 15, 2059, 2006.
94. Matsuda, N. et al., Diverse effects of pathogenic mutations of parkin that catalyze multiple monoubiquitylation *in vitro*, *J Biol Chem* 281, 3204, 2006.
95. Lim, K. L. et al., Parkin mediates nonclassical, proteasomal-independent ubiquitination of synphilin-1: Implications for Lewy body formation, *J Neurosci* 25, 2002, 2005.
96. Lim, K. L., Dawson, V. L., and Dawson, T. M., Parkin-mediated lysine 63-linked polyubiquitination: A link to protein inclusions formation in Parkinson's and other conformational diseases? *Neurobiol Aging* 27, 524, 2006.
97. Shimura, H. et al., Ubiquitination of a new form of alpha-synuclein by parkin from human brain: Implications for Parkinson's disease, *Science* 293, 263, 2001.
98. Chung, K. K. et al., Parkin ubiquitinates the alpha-synuclein-interacting protein, synphilin-1: Implications for Lewy-body formation in Parkinson disease, *Nat Med* 7, 1144, 2001.
99. Imai, Y. et al., An unfolded putative transmembrane polypeptide, which can lead to endoplasmic reticulum stress, is a substrate of parkin, *Cell* 105, 891, 2001.
100. Zhang, Y. et al., Parkin functions as an E2-dependent ubiquitin-protein ligase and promotes the degradation of the synaptic vesicle-associated protein, CDCrel-1, *Proc Natl Acad Sci U S A* 97, 13354, 2000.
101. Staropoli, J. F. et al., Parkin is a component of an SCF-like ubiquitin ligase complex and protects postmitotic neurons from kainate excitotoxicity, *Neuron* 37, 735, 2003.
102. Corti, O. et al., The p38 subunit of the aminoacyl-tRNA synthetase complex is a parkin substrate: Linking protein biosynthesis and neurodegeneration, *Hum Mol Genet* 12, 1427, 2003.
103. Huynh, D. P. et al., The autosomal recessive juvenile Parkinson disease gene product, parkin, interacts with and ubiquitinates synaptotagmin XI, *Hum Mol Genet* 12, 2587, 2003.
104. Ren, Y., Zhao, J., and Feng, J., Parkin binds to alpha/beta tubulin and increases their ubiquitination and degradation, *J Neurosci* 23, 3316, 2003.
105. Yamamoto, A. et al., Parkin phosphorylation and modulation of its E3 ubiquitin ligase activity, *J Biol Chem* 280, 3390, 2005.
106. Imai, Y. et al., CHIP is associated with parkin, a gene responsible for familial Parkinson's disease, and enhances its ubiquitin ligase activity, *Mol Cell* 10, 55, 2002.
107. Kalia, S. K. et al., BAG5 inhibits parkin and enhances dopaminergic neuron degeneration, *Neuron* 44, 931, 2004.
108. Fallon, L. et al., A regulated interaction with the UIM protein Eps15 implicates parkin in EGF receptor trafficking and PI(3)K-Akt signalling, *Nat Cell Biol* 8, 834, 2006.
109. Moore, D. J. et al., Parkin mediates the degradation-independent ubiquitination of Hsp70, *J Neurochem* 105, 1806, 2008.
110. Liu, Y. et al., The UCH-L1 gene encodes two opposing enzymatic activities that affect alpha-synuclein degradation and Parkinson's disease susceptibility, *Cell* 111, 209, 2002.
111. Nishikawa, K. et al., Alterations of structure and hydrolase activity of parkinsonism-associated human ubiquitin carboxyl-terminal hydrolase L1 variants, *Biochem Biophys Res Commun* 304, 176, 2003.
112. Choi, J. et al., Oxidative modifications and down-regulation of ubiquitin carboxyl-terminal hydrolase L1 associated with idiopathic Parkinson's and Alzheimer's diseases, *J Biol Chem* 279, 13256, 2004.
113. Chung, K. K., Dawson, T. M., and Dawson, V. L., Nitric oxide, S-nitrosylation and neurodegeneration, *Cell Mol Biol (Noisy-le-grand)* 51, 247, 2005.

114. Gu, Z. et al., Nitrosative and oxidative stress links dysfunctional ubiquitination to Parkinson's disease, *Cell Death Differ* 12, 1202, 2005.

115. Wang, C. et al., Stress-induced alterations in parkin solubility promote parkin aggregation and compromise parkin's protective function, *Hum Mol Genet* 14, 3885, 2005.

116. Wong, E. S. et al., Relative sensitivity of parkin and other cysteine-containing enzymes to stress-induced solubility alterations, *J Biol Chem*, 282, 12310, 2007.

117. LaVoie, M. J. et al., The effects of oxidative stress on parkin and other E3 ligases, *J Neurochem* 103, 2354, 2007.

118. LaVoie, M. J. et al., Dopamine covalently modifies and functionally inactivates parkin, *Nat Med* 11, 1214, 2005.

119. LaVoie, M. J., and Hastings, T. G., Peroxynitrite- and nitrite-induced oxidation of dopamine: Implications for nitric oxide in dopaminergic cell loss, *J Neurochem* 73, 2546, 1999.

120. Andrews, D. W., and Johnson, A. E., The translocon: More than a hole in the ER membrane? *Trends Biochem Sci* 21, 365, 1996.

121. Sidrauski, C., Chapman, R., and Walter, P., The unfolded protein response: An intracellular signalling pathway with many surprising features, *Trends Cell Biol* 8, 245, 1998.

122. Szegezdi, E. et al., Mediators of endoplasmic reticulum stress-induced apoptosis, *EMBO Rep* 7, 880, 2006.

123. Ellgaard, L., Molinari, M., and Helenius, A., Setting the standards: Quality control in the secretory pathway, *Science* 286, 1882, 1999.

124. Kaufman, R. J., Stress signaling from the lumen of the endoplasmic reticulum: Coordination of gene transcriptional and translational controls, *Genes Dev* 13, 1211, 1999.

125. Mori, K., Tripartite management of unfolded proteins in the endoplasmic reticulum, *Cell* 101, 451, 2000.

126. Patil, C., and Walter, P., Intracellular signaling from the endoplasmic reticulum to the nucleus: The unfolded protein response in yeast and mammals, *Curr Opin Cell Biol* 13, 349, 2001.

127. Yoshida, H. et al., XBP1 mRNA is induced by ATF6 and spliced by IRE1 in response to ER stress to produce a highly active transcription factor, *Cell* 107, 881, 2001.

128. Lyles, M. M., and Gilbert, H. F., Catalysis of the oxidative folding of ribonuclease A by protein disulfide isomerase: Dependence of the rate on the composition of the redox buffer, *Biochemistry* 30, 613, 1991.

129. Edman, J. C. et al., Sequence of protein disulphide isomerase and implications of its relationship to thioredoxin, *Nature* 317, 267, 1985.

130. Vuori, K. et al., Site-directed mutagenesis of human protein disulphide isomerase: Effect on the assembly, activity and endoplasmic reticulum retention of human prolyl 4-hydroxylase in *Spodoptera frugiperda* insect cells, *EMBO J* 11, 4213, 1992.

131. Ellgaard, L., and Ruddock, L. W., The human protein disulphide isomerase family: Substrate interactions and functional properties, *EMBO Rep* 6, 28, 2005.

132. Gruber, C. W. et al., Protein disulfide isomerase: The structure of oxidative folding, *Trends Biochem Sci* 31, 455, 2006.

133. Tian, G. et al., The crystal structure of yeast protein disulfide isomerase suggests cooperativity between its active sites, *Cell* 124, 61, 2006.

134. Conn, K. J. et al., Identification of the protein disulfide isomerase family member PDIp in experimental Parkinson's disease and Lewy body pathology, *Brain Res* 1022, 164, 2004.

135. Hetz, C. et al., The disulfide isomerase Grp58 is a protective factor against prion neurotoxicity, *J Neurosci* 25, 2793, 2005.

136. Hu, B. R. et al., Protein aggregation after transient cerebral ischemia, *J Neurosci* 20, 3191, 2000.

137. Rao, R. V., and Bredesen, D. E., Misfolded proteins, endoplasmic reticulum stress and neurodegeneration, *Curr Opin Cell Biol* 16, 653, 2004.

138. Atkin, J. D. et al., Induction of the unfolded protein response in familial amyotrophic lateral sclerosis and association of protein-disulfide isomerase with superoxide dismutase 1, *J Biol Chem* 281, 30152, 2006.

139. Tanaka, S., Uehara, T., and Nomura, Y., Up-regulation of protein-disulfide isomerase in response to hypoxia/brain ischemia and its protective effect against apoptotic cell death, *J Biol Chem* 275, 10388, 2000.

140. Ko, H. S., Uehara, T., and Nomura, Y., Role of ubiquilin associated with protein-disulfide isomerase in the endoplasmic reticulum in stress-induced apoptotic cell death, *J Biol Chem* 277, 35386, 2002.

141. Gotoh, T. et al., Nitric oxide-induced apoptosis in RAW 264.7 macrophages is mediated by endoplasmic reticulum stress pathway involving ATF6 and CHOP, *J Biol Chem* 277, 12343, 2002.

142. Oyadomari, S. et al., Nitric oxide-induced apoptosis in pancreatic beta cells is mediated by the endoplasmic reticulum stress pathway, *Proc Natl Acad Sci U S A* 98, 10845, 2001.
143. Forrester, M. T., Benhar, M., and Stamler, J. S., Nitrosative stress in the ER: A new role for S-nitrosylation in neurodegenerative diseases, *ACS Chem Biol* 1, 355, 2006.
144. Xu, L. et al., Activation of the cardiac calcium release channel (ryanodine receptor) by poly-S-nitrosylation, *Science* 279, 234, 1998.
145. Tao, L. et al., Nitrative inactivation of thioredoxin-1 and its role in postischemic myocardial apoptosis, *Circulation* 114, 1395, 2006.
146. Aracena-Parks, P. et al., Identification of cysteines involved in S-nitrosylation, S-glutathionylation, and oxidation to disulfides in ryanodine receptor type 1, *J Biol Chem* 281, 40354, 2006.
147. Martinez-Ruiz, A. et al., S-nitrosylation of Hsp90 promotes the inhibition of its ATPase and endothelial nitric oxide synthase regulatory activities, *Proc Natl Acad Sci U S A* 102, 8525, 2005.
148. Kakimura, J. et al., Microglial activation and amyloid-beta clearance induced by exogenous heat-shock proteins, *FASEB J* 16, 601, 2002.
149. Dou, F. et al., Chaperones increase association of tau protein with microtubules, *Proc Natl Acad Sci U S A* 100, 721, 2003.
150. Gray, D. A., Tsirigotis, M., and Woulfe, J., Ubiquitin, proteasomes, and the aging brain, *Sci Aging Knowledge Environ* 2003, RE6, 2003.
151. Paz Gavilan, M. et al., Cellular environment facilitates protein accumulation in aged rat hippocampus, *Neurobiol Aging* 27, 973, 2006.
152. Ditzler, K., Efficacy and tolerability of memantine in patients with dementia syndrome. A double-blind, placebo controlled trial, *Arzneimittelforschung* 41, 773, 1991.
153. Fleischhacker, W. W., Buchgeher, A., and Schubert, H., Memantine in the treatment of senile dementia of the Alzheimer type, *Prog Neuropsychopharmacol Biol Psychiatry* 10, 87, 1986.
154. Parsons, C. G., Danysz, W., and Quack, G., Memantine is a clinically well tolerated *N*-methyl-d-aspartate (NMDA) receptor antagonist — A review of preclinical data, *Neuropharmacology* 38, 735, 1999.
155. Chen, H. S. et al., Open-channel block of *N*-methyl-d-aspartate (NMDA) responses by memantine: Therapeutic advantage against NMDA receptor-mediated neurotoxicity, *J Neurosci* 12, 4427, 1992.
156. Chen, H. S. et al., Neuroprotective concentrations of the *N*-methyl-d-aspartate open-channel blocker memantine are effective without cytoplasmic vacuolation following post-ischemic administration and do not block maze learning or long-term potentiation, *Neuroscience* 86, 1121, 1998.
157. Lipton, S. A., Memantine prevents HIV coat protein-induced neuronal injury *in vitro*, *Neurology* 42, 1403, 1992.
158. Osborne, N. N., Memantine reduces alterations to the mammalian retina, in situ, induced by ischemia, *Vis Neurosci* 16, 45, 1999.
159. Lipton, S. A., and Wang, Y. F., NO-related species can protect from focal cerebral ischemia/reperfusion, in *Pharmacology of Cerebral Ischemia*, Krieglstein, J., and Oberpichler-Schwenk, H., Eds., Wissenschaftliche Verlagsgesellschaft, Stuttgart, Germany, 1996, pp. 183.
160. Zurakowski, D. et al., Nitrate therapy may retard glaucomatous optic neuropathy, perhaps through modulation of glutamate receptors, *Vision Res* 38, 1489, 1998.
161. Takahashi, H. et al., Hypoxia enhances *S*-nitrosylation-mediated NMDA receptor inhibition via a thiol oxygen sensor motif, *Neuron* 53, 53, 2007.

7 Cognitive and Behavioral Consequences of Iron Deficiency in Women of Reproductive Age

Laura E. Murray-Kolb

Center for Human Nutrition, Department of International Health, Bloomberg School of Public Health, Johns Hopkins University, Baltimore, Maryland, USA

CONTENTS

7.1 INTRODUCTION

Iron deficiency remains the greatest single nutrient deficiency worldwide despite widespread efforts to address this problem.[1] While it affects individuals of all ages and classes, it disproportionately affects infants, children, and women of reproductive age.[2] Global anemia prevalence rates in children range from 25.4% (school-aged) to 47.4% (preschool-aged).[2] In women, the global prevalence rates are 30.2% for non-pregnant women and 41.8% for pregnant women.[2] While the underlying causes of this anemia are not always immediately apparent, iron deficiency is thought to contribute to at least half of the anemia prevalence worldwide. In 2002, the World Health Organization listed iron deficiency anemia as one of the ten greatest global health risks with consequences of increasing morbidity and mortality in preschool-aged children and pregnant women.[3] Although the highest prevalence of iron deficiency is found in non-industrialized countries, it remains a problem even in industrialized areas of the world. In the United States, the latest National Health and Nutrition Examination Survey data reveal a prevalence of iron deficiency anemia of 7% for children between

1 and 2 years old and that of 9–16% for adolescent females and women of reproductive age.[4] By the end of pregnancy, it is estimated that 33% of women in the United States have iron deficiency anemia.[5] This prevalence rate has not changed since 1979 and is largely due to the enormous nutritional stressors on the body during pregnancy and lactation. Pregnant women require 50% more iron in their diets than non-pregnant women of reproductive age. The high demands coupled with inadequate intakes that are often documented in this age group frequently lead to iron depletion by the end of pregnancy and failure to recover stores postpartum.

Included among the many consequences of iron deficiency are changes in cognitive performance, emotions, and behavior.[6-11] Although the prevalence of iron deficiency in women of reproductive age is of public health significance, to date, there have been few studies examining the relationship between iron status and cognitive functioning in this age group, with most studies focusing on infants and young children instead. As a result, considerable evidence has accumulated over the last 35 years on the relationship between iron status, cognitive performance, and behavior in children.[6,7,12-15] Numerous studies have been conducted throughout the world, with most showing that iron deficiency in infants and young children has negative consequences for mental and motor development, as well as behavior. Findings from these studies suggest that iron-deficient infants are less attentive and more wary, clingy, and hesitant than their non-iron-deficient counterparts.[15-18] Particularly concerning are the results of long-term follow-up studies in individuals who were iron-deficient anemic as infants as they reveal persistent cognitive and behavioral changes into young adulthood.[19,20] The longest-term follow-up study, which was conducted by Lozoff et al., revealed an increase in the magnitude of the functional deficit with age between those who were iron-deficient anemic and those who were iron sufficient in infancy.[20] This speaks to the importance of preventing iron deficiency in infancy.

There are also a number of reports documenting the effects of iron deficiency in pre-adolescent and adolescent children.[21-26] These reports indicate that iron deficiency anemia results in lower scores on cognitive function tests and educational achievement tests in this age group. The reports by Bruner et al.[25] and Halterman et al.[26] provide evidence that the detrimental cognitive effects of iron deficiency occur before the deficiency has reached the stage of overt anemia. It is important to note that, in most cases, an intervention that normalizes the iron status in these age groups returns performance to levels similar to controls.

Although a vulnerable group for iron deficiency, women of reproductive age have received less attention in this area, with the literature regarding this relationship just beginning to emerge. The traditional lack of focus on women of reproductive age is due in large part to the long-held belief that the effects of iron status on cognition and behavior are developmentally dependent. While several of the brain alterations occurring as a result of iron deficiency are developmentally dependent (i.e., myelination), others are not (i.e., neurotransmitter alterations). Findings from animal studies indicate potential detrimental consequences even in adulthood.[7,27] As a result of such findings, researchers have begun to explore the cognitive and behavioral alterations resulting from iron deficiency in women of reproductive age.

7.2 FUNCTIONAL CONSEQUENCES OF IRON DEFICIENCY IN WOMEN OF REPRODUCTIVE AGE

For many years there has been anecdotal evidence that iron deficiency is related to irritability, apathy, fatigue, and depressive symptoms among women of reproductive age,[28] but the empirical evidence has been largely lacking. Recent interest in this topic has led to the emergence of studies reporting on findings in this area. These studies can be divided by design (observational vs. intervention studies) and primary outcomes of interest [mood/quality of life and cognition (with one study also reporting on mother-child interaction and child development as outcome variables)]. While the intervention studies have a stronger design than the observational ones, several of the intervention studies occur in populations with many other confounding variables (e.g., renal dialysis patients) and/or lack a control group, making causal inferences difficult. The remainder of this chapter will focus on reviewing the findings from these studies and the potential underlying neurobiology contributing to these findings. Although adolescents are of reproductive age, this chapter will focus on those studies that included subjects aged 18 years or older and who were premenopausal.

7.2.1 OBSERVATIONAL STUDIES

Nine observational studies have been conducted in this age group throughout the world. Table 7.1 summarizes the main outcome variables and findings from these studies.

7.2.1.1 Mood/Quality of Life

Of the studies examining the relation between iron status and mood/quality of life in women of reproductive age, the outcome of interest that has been most studied is depressive symptoms. Results of these studies are mixed. There are many possibilities for the mixed results, including the fact that the studies have been conducted in various settings throughout the world, used various scales to measure depression, used various measures to classify iron status, and used differing designs. Rangan et al.[29] and Fordy and Benton[30] conducted studies among college students and reported little evidence of a relation between iron status and mood except in women who were iron deficient (defined by low ferritin levels) and taking an oral contraceptive. Rangan et al.[29] reported significantly higher depressive symptoms, irritability, and difficulty concentrating in such women, while Fordy and Benton[30] reported higher depressive symptoms in these women. Hunt and Penland,[31] on the other hand, conducted a study in premenopausal women (age range of 20 to 45 years) and found no association between iron status and the depression scale on the Minnesota Multiphasic Personality Inventory. Two other studies[32,33] classified women as either having high depressive symptoms (depressed) or low depressive symptoms (healthy), and both found an association with iron status. The first study[32] was conducted in the United States during the early postpartum period and revealed that hemoglobin levels on Day 7 postpartum were negatively correlated with depressive symptoms on Day 28, as measured by the Center for Epidemiological Studies Depression Inventory. These data suggest a relation between postpartum anemia and depressive symptoms such that those women who are anemic exhibit significantly more depressive symptoms than those women who are not anemic. However, as this study was not originally designed to assess the relationship between iron status and depression, there are some limitations, including a small sample size and lack of data on indicators of iron status other than hemoglobin. It is likely that the anemia in this particular group of women was due to iron deficiency, but the lack of data on other iron status indicators precludes this conclusion from being made. The second study[33] was conducted recently in Iran among female medical students. Hemoglobin and ferritin levels were used as iron status indicators among other nutritional variables. Women who were anemic were excluded from the study. Each of the participants completed the Beck Depression Inventory (BDI) and was then classified as either depressed (BDI score ≥ 10) or healthy (BDI score < 10). Ferritin levels in those with depression were significantly lower than in those who were healthy. The authors concluded that low iron stores may be related to psychological changes in women even in the absence of anemia.

Rather than reporting specifically on depressive symptoms, some of the observational studies conducted in this area have used a general health quality-of-life questionnaire as the outcome variable of interest. Three studies examining the relation between iron status and general quality of life have been conducted in otherwise healthy populations.[34–36] One study was conducted in Australia, and the other two were conducted in France. The Australian study[34] was conducted among women between 18 and 50 years old and used the short-form Medical Outcomes Study (MOS SF-36) as the outcome measure. This measure yields a physical component score, a mental component score, and a vitality score. The women in this study reported whether or not their physician had ever told them that they had "low iron." There were differences

TABLE 7.1
Observation Studies

Reference	Location	Age	Iron Status Measure	Outcome Measure	Main Findings	Association between Iron Status and Outcome Measure?
Foley et al.[37] 1986	USA	19–36 years	ZPP	Cognitive performance	• Negative association between ZPP and 3 spatial tests	Yes
Fordy and Benton[30] 1994	Wales	College students	Ft	Cognitive performance, quality of life	• No association between Ft status and performance on cognitive tests • Females taking oral contraceptives and who had low Ft levels reported increased depressive symptoms	No, except for a subset of the population
Rangan et al.[29] 1998	Australia	15–30 years	Hb, MCV, TSAT, Ft, TfR	Quality of life	• Anemic women had increased psychological distress • Women taking oral contraceptives and who had low Ft levels reported increased depressive symptoms, irritability, and difficulty concentrating	No, except for a subset of the population
Hunt and Penland[31] 1999	USA	20–45 years	Hb, TSAT, Ft	Depressive symptoms	• No association between iron status and depressive symptoms	No
Patterson et al.[34] 2000	Australia	18–50 years	Self-report of "low iron"	Quality of life	• Those ever reporting low iron had decreased PCS, MCS, and VT scores	Yes
Corwin et al.[32] 2003	USA	18–35 years	Hb	Depressive symptoms	• Increased depressive symptoms in anemic women	Yes
Duport et al.[35] 2003	France	35–51 years	Ft	Quality of life	• No association between iron status and quality of life	No
Vahdat Shariatpanaahi et al.[33] 2007	Iran	24.5 years (M)	Hb, Ft, CRP	Depressive symptoms	• Increased depressive symptoms in those with low Ft	Yes
Grondin et al.[36] 2008	France	17–38 years	Ft	Quality of life	• Lower general health scores in those with low Ft	Yes

ZPP, zinc protoporphyrin; Ft, ferritin; Hb, hemoglobin MCV, mean corpuscular volume; TSAT, transferrin saturation; TfR, transferrin receptor; CRP, C-reactive protein; M, mean; PCS, physical component score (of MOS SF-36); MCS, mental component score (of MOS SF-36); VT, vitality score (of MOS SF-36).

on the physical component score, mental component score, and vitality score between those women who had been told they had low iron in the past and those who had never been told that they had low iron. The first French study[35] followed women for 2 years and classified them as either chronically iron deplete (ferritin < 15 µg/L at both baseline and the 2-year measurement) or chronically iron sufficient (ferritin between 30 and 80 µg/L at both baseline and the 2-year measurement). Women completed the Duke Health Profile as a measure of quality of life. No association was found between iron status and scores on the Duke Health Profile. The most recent study[36] enrolled females between 17 and 38 years old and classified them as iron deficient (ferritin < 15 µg/L), borderline deficient (ferritin = 15–20 µg/L), and iron replete (ferritin > 20 µg/L). Quality of life was assessed via a questionnaire adapted from the MOS SF-36. The authors reported lower general health scores in the iron-deficient women compared with the iron-replete women.

7.2.1.2 Cognition

There are only two observational studies[30,37] that have examined the relation between iron status and cognition in women of reproductive age. These studies used both male and female college students. The study by Foley et al.[37] used zinc protoporphyrin as the sole iron status measure, while Fordy and Benton[30] used ferritin levels as the sole iron status measure in their population. Foley et al. reported a negative correlation between zinc protoporphyrin and three spatial tests, while Fordy and Benton reported no association between ferritin levels and performance on cognitive tests.

7.2.1.3 Summary of Observational Studies

Of the nine observational studies conducted, seven examined depressive symptoms/quality of life as the main outcome, one examined cognitive performance as the main outcome, and one looked at both cognitive performance and quality of life. Five of the studies report an association between iron status and the outcome measures such that those with a higher iron status have more optimal outcomes (less depressive symptoms, better quality of life, and/or better performance on cognitive tests). Two of the studies report a lack of association between iron status and the outcome variables except for that in a subsample of the population (iron-deficient women taking oral contraceptives). Two studies report no association between iron status and the outcome variables. These results are clearly equivocal but provide evidence that iron status may be related to at least some facets of psychological functioning in women of reproductive age. As observational studies do not provide the data needed for causal inferences, the intervention studies may prove to be more informative.

7.2.2 INTERVENTION STUDIES

Some of the intervention studies conducted in this area are randomized controlled trials, others supplied iron supplements to all the iron-deficient women in their study, and one study used each subject as her own control. Additionally, some of the studies were conducted in patients on renal dialysis,[38–41] and it is therefore difficult to interpret the results. Given the many possible confounders in those studies, they will not be discussed here. Table 7.2 summarizes the main outcome variables and findings from the intervention studies conducted in generally healthy women of reproductive age.

7.2.2.1 Mood/Quality of Life

Like the observational studies, the outcome variables most often examined in the intervention studies are those dealing with mood/general quality-of-life variables. Beutler et al.[42] conducted an elegant study in 1960 in which they supplemented women of reproductive age who were chronically fatigued. Anemic women were excluded from the study. The women served as their own control

TABLE 7.2
Intervention Studies

Reference	Location	Age	Iron Status Measure	Length of Intervention	Outcome Measure	Main Findings	Association between Iron Status and Outcome Measure?
Beutler et al.[42] 1960	USA	College students	Hb, RBC, Hct, TSAT, bone marrow biopsy for stainable iron	3 months	Mood	• Iron-deficient women experienced a decrease in fatigue after treatment with iron	Yes
Groner et al.[48] 1986	USA	14–24 years	Hb, Hct, MCV, MCH, Ft	1 month	Cognitive performance	• Short-term memory and attention scores increased in iron-treated group	Yes
Risser et al.[45] 1988	USA	19.9 (M)	Hb, MCV, TSAT, Ft	1 intercollegiate sports season	Mood	• No association between iron status or treatment and mood	No
Kretsch et al.[49] 1998	USA	25–42 years	Hb, Hct, RBC, MCV, MCH, RDW, TSAT, Ft		Cognitive performance	• Sustained attention was positively correlated with Hb and TSAT levels	Yes
Patterson et al.[44] 2001	Australia	18–50 years	Hb, TSAT, Ft	4 months	Mood, quality of life	• Iron-deficient women had lower MCS and VT scores and higher PFS scores at baseline • Iron treatment improved MCS, VT, and PFS scores	Yes
Verdon et al.[43] 2003	Switzerland	18–55 years	Hb, Ft	1 month	Mood, quality of life	• Greater improvement in fatigue and anxiety scores in women with low Ft who received iron	Yes

Study	Country	Age	Measures	Duration	Domain	Findings	
Beard et al.[47] 2005 Perez et al.[55] 2005 Murray-Kolb and Beard[59] 2009	South Africa	18–30 years	Hb, MCV, TSAT, Ft, CRP	6.5 months	Mood, cognitive, performance, mother-child interaction, child development	• Less depressive symptoms and anxiety at endpoint in those treated with iron • Improvement in reasoning ability in those treated with iron • Improvement in mother-child interaction scores in those treated with iron • Developmental delays in children of mothers who were iron-deficient anemic; delay persisted after normalization of maternal iron status	Yes
Ando et al.[46] 2006	Japan	19–54 years	Hb, Ft	3 months	Quality of life	• Improvement in quality of life with iron supplements	Yes
Murray-Kolb and Beard[50] 2007	USA	18–35 years	CBC (including Hb and Hct), TSAT, Ft, TfR	4 months	Cognitive performance	• Lower scores on attention, memory, and learning domains in iron-deficient anemic women at baseline • Significant improvement in memory and attention scores in women who improved their iron status	Yes

RBC, red blood cell; Ft, ferritin; Hb, hemoglobin; Hct, hematocrit; MCV, mean corpuscular volume; MCH, mean corpuscular hemoglobin; RDW, red cell distribution width; TSAT, transferrin saturation; TfR, transferrin receptor; CRP, C-reactive protein; CBC, complete blood count; M, mean; MCS, mental component score (of MOS SF-36); VT, vitality score (of MOS SF-36); PFS, Piper Fatigue Scale.

and therefore consumed an iron supplement or placebo for 3 months, followed by 1 month of no supplementation and then 3 more months of supplementation (either iron or placebo, depending on what they received during the first 3-month period). At the time of this study, the measurement of ferritin as a marker of iron stores had not yet been identified; therefore, these women underwent bone marrow biopsy at baseline to determine whether or not they had any stainable iron. Several other measures of iron status were also determined (see Table 7.2). The women who were iron deficient at baseline experienced a significant decrease in fatigue after taking the iron supplements versus the placebo. More recently, Verdon et al.[43] used a similar population (women with unexplained fatigue) and a randomized controlled design to examine the relation between iron status, fatigue, depression, and anxiety. The women who had a ferritin level below 50 µg/L at baseline and received iron experienced a greater decrease in fatigue and anxiety compared with the women who received a placebo. The finding of lowered fatigue levels in women after iron supplementation was also reported in Australia.[44] Half of the iron-deficient women in this study were assigned to receive iron supplements, while the other half received counseling on consuming a high-iron diet and were given meal vouchers. There was also an iron-replete group who received no intervention. Fatigue and general quality of life were the outcome measures. At baseline, quality-of-life scores were lower and fatigue scores were higher in the iron-deficient women versus the iron-replete women. At endpoint, women in both treatment groups showed improvements in quality-of-life and fatigue scores that were similar in magnitude. The group receiving the iron supplement improved their iron status to a greater degree than the group receiving the high-iron diet counseling. Three other studies have examined the relation between iron status and mood/quality of life (although not fatigue) in women of reproductive age. One study provided iron treatment to all the subjects who were iron deficient and no treatment to those who were iron replete at baseline.[45] This study reported no impact of iron status or supplementation on mood. A recent Japanese study included only iron-deficient anemic premenopausal women who were all supplemented with iron.[46] At baseline, the scores on the general health questionnaire were significantly lower than the Japanese norms. After 1 month of supplementation, the women experienced a significant increase in nearly all the domain scores on the general health questionnaire; after 3 months of supplementation, scores on all the domains were comparable with or greater than the national norms. The most comprehensive study to date examining the relation between iron status and mood was conducted by Beard et al. in South Africa.[47] It was a randomized controlled trial including a group of iron-replete women receiving no intervention and a group of iron-deficient anemic women, of whom half received iron (plus vitamin C and folate) and the other half received a placebo (vitamin C and folate). Baseline testing was conducted at 10 weeks postpartum, and endpoint testing was conducted at 9 months postpartum. Multiple outcome measures were used, including mood variables (depression, anxiety, stress). At baseline, scores on the mood variables did not differ between the iron-deficient anemic groups and the iron-replete group. At endpoint, the women who received iron had lower depressive symptom scores as compared with the women who received the placebo and the iron-replete women. In this instance, the iron seems to have been protective against the increased depressive symptoms experienced by the groups who did not receive iron.

7.2.2.2 Cognition

There are only four intervention studies that have used cognitive variables as outcome measures in women of reproductive age. The first study was conducted by Groner et al.[48] among pregnant women. Half of the women received a prenatal vitamin, while the other half received a prenatal vitamin with iron. Outcome measures included short-term memory and attention span. After 1 month of iron supplementation, scores for both of these outcome measures increased in those women who received iron compared with those women who received only the prenatal supplement. Kretsch et al.[49] also found an association between iron status and attention. Their study was conducted in obese dieting women who were placed on a 50% calorie-restricted diet followed by a weight stabilization diet. Throughout the course of the weight loss, the women experienced a decrease in iron status. At the end of the study, hemoglobin and transferrin saturation levels were found to positively correlate with sustained attention. While both of these studies point to an association between iron and attention, Groner et al.'s study may be confounded by the fact that the participants were pregnant and Kretsch et al.'s study may be confounded by the fact that the participants were placed on a very restrictive diet and experienced significant weight loss during this period. A recent study conducted by Murray-Kolb and Beard was free of these confounding variables and used attention as an outcome variable (as well as memory and learning).[50] This study was conducted in women of reproductive age (age range of 18 to 35 years) and found a relation between iron and all three of the cognitive domains tested. After measurement of iron status parameters, women were classified as iron sufficient, iron deficient, or iron-deficient anemic. The women were then randomized, within groups, to receive iron or placebo for 16 weeks. Attention, memory, and learning were measured at both baseline and endpoint. At baseline, women in the iron-deficient anemic group performed significantly worse on tasks of attention, memory, and learning and took significantly longer to complete the tasks as compared with the iron-sufficient women. At endpoint, those women who experienced an improvement in their iron status also showed an improvement in memory and attention scores that was significantly greater than that occurring in women who did not experience an increase in iron status. As much as a seven-fold improvement in performance was noted in these women. In addition to these domains of cognition, Beard et al.[47] recently found an association between iron status and reasoning ability in South African women in the postpartum period. Iron-deficient anemic women who were supplemented with iron improved their scores on Raven's Coloured Progressive Matrices such that, at endpoint, they were not different from the iron-sufficient women and significantly different from the iron-deficient anemic women who received a placebo.

7.2.2.3 Mother-Child Interaction and Child Development

In addition to mood and cognitive outcomes, researchers have recently begun to hypothesize that mother-child interaction and child development may be negatively impacted as a result of iron deficiency in women of reproductive age. Many studies have examined the relationship between maternal iron status and child variables at birth, such as birth weight and infant iron status,[51–53] but child development outcome variables have been lacking. The possibility that mother-child interaction may be altered as a result of maternal iron deficiency is especially troubling, given the strong evidence for a relation between high-quality parenting and cognitive and social competencies in infants and older children.[54–57] The only study to report on these outcomes as a result of iron deficiency in the mother was conducted by Beard et al. in South Africa.[58,59] The study population was composed of black African women who were of lower socioeconomic status. Mother-infant interaction was assessed at 10 weeks postpartum, iron-deficient anemic mothers were then randomized to either an iron supplement or a placebo (iron-sufficient mothers received no intervention), and interaction was once again assessed at 9 months postpartum. Interactions were videotaped during 20 minutes of naturalistic free play and coded using the Parent/Caregiver Involvement Scale and the Emotional Availability Scales. At baseline, anemic mothers were less responsive and more controlling of their infants than the iron-sufficient moms as assessed with the Parent/Caregiver Involvement Scale. At 9 months, mothers in the placebo group had more negative statements and less goal setting and responsiveness than mothers in the iron-sufficient group. Mothers in the iron treatment group behaved like those in the iron-sufficient group. Results of scoring with the Emotional Availability Scales at baseline revealed more maternal sensitivity and child responsiveness in the dyads in which maternal iron status was normal (iron sufficient) compared with those in which the mother was iron-deficient anemic. At 9 months postpartum, the dyads in which the mother had been given iron no longer differed from the iron-sufficient dyads, and these two groups scored significantly better than those dyads in which mothers had received a placebo on measures of maternal sensitivity, structuring, non-hostility, and child responsiveness. In addition to mother-child interaction, child development was measured using the Griffiths Scale of Infant Development. At baseline, infants whose mothers were iron-deficient anemic were delayed in hand-eye movement and the overall development quotient as compared with those whose mothers were iron sufficient. These developmental delays were not diminished at 9 months even in those children whose mothers were treated with iron. While this study provides us with some evidence that maternal iron deficiency is detrimental to mother-child interaction, there were some significant limitations in the study. These include: no measurement of infant iron status; baseline measures did not occur until 10 weeks post-partum, after mothers and infants had substantial opportunities to interact; and the co-occurrence of poverty in this population makes it difficult to untangle the effects of iron deficiency from many other factors known to influence mother-child interactions and child development.

7.2.2.4   Summary of Intervention Studies

Of the nine intervention studies conducted, five examined mood/quality of life as the main outcomes, three examined cognitive performance as the main outcome, and one looked at both cognitive per-formance and mood, as well as mother-child interaction and child development. All the studies that measured mood/quality of life report more optimal outcomes as a result of iron supplementation, with the exception of one—this study[45] reports no improvement in mood with iron treatment. The studies examining cognition as an outcome variable all report more optimal scores with higher iron status. Finally, the one study reporting on mother-child interaction and child development showed an improvement in mother-child interaction after iron supplementation but no improvement in child development scores despite normalization of maternal iron status. Results of these studies are in better agreement than those from the observational studies and seem to indicate that supplement-ing iron-deficient women with iron may improve their cognitive performance, mood, and behavior.

7.3   POTENTIAL BIOLOGICAL UNDERPINNINGS

A considerable body of studies has established that appropriate levels of brain iron are necessary for optimal brain development and functioning.[7,60,61] Limitations in functioning of iron-deficient individuals are apparently related to deficits of iron in essential pools in muscle, immune cells, and the central nervous system. Although the neurological underpinnings of these limitations are not completely understood, various theories have been proposed as a result of studies done in animal models and, more recently, some human models. The brain alterations seen as a result of iron defi-ciency during development have recently been reviewed.[12] This section focuses on the findings that may be pertinent to the already developed brain, as is the case for women of reproductive age.

Animal models have repeatedly shown alterations in three primary areas: morphology, bio-chemistry, and bioenergetics.[12] Altered biochemistry is most likely the culprit of the psychological changes reported in iron-deficient women of reproductive age. Catecholamines are known to affect arousal and activation and therefore modulate the capacity for early and later processing of infor-mation.[62] Modulation of emotional processes also seems to be associated with the neurotransmit-ters.[63] Therefore, iron may affect cognition and mood through its action in the synthesis, as well as proper functioning of these compounds. Levels of monoamines change as a result of changes in brain iron status,[64–66] although specific mechanisms are still under investigation.[67] The effects of iron on dopamine have received much attention. Studies have consistently shown that dopaminergic tracts are sensitive to brain iron deficiency.[66–73] Studies have shown lower densities of dopamine D_2 receptors and the dopamine transporter in the striatum and nucleus accumbens of rats.[67,69,73,74] In vivo microdialysis studies showed that extracellular dopamine is elevated in the striatum of iron-deficient rats but returns to normal with iron repletion.[75] Alterations in dopamine metabolism could be related to altered perception and motivation as the ability to process environmental information is dependent on appropriate rates of dopamine clearance from the interstitial space. Dopamine mediates many forms of motivated behavior and motor function;[76,77] therefore, it is to be expected that alterations in dopamine metabolism as a consequence of iron deficiency will lead to alterations in motivation in animals and in humans.

Reports as to whether serotonergic and noradrenergic systems are also sensitive to changes in brain iron status are conflicting.[75,78] Lower densities of the serotonin transporter have been observed in the striatum of iron-deficient rats in one study, while an earlier study showed no difference between iron-deficient and non-iron-deficient animals.[71,73] In vivo microdialysis experiments in rats showed elevated interstitial levels of norepinephrine, but direct measurements in tissue homogenates showed no difference relative to control animals.[65,79,80]

A high amount of similarity in the brain regions that are high in iron and brain regions that receive input from γ-aminobutyric acid (GABA) fibers has been noted.[81] To look at the relationship between brain iron content and GABA metabolism, the author injected rats with an inhibitor of GABA transaminase.[82] She found that injection of the inhibitor into the striatum/pallidum resulted in a significant reduction in iron concentration in the pallidum, globus pallidus, and substantia nigra and concluded that this was evidence for a relationship between iron in the brain and utilization of GABA. The effects of iron concentrations on GABA metabolism were reported by Youdim et al.[79] They reported no effect of iron deficiency on amount of GABA, production rates, or receptor population. Other studies have reported decreases in glutamate decarboxylase, glutamate dehydrogenase, and GABA transaminase activities in rats.[83,84]

In addition to animal studies, one group of researchers has explored the neurophysiological changes occurring in adults as a result of iron deficiency. These studies were conducted more than 20 years ago by Tucker et al.[85–87] They utilized correlational and regression analyses to examine the strength of association between body iron levels, cognitive performance, and neural activity. Specifically, they found that iron status was related to electroencephalographic asymmetry.

7.4 SUMMARY

Numerous influences exist for cognition and behavior, with nutrition being one of the very important factors. Despite various studies conducted over several decades, much information is still lacking about the impact of nutrition on cognition and behavior, especially in women of reproductive age. While studies in this population are beginning to accumulate, there is a need for well-designed studies that systematically examine the relationship between iron status and specific measures of cognition, as well as mood and behavior. These studies should be designed in such a way that the effects of iron deficiency versus iron deficiency anemia may be teased apart. There is an especially critical need for studies that examine the impact of maternal iron status on mother-child interaction and child development. These studies should take maternal and child iron status into account in addition to environmental factors that are known to impact interaction and development. Also needed are studies that examine the association between maternal gestational iron status and subsequent child cognitive and behavioral development. To date, there is not enough data in this area to make any sound conclusions. Well-designed studies are needed to determine the effects of maternal gestational iron status on later child development, what, if any, "critical periods" exist during gestation with respect to these outcomes, and the severity of iron deficiency associated with negative outcomes. While animal studies can provide us with some clues for answering these questions, human studies are needed in order to aid policymakers in setting the appropriate recommendations with respect to iron supplementation of women of reproductive age. Finally, a need for studies in women of reproductive age that examine the underlying neurophysiological changes accompanying the changes seen in cognition and behavior exists. Recent advances in non-invasive techniques used to measure specific neurophysiological functioning should be utilized as we seek to better define the role iron has in cognitive functioning, mood, and behavior in women of reproductive age. By focusing on this age group, we have the ability to impact not only the health of these women but also that of their children.

REFERENCES

1. ACC/SCN, Fourth Report on the World Nutrition Situation, Geneva: ACC/SCN in collaboration with IFPRI, 2000.
2. de Benoist, B., et al., eds. Worldwide Prevalence of Anaemia 1993–2005, WHO Global Database on Anaemia, Geneva: World Health Organization, 2008.
3. The World Health Report 2002: Reducing Risks, Promoting Healthy Life.
4. Looker, A.C., Cogswell, M.E., and Gunter, E.W., Iron deficiency—United States, 1999–2000, CDC Surveillance Summaries, *MMWR* 51, 897, 2002.
5. Kim, L., et al., Pregnancy nutrition surveillance system—United States, 1979–1990, CDC Surveillance Summaries, *MMWR* 41, 25, 1992.
6. Lozoff, B., et al., Long-lasting neural and behavioral effects of iron deficiency in infancy, *Nutr Rev* 64, S34, 2006.
7. Beard, J.L. and Connor, J.R., Iron status and neural functioning, *Annu Rev Nutr* 23, 41, 2003.
8. UNICEF/WHO/UNU/MI, *Preventing Iron Deficiency in Women and Children: Technical Consensus on Key Issues and Resources for Programme Advocacy, Planning, and Implementation*, New York: UNICEF, 1998.
9. Beard, J.L., Iron biology in immune function, muscle metabolism, and neuronal functioning, *J Nutr* 131, 568S, 2001.
10. Hallberg, L. and Asp, N.G., *Iron Nutrition in Health and Disease*, London: John Libbey, 1996, 364.
11. Beard, J.L., Neuroendocrine alterations in iron deficiency, *Prog Food Nutr Sci* 14, 45, 1990.
12. Beard, J., Recent evidence from human and animal studies regarding iron status and infant development, *J Nutr*, 137, 524S, 2007.
13. Pollitt, E., The developmental and probabilistic nature of the functional consequences of iron-deficiency anemia in children, *J Nutr* 131, 669S, 2001.
14. Lozoff, B., Behavioral alterations in iron deficiency, *Adv Pediatr* 35, 331, 1988.
15. Lozoff, B., et al., Behavior of infants with iron-deficiency anemia, *Child Dev* 69, 24, 1998.
16. Lozoff, B., et al., Abnormal behavior and low developmental test scores in iron-deficient anemic infants, *J Dev Behav Pediatr* 6, 69, 1985.
17. Lozoff, B., Klein, N.K., and Prabucki, K.M., Iron-deficient anemic infants at play, *J Dev Behav Pediatr* 7, 152, 1986.
18. Honig, A.S. and Oski, F.A., Solemnity: A clinical risk index for iron deficient infants, *Early Child Dev Care* 16, 69, 1984.
19. Lozoff, B., et al., Poorer behavioral and developmental outcomes more than 10 years after treatment for iron deficiency in infancy, *Pediatrics* 105, E51, 2000.
20. Lozoff, B., Jimenez, E., and Smith, J.B., Double burden of iron deficiency in infancy and low socioeconomic status: A longitudinal analysis of cognitive test scores at age 19 years, *Arch Pediatr Adolesc Med* 160, 1108, 2006.
21. Soemantri, A.G., Pollitt, E., and Kim, I., Iron deficiency anemia and educational achievement, *Am J Clin Nutr* 42, 1221, 1985.
22. Pollitt, E., et al., Iron deficiency and educational achievement in Thailand, *Am J Clin Nutr* 50, 687, 1989.
23. Seshadri, S. and Gopaldas, T., Impact of iron supplementation on cognitive functions in preschool and school aged children: The Indian experience, *Am J Clin Nutr* 50, 675, 1989.
24. Ballin, A., et al., Iron state in female adolescents, *AJDC* 146, 803, 1992.
25. Bruner, A.B., et al., Randomized study of cognitive effects of iron supplementation in non-anemic iron deficient adolescent girls, *Lancet* 348, 992, 1996.
26. Halterman, J.S., et al., Iron deficiency and cognitive achievement among school-aged children and adolescents in the United States, *Pediatrics* 107, 1381, 2001.
27. Burhans, M.S., et al., Iron deficiency: Differential effects on monoamine transporters, *Neurosci* 8, 31, 2005.
28. Bothwell, T.H., et al., *Iron Metabolism in Man*, Oxford: Blackwell Scientific Publications, 1979.
29. Rangan, A.M., Blight, G.D., and Binns, C.W., Iron status and non-specific symptoms of female students, *J Am Col Nutr* 17, 351, 1998.
30. Fordy, J. and Benton, D., Does low iron status influence psychological functioning? *J Hum Nutr Diet* 7, 127, 1994.
31. Hunt, J.R. and Penland, J.G., Iron status and depression in premenopausal women: An MMPI study, *Behav Med* 25, 62, 1999.

32. Corwin, E.J., Murray-Kolb, L.E., and Beard, J.L., Low hemoglobin level is a risk factor for postpartum depression, *J Nutr* 133, 4139, 2003.

33. Vahdat Shariatpanaahi, M., The relationship between depression and serum ferritin level, *Eur J Clin Nutr* 61, 532, 2007.

34. Patterson, A.J., et al., Iron deficiency, general health and fatigue: Results from the Australian Longitudinal Study on Women's Health, *Qual Life Res* 9, 491, 2000.

35. Duport, N., et al., Consequences of iron depletion on health in menstruating women, *Eur J Clin Nutr* 57, 1169, 2003.

36. Grondin, M.-A., et al., Prevalences of iron deficiency and health-related quality of life among female students, *J Am Coll Nutr* 27, 337, 2008.

37. Foley, D., Hay, D.A., and Mitchell, R.J., Specific cognitive effects of mild iron deficiency and associations with blood polymorphisms in young adults, *Ann Hum Biol* 13, 417, 1986.

38. McMahon, L.P., et al., Effects of hemoglobin normalization on quality of life and cardiovascular parameters in end-stage renal failure, *Nephrol Dial Transplant* 15, 1425, 2000.

39. Stivelman, J.C., Benefits of anaemia treatment on cognitive function, *Nephrol Dial Transplant* 15, 29, 2000.

40. Temple, R.M., et al., Recombinant erythropoietin improves cognitive function in chronic haemodialysis patients, *Nephrol Dial Transplant* 7, 240, 1992.

41. Temple, R.M., Deary, I.J., and Winney, R.J., Recombinant erythropoietin improves cognitive function in patients maintained on chronic ambulatory peritoneal dialysis, *Nephrol Dial Transplant* 10, 1733, 1995.

42. Beutler, E., Larsh, S.E., and Gurney, C.W., Iron therapy in chronically fatigued, nonanemic women: A double-blind study, *Ann Intern Med* 52, 378, 1960.

43. Verdon, F., et al., Iron supplementation for unexplained fatigue in non-anaemic women: Double blind randomized placebo controlled trial, *BMJ* 326, 1124, 2003.

44. Patterson, A.J, Brown, W.J., and Roberts, D.C.K., Dietary and supplement treatment of iron deficiency results in improvements in general health and fatigue in Australian women of childbearing age, *J Am Coll Nutr* 20, 337, 2001.

45. Risser, W.L., et al., Iron deficiency in female athletes: Its prevalence and impact on performance, *Med Sci Sports Exerc*, 20, 116, 1988.

46. Ando, K., et al., Health-related quality of life among Japanese women with iron-deficiency anemia, *Qual Life Res* 15, 1559, 2006.

47. Beard, J.L., et al., Maternal iron deficiency anemia affects postpartum emotions and cognition, *J Nutr* 135, 267, 2005.

48. Groner, J.A., et al., A randomized trial of oral iron on tests of short-term memory and attention span in young pregnant women, *J Adolesc Health Care* 7, 44, 1986.

49. Kretsch, M.J., et al., Cognitive function, iron status, and hemoglobin concentration in obese dieting women, *Eur J Clin Nutr* 52, 512, 1998.

50. Murray-Kolb, L.E. and Beard, J.L., Iron treatment normalizes cognitive functioning in young adult women, *AJCN* 85, 778, 2007.

51. Bornstein, M.H., Tamis-LeMonda, C.S., and Haynes, O.M., First words in the second year: Continuity, stability, and models of concurrent and predictive correspondence in vocabulary and verbal responsiveness across age and context, *Infant Behav Dev* 22, 65, 1999.

52. Cole, P.M., Michel, M.K., and Teti, L.O., The development of emotion regulation and dysregulation: A clinical perspective, *Monogr Soc Res Child Dev* 59, 73, 1994.

53. Grolnick, W.S., et al., Mothers' strategies for regulating their toddlers' distress, *Infant Behav Dev* 21, 437, 1998.

54. Lamb, M.E., et al., Parent-child relationships: Development in the context of the family, in Bornstein, M.H. and Lamb, M.E. (eds.), *Developmental Psychology: An Advanced Textbook*, 4th ed., Mahwah, NJ: Lawrence Erlbaum Associates, 1999, 411.

55. Perez, E.M., et al., Mother-infant interactions and infant development are altered by maternal iron deficiency, *J Nutr* 135, 850, 2005.

59. Murray-Kolb, L.E. and Beard, J.L., Iron deficiency and child and maternal health, *AJCN* 89, 946S, 2009.

60. Beard, J.L., Connor, J.D., and Jones, B.C., Brain iron: Location and function, *Prog Food Nutr Sci* 17, 183, 1993.

61. Earley, C.J., et al., Abnormalities in CSF concentrations of ferritin and transferrin in restless legs syndrome, *Neurology* 54, 1698, 2000.

62. Izquierdo, I., Different forms of post-training memory processing, *Behav Neur Biol* 51, 171, 1989.

63. Izquierdo, I. and Medina, J.H., GABA receptor modulation of memory: The role of endogenous benzo-diazepines, *Trends Pharmacol Sci* 12, 260, 1991.
64. Ashkenazi, R., Ben-Shachar, D., and Youdim, M.B.H., Nutritional iron deficiency and dopamine binding sites in the rat brain, *Pharmacol Biochem Behav* 17, 43, 1982.
65. Kwik-Uribe, C.L., et al., Chronic marginal iron intakes during early development in mice result in persistent changes in dopamine metabolism and myelin composition, *J Nutr* 130, 2821, 2000.
66. Nelson, C., et al., Alterations in dopamine metabolism in iron deficient rats, *J Nutr* 12, 2282, 1997.
67. Erikson, K.M., Jones, B.C., and Beard, J.L., Iron deficiency alters dopamine transporter functioning in rat striatum, *J Nutr* 130, 2831, 2000.
68. Ben-Schachar, D., Finberg, J.P.M., and Youdim, M.B.H., The effect of iron chelators on dopamine D_2 receptors, *J Neurochem* 45, 999, 1985.
69. Erikson, K.M., et al., Altered function of dopamine D_1 and D_2 receptors in brains of iron deficient rats, *Physiol Pharmacol Behav* 69, 409, 2001.
70. Erikson, K.M., et al., Iron status and distribution of iron in the brain of developing rats, *J Nutr* 127, 2030, 1997.
71. Morse, A., Beard, J.L., and Jones, B., Behavioral and neurochemical alterations in iron deficient mice, *Proc Soc Exp Biol Med* 220, 147, 1999a.
72. Morse, A., Beard, J.L., and Jones, B., Sex and genetics are important cofactors in assessing the impact of iron deficiency on the developing mouse brain, *Nutr Neurosci* 2, 323, 1999b.
73. Yehuda, S., Neurochemical basis of behavioral effects of brain iron deficiency in animals, in Dobbing, J. (ed.), *Brain, Behavior and Iron in the Infant Diet*, London, UK: Springer-Verlag, 1990, 83.
74. Yehuda, S. and Youdim, M.B.H., Brain iron: A lesson from animal models, *Am J Clin Nutr* 50, 618, 1989.
75. Chen, Q., Beard, J.L., and Jones, B.C., Abnormal brain monoamine metabolism in iron deficiency anemia, *J Nutr Biochem* 6, 486, 1995.
76. Robbins, T.W. and Everitt, B.J., Functional studies of the central catecholamines, *Int Rev Neurobiol* 23, 303, 1982.
77. Izquierdo, I., Dopamine receptor in the caudate nucleus and memory processes, *Trends Neurosci* 13, 7, 1992.
78. Adhami, V.M., et al., Influence of iron deficiency and lead treatment on behavior and cerebellar and hippocampal polyamine levels in neonatal rats, *Neurochem Res* 21, 915, 1996.
79. Youdim, M.B.H., Ben-Schachar, D., and Yehuda, S., Putative biological mechanisms on the effects of iron deficiency on brain metabolism, *Am J Clin Nutr* 50, 607, 1989.
80. Youdim, M.B.H., et al., The effects of iron deficiency on brain biogenic monoamine biochemistry and function in rats, *Neuropharmacology* 19, 259, 1980.
81. Hill, J.M., The distribution of iron in the brain, in Youdim, M.B.H. (ed.), *Brain Iron: Neurochemistry and Behavioural Aspects*, London, UK: Taylor & Francis, 1988, 1.
82. Hill, J.M., et al., Transferrin receptors in rat brain: Neuropeptide-like pattern and relationship to iron distribution, *Proc Natl Acad Sci U S A* 82, 4553, 1985.
83. Li, D., Effects of iron deficiency on iron distribution and gamma-aminobutyric acid metabolism in young rat brain tissue, *Hokkaido Igaku Zasshi* 73, 215, 1998.
84. Taneja, V., Mishra, K., and Agarwal, K.N., Effect of early iron deficiency in the rat on gamma-aminobutyric acid shunt in brain, *J Neurochem* 46, 1670, 1986.
85. Tucker, D.M. and Sandstead, H.H., Spectral electroencephalographic correlates of iron status: Tired blood revisited, *Physiol Behav* 26, 439, 1981.
86. Tucker, D.M., et al., Longitudinal study of brain function and depletion of iron stores in individual subjects, *Physiol Behav* 29, 737, 1982.
87. Tucker, D.M., et al., Iron status and brain function: Serum ferritin levels associated with asymmetries of cortical electrophysiology and cognitive performance, *Am J Clin Nutr*, 39, 105 1984.

8 Micronutrient Needs of the Developing Brain
Priorities and Assessment

Anita J. Fuglestad[1], Sara E. Ramel[2], and Michael K. Georgieff[3]

[1]Institute of Child Development, University of
Minnesota, Minneapolis, Minnesota, USA

[2]Neonatal-Perinatal Medicine, Division of Neonatology, Department
of Pediatrics, University of Minnesota, Minneapolis, Minnesota, USA

[3]Center for Neurobehavioral Development, Section of Neonatology,
Department of Pediatrics, University of Minnesota School
of Medicine, Minneapolis, Minnesota, USA

CONTENTS

8.1 INTRODUCTION

Poor nutrition can have a significant impact on the physical, emotional, behavioral, and cognitive development of infants and young children. Although food insecurity does not occur to the same degree in the United States as in developing nations, the diets of many pregnant women and young children lack sufficient amounts of specific nutrients to support optimal functioning.[1] Although all nutrients are required to support neurodevelopment, research studies in animal models and human populations show that certain nutrients are particularly important during the first years of life to support rapidly developing brain systems. The neurodevelopmental effects of a given nutrient are based on that nutrient's metabolic role in regional brain development, the timing of the deficiency, and the degree and duration of the deficiency.[2,3]

Depending on the metabolic pathways and the structural components in which a nutrient is involved in the central nervous system (CNS), neuroanatomy, neurophysiology, and/or neurochemistry may be disrupted, each of which will result in altered neuronal function. Neuroanatomical changes include effects on cell proliferation, as well as cell differentiation, including synaptogenesis and dendritic arborization.[4,5] Neurophysiological changes often affect signal propagation or changes in metabolism, while neurochemical changes involve neurotransmitter and receptor synthesis.[6,7] These CNS changes may be region specific rather than global, as there is often regional sensitivity to a given deficiency.

The effect of a particular nutrient deficiency is also dependent on the timing of the deficit. Specifically, the neurodevelopmental processes that co-occur with a given deficiency may be disrupted. Thus, the most profound disruptions in the CNS will occur during periods of rapid development. Furthermore, since structures and systems in the CNS reach maturity at different ages, brain regions that are growing particularly rapidly during a nutritional deficiency will be most affected. The developing brain between 24 and 42 weeks of gestation is extremely vulnerable to nutritional insults, because many accelerated developmental events are occurring within this period. The human brain changes tremendously during this time. It begins as a smooth bilobed structure at the beginning of the third trimester, and it has become as morphologically complex as the adult brain by term.[8] This increase in complexity is reflective of extensive cortical neuronal growth and differentiation. Also, the process of synaptogenesis begins during this period and continues into the teenage years. The process of myelination begins prior to term and continues through 5–10 years of age. The sensory system develops prior to the regions of the brain that mediate cognition [e.g. prefrontal cortex (PFC)], which may not reach full maturity until adolescence.[9] Delays in maturation of the auditory and visual systems may also lead to further delays in cognitive development.[10] Thus, the same nutritional deficiency can have different neurodevelopmental effects depending on the developmental period during which it is experienced.

The duration of insult and the concentration of the nutrient determine the effect that the deficiency has on development.[3] Deficiency that occurs chronically will have more severe effects as sustained disruption of neurodevelopment may interfere with several processes across development. Furthermore, more severe deficiencies will result in greater neurodevelopmental perturbations and more detrimental effects on development. However, while a nutrient may support normal development at one concentration, it may be toxic at another, depending on its range of tolerance.[4] Thus, supplementation to correct a deficiency may improve development in those who are deficient, whereas supplementation in individuals who are already sufficient may have negative effects on brain development. Moreover, micronutrient deficiencies rarely occur in isolation, and their interactions may exponentially increase the brain effects. Furthermore, genetic (e.g., polymorphisms of transporter genes) and environmental (e.g., stress with activation of the hypothalamic-pituitary-adrenal axis, inhibitors of absorption) factors can alter nutrient availability in spite of adequate intake.

Although nutritional deficiencies may disrupt CNS development, the young brain is extremely amenable to repair and often able to recover from early insults. Accordingly, nutritional insults may cause temporary damage that resolves when the nutrient is repleted. On the other hand, if the insult occurs during a "critical" period of development, the damage may not be temporary and instead have long-lasting consequences.[10] Numerous studies of nutrient deficiency show that CNS dysfunction occurs during the deficiency and continues after nutrient repletion. Thus, it seems that its vulnerability to nutritional insults may outweigh its plasticity.[4]

While all nutrients are essential for normal brain development, this chapter focuses specifically on five micronutrients: iron, zinc, iodine, selenium, and copper. The importance of these nutrients in normal CNS development has been documented through nutrient deficit studies in animal models and humans, as well as through knowledge of their roles in various biochemical pathways.[4]

8.2 IRON

Iron, through iron-containing enzymes and hemoproteins, plays a key role in many basic neurodevelopmental processes. Specifically, iron-dependent enzymes are involved in cell division (ribonucleotide reductase), myelination (delta-9 desaturase), oxidative phosphorylation (cytochromes), and neurotransmitter synthesis, including tryptophan hydroxylase (serotonin) and tyrosine hydroxylase (norepinephrine and dopamine).[11] In addition to iron-containing enzymes, iron deficiency alters gene and protein profiles to compensate for and adapt to the deficiency.[12] The disruption of iron deficiency on neurodevelopment is dependent on which of these iron-dependent processes is occurring at the time of the deficiency and which neurobiological system is undergoing rapid changes.

Iron deficiency anemia is the most common single-nutrient deficiency disease in the world. Anemia is estimated by the World Health Organization to affect one to two billion people worldwide, including approximately 50% of pregnant women and children in developing countries. At least half of these cases are attributed to iron deficiency.[13] Children are particularly at risk of developing iron deficiency during three specific periods — the fetal-neonatal period, late infancy and early toddlerhood, and early adolescence for females following the onset of menarche.[2] This section focuses on the effects of deficiency during the first two periods, when brain growth and development are still quite rapid.

8.2.1 FETAL-NEONATAL IRON DEFICIENCY

The third trimester is a time of rapid iron accretion for the developing fetus, with more than 80% of the iron in a newborn accumulating during the final trimester.[14] Disruptions in iron accumulation during this period place infants at risk for iron deficiency. Infants born to mothers with severe iron deficiency anemia are subsequently at risk. While this is rare in the United States, severe maternal iron deficiency is more common in developing countries. Infants born prematurely are also at an increased risk for iron deficiency due to a shortened period of iron accretion during the third trimester.[14] Fetal iron deficiency may also be due to decreased placental transfer of iron. For instance, in pregnant women with hypertension, the placental vessels often undergo infarction and atheromatous plugging, leading to restriction of many nutrients, including iron, to the developing fetus. Finally, fetuses that are chronically hypoxic, and therefore polycythemic, are at an increased risk for iron deficiency. During limited iron availability, iron will first be used for hemoglobin synthesis, potentially placing other tissues, including the developing brain, at risk.[15,16]

Several populations are at particular risk for iron deficiency due to chronic hypoxia.[2,14,15] Infants born to mothers with pregnancy-induced hypertension associated with intrauterine growth restriction have lower cord ferritin levels and calculated iron stores, as well as a 33% decrease in neonatal brain iron concentration at autopsy.[17,18] In infants of diabetic mothers, fetal oxygen consumption rates are up to 30% higher because of chronic fetal hyperglycemia and hyperinsulinemia. The placenta is unable to completely compensate for these increased oxygen requirements, and the fetus becomes chronically hypoxic and therefore polycythemic.[15,19] As iron is prioritized to red cell production, the developing fetal brain is at risk of becoming iron deficient. Accordingly, brain iron deficiency occurs before anemia; thus, anemia is not always an accurate marker for brain iron deficiency.

*Fetal-neonatal iron deficiency is associated with cognitive and psychomotor delays. For instance, infants born to mothers with iron deficiency anemia during the peripartum period scored lower on tests of global development and hand-eye coordination at 10 weeks compared with infants born to iron-sufficient mothers. Delays in these infants persisted at 9 months of age, with lower locomotor performance and a trend for lower global developmental scores.[20] In another population, children with low cord ferritin levels at birth had lower scores on tests of language comprehension and fine motor skills at 5 years.[21]

Several studies on infants born to diabetic mothers have focused on hippocampus-based memory development. Compared with infants of diabetic mothers with normal serum ferritin levels, infants of diabetic mothers with low cord serum ferritin concentrations had electrophysiological abnormalities on an auditory recognition memory task during the neonatal period.[22] There is also evidence that such memory impairments are persistent. Infants of diabetic mothers, a group with reduced cord ferritin concentrations at birth, had poorer performance on a delayed recall task (elicited imitation) at 12 months compared to controls.[23]

Perinatal iron deficiency is also associated with affective and neurobehavioral alterations. One study found increased irritability in infants born to mothers with iron deficiency anemia,[24] and another reported infants with low iron status to have higher levels of negative emotionality and lower levels of alertness and soothability.[25] Finally, preterm infants with iron deficiency anemia have several abnormal reflexes.[26]

Iron deficiency in the perinatal period is clearly associated with adverse developmental outcomes, and several studies report persistent developmental effects of perinatal iron deficiency. However, it is not known whether such effects may be reversed with iron supplementation postpartum. Infants with reduced iron levels at birth are at an increased risk for iron deficiency later in infancy.[27] Thus, the long-term developmental outcomes of perinatal iron deficiency in these studies may not be solely attributed to the perinatal period.

8.2.2 Infancy and Toddlerhood Iron Deficiency

While there are relatively few studies evaluating the developmental outcomes of perinatal iron deficiency, a significant amount of research has focused on the effects of iron deficiency in late infancy and early toddlerhood. Iron deficiency most commonly occurs during infancy, between 6 and 24 months of age. Iron is essential to support the high growth rate during this period; however, children in this age range often have a low dietary intake of iron. Insufficient intake is due to the consumption of low-iron formula or cow's milk (a poor source of absorbable iron) and/or delayed introduction of iron-containing solids. Although iron in breast milk is better absorbed than iron in cow's milk, the concentrations in breast milk are relatively low, and after 6 months of age, the introduction of iron-containing solid foods is needed to minimize the risk for iron deficiency. Moreover, infants who experienced disruptions in fetal accumulation during the third trimester are at an increased risk for iron deficiencies during infancy due to decreased iron stores.

Numerous studies have revealed poorer cognitive and motor development in affected infants as compared with those without iron deficiency. Moreover, these impairments persist long term and are irreversible with supplementation.[15,28] In a review of multiple well-controlled studies, infants with iron deficiency anemia scored on average 6 to 15 points lower on developmental assessments compared with those without iron deficiency anemia.[28] Such studies also demonstrate significant decrements in motor achievement, with infants with iron deficiency anemia scoring on average 6 to 17 points lower on both gross and fine motor developmental assessments compared with infants without iron deficiency.[28]

Studies that have continued to follow infants after iron replacement therapy have revealed persisting differences even after 3 months of treatment.[15] Follow-up at 10 years of otherwise healthy Costa Rican infants with iron deficiency revealed persistent difficulties with arithmetic, writing, specific memory tasks, and motor functioning into adolescence, as well as a widening gap in cognitive scores.[29,30] Also, previously iron-deficient children in this cohort had an increased rate of grade repetition and need for special services, as well as social, behavioral, and attention difficulties.[29] Finally, long-term follow-up studies have shown effects on executive functioning, especially with respect to inhibition and planning.[31,32]

Iron deficiency during infancy and toddlerhood is associated with specific neurobehavioral outcomes that are consistent with iron's functions in the CNS. Physiological data reveal hippocampal perturbations in infants with iron deficiency. In a recognition memory study, compared with iron-sufficient infants, infants with iron deficiency anemia had electrophysiological responses associated with delayed hippocampal function.[33] However, whether this translates to altered memory-dependent behavior has not yet been investigated. Neurophysiology studies also suggest that iron deficiency results in hypomyelination. Infants with iron deficiency anemia have slower speeds of neural transmission in auditory brain stem responses and visual evoked potentials (VEPs), and these slower physiological responses persist at age 3 to 4 years despite iron repletion.[34,35] Rapid changes in sleep organization are occurring during the first two years of life and are also altered by iron deficiency.[36]

Infants with iron deficiency have behavioral alterations, including socioemotional behaviors and responses to novelty. Infants with iron deficiency anemia maintain closer proximity to their caregivers, have increased irritability, and have decreased positive affect.[37] Furthermore, such behavioral alterations are more prominent in novel settings.[38] In another study, preschoolers with iron deficiency anemia were more hesitant to touch a novel toy, and they displayed less positive affect and less social referencing to their mother in response to the novel toy.[39] Additionally, infants with iron deficiency anemia had decreased motor activity in an unfamiliar environment but comparable activity with nonanemic children in a familiar environment.[40]

8.2.3 ANIMAL MODELS OF IRON DEFICIENCY

Animal models offer the opportunity to study the effects of iron deficiency in a controlled environment without the variables that can complicate studies in human populations. There have been a number of rodent models that allow a comparative assessment of the effects of iron deficiency on the developing brain, based on timing, dose, and duration of the insult. Animal models also provide the opportunity to study the neurobiological systems that are disrupted by iron deficiency, which may explain the effects of iron deficiency in humans.

Iron deficiency can be experimentally induced in animals at varying stages of development to correlate with stages of human brain development. Many animal studies of iron deficiency are designed to model the most common periods of iron deficiency in humans, from late gestation through toddlerhood. In animal models, iron deficiency decreases brain iron concentrations, and these decreases have regional variation depending on the severity and the developmental timing of the deficiency. Although brain iron concentrations can be restored by iron therapy, the irreversibility of functional deficits from the deficiency depends on the timing of the deficiency and the timing of the normalization.[41–43] Animal models of early iron deficiency have focused on the neural and behavioral alterations in energy metabolism (primarily in the hippocampus), myelination, and monoamine neurotransmission, and recent studies have begun to investigate altered gene profiles.

To investigate the hippocampal effects of iron deficiency, animal studies have been developed to model iron deficiency during the third trimester in humans. These models correspond to the conditions of intrauterine growth restriction and diabetes during pregnancy, where there is moderate iron deficiency during late gestation followed by an iron-sufficient postnatal diet. Neurochemical alterations in this model suggest long-lasting effects in hippocampal energy metabolism, myelination, and neurotransmission.[7] Iron is required for energy metabolism through its role in cytochrome activity during oxidative phosphorylation. In the iron-deficient rat brain, the hippocampus and frontal lobes show the largest reductions in neuronal energy metabolism.[44] During the perinatal period in the rodent, the hippocampus undergoes a highly energy-dependent growth spurt involving dendritic growth and arborization and synaptogenesis,[45,46] and iron uptake into the hippocampus increases prior to this period of growth.[47,48] Perinatal iron deficiency disrupts this energy-dependent structural development, leading to long-term alterations in dendritic structure and dendritic extension and remodeling during synaptogenesis.[43,49,50] There is also some evidence for altered myelination in the hippocampus in the perinatal iron-deficient rat. Nuclear magnetic resonance spectroscopy results suggested myelination alterations in the hippocampus;[7] however, another study did not find reduced myelination in the hippocampus.[51] More recent research has also reported that early iron deficiency alters the expression of genes in the hippocampus, with some of these genetic alterations persisting despite iron repletion.[52] Genes that have been affected by iron deficiency include those involved in cell growth, energy metabolism, and cellular structure (e.g., dendrite structure and synaptic connectivity). These metabolic and structural alterations in the hippocampus are consistent with hippocampus-dependent performance in the iron-deficient rat. For instance, rats that were iron deficient with anemia during the perinatal period continued to have impaired performance on hippocampal spatial learning[53] and trace conditioning.[54–56] These hippocampal changes and behavioral impairments in animal models correspond with the studies of impaired recognition memory in infants of diabetic mothers.[23,24]

Another common condition in humans is chronic moderate iron deficiency, which exists from late gestation through toddlerhood. In a study that modeled this chronic iron deficiency, iron-deficient rodents had short-term and long-term impairments on hippocampally mediated tasks.[42] In humans, there is significant hippocampal development through toddlerhood, with the processes of myelination, dendritogenesis, and synaptogenesis occurring, and as in animal studies, these processes may be disrupted by insufficient iron.[29] This corresponds with the delayed electrophysiological responses on a hippocampal task in infants with iron deficiency anemia.[33]

Iron deficiency has been linked to many alterations in the process of myelination in rat models. These changes include decreases in myelin content that are long-lasting, as well as effects on individual myelin proteins and phospholipids involved in compaction.[57–61] Iron deficiency affects the proliferation of oligodendrocyte precursor cells and therefore the generation of oligodendrocytes. The long-lasting myelination defects that are not corrected with iron repletion may be due to the decreased numbers of oligodendrocytes[62] or to reduced metabolism during the process of myelination.[29] Consistent with the myelination effects of early iron deficiency, perinatal iron-deficient rats with reduced myelination have delayed neurological reflexes, which may be a marker for myelination-dependent speed of processing.[51] Hypomyelination is likely the cause of altered reflexes in the iron-deficient neonate[27] and the slower conduction in the auditory and visual systems found in human infants with iron deficiency.[35] Many other brain systems are also undergoing myelination during infancy and toddlerhood, the time of peak incidence of iron deficiency, and myelination defects may contribute to global delays, poor cognitive outcomes,[29] and motor impairments.[51]

Iron is essential for enzymes involved in monoamine neurotransmitter synthesis, including tyrosine hydroxylase (norepinephrine and dopamine) and tryptophan hydroxylase (serotonin).[11] These neurotransmitter systems are developing during early life when there is an increased risk for iron deficiency,[29] and the extent of neurotransmitter changes in iron-deficient animals correlates with the degree of brain iron deficiency.[41,63] Although most research studies on iron deficiency and neurotransmitter metabolism have focused on dopamine, iron deficiency induces alterations of all the monoamine transporters (dopamine, serotonin, and norepinephrine)[64] and in extracellular concentrations of dopamine and norepinephrine.[41,42,64,65]

The timing of iron deficiency has unique effects on the neurotransmitter systems and subsequent behaviors. For instance, iron deficiency during gestation in the rodent increases dopamine and its transmitter, whereas early postnatal deficiency decreases dopamine, dopamine receptors, and dopamine transporter.[41,66] The timing of the deficiency may also have unique effects on behaviors. Gestational iron deficiency in monkeys was associated with less fearful and more impulsive behaviors, whereas postnatal iron deficiency was associated with more inhibited and anxious behaviors.[67] Finally, there is regional specificity in neurotransmitter alterations dependent on the timing of the deficiency. For instance, dopamine alterations are seen in the nucleus accumbens, striatum, and PFC.[68] These regions undergo high rates of iron accretion at different periods of development, and each region will be vulnerable to the deficiency during its period of rapid accumulation.

Behaviors dependent on dopaminergic systems are altered in rodent models of iron deficiency and are consistent with behavioral alterations observed in children. Rodents with moderate early iron deficiency have impaired sensorimotor and grooming behaviors that are likely associated with the nigrostriatal and corticostriatal dopaminergic circuits.[29,41,42] Such motor alterations may explain the motor impairments reported in children with iron deficiency during infancy. Similar to young rats with dopamine depletion, rats with early iron deficiency have increased anxious behaviors and decreased exploratory behaviors in novel environments.[29,42,68] Such findings are consistent with infants with iron deficiency who show affective changes and who respond differently to context.[37,38,40] Alterations in dopamine metabolism may also explain the impairments in executive function years after iron deficiency.[29,33]

Finally, many of the CNS alterations associated with iron deficiency in animal models are persistent. More than 300 transcripts were altered in a model of moderate iron deficiency, with five genes remaining changed at 180 days.[12] A similar long-term gene effect is seen in the hippocampus.[52] The altered genes are involved in cytoskeletal stability and synaptic function and may be associated with some of the long-term behavioral changes in infants with early iron deficiency.

8.2.4 SUMMARY

Iron deficiency is the most common single-nutrient deficiency in the world. Children are especially at risk during the fetal-neonatal period, as well as during late infancy and early toddlerhood, a time when their brains are still undergoing rapid developmental changes. Rodent models have helped researchers gain a better understanding of the effects of iron deficiency on several processes in the CNS, including hippocampal metabolism, dendritogenesis, synaptogenesis, myelination, and neurotransmitter function. Studies involving human infants have shown long-term effects on cognition, motor functioning, social-emotional behaviors, and neurophysiological development.

8.3 ZINC

Zinc is a cofactor in enzymes that mediates nucleic acid and protein biochemistry[69,70] and thus is necessary for processes critical for brain growth, including cell replication and protein synthesis. Zinc may also have indirect effects on brain growth through its role in the growth hormone signaling pathway. Zinc-dependent DNA-binding proteins are essential for the genetic expression of insulin-like growth factor I and growth hormone receptor.[71] Accordingly, the effects of fetal zinc deficiency include reduced brain DNA, RNA, and protein concentrations in the rat pup[69,72] and postnatal zinc deficiency decreases insulin-like growth factor I and growth hormone receptor gene expression.[71]

Zinc deficiency has structural, chemical, and physiological effects in the CNS. Structural alterations are likely due to the timing of deficiency coinciding with rapid structural development and growth, while functional effects (e.g., neurochemical and neurophysiological) of zinc deficiency may occur during any period of development. Finally, the effects of zinc on neurodevelopment and function are not homogenous; zinc has regional effects in the CNS, with the autonomic nervous system (ANS), cerebellum, limbic system, and cerebral cortex particularly vulnerable.

Although severe zinc deficiency is rare, mild to moderate zinc deficiency may be common worldwide.[73] The prevalence of zinc deficiency has been difficult to estimate due to the lack of agreement on the indicators for zinc status, and there is no global estimate of the prevalence of zinc deficiency. However, using estimated dietary zinc availability in a country's food supply, the International Zinc Nutrition Consultative Group estimated that approximately one-third of the global population is zinc deficient, with regional prevalence rates ranging from 4% to 73%.[74] Populations at risk for deficiency include those who have inadequate zinc intake due primarily to low consumption of animal foods (the best source of zinc) and/or consume diets with low bioavailability due to high intakes of foods that interfere with zinc absorption (e.g., phytates or fiber). Secondary zinc deficiency may also occur despite adequate zinc intake. For instance, during the acute phase response, zinc is sequestered to the liver, decreasing its availability for other tissues.[75] Zinc deficiency may also occur due to increased losses during diarrhea.[76] Since zinc is critical for growth, the risk for zinc deficiency increases during periods of growth.

8.3.1 Fetal-Neonatal Zinc Deficiency

Pregnant women are vulnerable to zinc deficiency as fetal growth increases the requirements for zinc. It has been estimated that at least 80% of pregnant women worldwide have inadequate zinc intakes.[77] Severe zinc deficiency is associated with decreased fertility in women.[78] Severe zinc deficiency also interferes with neural crest development in animal models,[79] and there have been case reports of neural tube problems in infants of mothers with severe zinc deficiency.[80,81] More moderate maternal zinc deficiency may alter fetal ANS functioning as maternal zinc deficiency is associated with higher fetal heart rate and lower heart rate variability.[82,83] However, the association between zinc deficiency and ANS stability is based on behavioral measures without an understanding of specific biological mechanisms.

Research on mild to moderate maternal zinc deficiency suggests regional rather than global developmental effects. Although maternal zinc deficiency is associated with reduced brain DNA, RNA, and protein content in animal models,[69,72] there is a lack of human evidence for irreversible global developmental effects in infants born to mothers with zinc deficiency. However, mild to moderate maternal zinc deficiency is linked to cerebellum- and hippocampus-dependent infant behavior. In a village in Egypt where mild to moderate undernutrition was common, a cohort of pregnant women was followed through pregnancy until 6 months of lactation.[84] Although maternal zinc nutriture was not related to global developmental measures, both cerebellum- and hippocampus-dependent behaviors were altered. Infants born to mothers with higher intakes of plant zinc, fiber, and phytate (markers of low bioavailable zinc) were more likely to have lower scores on the psychomotor index of the Bayley Scales of Infant Development (BSID) at 6 months. However, the dietary intake measure included intakes during pregnancy through 6 months of lactation, and the link between maternal zinc intake and infant motor development cannot be attributed solely to prenatal zinc. Maternal zinc intake was also related to the habituation cluster of the Brazelton Neonatal Assessment Scale; specifically, the amount of available zinc in the maternal diet during the second and third trimesters was associated with better ability of the neonate to habituate to repeated stimulation.[84] These data provide a link between maternal zinc status and neonatal hippocampal development; however, other nutrients (e.g., iron) were also associated with habituation performance and, as the authors mentioned, this association between zinc and habituation could be due to an interaction of zinc and iron deficiencies. The hippocampal effects of zinc deficiency may be amplified by concurrent iron deficiency.

Despite the association between maternal zinc status and infant motor and hippocampus-dependent behavior, these findings rely on correlational data and experimental studies of maternal mild to moderate zinc deficiency and infant behavior will be important in future research. There is a need for both randomized controlled trials of zinc supplementation and long-term follow-up studies to define the relationship between maternal zinc status and infants' motor and cognitive abilities. Additionally, the form of zinc supplementation to promote optimal development needs to be determined. Prenatal zinc supplementation may even hinder global development. In one study, infants of Bangladeshi women who were given elemental zinc supplements during pregnancy performed worse on both the mental and motor scales of the BSID than infants of mothers who were given a placebo.[85] The authors caution that giving undernourished pregnant women zinc alone, rather than as part of a micronutrient mixture, may interfere with the metabolism of other nutrients that are critical for neurodevelopment. For example, when given in large quantities, zinc may limit the absorption of iron and copper from the intestine,[85,86] leading to secondary iron or copper deficiency.

8.3.2   Infancy and Childhood Zinc Deficiency

During postnatal development, periods of rapid growth (e.g., infancy and adolescence) place infants and children at risk for zinc deficiency due to the zinc requirements of growth. Zinc concentrations in breast milk decrease after 6 months and no longer supply adequate zinc, increasing the risk for dietary zinc deficiency in infants older than 6 months.[74,87] Although there is no estimate of the prevalence of infant zinc deficiency, dietary reports from middle-income families in the United States indicate that marginal zinc deficiency may be fairly widespread.[88] During childhood and adolescence, children relying on diets with limited bioavailable zinc (i.e., plant-based diets) during periods of growth are also at risk for zinc deficiency.

Similar to the prenatal period, zinc deficiency during infancy and childhood is associated with regional rather than global effects, localized to the cerebellum, hippocampus, and cerebral cortex. There is little human evidence that infants with moderate zinc deficiency have global delays. Several well-controlled studies in undernourished populations have found no mean differences on tests of global cognitive development in zinc-supplemented infants, and these findings are consistent whether zinc was given alone[89] or as part of a micronutrient mixture.[90] One study however reported a small effect of zinc supplementation on global cognitive development in a group of Chilean infants. Although there were no mean differences on the mental developmental index of the BSID between the supplemented group and the placebo group, a smaller proportion of the supplemented group scored below the mean standardized score of 100.[91]

There is some evidence for the role of zinc in motor development and that zinc supplementation promotes motor activity during infancy. Zinc supplementation (zinc-copper supplement) given during infancy is associated with improved motor development in infants with a very low birth weight.[92] Additionally, in the Chilean cohort of term infants mentioned above, a lower proportion of the supplemented group had motor dysfunction (in the areas of gross motor, fine motor, and control of movement) compared with the placebo group despite no mean difference on standardized scores of motor development.[91] On the other hand, several supplementation trials have failed to find any improvements in motor development in other populations of infants.[89,93,94] Two of these studies did however find differences in motor activity. For instance, supplemented infants spent more time sitting up, less time lying down, and more time in play compared with the infants administered a placebo despite no difference in the number of motor milestones (e.g., walking) attained between the groups.[93] A similar study found that children supplemented with zinc were more active and expended more energy.[94]

Although less studied than that in the prenatal and infancy periods, zinc deficiency during later childhood is associated with decreased abilities dependent on the PFC. In three Chinese cities, children between 6 and 9 years old who were supplemented with a micronutrient mixture with zinc performed better on a complex reasoning task involving abstract reasoning and concept formation compared with those who were supplemented with a micronutrient mixture without zinc.[95] However, there are no structural or functional [e.g., magnetic resonance imaging (MRI)] data on zinc-deficient human populations to connect the prefrontal-dependent behavioral data with underlying CNS functioning.

8.3.3   Animal Models of Zinc Deficiency

The human data on cerebellar/motor activity, hippocampal, and cerebral cortical effects are corroborated by animal models. Zinc has several roles in neurotransmission consistent with these effects. Zinc-containing neurons are present in the limbic and cortical regions.[96] In the hippocampus, zinc-containing neurons release zinc from presynaptic boutons into the synapse, which may modulate glutamate neurotransmission and neuron excitability.[97] Zinc deficiency also inhibits γ-aminobutyric acid-stimulated Cl influx into hippocampal, cortical, and cerebellar neurons, with the greatest inhibition in the hippocampus.[98] Neurophysiology in the hippocampus is also affected by zinc, as zinc-deficient rats have abnormal evoked responses in hippocampal neurons.[99] In the cerebral cortex, zinc deficiency disrupts opioid receptor function.[100]

Consistent with the effects of prenatal zinc deficiency on motor development, prenatal zinc deficiency in rodents alters cerebellar dendritic arborization[101] and results in reduced cerebellar volume.[101–103] Similar to the reduced activity associated with zinc deficiency in humans, several animal models of zinc deficiency show reduced spontaneous motor activity.[104,105] Future animal models of moderate zinc deficiency will be important to clarify brain-behavior relationships due to zinc deficiency (e.g., to determine whether delayed motor development is associated with altered cerebellar structure) and to determine whether there are "sensitive" periods when motor development is susceptible to zinc deficiency.

Animal models of moderate zinc deficiency during adolescence also support the role of zinc in hippocampal and cortical activity. Consistent with its role in hippocampal neurotransmission, animal models of zinc deficiency during adolescence and adulthood show altered hippocampus-dependent behaviors. Adolescent rhesus monkeys performed worse on hippocampus-dependent memory tasks (delayed match-to-sample test) when fed a diet that is moderately zinc deficient compared with a zinc-sufficient diet.[104] Additionally, zinc-deficient rats performed poorly in hippocampus-dependent spatial working memory tasks compared with zinc-sufficient rats.[106] Although hippocampal effects in humans have been observed in the neonate, hippocampal effects of zinc deficiency during adolescence and adulthood in animal models support the essentiality of zinc for normal hippocampal function beyond infancy. Cortical effects were also observed in the adolescent rhesus monkeys that were fed a diet that is moderately zinc deficient. Diets that are moderately zinc deficient were associated with impaired performance on a continuous performance test, a measure of attention and inhibition.[104]

8.3.4 SUMMARY

There is evidence for the essentiality of zinc in normal CNS development and function from fetal life through childhood, and the effects appear to be more regional than global. Animal models and behavioral human research suggest localized effects of early zinc deficiency on the developing ANS, cerebellum, hippocampus, and cortex. The small amount of behavioral evidence is compatible with specific CNS alterations found in animal models; however, there is a lack of data on the specific brain-behavior pathology due to zinc deficiency.

8.4 IODINE

Iodine is a component of thyroid hormones, with insufficient iodine leading to hypothyroidism. Iodine does not have direct roles in the CNS, but it alters brain development and function through thyroid activity. Thyroid hormones regulate cellular metabolism and cell differentiation and development. In the CNS, thyroid hormones regulate metabolic rate, affecting mitochondrial enzymatic activity[107] and playing a role in glucose transport in astrocytes.[108] Thyroid hormone is also essential for cell differentiation and growth, during which it acts to regulate metabolism to meet the cellular activity and energy requirements and to regulate structural proteins and growth factors.[109] Additionally, thyroid hormones affect the genetic expression of several myelination proteins.[110,111] Finally, thyroid hormones have localized gene targets in the brain; regions particularly affected include the developing cortex, hippocampus, striatum, and cerebellum.[109,112]

Iodine is present in the soil, and populations who depend on plant crops and animals raised on pasture grasses grown in soils with low levels of iodine are at risk for deficiency. Approximately one-third of the world's population (nearly two billion individuals) is estimated to have insufficient iodine intakes, with the highest proportions found in regions of Southeast Asia and Europe.[113] The severity of iodine deficiency is correlated with the availability of iodine in the food supply.[114] Universal salt iodization is recognized as the most effective method of preventing iodine deficiency.[113]

Iodine deficiency is the leading cause of preventable brain damage worldwide.[113] For individuals living in endemic regions, iodine deficiency may occur during any developmental period; however, the neurodevelopmental effects are critically dependent on the severity and timing of the deficiency. Iodine deficiency that occurs during the second trimester of pregnancy through the third postnatal year is associated with the greatest risk for neurodevelopmental impairments;[113] however, fetal effects are the most profound. Severe iodine deficiency during early pregnancy leads to the syndrome of cretinism, characterized by irreversible mental retardation, deaf mutism, spastic diplegia, and dwarfism.[115,116] While iodine deficiency that occurs later in pregnancy or is mild to moderate does not cause cretinism, it is associated with reduced head size and irreversible global deficits.

8.4.1 Prenatal Iodine Deficiency

One study examined the timing of iodine deficiency in an endemic region of China by giving pregnant women and young children iodine supplements during specific developmental periods.[115] The most positive outcomes were associated with supplementation before the end of the second trimester. Specifically, iodine treatment during the first two trimesters reduced the incidence of neurological symptoms. Moreover, supplementation during the second trimester increased head size and improved cognitive and motor developmental scores at 2 years of age. Iodine treatment during the third trimester, during the neonatal period, or between the ages of 3 and 12 months did not reduce neurological symptoms but did slightly improve head growth and developmental scores at 2 years of age compared with the untreated controls. Furthermore, when tested at school age, children whose mothers were supplemented early in pregnancy had improved visual perception and fine motor skills compared with children who were supplemented either during the third trimester or at 2 years of age.[116] These findings suggest that the developmental effects of iodine deficiency during early gestation are irreversible with iodine repletion.

8.4.2 Childhood Iodine Deficiency

Although the most severe developmental effects are due to prenatal iodine deficiency, childhood iodine deficiency is also associated with cognitive and motor impairments, and there is evidence for improvements with iodine treatment. Several studies have compared children from iodine-deficient regions with children living in similar regions with higher levels of iodine in the soil. Children living in the iodine-deficient regions had more cognitive impairments compared with children from the iodine-sufficient regions,[117] and the degree of iodine deficiency of a region was associated with the severity of impairments in motor and IQ assessments.[118] Additionally, simple reaction times, used to assess nervous transmission and information processing efficiency, were delayed in children living in iodine-deficient regions compared with those in iodine-sufficient regions.[119] Iodine supplementation in school-age children improves cognitive and motor performance, emphasizing the significance of iodine for normal brain function. Specifically, moderately deficient children between the ages of 10 and 12 years who received iodine supplements performed better on several cognitive measures of information processing and visual problem solving, as well as fine motor tasks, compared with those given a placebo.[120] The cognitive and motor improvements in these studies suggest a metabolic effect that is reversible with iodine repletion rather than an irreversible anatomical effect of iodine.

8.4.3 Animal Models of Iodine Deficiency

Hypothyroidism in animals has histological effects in the CNS, including alterations in neuronal differentiation, neural process formation, and synaptogenesis.[115,121] Such structural changes are consistent with thyroid's role in cell growth and development. Additionally, thyroid hormones alter

CNS-specific structural proteins involved in cytoskeletal structure and neurite outgrowth, which may affect synaptic connectivity.[109] Specifically, thyroid hormones have been shown to alter microtubule-associated proteins and neurotubule development, especially during early development.[122,123] Thyroid hormones are also involved in the regulation of myelination proteins, myelin basic protein, and myelin-associated glycoprotein.[110,111]

Animal studies support the significance of the timing of the deficiency and the subsequent neurobehavioral effects reported in humans. Neuronal proliferation of the structures that are affected in cretinism (e.g., cerebral cortex, cochlea, and basal ganglia) occurs primarily during the second trimester.[109,115] During the third trimester, cell differentiation takes place, rapidly increasing brain weight and DNA and protein content. This period of increased brain weight corresponds to the finding of improved head growth in infants whose mothers were supplemented with iodine during late pregnancy.[115] Thyroid deficiency or iodine deficiency results in reduced dendritic arborization, synaptic counts, and myelination.[121,124,125] These processes are occurring during the first 3 years in humans and are consistent with the developmental deficits associated with iodine deficiency during early life. Accordingly, delayed myelination and/or impaired synaptogenesis may be the cause for the decreased reaction time observed in iodine-deficient children.[119]

Much of the research on iodine deficiency in humans has focused on global developmental outcomes, such as intelligence or developmental quotients (e.g., BSID). However, animal research demonstrates that iodine deficiency likely has regional effects with specific impairments consistent with the localized gene targets of thyroid hormone. Thyroid hormone alters the genetic expression of neurogranin, a dendritic protein, in the cerebral cortex, hippocampus, and striatum.[112] In hypothyroid rats, neurotrophin genes are altered. Specifically, nerve growth factor is reduced in the cortex, hippocampus, and cerebellum, and its receptor is reduced in the striatum.[126]

Thyroid hormones have several other functions specific to the hippocampus. Hypothyroidism is associated with delayed maturation of glial cells in the hippocampus, which may impair neuronal migration during midgestation.[127] Neurogranin, which is reduced by hypothyroidism in the hippocampus, is necessary for postsynaptic calcium events, including long-term potentiation, a process involved in memory and learning.[112] Acetylcholine is also involved in long-term potentiation in the hippocampus, and thyroid hormones are involved in cholinergic neurotransmission. For instance, hypothyroidism reduces choline acetyltransferase activity necessary for acetylcholine synthesis.[128] Accordingly, in adult rats, thyroid hormone supplementation improved performance on a spatial learning task compared with nonsupplemented rats, and this effect seemed to be due to thyroid hormone alteration of acetylcholine neurotransmission in the hippocampus.[109] Such animal studies suggest that thyroid deficiency hampers hippocampus-dependent memory performance; however, the specific effects of iodine deficiency on memory and hippocampal development in children have yet to be investigated.

Localized gene targets of thyroid hormone in the striatum and cerebellum may also be responsible for motor impairments reported in iodine-deficient humans. In addition to the alterations in neurogranin[112] and neurotrophin receptors in the striatum,[126] thyroid hormones regulate Purkinje-specific genes in the cerebellum.[129]

8.4.4 SUMMARY

The neurobehavioral effects of iodine deficiency are critically dependent on the timing and severity of the deficiency. Iodine deficiency, which occurs early in development, results in the most severe and irreversible developmental impairment. Although less studied, iodine supplementation during childhood improves cognitive and motor performance, highlighting that iodine is important for normal CNS functioning. Most research has focused on global development, but as animal research studies on hypothyroidism continue to identify the gene targets under control of thyroid hormones, investigations into particular domains will be important to link these genetic effects of thyroid hormones with the behavioral effects in humans. Furthermore, animal studies of thyroid hormone do not address iodine deficiency directly, but they can guide research and further the understanding on iodine's effects on brain development. Future studies will be important to connect the severity of iodine deficiency with the subsequent thyroid dysfunction in the CNS.

8.5 SELENIUM

Selenium is required for the synthesis of selenoproteins, during which selenium as selenocysteine is incorporated into selenoproteins during translation. Several roles of selenoproteins relevant to neurodevelopment include thyroid metabolism and antioxidant metabolism. Selenium is necessary for normal thyroid function as selenoproteins activate thyroid hormone, deiodinizing thyroxine (T4) to triiodothyronine (T3). Thus, like iodine, selenium deficiency may impair thyroid activity. Several selenoproteins are involved in antioxidant metabolism necessary for protecting against free radical damage and heavy metal toxicity (e.g., mercury), which can lead to cognitive impairments.[130] Specifically, selenium deficiency is associated with reduced glutathione peroxidase activity in the brain, whereas thioredoxin reductase does not seem to be affected in the CNS.[131] Finally, the roles of selenium protein P in the CNS are also being investigated.[130,132]

Like iodine, selenium is present in the soil, and populations who depend on food raised in soils with low levels of selenium are at risk for deficiency. The highest rates of selenium deficiency are found in regions throughout China with low soil content.[133] New Zealand and parts of the United Kingdom, among several other regions, also have low soil content. Preterm infants are at a relatively high risk for selenium deficiency due to their lower body stores and generally poorer antioxidant status. Phenylketonuria (PKU) patients may also be at risk for decreased selenium status due to restricted protein intake that may limit their dietary selenium.[134] However, during selenium deficiency, the brain is highly protected, and selenium levels in the brain are the last to be depleted.[135,136] Thyroid function also seems to be preferentially preserved when there is marginal selenium;[137] however, during severe selenium deficiency, which likely occurs in regions with low soil content, thyroid metabolism may be affected.[138]

8.5.1 Selenium Deficiency in Humans

Research on the neurodevelopmental effects of selenium deficiency in children is limited. Due to its role in thyroid metabolism, selenium may have effects on brain development (see Section 8.4) similar to those of iodine. Additionally, in children and adolescents with PKU, poor attention abilities are associated with selenium status.[134] In these children, thyroid function was normal, suggesting that altered antioxidant metabolism is a likely cause. Research studies in adults indicate that selenium deficiency is associated with depression.[139] Also, in adults, selenium deficiency, likely through reduced antioxidant metabolism, has been found to be associated with aging and cognitive decline.[134,140]

8.5.2 Animal Models of Selenium Deficiency

Although research on selenium deficiency and neurodevelopment in children is sparse, animal models support the essentiality of selenium in the CNS.[141] Thyroid hormones are necessary for myelination (see Section 8.4). In rodent models, the combined deficiencies of iodine and selenium resulted in myelination delays greater than those from the individual deficiencies.[142] Selenium promotes neuronal survival,[130] and its deficiency reduces brain-derived neurotrophic factor expression.[142] Additionally, selenium protein P seems to have a role in neuronal survival.[143] Selenium (e.g., glutathione peroxidases) is also important in protection against neuronal damage from lipid peroxidation of cell membranes from oxidative stress. For instance, oxidative metabolism of dopamine may produce free radicals, and several dopaminergic systems (e.g., PFC, nigrostriatal pathway) are vulnerable to oxidative damage during selenium deficiency.[144] Animal studies on selenium's role in antioxidant metabolism have been used to explain neurodegeneration, and whether these findings translate to neurodevelopment is unknown. However, a study in children with PKU who had impaired attention and inhibition, behaviors that are highly dependent on PFC, suggested that decreased glutathione peroxidase activity from selenium deficiency in combination with increased dopamine turnover rate leads to neuronal damage in the PFC and may explain the attention deficits.[134] Alterations in neurotransmitter systems may also be associated with depressed mood reported in humans.[139]

8.5.3 SUMMARY

Selenium has several roles in the CNS, primarily through thyroid and antioxidant metabolism. Research on the role of selenium deficiency on neurodevelopment in children is lacking; however, animal research provides evidence for the role of selenium in brain development and function.

8.6 COPPER

There are relatively high levels of copper present throughout the CNS compared with other extrahepatic tissues. Both animal and human studies support the importance of adequate copper for normal brain development. Clinical awareness of copper deficiency has been well recognized in the hematological profile since the 1840s.[145] However, it was not until the 1930s, when copper-deficient ewes were noted to give birth to ataxic offspring, that the role of copper in brain development was appreciated.[146] Since that time, the effect of perinatal copper deficiency on the developing brain has been documented in laboratory animals and humans.

Copper is present as a cofactor in enzymes (cuproenzymes) that serves a diversity of functions, and inadequate copper likely affects normal brain development through reduced activity of these enzymes.[147] Although cuproenzymes are present throughout the brain, copper deficiency seems to have regional specificity, particularly affecting the occipital and parietal cortex, striatum, cerebellum, and hippocampus.[148]

Clinical copper deficiency in the general human population is less common than iron and zinc deficiencies; however, several groups, including infants, toddlers, and adolescents, have an average intake less than the recommended dietary allowance and are at risk for marginal copper deficiency.[149] Copper is accumulated during the last trimester, and normal brain development may not be supported if copper accumulation is insufficient during this time. Accordingly, infants born prematurely are at risk for copper deficiency. This is of particular concern for teen pregnancies, when the mother's own copper needs are high because of her continued development. Also, the copper needs of pregnant or lactating women are far greater than those needed to support adult homeostasis,[150] which may also place the developing infant at risk. Additionally, excessive use of supplements that interfere with copper absorption, such as zinc for the common cold, is becoming increasingly popular and increases the risk for copper deficiency.[147]

8.6.1 EARLY COPPER DEFICIENCY

The role of copper in early brain development is evident from infants with the Menkes disease, an X-linked defect in a copper transporter gene that results in failure to transport copper across the intestinal basal lateral surface, leading to copper accumulation in gut epithelial cells and a deficiency in all other tissues. Patients with this disease have marked neurodegeneration and face death at about 2 to 4 years of age when untreated. Much less is known about the neurodevelopmental effects of marginal copper deficiency in humans, in part because of the difficulty in assessing copper status. Currently, there are methods of assessing copper status, including serum copper and ceruloplasmin concentrations. Unfortunately, while these markers are able to detect moderate to severe copper deficiency, they are much less sensitive to marginal deficiency. Cu,Zn-superoxide dismutase and copper chaperone CCS may be useful for assessing marginal copper status in humans.[151]

8.6.2 ANIMAL MODELS OF COPPER DEFICIENCY

Although the literature on copper deficiency and human neurodevelopment is scarce, laboratory animals with copper deficiency have severe developmental and morphological abnormalities. Rodents with perinatal copper deficiency have focal lesions in the occipital and parietal parts of the cerebral cortex, as well as alterations in the corpus striatum.[152] The developing cerebellum also seems to be at high risk in models of copper restriction during gestation, with resultant permanent impairment to motor function, balance, and coordination that persists even after long-term recovery from the deficiency.[145] Hippocampal development appears to be altered by early copper deficiency as well,[153] while the hypothalamus is spared.[154] Although these effects are reported in animal models, the extent of copper deficiency in most animal studies is severe and may not translate to the potential neurodevelopmental effects of moderate copper deficiency in humans.[147]

Over the past decades, the goal of research on copper deficiency has been to link these CNS effects with the known functions of cuproenzymes.[147] Prohaska[147] has proposed four hypotheses. The first is that limited cytochrome c oxidase, involved in energy metabolism, may impair protein and lipid synthesis during periods of increased brain growth. The second hypothesis is that limited Cu,Zn-superoxide dismutase, involved in antioxidant metabolism, may increase lipid peroxidation and membrane degeneration (e.g., hypo- and demyelination) and alter the trophic chemical environment. Third, limited dopamine B monooxygenase, which converts dopamine to norepinephrine, would alter neurotransmission. Finally, limited peptidyl α-amidating monooxygenase activity would limit peptide synthesis, as half of all peptides are α-amidated by this enzyme. Consistent with these cuproenzyme functions, copper deficiency has been linked specifically to aberrant energy metabolism, myelination defects, increased susceptibility to reactive oxygen toxicity, and altered dopamine and norepinephrine metabolism.[147,151,152,155–157] Finally, the cuproenzyme ceruloplasmin is involved in iron metabolism, and copper deficiency is associated with decreased iron accretion in the fetal and neonatal brain[157] and may thus have effects on neurodevelopment through secondary iron deficiency.

8.6.3 SUMMARY

While severe copper deficiency is not very common in humans, there are specific subgroups that are at risk. Studies in laboratory animals have confirmed that copper deficiency can lead to alterations in the structure and function of the developing CNS. It has also been shown that these alterations can lead to long-term difficulties with motor function, balance, and coordination.

8.7 CIRCUIT-SPECIFIC ASSESSMENTS

Multilevel analyses in infants and children at the neurostructural, neurofunctional, and neurobehavioral levels are important to address the roles of nutrients in specific neural systems, especially in conjunction with animal studies that can guide circuit-specific research in humans. Systems that are developing rapidly have increased metabolic requirements of nutrient substrates, and insufficient nutrition may disrupt the anatomy of the structures that are developing rapidly at that time. Volume measurements, such as head size, computerized axial tomography, or MRI scans, are important to estimate structural development. In the infant, occipitofrontal circumference is used as gross measurement of brain size. Circuit-specific structural estimates include regional brain volumes using structural MRI and neural connectivity using diffusion tensor imaging (DTI).[158] By about 6 years of age, MRI does not require sedation and is therefore a more feasible instrument in older children than in infants. Although structural estimates are useful in determining the role of any given nutrient in developing neural systems, it is vital to link structural effects to functional and neurobehavioral development to fully explicate the circuit-specific effects of that nutrient. Functional assessments include electrophysiology measures, such as electroencephalography, auditory brain stem evoked response (ABR), and event-related potentials (ERPs).[159] ERP studies are used to link electrophysiological response to specific stimuli and events and are particularly useful in prever-bal infants. In older children, functional imaging (e.g., fMRI) is valuable for linking structure and function.[160] The use of newer techniques, including magnetoencephalography and near-infrared spectroscopy, will likely be useful in measuring cortical functioning in younger populations.[161,162] Additionally, functioning of the ANS and hypothalamic-pituitary-adrenal axis in response to stress may be measured in infants and children.[163] Finally, neurobehavioral assessments are necessary to link the structural and functional effects of nutrient deficiencies to cognitive and behavioral development in children.

There are several systems that are particularly vulnerable to nutritional insults and may be assessed using this multilevel approach in infants and children. Hippocampal circuitry is involved in explicit memory formation.[164] In infants, this system can be evaluated through ERP studies. For instance, an electrophysiological response specific to hippocampal function is recorded during a recognition memory paradigm during which infants are presented with familiar and novel stimuli.[33] By age 4 to 6 months, behavioral measures are also useful to investigate this system. About this age, the preferential looking test[165] is effective; in older infants and toddlers, beginning at about 9 to 12 months, elicited imitation is useful in measuring hippocampus-dependent memory.[166,167] Moreover, behavioral measures in conjunction with ERP are important in connecting behaviors with hippocampal functioning.[167]

Development of the PFC and its connectivity to related structures are associated with the acquisition of tasks of working memory and inhibitory control and executive function. Frontal lobe development may be testable during infancy using behavioral measures, such as the A not B task, which requires the ability to hold a representation in memory over time and to inhibit a motor response.[168] Electrophysiology studies are useful in linking these behaviors to CNS functioning. For instance, successful performance on the A not B task following a delay was associated with frontal region electroencephalography activity in infants.[169,170] At about 4 years of age, the computerized battery Cambridge Neuropsychological Test Automated Battery is useful in investigating cognitive performance in specific domains, such as working memory and executive function.[171] Also, at about 6 years of age, fMRI is useful in structure-function associations on numerous cognitive tasks dependent on the frontal lobes, as well as functions specific to other systems (e.g., explicit and implicit memory systems).[160]

Synaptic efficacy and myelination can be effectively evaluated in infants and children using speed-of-processing measures with ABR and ERP (e.g., VEPs).[35] Neonatal behaviors during a neurological examination, including neurological reflexes, can also provide limited information on speed of processing.[27] In older children, DTI studies are valuable in measuring connectivity and associated behaviors. For instance, DTI studies have shown that age-related changes in white matter (e.g., in PFC connections) throughout childhood and adolescence are associated with improved cognition.[172,173]

Socioemotional behaviors that are dependent on monoaminergic neurotransmission can be directly observed and coded during infancy and childhood.[38,40] Furthermore, as in cognitive systems, functional and structural imaging techniques in older children are important to link underlying structures and associated neural systems with socioemotional behaviors.

Many studies have used general cognitive outcomes such as developmental and intelligence quotients (e.g., BSID) to investigate the effects of nutrient deficiencies on development. Although such measures are practical and demonstrate that specific nutrients are necessary for typical neurodevelopment, these general measures are not specific to the neural systems that may be disrupted by a given nutrient. Additionally, general measures may not be sensitive enough to identify the effects of mild to moderate nutrient deficiencies. However, it is important to note that neural systems may compete with, cooperate with, and/or compensate for activity of other neural systems.[174,175] Thus, while specific hypotheses can be made on the neurodevelopmental effects of a nutrient deficiency based on that nutrient's roles in a particular neural system, neural systems do not function in isolation.

Examples of published studies on micronutrient deficiencies that have used circuit-specific assessments are listed in Table 8.1. Continued investigations into the circuit-specific effects of nutrient

TABLE 8.1
Micronutrient Deficiency Effects on Developing Brain Circuitry: Published Studies Describing Circuit-Specific Assessments in Humans

Micronutrient	Processes Affected	Principal Brain Region or Circuitry Affected	Circuit-Specific Assessments
Iron	Myelination, monoamine synthesis, and oxidative phosphorylation	White matter, frontal, striatum, and hippocampus	Neurological examination,[26] ABR and VEPs,[34,35] elicited imitation,[23] ERP for hippocampus-dependent memory,[22,33] executive functioning tasks,[31,32] and socioemotional behavior coding[24,25,37–39]
Zinc	Neurotransmission, cell replication, and protein synthesis	ANS, cerebellum, hippocampus, and cortex	Heart rate variability,[82,83] standardized motor assessment,[84,91,92] motor activity,[93,94] habituation,[84] and complex reasoning tasks[95]
Iodine	Thyroid metabolism (myelination, energy metabolism, neuronal proliferation, cell differentiation, acetylcholine synthesis)	Cortex, striatum, cerebellum, and hippocampus	Reaction time,[119] motor assessment,[116,118,120] and information processing and working memory tasks[120]
Selenium	Antioxidant metabolism and thyroid metabolism	PFC and striatum	Attention and inhibition tasks[134]
Copper	Energy metabolism, antioxidant metabolism, myelination, and monoamine synthesis	Cerebellum	No study available

status will be important for expanding our basic understanding of the typical development of neural systems, determining the role of particular nutrients in the development of these systems, and developing effective early intervention programs for populations at risk for early nutrient deficiencies.

REFERENCES

1. Tanner, E.M., and Finn-Stevenson, M., Nutrition and brain development: Social policy implications, *Am J Orthopsychiatry* 72, 182, 2002.
2. Fuglestad, A.J., Rao, R., and Georgieff, M.K., The role of nutrition in cognitive development, in *Handbook of Developmental Cognitive Neuroscience*, 2nd ed., Nelson, C.A., and Luciana, M., Eds., MIT Press, Cambridge, MA, 2008, 623.
3. Kretchmer, N., Beard, J.L., and Carlson, S., The role of nutrition in the development of normal cognition, *Am J Clin Nutr* 63, 997S, 1996.
4. Georgieff, M.K., Nutrition and the developing brain: Nutrient priorities and measurement, *Am J Clin Nutr* 85, 614S, 2007.
5. Winick, M., and Rosso, P., The effect of severe early malnutrition on cellular growth in the human brain, *Pediatr Res* 3, 181, 1969.
6. Beard, J.L., and Connor, J.R., Iron status and neural functioning, *Annu Rev Nutr* 23, 31, 2003.
7. Rao, R. et al., Perinatal iron deficiency alters the neurochemical profile of the developing rat hippocampus, *J Nutr* 133, 3215, 2003.
8. Pomeroy, S.L., and Ullrich, N.J., Development of the nervous system, in *Fetal and Neonatal Physiology*, 3rd ed., Polin, R., Fox, W., and Abman, S., Eds., Saunders, Philadelphia, PA, 2004, 1675.
9. Thompson, R.A., and Nelson, C.A., Developmental science and the media: Early brain development, *Am Psychol* 56, 5, 2001.

10. Yehuda, S., Rabinovitz, S., and Mostofsky, D.I., Nutritional deficiencies in learning and cognition, *J Pediatr Gastroenterol Nutr* 43, S22, 2006.

11. Wigglesworth, J.M., and Baum, H., Iron dependent enzymes in the brain, in *Brain Iron: Neurochemical and Behavioural Aspects*, Youdim, M.B.H., Ed., Taylor and Francis, New York, NY, 1988, 25.

12. Clardy, S.L. et al., Acute and chronic effects of developmental iron deficiency on mRNA expression patterns in the brain, in *Oxidative Stress and Neuroprotection*, Vol. 71, Springer, Vienna, 2006, 173.

13. Stoltzfus, R.J., Defining iron-deficiency anemia in public health terms: A time for reflection, *J Nutr* 131, 565, 2001.

14. Rao, R., and Georgieff, M.K., Perinatal aspects of iron metabolism, *Acta Paediatr Suppl* 91, 124, 2002.

15. Lozoff, B., and Georgieff, M.K., Iron deficiency and brain development, *Semin Pediatr Neurol* 13, 158, 2006.

16. Georgieff, M.K. et al., Fetal iron and cytochrome *c* status after intrauterine hypoxemia and erythropoietin administration, *Am J Physiol* 262, R485, 1992.

17. Chockalingham, U.M. et al., Cord transferrin and ferritin values in newborn infants at risk for prenatal uteroplacental insufficiency and chronic hypoxia, *J Pediatr* 111, 283, 1987.

18. Georgieff, M.K. et al., Liver and brain iron deficiency in infants with bilateral renal agenesis (Potter's syndrome), *Pediatr Pathol Lab Med* 16, 509, 1996.

19. Georgieff, M.K. et al., Abnormal iron distribution in infants of diabetic mothers: Spectrum and maternal antecedents, *J Pediatr* 117, 455, 1990.

20. Perez, E.M. et al., Mother-infant interactions and infant development are altered by maternal iron deficiency anemia, *J Nutr* 135, 850, 2005.

21. Tamura, T. et al., Cord serum ferritin concentrations and mental and psychomotor development of children at 5 years of age, *J Pediatr* 140, 165, 2002.

22. Siddappa, A.M. et al., Iron deficiency alters auditory recognition memory, *Pediatr Res* 55, 1034, 2004.

23. DeBoer, T. et al., Explicit memory performance in infants of diabetic mothers at 1 year of age, *Dev Med Child Neurol* 47, 525, 2005.

24. Vaughn, J., Brown, J., and Carter, J.P., The effects of maternal anemia on infant behavior, *J Natl Med Assoc* 78, 963, 1986.

25. Wachs, T.D. et al., Relation of neonatal iron status to individual variability in neonatal temperament, *Dev Psychobiol* 46, 141, 2005.

26. Armony-Sivan, R. et al., Iron status and neurobehavioral development of premature infants, *J Perinat* 24, 757, 2004.

27. Georgieff, M.K. et al., Iron status at 9 months of infants with low iron stores at birth, *J Ped* 141, 405, 2002.

28. Lozoff, B. et al., Long-lasting neural and behavioral effects of iron deficiency in infancy, *Nutr Rev* 64, S34, 2006.

29. Lozoff, B. et al., Poorer behavioral and developmental outcome more than 10 years after treatment for iron deficiency in infancy, *Pediatrics* 105, 51, 2000.

30. Lozoff, B., Jimenez, E., and Smith, J.B., Double burden of iron deficiency in infancy and low socioeconomic status, *Arch Pediatr Adolesc Med* 160, 1108, 2006.

31. Peirano, P.D. et al., Cerebral executive function in preadolescents is affected by iron deficiency in infancy, *Pediatr Res* 55, 279A, 2004.

32. Burden, M., Koss, M., and Lozoff, B., Neurocognitive differences in 19-year-olds treated for iron deficiency in infancy, *Pediatr Res* 55, 279A, 2004.

33. Burden, M.J. et al., An event-related potential study of attention and recognition memory in infants with iron-deficiency anemia, *Pediatrics* 120, e336, 2007.

34. Roncagliolo, M. et al., Evidence of altered central nervous system development in infants with iron deficiency anemia at 6 mo: Delayed maturation of auditory brainstem responses, *Am J Clin Nutr* 68, 683, 1998.

35. Algarin, C. et al., Iron deficiency anemia in infancy: Long-lasting effects on auditory and visual system functioning, *Pediatr Res* 53, 217, 2003.

36. Peirano, P.D. et al., Iron deficiency anemia in infancy is associated altered temporal organization of sleep states in childhood, *Pediatr Res* 62, 715, 2007.

37. Lozoff, B. et al., Behavior of infants with iron-deficiency anemia, *Child Dev* 69, 24, 1998.

38. Lozoff, B. et al., Dose-response relationships between iron deficiency with or without anemia and infant social-emotional behavior, *J Ped* 152, 696, 2008.

39. Lozoff, B. et al., Preschool-aged children with iron deficiency anemia show altered affect and behavior, *J Nutr*, 137, 683, 2007.

40. Angulo-Kinzler, R.M. et al., Twenty-four-hour motor activity in human infants with and without iron deficiency anemia, *Early Hum Dev*, 70, 85, 2002.

41. Beard, J.L. et al., Moderate iron deficiency in infancy: Biology and behavior in young rats, *Behav Brain Res* 170, 224, 2006.

42. Felt, B.T. et al., Persistent neurochemical and behavioral abnormalities in adulthood despite early iron supplementation for perinatal iron deficiency anemia in rats, *Brain Behav Res* 171, 261, 2006.

43. Jorgenson, L.A. et al., Fetal iron deficiency disrupts the maturation of synaptic function and efficacy in area CA1 of the developing rat hippocampus, *Hippocampus* 15, 1094, 2005.

44. DeUngria, M. et al., Perinatal iron deficiency decreases cytochrome *c* oxidase (CytOx) activity in selected regions of neonatal rat brain, *Pediatr Res* 48, 169, 2000.

45. Pokorný, J., and Yamamoto, T., Postnatal ontogenesis of hippocampal CA1 area in rats: I. Development of dendritic arborisation in pyramidal neurons, *Brain Res Bull* 7, 113, 1981.

46. Steward, O., and Falk, P.M., Selective localization of polyribosomes beneath developing synapses: A quantitative analysis of the relationships between polyribosomes and developing synapses in the hippocampus and dentate gyrus, *J Comp Neurol* 314, 545, 1991.

47. Roskams, A.J., and Connor, J.R., Iron, transferrin, and ferritin in the rat brain during development and aging, *J Neurochem* 63, 709, 1994.

48. Siddappa, A.J. et al., Developmental changes in the expression of iron regulatory proteins and iron transport proteins in the perinatal rat brain, *J Neurosci Res* 68, 761, 2002.

49. Jorgenson, L.A., Wobken, J.D., and Georgieff, M.K., Perinatal iron deficiency alters apical dendritic growth in hippocampal CA1 pyramidal neurons, *Dev Neurosci* 25, 412, 2003.

50. Yoo, Y. et al., Iron enhances NGF-induced neurite outgrowth in PC12 cells, *Mol Cells* 17, 340, 2004.

51. Wu, L.-L. et al., Effect of perinatal iron deficiency on myelination and associated behaviors in rat pups, *Behav Brain Res* 188, 263, 2008.

52. Carlson, E.S. et al., Perinatal iron deficiency results in altered developmental expression of genes mediating energy metabolism and neuronal morphogenesis in hippocampus, *Hippocampus* 17, 679, 2007.

53. Felt, B.T., and Lozoff, B., Brain iron and behavior of rats are not normalized by treatment of iron deficiency anemia during early development, *J Nutr* 126, 693, 1996.

54. Schmidt, A.T. et al., Dissociating the long-term effects of fetal/neonatal iron deficiency on neonatal iron deficiency on three types of learning in the rat, *Behav Neurosci* 121, 475, 2007.

55. McEchron, M.D. et al., Perinatal nutritional iron deficiency impairs hippocampus-dependent trace eyeblink conditioning in rats, *Dev Neurosci* 30, 243, 2008.

56. McEchron, M.D. et al., Perinatal nutritional iron deficiency permanently impairs hippocampus-dependent trace fear conditioning in rats, *Nutr Neurosci* 8, 195, 2005.

57. Yu, G.S. et al., Effect of prenatal iron deficiency on myelination in rat pups, *Am J Pathol* 125, 620, 1986.

58. Oloyede, O.B., Folayan, A.T., and Odutuga, A.A., Effects of low-iron status and deficiency of essential fatty acids on some biochemical constituents of rat brain, *Biochem Int* 27, 913, 1992.

59. Kwik-Uribe, C.L. et al., Chronic marginal iron intakes during early development in mice result in persistent changes in dopamine metabolism and myelin composition, *J Nutr*, 130, 2821 2000.

60. Beard, J.L., Wiesinger, J.A., and Connor, J.R., Pre- and postweaning iron deficiency alters myelination in Sprague-Dawley rats, *Dev Neurosci* 25, 308, 2003.

61. Ortiz, E. et al., Effect of manipulation of iron storage, transport, or availability on myelin composition and brain iron content in three different animal models, *J Neurosci Res* 77, 681, 2004.

62. Morath, D.J., and Mayer-Pröschel, M., Iron deficiency during embryogenesis and consequences for oligodendrocyte generation in vivo, *Dev Neurosci* 24, 197, 2002.

63. Beard, J.L., and Connor, J.R., Iron status and neural functioning, *Annu Rev Nutr*, 23, 41, 2003.

64. Burhans, M.S. et al., Iron deficiency: Differential effects on monoamine transporters, *Nutr Neurosci* 8, 31, 2005.

65. Morse, A., Beard, J.L., and Jones, B., Behavioral and neurochemical alterations in iron deficient mice, *Proc Soc Exp Biol Med* 220, 147, 1999.

66. Youdim, M.B., Ben-Sacher, D., and Yehuda, S., Putative biological mechanisms of the effect of iron deficiency on brain biochemistry and behavior, *Am J Clin Nutr* 50, 607, 1989.

67. Golub, M.S. et al., Behavioral consequences of developmental iron deficiency in infant rhesus monkeys, *Neurotoxicol Teratol* 28, 3, 2006.

68. Beard, J., Erikson, K.M., and Jones, B.C., Neonatal iron deficiency results in irreversible changes in dopamine function in rats, *J Nutr* 133, 1174, 2003.

69. Sandstead, H.H., Zinc: Essentiality for brain development and function, *Nutr Rev* 43, 129, 1985.

70. Terhune, M.W., and Sandstead, H.H., Decreased RNA polymerase activity in mammalian zinc deficiency, *Science* 177, 68, 1972.

71. McNall, A.D., Etherton, T.D., and Fosmire, G.J., The impaired growth induced by zinc deficiency in rats is associated with decreased expression of the hepatic insulin-like growth factor I and growth hormone receptor genes, *J Nutr* 125, 874, 1995.

72. Duncan, J., and Hurley, L., Thymidine kinase and DNA polymerase activity in normal and zinc deficient developing rat embryos, *Proc Soc Exp Biol Med* 159, 39, 1978.

73. Sandstead, H.H., Is zinc deficiency a public health problem? *Nutrition* 11, 87, 1995.

74. Caulfield, L.E., and Black, R.E., Zinc deficiency, in *Comparative Quantification of Health Risks: Global and Regional Burden of Disease Attributable to Selected Major Risk Factors*, Vol. 1, Ezzati, M., Lopez, A.D., Rodgers, A., and Murray, C.J.L., Eds., World Health Organization, Geneva, 2004, 257.

75. Keen, C.L. et al., Primary and secondary zinc deficiency as factors underlying abnormal CNS development, *Ann N Y Acad Sci* 678, 37, 1993.

76. WHO, Zinc, in *Trace Elements in Human Nutrition and Health*, World Health Organization, Geneva, 1996.

77. Caulfield, L.E. et al., Potential contribution of maternal zinc supplementation during pregnancy to maternal and child survival, *Am J Clin Nutr* 68, 488S, 1998.

78. Jameson, S., Zinc status in pregnancy: The effect of zinc therapy on perinatal mortality, prematurity, and placental ablation, *Ann N Y Acad Sci* 678, 178, 1993.

79. Rogers, J.M. et al., Zinc deficiency causes apoptosis but not cell cycle alterations in organogenesis-stage rat embryos: Effect of varying duration of deficiency, *Teratology* 52, 149, 1995.

80. Cavdar, A.O. et al., Effect of zinc supplementation in a Turkish woman with two previous anencephalic infants, *Gynecol Obstet Invest* 32, 123, 1991.

81. Mambidge, K.M., Neldner, K.H., and Walravens, P.A., Letter: Zinc, acrodermatitis enteropathica, and congenital malformations, *Lancet* 1, 577, 1975.

82. Merialdi, M. et al., Adding zinc to prenatal iron and folate tablets improves fetal neurobehavioral development, *Am J Obstet Gynecol* 180, 483, 1999.

83. Merialdi, M. et al., Randomized controlled trial of prenatal zinc supplementation and the development of fetal heart rate, *Am J Obstet Gynecol* 190, 1106, 2004.

84. Kirksey, A. et al., Relation of maternal zinc nutriture to pregnancy outcome and infant development in an Egyptian village, *Am J Clin Nutr* 60, 782, 1994.

85. Hamadani, J.D. et al., Zinc supplementation during pregnancy and effects on mental development and behaviour of infants: A follow-up study, *Lancet* 360, 290, 2002.

86. Joint FAO/WHO Expert Consultation, Zinc, in *Vitamin and Mineral Requirements in Human Nutrition*, 2nd ed., World Health Organization, Geneva, 2004, 230.

87. Krebs, N.F., Dietary zinc and iron sources, physical growth and cognitive development of breastfed infants, *J Nutr* 130, 358S, 2000.

88. Skinner, J.D. et al., Longitudinal study of nutrient and food intakes of infants aged 2 to 24 months, *J Am Diet Assoc* 97, 496, 1997.

89. Ashworth, A. et al., Zinc supplementation, mental development and behavior in low birth weight term infants in northeast Brazil, *Eur J Clin Nutr* 52, 223, 1998.

90. Black, M.M. et al., Cognitive and motor development among small-for-gestational-age infants: Impact of zinc supplementation, birth weight, and caregiving practices, *Pediatrics* 113, 1297, 2004.

91. Castillo-Duran, C. et al., Effect of zinc supplementation on development and growth of Chilean infants, *J Ped* 138, 229, 2001.

92. Friel, J.K. et al., Zinc supplementation in very-low-birth-weight infants, *J Pediatr Gastroenterol Nutr* 17, 97, 1993.

93. Bentley, M.E. et al., Zinc supplementation affects the activity patterns of rural Guatemalan infants, *J Nutr* 127, 1333, 1997.

94. Sazawal, S. et al., Effect of zinc supplementation on observed activity in low socioeconomic Indian preschool children, *Pediatrics* 98, 1132, 1996.

95. Penland, J.G., Behavioral data and methodology issues in studies in zinc nutrition in humans, *J Nutr* 130, 361S, 2000.

96. Frederickson, C., and Danscher, G., Zinc-containing neurons in hippocampus and related CNS structures, *Prog Brain Res* 83, 71, 1990.

97. Sandstead, H.H., Frederickson, C.J., and Penland, J.G., History of zinc as related to brain function, *J Nutr* 130, 496, 2000.

98. Li, M., Rosenberg, H., and Chiu, T., Zinc inhibition of GABA-stimulated Cl⁻ flux in rat brain regions is unaffected by acute or chronic benzodiazepine, *Pharmacol Biochem Behav* 49, 477, 1994.

99. Hesse, G., Chronic zinc deficiency alters neuronal function of hippocampal mossy fibers, *Science* 205, 1005, 1979.

100. Tejwani, G., and Hanissian, S., Modulation of mu, delta and kappa opioid receptors in rat brain by metal ions and histidine, *Neuropharmacology* 29, 445, 1990.

101. Dvergsten, C.L., Johnson, L.A., and Sandstead, H.H., Alterations in the postnatal development of the cerebellar cortex due to zinc deficiency: III. Impaired dendritic differentiation of basket and stellate cells, *Brain Res* 318, 21, 1984.

102. Dvergsten, C.L. et al., Alterations in the postnatal development of the cerebellar cortex due to zinc deficiency: I. Impaired acquisition of granule cells, *Brain Res* 271, 217, 1983.

103. Dvergsten, C.L. et al., Alterations in the postnatal development of the cerebellar cortex due to zinc deficiency: II. Impaired maturation of Purkinje cells, *Brain Res* 318, 11, 1984.

104. Golub, M.S. et al., Modulation of behavioral performance of prepubertal monkeys by moderate dietary zinc deprivation, *Am J Clin Nutr* 60, 238, 1994.

105. Golub, M.S. et al., Activity and attention in zinc-deprived adolescent monkeys, *Am J Clin Nutr*, 64 908, 1996.

106. Frederickson, R.E., Frederickson, C.J., and Danscher, G., *In situ* binding of bouton zinc reversibly disrupts performance on a spatial memory task, *Behav Brain Res* 38, 25, 1992.

107. Dembri, A. et al., Effects of short- and long-term thyroidectomy on mitochondrial and nuclear activity in adult rat brain, *Mol Cell Endo*, 33 211, 1983.

108. Roder, L.M., Williams, I.B., and Tildon, J.T., Glucose transport in astrocytes: Regulation by thyroid hormone, *J Neurochem* 45, 1653, 1985.

109. Smith, J.W. et al., Thyroid hormones, brain function and cognition: A brief review, *Neurosci Biobehav Rev* 26, 45, 2002.

110. Farsetti, A. et al., Characterization of myelin basic protein thyroid hormone response element and its function in the context of native and heterologous promoter, *J Biol Chem* 267, 15784, 1992.

111. Rodriquez-Pena, A., Oligodendrocyte development and thyroid hormone, *J Neurobiol* 40, 497, 1999.

112. Iniguez, M.A. et al., Cell-specific effects of thyroid hormone on RC3/neurogranin expression in rat brain, *Endocrinology* 137, 1032, 1996.

113. WHO, *Assessment of Iodine Deficiency Disorders and Monitoring Their Elimination: A Guide for Programme Managers*, WHO Press, Geneva, 2007.

114. Hetzel, B.S., and Mano, M.T., A review of experimental studies of iodine deficiency during fetal development, *J Nutr* 119, 145, 1989.

115. Cao, X.-Y. et al., Timing of vulnerability of the brain to iodine deficiency in endemic cretinism, *N Engl J Med* 331, 1739, 1994.

116. O'Donnell, K. et al., Effects of iodine supplementation during pregnancy on child growth and development at school age, *Dev Med Child Neurol* 44, 76, 2002.

117. Vermiglio, F. et al., Defective neuromotor and cognitive ability in iodine-deficient schoolchildren of an endemic goiter region in Sicily, *J Clin Endocrinol Metab* 70, 379, 1990.

118. Azizi, F. et al., Impairment of neuromotor and cognitive development in iodine-deficient schoolchildren with normal physical growth, *Acta Endocrinol* 129, 497, 1993.

119. Aghini-Lombardi, F.A. et al., Mild iodine deficiency during fetal/neonatal life and neuropsychological impairment in Tuscany, *J Endocrinol Invest* 18, 57, 1995.

120. Zimmermann, M.B. et al., Iodine supplementation improves cognition in iodine-deficient schoolchildren in Albania: A randomized, controlled, double-blind study, *Am J Clin Nutr* 83, 108, 2006.

121. Eayrs, J.T., Influence of the thyroid on the central nervous system, *Br Med Bull* 16, 122, 1960.

122. Chaudhury, S., Chatterjee, D., and Sarkar, P.K., Induction of brain tubulin by triiodothyronine: Dual effect of the hormone on the synthesis and turnover of the protein, *Brain Res* 339, 191, 1985.

123. Francon, J. et al., Is thyroxine a regulatory signal for neurotubule assembly during brain development? *Nature* 266, 188, 1977.

124. Potter, B.J. et al., Restoration of brain growth in fetal sheep after iodized oil administration to pregnant iodine-deficient ewes, *J Neurol Sci* 66, 15, 1984.

125. Scthi, V., and Kapil, U., Iodine deficiency and development of the brain, *Indian J Pediatr* 71, 325, 2004.

126. Alvarez-Dolado, M. et al., Expression of neurotrophins and the trk family of neurotrophin receptors in normal and hypothyroid rat brain, *Mol Brain Res* 27, 249, 1994.

127. Martínez-Galán, J.R. et al., Early effects of iodine deficiency on radial glial cells of the hippocampus of the rat fetus. A model of neurological cretinism, *J Clin Invest* 99, 2701, 1997.

128. Patel, A.J., Hayashi, M., and Hunt, A., Selective persistent reduction in choline acetyltransferase activity in basal forebrain of the rat after thyroid deficiency during early life, *Brain Res* 422, 182, 1987.
129. Zou, L. et al., Identification of thyroid hormone response elements in rodent Pcp-2, a developmentally regulated gene of cerebellar Purkinje cells, *J Biol Chem* 269, 13346, 1994.
130. Chen, J., and Berry, M.J., Selenium and selenoproteins in the brain and brain diseases, *J Neurochem* 86, 1, 2003.
131. Whanger, P.D., Selenium and the brain: A review, *Nutr Neurosci* 4, 81, 2001.
132. Scharpf, M. et al., Neuronal and ependymal expression of selenoprotein P in the human brain, *J Neural Transm* 114, 877, 2007.
133. FAO/WHO, *Vitamin and Mineral Requirements in Human Nutrition*, 2nd ed., WHO Press, Geneva, 2004.
134. Gassio´, R. et al., Cognitive functions and the antioxidant system in phenylketonuric patients, *Neuropsychology* 22, 426, 2008.
135. Brown, D.G., and Burk, R.F., Selenium retention in tissues and sperm of rats fed a Torula yeast diet, *J Nutr* 102, 102, 1972.
136. Prohaska, J.R., and Ganther, H.E., Selenium and glutathione peroxidase in developing rat brain, *J Neurochem* 27, 1379, 1976.
137. Schomburg, L. et al., Synthesis and metabolism of thyroid hormones is preferentially maintained in selenium-deficient transgenic mice, *Endocrinology* 147, 1306, 2006.
138. Rayman, M.P., The importance of selenium to human health, *Lancet* 356, 233, 2000.
139. Benton, D., Selenium intake, mood and other aspects of psychological functioning, *Nutr Neurosci* 5, 363, 2002.
140. Berr, C. et al., Systematic oxidative stress and cognitive performance in the population-based EVA study, *Free Radic Biol Med* 24, 1202, 1998.
141. Zhang, Y. et al., Comparative analysis of selenocysteine machinery and selenoproteome gene expression in mouse brain identifies neurons as key functional sites of selenium in mammals, *J Biol Chem* 283, 2427, 2008.
142. Mitchell, J.H. et al., Selenoprotein expression and brain development in preweanling selenium- and iodine-deficient rats, *J Mol Endocrinol* 20, 203, 1998.
143. Yan, J., and Barrett, J.N., Purification from bovine serum of a survival-promoting factor for cultured central neurons and its identification as selenoprotein P, *J Neurosci* 18, 8682, 1998.
144. Castaño, A. et al., Low selenium diet increases the dopamine turnover in prefrontal cortex of the rat, *Neurochem Int* 30, 549, 1997.
145. Penland, J.G., and Prohaska, J.R., Abnormal motor function persists following recovery from perinatal copper deficiency in rats, *J Nutr* 134, 1984, 2004.
146. Bennets, H.W., and Chapman, F.E., Copper deficiency in sheep in Western Australia: A preliminary account of the etiology of enzootic ataxia of lambs and an anemia of ewes, *Aust Vet J* 13, 138, 1937.
147. Prohaska, J.R., Long-term functional consequences of malnutrition during brain development: Copper, *Nutrition* 16, 502, 2000.
148. Prohaska, J.R., and Bailey, W.R., Persistent regional changes in brain copper, cuproenzymes and catecholamines following perinatal copper deficiency in mice, *J Nutr* 123, 1226, 1993.
149. Hunt, C.D., and Meacham, S.L., Aluminum, boron, calcium, copper, iron, magnesium, manganese, molybdenum, phosphorus, potassium, sodium, and zinc: Concentrations in common Western foods and estimated daily intakes by infants; toddlers; and male and female adolescents, adults, and seniors in the United States, *J Am Diet Assoc* 101, 1058, 2001.
150. Prohaska, J.R., and Brokate, B., The timing of perinatal copper deficiency in mice influences offspring survival, *J Nutr* 132, 3142, 2002.
151. West, E.C., and Prohaska, J.R., Cu,Zn-superoxide dismutase is lower and copper chaperone CCS is higher in erythrocytes of copper-deficient rats and mice, *Exp Biol Med* 229, 756, 2004.
152. Gybina, A.A., and Prohaska, J.R., Increased rat brain cytochrome *C* correlates with degree of perinatal copper deficiency rather than apoptosis, *J Nutr* 133, 3361, 2003.
153. Hunt, C.D., and Idso, J.P., Moderate copper deprivation during gestation and lactation affects dentate gyrus and hippocampal maturation in immature male rats, *J Nutr* 125, 2700, 1995.
154. Prohaska, J.R., Baily, W.R., and Lear, P.M., Copper deficiency alters rat peptidylglycine alpha-amidating monooxygenase activity, *J Nutr* 125, 1447, 1995.
155. Prohaska, J.R., and Brokate, B., Dietary copper deficiency alters protein levels of rat β-monooxygenase and tyrosine monooxygenase, *Exp Biol Med* 226, 199, 2001.

156. Prohaska, J.R., and Wells, W.W., Copper deficiency in the developing rat brain: A possible model for Menkes' steely-hair disease, *J Neurochem* 23, 91, 1974.
157. Prohaska, J.R., and Gybina, A.A., Rat brain iron concentration is lower following perinatal copper deficiency, *J Neurochem* 93, 698, 2005.
158. Wozniak, J.R., Mueller, B.A., and Lim, K.O., Diffusion tensor imaging, in *Handbook of Developmental Cognitive Neuroscience*, 2nd ed., Nelson, C.A., and Luciana, M., Eds., MIT Press, Cambridge, MA, 2008, 301.
159. Csibra, G., Kushnerenko, E., and Grossman, T., Electrophysiological methods in studying infant cognitive development, in *Handbook of Developmental Cognitive Neuroscience*, 2nd ed., Nelson, C.A., and Luciana, M., Eds., MIT Press, Cambridge, MA, 2008, 247.
160. Thomas, K.M., and Tseng, A., Functional MRI methods in developmental cognitive neuroscience, in *Handbook of Developmental Cognitive Neuroscience*, 2nd ed., Nelson, C.A., and Luciana, M., Eds., MIT Press, Cambridge, MA, 2008, 311.
161. Yanowitz, T.D. et al., Variability in cerebral oxygen delivery is reduced in premature neonates exposed to chorioamnionitis, *Pediatr Res* 59, 299, 2006.
162. Pihko, E., and Lauronen, L., Somatosensory processing in healthy newborns, *Exp Neurol* 190, 2, 2004.
163. Davis, E.P. et al., Effects of prenatal betamethasone exposure on regulation of stress physiology in healthy premature infants, *Psychoneuroendocrinology* 29, 1028, 2004.
164. Nelson, C.A., The ontogeny of human memory: A cognitive neuroscience perspective, *Dev Psychol* 31, 723, 1995.
165. Fagan III, J.F. et al., Selective screening device for the early detection of normal or delayed cognitive development in infants at risk for later mental retardation, *Pediatrics* 78, 1021, 1986.
166. Bauer, P.J., Recalling past events: From infancy to early childhood, *Ann Child Dev* 11, 25, 1995.
167. Bauer, P.J. et al., Electrophysiological indexes of encoding and behavioral indexes of recall: Examining relations and developmental change late in the first year of life, *Dev Neuropsychol* 29, 293, 2006.
168. Diamond, A., Development of the ability to use recall to guide action as indicated by infants' performance on A not B, *Child Dev* 56, 868, 1985.
169. Bell, M.A., and Fox, N.A., The relations between frontal brain electrical activity and cognitive development during infancy, *Child Dev* 63, 1142, 1992.
170. Bell, M.A., and Wolfe, C.D., Changes in brain functioning from infancy to early childhood: Evidence from EEG power and coherence working memory tasks, *Dev Neuropsychol* 31, 21, 2007.
171. Luciana, M., and Nelson, C.A., Assessment of neuropsychological function through use of the Cambridge Neuropsychological Testing Automated Battery: Performance in 4- to 12-year-old children, *Dev Neuropsychol* 22, 595, 2002.
172. Barnea-Goraly, N. et al., White matter development during childhood and adolescence: A cross-sectional diffusion tensor imaging study, *Cereb Cortex* 15, 1848, 2005.
173. Olesen, P.J. et al., Combined analysis of DTI and fMRI data reveals a joint maturation of white and grey matter in a fronto-parietal network, *Cogn Brain Res* 18, 48, 2003.
174. White, N.M., and McDonald, R.J., Multiple parallel memory systems in the brain of the rat, *Neurobiol Learn Mem* 77, 125, 2002.
175. Kim, J.J., and Baxter, M.G., Multiple brain-memory systems: The whole does not equal the sum of its parts, *Trends Neurosci* 24, 324, 2001.

9 Therapeutics of Alzheimer's Disease Based on Metal Bioavailability

Su San Mok[1,2] and Ashley I. Bush[1–3]

[1]Oxidation Biology Laboratory, Mental Health Research Institute of Victoria, Parkville, Victoria, Australia

[2]Department of Pathology, University of Melbourne, Parkville, Victoria, Australia

[3]Department of Psychiatry, Massachusetts General Hospital East, Charlestown, Massachusetts, USA

CONTENTS

9.1   INTRODUCTION

Alzheimer's disease (AD) is characterized by progressive dementia resulting in loss of cognitive function. It is the most common form of dementia and is significantly associated with age. Risks associated with the disease are thus associated with increasing age. Approximately 1% of individuals aged between 65 and 70 years are affected by AD, and the incidence increases to 6–8% for individuals older than 85 years [1]. Familial mutations cause both early-onset (before the age of 65 years) and late-onset (over 65 years) AD. In 2006, 26.6 million people worldwide were affected by AD. This number is expected to quadruple by 2050. As the world population ages and life expectancy increases, significant resources will be required to adequately care for those afflicted with this disease.

9.2   METALS AND AGING

Metals such as copper and iron play biologically important roles in many cellular processes. Many proteins require metal ions to perform biological functions. The importance of transition metals in the body is thus encompassed by their many functions within the living organism (for a review, see Reference. [2]). The levels of such transition metals as manganese, iron, copper, and zinc in brains and plasma in both humans and animals have been determined [3–8]. One of the key cellular processes known to be affected by age is metal metabolism, and changes in metal bioavailability may thus play a significant role in the aging process. Studies show that aging is associated with elevated levels of plasma copper [9–14] and decreased levels of plasma zinc [9,15–19]. As the concentrations of these metals are relatively high within the brain (ranging from approximately 0.1 to 1 mM), alterations in plasma levels may thus reflect changes of these metal ion levels within the brain. As a consequence, impairment of metal homeostasis may also result in changes in neurological function [20].

The brain itself contains significantly high total concentrations of different transition metal ions, such as iron, copper, and zinc, which accumulate as a consequence of normal aging [6,21–23]. In conjunction with the high levels of oxidative stress in this tissue, the potential for significant oxidative damage to occur is therefore extremely high. As such, the brain must be capable of efficiently and strictly regulating ionic metal levels by different homeostatic mechanisms to prevent damage. In accordance, it would be expected that any dysregulation of these mechanisms may result in increased levels of free metal, allowing the subsequent development and progression of oxidative injury.

9.3   OXIDATIVE STRESS

As the brain requires a continuous and substantial amount of energy for its metabolism, it is constantly exposed to free radicals [in particular, reactive oxygen species (ROS)] generated within this process. In order to combat the harmful effects of ROS, normal cellular function includes several antioxidant defense mechanisms that can inhibit or repair oxidative damage. These include the antioxidant enzymes glutathione peroxidase and glutathione reductase, catalase and superoxide dismutase (SOD) [2,24], and such antioxidants as glutathione and ascorbate. Oxidative stress plays a major role in neurodegenerative disorders, and the AD brain is under increased oxidative load. As a consequence, there is substantial evidence for oxidative injury in AD resulting in lipid peroxidation causing lipid peroxidation adducts, damage to deoxyribonucleic acid resulting in base modification, cross-linkages and strand breakages, and damage to proteins generating protein carbonyl modifications and dityrosine cross-linking [25–27], all of which are modifications that have been demonstrated in the amyloid beta (Aβ) peptide extracted from amyloid plaques from postmortem AD brain tissues.

9.4   THE NEUROCHEMISTRY OF TRANSITION METAL IONS

It is commonly misunderstood that the neurological syndromes in which metals are implicated are hypothetically caused by toxicological exposure to copper, iron, zinc, and manganese. It is often thought that ingestion or exposure to these metals causes an abnormal protein interaction, which then causes the disease. This is now known to not be really accurate. The brain itself has total concentrations of metal ions, such as copper, iron, zinc, and manganese, high enough to be able to cause damage and dysregulate normal metabolic pathways on its own accord. For example, during neurotransmission, approximately 300 μM Zn^{2+} is released. This is a particularly high concentration and can cause toxicity to neuronal cells in culture [28]. This toxicity however does not occur within the brain, indicating that the brain must have effective homeostatic mechanisms and buffers to prevent any abnormal metal ion compartmentalization. In addition, the blood-brain barrier (BBB) is relatively impermeable to changing levels of metal ions in the plasma.

Biometals such as iron, copper, and zinc are generally bound to ligands (e.g., transferrin) and not found as free species. Recent data however demonstrate the release of free ionic or exchangeable zinc and copper in the synaptic cleft. Zinc has also been shown to act like calcium as a new class of second messenger [29]. It is well established that the intracellular pool of free iron (the labile iron pool) is able to modulate the expression of a number of proteins, including the amyloid precursor protein (APP) [30]. Much intracellular copper is considered to be bound but is clearly exchangeable and transferred from protein to protein (e.g., by CCS1—the copper chaperone of SOD1) [31].

Several important basic discoveries have been made about copper and zinc release and flux at the glutamatergic synapse in the cortex and hippocampus. As the site of long-term potentiation, a physical substrate of memory formation, it is also the first site of Aβ deposition in AD, making it probably the most important site for dysfunction in this disorder [32]. The presence of zinc and copper released by hippocampal tissue has been of interest for about the last 20 years (reviewed by Frederickson et al. [29]). Zinc is reported to be released either as a free or an exchangeable ionic species into the extracellular space [33]. Zinc transporter protein-3, found within the membrane of glutamatergic vesicles, is responsible for this pool of vesicular zinc. Zinc is released possibly with glutamate during neurotransmission [34] and seems to suppress the response of the NMDA receptor, likely preventing seizure activity.

It has recently been observed that postsynaptic NMDA neurites release free ionic copper upon NMDA activation [35]. In hippocampal neurons, activation of synaptic NMDA receptors results in trafficking of the Menkes ATPase and an associated efflux of copper [35]. Catalytic amounts of copper promote the reaction of nitric oxide with thiols by functioning as electron acceptors. The release of copper may function as a molecular switch to control extracellular S-nitrosylation of the NMDA receptor, a posttranslational mechanism critical for modulating receptor function [35]. Copper has a protective effect against NMDA-mediated excitotoxic cell death in primary hippocampal neurons dependent on endogenous nitric oxide production in these cells [36].

The glutamatergic synapse is therefore a site where chemically exchangeable Zn and copper converge, a uniquely occurring situation in the body. This may explain why Aβ, with its tendency to precipitate and cross-link when induced by these metal ions, initially precipitates in this site in AD. Another component present within this vicinity that could modulate the availability of zinc and copper ions to bind to the NMDA receptor or to Aβ is the release of metallothionein-3 (MT3) [or GIF (growth inhibitory factor)] by the neighboring astrocytes [37]. Levels of this protein are decreased in AD [38].

9.5   METALS AND ALZHEIMER'S DISEASE

The metal ion content of the brain is tightly regulated. Movement of metals across the BBB is highly regulated such that no passive flux of metals occurs from the circulation to the brain. The breakdown in homeostatic mechanisms that compartmentalize and regulate metals plays a major role in the interactions of metals such as iron, copper, and zinc with the major protein components of neurodegenerative disease.

Transition metals have been demonstrated to be important in AD pathogenesis. Changes in the levels of these metals in both the brain and the cerebrospinal fluid (CSF) in AD subjects strongly suggest an imbalance of metal homeostasis. Studies have demonstrated increases in copper, iron, and zinc in AD patients compared with age-matched controls [39]. The concentrations of copper (0.39 mM), iron (0.94 mM), and zinc (1.05 mM) are all increased in plaques of AD brains. In contrast, normal levels in age-matched controls were found to be 0.07, 0.35, and 0.34 mM for copper, zinc, and iron, respectively [39]. Metal levels have been measured in both CSF and serum from AD patients with differing results. Some studies have shown a significant decrease in the level of zinc in the CSF of AD patients with no difference in the levels of iron, copper, and manganese and no change in the level of any of these metals in serum [40]. In other studies, copper levels have been shown to be increased, while iron and zinc levels were decreased [41,42].

Aging has also been identified as a dominant risk factor not only for AD but also for neurodegenerative diseases. Studies in animals and humans have reported increases in copper levels in the brain from youth to adulthood (reviewed by Adlard and Bush [43]). However, from middle age onward, copper levels drop substantially, accompanied by a loss of copper-dependent enzyme activities (e.g., cytochrome c oxidase, SOD1, and ceruloplasmin) [44]. Age-related increases in brain iron have been observed in humans [45,46], primates [47], rodents [6,48–50], and *Drosophila* [51]. The brain is susceptible to abnormal iron regulation, observed when ferroxidases, such as ceruloplasmin, ferritin [52], and frataxin [53], fail to function properly, leading to neurodegenerative disorders [54]. The breakdown in metal ion regulation in the glutamatergic synapse, possibly by inhibition of reuptake, raises the average concentrations of zinc and copper in the cleft, leading to Aβ aggregation and synaptotoxicity.

As mentioned previously, changes in the levels of metal ions may be a reflection of alterations in homeostasis and function of different proteins that bind, store, and transport these metals. Levels of ceruloplasmin are demonstrated to be increased in CSF and brains in AD cases [55,56]. Changes have also been observed in iron-binding proteins. In AD cases, the iron regulatory protein-2 was found to be localized in neurofibrillary tangles (NFTs), senile plaques, and neuropil threads [57], suggesting that iron regulatory protein-2 changes in the disease may be directly associated with dysregulation of iron homeostasis [58]. Other examples of altered metal binding or transport proteins in AD include the MTs, which may be increased (MT1 and MT2) or decreased (MT3) in AD [38,59,60], and zinc transporter protein-1, which transports zinc into the extracellular space [61].

SOD+/−x APP transgenic mice display an interesting feature of the mechanism of increased AD pathology demonstrating decreased brain copper, zinc, and iron levels due to the mitochondrial lesion [62]. This reflects a similar observation in AD pathology, where copper levels decrease with advanced pathology [44]. Amyloid pathology improves in APP transgenic mice subject to dietary and genetic manipulations that increase brain copper levels [63,64]. Conversely, studies also show that exposure to copper in combination with a high-fat diet increases the risk for AD [65], also demonstrated in studies of rabbits exposed to copper and cholesterol [66,67]. In contrast, the increase in zinc levels in advanced AD correlates with brain Aβ burden in humans but not in APP transgenic mice [44]. Nutritional deficiency in zinc is common in aging, and it has been recently reported that zinc-deficient APP transgenic mice demonstrate an increased volume of amyloid plaques [68]. Metals also play significant roles in other metabolic pathways, including enzymatic and mitochondrial function [69,70], immunity [71], myelination [72–74], and learning and memory [75–77].

9.6 INTERACTION OF THE AMYLOID PRECURSOR PROTEIN WITH METALS

The APP is a membrane-bound glycoprotein that contains a large extracellular N-terminal ectodomain and a short cytoplasmic domain. It is ubiquitously expressed and exists in three isoforms—neurons produce APP695, while the APP751/770 isoforms are expressed peripherally. APP is cleaved by "secretases," resulting in the generation of different fragments. α-Secretase processing generates a large secreted soluble APP ectodomain (sAPPα), while β-secretase (BACE-1) and γ-secretase cleavage results in the generation of Aβ peptides that are observed in plaques in the brain [78–80]. BACE-1 interacts with domain 1 of CCS1 through a copper binding site in its C-terminal cytoplasmic domain [81]. While the precise functional implications of this interaction are still unknown, it suggests that copper levels may affect the generation of Aβ. A recent report has demonstrated that γ-secretase activity can be inhibited by low concentrations of Zn^{2+}, but again, the physiological implications are unclear [82].

To date, the biological function of APP is still unclear. Studies show that the APP ectodomain has neurotrophic function, including roles in neurite outgrowth, synaptogenesis and plasticity, and cell adhesion. APP has also been shown to play a role in maintaining metal ion homeostasis (reviewed by Inestrosa et al. [83] and Cerpa et al. [84]) and has metal binding sites for copper [85–90] and zinc [91–95]. In vivo studies have shown that APP expression can modulate copper homeostasis in neurons

[96]. Increasing copper concentrations also play a role in modulating APP processing, and APP is sensitive to changes in copper levels [93,97]. A copper binding domain has been identified at the amino-terminus of the molecule, and APP expression promotes the export of neuronal copper [98]. Studies show that cellular copper drives the expression of APP mRNA [97,99], lending support for a functional role of APP in copper homeostasis. In AD, copper accumulates in amyloid plaques, while a deficiency of copper is observed in neighboring cells, indicating abnormal brain copper distribution. Excess copper inhibited Aβ production from APP-transfected Chinese hamster ovary cells [100], but the effects of deficiency were not studied. Genetically manipulated human fibroblasts, which accumulate copper, secreted higher levels of sAPPα into the medium, while copper-deficient genetically manipulated fibroblasts secreted predominantly sAPPβ, producing more of the amyloidogenic C-termini of APP (C99) [101]. APP was found to be processed simultaneously by both α- and β-secretases in this system, as copper-deficient fibroblasts produced α-cleaved C83 but secreted sAPPβ exclusively [101]. A decrease in copper also significantly reduced the steady-state level of APP mRNA, while a constant level of APP protein was maintained, indicating that copper deficiency may accelerate APP translation [101]. Copper-deficient human neuroblastoma cells significantly increased levels of secreted Aβ without affecting cleavage of APP itself. These studies indicate that copper deficiency markedly alters APP metabolism and can increase Aβ secretion by either affecting APP cleavage or by inhibiting its degradation, with the mechanism dependent on cell type [101].

Aβ is degraded extracellularly by the action of enzymes that include zinc metalloproteinases, such as neprilysin, insulin-degrading enzyme, and matrix metalloproteinases (reviewed by Adlard and Bush [43]). This may explain the inverse correlation between CSF zinc and copper levels and CSF Aβ1–42 levels in normal humans [102]. This possibility is further supported by the observation that addition of low micromolar concentrations of zinc or copper to ex vivo CSF samples accelerates Aβ degradation [102].

A subdomain within the extracellular portion of APP contains a copper binding domain at amino acids 124–189, which consists of four ligands in amino acids: His147, His151, Tyr168, and Met170 [86]. This region has an exposed surface and has structural folds similar to those of other copper chaperone proteins, such as Menkes, Atox-1, and CCS1 [86]. A second binding site for copper lies within the Aβ sequence of APP [103]. Zinc binding to APP involves a region within amino acids 170–188 of the molecule [91,92]. This region contains two cysteines at positions 186 and 187, in addition to several other potential ligands that are believed to be crucial for zinc binding [104]. Binding of zinc to APP may cause dimerization of the protein, which may be important for its function [104]. In addition, secreted APP aggregates in the presence of zinc [105]. APP translation is mediated by an iron-regulatory element (IRE type II) present in its 5′-untranslated region [106]. This translation is up-regulated upon exposure to iron and down-regulated in response to intracellular metal chelation [30]. These studies demonstrate that APP expression can be regulated by metals that in turn play a role in the self-regulation of the molecule. Thus, the direct effect of metals on APP may affect its processing and, subsequently, levels of the Aβ peptide.

9.7   INTERACTION OF THE AMYLOID BETA PEPTIDE WITH METALS

The 39- to 42-aa Aβ peptide is generated from the processing of its larger precursor, APP. This peptide was identified as a constituent of extracellular amyloid plaques in the mid-1980s [107,108]. Pathological mutations near or within the Aβ sequence result in higher levels of the peptide and are associated with familial forms of AD [109]. A positive correlation has been identified between the amount of Aβ and decline in cognition in AD [110]. Initial studies were focused on the toxic effects of extracellular Aβ. However, subsequent studies have shown that soluble Aβ oligomers and protofibrils exert more neurotoxic effects [111–115] and that intracellular Aβ can also be neurotoxic [116–118].

Aβ peptides have been shown to interact with zinc and copper, resulting in the aggregation of the peptide [43,85,93,105,119–124]. It has been demonstrated that these Aβ-metal aggregates may seed the formation of amyloid plaques. While the amount of available zinc and copper in the brain varies, in most brain regions, the extracellular concentrations of both Cu^{2+} and Zn^{2+} are usually very low [125,126]. However, levels of Zn^{2+} and Cu^{2+} have been found to be increased in amyloid plaques [39,127] and the homeostatic regulation of these metals is demonstrated to be disrupted in the AD brain [43].

Cu^{2+} and Fe^{3+} can also bind to Aβ *in vitro*, potentiating aggregation but only under mildly acidic conditions (pH 6.8–7.0), similar to conditions present in AD brains. Cu^{2+} precipitates Aβ more rapidly than Fe^{3+}. Monomeric Aβ binds copper ions via three histidine residues and one tyrosine residue, while aggregated Aβ binds via a bridging histidine [128,129]. Spectrophotometric studies show that the interaction between Aβ and copper *in vitro* induces the production of hydrogen peroxide (H_2O_2) as Cu^{2+} is reduced to Cu^+ by the peptide [85,130,131]. H_2O_2 can then diffuse across cell membranes and mediate oxidative damage.

Aβ has high- and low-affinity binding sites for Cu^{2+} [85,103]. While both Aβ1–40 and Aβ1–42 have a similar low-affinity binding for copper (approximately 5.0×10^{-9} M), the high-affinity binding site of Aβ1–40 is 5.0×10^{-11} M and that of Aβ1–42 is much higher at 7.0×10^{-18} M [103]. This higher affinity of Aβ1–42 for Cu^{2+} correlates with increased precipitation and SDS-resistant dimerization of the peptide by Cu^{2+} [85,103]. In addition, Cu^{2+}/Aβ1–42 complexes also demonstrate increased redox activity.

The binding of iron to Aβ *in vitro* also results in aggregation of the peptide [132]. As seen with copper, the binding of Aβ reduces Fe^{3+} to Fe^{2+} and causes the formation of H_2O_2 via Fenton chemistry [133–135]. This reaction with iron occurs to a lesser degree than that with copper, and the iron-reducing ability of Aβ1–42 is greater than that of Aβ1–40 [136]. Thus, iron is also recognized to have a role in AD pathogenesis.

Zinc is increased in amyloid plaques both in AD brains [39,127,129] and in APP-overexpressing (Tg2576) transgenic mice [137]. Zinc levels are highest in the hippocampus, amygdala, and cortex, regions most affected in AD [127]. Initial studies in our laboratory demonstrated that Aβ could be rapidly precipitated by Zn^{2+} [93,94,138]. Aβ1–40 has both high and low affinity for zinc, mediated by histidine residues, particularly histidine 13 [85,93,138]. Unlike copper and iron, zinc precipitates Aβ at a physiological pH (7.4) *in vitro*. High-affinity zinc binding K_d is reported to be 100 nM; low-affinity K_d, 5 μM [93,94]. The ability of zinc to precipitate Aβ at a physiological pH means that these aggregates are less able to be solubilized and, as a result, less likely to be cleared and eventually degraded [139].

These oxidation injury processes are reported to be mediated by H_2O_2 and peroxidation products. H_2O_2 freely diffuses across tissues, reacts with reduced metal ions such as Fe^{2+} and Cu^+, and generates hydroxyl radicals (OH•) by Fenton chemistry. Metal reduction and OH• and H_2O_2 formation are highest for Aβ42$_{human}$, with the order being Aβ42$_{human}$ > Aβ40$_{human}$ >> Aβ40$_{mouse}$ ≈ 0 [131], agreeing with observations that Aβ1–42 is the major species observed in amyloid plaques in AD and is generated in mutations that cause familial forms of AD. This relationship is also observed in neuronal cell culture, where neurotoxicity corresponds with the ability of the respective peptides to produce redox activity and where the Cu^{2+}/Aβ interaction plays a major role [136,140]. Binding of Aβ to Cu^{2+} and Zn^{2+} increases its interaction with cell membranes [128,141]. In culture, this interaction can be reversed by copper and iron chelators, such as TETA and clioquinol (CQ), which inhibit these reactions and attenuate the Aβ toxicity [142,143]. Binding of Aβ to copper results in ROS generation when Cu^{2+} is converted to Cu^{1+}. If this reduction process is not accompanied by the oxidation of other moieties, such as cholesterol [143], the side chains of Aβ itself will be oxidized. This results in the formation of different species of oxidized Aβ and oligomerization of the peptide. Many adducts can be produced as a result of this copper-mediated redox reaction. These include Aβ methionine sulfoxide [144], 2-oxo-histidine adducts of Aβ extracted from AD plaques [145], $N3$-pyroglutamate-modified forms of Aβ [146], aldehyde adducts to the lysine residues [147], and tyrosine residues modified with adducts such as DOPA, dopamine, dopamine quinine, dihydroxyindol, and isodityrosine [144,148]. Tyrosine residues are especially susceptible to free radical attack due to the presence of the conjugated aromatic ring. In AD plaques, increased levels of dityrosine and 3-nitrotyrosine have been reported. In the presence of Cu^{2+} and H_2O_2, Aβ42 forms dityrosine cross-linked oligomers *in vitro*, a modification that is resistant to proteolysis [149]. Dityrosine-linked Aβ results in further peptide aggregation, and subsequent formation of higher-order oligomers [150]. Aβ radicals formed after reduction of copper can also form covalent adducts onto other proteins. As an example, levels of the peroxidase cyclooxygenase-2 covalently complexed to Aβ are elevated in the AD brain, due to formation of dityrosine bridges between the two molecules [151].

9.8 METALS AND TAU PATHOLOGY

The paired helical filaments in NFTs, the other major hallmark of AD, are composed predominantly of the microtubule-associated protein tau. Tau is also present in other neurodegenerative disorders, such as frontotemporal dementia. It is still unclear how the Aβ and tau lesions may be linked by metal interactions, and current evidence linking tau to metal neurochemistry is not as advanced as the literature about APP and Aβ interactions with metals. However, several reports indicate that tau may play a role in the metal-related abnormalities observed in AD. Tau and two peptides corresponding to the second and third repeat regions of the tau microtubule binding domain bind copper (Cu^{2+}), dependent on both pH level and stoichiometry. This binding results in a conformational change in tau that may be important in the formation of paired helical filaments [152–154]. Earlier studies demonstrated that NFTs are capable of binding adventitious copper and iron in a redox-competent manner, creating a source for ROS within the neuron [155]. Recent data indicate that the interaction of tau with Cu^{2+} can mediate H_2O_2 generation similar to that mediated by Aβ [156].

Tau is phosphorylated by many kinases (causing hyperphosphorylation of the tau molecule and subsequent dissociation from the microtubules), and, interestingly, many of these kinases, such as the extracellular signal-regulated kinase (ERK1/2), are induced by metal ions such as zinc [157,158]. Tau hyperphosphorylation has been induced in both SH-SY5Y and N2a cells by zinc [159]. These studies underscore the importance of zinc as levels of this metal rise dramatically in AD-affected neocortex [39,44,160,161] and tangle-bearing neurons fill with zinc [162]. Additionally, H_2O_2 generated by Cu/Fe-Aβ complexes [131,136] may also inhibit phosphatases [163]. In contrast, hippocampal neuron cultures treated with iron citrate demonstrate a decrease in tau phosphorylation at AD-related epitopes, possibly due to a decrease in the activity of the Cdk5-p25 complex, where p25 may regulate the activity of the Cdk5 kinase [164]. Furthermore, Fe^{3+}, but not Fe^{2+}, induces aggregation of hyperphosphorylated tau, which can be reversed when Fe^{3+} is reduced to Fe^{2+} [165]. When tau aggregates from the AD brain are treated with reducing agents, resolubilization of the tau molecule and release of iron (II) are observed, further demonstrating the potential role of iron in NFT formation [165].

Other cytoskeletal components found within NFTs are also known to interact with metals. Neurofilaments (NF), which are found within the NFTs and in association with the dystrophic neurites surrounding amyloid plaques, show phosphorylation-dependent alterations very early in the AD cascade [166]. Purified NFs stoichiometrically bind at least 1 mol copper and 4 mol zinc [167]. The assembly of NFs may be partly mediated by copper, fostering the assembly of the light NF subunit. NFs are also phosphorylated by a number of the kinases that phosphorylate tau and as such can be modulated by metal ions. Indeed, zinc can also induce the phosphorylation of NFs at various sites [159].

These studies highlight the importance of regulation of biometals in AD and that the biochemistry of Aβ-metal complexes is pathophysiologically relevant. Thus, any dysregulation resulting in changes in normal metal ion homeostasis will contribute to alterations in the disease. It is generally agreed now that in AD, zinc and copper are enriched in amyloid, where they coordinate Aβ, there is strong evidence of functional copper deficiency, and iron is enriched in the tissue and involved in neuritic pathology. Therefore, pharmacotherapy that targets abnormal Aβ metallation should ideally release the metals trapped by Aβ and return them to normal metabolism. As such, the metal theory of AD may aid in explaining AD pathology, which remains unexplained by the amyloid cascade hypothesis. Blocking these metal-mediated reactions is thus a useful pharmacological intervention.

9.9   THERAPEUTIC INVENTIONS BASED ON METAL THEORY

Current drug therapies for AD target symptomatic relief. More recent approaches target the underlying disease. Drugs currently approved for the treatment of AD include cholinesterase inhibitors, such as donepezil and galantamine, and the NMDA receptor antagonist memantine, all of which are mainly used for symptomatic relief. In addition, other candidates targeting different therapeutic strategies are in development and many have progressed into clinical trials. These include modifiers for the enzymes (secretases) responsible for the processing of APP and the use of Aβ in passive and active immunotherapies. As oxidative stress is known to play a role in AD, antioxidants such as vitamins C and E have also been trialed as potential therapies but have not yielded useful benefit. More recently, small peptide modifiers, such as glycosaminoglycans, which can inhibit or disrupt Aβ aggregation, have been studied.

Chelation, the removal of metal ions from tissue, is approved for use in several genuine overexposure situations (e.g., Wilson's disease or lead toxicity) and in rheumatoid arthritis. There is currently no evidence that AD will respond to traditional medical chelation. In removing essential metal ions, the risk for subsequent side effects (e.g., iron-deficiency anemia) is inevitable. These issues can be partially overcome by engineering small molecules to target specific compartments or organelles. Nevertheless, complex situations, such as the pooling of metals in plaques and their relative deficiency within neighboring cells, require the development of small molecules with sophisticated properties, such as ionophores.

The aim of these neurotherapeutic small molecules is to target the initiating event in the generation of free radicals. These metal-complexing agents can be used to prevent the metal ions from participating in redox chemistry. In addition, these potential neurodegenerative therapeutics should have the ability to cross the BBB. This requirement excludes many common metal chelators due to their hydrophilic nature. In one study, AD patients treated twice daily with intramuscular desferrioxamine were reported to benefit from the treatment over a 2-year period [168]. This treatment resulted in a significant reduction in the rate of decline of daily living skills, which the researchers originally attributed to chelation of aluminum. However, desferrioxamine also chelates zinc, iron, and copper. In another small double-blind trial, thirty-four AD subjects treated with d-penicillamine or placebo reported a decrease in serum oxidative markers over a 6-month period with unchanged cognitive decline [169]. As there was a large dropout rate in the study, the results are inconclusive.

A range of metal-complexing agents have been tested in a variety of preclinical systems for AD. The bicyclam analogue JKL169 (1,1′-xylyl bis-1,4,8,11 tetraazacyclotetradecane) reduced copper levels in brain cortex while maintaining normal levels of copper in the blood, CSF, and corpus callosum of rats [170]. In a mouse model, the lipophilic chelator DP109 was able to reduce levels of aggregated insoluble Aβ and increased soluble Aβ [171]. Treatment of transgenic APP (Tg2576) mice orally with the chelator CQ (5-chloro-7-iodo-8-hydroxyquinoline) resulted in a 49% reduction of amyloid deposition in the cortex, accompanied by an improvement of stability of the general health and weight in these animals compared with untreated mice [172]. This quinoline compound was able to cross the BBB and increase brain copper and zinc levels in the treated mice. CQ has nanomolar affinity for Cu^{2+} and Zn^{2+} and dissociates metal ions from the low-affinity binding sites of Aβ [161]. CQ was demonstrated to cross the BBB in Tg2576 mice given peripheral doses of the compound. CQ was found to complex with amyloid plaques and with Zn^{2+}-metallated Aβ in brain samples of postmortem AD specimens [161]. In a following pilot Phase II human trial, CQ was administered orally for 36 weeks in moderately severe AD patients. The results demonstrated a slower rate of cognitive decline (significant at Weeks 4 and 16, with a trend to improvement at other intervals) and a reduction in plasma Aβ1–42 levels in treated subjects compared with the placebo controls [173]. CQ may modulate metal levels as it has strong ionophore activity, and the CQ-Cu complex has been shown to mediate copper transport into cells, resulting in the activation of matrix metalloproteases (MMPs) and subsequent degradation of Aβ [174]. This demonstrates that compounds targeting the Aβ-metal interaction and/or metal homeostasis have genuine therapeutic potential. The main objective of using this class of metal-protein-attenuating compound is to remove the essential metals (copper and zinc) from where they may be harmful and cause damage (i.e., coordinated to Aβ) and relocate them to a place where they will be beneficial (i.e., restore dysregulated metal homeostasis). CQ has also been tested in other neurodegenerative disease models and shown to be effective in both Parkinson's disease [175] and Huntington's disease [176] animal models. Interestingly, both of these diseases have been associated with iron overload leading to oxidative stress.

APP transgenic mouse models treated with PBT-2, the second-generation 8-hydroxyquinoline derivative of CQ, demonstrated marked improvement within days of commencement of treatment [177]. This compound has greater BBB penetration than CQ. Treated animals had significantly decreased levels of Aβ in the brain accompanied by reversed cognitive deficit in the Morris water maze task. These studies provided the rationale for then studying more mildly affected AD cases with this compound. The first double-blind, placebo-controlled Phase II clinical trial was recently completed for PBT-2. The trial was performed in 78 subjects over 12 weeks for the treatment of early AD. The results revealed that the drug was safe and well tolerated at 50- and 250-mg daily doses for the 12 weeks. In addition, Aβ levels in the CSF were significantly lowered by the 250-mg dose at 12 weeks, and significant improvement above baseline in performance on executive tests of the Neuropsychological Test Battery was observed at 12 weeks [178]. These results are encouraging and appear to be the basis for proceeding with further Phase IIb or III testing of what may be a promising disease-modifying drug based on the metal theory.

However, the mechanisms of action of CQ and PBT-2 are still to be confirmed. Studies in mice suggest that CQ enters the brain, combines with metallated Aβ in plaques, and possibly diffuses the Aβ [161]. CQ modestly increased brain zinc and copper levels in treated transgenic mice [172] and significantly increased the plasma zinc levels in AD patients in the cited Phase II clinical trial (normalized from a baseline of deficiency) [173]. These studies show that CQ does not just act as a simple chelator. *In vitro*, CQ-Cu complexes enter cells, inhibiting Aβ secretion via degradation of the peptide through up-regulation of MMP-2 and MMP-3. Increase in activity of these MMPs was through activation of phosphoinositol 3-kinase and Jun N-terminal kinase. The CQ-Cu complex was also able to promote phosphorylation of glycogen synthase kinase-3, leading to activation of Jun N-terminal kinase and degradation of Aβ1–40 [174]. A proposed mechanism of action for AD could be that when CQ or PBT-2 enters the brain, the compounds are attracted to the extracellular pool of metals that are in a dissociable equilibrium in amyloid. Binding of the molecule to zinc or copper in the amyloid possibly forms a ternary complex with Aβ, whereby the drug-metal complex then enters the cell. This allows activation of MMPs and facilitates the clearance of Aβ in the synapse. The oxidative oligomerization of Aβ and the toxic redox activity of Aβ oligomers are concomitantly blocked by the drug.

Iron chelation therapy has also been used as an approach to metal-based therapeutics for AD. Targeting the increase in brain iron in the disease, this involves the utilization of molecules that pass the BBB and are designed to be multifunctional. These include the inhibition of acetylcholinesterase through the attachment of a propargylamine moiety and by using compounds that possess antioxidant or monoamine oxidase inhibitor activity [179–181]. In these studies, the labile iron pool is decreased by the drugs; as a consequence, APP translation and Aβ generation are also expected to decrease [182]. Inhibition of hypoxia-inducible factor prolyl 4-hydroxylases also depletes iron, resulting in neuroprotection [183]. Despite attempts to achieve complete metal ion specificity, it is still likely that molecules that are thought to target iron will also interact with copper, zinc, and other metal ions. While this may have advantages in the dissolution of Aβ aggregates, excess deple-tion of copper and zinc may paradoxically exaggerate AD pathology. Only empirical testing can determine whether that will be of value in the clinical situation. It thus seems likely that future phar-macotherapeutic approaches to AD will involve combination therapies as there currently appears to be no biochemical barrier in utilizing the combination of metal-complexing agents with other potentially disease-modifying interventions, such as cholinergic modulators, secretase inhibitors, and immunotherapy.

9.10   SUMMARY

There is currently no effective therapy for AD. The disease is characterized by an increase in tissue iron and accumulation of copper and zinc in amyloid plaques. These are associated with a perturbation in metal homeostasis. Therapeutic agents that modulate metal bioavailability are an approach to correct these metal abnormalities. Small molecules utilizing metal-based thera-peutics have shown promise in the treatment of the disease and are currently progressing through clinical trials.

ACKNOWLEDGMENTS

This work was supported by funds from the National Health and Medical Research Council of Australia and the Australian Research Council.

REFERENCES

1. Ferri, R.T. and Levitt, P., Regulation of regional differences in the differentiation of cerebral cortical neurons by EGF family-matrix interactions. *Development* 121: 1995.
2. Valko, M., Morris, H., and Cronin, M.T., Metals, toxicity and oxidative stress. *Curr Med Chem* 12: 2005.
3. Tarohda, T., Yamamoto, M., and Amamo, R., Regional distribution of manganese, iron, copper, and zinc in the rat brain during development. *Anal Bioanal Chem* 380: 2004.
4. Madaric, A., Ginter, E., and Kadrabova, J., Serum copper, zinc and copper/zinc ratio in males: Influence of aging. *Physiol Res* 43: 1994.
5. Martinez Lista, E., et al., Changes in plasma copper and zinc during rat development. *Biol Neonate*, 64: 1993.
6. Maynard, C.J., et al., Overexpression of Alzheimer's disease amyloid-beta opposes the age-dependent elevations of brain copper and iron. *J Biol Chem* 277: 2002.
7. Ahluwalia, N., et al., Iron status and stores decline with age in Lewis rats. *J Nutr* 130: 2000.
8. Ke, Y., et al., Age-dependent and iron-independent expression of two mRNA isoforms of divalent metal transporter 1 in rat brain. *Neurobiol Aging* 26: 2005.
9. Ekmekcioglu, C., The role of trace elements for the health of elderly individuals. *Nahrung* 45: 2001.
10. Iskra, M., Patelski, J., and Majewski, W., Concentrations of calcium, magnesium, zinc and copper in rela-tion to free fatty acids and cholesterol in serum of atherosclerotic men. *J Trace Elem Electrolytes Health Dis* 7: 1993.
11. McMaster, D., et al., Serum copper and zinc in random samples of the population of Northern Ireland. *Am J Clin Nutr* 56: 1992.
12. Menditto, A., et al., Association of serum copper and zinc with serum electrolytes and with selected risk factors for cardiovascular disease in men aged 55–75 years. NFR Study Group. *J Trace Elem Electrolytes Health Dis* 7: 1993.
13. Milne, D.B. and Johnson, P.E., Assessment of copper status: Effect of age and gender on reference ranges in healthy adults. *Clin Chem* 39: 1993.
14. Prasad, A.S., et al., Zinc deficiency in elderly patients. *Nutrition* 9: 1993.
15. Bunker, V.W., et al., Metabolic balance studies for zinc and copper in housebound elderly people and the relationship between zinc balance and leukocyte zinc concentrations. *Am J Clin Nutr* 46: 1987.

16. Lindeman, R.D., Clark, M.L., and Colmore, J.P., Influence of age and sex on plasma and red-cell zinc concentrations. *J Gerontol* 26: 1971.
17. Monget, A.L., et al., Micronutrient status in elderly people. Geriatrie/Min. Vit. Aux Network. *Int J Vitam Nutr Res* 66: 1996.
18. Munro, H.N., Suter, P.M., and Russell, R.M., Nutritional requirements of the elderly. *Annu Rev Nutr* 7: 1987.
19. Ravaglia, G., et al., Blood micronutrient and thyroid hormone concentrations in the oldest-old. *J Clin Endocrinol Metab* 85: 2000.
20. Keen, C.L., et al., Developmental consequences of trace mineral deficiencies in rodents: Acute and long-term effects. *J Nutr* 133: 2003.
21. Massie, H.R., Aiello, V.R., and Iodice, A.A., Changes with age in copper and superoxide dismutase levels in brains of C57BL/6J mice. *Mech Ageing Dev* 10: 1979.
22. Morita, A., Kimura, M., and Itokawa, Y., The effect of aging on the mineral status of female mice. *Biol Trace Elem Res* 42: 1994.
23. Takahashi, S., et al., Age-related changes in the concentrations of major and trace elements in the brain of rats and mice. *Biol Trace Elem Res* 80: 2001.
24. Rae, T.D., et al., Undetectable intracellular free copper: The requirement of a copper chaperone for super-oxide dismutase. *Science* 284: 1999.
25. Markesbery, W.R. and Lovell, M.A., Damage to lipids, proteins, DNA, and RNA in mild cognitive impairment. *Arch Neurol* 64: 2007.
26. Smith, M.A., et al., Oxidative damage in Alzheimer's. *Nature* 382: 1996.
27. Smith, M.A., et al., Widespread peroxynitrite-mediated damage in Alzheimer's disease. *J Neurosci* 17: 1997.
28. Frederickson, C.J., Neurobiology of zinc and zinc-containing neurons. *Int Rev Neurobiol* 31: 1989.
29. Frederickson, C.J., Koh, J.Y., and Bush, A.I., The neurobiology of zinc in health and disease. *Nat Rev Neurosci* 6: 2005.
30. Rogers, J.T., et al., An iron-responsive element type II in the 5¢-untranslated region of the Alzheimer's amyloid precursor protein transcript. *J Biol Chem* 277: 2002.
31. Son, M., et al., Overexpression of CCS in G93A-SOD1 mice leads to accelerated neurological deficits with severe mitochondrial pathology. *Proc Natl Acad Sci U S A* 104: 2007.
32. Terry, R.D., et al., Physical basis of cognitive alterations in Alzheimer's disease: Synapse loss is the major correlate of cognitive impairment. *Ann Neurol* 30: 1991.
33. Frederickson, C.J., et al., Synaptic release of zinc from brain slices: Factors governing release, imaging, and accurate calculation of concentration. *J Neurosci Methods* 154: 2006.
34. Danscher, G. and Stoltenberg, M., Zinc-specific autometallographic in vivo selenium methods: Tracing of zinc-enriched (ZEN) terminals, ZEN pathways, and pools of zinc ions in a multitude of other ZEN cells. *J Histochem Cytochem* 53: 2005.
35. Schlief, M.L., Craig, A.M., and Gitlin, J.D., NMDA receptor activation mediates copper homeostasis in hippocampal neurons. *J Neurosci* 25: 2005.
36. Schlief, M.L., et al., Role of the Menkes copper-transporting ATPase in NMDA receptor-mediated neuronal toxicity. *Proc Natl Acad Sci U S A* 103: 2006.
37. Uchida, Y., et al., Growth inhibitory factor prevents neurite extension and the death of cortical neurons caused by high oxygen exposure through hydroxyl radical scavenging. *J Biol Chem* 277: 2002.
38. Uchida, Y., et al., The growth inhibitory factor that is deficient in the Alzheimer's disease brain is a 68 amino acid metallothionein-like protein. *Neuron* 7: 1991.
39. Lovell, M.A., et al., Copper, iron and zinc in Alzheimer's disease senile plaques. *J Neurol Sci* 158: 1998.
40. Molina, J.A., et al., Cerebrospinal fluid levels of transition metals in patients with Alzheimer's disease. *J Neural Transm* 105: 1998.
41. Basun, H., et al., Metals and trace elements in plasma and cerebrospinal fluid in normal aging and Alzheimer's disease. *J Neural Transm Park Dis Dement Sect* 3: 1991.
42. Squitti, R., et al., Elevation of serum copper levels in Alzheimer's disease. *Neurology* 59: 2002.
43. Adlard, P.A. and Bush, A.I., Metals and Alzheimer's disease. *J Alzheimers Dis* 10: 2006.
44. Religa, D., et al., Elevated cortical zinc in Alzheimer disease. *Neurology* 67: 2006.
45. Bartzokis, G., et al., In vivo MR evaluation of age-related increases in brain iron. *AJNR Am J Neuroradiol* 15: 1994.
46. Hallgren, B. and Sourander, P., The effect of age on the non-haemin iron in the human brain. *J Neurochem* 3: 1958.

47. Hardy, P.A., et al., Correlation of R^2 with total iron concentration in the brains of rhesus monkeys. *J Magn Reson Imaging* 21: 2005.

48. Roskams, A.J. and Connor, J.R., Iron, transferrin, and ferritin in the rat brain during development and aging. *J Neurochem* 63: 1994.

49. Sohal, R.S., et al., Effect of age and caloric restriction on bleomycin-chelatable and nonheme iron in different tissues of C57BL/6 mice. *Free Radic Biol Med* 27: 1999.

50. Suh, J.H., et al., Dietary supplementation with (R)-alpha-lipoic acid reverses the age-related accumulation of iron and depletion of antioxidants in the rat cerebral cortex. *Redox Rep* 10: 2005.

51. Massie, H.R., Aiello, V.R., and Williams, T.R., Iron accumulation during development and ageing of *Drosophila. Mech Ageing Dev* 29: 1985.

52. Chinnery, P.F., et al., Clinical features and natural history of neuroferritinopathy caused by the FTL1 460InsA mutation. *Brain* 130: 2007.

53. Mantovan, M.C., et al., Exploring mental status in Friedreich's ataxia: A combined neuropsychological, behavioral and neuroimaging study. *Eur J Neurol* 13: 2006.

54. Zecca, L., et al., Iron, brain ageing and neurodegenerative disorders. *Nat Rev Neurosci* 5: 2004.

55. Castellani, R.J., et al., Contribution of redox-active iron and copper to oxidative damage in Alzheimer disease. *Ageing Res Rev* 3: 2004.

56. Moreira, P.I., et al., Oxidative stress mechanisms and potential therapeutics in Alzheimer disease. *J Neural Transm* 112: 2005.

57. Smith, M.A., et al., Abnormal localization of iron regulatory protein in Alzheimer's disease. *Brain Res* 788: 1998.

58. Bishop, G.M., et al., Iron: A pathological mediator of Alzheimer disease? *Dev Neurosci* 24: 2002.

59. Yu, W.H., et al., Metallothionein III is reduced in Alzheimer's disease. *Brain Res* 894: 2001.

60. Adlard, P.A., West, A.K., and Vickers, J.C., Increased density of metallothionein I/II-immunopositive cortical glial cells in the early stages of Alzheimer's disease. *Neurobiol Dis* 5: 1998.

61. Lovell, M.A., et al., Alterations in zinc transporter protein-1 (ZnT-1) in the brain of subjects with mild cognitive impairment, early, and late-stage Alzheimer's disease. *Neurotox Res* 7: 2005.

62. Melov, S., et al., Mitochondrial oxidative stress causes hyperphosphorylation of tau. *PLoS ONE* 2: 2007.

63. Bayer, T.A., et al., Dietary Cu stabilizes brain superoxide dismutase 1 activity and reduces amyloid Abeta production in APP23 transgenic mice. *Proc Natl Acad Sci U S A* 100: 2003.

64. Phinney, A.L., et al., In vivo reduction of amyloid-beta by a mutant copper transporter. *Proc Natl Acad Sci U S A* 100: 2003.

65. Morris, M.C., et al., Dietary copper and high saturated and trans fat intakes associated with cognitive decline. *Arch Neurol* 63: 2006.

66. Sparks, D.L. and Schreurs, B.G., Trace amounts of copper in water induce beta-amyloid plaques and learning deficits in a rabbit model of Alzheimer's disease. *Proc Natl Acad Sci U S A* 100: 2003.

67. Sparks, D.L., et al., Trace copper levels in the drinking water, but not zinc or aluminum influence CNS Alzheimer-like pathology. *J Nutr Health Aging* 10: 2006.

68. Stoltenberg, M., et al., Immersion autometallographic tracing of zinc ions in Alzheimer beta-amyloid plaques. *Histochem Cell Biol* 123: 2005.

69. Yamaguchi, M., Kura, M., and Okada, S., Role of zinc as an activator of mitochondrial function in rat liver. *Biochem Pharmacol* 31: 1982.

70. Rossi, L., et al., Mitochondrial dysfunction in neurodegenerative diseases associated with copper imbalance. *Neurochem Res* 29: 2004.

71. Percival, S.S., Copper and immunity. *Am J Clin Nutr* 67: 1998.

72. Wegner, M., Transcriptional control in myelinating glia: The basic recipe. *Glia* 29: 2000.

73. Ortiz, E., et al., Effect of manipulation of iron storage, transport, or availability on myelin composition and brain iron content in three different animal models. *J Neurosci Res* 77: 2004.

74. Hara, A. and Taketomi, T., Cerebral lipid and protein abnormalities in Menkes' steely-hair disease. *Jpn J Exp Med* 56: 1986.

75. Bhatnagar, S. and Taneja, S., Zinc and cognitive development. *Br J Nutr* 85: 2001.

76. Takeda, A., Zinc homeostasis and functions of zinc in the brain. *Biometals* 14: 2001.

77. Youdim, M.B. and Yehuda, S., The neurochemical basis of cognitive deficits induced by brain iron deficiency: Involvement of dopamine-opiate system. *Cell Mol Biol (Noisy-le-grand)* 46: 2000.

78. Tanzi, R.E. and Bertram, L., Twenty years of the Alzheimer's disease amyloid hypothesis: A genetic perspective. *Cell* 120: 2005.

79. Haass, C. and Selkoe, D.J., Soluble protein oligomers in neurodegeneration: Lessons from the Alzheimer's amyloid beta-peptide. *Nat Rev Mol Cell Biol* 8: 2007.

80. Blennow, K., de Leon, M.J., and Zetterberg, H., Alzheimer's disease. *Lancet* 368: 2006.

81. Angeletti, B., et al., BACE1 cytoplasmic domain interacts with the copper chaperone for superoxide dismutase-1 and binds copper. *J Biol Chem* 280: 2005.

82. Hoke, D.E., et al., *In vitro* gamma-secretase cleavage of the Alzheimer's amyloid precursor protein correlates to a subset of presenilin complexes and is inhibited by zinc. *FEBS J* 272: 2005.

83. Inestrosa, N.C., Cerpa, W., and Varela-Nallar, L., Copper brain homeostasis: Role of amyloid precursor protein and prion protein. *IUBMB Life* 57: 2005.

84. Cerpa, W., et al., Is there a role for copper in neurodegenerative diseases? *Mol Aspects Med* 26: 2005.

85. Atwood, C.S., et al., Dramatic aggregation of Alzheimer abeta by Cu(II) is induced by conditions representing physiological acidosis. *J Biol Chem* 273: 1998.

86. Barnham, K.J., et al., Structure of the Alzheimer's disease amyloid precursor protein copper binding domain. A regulator of neuronal copper homeostasis. *J Biol Chem* 278: 2003.

87. Hesse, L., et al., The beta A4 amyloid precursor protein binding to copper. *FEBS Lett* 349: 1994.

88. Simons, A., et al., Evidence for a copper-binding superfamily of the amyloid precursor protein. *Biochemistry* 41: 2002.

89. Valensin, D., et al., Identification of a novel high affinity copper binding site in the APP(145–155) fragment of amyloid precursor protein. *Dalton Trans* (1): 2004.

90. Kong, G.K., et al., Copper binding to the Alzheimer's disease amyloid precursor protein. *Eur Biophys J* 37: 2008.

91. Bush, A.I., et al., A novel zinc(II) binding site modulates the function of the beta A4 amyloid protein precursor of Alzheimer's disease. *J Biol Chem* 268: 1993.

92. Bush, A.I., et al., The amyloid beta-protein precursor and its mammalian homologues. Evidence for a zinc-modulated heparin-binding superfamily. *J Biol Chem* 269: 1994.

93. Bush, A.I., et al., Rapid induction of Alzheimer A beta amyloid formation by zinc. *Science* 265: 1994.

94. Bush, A.I., et al., Modulation of A beta adhesiveness and secretase site cleavage by zinc. *J Biol Chem* 269: 1994.

95. Multhaup, G., et al., Interaction between the zinc (II) and the heparin binding site of the Alzheimer's disease beta A4 amyloid precursor protein (APP). *FEBS Lett* 355: 1994.

96. White, A.R., et al., Copper levels are increased in the cerebral cortex and liver of APP and APLP2 knockout mice. *Brain Res* 842: 1999.

97. Armendariz, A.D., et al., Gene expression profiling in chronic copper overload reveals upregulation of Prnp and App. *Physiol Genomics* 20: 2004.

98. Bellingham, S.A., et al., Gene knockout of amyloid precursor protein and amyloid precursor-like protein-2 increases cellular copper levels in primary mouse cortical neurons and embryonic fibroblasts. *J Neurochem* 91: 2004.

99. Bellingham, S.A., et al., Copper depletion down-regulates expression of the Alzheimer's disease amyloid-beta precursor protein gene. *J Biol Chem* 279: 2004.

100. Borchardt, T., et al., Copper inhibits beta-amyloid production and stimulates the non-amyloidogenic pathway of amyloid-precursor-protein secretion. *Biochem J* 344: 1999.

101. Cater, M.A., et al., Intracellular copper deficiency increases amyloid-beta secretion by diverse mechanisms. *Biochem J* 412: 2008.

102. Strozyk, D., et al., Zinc and copper modulate Alzheimer Abeta levels in human cerebrospinal fluid. *Neurobiol Aging* 2007.

103. Atwood, C.S., et al., Characterization of copper interactions with Alzheimer amyloid beta peptides: Identification of an attomolar-affinity copper binding site on amyloid beta1–42. *J Neurochem* 75: 2000.

104. Ciuculescu, E.D., Mekmouche, Y., and Faller, P., Metal-binding properties of the peptide APP170–188: A model of the ZnII-binding site of amyloid precursor protein (APP). *Chemistry* 11: 2005.

105. Brown, A.M., et al., Selective aggregation of endogenous beta-amyloid peptide and soluble amyloid precursor protein in cerebrospinal fluid by zinc. *J Neurochem* 69: 1997.

106. Venti, A., et al., The integrated role of desferrioxamine and phenserine targeted to an iron-responsive element in the APP-mRNA 5¢-untranslated region. *Ann N Y Acad Sci* 1035: 2004.

107. Glenner, G.G. and Wong, C.W., Alzheimer's disease: Initial report of the purification and characterization of a novel cerebrovascular amyloid protein. *Biochem Biophys Res Commun* 120: 1984.

108. Masters, C.L., et al., Amyloid plaque core protein in Alzheimer disease and Down syndrome. *Proc Natl Acad Sci U S A* 82: 1985.

109. Selkoe, D.J., Amyloid beta-protein and the genetics of Alzheimer's disease. *J Biol Chem* 271: 1996.
110. McLean, C.A., et al., Soluble pool of Abeta amyloid as a determinant of severity of neurodegeneration in Alzheimer's disease. *Ann Neurol* 46: 1999.
111. Hartley, D.M., et al., Protofibrillar intermediates of amyloid beta-protein induce acute electrophysiological changes and progressive neurotoxicity in cortical neurons. *J Neurosci* 19: 1999.
112. Lambert, M.P., et al., Diffusible, nonfibrillar ligands derived from Abeta1–42 are potent central nervous system neurotoxins. *Proc Natl Acad Sci U S A* 95: 1998.
113. Lambert, M.P., et al., Vaccination with soluble Abeta oligomers generates toxicity-neutralizing antibodies. *J Neurochem* 79: 2001.
114. Walsh, D.M., et al., Amyloid beta-protein fibrillogenesis. Structure and biological activity of protofibrillar intermediates. *J Biol Chem* 274: 1999.
115. Walsh, D.M., et al., Naturally secreted oligomers of amyloid beta protein potently inhibit hippocampal long-term potentiation in vivo. *Nature* 416: 2002.
116. Crouch, P.J., et al., Copper-dependent inhibition of human cytochrome *c* oxidase by a dimeric conformer of amyloid-beta1–42. *J Neurosci* 25: 2005.
117. Yan, S.D., et al., An intracellular protein that binds amyloid-beta peptide and mediates neurotoxicity in Alzheimer's disease [see comments]. *Nature* 389: 1997.
118. Yan, S.D., et al., Role of ERAB/l-3-hydroxyacyl-coenzyme A dehydrogenase type II activity in Abeta-induced cytotoxicity. *J Biol Chem* 274: 1999.
119. Bush, A.I., Masters, C.L., and Tanzi, R.E., Copper, beta-amyloid, and Alzheimer's disease: Tapping a sensitive connection. *Proc Natl Acad Sci U S A* 100: 2003.
120. Bush, A.I., The metallobiology of Alzheimer's disease. *Trends Neurosci* 26: 2003.
121. Cuajungco, M.P. and Faget, K.Y., Zinc takes the center stage: Its paradoxical role in Alzheimer's disease. *Brain Res Brain Res Rev* 41: 2003.
122. Esler, W.P., et al., Zinc-induced aggregation of human and rat beta-amyloid peptides *in vitro*. *J Neurochem* 66: 1996.
123. Gaggelli, E., et al., Copper homeostasis and neurodegenerative disorders (Alzheimer's, prion, and Parkinson's diseases and amyotrophic lateral sclerosis). *Chem Rev* 106: 2006.
124. Mantyh, P.W., et al., Aluminum, iron, and zinc ions promote aggregation of physiological concentrations of beta-amyloid peptide. *J Neurochem* 61: 1993.
125. Masuoka, J. and Saltman, P., Zinc(II) and copper(II) binding to serum albumin. A comparative study of dog, bovine, and human albumin. *J Biol Chem* 269: 1994.
126. Frederickson, C.J., et al., Concentrations of extracellular free zinc (pZn)e in the central nervous system during simple anesthetization, ischemia and reperfusion. *Exp Neurol* 198: 2006.
127. Miller, L.M., et al., Synchrotron-based infrared and X-ray imaging shows focalized accumulation of Cu and Zn co-localized with beta-amyloid deposits in Alzheimer's disease. *J Struct Biol* 155: 2006.
128. Curtain, C.C., et al., Alzheimer's disease amyloid-beta binds copper and zinc to generate an allosterically ordered membrane-penetrating structure containing superoxide dismutase-like subunits. *J Biol Chem* 276: 2001.
129. Dong, J., et al., Metal binding and oxidation of amyloid-beta within isolated senile plaque cores: Raman microscopic evidence. *Biochemistry* 42: 2003.
130. Behl, C., et al., Hydrogen peroxide mediates amyloid beta protein toxicity. *Cell* 77: 1994.
131. Huang, X., et al., The A beta peptide of Alzheimer's disease directly produces hydrogen peroxide through metal ion reduction. *Biochemistry* 38: 1999.
132. Kuroda, Y. and Kawahara, M., Aggregation of amyloid beta-protein and its neurotoxicity: Enhancement by aluminum and other metals. *Tohoku J Exp Med* 174: 1994.
133. Kanski, J., et al., Role of glycine-33 and methionine-35 in Alzheimer's amyloid beta-peptide 1–42-associated oxidative stress and neurotoxicity. *Biochim Biophys Acta* 1586: 2002.
134. Rottkamp, C.A., et al., Redox-active iron mediates amyloid-beta toxicity. *Free Radic Biol Med* 30: 2001.
135. Smith, M.A., et al., Iron accumulation in Alzheimer disease is a source of redox-generated free radicals. *Proc Natl Acad Sci U S A* 94: 1997.
136. Opazo, C., et al., Metalloenzyme-like activity of Alzheimer's disease beta-amyloid. Cu-dependent catalytic conversion of dopamine, cholesterol, and biological reducing agents to neurotoxic H(2)O(2). *J Biol Chem* 277: 2002.
137. Lee, J.Y., Mook-Jung, I., and Koh, J.Y., Histochemically reactive zinc in plaques of the Swedish mutant beta-amyloid precursor protein transgenic mice. *J Neurosci* 19: 1999.

138. Ha, C., Ryu, J., and Park, C.B., Metal ions differentially influence the aggregation and deposition of Alzheimer's beta-amyloid on a solid template. *Biochemistry* 46: 2007.

139. Moir, R.D., et al., Differential effects of apolipoprotein E isoforms on metal-induced aggregation of A beta using physiological concentrations. *Biochemistry* 38: 1999.

140. Huang, X., et al., Cu(II) potentiation of Alzheimer abeta neurotoxicity. Correlation with cell-free hydrogen peroxide production and metal reduction. *J Biol Chem* 274: 1999.

141. Curtain, C.C., et al., Metal ions, pH, and cholesterol regulate the interactions of Alzheimer's disease amyloid-beta peptide with membrane lipid. *J Biol Chem* 278: 2003.

142. Abramov, A.Y., Canevari, L., and Duchen, M.R., Changes in intracellular calcium and glutathione in astrocytes as the primary mechanism of amyloid neurotoxicity. *J Neurosci* 23: 2003.

143. Puglielli, L., et al., Alzheimer disease beta-amyloid activity mimics cholesterol oxidase. *J Clin Invest* 115: 2005.

144. Ali, F.E., et al., Methionine regulates copper/hydrogen peroxide oxidation products of Abeta. *J Pept Sci* 11: 2005.

145. Metodiewa, D., Molecular mechanisms of cellular injury produced by neurotoxic amino acids that generate reactive oxygen species. *Amino Acids* 14: 1998.

146. Maeda, J., et al., Longitudinal, quantitative assessment of amyloid, neuroinflammation, and anti-amyloid treatment in a living mouse model of Alzheimer's disease enabled by positron emission tomography. *J Neurosci* 27: 2007.

147. Chen, K., Kazachkov, M., and Yu, P.H., Effect of aldehydes derived from oxidative deamination and oxidative stress on beta-amyloid aggregation; pathological implications to Alzheimer's disease. *J Neural Transm* 114: 2007.

148. Ali, F.E., et al., Dimerisation of *N*-acetyl-l-tyrosine ethyl ester and Abeta peptides via formation of dityrosine. *Free Radic Res* 40: 2006.

149. Atwood, C.S., et al., Copper mediates dityrosine cross-linking of Alzheimer's amyloid-beta. *Biochemistry* 43: 2004.

150. Barnham, K.J., et al., Tyrosine gated electron transfer is key to the toxic mechanism of Alzheimer's disease beta-amyloid. *FASEB J* 18: 2004.

151. Nagano, S., et al., Peroxidase activity of cyclooxygenase-2 (COX-2) cross-links beta-amyloid (Abeta) and generates Abeta-COX-2 hetero-oligomers that are increased in Alzheimer's disease. *J Biol Chem* 279: 2004.

152. Ma, Q., et al., Copper binding properties of a tau peptide associated with Alzheimer's disease studied by CD, NMR, and MALDI-TOF MS. *Peptides* 27: 2006.

153. Ma, Q.F., et al., Binding of copper (II) ion to an Alzheimer's tau peptide as revealed by MALDI-TOF MS, CD, and NMR. *Biopolymers* 79: 2005.

154. Zhou, L.X., et al., Copper (II) modulates *in vitro* aggregation of a tau peptide. *Peptides* 28: 2007.

155. Sayre, L.M., et al., In situ oxidative catalysis by neurofibrillary tangles and senile plaques in Alzheimer's disease: A central role for bound transition metals. *J Neurochem* 74: 2000.

156. Su, X.Y., et al., Hydrogen peroxide can be generated by tau in the presence of Cu(II). *Biochem Biophys Res Commun* 358: 2007.

157. An, W.L., et al., Mechanism of zinc-induced phosphorylation of p70 S6 kinase and glycogen synthase kinase 3beta in SH-SY5Y neuroblastoma cells. *J Neurochem* 92: 2005.

158. Harris, F.M., et al., Increased tau phosphorylation in apolipoprotein E4 transgenic mice is associated with activation of extracellular signal-regulated kinase: Modulation by zinc. *J Biol Chem*, 279: 2004.

159. Bjorkdahl, C., et al., Zinc induces neurofilament phosphorylation independent of p70 S6 kinase in N2a cells. *NeuroReport* 16: 2005.

160. Danscher, G., et al., Increased amount of zinc in the hippocampus and amygdala of Alzheimer's diseased brains: A proton-induced X-ray emission spectroscopic analysis of cryostat sections from autopsy material. *J Neurosci Methods* 76: 1997.

161. Opazo, C., et al., Radioiodinated clioquinol as a biomarker for beta-amyloid: Zn complexes in Alzheimer's disease. *Aging Cell* 5: 2006.

162. Suh, S.W., et al., Histochemically-reactive zinc in amyloid plaques, angiopathy, and degenerating neurons of Alzheimer's diseased brains. *Brain Res* 852: 2000.

163. Denu, J.M. and Tanner, K.G., Specific and reversible inactivation of protein tyrosine phosphatases by hydrogen peroxide: Evidence for a sulfenic acid intermediate and implications for redox regulation. *Biochemistry* 37: 1998.

164. Egana, J.T., et al., Iron-induced oxidative stress modify tau phosphorylation patterns in hippocampal cell cultures. *Biometals* 16: 2003.

165. Yamamoto, A., et al., Iron (III) induces aggregation of hyperphosphorylated tau and its reduction to iron (II) reverses the aggregation: Implications in the formation of neurofibrillary tangles of Alzheimer's disease. *J Neurochem* 82: 2002.

166. King, C.E., et al., Neuronal response to physical injury and its relationship to the pathology of Alzheimer's disease. *Clin Exp Pharmacol Physiol* 27: 2000.

167. Pierson, K.B. and Evenson, M.A., 200 K_d neurofilament protein binds Al, Cu and Zn. *Biochem Biophys Res Commun* 152: 1988.

168. Crapper McLachlan, D.R., et al., Intramuscular desferrioxamine in patients with Alzheimer's disease. *Lancet* 337: 1991.

169. Squitti, R., et al., d-penicillamine reduces serum oxidative stress in Alzheimer's disease patients. *Eur J Clin Invest* 32: 2002.

170. Moret, V., et al., 1,1¢-Xylyl bis-1,4,8,11-tetraaza cyclotetradecane: A new potential copper chelator agent for neuroprotection in Alzheimer's disease. Its comparative effects with clioquinol on rat brain copper distribution. *Bioorg Med Chem Lett* 16: 2006.

171. Lee, J.Y., et al., The lipophilic metal chelator DP-109 reduces amyloid pathology in brains of human beta-amyloid precursor protein transgenic mice. *Neurobiol Aging* 25: 2004.

172. Cherny, R.A., et al., Treatment with a copper-zinc chelator markedly and rapidly inhibits beta-amyloid accumulation in Alzheimer's disease transgenic mice. *Neuron* 30: 2001.

173. Ritchie, C.W., et al., Metal-protein attenuation with iodochlorhydroxyquin (clioquinol) targeting Abeta amyloid deposition and toxicity in Alzheimer disease: A pilot Phase 2 clinical trial. *Arch Neurol* 60: 2003.

174. White, A.R., et al., Degradation of the Alzheimer disease amyloid beta-peptide by metal-dependent up-regulation of metalloprotease activity. *J Biol Chem* 281: 2006.

175. Kaur, D., et al., Genetic or pharmacological iron chelation prevents MPTP-induced neurotoxicity in vivo: A novel therapy for Parkinson's disease. *Neuron* 37: 2003.

176. Nguyen, T., Hamby, A., and Massa, S.M., Clioquinol down-regulates mutant huntingtin expression *in vitro* and mitigates pathology in a Huntington's disease mouse model. *Proc Natl Acad Sci U S A* 102: 2005.

177. Adlard, P.A., et al., Rapid restoration of cognition in Alzheimer's transgenic mice with 8-hydroxy quinoline analogs is associated with decreased interstitial Abeta. *Neuron* 59: 2008.

178. Lannfelt, L., et al., Safety, efficacy, and biomarker findings of PBT2 in targeting Abeta as a modifying therapy for Alzheimer's disease: A Phase IIa, double-blind, randomised, placebo-controlled trial. *Lancet Neurol* 7: 2008.

179. Yogev-Falach, M., et al., A multifunctional, neuroprotective drug, ladostigil (TV3326), regulates holo-APP translation and processing. *FASEB J* 20: 2006.

180. Avramovich-Tirosh, Y., et al., Neurorescue activity, APP regulation and amyloid-beta peptide reduction by novel multi-functional brain permeable iron- chelating-antioxidants, M-30 and green tea polyphenol, EGCG. *Curr Alzheimer Res* 4: 2007.

181. Avramovich-Tirosh, Y., et al., Therapeutic targets and potential of the novel brain-permeable multifunctional iron chelator-monoamine oxidase inhibitor drug, M-30, for the treatment of Alzheimer's disease. *J Neurochem* 100: 2007.

182. Reznichenko, L., et al., Reduction of iron-regulated amyloid precursor protein and beta-amyloid peptide by (-)-epigallocatechin-3-gallate in cell cultures: Implications for iron chelation in Alzheimer's disease. *J Neurochem* 97: 2006.

183. Siddiq, A., et al., Hypoxia-inducible factor prolyl 4-hydroxylase inhibition. A target for neuroprotection in the central nervous system. *J Biol Chem* 280: 2005.

10 Lipoic Acid as a Novel Treatment for Mild Cognitive Impairment and Early-Stage Alzheimer's Disease

Annette E. Maczurek[1,2], Martina Krautwald[1], Megan L. Steele[1,2], Klaus Hager[3], Marlene Kenklies[3], Matt Sharman[4], Ralph Martins[4], Jürgen Engel[5], David Carlson[6], Gerald Münch[1]

[1]Department of Pharmacology, School of Medicine, University of Western Sydney, Penrith South DC, New South Wales, Australia

[2]Department of Biochemistry and Molecular Biology, James Cook University, Townsville, Queensland, Australia

[3]Klinik für Medizinische Rehabilitation und Geriatrie der Henriettenstiftung, Hannover, Germany

[4]Centre of Excellence for Alzheimer's Disease Research and Care, Edith Cowan University, Perth, Western Australia, Australia

[5]Zentaris GmbH, Frankfurt am Main, Germany

[6]GeroNova Research, Inc., Reno, Nevada, USA

CONTENTS

10.1 EPIDEMIOLOGY AND PATHOBIOCHEMISTRY OF ALZHEIMER'S DISEASE

Alzheimer's disease (AD) is a progressive neurodegenerative brain disorder that gradually destroys a patient's memory and ability to learn, make judgments, communicate within the social environment, and carry out daily activities. In the course of the disease, short-term memory is affected first, caused by neuronal dysfunction and degeneration in the hippocampus and amygdala. As the disease progresses further, neurons also degenerate and die in other cortical regions of the brain.[1] At this stage, sufferers often experience dramatic changes in personality and behavior, such as anxiety, suspiciousness and agitation, as well as delusions and hallucinations.[2] AD prevalence rates are 1% for individuals between 65 and 69 years old, 3% for those between 70 and 74 years old, 6% for those between 75 and 79 years old, 12% for those between 80 and 84 years old, and 25% for those aged 85 years and older. AD is further characterized by two major neuropathological hallmarks: The deposition of neuritic, β-amyloid (Aβ) peptide-containing senile plaques in hippocampal and cerebral cortical regions of AD patients is accompanied by the presence of intracellular neurofibrillary tangles that occupy much of the cytoplasm of pyramidal neurons. Inflammation, as evidenced by the activation of microglia and astroglia, is another hallmark of AD. Inflammation, including superoxide production ("oxidative burst"), is an important source of oxidative stress in AD patients.[3,4] The inflammatory process occurs mainly around the amyloid plaques and is characterized by proinflammatory substances released from activated microglia and astroglia.[5] Glia-produced cytokines, including interleukin-1β (IL-1β), IL-6, macrophage colony-stimulating factor, and tumor necrosis factor-α (TNF-α),[6] are prominent molecules in the inflammatory process.

Besides morphological alterations, AD is also associated with a markedly impaired cerebral glucose metabolism, as detected by reduced cortical ^{18}F-labeled desoxyglucose utilization in positron emission tomography of AD patients.[7]

10.2 THE CHOLINERGIC DEFICIT IN ALZHEIMER'S DISEASE

AD patients show a progressive neuronal cell loss that is associated with region-specific brain atrophy. In particular, the cholinergic projection pathway from the nucleus basalis of the Meynert area to areas of the cerebral cortex is most severely affected in brains of AD patients.[8] Loss of basal forebrain cholinergic neurons is demonstrated by reductions in the number of cholinergic markers, such as choline acetyltransferase (ChAT), muscarinic and nicotinic acetylcholine (ACh) receptor binding, and levels of the neurotransmitter ACh itself.[9] These changes are highly correlated with the degree of dementia in AD. ACh is derived from choline and acetyl-coenzyme A (CoA), the final product of the glycolytic pathway. Pyruvate derived from glycolytic metabolism serves as an important energy source in neurons. Therefore, the inhibition of pyruvate production (e.g., by glucose depletion) is considered a crucial factor that leads to acetyl-CoA deficits in AD brains. Based on the findings that (a) AD patients have reduced levels of ChAT and ACh compared with healthy elderly people and (b) ACh is hydrolyzed by ACh esterase (AChE), AChE inhibitors (AChIs) were the first drug class successfully introduced for the treatment of AD patients.

10.3 ALZHEIMER'S DISEASE—CURRENT TREATMENT STRATEGIES

Presently, only symptomatic treatments with AChIs are approved for treatment of mild to moderate forms of AD. The need for a "cholinergic + proenergetic + neuroprotective" therapy for early-stage AD is urgent, and we have proposed that lipoic acid (LA) may be a promising candidate for such treatment.[10]

10.4 LIPOIC ACID—A MULTIMODAL DRUG FOR THE TREATMENT OF ALZHEIMER'S DISEASE

10.4.1 LIPOIC ACID—A NATURAL MOLECULE

The chemical name for LA is 1,2-dithiolane-3-pentanoic acid. LA contains two thiol (sulfur) groups, which may be oxidized or reduced. The reduced form is known as dihydrolipoic acid (DHLA), while the oxidized form is known as LA. α-LA also contains an asymmetric carbon, meaning there are two possible optical isomers that are mirror images of each other (RLA and SLA). Only the *R*-isomer is endogenously synthesized, and it is an essential cofactor in aerobic metabolism, specifically as a cofactor of the pyruvate dehydrogenase (PDH) complex.

10.4.2 POSSIBLE MODES OF ACTION OF LIPOIC ACID INTERFERING WITH ALZHEIMER'S DISEASE-SPECIFIC DEGENERATION

In vitro and *in vivo* studies suggest that LA acts as a powerful micronutrient with diverse pharmacological and antioxidant properties.[11] In brief, LA has been suggested to have the following anti-dementia/anti-AD properties:

A. To increase ACh production by activation of ChAT
B. To increase glucose uptake, supplying more acetyl-CoA for the production of ACh
C. To chelate redox-active transition metals, inhibiting the formation of hydrogen peroxide and hydroxyl radicals
D. To increase the level of reduced glutathione (GSH) by scavenging reactive oxygen species (ROS)
E. To down-regulate inflammatory processes by scavenging ROS
F. To scavenge lipid peroxidation products
G. To induce the enzymes of GSH synthesis and other antioxidative enzymes

These diverse actions suggest that LA acts through multiple physiological and pharmacological mechanisms, many of which are being explored only now. It has been initially proposed that the reduced form of LA, DHLA, is responsible for many of its *pharmacological* benefits. However, more and more evidence suggest that many of the "antioxidant" in vivo effects of LA are mediated by an indirect effect in which LA acts as a pro-oxidant and activates the nuclear factor erythroid 2-related factor 2 (Nrf2), which subsequently up-regulates the expression of phase II detoxification enzymes and antioxidative acting proteins, including GSH *S*-transferase, NAD(P)H:quinone oxidoreductase-1, γ-glutamylcysteine synthase [GCS; also termed glutamate cysteine ligase (GCL)], ferritin, and heme oxygenase-1.[12,13]

10.4.3 LIPOIC ACID—AN ACTIVATOR OF CHOLINE ACETYLTRANSFERASE

Haugaard and Levin demonstrated that DHLA, which is formed by the PDH complex-catalyzed reduction of LA, increases the activity of ChAT.[14] Removal of DHLA by dialysis from purified ChAT causes complete disappearance of enzyme activity. By addition of DHLA (but not reduced ascorbic acid or reduced nicotinamide adenine dinucleotide), enzymatic activity could be restored to normal levels.[15] The authors concluded that DHLA serves an essential function in the action of this enzyme and that the ratio of reduced to oxidized LA plays an important role in ACh synthesis. It was furthermore suggested that DHLA (a) may act as a coenzyme in the ChAT reaction or (b) is able to reduce an essential functional cysteine residue in ChAT that cannot be reduced by any other physiological antioxidant, including reduced GSH.[15]

10.4.4   Lipoic Acid—A Potent Metal Chelator

The Aβ peptide, the main component of amyloid plaques in the AD brain, undergoes an age-dependent reaction with excess brain metal ions (Cu, Fe, and Zn) that induces peptide precipitation and plaque formation. Furthermore, combination of Aβ peptides with Cu or Fe ions induces the production of hydrogen peroxide from molecular oxygen,[16] which subsequently generates the cytotoxic hydroxyl radical by Fenton or Haber-Weiss reactions. As LA is a potent chelator of divalent metal ions *in vitro*, it was investigated whether feeding RLA could lower cortical iron levels and improve antioxidant status in rats. Results show that cerebral iron levels in old animals fed LA were lower when compared to controls and were similar to levels seen in young rats.[17] Since amyloid aggregates have been shown to be stabilized by transition metals such as iron and copper, it was speculated that LA could also inhibit the formation of these aggregates or maybe even be able to dissolve existing amyloid deposits. Fonte et al. successfully resolubilized Aβ aggregations with transition metal ion chelators and showed that LA enhanced the extraction of Aβ peptides from the frontal cortex of amyloid precursor protein-overexpressing transgenic mice, suggesting that, like other metal chelators, it could reduce amyloid plaque burden in AD patients.[18] In another published study, the effects of LA were tested on Tg2576 mice, a transgenic model of AD. Ten-month-old Tg2576 and wild-type mice were fed LA for 6 months, and the influence of this diet on memory and neuropathology was assessed. The authors could demonstrate that LA-treated Tg2576 mice exhibited significantly improved learning and memory retention in the Morris water maze task as compared with untreated Tg2576 mice. However, assessment of brain soluble and insoluble Aβ peptide and nitrotyrosine levels revealed no difference between LA-treated and untreated Tg2576 mice. These data suggest that chronic LA supplementation can reduce hippocampus-dependent memory deficits in Tg2576 mice without affecting Aβ peptide levels or plaque deposition.[19] A potential side effect of a long-term therapy with high doses of a metal chelator, such as LA, could be the inhibition of metal-containing enzymes, such as insulin-degrading enzyme and superoxide dismutase. Suh et al. investigated whether LA or DHLA could remove Cu or Fe from the active site of enzymes, such as Cu/Zn-superoxide dismutase and the Fe-containing enzyme aconitase. It was however found that even at millimolar concentrations, neither LA nor DHLA altered the activity of these enzymes.[20]

10.4.5   Lipoic Acid—An Anti-Inflammatory Antioxidant
and Modulator of Redox-Sensitive Signaling

AD is characterized by a chronic inflammatory process surrounding amyloid plaques, the activation of microglia and astroglia and increased levels of radicals and proinflammatory cytokines, such as inducible nitric oxide synthase (iNOS), IL-1β, IL-6, and TNF.[6] AD patients also show increased cytokine levels (e.g., IL-1β and TNF) in the cerebrospinal fluid, with TNF being a good predictor for the progression of mild cognitive impairment to AD. Recently, much attention has been paid to ROS as mediators in signaling processes, termed "redox-sensitive signal transduction." ROS modulate the activity of cytoplasmic signal-transducing enzymes by at least two mechanisms: oxidation of cysteine residues or reaction with iron-sulfur clusters. One widely investigated sensor protein is the small G-protein p21 (Ras).[21] Activation of Ras by oxidants is caused by modification of a specific cysteine (Cys118). Ras interacts with PI3 kinase, protein kinase C, diacylglycerol kinase, and MAP kinase kinase kinase, thereby regulating the expression of IL-1, IL-6, and iNOS. LA can scavenge intracellular free radicals (which are acting as second messengers) and down-regulate proinflammatory redox-sensitive signal transduction processes, including nuclear factor-κB translocation. As a result, LA attenuates the release of more free radicals and cytotoxic cytokines.[22,23] In addition, LA induces a scope of cellular antioxidants and phase II enzymes, including catalase, reduced GSH, GSH reductase, GSH S-transferase, and NAD(P)H:quinone oxidoreductase-1.[24] GSH content significantly declines during aging, with GCS being the rate-limiting enzyme in GSH synthesis.[25] With age, the expression of both the catalytic (GCLC) and the modulatory (GCLM) subunits of GCS and the overall enzyme activity decline by approximately 50%. Because Nrf2 governs basal and inducible GCLC and GCLM expression by means of the antioxidant response element, Suh et al. hypothesized that aging results in the dysregulation of Nrf2-mediated GCS expression. They observed an approximately 50% age-related loss in total and nuclear Nrf2 levels, which suggests attenuation in Nrf2-dependent gene transcription. However, when old rats were treated with RLA, nuclear Nrf2 increased and, consequently, higher GCLC levels and GCS activity were observed as early as 24 h after injection of RLA.[26]

10.4.6 LIPOIC ACID—A CARBONYL SCAVENGER

Cell and mitochondrial membranes contain a significant amount of arachidonic acid and linoleic acid, precursors of the lipid peroxidation products 4-hydroxynonenal and 2-propen-1-al (acrolein) that are extremely reactive. Acrolein decreases PDH and KGDH activities by covalently binding to LA, a cofactor of both the PDH and KGDH complexes, most likely explaining the loss of enzymatic activity in AD brain. This action of acrolein might be partially responsible for the dysfunction of mitochondria and energy loss found in AD brain, potentially contributing to the neurodegeneration in this disorder.[27] In a further study, levels of lipid peroxidation, oxidized GSH (GSSG), and nonenzymatic antioxidants and the activities of mitochondrial enzymes were measured in liver and kidney mitochondria of young and aged rats before and after LA supplementation. In both tissues, a decrease in mitochondrial enzymatic activity was observed in aged rats. LA-fed aged rats showed a decrease in lipid peroxidation levels and restored activities of mitochondrial enzymes, such as isocitrate dehydrogenase, KGDH, succinate dehydrogenase, NADH dehydrogenase, and cytochrome C oxidase. The authors concluded that LA reverses the age-associated decline in mitochondrial enzyme activity and therefore lowers oxidative damage during aging.[28]

10.4.7 LIPOIC ACID—A STIMULATOR OF GLUCOSE UPTAKE AND UTILIZATION ("INSULINOMIMETIC")

Increased prevalence of insulin abnormalities and insulin resistance in AD may contribute to the disease pathophysiology and clinical symptoms.[29] Insulin and insulin receptors are densely but selectively expressed in the brain, including the medial temporal regions that support memory formation. It has recently been demonstrated that insulin-sensitive glucose transporters are localized in the same regions and that insulin plays a role in memory functions. Collectively, these findings suggest that insulin contributes to normal cognitive function and that insulin abnormalities may exacerbate cognitive impairments, such as those associated with AD.[30] This view is further supported by the finding that higher fasting plasma insulin levels and reduced cerebrospinal fluid-to-plasma insulin ratios, suggestive of insulin resistance, have also been observed in patients with AD. When AD sufferers were treated with insulin in a glucose clamp approach (plasma level raised to 85 μU/mL), a marked enhancement in memory was observed, whereas normal adults' memory was unchanged.[31] As described previously, AD is also associated with a significantly impaired cerebral glucose metabolism in affected regions. Impaired glucose uptake (partially mediated by insulin resistance) in vulnerable neuronal populations compromises production of ACh but also renders neurons vulnerable to excitotoxicity and apoptosis. Evidence shows that LA can ameliorate insulin resistance and impaired glucose metabolism in the periphery of type II diabetes mellitus patients. One study examined the beneficial effects of LA on glucose uptake using soleus muscles derived from nonobese, insulin-resistant Goto-Kakizaki rats. In this model, chronic administration of LA partly ameliorated diabetes-related deficiency in glucose metabolism and decreased protein oxidation but also induced the insulin-stimulated activation of various steps of the insulin signaling pathway, including the enzymes Akt/PKB and PI3 kinase.[32] In a further study, the incorporation of ^{14}C-labeled 2-deoxyglucose (2DG) into areas of basal ganglia was investigated in rats treated acutely or for 5 days with RLA or SLA. Following acute administration, RLA was more effective than SLA in increasing 2DG incorporation. In the substantia nigra, for example, acute administration of RLA caused an approximately 40% increase in 2DG incorporation, while SLA showed no effect. Following subacute administration, the pattern of 2DG incorporation was altered, and both isomers were equally effective. The effects of RLA were largely maintained with increasing animal age, but the ability of the *S*-isomer to alter incorporation was lost by 30 months. The authors concluded that RLA has the ability to increase glucose utilization *in vivo*, which may be relevant to the treatment of neurodegenerative disorders.[33]

In addition, there is evidence that LA improves mitochondrial function *in vivo*. A combined dietary supplementation with acetyl-l-carnitine (ALCAR) and LA in 21-month-old rats increased the proliferation of intact mitochondria and reduced the density of mitochondria associated with vacuoles and lipofuscin in neurons, as compared with control animals. These results suggest that feeding a combination of ALCAR and LA may ameliorate age-associated mitochondrial ultrastructural decay, which might be the limiting factor for glucose metabolism.[34]

Based on this and similar studies, it is quite conceivable that LA might increase glucose uptake in insulin-resistant neurons and thus provide more glycolytic metabolites, including acetyl-CoA, for these neurons. Since ACh synthesis depends on the availability of acetyl-CoA (provided from glucose breakdown) and insulin, which controls ChAT activity, LA additionally might be able to increase the substrate acetyl-CoA concentration for ACh synthesis directly.[35]

10.4.8 PROTECTION OF CULTURED NEURONS AGAINST TOXICITY OF AMYLOID, IRON, AND OTHER NEUROTOXINS BY LIPOIC ACID

Neurotoxicity of Aβ peptides contributes to neuronal degeneration in AD by stimulating formation of free radicals. Zhang et al.[36] have investigated potential neuroprotective effects of LA against cytotoxicities induced by either Aβ (30 μM) or hydrogen peroxide (100 μM) in primary neurons of the rat cerebral cortex and found that treatment with LA protected cortical neurons against both cytotoxins. A similar effect was observed by Lovell et al.,[37] who investigated the effects of LA and DHLA on neurons (hippocampal cultures) treated with Aβ (25–35 μM) and/or iron/hydrogen peroxide (Fe/H_2O_2). In a further study, Muller and Krieglstein[38] tested whether pretreatment with LA can protect cultured neurons against injury caused by cyanide, glutamate, or iron ions. Neuroprotective effects were only significant when the pretreatment with LA exceeded 24 h. The authors concluded that neuroprotection occurs only after prolonged pretreatment with LA and is probably due to the radical scavenger properties of endogenously formed DHLA.[38]

In summary, data from these studies suggest that pretreatment of neurons with LA before exposure to Aβ or Fe/H_2O_2 (or direct application of DHLA at the time point of the insult) significantly reduces oxidative stress and increases cell survival.

10.4.9 PROTECTIVE EFFECTS OF LIPOIC ACID AGAINST AGE-RELATED COGNITIVE DEFICITS IN AGED RATS AND MICE

Protective effects of LA against cognitive deficits have been shown in several studies in aged rats and mice. In one study, a diet supplemented with RLA was fed to aged rats to determine its efficacy in reversing the decline in metabolism seen with age. Young (3–5 months old) and aged (24–26 months old) rats were fed for 2 weeks. Ambulatory activity, a measure of general metabolic activity, was almost three-fold lower in untreated old rats versus controls (untreated young rats) (untreated rats are usually controls!), but this decline was reversed in old rats fed RLA.[39] In a combination treatment study, the effects on cognitive function, brain mitochondrial structure, and biomarkers of oxidative damage were studied after feeding old rats a combination of ALCAR and RLA. Spatial memory was assessed using the Morris water maze task, and temporal memory was tested using the peak procedure (a time-discrimination procedure). Dietary supplementation with ALCAR and RLA improved memory, with the combination being the most effective in different tests of spatial memory and temporal memory. The authors suggested that a combined diet of ALCAR and RLA to old rats improves performance on memory tasks by lowering oxidative damage and improving mitochondrial function. Feeding a combination of ALCAR and RLA restores the velocity of the reaction for ALCAR transferase and mitochondrial function. The principle appears to be that, with age, increased protein oxidation causes a deformation of the structure of key enzymes with a consequent lessening of affinity (K_m) for the enzyme substrate.[40]

Similar experiments were performed in the senescence-accelerated-prone mouse strain 8 (SAMP8), which exhibits age-related deterioration in memory and learning as well as increased oxidative markers and therefore provides a good model for disorders with age-related cognitive impairment. In one study, the ability of LA to reverse the cognitive deficits found in the SAMP8 mice was investigated. Chronic administration of LA improved cognition of 12-month-old SAMP8 mice in the T-maze footshock avoidance paradigm and the lever-press appetitive task. It furthermore reversed several markers of oxidative stress. These results provide further support for a therapeutic role for LA in age- and oxidative stress-mediated cognitive impairment, including that of AD.[41]

10.4.10 Clinical Trials with Lipoic Acid in Patients with Alzheimer's Disease

Although LA has been used for the treatment of diabetic polyneuropathy in Germany for more than 40 years, no epidemiological study has taken advantage of this large patient population and investigated the incidence of AD in the patient group. Therefore, the first indication for a beneficial effect of LA in AD and related dementias came from the study of an untypical AD case. In 1997, a 74-year-old patient presented herself at the Henriettenstiftung Hannover Department of Medical Rehabilitation and Geriatrics with signs of cognitive impairment. Diabetes mellitus and a mild form of polyneuropathy were her main concomitant diseases. With clinical criteria of *Diagnostic and Statistical Manual of Mental Disorders, Third Edition, Revised*, deficits in the neuropsychological tests, a magnetic resonance image without signs of ischemia, and a typical SPECT showing a decreased bitemporal and biparietal perfusion, early-stage AD was diagnosed. A treatment with AChIs was initiated in addition to the 600 mg of LA taken daily by the patient for the treatment of diabetic polyneuropathy. Since 1997, several retests have been performed, which showed no substantial decline of the cognitive functions. Therefore, the diagnosis of mild AD was reevaluated several times, but the diagnostic features did not change and the neuropsychological tests showed an unusually slow progress of cognitive impairment. This observation inspired an open pilot trial at the Henriettenstiftung Hospital in Hannover in 1999. A total of 600 mg of LA was given once daily (in the morning, 30 min before breakfast) to nine patients with probable AD [age = 67 ± 9 years; Mini Mental State Examination (MMSE) score at first visit/start of AChI therapy = 23 ± 2 points) receiving a standard treatment with AChIs over an observation period of 337 ± 80 days. The cognitive performance of the patients before and that after addition of LA to their standard medication was compared. A steady decrease in cognitive performance [a decrease of two points per year in scores in the MMSE and an increase of four points per year increase in the AD Assessment Scale-Cognitive Subscale (ADAS-cog)] was observed before initiation of the LA regimen. Treatment with LA led to a stabilization of cognitive functions, demonstrated by constant scores in two neuropsychological tests for nearly 1 year.[42] This study was continued, finally

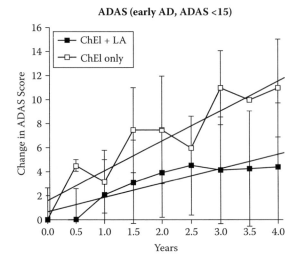

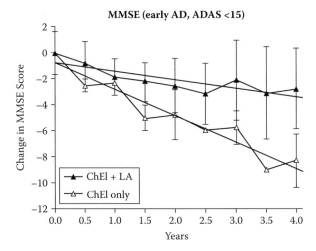

FIGURE 10.1 Time-dependent changes in (a) ADAS and (b) MMSE scores in early-stage AD patients. Early-stage AD patients (ADAS, <15) were divided into a control group (receiving cholinesterase inhibitors only) and a treatment group (receiving LA as add-on therapy) and treated over an observation period of up to 4 years. Scores in the two cognitive tests ADAS (a) and MMSE (b) were used to assess disease progression. Data are presented as mean ± SEM.

including 43 patients who were followed for an observation period of up to 48 months.[43] In patients with mild dementia (ADAS-cog, <15), the disease progressed extremely slowly (ADAS-cog, +1.2 points/year; MMSE, −0.6 points/year) (Figure 10.1a and b). However, this study was small and not randomized. In addition, patients were diagnosed with "probable" AD, and the diagnoses were not confirmed by neuropathological postmortem analysis. Therefore, a double-blind, placebo-controlled phase II trial is urgently needed before LA can be confidently recommended as a therapy for AD and related dementias.

10.4.11 PHARMACOKINETICS OF LIPOIC ACID

The correlation of pharmacokinetic (PK) parameters with therapeutic efficacy provides important basic data for the rational design of preclinical and clinical studies. The PKs of a compound may constitute a limitation to its clinical use, if pharmacological concentrations cannot be achieved or maintained long enough to achieve a therapeutic response. To date, it has not been possible to positively correlate the PKs and pharmacodynamics of LA,[44] indicating that the therapeutic effects may be more dependent on the maximum plasma concentration (C_{max}) and the area under the curve (AUC) than on the time to maximum concentration (T_{max}), the elimination half-life ($T_{1/2}$), or the mean residence time in plasma. This contention is supported by a recent PK study with multiple sclerosis patients in which the therapeutic response (reductions in matrix metallopeptidase 9 and the soluble intercellular adhesion molecule) was positively correlated with C_{max}.[45] Several PK studies utilizing 600 mg of rac-LA either orally or intravenously (the most frequent dose and form of LA used clinically to date and the dose used in the AD study; Section 10.4.9) have been reported. The mean values from eight human PK studies utilizing 600 mg of rac-LA indicate that the effective dose (ED_{50}) of LA is equal to the C_{max} of 4–5 µg/mL (~20–25 µM) and that the AUC is equal to 2.85 µg h/mL.[44,46] Recently, the PK profile of RLA, administered as an aqueous solution of Na-RLA, was reported in a study with 12 healthy subjects.[47] The average dose was 8.25 mg/kg, generating a $C_{max,mean}$ of 16.03 µg/mL (range = 10.6–33.8 µg/mL), a $T_{max,median}$ of 15 min (range = 10–20 min), and an AUC_{mean} of 441.59 µg min/mL (7.36 µg h/mL).

In a comparison of the $C_{max,mean}$ and the AUC_{mean} of eight published PK trials using 600 mg of rac-LA and the Na-RLA study, it was found that the average plasma C_{max} levels are four times and the bioavailability (AUC) is three times greater for RLA than for rac-LA, indicating that Na-RLA would be the preferred formulation for chronic administration. Therefore, it should be possible to achieve comparable plasma concentrations with 150 mg of RLA (as Na-RLA oral solution) as with 600 mg of rac-LA, which is generally well tolerated during chronic treatment (Carlson, unpublished data). It has been suggested that the naturally occurring enantiomer RLA has its therapeutic window at higher plasma concentrations of 10–20 µg/mL (~50–100 µM).[44] Based on the tolerability of LA, the upper limit of the human therapeutic window is at a concentration of ~50 µg/mL (~250 µM).[48] The use of 600 mg of rac-LA for the treatment of diabetic complications achieved a rational basis by demonstration of a reduction in the dose necessary to produce an improvement in the glucose challenge and insulin clamp tests in diabetic patients.[49] Additionally, several clinical trials have indicated that this dose is effective for the treatment of diabetic neuropathy and subsequently causes fewer side effects than treatment with 1200 or 1800 mg of LA.[50] Intravenous doses of 600–1000 mg of rac-LA for 10 days were reduced to 500 mg and achieved the same result (improved glucose metabolic clearance rate and improved insulin sensitivity index) as with oral dosages of 600–1200 mg of rac-LA daily for 4 weeks. The oral treatment led to similar reductions in the effective doses.[44] A preliminary evaluation was made on the bioavailability of R-DHLA in human plasma. R-DHLA, when administered as such, produced insignificant increases in RLA and R-DHLA levels and is therefore not bioequivalent to RLA. This indicates that for R-DHLA to have clinical efficacy, it must be either administered intravenously or obtained by endogenous reduction of administered RLA. Any therapeutic potential of R-DHLA must consider its short biological half-life since it is rapidly metabolized.[31]

The three forms of LA (RLA, SLA, and rac-LA) produce different PK values when administered as sodium salts. Recently, Na-RLA was compared with Na-rac-LA and Na-SLA in humans using a simple three-period crossover design. The mean values for C_{max} and AUC showed that RLA values were considerably greater than those for either rac-LA or SLA [RLA C_{max} = 15.67 µg/mL (78 µM), RLA AUC = 6.86 µg h/mL; SLA C_{max} = 6.43 µg/mL, SLA AUC = 3.89 µg h/mL; rac-LA C_{max} = 6.37 µg/mL; rac-LA AUC = 2.69 µg h/mL]. The values for T_{max} reveal that SLA > RLA > rac-LA (18.33, 13.33, and 10.00 min), and those for $T_{1/2}$ reveal that SLA > rac-LA > RLA (28.87, 20.7, and 12.5 min). The diminished peak plasma concentration of rac-LA dosages compared with RLA and SLA suggests not only a stereoselective transporter favoring RLA but also that the presence of SLA in the racemate may limit the overall bioavailability of RLA (Carlson, unpublished data).

10.4.12   Combination Treatment of Lipoic Acid with Nutraceuticals

Since AD is a multifactorial disease, it has been suggested that a combination rather than a single drug treatment might be most beneficial for AD patients. Among many suggested add-on treatments to LA, nutraceuticals with antioxidant and anti-inflammatory properties might be quite promising candidates.[52] Nutraceuticals may be broadly defined as any food substance that is considered to offer health or medical benefits.[53] Given that plant foods are derived from biological systems, they contain many compounds, in addition to traditional nutrients, that can elicit biologic responses. Therefore, they are also known as phytonutrients. Polyphenols are one of the largest groups of phytonutrients that may confer beneficial health effects. Over the past decade, polyphenols, which are abundant in fruits and vegetables, have gained recognition for their antioxidant properties and their roles in protection against chronic diseases such as cancer and cardiovascular diseases.[54] Consequently, diet is now considered to be an important environmental factor in the development of late-onset AD.[55] Polyphenols are therefore beginning to attract increasing interest, with numerous epidemiological studies suggesting a positive association between consumption of polyphenol-rich foods and the prevention of diseases. A recent epidemiological study reported that consumption of fruit and vegetable juices high in polyphenols more than three times a week resulted in a 76% reduction in the risk of developing probable AD over a 9-year period.[56] Another epidemiological study of 1010 subjects with an age range of 60 to 93 years reported that individuals who consumed curry (containing curcumin) "often" and "very often" had significantly better cognitive test scores as measured via MMSE.[57] As antioxidants, polyphenols may protect cells against oxidative damage, thereby limiting the risk for AD, which is associated with oxidative stress. Converging epidemiological data also suggest that a low dietary intake of omega-3 essential fatty acids is a candidate risk factor for AD.[58] Docosahexaenoic acid (DHA) is one of the major omega-3 fatty acids in the brain, where it is enriched in neurons and synapses. DHA is associated with learning memory and is also required for the structure and function of brain cell membranes. In the AD brain, DHA is known to be decreased,[59] while people who ingest higher levels of DHA are less likely to develop AD.[60] As many Western diets have been reported to be deficient in DHA and low in polyphenolic content, supplementation with DHA and polyphenols may offer potential preventative treatments for AD. Several of these nutraceuticals/phytonutrients [e.g., (–)-epigallocatechin gallate (EGCG) from green tea, curcumin from the curry spice turmeric, and omega-3 DHA from fish oils] have shown promising results when used as single therapies in animal studies.

EGCG has previously been shown to prevent neuronal cell death caused by Aβ in cell cultures.[61] A study by Rezai-Zadeh et al.[62] reported that EGCG not only reduced Aβ generation *in vitro* in a neuronal cell line and primary neuronal cultures from Tg2576 mice but also promoted the nonamyloidogenic α-secretase proteolytic pathway. Twelve-month-old Tg2576 mice were intraperitoneally injected with 20 mg/kg of EGCG for 60 days to validate these findings. The mice showed decreased Aβ levels and plaque load in the brain, as well as an increase in the α-secretase pathway.[62] Recent studies have furthermore shown that EGCG promotes APP processing by the nonamyloidogenic α-secretase pathway via protein kinase C-dependent activation.[61,63] Curcumin has been reported to be a several times more potent free-radical scavenger than vitamin E,[64] and there is also increasing evidence showing that curcumin can inhibit Aβ aggregation.[65] In a study by Lim et al.,[66] curcumin was tested for its ability to inhibit the combined inflammatory and oxidative damage in Tg2576 transgenic mice. In this study, 10-month-old Tg2576 mice were fed a curcumin diet (160 ppm) for 6 months. The results showed that the curcumin diet significantly lowered the levels of not only oxidized proteins, such as IL-1β, the astrocyte marker, glial fibrillary acidic protein, soluble Aβ, and insoluble Aβ, but also the plaque burden.[66] Following on from this work, Yang et al.[65] evaluated the effect of a curcumin diet (500 ppm) in 17-month-old Tg2576 mice for 6 months. When fed to the aged Tg2576 mice with advanced amyloid accumulation, curcumin resulted in reduced soluble amyloid levels and plaque burden. These data raise the possibility that dietary supplementation with curcumin may provide a potential preventative treatment for AD not only by decreasing Aβ levels and plaque load via inhibition of Aβ oligomer formation and fibrillation but also by decreasing oxidative stress and inflammation.

A recent study by Florent et al.[67] demonstrated that DHA provides cortical neurons *in vitro* with a higher resistance level against the cytotoxic effects induced by soluble Aβ oligomers. Lukiw et al.[68] also demonstrated that DHA decreased $Aβ_{40}$ and $Aβ_{42}$ secretion from aged human neuronal cells. A study by Calon et al.[69] showed that a reduction of dietary omega-3 polyunsaturated fatty acids in Tg2576 transgenic mice resulted in loss of postsynaptic proteins and behavioral deficits, while a DHA-enriched diet prevented these effects. Other studies have shown that DHA protects neurons from Aβ accumulation and toxicity and ameliorates cognitive impairment in rodent models of AD.[70] Thus, dietary supplementation with DHA may also provide a potential preventative treatment for AD via prevention of cognitive deficits and decreased Aβ accumulation and oxidative stress.

The discussed nutraceuticals have all demonstrated to possess varying mechanisms of action relating to decreasing cognitive deficits, oxidative stress, inflammation, and Aβ levels. Therefore, a combination of polyphenols (e.g., EGCG and curcumin), omega-3 essential fatty acids (e.g., DHA), and LA might have the potential to provide nutritional supplement therapies for the prevention of AD.

10.4.13 SUMMARY

AD is a progressive neurodegenerative disorder that destroys patient memory and cognition, communication ability within the social environment, and the ability to carry out daily activities. Despite extensive research into the pathogenesis of AD, a neuroprotective treatment—particularly for the early stages of the disease—remains unavailable for clinical use. In this review, we advance the suggestion that LA may fulfill this therapeutic need. A naturally occurring cofactor for the mitochondrial enzymes PDH and α-ketoglutarate dehydrogenase, LA has been shown to possess a variety of properties that can interfere with the pathogenesis or progression of AD. For example, LA increases ACh production by activation of ChAT and increases glucose uptake, thus supplying more acetyl-CoA for the production of ACh. LA chelates redox-active transition metals, thus inhibiting the formation of hydroxyl radicals, but also scavenges ROS, leading to an increase in reduced GSH levels. In addition, LA down-regulates the expression of redox-sensitive proinflammatory proteins such as TNF and iNOS. Furthermore, LA can scavenge lipid peroxidation products such as 4-hydroxynonenal and acrolein. In human plasma, LA exists in an equilibrium of free and plasma protein-bound forms. At levels of up to 150 μM, LA is bound most likely to high-affinity fatty acid sites on human serum albumin, suggesting that one large dose rather than continuous low doses (as provided by "slow-release" LA) will be beneficial for delivery of LA to the brain. Evidence for a clinical benefit for LA in dementia is as yet limited. There are only two published studies in which 600 mg of LA was given daily to 43 patients with AD (receiving a standard treatment with cholinesterase inhibitors) in an open-label study over an observation period of up to 48 months. Whereas the improvement in patients with moderate dementia was not significant, the disease progressed extremely slowly (change in ADAS-cog, 1.2 points/year; MMSE, −0.6 points/year) in patients with mild dementia (ADAS-cog, <15). Data from cell culture and animal models suggest that LA could be combined with nutraceuticals such as curcumin, EGCG (from green tea), and DHA (from fish oil) to synergistically decrease oxidative stress, inflammation, Aβ levels, and Aβ plaque load and thus provide a combined benefit in the treatment of AD.

ACKNOWLEDGMENTS

This work was supported by the J.O. and J.R Wicking Foundation, Alzheimer's Australia, and the National Health and Medical Research Council.

REFERENCES

1. Stuchbury, G., and Münch, G., Alzheimer's associated inflammation, potential drug targets and future therapies, *J Neural Transm* 112(3), 429–53, 2005.
2. Cummings, J. L., Treatment of Alzheimer's disease: Current and future therapeutic approaches, *Rev Neurol Dis* 1(2), 60–9, 2004.
3. Retz, W., Gsell, W., Münch, G., Rosler, M., and Riederer, P., Free radicals in Alzheimer's disease, *J Neural Transm Suppl* 54, 221–36, 1998.
4. Münch, G., Schinzel, R., Loske, C., Wong, A., Durany, N., Li, J. J., Vlassara, H., Smith, M. A., Perry, G., and Riederer, P., Alzheimer's disease—Synergistic effects of glucose deficit, oxidative stress and advanced glycation endproducts, *J Neural Transm* 105(4–5), 439–61, 1998.
5. Wong, A., Luth, H. J., Deuther-Conrad, W., Dukic-Stefanovic, S., Gasic-Milenkovic, J., Arendt, T., and Münch, G., Advanced glycation endproducts co-localize with inducible nitric oxide synthase in Alzheimer's disease, *Brain Res* 920(1–2), 32–40, 2001.
6. Griffin, W. S., Sheng, J. G., Roberts, G. W., and Mrak, R. E., Interleukin-1 expression in different plaque types in Alzheimer's disease: Significance in plaque evolution, *J Neuropathol Exp Neurol* 54(2), 276–81, 1995.
7. Ishii, K., and Minoshima, S., PET is better than perfusion SPECT for early diagnosis of Alzheimer's disease, *Eur J Nucl Med Mol Imaging* 32(12), 1463–5, 2005.
8. Nordberg, A., Nyberg, P., Adolfsson, R., and Winblad, B., Cholinergic topography in Alzheimer brains: A comparison with changes in the monoaminergic profile, *J Neural Transm* 69(1–2), 19–32, 1987.
9. Nordberg, A., and Winblad, B., Reduced number of [^{3}H]nicotine and [^{3}H]acetylcholine binding sites in the frontal cortex of Alzheimer brains, *Neurosci Lett* 72(1), 115–9, 1986.
10. Holmquist, L., Stuchbury, G., Berbaum, K., Muscat, S., Young, S., Hager, K., Engel, J., and Münch, G., Lipoic acid as a novel treatment for Alzheimer's disease and related dementias, *Pharmacol Ther* 113(1), 154–64, 2007.
11. Packer, L., Witt, E. H., and Tritschler, H. J., alpha-Lipoic acid as a biological antioxidant, *Free Radic Biol Med* 19(2), 227–50, 1995.
12. Nguyen, T., Yang, C. S., and Pickett, C. B., The pathways and molecular mechanisms regulating Nrf2 activation in response to chemical stress, *Free Radic Biol Med* 37(4), 433–41, 2004.
13. Chen, X. L., and Kunsch, C., Induction of cytoprotective genes through Nrf2/antioxidant response element pathway: A new therapeutic approach for the treatment of inflammatory diseases, *Curr Pharm Des* 10(8), 879–91, 2004.
14. Haugaard, N., and Levin, R. M., Regulation of the activity of choline acetyl transferase by lipoic acid, *Mol Cell Biochem* 213(1–2), 61–3, 2000.
15. Haugaard, N., and Levin, R. M., Activation of choline acetyl transferase by dihydrolipoic acid, *Mol Cell Biochem* 229(1–2), 103–6, 2002.
16. Huang, X., Atwood, C. S., Hartshorn, M. A., Multhaup, G., Goldstein, L. E., Scarpa, R. C., Cuajungco, M. P., Gray, D. N., Lim, J., Moir, R. D., Tanzi, R. E., and Bush, A. I., The A beta peptide of Alzheimer's disease directly produces hydrogen peroxide through metal ion reduction, *Biochemistry* 38(24), 7609–16, 1999.
17. Suh, J. H., Moreau, R., Heath, S. H., and Hagen, T. M., Dietary supplementation with (*R*)-alpha-lipoic acid reverses the age-related accumulation of iron and depletion of antioxidants in the rat cerebral cortex, *Redox Rep* 10(1), 52–60, 2005.
18. Fonte, J., Miklossy, J., Atwood, C., and Martins, R., The severity of cortical Alzheimer's type changes is positively correlated with increased amyloid-beta levels: Resolubilization of amyloid-beta with transition metal ion chelators, *J Alzheimers Dis* 3(2), 209–19, 2001.
19. Quinn, J. F., Bussiere, J. R., Hammond, R. S., Montine, T. J., Henson, E., Jones, R. E., and Stackman, R. W., Jr., Chronic dietary alpha-lipoic acid reduces deficits in hippocampal memory of aged Tg2576 mice, *Neurobiol Aging* 28(2), 213–25, 2007.
20. Suh, J. H., Zhu, B. Z., deSzoeke, E., Frei, B., and Hagen, T. M., Dihydrolipoic acid lowers the redox activity of transition metal ions but does not remove them from the active site of enzymes, *Redox Rep* 9(1), 57–61, 2004.
21. Lander, H. M., Tauras, J. M., Ogiste, J. S., Hori, O., Moss, R. A., and Schmidt, A. M., Activation of the receptor for advanced glycation end products triggers a p21(ras)-dependent mitogen-activated protein kinase pathway regulated by oxidant stress, *J Biol Chem* 272(28), 17810–4, 1997.
22. Wong, A., Dukic-Stefanovic, S., Gasic-Milenkovic, J., Schinzel, R., Wiesinger, H., Riederer, P., and Münch , G., Anti-inflammatory antioxidants attenuate the expression of inducible nitric oxide synthase mediated by advanced glycation endproducts in murine microglia, *Eur J Neurosci* 14(12), 1961–7, 2001.

23. Bierhaus, A., Chevion, S., Chevion, M., Hofmann, M., Quehenberger, P., Illmer, T., Luther, T., Berentshtein, E., Tritschler, H., Muller, M., Wahl, P., Ziegler, R., and Nawroth, P. P., Advanced glycation end product-induced activation of NF-kappaB is suppressed by alpha-lipoic acid in cultured endothelial cells, *Diabetes* 46(9), 1481–90, 1997.

24. Cao, Z., Tsang, M., Zhao, H., and Li, Y., Induction of endogenous antioxidants and phase 2 enzymes by alpha-lipoic acid in rat cardiac H9C2 cells: Protection against oxidative injury, *Biochem Biophys Res Commun* 310(3), 979–85, 2003.

25. Suh, J. H., Shenvi, S. V., Dixon, B. M., Liu, H., Jaiswal, A. K., Liu, R. M., and Hagen, T. M., Decline in transcriptional activity of Nrf2 causes age-related loss of glutathione synthesis, which is reversible with lipoic acid, *Proc Natl Acad Sci U S A* 101(10), 3381–6, 2004.

26. Suh, J. H., Wang, H., Liu, R. M., Liu, J., and Hagen, T. M., (*R*)-alpha-lipoic acid reverses the age-related loss in GSH redox status in post-mitotic tissues: Evidence for increased cysteine requirement for GSH synthesis, *Arch Biochem Biophys* 423(1), 126–35, 2004.

27. Pocernich, C. B., and Butterfield, D. A., Acrolein inhibits NADH-linked mitochondrial enzyme activity: Implications for Alzheimer's disease, *Neurotox Res* 5(7), 515–20, 2003.

28. Arivazhagan, P., Ramanathan, K., and Panneerselvam, C., Effect of dl-alpha-lipoic acid on glutathione metabolic enzymes in aged rats, *Exp Gerontol* 37(1), 81–7, 2001.

29. Hoyer, S., Is sporadic Alzheimer disease the brain type of non-insulin dependent diabetes mellitus? A challenging hypothesis, *J Neural Transm* 105(4–5), 415–22, 1998.

30. Watson, G. S., and Craft, S., The role of insulin resistance in the pathogenesis of Alzheimer's disease: Implications for treatment, *CNS Drugs* 17(1), 27–45, 2003.

31. Craft, S., Asthana, S., Cook, D. G., Baker, L. D., Cherrier, M., Purganan, K., Wait, C., Petrova, A., Latendresse, S., Watson, G. S., Newcomer, J. W., Schellenberg, G. D., and Krohn, A. J., Insulin dose-response effects on memory and plasma amyloid precursor protein in Alzheimer's disease: Interactions with apolipoprotein E genotype, *Psychoneuroendocrinology* 28(6), 809–22, 2003.

32. Bitar, M. S., Wahid, S., Pilcher, C. W., Al-Saleh, E., and Al-Mulla, F., Alpha-lipoic acid mitigates insulin resistance in Goto-Kakizaki rats, *Horm Metab Res* 36(8), 542–9, 2004.

33. Seaton, T. A., Jenner, P., and Marsden, C. D., The isomers of thioctic acid alter C-deoxyglucose incorporation in rat basal ganglia, *Biochem Pharmacol* 51(7), 983–6, 1996.

34. Aliev, G., Liu, J., Shenk, J. C., Fischbach, K., Pacheco, G. J., Chen, S. G., Obrenovich, M. E., Ward, W. F., Richardson, A. G., Smith, M. A., Gasimov, E., Perry, G., and Ames, B. N., Neuronal mitochondrial amelioration by feeding acetyl-l-carnitine and lipoic acid to aged rats, *J Cell Mol Med*, 320–33, 2008.

35. Hoyer, S., Memory function and brain glucose metabolism, *Pharmacopsychiatry* 36 Suppl 1, S62–7, 2003.

36. Zhang, L., Xing, G. Q., Barker, J. L., Chang, Y., Maric, D., Ma, W., Li, B. S., and Rubinow, D. R., Alpha-lipoic acid protects rat cortical neurons against cell death induced by amyloid and hydrogen peroxide through the Akt signalling pathway, *Neurosci Lett* 312(3), 125–8, 2001.

37. Lovell, M. A., Xie, C., Xiong, S., and Markesbery, W. R., Protection against amyloid beta peptide and iron/hydrogen peroxide toxicity by alpha lipoic acid, *J Alzheimers Dis* 5(3), 229–39, 2003.

38. Muller, U., and Krieglstein, J., Prolonged pretreatment with alpha-lipoic acid protects cultured neurons against hypoxic, glutamate-, or iron-induced injury, *J Cereb Blood Flow Metab* 15(4), 624–30, 1995.

39. Hagen, T. M., Ingersoll, R. T., Lykkesfeldt, J., Liu, J., Wehr, C. M., Vinarsky, V., Bartholomew, J. C., and Ames, A. B., (*R*)-alpha-lipoic acid-supplemented old rats have improved mitochondrial function, decreased oxidative damage, and increased metabolic rate, *FASEB J* 13(2), 411–8, 1999.

40. Liu, J., Killilea, D. W., and Ames, B. N., Age-associated mitochondrial oxidative decay: Improvement of carnitine acetyltransferase substrate-binding affinity and activity in brain by feeding old rats acetyl-l-carnitine and/or *R*-alpha-lipoic acid, *Proc Natl Acad Sci U S A* 99(4), 1876–81, 2002.

41. Farr, S. A., Poon, H. F., Dogrukol-Ak, D., Drake, J., Banks, W. A., Eyerman, E., Butterfield, D. A., and Morley, J. E., The antioxidants alpha-lipoic acid and *N*-acetylcysteine reverse memory impairment and brain oxidative stress in aged SAMP8 mice, *J Neurochem* 84(5), 1173–83, 2003.

42. Hager, K., Marahrens, A., Kenklies, M., Riederer, P., and Münch, G., Alpha-lipoic acid as a new treatment option for Alzheimer type dementia, *Arch Gerontol Geriatr* 32(3), 275–282, 2001.

43. Hager, K., Kenklies, M., McAfoose, J., Engel, J., and Münch, G., Alpha-lipoic acid as a new treatment option for Alzheimer's disease—A 48 months follow-up analysis, *J Neural Transm Suppl* (72), 189–93, 2007.

44. Krone, D., The pharmacokinetics and pharmacodynamics of *R*-(+)-alpha lipoic acid, Ph.D. thesis, Johann Wolfgang Goethe University, Frankfurt, Germany, 2002.

45. Yadav, V., Marracci, G., Lovera, J., Woodward, W., Bogardus, K., Marquardt, W., Shinto, L., Morris, C., and Bourdette, D., Lipoic acid in multiple sclerosis: A pilot study, *Mult Scler* 11, 159–65, 2005.

46. Chen, J., Jiang, W., Cai, J., Tao, W., Gao, X., and Jiang, X., Quantification of lipoic acid in plasma by high-performance liquid chromatography-electrospray ionization mass spectrometry, *J Chromatogr B Analyt Technol Biomed Life Sci* 824(1–2), 249–57, 2005.

47. Carlson, D. A., Smith, A. R., Fischer, S. J., Young, K. L., and Packer, L., The plasma pharmacokinetics of *R*-(+)-lipoic acid administered as sodium *R*-(+)-lipoate to healthy human subjects, *Altern Med Rev* 12(4), 343–51, 2007.

48. Hermann, R., Niebch, G., and Borbe, H., Enantioselective pharmacokinetics and bioavailability of different racemic α-lipoic acid formulations in healthy volunteers, *Eur J Pharm Sci* 4, 167–174, 1996.

49. DeFronzo, R. A., Tobin, J. D., and Andres, R., Glucose clamp technique: A method for quantifying insulin secretion and resistance, *Am J Physiol* 237(3), E214–23, 1979.

50. Ziegler, D., Thioctic acid for patients with symptomatic diabetic polyneuropathy: A critical review, *Treat Endocrinol* 3(3), 173–89, 2004.

51. Carlson, D., Young, K., Fischer, S., and Ulrich, H., *An Evaluation of the Stability and Plasma Pharmacokinetics of R-Lipoic Acid (RLA) and R-Dihydrolipoic Acid (R-DHLA) Dosage Forms in Human Plasma from Healthy Volunteers*, Taylor & Francis, Boca Raton, FL, 2008.

52. Steele, M., Stuchbury, G., and Münch, G., The molecular basis of the prevention of Alzheimer's disease through healthy nutrition, *Exp Gerontol* 42(1–2), 28–36, 2007.

53. Ferrari, C. K., Functional foods, herbs and nutraceuticals: Towards biochemical mechanisms of healthy aging, *Biogerontology* 5(5), 275–89, 2004.

54. Liu, R. H., Potential synergy of phytochemicals in cancer prevention: Mechanism of action, *J Nutr* 134(12), 3479S–85S, 2004.

55. Solfrizzi, V., Panza, F., and Capurso, A., The role of diet in cognitive decline, *J Neural Transm* 110(1), 95–110, 2003.

56. Dai, Q., Borenstein, A. R., Wu, Y., Jackson, J. C., and Larson, E. B., Fruit and vegetable juices and Alzheimer's disease: The Kame Project, *Am J Med* 119(9), 751–9, 2006.

57. Ng, T. P., Chiam, P. C., Lee, T., Chua, H. C., Lim, L., and Kua, E. H., Curry consumption and cognitive function in the elderly, *Am J Epidemiol* 164(9), 898–906, 2006.

58. Maclean, C. H., Issa, A. M., Newberry, S. J., Mojica, W. A., Morton, S. C., Garland, R. H., Hilton, L. G., Traina, S. B., and Shekelle, P. G., Effects of omega-3 fatty acids on cognitive function with aging, dementia, and neurological diseases, *Evid Rep Technol Assess (Summ)* (114), 1–3, 2005.

59. Prasad, M. R., Lovell, M. A., Yatin, M., Dhillon, H., and Markesbery, W. R., Regional membrane phospholipid alterations in Alzheimer's disease, *Neurochem Res* 23(1), 81–8, 1998.

60. Tully, A. M., Roche, H. M., Doyle, R., Fallon, C., Bruce, I., Lawlor, B., Coakley, D., and Gibney, M. J., Low serum cholesteryl ester-docosahexaenoic acid levels in Alzheimer's disease: A case–control study, *Br J Nutr* 89(4), 483–9, 2003.

61. Levites, Y., Amit, T., Mandel, S., and Youdim, M. B., Neuroprotection and neurorescue against Abeta toxicity and PKC-dependent release of nonamyloidogenic soluble precursor protein by green tea polyphenol (−)-epigallocatechin-3-gallate, *FASEB J* 17(8), 952–4, 2003.

62. Rezai-Zadeh, K., Shytle, D., Sun, N., Mori, T., Hou, H., Jeanniton, D., Ehrhart, J., Townsend, K., Zeng, J., Morgan, D., Hardy, J., Town, T., and Tan, J., Green tea epigallocatechin-3-gallate (EGCG) modulates amyloid precursor protein cleavage and reduces cerebral amyloidosis in Alzheimer transgenic mice, *J Neurosci* 25(38), 8807–14, 2005.

63. Mandel, S., Weinreb, O., Amit, T., and Youdim, M. B., Cell signaling pathways in the neuroprotective actions of the green tea polyphenol (−)-epigallocatechin-3-gallate: Implications for neurodegenerative diseases, *J Neurochem* 88(6), 1555–69, 2004.

64. Zhao, B. L., Li, X. J., He, R. G., Cheng, S. J., and Xin, W. J., Scavenging effect of extracts of green tea and natural antioxidants on active oxygen radicals, *Cell Biophys* 14(2), 175–85, 1989.

65. Yang, F., Lim, G. P., Begum, A. N., Ubeda, O. J., Simmons, M. R., Ambegaokar, S. S., Chen, P. P., Kayed, R., Glabe, C. G., Frautschy, S. A., and Cole, G. M., Curcumin inhibits formation of amyloid beta oligomers and fibrils, binds plaques, and reduces amyloid in vivo, *J Biol Chem* 280(7), 5892–901, 2005.

66. Lim, G. P., Chu, T., Yang, F., Beech, W., Frautschy, S. A., and Cole, G. M., The curry spice curcumin reduces oxidative damage and amyloid pathology in an Alzheimer transgenic mouse, *J Neurosci* 21(21), 8370–7, 2001.

67. Florent, S., Malaplate-Armand, C., Youssef, I., Kriem, B., Koziel, V., Escanye, M. C., Fifre, A., Sponne, I., Leininger-Muller, B., Olivier, J. L., Pillot, T., and Oster, T., Docosahexaenoic acid prevents neuronal apoptosis induced by soluble amyloid-beta oligomers, *J Neurochem* 96(2), 385–95, 2006.

68. Lukiw, W. J., Cui, J. G., Marcheselli, V. L., Bodker, M., Botkjaer, A., Gotlinger, K., Serhan, C. N., and Bazan, N. G., A role for docosahexaenoic acid-derived neuroprotectin D_1 in neural cell survival and Alzheimer disease, *J Clin Invest* 115(10), 2774–83, 2005.

69. Calon, F., Lim, G. P., Yang, F., Morihara, T., Teter, B., Ubeda, O., Rostaing, P., Triller, A., Salem, N., Jr., Ashe, K. H., Frautschy, S. A., and Cole, G. M., Docosahexaenoic acid protects from dendritic pathology in an Alzheimer's disease mouse model, *Neuron* 43(5), 633–45, 2004.

70. Lim, G. P., Calon, F., Morihara, T., Yang, F., Teter, B., Ubeda, O., Salem, N., Jr., Frautschy, S. A., and Cole, G. M., A diet enriched with the omega-3 fatty acid docosahexaenoic acid reduces amyloid burden in an aged Alzheimer mouse model, *J Neurosci* 25(12), 3032–40, 2005.

11 Zinc and the Cytoskeleton in Neuronal Signaling

Gerardo G. Mackenzie[1,2] and Patricia I. Oteiza[1,2]

Departments of [1]Environmental Toxicology and [2]Nutrition, University of California–Davis, Davis, California, USA

CONTENTS

11.1 ZINC AND THE NERVOUS SYSTEM

Zinc is essential for the normal development and function of the nervous system. In humans, and depending on the developmental stage, zinc deficiency affects infant behavior, cognition, and motor performance [1–6]. In this regard, low plasma zinc values were found to be associated with altered cognition and emotional functioning in three rural communities of low-income 3- to 5-year-old children in the United States [7] and with pregnancy complications and altered child psychomotor development in a population from central Russia [8]. Low plasma zinc concentrations are also correlated with the occurrence of inattentive symptoms in children with attention/hyperactivity disorder [9]. Stressing the relevance of an adequate zinc nutrition for normal brain physiology, several studies have shown that zinc supplementation of undernourished children improves developmental quotients, activity patterns, and neuropsychological functions [2,3,6,10–12]. Given these data and the fact that the risk for zinc undernutrition is high throughout the world [13,14], the potential deleterious impact of zinc deficiency on brain development and function has to be considered of significant health concern in human populations.

The findings that severe gestational zinc deficiency causes brain malformations in the offspring [15] stress the relevance of this metal in key neurodevelopmental events, including neuronal proliferation, migration, differentiation, and survival. In fact, in rats fed zinc-deficient diet postnatally, several neuronal alterations were observed in the cerebellum. These alterations included a decreased number of cerebellum granule cells (CGNs) and altered dendritic differentiation of basket, stellate, and Purkinje cells [16–18]. In human IMR-32 neuroblastoma cells, which differentiate *in vitro* to a cortical neuron phenotype, we observed that decreased zinc availability causes oxidative stress [19], alterations in cytoskeleton structure and dynamics [20,21], deregulation of oxidant sensitive signaling cascades [20–23], apoptotic cell death [20–24], and cell cycle arrest.

Although the adverse effects of zinc deprivation on the cytoskeleton have been known for many years [25–29], the underlying mechanisms and the functional impact of these alterations are poorly

149

characterized. This chapter addresses the relevance of zinc and the cytoskeleton, particularly of microtubules, in the modulation of transcription factors NFAT (nuclear factor of activated T cells) and NF-κB. It is worth mentioning that in neurons, the long-distance intracellular trafficking of transcription factors that are activated in synapse/cytosol and have to be transported into the nucleus to regulate transcription could be particularly affected by cytoskeletal alterations.

11.2   REGULATION OF NF-κB AND NFAT

Rel/NF-κB is a ubiquitous transcription factor that has been proposed to play an important role in brain development and function. The Rel/NF-κB family of proteins shares a Rel homology domain and includes c-Rel, RelB, RelA (p65), p50, and p52. Through the interactions with inhibitory IκB proteins, NF-κB dimers are kept in the cytosol in an inactive form. Upon different stimuli, IκB kinases phosphorylate two conserved serines in IκB, leading to the subsequent ubiquitination and degradation of IκB [30,31]. Once IκB is degraded, the nuclear localization signal (NLS) is unmasked and the active dimer interacts with karyopherin α [32], with the subsequent recognition by karyopherin β, which drives the transport of NF-κB into the nucleus.

In the nervous system, NF-κB is widely distributed and can be involved in processes that are central to neurodevelopment, including neuronal proliferation, migration, differentiation, and survival [33–36]. Furthermore, NF-κB can modulate neural synaptic plasticity and memory (reviewed by Romano et al. [37]). NF-κB also regulates the expression of pro-survival genes and protects neurons from different pro-apoptotic stimuli (reviewed by Mattson et al. [34]). Accordingly, NF-κB inactivation by proteasome inhibitors triggered apoptosis in different areas of the central nervous system [38] and in IMR-32 cells [21]. In CGNs, survival is dependent on continuous NF-κB activation. Serum deprivation and potassium deprivation lead to the inactivation of NF-κB [39,40], to a decreased expression of NF-κB-regulated antiapoptotic proteins [41], and to CGN apoptosis. In addition, the presence of NF-κB during development and in the mature mouse brain in areas of active neurogenesis supports the participation of NF-κB in neuronal proliferation and migration [42].

NFAT can also be involved in the development of the nervous system [43]. At this time, five members of the NFAT family of proteins have been described. NFAT1 (also named NFATc2 or NFATp), NFAT2 (NFATc1 or NFATc), NFAT3 (NFATc4), NFAT4 (NFATc3 or NFATx), and NFAT5 (TonEBP) (reviewed by Macian et al. [44]). NAFTc1–c4 contain a regulatory domain that determines the cellular distribution of NFAT and controls the transcriptional activation. NFATc1–c4 are activated by calcineurin, a calcium/calmodulin-dependent phosphatase [45]. The elevation of intracellular calcium triggers the binding of calcineurin to NFATc1–c4, leading to NFAT dephosphorylation. The unmasking of an NLS drives the translocation of NFAT to the nucleus [46], where it regulates transcription.

Calcineurin-NFAT signaling can have important regulatory roles in neuronal development and function, particularly on the modulation of synaptic plasticity. Mice bearing mutations in NFATc2, NFATc3, and NFATc4, although not presenting alterations in neuronal differentiation and survival, showed a remarkable impairment in neurotrophins- or netrins-induced axonal outgrowth [47]. In addition, NFAT regulates the transcription of several genes, including the inositol 1,4,5-triphosphate receptor in CGNs [48] and hippocampal neurons [49]. Furthermore, the transcription of brain-derived neurotrophic factor, which has a critical role in neuronal survival and proliferation, is regulated by NFATc4 in hippocampal neurons [50]. Accordingly, NFATc4/NFAT3 protects CGNs from apoptotic death [51]. Overall, current evidence supports a role for the NFAT signaling cascade in neuronal survival and plasticity during development and in the mature brain [43,52].

11.3   ZINC AND THE NEURONAL REGULATION OF NF-κB AND NFAT

Decreased zinc availability affects different aspects of NF-κB modulation. In IMR-32 cells incubated in media with low-zinc content, an overall activation of NF-κB is observed. NF-κB-DNA binding in total cell extracts is higher in the zinc-deficient IMR-32 cells than in controls (Figure 11.1A) [21].

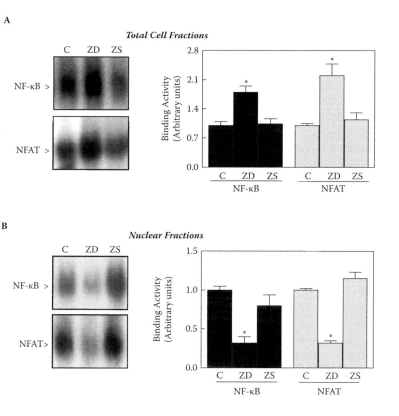

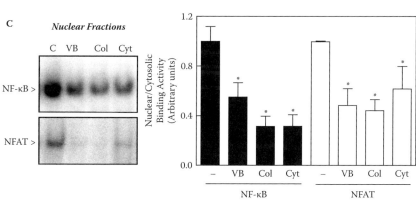

FIGURE 11.1   Zinc deficiency and cytoskeleton-disrupting drugs modulate transcription factors NF-κB and NFAT in human neuroblastoma IMR-32 cells. To investigate the influence of zinc on NF-κB and NFAT modulation in neuronal cells, we isolated total and nuclear fractions after 24 h of incubating IMR-32 cells in control (C), zinc-deficient (ZD), or zinc-supplemented (ZS) media. (A) EMSA for NF-κB and NFAT in total fractions. (B) EMSA for NF-κB and NFAT in nuclear fractions isolated from cells incubated for 24 h in the corresponding media. After EMSA, the bands of the transcription factor-DNA complexes were quantitated, and the results are shown as the mean ± SEM of five independent experiments. *Significantly different compared with the C and ZS groups ($p < .05$, one-way analysis of variance test). (C) To investigate the effects of cytoskeleton-disrupting drugs, we isolated nuclear and cytosolic fractions after incubating IMR-32 cells for 24 h in control (C) media incubated without (–) or with 0.5 μM vinblastine (VB), 0.5 μM colchicine (Col), or 0.5 μM cytochalasin D (Cyt). Left panel: EMSA for NF-κB or NFAT in nuclear fractions. Right panel: After the EMSA assays, bands were quantitated and values were expressed as the nuclear/cytosolic binding activity ratio. Results are shown as the mean ± SEM of six independent experiments. *Significantly different compared with the C group ($p < .01$, one-way analysis of variance test). This figure is an adaptation of results previously published [20,22].

The cytosolic events in the NF-κB cascade (IκBα phosphorylation and degradation) are activated in zinc-deficient cells [21], and this activation is triggered by an increase in cell oxidants [19]. However, NF-κB-DNA binding in nuclear extracts is lower in zinc-deficient IMR-32 cells compared with control and zinc-supplemented cells (Figure 11.1B) [21]. Similarly, a low nuclear NF-κB-DNA binding activity has been described for different cell lines, including 3T3 fibroblasts [53], C6 rat glioma cells [54], and a T-lymphoblastoid cell line (HUT-78) [55]. The low NF-κB-DNA binding translates in lower transactivation of both reporter and endogenous NF-κB-dependent genes. After 24 h of incubation in zinc-deficient media, a decreased expression of p105, the precursor for p50, and that of the antiapoptotic protein bcl-2 were observed as compared with controls [20]. Given that NF-κB is generally a pro-survival signal and recent findings (unpublished results) that zinc deficiency downregulates several antiapoptotic proteins in neurons, the inhibition of NF-κB when cellular zinc decreases could in part underlie zinc deficiency-induced neuronal apoptosis [23,24].

Similarly, zinc deficiency affects NFAT modulation in neuronal cells at different steps of this signaling pathway [22]. When measured in total cell fractions, NFAT-DNA binding was high in IMR-32 cells incubated in low-zinc media (Figure 11.1A). Given that NFAT activation is dependent on calcineurin, a calcium/calmodulin-dependent phosphatase, the possible involvement of calcium in the observed activation was investigated. Cellular calcium levels increased in the cells incubated in zinc-deficient media [22], suggesting that the initial events of NFAT activation are triggered by changes in calcium fluxes associated to the decrease in extracellular zinc concentrations [22]. Furthermore, the simultaneous incubation of zinc-deficient cells with the antioxidants α-lipoic acid and N-acetyl cysteine prevents NFAT activation, suggesting that oxidants are also involved in the initial activation of NFAT in neuronal cells when cellular zinc decreases [22]. Similarly to that observed for NF-κB, the nuclear translocation of NFAT is impaired in IMR-32 (Figure 11.1B) and PC-12 cells, and, as a consequence, NFAT-dependent gene expression is reduced in the zinc-deficient IMR-32 cells [22]. The stabilization of microtubules with taxol prevents zinc deficiency-induced alterations in NFAT nuclear translocation [22].

Both NF-κB and NFAT exist in the cytosol in an inactive form, and they are translocated to the nucleus to modulate transcription upon activation and exposure of the NLS. Given that zinc deficiency is known to affect tubulin polymerization in the nervous system, we hypothesized that zinc deficiency-induced alteration of the cytoskeleton is one mechanism affecting both NF-κB and NFAT transport from the synapse/cytosol into the nucleus. Thus, we studied the importance of a functional cytoskeleton in neuronal NF-κB and NFAT transport and nuclear translocation.

11.4   ZINC AND THE CYTOSKELETON

Alterations in tubulin polymerization have been described in association with zinc deficiency in fetal and adult rat brains and in IMR-32 cells [20,21,25,27,29,56]. Zinc deficiency affects the kinetics of brain tubulin polymerization, with a lower initial velocity (Figure 11.2A) and a longer lag period in tubulin assembly being observed in brain supernates obtained from zinc-deficient rats compared with controls [21,25,27]. Similarly, zinc deficiency affects the dynamics, the structure of microtubules, and the amount of polymerized tubulin in IMR-32 cells [20]. The kinetics of *in vitro* tubulin polymerization showed lower rates of polymerization in cells incubated in zinc-deficient media (Figure 11.2A) [21]. *In vitro* tubulin polymerization rates are markedly lower in the brain of gestation day 19 fetuses from dams fed zinc-deficient diets throughout gestation compared with those given a control diet ad libitum and those given the control diet at a restricted intake (Figure 11.2A). As evaluated by fluorescence microscopy, the structure of microtubules is altered in the zinc-deficient cells, and this is prevented by supplementing the media with zinc (Figure 11.2B).

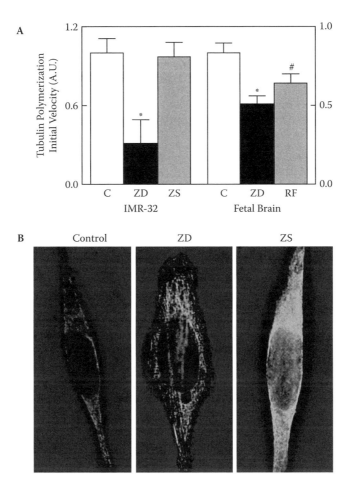

FIGURE 11.2 (See color insert following page 234.) Zinc deficiency affects microtubule structure and polymerization in human neuroblastoma IMR-32 cells and in fetal brain supernatants. (A) For the cell study, IMR-32 cells were incubated for 24 h in control (C), zinc-deficient (ZD), or zinc-supplemented (ZS) media. For the animal study, Sprague-Dawley rats were fed from day 0 until day 19 of gestation a control diet (25 µg Zn/g) given ad libitum (C), a zinc-deficient diet containing 0.5 µg Zn/g (ZD), or the control diet at a restricted intake equal to that consumed by the ZD group (RF). *In vitro* tubulin polymerization kinetics in cells or in fetal brain supernatants were assessed as previously described [21]. The slope in the linear portion of the curves was calculated. Results are shown as the mean ± SEM. (*) and (#) indicate, significantly lower compared with the C and ZS groups (*) or with the C group (#) (p < 0.05, one-way analysis of variance test). (B) Differentiated IMR-32 cells were incubated for 24 h in control (C), zinc-deficient (ZD), or zinc-supplemented (ZS) media. β-tubulin was immunodetected by using a specific anti-β-tubulin antibody, followed by an FITC-labeled secondary antibody. Images of cells incubated in control (C), zinc-deficient (ZD), or zinc-supplemented (ZS) media are shown. This figure is an adaptation of results previously published [20,21,27].

The altered pattern of tubulin distribution observed in the zinc-deficient cells resembled that of neuronal cells treated with colchicine, an inhibitor of tubulin polymerization [20].

Although the evidence clearly demonstrates that decreased zinc availability alters the microtubule network in the nervous system, the underlying mechanisms and the functional consequences are not characterized. A decrease in available zinc could per se cause tubulin disruption. Tubulin contains zinc binding sites, and zinc stimulates *in vitro* tubulin polymerization [26]. Given that zinc deficiency is associated with an increase in neuronal oxidants [19] and altered antioxidant defenses [57], tubulin oxidation could also underlie the impaired tubulin polymerization when neuronal zinc decreases. Oxidative modifications of tubulin and tubulin-associated proteins can lead to decreased tubulin polymerization. Tubulin has twenty cysteine residues that can be oxidized to cystine. Oxidants that react with cysteines, such as peroxynitrite (ONOO–), 4-hydroxynonenal, H_2O_2, and thiol-reactive agents, inhibit tubulin polymerization [58–61]. However, while zinc deficiency seems to exert a controlled and specific deregulation of tubulin polymerization, other conditions of oxidative stress lead to tubulin fragmentation and loss of tubulin protein [62,63]. In zinc deficiency, marked alterations in tubulin polymerization occur without affecting tubulin content [20,27]. Although additional research is necessary to further understand the underlying mechanism, a decrease in zinc availability and an increase in tubulin oxidation can act jointly to impair microtubule assembly in zinc-deficient neurons.

11.5   THE ROLE OF THE CYTOSKELETON IN THE NUCLEAR TRANSPORT OF NF-κB AND NFAT

The extended morphology of neurons creates a particular challenge to intracellular signal transduction. Signals generated at the synapse must be transported long distances to the nucleus in order to induce changes in gene expression. The most efficient mechanism for intracellular long-distance transport involves the association of a transcription factor with molecular motors that move along the cytoskeleton [64]. Several reports indicate that microtubules are involved in dynein-mediated transport of NLS-containing protein cargos [64,65]. In *Aplysia* neurons, microtubules are required for retrograde transport of NLS-labeled proteins along the axon [66]. In mammalian neurons, microtubules are involved in retrograde transport of nuclear import complexes within the axon after injury [67]. Salman et al. showed that NLS-containing peptides induced molecular delivery along microtubules in an extract of *Xenopus laevis* eggs [68]. The nuclear transport of the parathyroid hormone-related protein was reduced in cells with microtubules depolymerized by nocodazole [69]. The tumor suppressor protein p53 was found to be associated with microtubules and to require dynein for its nuclear import [70]. Given these data, we used drugs that perturb the normal assembly and/or structure of microtubules and actin filaments to investigate the role of the cytoskeleton in the neuronal transport of NF-κB and NFAT.

As described in the previous section, alterations in tubulin polymerization occur in association with an impaired nuclear translocation of NF-κB and NFAT in zinc-deficient neurons. These findings, together with the above-mentioned evidence of the involvement of cytoskeleton in long-distance protein transport, suggest that the cytoskeleton could participate in neuronal NF-κB and NFAT transport and nuclear translocation.

11.5.1   NF-κB

We investigated the requirement of a functional cytoskeleton for the nuclear transport of NF-κB in neuronal cells. By electrophoretic mobility shift assay (EMSA), we observed decreased NF-κB-DNA binding in nuclear fractions isolated from cells exposed to cytoskeleton-disrupting drugs. The nuclear/cytosolic DNA binding ratio of NF-κB and NFAT is significantly lower in IMR-32 cells treated with inhibitors of tubulin (vinblastine and colchicine) or actin (cytochalasin) polymerization (Figure 11.1C). That this finding was due to a lower nuclear transport of the active NF-κB is supported by the observed patterns of RelA and p50 nuclear cytosolic distribution in the treated cells. RelA accumulates in the cytoplasm when human neuroblastoma IMR-32 cells are incubated with vinblastine and colchicine and in the zinc-deficient cells, compared with control and zinc-supplemented cells [20]. Similar findings were observed in primary cultures of rat CGNs and cortical neurons [20]. Accordingly, zinc deficiency and treatments with vinblastine or colchicine reduced NF-κB-dependent gene transactivation [20]. The observation that supplementation with zinc prevents both the tubulin polymerization alterations and NF-κB nuclear transport disruption in the cells cultured in zinc-deficient media strongly supports the association between both events. Indicating the particular relevance of microtubules in NF-κB transport, the stabilization of microtubules with taxol prevents zinc deficiency-induced alterations in NF-κB nuclear translocation [20].

In studies using photobleaching techniques, Meffert et al. showed that NF-κB movement within dendrites is directed in the retrograde direction [71]. Gene delivery vectors containing repetitive binding sites for NF-κB were transported along microtubules in a dynein-dependent manner [72]. Furthermore, hippocampal neurons exposed to vincristine or colchicine showed reduced mobility of activated NF-κB toward the nucleus and lower NF-κB-dependent transcription activity after glutamate stimulus [73]. Interestingly, dynein intermediate chain (IC74) and two subunits of dynactin, dynamitin and p150[Glued], are present in complex with NF-κB. Additionally, with the use of *in vitro* protein pull-down experiments, it was demonstrated that the formation of this complex requires a functional NLS, which is present in NF-κB p65 [73]. In contrast, non-neuronal tumor cell lines do not require the cytoskeleton to transport activated NF-κB toward the nucleus [74], suggesting that this could be an exclusive mechanism for neuronal cells. The finding that p65-GFP fusion protein redistributes from distal processes to the nucleus after glutamate or kainate stimulation in hippocampal neurons in a retrograde way further stresses the importance of the retrograde transport for NF-κB in neurons [75].

The actin cytoskeleton can also regulate gene expression. Neuroblastoma IMR-32 cells treated with the inhibitor of actin polymerization cytochalasin show an impaired NF-κB nuclear translocation (Figure 11.1C) with a subsequent inhibition in the expression of NF-κB-regulated genes. The RelA distribution pattern is altered as observed by immunocytochemistry in the cells incubated with cytochalasin compared with control untreated cells. Jasplakinolide, a drug that stabilizes F-actin and promotes actin polymerization, did not restore NF-κB nuclear translocation. This could be due to the fact that alterations in the actin network structure may occur secondarily to the initial disruption of the microtubule network in zinc deficiency. A requirement of functional microfilaments for NF-κB transport in neurons can be explained by the close interrelationship between microtubules and microfilaments in sustaining the normal structure and function of the cell cytoskeleton.

The nuclear transport of proteins containing the NLS involves the recognition and binding of the NLS protein with karyopherin α, the subsequent interaction with karyopherin β, the binding of the complex to the nuclear pore, and the translocation of the NLS protein through the pore in an energy-dependent process mediated by the Ran guanosine 5′-triphosphate-binding proteins. Karyopherins are found in the neuronal axoplasm and are involved in the retrograde transport of NLS-containing proteins [67]. In the microtubule-dependent retrograde axonal transport, the motor protein dynein plays a central role as the motor force for the mobilization of cargos through microtubules [76]. It is now clear that an association between karyopherin α, dynein, and microtubules is necessary for the nuclear translocation of NLS-containing proteins [67]. Given these data, we investigated if the altered NF-κB nuclear translocation associated with microtubule depolymerization could affect the formation of a tubulin-dynein-karyopherin α-p50 complex. In IMR-32 cells, immunoblotting assays demonstrated that dynein, karyopherin α1, and p50 coprecipitate with β-tubulin [20]. Cells incubated with vinblastine or colchicine or in zinc-deficient media show a markedly reduced association between β-tubulin and dynein, karyopherin α1, and p50. Of particular relevance is the finding that a decrease in neuronal zinc, which affects microtubule polymerization and structure, causes an alteration in the formation of the β-tubulin-dynein-karyopherin α1-p50 complex that is similar to that induced by cytoskeleton-disrupting drugs. Taken together, these results stress the importance of zinc and that of a functional microtubule network in the formation of the complex and in the subsequent neuronal transport and nuclear translocation of NF-κB. In support of this, *in vitro* studies have previously shown an association between karyopherin α and the cytoskeleton in tobacco protoplasts [77]. It is proposed that this association is necessary for the transport of NLS-containing proteins from the cytosol into the nucleus [77].

11.5.2   NFAT

NFAT, similar to NF-κB, is an NLS-bearing transcription factor that is sequestered inactive in the cytosol and translocates into the nucleus once it is activated. To investigate the possible requirement of an organized microtubule network in the nuclear transport of NFAT that could explain the observed alterations in NFAT nuclear translocation in zinc-deficient cells, we investigated the effects of cytoskeleton disruptors.

The inhibition of tubulin polymerization by colchicine and vinblastine and that of actin polymerization by cytochalasin inhibit the nuclear translocation of the active NFAT as evidenced by EMSA (Figure 11.1C) and by the nuclear/cytosolic distribution of the NFATc4 protein by Western blot. As observed by immunocytochemistry, tubulin and actin polymerization inhibitors lead to the formation of NFATc4 aggregates in the cytoplasm that are similar to those observed in the zinc-deficient cells [22]. The depolymerization of microtubules by cold also leads to the accumulation of NFAT in the cytosol [22], and microtubule repolymerization by warming cells at 37°C restores the translocation of NFAT to the nucleus. The above-mentioned results show that a functional microtubule and actin cytoskeleton is required for the transport of the active NFAT into the nucleus. As a consequence of altered microtubule dynamics, a decrease in cellular zinc also affects NFAT nuclear transport and transcriptional activity.

11.6   SUMMARY

Zinc is required for the normal function of the cytoskeleton and the regulation of transcription factors NF-κB and NFAT (summarized in Figure 11.3). The obtained evidence shows that the cytoskeleton, and particularly microtubules, is necessary for the neuronal long-distance transport of activated NF-κB and NFAT into the nucleus. During development, as well as in the adult brain, conditions that affect the normal structure and function of the cytoskeleton, such as zinc deficiency, can affect neuronal proliferation, differentiation, synaptic plasticity, and neuronal survival through the impairment of NF-κB/NFAT nuclear translocation and NF-κB/NFAT-dependent gene regulation. In this regard, the decreased expression of NF-κB- and NFAT-dependent genes could be one of the underlying causes of the decreased neuronal proliferation and increased apoptotic death associated with zinc deficiency. Variations of neuronal zinc levels could, through the modulation of NF-κB and NFAT, constitute a signal involved in the modulation of neuronal development.

ACKNOWLEDGMENTS

This work was supported by grants from the U.S. National Institutes of Health (grant no. HD 01743), the University of California, Davis, and the University of Buenos Aires and CONICET (Argentina).

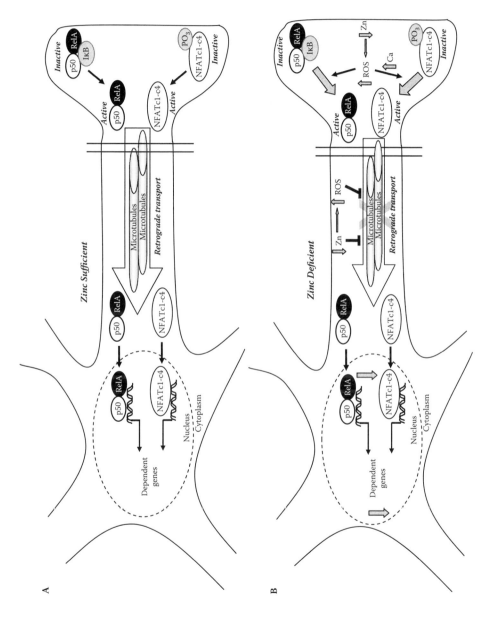

FIGURE 11.3 Cellular events triggered by a decrease in zinc availability in neuronal cells. (A) In a physiological (zinc-sufficient) condition, both NF-κB and NFAT located at the synapse/cytosol of neurons are activated, transported by retrograde transport, and translocate into the nucleus, where they regulate several genes, including those involved in cell proliferation, differentiation, and apoptosis. (B) A decrease in zinc availability (zinc-deficient scenario) is associated with an increase in cellular calcium, a condition of oxidative stress, including elevated levels of H_2O_2 and alterations in the cytoskeleton (microtubules and microfilaments) structure and dynamics. The elevation of oxidant species leads to the initial activation of NF-κB and NFAT. However, microtubule disruption impairs NF-κB/NFAT retrograde transport, NF-κB/NFAT nuclear translocation, and NF-κB/NFAT-dependent gene expression.

REFERENCES

1. Arnold, L. E., and DiSilvestro, R. A. (2005). Zinc in attention-deficit/hyperactivity disorder. *J Child Adolesc Psychopharmacol* **15**, 619–627.
2. Bentley, M. E., Caulfield, L. E., Ram, M., Santizo, M. C., Hurtado, E., Rivera, J. A., Ruel, M. T., and Brown, K. H. (1997). Zinc supplementation affects the activity patterns of rural Guatemalan infants. *J Nutr* **127**, 1333–1338.
3. Gardner, J. M., Powell, C. A., Baker-Henningham, H., Walker, S. P., Cole, T. J., and Grantham-McGregor, S. M. (2005). Zinc supplementation and psychosocial stimulation: Effects on the development of under-nourished Jamaican children. *Am J Clin Nutr* **82**, 399–405.
4. Golub, M. S., Keen, C. L., and Gershwin, M. E. (2000). Moderate zinc-iron deprivation influences behavior but not growth in adolescent rhesus monkeys. *J Nutr* **130**, 354S–357S.
5. Kirksey, A., Wachs, T. D., Yunis, F., Srinath, U., Rahmanifar, A., McCabe, G. P., Galal, O. M., Harrison, G. G., and Jerome, N. W. (1994). Relation of maternal zinc nutriture to pregnancy outcome and infant development in an Egyptian village. *Am J Clin Nutr* **60**, 782–792.
6. Penland, J. G., Sandstead, H. H., Alcock, N. W., Dayal, H. H., Chen, X. C., Li, J. S., Zhao, F., and Yang, J. J. (1997). A preliminary report: Effects of zinc and micronutrient repletion on growth and neuropsychological function of urban Chinese children. *J Am Coll Nutr* **16**, 268–272.
7. Hubbs-Tait, L., Kennedy, T. S., Droke, E. A., Belanger, D. M., and Parker, J. R. (2007). Zinc, iron, and lead: Relations to head start children's cognitive scores and teachers' ratings of behavior. *J Am Diet Assoc* **107**, 128–133.
8. Scheplyagina, L. A. (2005). Impact of the mother's zinc deficiency on the woman's and newborn's health status. *J Trace Elem Med Biol* **19**, 29–35.
9. Arnold, L. E., Bozzolo, H., Hollway, J., Cook, A., DiSilvestro, R. A., Bozzolo, D. R., Crowl, L., Ramadan, Y., and Williams, C. (2005). Serum zinc correlates with parent- and teacher-rated inattention in children with attention-deficit/hyperactivity disorder. *J Child Adolesc Psychopharmacol* **15**, 628–636.
10. Brown, K. H., Peerson, J. M., Rivera, J., and Allen, L. H. (2002). Effect of supplemental zinc on the growth and serum zinc concentrations of prepubertal children: A meta-analysis of randomized controlled trials. *Am J Clin Nutr* **75**, 1062–1071.
11. Bhutta, Z. A., Black, R. E., Brown, K. H., Gardner, J. M., Gore, S., Hidayat, A., Khatun, F., Martorell, R., Ninh, N. X., Penny, M. E., Rosado, J. L., Roy, S. K., Ruel, M., Sazawal, S., and Shankar, A. (1999). Prevention of diarrhea and pneumonia by zinc supplementation in children in developing countries: Pooled analysis of randomized controlled trials. Zinc Investigators' Collaborative Group. *J Pediatr* **135**, 689–697.
12. Sazawal, S., Black, R. E., Menon, V. P., Dinghra, P., Caulfield, L. E., Dhingra, U., and Bagati, A. (2001). Zinc supplementation in infants born small for gestational age reduces mortality: A prospective, randomized, controlled trial. *Pediatrics* **108**, 1280–1286.
13. Briefel, R. R., Bialostosky, K., Kennedy-Stephenson, J., McDowell, M. A., Ervin, R. B., and Wright, J. D. (2000). Zinc intake of the U.S. population: Findings from the Third National Health and Nutrition Examination Survey, 1988–1994. *J Nutr* **130**, 1367S–1373S.
14. Wuehler, S. E., Peerson, J. M., and Brown, K. H. (2005). Use of national food balance data to estimate the adequacy of zinc in national food supplies: Methodology and regional estimates. *Public Health Nutr* **8**, 812–819.
15. Hurley, L. S., and Swenerton, H. (1966). Congenital malformations resulting from zinc deficiency in rats. *Proc Soc Exp Biol Med* **123**, 692–696.
16. Dvergsten, C. L., Fosmire, G. J., Ollerich, D. A., and Sandstead, H. H. (1983). Alterations in the postnatal development of the cerebellar cortex due to zinc deficiency: I. Impaired acquisition of granule cells. *Brain Res* **271**, 217–226.
17. Dvergsten, C. L., Fosmire, G. J., Ollerich, D. A., and Sandstead, H. H. (1984). Alterations in the postnatal development of the cerebellar cortex due to zinc deficiency: II. Impaired maturation of Purkinje cells. *Brain Res* **318**, 11–20.
18. Dvergsten, C. L., Johnson, L. A., and Sandstead, H. H. (1984). Alterations in the postnatal development of the cerebellar cortex due to zinc deficiency: III. Impaired dendritic differentiation of basket and stellate cells. *Brain Res* **318**, 21–26.
19. Mackenzie, G. G., Zago, M. P., Erlejman, A. G., Aimo, L., Keen, C. L., and Oteiza, P. I. (2006). alpha-Lipoic acid and *N*-acetyl cysteine prevent zinc deficiency-induced activation of NF-kappaB and AP-1 transcription factors in human neuroblastoma IMR-32 cells. *Free Radic Res* **40**, 75–84.

20. Mackenzie, G. G., Keen, C. L., and Oteiza, P. I. (2006). Microtubules are required for NF-kappaB nuclear translocation in neuroblastoma IMR-32 cells: Modulation by zinc. *J Neurochem* **99**, 402–415.
21. Mackenzie, G. G., Zago, M. P., Keen, C. L., and Oteiza, P. I. (2002). Low intracellular zinc impairs the translocation of activated NF-kappa B to the nuclei in human neuroblastoma IMR-32 cells. *J Biol Chem* **277**, 34610–34617.
22. Mackenzie, G. G., and Oteiza, P. I. (2007). Zinc and the cytoskeleton in the neuronal modulation of transcription factor NFAT. *J Cell Physiol* **210**, 246–256.
23. Zago, M. P., Mackenzie, G. G., Adamo, A. M., Keen, C. L., and Oteiza, P. I. (2005). Differential modulation of MAP kinases by zinc deficiency in IMR-32 cells: Role of H(2)O(2). *Antioxid Redox Signal* **7**, 1773–1782.
24. Verstraeten, S. V., Zago, M. P., MacKenzie, G. G., Keen, C. L., and Oteiza, P. I. (2004). Influence of zinc deficiency on cell-membrane fluidity in Jurkat, 3T3 and IMR-32 cells. *Biochem J* **378**, 579–587.
25. Hesketh, J. E. (1981). Impaired microtubule assembly in brain from zinc-deficient pigs and rats. *Int J Biochem* **13**, 921–926.
26. Hesketh, J. E. (1982). Zinc-stimulated microtubule assembly and evidence for zinc binding to tubulin. *Int J Biochem* **14**, 983–990.
27. Oteiza, P. I., Cuellar, S., Lonnerdal, B., Hurley, L. S., and Keen, C. L. (1990). Influence of maternal dietary zinc intake on *in vitro* tubulin polymerization in fetal rat brain. *Teratology* **41**, 97–104.
28. Oteiza, P. I., Hurley, L. S., Lonnerdal, B., and Keen, C. L. (1990). Effects of marginal zinc deficiency on microtubule polymerization in the developing rat brain. *Biol Trace Elem Res* **24**, 13–23.
29. Oteiza, P. I., Hurley, L. S., Lonnerdal, B., and Keen, C. L. (1988). Marginal zinc deficiency affects maternal brain microtubule assembly in rats. *J Nutr* **118**, 735–738.
30. Baeuerle, P. A., and Baltimore, D. (1988). I kappa B: A specific inhibitor of the NF-kappa B transcription factor. *Science* **242**, 540–546.
31. Karin, M. (1999). How NF-kappaB is activated: The role of the IkappaB kinase (IKK) complex. *Oncogene* **18**, 6867–6874.
32. Cunningham, M. D., Cleaveland, J., and Nadler, S. G. (2003). An intracellular targeted NLS peptide inhibitor of karyopherin alpha:NF-kappa B interactions. *Biochem Biophys Res Commun* **300**, 403–407.
33. Albensi, B. C., and Mattson, M. P. (2000). Evidence for the involvement of TNF and NF-kappaB in hippocampal synaptic plasticity. *Synapse* **35**, 151–159.
34. Mattson, M. P., Culmsee, C., Yu, Z., and Camandola, S. (2000). Roles of nuclear factor kappaB in neuronal survival and plasticity. *J Neurochem* **74**, 443–456.
35. Mattson, M. P. (2005). NF-kappaB in the survival and plasticity of neurons. *Neurochem Res* **30**, 883–893.
36. Mattson, M. P., and Meffert, M. K. (2006). Roles for NF-kappaB in nerve cell survival, plasticity, and disease. *Cell Death Differ* **13**, 852–860.
37. Romano, A., Freudenthal, R., Merlo, E., and Routtenberg, A. (2006). Evolutionarily-conserved role of the NF-kappaB transcription factor in neural plasticity and memory. *Eur J Neurosci* **24**, 1507–1516.
38. Taglialatela, G., Kaufmann, J. A., Trevino, A., and Perez-Polo, J. R. (1998). Central nervous system DNA fragmentation induced by the inhibition of nuclear factor kappa B. *NeuroReport* **9**, 489–493.
39. Piccioli, P., Porcile, C., Stanzione, S., Bisaglia, M., Bajetto, A., Bonavia, R., Florio, T., and Schettini, G. (2001). Inhibition of nuclear factor-kappaB activation induces apoptosis in cerebellar granule cells. *J Neurosci Res* **66**, 1064–1073.
40. Lilienbaum, A., and Israel, A. (2003). From calcium to NF-kappa B signaling pathways in neurons. *Mol Cell Biol* **23**, 2680–2698.
41. Kovacs, A. D., Chakraborty-Sett, S., Ramirez, S. H., Sniderhan, L. F., Williamson, A. L., and Maggirwar, S. B. (2004). Mechanism of NF-kappaB inactivation induced by survival signal withdrawal in cerebellar granule neurons. *Eur J Neurosci* **20**, 345–352.
42. Denis-Donini, S., Caprini, A., Frassoni, C., and Grilli, M. (2005). Members of the NF-kappaB family expressed in zones of active neurogenesis in the postnatal and adult mouse brain. *Brain Res Dev Brain Res* **154**, 81–89.
43. Nguyen, T., and Di Giovanni, S. (2008). NFAT signaling in neural development and axon growth. *Int J Dev Neurosci* **26**, 141–145.
44. Macian, F., López-Rodriguez, C., and Rao, A. (2001). Partners in transcription: NFAT and AP-1. *Oncogene* **20**, 2476–2489.
45. Dolmetsch, R. E., Lewis, R. S., Goodnow, C. G., and Healy, J. I. (1997). Differential activation of transcription factors induced by Ca2+ response amplitude and duration. *Nature* **386**, 855–858.

46. Beals, C. R., Sheridan, C. M., Turck, C. W., Gardner, P., and Crabtree, G. R. (1997). Nuclear export of NF-ATc enhanced by glycogen synthase kinase-3. *Science* **275**, 1930–1934.

47. Graef, I. A., Wang, F., Charron, F., Chen, L., Neilson, J., Tessier-Lavigne, M., and Crabtree, G. R. (2003). Neurotrophins and netrins require calcineurin/NFAT signaling to stimulate outgrowth of embryonic axons. *Cell* **113**, 657–670.

48. Kramer, D., Fresu, L., Ashby, D. S., Freeman, T. C., and Genazzani, A. A. (2003). Calcineurin controls the expression of numerous genes in cerebellar granule cells. *Mol Cell Neurosci* **23**, 325–330.

49. Graef, I. A., Mermelstein, P. G., Stankunas, K., Neilson, J. R., Deisseroth, K., Tsien, R. W., and Crabtree, G. R. (1999). L-type calcium channels and GSK-3 regulate the activity of NF-ATc4 in hippocampal neurons. *Nature* **401**, 703–708.

50. Groth, R. D., and Mermelstein, P. G. (2003). Brain-derived neurotrophic factor activation of NFAT (nuclear factor of activated T-cells)-dependent transcription: A role for the transcription factor NFATc4 in neurotrophin-mediated gene expression. *J Neurosci* **23**, 8125–8134.

51. Benedito, A. B., Lehtinen, M., Massol, R., Lopes, U. G., Kirchhausen, T., Rao, A., and Bonni, A. (2005). The transcription factor NFAT3 mediates neuronal survival. *J Biol Chem* **280**, 2818–2825.

52. Graef, I. A., Chen, F., and Crabtree, G. R. (2001). NFAT signaling in vertebrate development. *Curr Opin Genet Dev* **11**, 505–512.

53. Oteiza, P. I., Clegg, M. S., Zago, M. P., and Keen, C. L. (2000). Zinc deficiency induces oxidative stress and AP-1 activation in 3T3 cells. *Free Radic Biol Med* **28**, 1091–1099.

54. Ho, E., and Ames, B. N. (2002). Low intracellular zinc induces oxidative DNA damage, disrupts p53, NFkappa B, and AP1 DNA binding, and affects DNA repair in a rat glioma cell line. *Proc Natl Acad Sci U S A* **99**, 16770–16775.

55. Prasad, A. S., Bao, B., Beck, F. W., and Sarkar, F. H. (2002). Zinc enhances the expression of interleukin-2 and interleukin-2 receptors in HUT-78 cells by way of NF-kappaB activation. *J Lab Clin Med* **140**, 272–289.

56. Mackenzie, G. G., Keen, C. L., and Oteiza, P. I. (2002). Zinc status of human IMR-32 neuroblastoma cells influences their susceptibility to iron-induced oxidative stress. *Dev Neurosci* **24**, 125–133.

57. Mackenzie, G. G., Zago, M. P., Aimo, L., and Oteiza, P. I. (2007). Zinc deficiency in neuronal biology. *IUBMB Life* **59**, 299–307.

58. Landino, L. M., Hasan, R., McGaw, A., Cooley, S., Smith, A. W., Masselam, K., and Kim, G. (2002). Peroxynitrite oxidation of tubulin sulfhydryls inhibits microtubule polymerization. *Arch Biochem Biophys* **398**, 213–220.

59. Neely, M. D., Sidell, K. R., Graham, D. G., and Montine, T. J. (1999). The lipid peroxidation product 4-hydroxynonenal inhibits neurite outgrowth, disrupts neuronal microtubules, and modifies cellular tubulin. *J Neurochem* **72**, 2323–2333.

60. Luduena, R. F., Roach, M. C., Jordan, M. A., and Murphy, D. B. (1985). Different reactivities of brain and erythrocyte tubulins toward a sulfhydryl group-directed reagent that inhibits microtubule assembly. *J Biol Chem* **260**, 1257–1264.

61. Mellon, M. G., and Rebhun, L. I. (1976). Sulfhydryls and the *in vitro* polymerization of tubulin. *J Cell Biol* **70**, 226–238.

62. Bernhard, D., Csordas, A., Henderson, B., Rossmann, A., Kind, M., and Wick, G. (2005). Cigarette smoke metal-catalyzed protein oxidation leads to vascular endothelial cell contraction by depolymerization of microtubules. *FASEB J* **19**, 1096–1107.

63. Sponne, I., Fifre, A., Drouet, B., Klein, C., Koziel, V., Pincon-Raymond, M., Olivier, J. L., Chambaz, J., and Pillot, T. (2003). Apoptotic neuronal cell death induced by the non-fibrillar amyloid-beta peptide proceeds through an early reactive oxygen species-dependent cytoskeleton perturbation. *J Biol Chem* **278**, 3437–3445.

64. Hanz, S., and Fainzilber, M. (2004). Integration of retrograde axonal and nuclear transport mechanisms in neurons: Implications for therapeutics. *Neuroscientist* **10**, 404–408.

65. Campbell, E. M., and Hope, T. J. (2003). Role of the cytoskeleton in nuclear import. *Adv Drug Deliv Rev* **55**, 761–771.

66. Ambron, R. T., Schmied, R., Huang, C. C., and Smedman, M. (1992). A signal sequence mediates the retrograde transport of proteins from the axon periphery to the cell body and then into the nucleus. *J Neurosci* **12**, 2813–2818.

67. Hanz, S., Perlson, E., Willis, D., Zheng, J. Q., Massarwa, R., Huerta, J. J., Koltzenburg, M., Kohler, M., van-Minnen, J., Twiss, J. L., and Fainzilber, M. (2003). Axoplasmic importins enable retrograde injury signaling in lesioned nerve. *Neuron* **40**, 1095–1104.

68. Salman, H., Abu-Arish, A., Oliel, S., Loyter, A., Klafter, J., Granek, R., and Elbaum, M. (2005). Nuclear localization signal peptides induce molecular delivery along microtubules. *Biophys J* **89**, 2134–2145.
69. Lam, M. H., Thomas, R. J., Loveland, K. L., Schilders, S., Gu, M., Martin, T. J., Gillespie, M. T., and Jans, D. A. (2002). Nuclear transport of parathyroid hormone (PTH)-related protein is dependent on microtubules. *Mol Endocrinol* **16**, 390–401.
70. Giannakakou, P., Sackett, D. L., Ward, Y., Webster, K. R., Blagosklonny, M. V., and Fojo, T. (2000). p53 is associated with cellular microtubules and is transported to the nucleus by dynein. *Nat Cell Biol* **2**, 709–717.
71. Meffert, M. K., Chang, J. M., Wiltgen, B. J., Fanselow, M. S., and Baltimore, D. (2003). NF-kappa B functions in synaptic signaling and behavior. *Nat Neurosci* **6**, 1072–1078.
72. Mesika, A., Kiss, V., Brumfeld, V., Ghosh, G., and Reich, Z. (2005). Enhanced intracellular mobility and nuclear accumulation of DNA plasmids associated with a karyophilic protein. *Hum Gene Ther* **16**, 200–208.
73. Mikenberg, I., Widera, D., Kaus, A., Kaltschmidt, B., and Kaltschmidt, C. (2007). Transcription factor NF-kappaB is transported to the nucleus via cytoplasmic dynein/dynactin motor complex in hippocampal neurons. *PLoS ONE* **2**, e589.
74. Mikenberg, I., Widera, D., Kaus, A., Kaltschmidt, B., and Kaltschmidt, C. (2006). TNF-alpha mediated transport of NF-kappaB to the nucleus is independent of the cytoskeleton-based transport system in non-neuronal cells. *Eur J Cell Biol* **85**, 529–536.
75. Wellmann, H., Kaltschmidt, B., and Kaltschmidt, C. (2001). Retrograde transport of transcription factor NF-kappa B in living neurons. *J Biol Chem* **276**, 11821–11829.
76. Guzik, B. W., and Goldstein, L. S. (2004). Microtubule-dependent transport in neurons: Steps towards an understanding of regulation, function and dysfunction. *Curr Opin Cell Biol* **16**, 443–450.
77. Smith, H. M., and Raikhel, N. V. (1998). Nuclear localization signal receptor importin alpha associates with the cytoskeleton. *Plant Cell* **10**, 1791–1799.

12 Tocotrienol Neuroprotection
The Most Potent Biological Function of All Natural Forms of Vitamin E

Chandan K. Sen, Savita Khanna, and Sashwati Roy

Laboratory of Molecular Medicine, Department of
Surgery, Davis Heart & Lung Research Institute, Ohio State
University Medical Center, Columbus, Ohio, USA

CONTENTS

12.1 INTRODUCTION

The brain and neural cells are highly susceptible to oxidative damage because they (1) contain high amounts of polyunsaturated (20:4 and 22:6) fatty acids that are susceptible to lipid peroxidation, (2) have high oxygen consumption, (3) are relatively deficient in certain antioxidant enzymes, and (4) are rich in iron concentration. Thus, antioxidant defenses are critically important to protect the brain and neural tissues from oxidative damage. Numerous pathophysiological conditions have indeed been associated with increased levels of oxidative stress indices [1]. Neuroprotection by antioxidants has therefore drawn much interest. Much of the available research on the role of antioxidant nutrients in neurological function and disease has focused on vitamin E. Vitamin E is essential for normal neurological function [2,3]. It represents a family of major lipid-soluble, chain-breaking antioxidants in the body protecting the integrity of membranes by inhibiting lipid peroxidation. Mostly based on symptoms of primary vitamin E deficiency, research has demonstrated that vitamin E has a central role in maintaining neurological structure and function [3]. Most of the vitamin E-sensitive neurological disorders are associated with elevated levels of oxidative damage markers. During the last decade, this led to the popular hypothesis stating that the neuroprotective effects of vitamin E are mediated by its antioxidant property [4].

The natural vitamin E family includes eight chemically distinct molecules: the α-, β-, γ-, and δ-tocopherols and the α-, β-, γ-, and δ-tocotrienols. Tocochromanols contain a polar chromanol head group with a long isoprenoid side chain. Depending on the nature of the isoprenoid chain, tocopherols (containing a phytyl chain), or tocotrienols (geranylgeranyl chain) can be distinguished [5]. A striking asymmetry in our understanding of the eight-membered natural vitamin E tocol family has deprived us of the full complement of benefits offered by the natural vitamin E molecules. Approximately only 1% of the entire literature on vitamin E addresses tocotrienols. A review of the National Institutes of Health CRISP (Computer Retrieval of Information on Scientific Projects) database shows that funding for tocotrienol research represents less than 1% of all vitamin E research during the last 30+ years. Within the tocopherol literature, the non-α forms remain poorly studied [6–8]. This represents a major void in vitamin E research. The significance of the void is substantially enhanced by the observation that the biological functions of the different homologs of natural vitamin E are not identical. Thus, drawing conclusions against the efficacy of vitamin E as a whole based on the study of tocopherols is misleading [9]. With reference to work by our laboratory, a recent issue of *Harvard Health Publications* recognized vitamin E to be "separate" and "unequal," acknowledging that lumping all the varieties of vitamin E together has glossed over some important distinctions between the different types that could put vitamin E in a better light [10]. During the past 5 years, tocotrienol research has gained substantial momentum. More than three-fourths of the entire PubMed literature on tocotrienols was published in or after 2000. This represents a major swing in the overall direction of vitamin E research. In 2005, the definition of vitamin E in the *Merck Manual* had no reference to tocotrienol. Today, the manual recognizes that "Vitamin E is a group of compounds (including tocopherols and tocotrienols) that have similar biologic activities. The most biologically active is α-tocopherol, but β-, γ-, and δ-tocopherols, four tocotrienols, and several stereoisomers may also have important biologic activity. This chapter highlights the potential significance of the tocotrienol half of the vitamin E family as a natural neuroprotective agent.

12.2 VITAMIN E: A FAMILY OF SEPARATE AND UNEQUAL MEMBERS

All eight tocols in the natural vitamin E family share close structural homology and hence possess comparable antioxidant efficacy. However, current studies of the biological functions of vitamin E continue to indicate that members of the vitamin E family possess unique biological functions often not shared by other family members (Figure 12.1). One of the earliest observations suggesting that α-tocopherol may have functions independent of its antioxidant property came from the observation that α-tocopherol strongly inhibited platelet adhesion. The antiadhesive effect of α-tocopherol appeared to be related to a reduction in the number and size of pseudopodia upon platelet activation, and this finding led to the

	Tocopherol	Tocotrienol
Cholesterol lowering	−	+
Mean life span increase	−	+
Tumor latency	−	+
Human breast cancer cell growth arrest	−	+
Inhibition of telomerase activity in cancer cells	−	+
Restricts proliferation of Tenon's capsule fibroblast	−	+
Activate steroid and xenobiotic receptor	−	+
Prevents side-effect of long-term glucocorticoid use	−	+
Preventsstress-induced gastric acidity	−	+
Anti-angiogenic	−	+
nM Neuroprotection	−	+
Blood pressure lowering	+	++
Antioxidant	+	++
Inhibition of cell adhesion	+	10x
Uptake by endothelial cells	+	25–95x
Uptake of topical by ocular tissue	+	++
Membrane incorporation	+	++

FIGURE 12.1 Tocotrienols and tocopherols are separate and unequal in function. For literature support, see the review article by Sen et al. [11].

hypothesis that within the human body vitamin E may exert functions beyond its antioxidant property [12]. That members of the tocopherol family may have functions independent of their antioxidant properties gained more prominence when vitamin E molecules with comparable antioxidant properties exhibited contrasting biological effects [13]. At the posttranslational level, α-tocopherol inhibits protein kinase C, 5-lipoxygenase, and phospholipase A2, and activates protein phosphatase 2A and diacylglycerol kinase. Some genes [e.g., scavenger receptors, α-tocopherol transfer protein (TTP), α-tropomyosin, matrix metalloproteinase-19, and collagenase] are specifically modulated by α-tocopherol at the transcriptional level. α-Tocopherol also inhibits cell proliferation, platelet aggregation and monocyte adhesion. These effects have been characterized to be unrelated to the antioxidant activity of vitamin E and possibly reflect specific interactions of α-tocopherol with enzymes, structural proteins, lipids and transcription factors [14]. γ-Tocopherol represents the major form of vitamin E in the diet in the United States, but not in Europe. Desmethyl tocopherols, such as γ-tocopherol and specific tocopherol metabolites, most notably the carboxyethyl-hydroxychroman products, exhibit functions that are not shared by α-tocopherol. The activities of these other tocopherols do not map directly to their chemical antioxidant behavior but rather reflect anti-inflammatory, antineoplastic, and natriuretic functions possibly mediated through specific binding interactions [7]. Metabolites of γ-tocopherol [2,7,8-trimethyl-2-(β-carboxyethyl)-6-hydroxychroman], but not those of α-tocopherol, provide natriuretic activity. Moreover, a nascent body of epidemiological data suggests that γ-tocopherol is a better negative risk factor for certain types of cancer and myocardial infarction than α-tocopherol [15].

Structurally, tocotrienols differ from tocopherols by the presence of three *trans* double bonds in the hydrocarbon tail. Because of these unsaturations in the isoprenoid side chain, tocotrienols are thought to assume a unique conformation [16]. Indeed, α-tocotrienol possesses numerous functions that are not shared by α-tocopherol [17]. For example, nanomolar concentrations of α-tocotrienol uniquely prevent inducible neurodegeneration by regulating specific mediators of cell death [18–20]. Oral supplementation of tocotrienol protects against stroke [21,22]. Micromolar amounts of tocotrienol suppress the activity of HMG-CoA (3-hydroxy-3-methylglutaryl coenzyme A) reductase, the hepatic enzyme responsible for cholesterol synthesis [23,24]. Tocopherols do not share the cholesterol-lowering properties of tocotrienols [25,26]. Sterol-regulated ubiquitination marks HMG-CoA reductase for endoplasmic reticulum-associated degradation by 26S proteasomes. This degradation, which results from sterol-induced binding of reductase to endoplasmic reticulum membrane proteins called Insigs, contributes to the complex and multivalent feedback regulation of the enzyme. It has recently been demonstrated that δ-tocotrienol stimulates ubiquitination and degradation of reductase and blocks processing of sterol regulatory element binding proteins (SREBPs), another sterol-mediated action of Insigs. The γ-tocotrienol analog is more selective in enhancing reductase ubiquitination and degradation than in blocking SREBP processing. Other forms of vitamin E neither accelerate reductase degradation nor block SREBP processing [27].

Tocotrienol, not tocopherol, administration reduces oxidative protein damage and extends the mean lifespan of *Caenorhabditis elegans* [28]. Tocotrienols are thought to have more potent antioxidant properties than α-tocopherol [29,30]. Reportedly, the unsaturated side chain of tocotrienol allows for more efficient penetration into tissues that have saturated fatty layers, such as the brain and liver [31]. Experimental research examining the antioxidant, free radical scavenging effects of tocopherols and tocotrienols revealed that tocotrienols appear superior due to their better distribution in the fatty layers of the cell membrane [31]. Furthermore, tocotrienol but not tocopherol, suppresses growth of human breast cancer cells [32].

12.3 TOCOTRIENOL: THE NON-GREEN SEED VITAMIN E

Tocotrienols are the primary form of vitamin E in the seed endosperm of most monocots, including agronomically important cereal grains such as wheat, rice, and barley. Palm oil contains significant quantities of tocotrienol [33]. Tocotrienols are also found in the seed endosperm of a limited number of dicots, including *Apiaceae* species and certain *Solanaeceae* species, such as tobacco. These molecules are found only rarely in vegetative tissues of plants. Crude palm oil extracted from the fruits of *Elaeis guineensis* particularly contains a high amount of tocotrienols (up to 800 mg/kg), mainly consisting of γ-tocotrienol and α-tocotrienol. Compared with tocopherols, tocotrienols are considerably less widespread in the plant kingdom [34]. In 80 plant species studied, 24 were found to contain significant amounts of tocotrienols. No taxonomic relation was apparent among the 16 dicotyledonous species that were found to contain tocotrienol. Monocotyledonous species (eight species) belonged either to the *Poaceae* (six species) or the *Aracaceae* (two species). A more detailed analysis of tocotrienol accumulation revealed the presence of this natural vitamin E in several nonphotosynthetic tissues and organs (i.e., seeds, fruits) and in latex. No tocotrienol could be detected in mature photosynthetic tissues. Transient accumulation of low levels of tocotrienols is found in the young coleoptiles of plant species whose seeds contained tocotrienols. No measurable tocotrienol biosynthesis was apparent in coleoptiles or in chloroplasts isolated from coleoptiles. Tocotrienol accumulation in coleoptiles was not associated with chloroplasts. Tocotrienols seem to be transiently present in photosynthetically active tissues, but it remains to be proven whether they are biosynthesized in such tissues or imported from elsewhere in the plant [34].

In contrast to tocotrienols, tocopherols occur ubiquitously in plant tissues and are the exclusive form of vitamin E in leaves of plants and seeds of most monocot plants. Transgenic expression of the barley HGGT (homogentisic acid transferase, which catalyzes the committed step of tocotrienol biosynthesis) in *Arabidopsis thaliana* leaves resulted in accumulation of tocotrienols, which were absent from leaves of nontransformed plants, and a 10- to 15-fold increase in total vitamin E antioxidants (tocotrienols plus tocopherols). Overexpression of the barley HGGT in corn seeds increased tocotrienol and tocopherol content by as much as six-fold. These results provide insight into the genetic basis for tocotrienol biosynthesis in plants and demonstrate the ability to enhance the antioxidant content of crops by introduction of an enzyme that redirects metabolic flux [35]. More recently, another strategy involving genetic engineering of metabolic pathways in plants has proved to be efficient in bolstering tocotrienol biosynthesis [36]. In plants, phenylalanine is the precursor of a myriad of secondary compounds termed phenylpropanoids. In contrast, much less carbon is incorporated into tyrosine, which provides *p*-hydroxyphenylpyruvate and homogentisate, the aromatic precursors of vitamin E. The flux of these two compounds has been upregulated by deriving their synthesis directly at the level of prephenate. This was achieved by the expression of the yeast prephenate dehydrogenase gene in tobacco plants that already overexpress the *Arabidopsis* *p*-hydroxyphenylpyruvate dioxygenase coding sequence. Massive accumulation of tocotrienols was observed in leaves. These molecules, which were undetectable in wild-type leaves, became the major forms of vitamin E in the leaves of the transgenic lines. An increased resistance of the transgenic plants toward the herbicidal *p*-hydroxyphenylpyruvate dioxygenase inhibitor diketonitril was also observed. Thus, the synthesis of *p*-hydroxyphenylpyruvate is a limiting step for the accumulation of vitamin E in plants [36].

Palm oil represents one of the most abundant natural sources of tocotrienols [37]. The distribution of vitamin E in palm oil is 30% tocopherols and 70% tocotrienols [33]. The oil palm (*E. guineensis*) is native to many West African countries, where local populations have used its oil for culinary and other purposes. Large-scale plantations, established principally in tropical regions of Asia, Africa, and Latin America, are mostly aimed at the production of oil [38], which is extracted from the fleshy mesocarp of the palm fruit, and endosperm or kernel oil. Palm oil is different from other plant and animal oils in that it contains 50% saturated fatty acids, 40% unsaturated fatty acids and 10% polyunsaturated fatty acids. Because of its high saturated fat content, palm oil has not been very popular in the United States. Hydrogenated fats contain high levels of trans-fatty acids, which are now thought to have adverse health effects. The U.S. Food and Drug Administration's final ruling on trans-fatty acid labeling issued in 2003 has caused a rapid transformation in the fat and oil industries [39]. Palm oil is free of trans-fatty acid and is rapidly gaining wider acceptance by the food industry in the country. Primary applications include bakery products, breakfast cereals, wafers, and candies.

Rice bran oil, a by-product of the rice milling industry, is a major natural source of γ-tocotrienol but a poor source of α-tocotrienol. In addition, rice bran oil provides desmethyl tocotrienols. Two novel tocotrienols were isolated from stabilized and heated rice bran, apart from the known α-, β-, γ-, and δ-tocopherols and tocotrienols. These new tocotrienols are known as desmethyl tocotrienol [3,4-dihydro-2-methyl-2-(4,8,12-trimethyltrideca-3′(E),7′(E),11′-trienyl)-2H-1-benzopyran-6-ol] and didesmethy tocotrienol [3,4-dihydro-2-(4,8,12-trimethyltrideca-3′(E),7′(E),11′-trienyl)-2H-1-benzopyran-6-ol] [40]. Although scientific evidence is relatively limited, rice bran oil is believed to be a healthy vegetable oil in Asian countries [41].

Cereals such as oat, rye, and barley contain small amounts of tocotrienol in them. α-Tocotrienol is the predominant form of tocotrienol in oat (*Avena sativa* L.) and barley (56 and 40 mg/kg of dry weight, respectively). β-Tocotrienol is the major form of tocotrienol found in hulled and dehulled wheats (from 33 to 43 mg/kg of dry weight) [42]. Steaming and flaking of dehulled oat groats result in moderate losses of tocotrienols but not of tocopherols [43]. Although tocotrienols are present in edible natural products, it is questionable whether these dietary sources could provide sufficient amounts of tocotrienol to humans. For example, the processing of 1000 kg of crude palm oil is necessary to derive 1 kg of the commercial product Tocomin 50% (Carotech, NJ). Roughly, one would have to consume 100–200 g of palm/rice-bran oil or 1.5–4 kg of wheat germ, barley, or oat to achieve doses that have been published to be effective biologically. With this consideration in mind, appropriately configured dietary supplements seem to be a prudent choice.

12.4 ORAL SUPPLEMENTS EFFECTIVELY DELIVER TOCOTRIENOLS

During the last two decades, efforts to understand how dietary vitamin E is transported to the tissues have focused on α-tocopherol transport [44–47]. TTP has been identified to mediate α-tocopherol secretion into the plasma, while other tocopherol-binding proteins seem to play a less important role [46]. Tocotrienols have been known for decades, but why have they not been studied as well as α-tocopherol? Although there does not seem to be a straightforward rational answer to this question, one contributing factor is whether tocotrienol taken orally reaches vital organs of the body. This concern was primarily based on a 1997 finding that the transport system, TTP, responsible to carry α-tocopherol to vital organs has a poorer efficiency to transport tocotrienols to tissues [48]. The lack of relative specific affinity of TTP for tocotrienols led to the urban legend that availability of dietary tocotrienol to vital organs is negligible.

Although TTP is known to bind to α-tocotrienol with 8.5-fold lower affinity than that for α-tocopherol [48], it has not been clear whether, or to what extent, the delivery of orally supplemented α-tocotrienol to vital organs is dependent on TTP. It has been reported that TTP-deficient females are infertile presumably because of vitamin E deficiency [49]. This important observation was confirmed in a lineage of TTP-deficient mice. Placentas of pregnant TTP-deficient females were severely impaired with marked reduction of labyrinthine trophoblasts, and the embryos died at mid-gestation even when fertilized eggs of TTP-containing wild-type mice were transferred into TTP-deficient recipients [50]. Even in the presence of dietary α-tocopherol, TTP knockout mice are known to suffer from α-tocopherol deficiency [49,50]. It has recently been noted that oral supplementation of female mice with α-tocotrienol restored fertility of TTP knockout mice, suggesting that tocotrienol was successfully delivered to the relevant tissues and that tocotrienol supported reproductive function under conditions of α-tocopherol deficiency [51]. This observation was consistent with another line of evidence from rats where tocotrienol supplementation spared loss of fertility caused by long-term vitamin E deficiency in the diet [51]. TTP continues to be a key transport mechanism for the delivery of α-tocopherol to tissues. The significance of TTP in the transport of other forms of vitamin E remains unclear at present. It is clear, however, that natural isomers of vitamin E do get transported to vital organs even in the absence of TTP. Identification and characterization of TTP-independent vitamin E transport mechanisms *in vivo* are warranted.

Current findings support that oral tocotrienol (Carotech) not only reaches the brain [21,51,52] but also does so in amounts sufficient to protect against stroke [21]. The standard laboratory chow contains excessive amounts of α-tocopherol [53,54] but negligible amounts of tocotrienol. Long-term lack of tocotrienol in the diet may repress any putative tocotrienol transport mechanism *in vivo*. Thus, long-term supplementation studies are needed. In light of the knowledge that natural analogs of vitamin E may compete for specific transporting mechanisms [48], it is important that tocotrienol supplementation be performed under conditions of minimized copresence of tocopherols. Another related consideration is that although incorporation of orally supplemented vitamin E into tissues is a slow and progressive process, rapid incorporation of the supplement into tissues of newborns may occur in response to gavaging of pregnant mother rats [52]. Thus, an experimental design incorporating long-term tocotrienol supplementation under conditions of minimal dietary copresence of tocopherols and breeding of the supplemented colony would be a valuable approach to generate proof of principle testing whether dietary α-tocotrienol is capable of being transported to vital organs in vivo. In a recent study, rats were maintained on a vitamin E–deficient diet and gavaged with α-tocotrienol alone, α-tocopherol alone, or both. Five generations of rats were studied over sixty weeks [51]. Skin, adipose, heart, lungs, skeletal muscle brain, spinal cord, liver, and blood were studied. Oral tocotrienol was delivered to all vital organs. In some tissues, the level of tocotrienols exceeded that of tocopherols, indicating the presence of an efficient tocotrienol transport system in vivo. Baseline levels of α-tocotrienol in the skin of tocopherol-fed rats that never received any tocotrienol supplementation were negligible. Orally supplemented tocotrienol was rapidly taken up by the skin. Already in second-generation rats, α-tocotrienol levels in the skin of tocotrienol-supplemented rats exceeded twice the α-tocopherol levels in this organ. Of note, the α-tocotrienol level in the skin matched the α-tocotrienol level in the skin of rats fed with a comparable amount of tocopherol. When tocotrienol and tocopherol were cosupplemented, the uptake of α-tocotrienol by the skin was clearly blunted. In this group, α-tocotrienol levels were lower than α-tocotrienol levels in the skin, suggesting a direct competition between orally taken tocotrienol and tocopherol for delivery to the skin. Longer supplementation resulted in a marked increase in the α-tocotrienol levels in the skin of tocotrienol-fed rats, indicating a build-up of α-tocotrienol over time. Interestingly, the levels of α-tocotrienol in the skin of these rats were folds higher than the α-tocopherol levels in the skin of tocopherol-fed rats. This observation suggests the presence of an effective transport mechanism delivering α-tocotrienol to the skin and efficient retention of α-tocotrienol in the skin over time. Cosupplementation of tocotrienol and tocopherol demonstrated favorable uptake of α-tocopherol over α-tocotrienol. Adipose tissue serves as storage organ for vitamin E [55]. Analysis of adipose tissue vitamin E content of fifth-generation rats revealed substantially more accumulation of α-tocotrienol in that tissue than α-tocopherol.

In the case of tocotrienol and tocopherol feeding, results from third- and fifth-generation rats indicate higher levels of vitamin E in the skin of female rats compared with that of male rats. This gender-specific effect suggesting better transport of tocotrienol in females than in males was noted as a general trend across all organs studied. Gender-based differences in the transport of dietary vitamins are known to exist in specific cases [56]. Although the effect of several physiological factors on vitamin E transport has been studied, the gender factor remains to be specifically addressed [57]. It has recently been demonstrated that γ-tocopherol is more rapidly metabolized in women than in men [58]. The level of α-tocotrienol in the ovary was over five-fold higher than that in the testes from the corresponding male rats [51]. In the ovary, tocopherol is known to accumulate via a lipoprotein receptor-dependent mechanism [59]. Whether tocotrienol shares that mechanism remains to be tested.

Vitamin E enters the circulation from the intestine in chylomicrons. The conversion of chylomicrons to remnant particles results in the distribution of newly absorbed vitamin E to all the circulating lipoproteins and ultimately to tissues. This enrichment of lipoproteins with vitamin E is a key mechanism by which vitamin E is delivered to tissues [44]. In the liver, newly absorbed dietary lipids are incorporated into nascent very low density lipoproteins. The liver is responsible for the control and release of α-tocopherol into blood plasma. In the absence of TTP, α-tocopherol is not secreted back into the plasma. Excess vitamin E is not accumulated in the liver, but it is excreted, mostly in bile [44]. It has recently been noted that α-tocotrienol levels in the liver of rats and that of TTP-deficient mice were much lower than the levels of this vitamin E isoform in most peripheral tissues studied [51]. Such observation argues against a central role of the liver in delivering oral α-tocotrienol to peripheral tissues. TTP has the ability to bind to both α-tocopherol and α-tocotrienol. The affinity to bind α-tocopherol is several folds higher than that for α-tocotrienol [48]. Thus, under conditions of coexistence, α-tocopherol is expected to out-compete α-tocotrienol for binding. Although studies with TTP-deficient mice [51] indicate the existence of a TTP-independent mechanism for the tissue delivery of oral α-tocotrienol, observations in the rat [51] indicate that the mechanisms for transporting α-tocopherol and α-tocotrienol seem to compete such that transport of α-tocopherol is favored. Thus, cosupplementation of α-tocopherol and α-tocotrienol is likely to compromise tissue delivery of α-tocotrienol [51].

Few studies have specifically looked at the fate of oral tocotrienol supplementation in humans. In a study investigating the pharmacokinetics and bioavailability of α-, γ-, and δ-tocotrienols under fed and fasted conditions in eight healthy volunteers, subjects were administered a single 300-mg oral dose of mixed tocotrienols under fed or fasted conditions. The peak concentration of α-tocotrienol in the blood plasma was just over 1 μM [60]. The fed state increased the onset and extent of absorption of tocotrienols by more than two-fold. In addition, the mean apparent elimination half-lives of α-, γ-, and δ-tocotrienols were estimated to be 4.4, 4.3, and 2.3 h, respectively, being between 4.5- and 8.7-fold shorter than the half-life reported for α-tocopherol [60]. In another study, human subjects took tocotrienyl acetate supplements (250 mg/day) for 8 weeks while being on low-fat diet. In response to supplementation, the concentrations of tocotrienol in the mean blood plasma were as follows: α-tocotrienol, 0.98 μM; γ-tocotrienol, 0.54 μM; and δ-tocotrienol, 0.09 μM [6]. Thus, tocotrienyl acetate supplements were observed to be hydrolyzed, absorbed, and detectable in human plasma. A novel formulation for improved absorption of tocotrienols has recently been developed [61]. Emulsions are known to increase absorption of fat-soluble drugs. This invention is based on SEDDS (self-emulsifying drug delivery systems) technology [62–64]. Soft-gelatin capsules (Tocovid Suprabio™, Carotech) containing tocotrienol have been produced. Once ingested, the tocotrienols form emulsion when the contents are released and mixed with human gastrointestinal fluid. In a recent study using Tocovid Suprabio™, the postabsorptive fate of tocotrienol isomers and their association with lipoprotein subfractions were examined in humans [54]. The peak α-tocotrienol concentrations in supplemented individuals averaged approximately 3 μM in blood plasma, 1.7 μM in low-density lipoprotein, 0.9 μM in triglyceride-rich lipoprotein, and 0.5 μM in high-density lipoprotein. These peak plasma concentrations of α-tocotrienol are two to three times higher than the peak concentrations reported in previous studies using generic supplements not based on SEDDS [6,60].

12.5 NEUROPROTECTIVE EFFECTS OF NANOMOLAR TOCOTRIENOL REPRESENT THE MOST POTENT BIOLOGICAL FUNCTION OF ALL NATURAL FORMS OF VITAMIN E

On a concentration basis, the neuroprotective effects of nanomolar tocotrienol represent the most potent biological function of all natural forms of vitamin E. Glutamate toxicity is a major contributor to neurodegeneration. It includes excitotoxicity and an oxidative stress component also known as oxytosis [65,66]. Murine HT hippocampal neuronal cells, which lack an intrinsic excitotoxicity pathway, have been used as a standard model to characterize the oxidant-dependent component of glutamate toxicity. In 1999, we conducted a side-by-side comparison of all eight forms of natural vitamin E in a model of glutamate-induced neurodegeneration of HT neural cells. In subsequent experiments, it was observed that the neuroprotective property of tocotrienol applies not only to neural cell lines but also to primary cortical neurons. This line of experimentation led to an observation that eventually turned out to be the most potent function of any natural form of vitamin E on a concentration basis reported. Until then, all biological functions of vitamin E studied *in vitro* were observed at micromolar concentrations. Our studies led to the first evidence that α-tocotrienol was the most potent neuroprotective form of vitamin E in glutamate-induced degeneration of HT4 hippocampal neural cells [19]. What was striking in this study was the observation that nanomolar concentrations of α-tocotrienol, not α-tocopherol, provide complete neuroprotection. At such a low dose, tocotrienol was not protective against direct oxidant insult, suggesting that the observed neuroprotective effects of nanomolar tocotrienol were not dependent on the widely known antioxidant property of vitamin E. That tocotrienol-dependent neuroprotection includes a significant antioxidant-independent mechanism has now been established [20]. The neuroprotective property of tocotrienol holds good not only in response to glutamate challenge but also in response to other insults such as homocysteic acid, glutathione deficiency, and linoleic acid-induced oxidative stress [19,20]. It is now evident that tocotrienol at micromolar concentrations protects neural cells by virtue of its antioxidant property. At nanomolar concentrations, however, tocotrienol regulates specific neurodegenerative signaling processes.

The major tocotrienol-sensitive signaling pathways that are known to be involved in glutamate-induced neurodegeneration include c-Src and 12-lipoxygenase (12-Lox) [18–21,67]. In our initial search for signaling pathways that are sensitive to tocotrienol and play a decisive role in neurodegeneration, we were led to c-Src kinase [19]. c-Src and the structurally related members of the Src family are nonreceptor tyrosine kinases that reside within the cell associated with cell membranes and appear to transduce signals from transmembrane receptors to the cell interior. SH2 and SH3 domains are known to play a central role in regulating the catalytic activity of src protein tyrosine kinase. High-resolution crystal structures of human Src, in their repressed state, have provided a structural explanation for how intramolecular interactions of the SH3 and SH2 domains stabilize the inactive conformation of Src [68].

Our hypothesis that tocotrienol prevents neurodegeneration by regulating specific signaling processes involved in neurotoxicity led to screening for potential tocotrienol-sensitive candidate death pathways in HT4 cells. During such screening studies, inhibitors of the protein tyrosine kinase activity completely prevented glutamate-induced cell death. Herbimycin and geldanamycin potently inhibit c-Src tyrosine kinase activity [69,70], whereas lavendustin A is an inhibitor of extracellular growth factor receptor protein tyrosine kinase activity [71]. The observation that herbimycin and geldanamycin but not lavendustin A prevented glutamate-induced death of HT4 neuronal cells hinted the involvement of c-Src kinase activity in the death pathway. Immunoprecipitation of tyrosine-phosphorylated protein from cellular extracts confirmed that protein tyrosine phosphorylation reactions were indeed triggered by exposure of cells to elevated levels of glutamate and that such reactions were inhibited by nanomolar concentrations of α-tocotrienol [19]. These results however did not provide any information regarding the specific kinases involved. The involvement of c-Src kinase activity in the death pathway was verified by experiments involving the overexpression of catalytically active or inactive Src kinase. Indeed, overexpression of catalytically active Src kinase markedly sensitized the cells to HT4-induced death. Tocotrienol treatment completely prevented glutamate-induced death even in active c-Src kinase overexpressing cells, indicating that it either inhibited c-Src kinase activity or regulated one or more events upstream of c-Src kinase activation. Further evidence supporting this contention was provided by results obtained from the determination of c-Src kinase activity in HT4 cells. Glutamate treatment resulted in marked enhancement of c-Src kinase activity, and this change was completely blocked in cells treated with nanomolar amounts of α-tocotrienol. Further evidence establishing that signal transduction processes related to the cell death pathway are involved in glutamate-induced cytotoxicity was obtained from the study of ERK1 (extracellular signal-regulated kinase 1) and ERK2 activation. MEKK (mitogen-activated/extracellular response kinase kinase kinase) is a serine-threonine kinase that regulates sequential protein phosphorylation pathways, leading to the activation of mitogen-activated protein kinases (MAPKs), including members of the ERKs. MEKK selectively regulates signal transduction pathways that contribute to the apoptotic response [72]. When activated, p44 and p42 MAPKs (ERK1 and ERK2) are phosphorylated at specific threonine and tyrosine residues. ERK has been implicated in mediating the signaling events that precede apoptosis. ERK2 plays an active role in mediating anti-immunoglobulin M-induced apoptosis of B cells [73]. It has also been shown that H_2O_2 induces the activation of multiple MAPKs in oligodendrocyte progenitors and that the activation of ERK is associated with oxidant-mediated cytotoxicity [74]. Our studies showed that ERK1 and ERK2 are sensitive to elevated levels of extracellular glutamate. Rapid activation of ERK, particularly ERK2, was observed in response to glutamate treatment. Such response of ERK was completely inhibited in cells treated with α-tocotrienol, suggesting that α-tocotrienol influences an early event in the glutamate-induced death pathway [19]. In some cases, Src kinase activity is known to be required for the activation of ERK [75]. Thus, it is likely that tocotrienol inhibits inducible ERK activation by down-regulating Src kinase activity [19].

c-Src is heavily expressed in the brain [76] and in human neural tissues [77]. Differentiating rodent neurons are known to express high levels of c-Src. In neurons and astrocytes, c-Src is present at 15–20 times higher levels than those found in fibroblasts. The specific activity of the c-Src protein from neuronal cultures is 6–12 times higher than that from the astrocyte cultures, suggesting a key function of this protein in neurons [78]. Initially, c-Src was identified as being important in growth cone-mediated neurite extension and synaptic plasticity [79] and in neuronal differentiation [80]. Targeted disruption of c-Src, however, did not cause any abnormality in the brain [76]. Our pursuit for the neuroprotective mechanisms of tocotrienols led to the first evidence demonstrating that rapid c-Src activation [19,81] plays a central role in executing neurodegeneration. Consistently, it was demonstrated in a subsequent report that Src deficiency or blockade of Src activity in mice provides cerebral protection following stroke [82]. Further support of our claim that c-Src is a key player in neurodegeneration is provided by observation that the Src family kinase inhibitor PP2 reduces focal ischemic brain injury [83]. Our observation that tocotrienol-dependent inhibition of c-Src is beneficial for neuroprotection has now been extended to the heart. A recent study showed that c-Src mediates postischemic cardiac injury and dysfunction. Tocotrienol supplementation inhibited c-Src activation and protected the heart [84]. Many intracellular pathways can be stimulated upon Src activation, and a variety of cellular consequences can result. High c-Src is tightly associated with carcinogenesis. c-Src inhibitors are being actively studied for cancer therapy [85–88]. Based on the inducible c-Src inhibitory properties of tocotrienol, one may postulate that tocotrienol has anticancer properties. The anticancer properties of tocotrienol are discussed in a separate section below.

GSH is the major cellular thiol present in mammalian cells and is critical for maintenance of redox homeostasis [89]. GSH is a key survival factor in cells of the nervous system, and lowered [GSH]i is one of the early markers of neurotoxicity induced by a variety of agonists [90,91]. We observed that α-tocotrienol clearly protects primary cortical neurons against a number of GSH-lowering neurotoxins [18]. Of interest, the neurons survived even in the face of GSH loss. These observations led to the hypothesis that loss of [GSH]i alone is not lethal [18]. Given that pro-GSH agents are known to be neuroprotective in a variety of scenarios [91–93], it becomes reasonable to hypothesize that glutamate-induced lowering of [GSH]i triggers downstream responses that execute cell death. Our work led to the identification of 12-Lox as a key tocotrienol-sensitive mediator of neurodegeneration [18]. Specific inhibition of 12-Lox by BL15 protected neurons from glutamate-induced degeneration, although [GSH]i is compromised by 80%. Similar protective effects of BL15 were noted when BSO, a specific inhibitor of GSH synthesis, was used as the agonist. Importantly, neurons isolated from mice lacking the 12-Lox gene were observed to be resistant to glutamate-induced loss of viability [18]. This key piece of evidence established that 12-Lox indeed represents a critical checkpoint in glutamate-induced neurodegeneration.

Understanding the intracellular regulation of 12-Lox requires knowledge of the distribution of both enzyme protein and its activity. For example, in human erythroleukemia cells, the membrane fraction contains about 90% of the total cellular 12-Lox activity, whereas only 10% of 12-Lox activity resides in the cytosol. However, the majority of cellular 12-Lox protein is found in the cytosol [94]. Upon activation, 12-Lox may translocate to the membrane [94]. Consistently, we have observed the decreased presence of 12-Lox in the cytosol and its increased presence in the membrane of glutamate-treated cells. For 5-lipoxygenase, both catalytic function and translocation of the enzyme from the cytosol to the membrane are known to be regulated by tyrosine kinases [95]. We have recently noted that 12-Lox is subject to rapid tyrosine phosphorylation in neuronal cells challenged with glutamate or GSH-lowering agents. Such rapid phosphorylation coincides with the timeline of c-Src activation [21]. Inhibitors of c-Src abrogated inducible 12-Lox tyrosine phosphorylation, supporting the notion that c-Src may directly phosphorylate 12-Lox in challenged neurons. To test this hypothesis, we utilized genetic approaches of overexpressing kinase-active, kinase-dead, or dominant-negative c-Src in neuronal cells. Findings from cell biology studies and from the study of c-Src and 12-Lox in cell-free systems indicate that c-Src is rapidly activated and phosphorylates 12-Lox in response to challenge by glutamate or GSH-lowering agents [21].

Neurons and the brain are rich in arachidonic acid (AA; 20:4ω-6). Massive amounts of AA are released from the membranes in response to brain ischemia or trauma [96–100]. Subsequent work has established that AA and its metabolites may be neurotoxic. There are three major pathways of AA metabolism: lipoxygenases, cyclooxygenases, and cytochrome P450. The cyclooxygenase pathway has been preliminarily ruled out from being a contributor to neurodegeneration [101]. In the lipoxygenase pathway, metabolites of 12-Lox seem to be the major metabolite of AA in the brain [102,103] and in cultured cortical neurons [104–106]. Lipoxygenases, mainly 5-, 12-, and 15-Lox, are named for their ability to insert molecular oxygen at the 5-, 12-, and carbon atoms of AA, forming a distinct hydroperoxyeicosatetraenoic (HPETE) acid [107]. 12-Lox produces 12(S)-HPETE, which is further metabolized into four distinct products: an alcohol [12(S)-HETE], a ketone (12-keto-eicosatetraenoic acid), or two epoxy alcohols (hepoxilins A3 and B3). Immunohistochemical studies revealed the occurrence of 12-Lox in neurons—particularly in hippocampus, striatum, olivary nucleus, as well as in glial and in cerebral endothelial cells [108,109]. With the use of immature cortical neurons and HT cells, it has been shown that a decrease in [GSH]i triggers the activation of neuronal 12-Lox, which leads to the production of peroxides, the influx of Ca^{2+}, and ultimately to cell death [65,110]. The 12-Lox metabolite 12-HPETE proved to be capable of causing cell death [111]. Inhibition of 12-Lox protected cortical neurons from β-amyloid–induced toxicity [112]. Intracellular calcium chelation delayed cell death by lipoxygenase-mediated free radicals in mouse cortical cultures [113]. In summary, 12-Lox poses a clear threat to neuronal survival especially under GSH-deficient conditions.

Lipoxygenase activity is sensitive to vitamin E. α-Tocopherol strongly inhibits purified 5-Lox with an IC_{50} of 5 μM. The inhibition is independent of the antioxidant property of tocopherol. Tryptic digestion and peptide mapping of the 5-Lox-tocopherol complex indicated that tocopherol binds strongly to a single peptide [114]. Another study reported inhibition of 15-Lox by tocopherol via specific interaction with the enzyme protein [115]. Of interest, inhibitors specific for cyclooxygenase or 5-Lox are not effective in protecting neuronal cells against glutamate-induced death, suggesting a specific role of 12-Lox in glutamate-induced death [18,21]. Our studies addressing the effect of α-tocotrienol on pure 12-Lox indicate that α-tocotrienol directly interacts with the enzyme to suppress AA metabolism. In silico studies examining possible docking sites of α-tocotrienol to 12-Lox support the presence of an α-tocotrienol binding solvent cavity close to the active site. It has been demonstrated in 15-Lox that the COOH terminal of AA enters this solvent cavity while accessing the catalytic site [116]. It is therefore plausible that the binding position of α-tocotrienol prevents access of the natural substrate AA to the active site of 12-Lox [18]. Does 12-Lox have a tangible impact on neurodegenerative processes in vivo? In 1992, it was reported that a mixed lipoxygenase/cyclooxygenase inhibitor SK&F 105809 reduced cerebral edema after closed head injury in rat [117]. We noted that mice deficient in 12-Lox, but not 5-Lox [118], were significantly protected against stroke-related injury of the brain [21]. The case for 12-Lox as an important mediator of neurodegeneration in vivo is gaining additional support from independent studies [119]. 12-Lox has been also implicated in the pathogenesis of Alzheimer's disease [120]. α-Tocotrienol is capable of resisting neurodegeneration *in vivo* by opposing the c-Src and 12-Lox pathways.

Other than stroke, another area where tocotrienol-based therapeutics shows promise is in the treatment of familial dysautonomia (FD), a neurodegenerative genetic disorder primarily affecting individuals of Ashkenazi Jewish descent. FD is a neurodevelopmental genetic disorder within the larger classification of hereditary sensory and autonomic neuropathies, each caused by a different genetic error. Clinical features reflect widespread involvement of sensory and autonomic neurons. The FD gene has been identified as IKBKAP, a gene that encodes the IκB kinase complex-associated protein (IKAP). Mutations result in tissue-specific expression of mutant (IKAP). The genetic error probably affects development, as well as maintenance, of neurons because there is neuropathological and clinical progression. Pathological alterations consist of decreased unmyelinated and small-fiber neurons. In 2003, it was first reported that tocotrienols may induce the transcription of IKAP mRNA in FD-derived cells such that expression of correctly spliced transcript and normal protein is increased [121]. These findings suggested that *in vivo* supplementation with tocotrienols may elevate IKBKAP gene expression and in turn increase the amount of functional IKAP protein produced in FD patients. FD is associated with elevated plasma levels of norepinephrine and dopamine. These changes are observed during autonomic crises. Fetal tissue homozygous for the common FD-causing mutation and peripheral blood cells of individuals with FD have reduced monoamine oxidase (MAO) A mRNA levels. FD-derived cells, stimulated with tocotrienols produced at increased levels of functional IKAP, express increased amounts of MAO A mRNA transcript and protein. Administration of tocotrienol to individuals with FD resulted in increased expression of both functional IKAP and MAO A transcripts in their peripheral blood cells. As ingestion of tocotrienols elevates IKAP and MAO A in FD patients, a recent study examined their impact on the frequency of hypertensive crises and cardiac function. After 3 to 4 months of tocotrienol ingestion, approximately 80% of patients reported a significant (≥50%) decrease in the number of crises. In a smaller group of patients, a postexercise increase in heart rate and a decrease in the QT interval were observed in the majority of participants [122]. Thus, tocotrienol therapy may improve the long-term clinical outlook and survival of individuals with FD. Randomized, placebo-controlled clinical studies are warranted.

12.6 SUMMARY

Members of the natural vitamin E family possess overlapping and unique functional properties. Knowledge about their biology and significance is limited but rapidly expanding. Among the natural vitamin E molecules, d-α-tocopherol (RRR-α-tocopherol) has the highest bioavailability and is the standard against which all the others are compared. However, it is only one of eight natural forms of vitamin E. Interestingly, symptoms caused by α-tocopherol deficiency can be alleviated by tocotrienols. Thus, tocotrienols may be viewed as being members of the natural vitamin E family not only structurally but also functionally. Disappointments with outcomes-based studies investigating α-tocopherol [123,124] need to be cautiously handled, recognizing the untapped opportunities offered by the other forms of natural vitamin E. It has recently been suggested that the safe dose of various tocotrienols for human consumption is 200–1000 mg/day [125]. Vitamin E represents one of the most fascinating natural resources that have the potential to influence a broad range of mechanisms underlying human health and disease. The current state of knowledge warrants strategic investment into the lesser-known forms of vitamin E with emphasis on uncovering the specific conditions that govern the function of vitamin E molecules *in vivo*. Outcome studies designed in light of such information have the clear potential of returning dividends that are more lucrative than findings of the current clinical outcomes studies.

ACKNOWLEDGMENT

Tocotrienol research in the laboratory was supported by the National Institutes of Health through grant RO1NS42617.

REFERENCES

1. Lehtinen, M. K., Bonni, A., Modeling oxidative stress in the central nervous system, *Curr Mol Med* 6: 871–881, 2006.
2. Muller, D. P., Goss-Sampson, M. A., Role of vitamin E in neural tissue, *Ann N Y Acad Sci* 570: 146–155, 1989.
3. Muller, D. P., Goss-Sampson, M. A., Neurochemical, neurophysiological, and neuropathological studies in vitamin E deficiency, *Crit Rev Neurobiol* 5: 239–263, 1990.
4. Vatassery, G. T., Vitamin E and other endogenous antioxidants in the central nervous system, *Geriatrics* 53 Suppl 1: S25–S27, 1998.
5. Dormann, P., Functional diversity of tocochromanols in plants, *Planta* 225: 269–276, 2007.
6. O'Byrne, D., Grundy, S., Packer, L., Devaraj, S., Baldenius, K., Hoppe, P. P., Kraemer, K., Jialal, I., Traber, M. G., Studies of LDL oxidation following alpha-, gamma-, or delta-tocotrienyl acetate supplementation of hypercholesterolemic humans, *Free Radic Biol Med* 29: 834–845, 2000.
7. Hensley, K., Benaksas, E. J., Bolli, R., Comp, P., Grammas, P., Hamdheydari, L., Mou, S., Pye, Q. N., Stoddard, M. F., Wallis, G., Williamson, K. S., West, M., Wechter, W. J., Floyd, R. A., New perspectives on vitamin E: gamma-Tocopherol and carboxyelthylhydroxychroman metabolites in biology and medicine, *Free Radic Biol Med* 36: 1–15, 2004.
8. Dietrich, M., Traber, M. G., Jacques, P. F., Cross, C. E., Hu, Y., Block, G., Does {gamma}-tocopherol play a role in the primary prevention of heart disease and cancer? A review, *J Am Coll Nutr* 25: 292–299, 2006.
9. Kamat, C. D., Gadal, S., Mhatre, M., Williamson, K. S., Pye, Q. N., Hensley, K., Antioxidants in central nervous system diseases: Preclinical promise and translational challenges, *J Alzheimers Dis* 15: 473–493, 2008.
10. Harvard Health Publications, Vitamin E: Separate and Unequal. http://www.gather.com/viewArticle.jsp?memberId=-1&articleId=281474977262406. *Harvard Health Publications*, 2008.
11. Sen, C. K., Khanna, S., Rink, C., Roy, S., Tocotrienols: The emerging face of natural vitamin E, *Vitam Horm* 76: 203–261, 2007.
12. Steiner, M., Vitamin E: More than an antioxidant, *Clin Cardiol* 16: I16–I18, 1993.
13. Boscoboinik, D., Szewczyk, A., Hensey, C., Azzi, A., Inhibition of cell proliferation by alpha-tocopherol. Role of protein kinase C, *J Biol Chem* 266: 6188–6194, 1991.
14. Zingg, J. M., Azzi, A., Non-antioxidant activities of vitamin E, *Curr Med Chem* 11: 1113–1133, 2004.
15. Wagner, K. H., Kamal-Eldin, A., Elmadfa, I., Gamma-tocopherol—An underestimated vitamin? *Ann Nutr Metab* 48: 169–188, 2004.
16. Atkinson, J., Epand, R. F., Epand, R. M., Tocopherols and tocotrienols in membranes: A critical review, *Free Radic Biol Med* 44: 739–764, 2008.
17. Sen, C. K., Khanna, S., Roy, S., Tocotrienols: Vitamin E beyond tocopherols, *Life Sci* 78: 2088–2098, 2006.
18. Khanna, S., Roy, S., Ryu, H., Bahadduri, P., Swaan, P. W., Ratan, R. R., Sen, C. K., Molecular basis of vitamin E action: Tocotrienol modulates 12-lipoxygenase, a key mediator of glutamate-induced neurodegeneration, *J Biol Chem* 278: 43508–43515, 2003.
19. Sen, C. K., Khanna, S., Roy, S., Packer, L., Molecular basis of vitamin E action. Tocotrienol potently inhibits glutamate-induced pp60(c-Src) kinase activation and death of HT4 neuronal cells, *J Biol Chem* 275: 13049–13055, 2000.
20. Khanna, S., Roy, S., Parinandi, N. L., Maurer, M., Sen, C. K., Characterization of the potent neuroprotective properties of the natural vitamin E alpha-tocotrienol, *J Neurochem*, 98: 1474–1486, 2006.
21. Khanna, S., Roy, S., Slivka, A., Craft, T. K., Chaki, S., Rink, C., Notestine, M. A., DeVries, A. C., Parinandi, N. L., Sen, C. K., Neuroprotective properties of the natural vitamin E alpha-tocotrienol, *Stroke* 36: 2258–2264, 2005.
22. Cherubini, A., Ruggiero, C., Morand, C., Lattanzio, F., Dell'aquila, G., Zuliani, G., Di Iorio, A., Andres-Lacueva, C., Dietary antioxidants as potential pharmacological agents for ischemic stroke, *Curr Med Chem* 15: 1236–1248, 2008.
23. Pearce, B. C., Parker, R. A., Deason, M. E., Dischino, D. D., Gillespie, E., Qureshi, A. A., Volk, K., Wright, J. J., Inhibitors of cholesterol biosynthesis: 2. Hypocholesterolemic and antioxidant activities of benzopyran and tetrahydronaphthalene analogues of the tocotrienols, *J Med Chem* 37: 526–541, 1994.

24. Pearce, B. C., Parker, R. A., Deason, M. E., Qureshi, A. A., Wright, J. J., Hypocholesterolemic activity of synthetic and natural tocotrienols, *J Med Chem* 35: 3595–3606, 1992.

25. Qureshi, A. A., Burger, W. C., Peterson, D. M., Elson, C. E., The structure of an inhibitor of cholesterol biosynthesis isolated from barley, *J Biol Chem* 261: 10544–10550, 1986.

26. Qureshi, A. A., Sami, S. A., Salser, W. A., Khan, F. A., Dose-dependent suppression of serum cholesterol by tocotrienol-rich fraction (TRF25) of rice bran in hypercholesterolemic humans, *Atherosclerosis* 161: 199–207, 2002.

27. Song, B. L., Debose-Boyd, R. A., Insig-dependent ubiquitination and degradation of 3-hydroxy-3-methylglutaryl coenzyme A reductase stimulated by delta- and gamma-tocotrienols, *J Biol Chem* 267: 9080–9086, 2006.

28. Adachi, H., Ishii, N., Effects of tocotrienols on life span and protein carbonylation in *Caenorhabditis elegans*, *J Gerontol A Biol Sci Med Sci* 55: B280–B285, 2000.

29. Serbinova, E., Kagan, V., Han, D., Packer, L., Free radical recycling and intramembrane mobility in the antioxidant properties of alpha-tocopherol and alpha-tocotrienol, *Free Radic Biol Med* 10: 263–275, 1991.

30. Serbinova, E. A., Packer, L., Antioxidant properties of alpha-tocopherol and alpha-tocotrienol, *Methods Enzymol* 234: 354–366, 1994.

31. Suzuki, Y. J., Tsuchiya, M., Wassall, S. R., Choo, Y. M., Govil, G., Kagan, V. E., Packer, L., Structural and dynamic membrane properties of alpha-tocopherol and alpha-tocotrienol: Implication to the molecular mechanism of their antioxidant potency, *Biochemistry* 32: 10692–10699, 1993.

32. Nesaretnam, K., Guthrie, N., Chambers, A. F., Carroll, K. K., Effect of tocotrienols on the growth of a human breast cancer cell line in culture, *Lipids* 30: 1139–1143, 1995.

33. Sundram, K., Sambanthamurthi, R., Tan, Y. A., Palm fruit chemistry and nutrition, *Asia Pac J Clin Nutr* 12: 355–362, 2003.

34. Horvath, G., Wessjohann, L., Bigirimana, J., Jansen, M., Guisez, Y., Caubergs, R., Horemans, N., Differential distribution of tocopherols and tocotrienols in photosynthetic and non-photosynthetic tissues, *Phytochemistry* 67: 1185–1195, 2006.

35. Cahoon, E. B., Hall, S. E., Ripp, K. G., Ganzke, T. S., Hitz, W. D., Coughlan, S. J., Metabolic redesign of vitamin E biosynthesis in plants for tocotrienol production and increased antioxidant content, *Nat Biotechnol* 21: 1082–1087, 2003.

36. Rippert, P., Scimemi, C., Dubald, M., Matringe, M., Engineering plant shikimate pathway for production of tocotrienol and improving herbicide resistance, *Plant Physiol* 134: 92–100, 2004.

37. Elson, C. E., Tropical oils: Nutritional and scientific issues, *Crit Rev Food Sci Nutr* 31: 79–102, 1992.

38. Solomons, N. W., Orozco, M., Alleviation of vitamin A deficiency with palm fruit and its products, *Asia Pac J Clin Nutr* 12: 373–384, 2003.

39. Tarrago-Trani, M. T., Phillips, K. M., Lemar, L. E., Holden, J. M., New and existing oils and fats used in products with reduced trans-fatty acid content, *J Am Diet Assoc* 106: 867–880, 2006.

40. Qureshi, A. A., Mo, H., Packer, L., Peterson, D. M., Isolation and identification of novel tocotrienols from rice bran with hypocholesterolemic, antioxidant, and antitumor properties, *J Agric Food Chem* 48: 3130–3140, 2000.

41. Sugano, M., Koba, K., Tsuji, E., Health benefits of rice bran oil, *Anticancer Res* 19: 3651–3657, 1999.

42. Panfili, G., Fratianni, A., Irano, M., Normal phase high-performance liquid chromatography method for the determination of tocopherols and tocotrienols in cereals, *J Agric Food Chem* 51: 3940–3944, 2003.

43. Bryngelsson, S., Dimberg, L. H., Kamal-Eldin, A., Effects of commercial processing on levels of antioxidants in oats (*Avena sativa* L.), *J Agric Food Chem* 50: 1890–1896, 2002.

44. Traber, M. G., Burton, G. W., Hamilton, R. L., Vitamin E trafficking, *Ann N Y Acad Sci* 1031: 1–12, 2004.

45. Traber, M. G., Arai, H., Molecular mechanisms of vitamin E transport, *Ann Rev Nutr* 19: 343–355, 1999.

46. Kaempf-Rotzoll, D. E., Traber, M. G., Arai, H., Vitamin E and transfer proteins, *Curr Opin Lipidol* 14: 249–254, 2003.

47. Blatt, D. H., Leonard, S. W., Traber, M. G., Vitamin E kinetics and the function of tocopherol regulatory proteins, *Nutrition* 17: 799–805, 2001.

48. Hosomi, A., Arita, M., Sato, Y., Kiyose, C., Ueda, T., Igarashi, O., Arai, H., Inoue, K., Affinity for alpha-tocopherol transfer protein as a determinant of the biological activities of vitamin E analogs, *FEBS Lett* 409: 105–108, 1997.

49. Terasawa, Y., Ladha, Z., Leonard, S. W., Morrow, J. D., Newland, D., Sanan, D., Packer, L., Traber, M. G., Farese, R. V., Jr., Increased atherosclerosis in hyperlipidemic mice deficient in alpha-tocopherol transfer protein and vitamin E, *Proc Natl Acad Sci U S A* 97: 13830–13834, 2000.

50. Jishage, K., Arita, M., Igarashi, K., Iwata, T., Watanabe, M., Ogawa, M., Ueda, O., Kamada, N., Inoue, K., Arai, H., Suzuki, H., Alpha-tocopherol transfer protein is important for the normal development of placental labyrinthine trophoblasts in mice, *J Biol Chem* 276: 1669–1672, 2001.

51. Khanna, S., Patel, V., Rink, C., Roy, S., Sen, C. K., Delivery of orally supplemented alpha-tocotrienol to vital organs of rats and tocopherol-transport protein deficient mice, *Free Radic Biol Med* 39: 1310–1319, 2005.

52. Roy, S., Lado, B. H., Khanna, S., Sen, C. K., Vitamin E sensitive genes in the developing rat fetal brain: A high-density oligonucleotide microarray analysis, *FEBS Lett* 530: 17–23, 2002.

53. van der Worp, H. B., Bar, P. R., Kappelle, L. J., de Wildt, D. J., Dietary vitamin E levels affect outcome of permanent focal cerebral ischemia in rats, *Stroke* 29: 1002–1005, discussion 1005–1006, 1998.

54. Khosla, P., Patel, V., Whinter, J. M., Khanna, S., Rakhkovskaya, M., Roy, S., Sen, C. K., Postprandial levels of the natural vitamin E tocotrienol in human circulation, *Antioxid Redox Signal* 8: 1059–1068, 2006.

55. Adachi, K., Miki, M., Tamai, H., Tokuda, M., Mino, M., Adipose tissues and vitamin E, *J Nutr Sci Vitaminol (Tokyo)* 36: 327–337, 1990.

56. Garry, P. J., Hunt, W. C., Bandrofchak, J. L., VanderJagt, D., Goodwin, J. S., Vitamin A intake and plasma retinol levels in healthy elderly men and women, *Am J Clin Nutr* 46: 989–994, 1987.

57. Lodge, J. K., Hall, W. L., Jeanes, Y. M., Proteggente, A. R., Physiological factors influencing vitamin E biokinetics, *Ann N Y Acad Sci* 1031: 60–73, 2004.

58. Leonard, S. W., Paterson, E., Atkinson, J. K., Ramakrishnan, R., Cross, C. E., Traber, M. G., Studies in humans using deuterium-labeled alpha- and gamma-tocopherols demonstrate faster plasma gamma-tocopherol disappearance and greater gamma-metabolite production, *Free Radic Biol Med* 38: 857–866, 2005.

59. Aten, R. F., Kolodecik, T. R., Behrman, H. R., Ovarian vitamin E accumulation: Evidence for a role of lipoproteins, *Endocrinology* 135: 533–539, 1994.

60. Yap, S. P., Yuen, K. H., Wong, J. W., Pharmacokinetics and bioavailability of alpha-, gamma- and delta-tocotrienols under different food status, *J Pharm Pharmacol* 53: 67–71, 2001.

61. Ho, D., Yuen, K. H., Yap, S. P., United States Patent No. 6596306, 2003.

62. Hong, J. Y., Kim, J. K., Song, Y. K., Park, J. S., Kim, C. K., A new self-emulsifying formulation of itraconazole with improved dissolution and oral absorption, *J Control Release* 110: 332–338, 2006.

63. Gao, P., Morozowich, W., Development of supersaturatable self-emulsifying drug delivery system formulations for improving the oral absorption of poorly soluble drugs, *Expert Opin Drug Deliv* 3: 97–110, 2006.

64. Araya, H., Tomita, M., Hayashi, M., The novel formulation design of self-emulsifying drug delivery systems (SEDDS) type O/W microemulsion: III. The permeation mechanism of a poorly water soluble drug entrapped O/W microemulsion in rat isolated intestinal membrane by the Ussing chamber method, *Drug Metab Pharmacokinet* 21: 45–53, 2006.

65. Tan, S., Schubert, D., Maher, P., Oxytosis: A novel form of programmed cell death, *Curr Top Med Chem* 1: 497–506, 2001.

66. Schubert, D., Piasecki, D., Oxidative glutamate toxicity can be a component of the excitotoxicity cascade, *J Neurosci* 21: 7455–7462, 2001.

67. Sen, C. K., Khanna, S., Roy, S., Tocotrienol: The natural vitamin E to defend the nervous system? *Ann N Y Acad Sci* 1031: 127–142, 2004.

68. Thomas, S. M., Brugge, J. S., Cellular functions regulated by Src family kinases, *Annu Rev Cell Dev Biol* 13: 513–609, 1997.

69. Yoneda, T., Lowe, C., Lee, C. H., Gutierrez, G., Niewolna, M., Williams, P. J., Izbicka, E., Uehara, Y., Mundy, G. R., Herbimycin A, a pp60c-src tyrosine kinase inhibitor, inhibits osteoclastic bone resorption *in vitro* and hypercalcemia *in vivo*, *J Clin Invest* 91: 2791–2795, 1993.

70. Hall, T. J., Schaeublin, M., Missbach, M., Evidence that c-src is involved in the process of osteoclastic bone resorption, *Biochem Biophys Res Commun* 199: 1237–1244, 1994.

71. Hsu, C. Y., Persons, P. E., Spada, A. P., Bednar, R. A., Levitzki, A., Zilberstein, A., Kinetic analysis of the inhibition of the epidermal growth factor receptor tyrosine kinase by lavendustin-A and its analogue, *J Biol Chem* 266: 21105–21112, 1991.

72. Johnson, N. L., Gardner, A. M., Diener, K. M., Lange-Carter, C. A., Gleavy, J., Jarpe, M. B., Minden, A., Karin, M., Zon, L. I., Johnson, G. L., Signal transduction pathways regulated by mitogen-activated/extracellular response kinase kinase kinase induce cell death, *J Biol Chem* 271: 3229–3237, 1996.

73. Lee, J. R., Koretzky, G. A., Extracellular signal-regulated kinase-2, but not c-Jun NH2-terminal kinase, activation correlates with surface IgM-mediated apoptosis in the WEHI 231 B cell line, *J Immunol* 161: 1637–1644, 1998.

74. Bhat, N. R., Zhang, P., Hydrogen peroxide activation of multiple mitogen-activated protein kinases in an oligodendrocyte cell line: Role of extracellular signal-regulated kinase in hydrogen peroxide-induced cell death, *J Neurochem* 72: 112–119, 1999.

75. Aikawa, R., Komuro, I., Yamazaki, T., Zou, Y., Kudoh, S., Tanaka, M., Shiojima, I., Hiroi, Y., Yazaki, Y., Oxidative stress activates extracellular signal-regulated kinases through Src and Ras in cultured cardiac myocytes of neonatal rats, *J Clin Invest* 100: 1813–1821, 1997.

76. Soriano, P., Montgomery, C., Geske, R., Bradley, A., Targeted disruption of the c-src proto-oncogene leads to osteopetrosis in mice, *Cell* 64: 693–702, 1991.

77. Pyper, J. M., Bolen, J. B., Neuron-specific splicing of C-SRC RNA in human brain, *J Neurosci Res* 24: 89–96, 1989.

78. Brugge, J. S., Cotton, P. C., Queral, A. E., Barrett, J. N., Nonner, D., Keane, R. W., Neurones express high levels of a structurally modified, activated form of pp60c-src, *Nature* 316: 554–557, 1985.

79. Maness, P. F., Aubry, M., Shores, C. G., Frame, L., Pfenninger, K. H., c-src gene product in developing rat brain is enriched in nerve growth cone membranes, *Proc Natl Acad Sci U S A* 85: 5001–5005, 1988.

80. Ingraham, C. A., Cox, M. E., Ward, D. C., Fults, D. W., Maness, P. F., c-src and other proto-oncogenes implicated in neuronal differentiation, *Mol Chem Neuropathol* 10: 1–14, 1989.

81. Khanna, S., Venojarvi, M., Roy, S., Sen, C. K., Glutamate-induced c-Src activation in neuronal cells, *Methods Enzymol* 352: 191–198, 2002.

82. Paul, R., Zhang, Z. G., Eliceiri, B. P., Jiang, Q., Boccia, A. D., Zhang, R. L., Chopp, M., Cheresh, D. A., Src deficiency or blockade of Src activity in mice provides cerebral protection following stroke, *Nat Med* 7: 222–227, 2001.

83. Lennmyr, F., Ericsson, A., Gerwins, P., Akterin, S., Ahlstrom, H., Terent, A., Src family kinase-inhibitor PP2 reduces focal ischemic brain injury, *Acta Neurol Scand* 110: 175–179, 2004.

84. Das, S., Powell, S. R., Wang, P., Divald, A., Nesaretnam, K., Tosaki, A., Cordis, G. A., Maulik, N., Das, D. K., Cardioprotection with palm tocotrienol: Antioxidant activity of tocotrienol is linked with its ability to stabilize proteasomes, *Am J Physiol Heart Circ Physiol* 289: H361–H367, 2005.

85. Alper, O., Bowden, E. T., Novel insights into c-Src, *Curr Pharm Des* 11: 1119–1130, 2005.

86. Ishizawar, R., Parsons, S. J., c-Src and cooperating partners in human cancer, *Cancer Cell* 6: 209–214, 2004.

87. Lau, A. F., c-Src: Bridging the gap between phosphorylation- and acidification-induced gap junction channel closure, *Sci STKE* 2005: pe33, 2005.

88. Shupnik, M. A., Crosstalk between steroid receptors and the c-Src-receptor tyrosine kinase pathways: Implications for cell proliferation, *Oncogene* 23: 7979–7989, 2004.

89. Sun, X., Shih, A. Y., Johannssen, H. C., Erb, H., Li, P., Murphy, T. H., Two-photon imaging of glutathione levels in intact brain indicates enhanced redox buffering in developing neurons and cells at the cerebro-spinal fluid and blood-brain interface, *J Biol Chem* 281: 17420–17431, 2006.

90. Dringen, R., Gutterer, J. M., Hirrlinger, J., Glutathione metabolism in brain metabolic interaction between astrocytes and neurons in the defense against reactive oxygen species, *Eur J Biochem* 267: 4912–4916, 2000.

91. Bains, J. S., Shaw, C. A., Neurodegenerative disorders in humans: The role of glutathione in oxidative stress-mediated neuronal death, *Brain Res - Brain Res Rev* 25: 335–358, 1997.

92. Han, D., Sen, C. K., Roy, S., Kobayashi, M. S., Tritschler, H. J., Packer, L., Protection against glutamate-induced cytotoxicity in C6 glial cells by thiol antioxidants, *Am J Physiol* 273: R1771–R1778, 1997.

93. Schulz, J. B., Lindenau, J., Seyfried, J., Dichgans, J., Glutathione, oxidative stress and neurodegeneration, *Eur J Biochem* 267: 4904–4911, 2000.

94. Hagmann, W., Kagawa, D., Renaud, C., Honn, K. V., Activity and protein distribution of 12-lipoxygenase in HEL cells: Induction of membrane-association by phorbol ester TPA, modulation of activity by glutathione and 13-HPODE, and Ca(2+)-dependent translocation to membranes, *Prostaglandins* 46: 471–477, 1993.

95. Lepley, R. A., Muskardin, D. T., Fitzpatrick, F. A., Tyrosine kinase activity modulates catalysis and translocation of cellular 5-lipoxygenase, *J Biol Chem* 271: 6179–6184, 1996.

96. Bazan, N. G., Jr., Effects of ischemia and electroconvulsive shock on free fatty acid pool in the brain, *Biochim Biophys Acta* 218: 1–10, 1970.

97. Bazan, N. G., Jr., Changes in free fatty acids of brain by drug-induced convulsions, electroshock and anaesthesia, *J Neurochem* 18: 1379–1385, 1971.

98. Bazan, N. G., Jr., Phospholipases A1 and A2 in brain subcellular fractions, *Acta Physiol Lat Am* 21: 101–106, 1971.

99. Bazan, N. G., Free arachidonic acid and other lipids in the nervous system during early ischemia and after electroshock, *Adv Exp Med Biol* 72: 317–335, 1976.

100. Bazan, N. G., Jr., Rakowski, H., Increased levels of brain free fatty acids after electroconvulsive shock, *Life Sci* 9: 501–507, 1970.

101. Kwon, K. J., Jung, Y. S., Lee, S. H., Moon, C. H., Baik, E. J., Arachidonic acid induces neuronal death through lipoxygenase and cytochrome P450 rather than cyclooxygenase, *J Neurosci Res* 81: 73–84, 2005.

102. Adesuyi, S. A., Cockrell, C. S., Gamache, D. A., Ellis, E. F., Lipoxygenase metabolism of arachidonic acid in brain, *J Neurochem* 45: 770–776, 1985.

103. Carlen, P. L., Gurevich, N., Zhang, L., Wu, P. H., Reynaud, D., Pace-Asciak, C. R., Formation and electrophysiological actions of the arachidonic acid metabolites, hepoxilins, at nanomolar concentrations in rat hippocampal slices, *Neuroscience* 58: 493–502, 1994.

104. Miyamoto, T., Lindgren, J. A., Hokfelt, T., Samuelsson, B., Regional distribution of leukotriene and mono-hydroxyeicosatetraenoic acid production in the rat brain. Highest leukotriene C4 formation in the hypothalamus, *FEBS Lett* 216: 123–127, 1987.

105. Miyamoto, T., Lindgren, J. A., Hokfelt, T., Samuelsson, B., Formation of lipoxygenase products in the rat brain, *Adv Prostaglandin Thromboxane Leukot Res* 17B: 929–933, 1987.

106. Ishizaki, Y., Murota, S., Arachidonic acid metabolism in cultured astrocytes: Presence of 12-lipoxygenase activity in the intact cells, *Neurosci Lett* 131: 149–152, 1991.

107. Yamamoto, S., Mammalian lipoxygenases: Molecular structures and functions, *Biochim Biophys Acta* 1128: 117–131, 1992.

108. Nishiyama, M., Watanabe, T., Ueda, N., Tsukamoto, H., Watanabe, K., Arachidonate 12-lipoxygenase is localized in neurons, glial cells, and endothelial cells of the canine brain, *J Histochem Cytochem* 41: 111–117, 1993.

109. Nishiyama, M., Okamoto, H., Watanabe, T., Hori, T., Hada, T., Ueda, N., Yamamoto, S., Tsukamoto, H., Watanabe, K., Kirino, T., Localization of arachidonate 12-lipoxygenase in canine brain tissues, *J Neurochem* 58: 1395–1400, 1992.

110. Li, Y., Maher, P., Schubert, D., A role for 12-lipoxygenase in nerve cell death caused by glutathione depletion, *Neuron* 19: 453–463, 1997.

111. Gu, J., Liu, Y., Wen, Y., Natarajan, R., Lanting, L., Nadler, J. L., Evidence that increased 12-lipoxygenase activity induces apoptosis in fibroblasts, *J Cell Physiol* 186: 357–365, 2001.

112. Lebeau, A., Esclaire, F., Rostene, W., Pelaprat, D., Baicalein protects cortical neurons from beta-amyloid (25–35) induced toxicity, *NeuroReport* 12: 2199–2202, 2001.

113. Wie, M. B., Koh, J. Y., Won, M. H., Lee, J. C., Shin, T. K., Moon, C. J., Ha, H. J., Park, S. M., Kim, H. C., BAPTA/AM, an intracellular calcium chelator, induces delayed necrosis by lipoxygenase-mediated free radicals in mouse cortical cultures, *Prog Neuropsychopharmacol Biol Psychiatry* 25: 1641–1659, 2001.

114. Reddanna, P., Rao, M. K., Reddy, C. C., Inhibition of 5-lipoxygenase by vitamin E, *FEBS Lett* 193: 39–43, 1985.

115. Grossman, S., Waksman, E. G., New aspects of the inhibition of soybean lipoxygenase by alpha-tocopherol. Evidence for the existence of a specific complex, *Int J Biochem* 16: 281–289, 1984.

116. Borngraber, S., Browner, M., Gillmor, S., Gerth, C., Anton, M., Fletterick, R., Kuhn, H., Shape and specificity in mammalian 15-lipoxygenase active site. The functional interplay of sequence determinants for the reaction specificity, *J Biol Chem* 274: 37345–37350, 1999.

117. Shohami, E., Glantz, L., Nates, J., Feuerstein, G., The mixed lipoxygenase/cyclooxygenase inhibitor SK&F 105809 reduces cerebral edema after closed head injury in rat, *J Basic Clin Physiol Pharmacol* 3: 99–107, 1992.

118. Kitagawa, K., Matsumoto, M., Hori, M., Cerebral ischemia in 5-lipoxygenase knockout mice, *Brain Res* 1004: 198–202, 2004.

119. Musiek, E. S., Breeding, R. S., Milne, G. L., Zanoni, G., Morrow, J. D., McLaughlin, B., Cyclopentenone isoprostanes are novel bioactive products of lipid oxidation which enhance neurodegeneration, *J Neurochem* 97: 1301–1313, 2006.

120. Yao, Y., Clark, C. M., Trojanowski, J. Q., Lee, V. M., Pratico, D., Elevation of 12/15 lipoxygenase products in AD and mild cognitive impairment, *Ann Neurol* 58: 623–626, 2005.

121. Anderson, S. L., Qiu, J., Rubin, B. Y., Tocotrienols induce IKBKAP expression: A possible therapy for familial dysautonomia, *Biochem Biophys Res Commun* 306: 303–309, 2003.

122. Rubin, B. Y., Anderson, S. L., Kapas, L., Can the therapeutic efficacy of tocotrienols in neurodegenerative familial dysautonomia patients be measured clinically? *Antioxid Redox Signal* 10: 837–841, 2008.

123. Greenberg, E. R., Vitamin E supplements: Good in theory, but is the theory good? *Ann Intern Med* 142: 75–76, 2005.

124. Friedrich, M. J., To "E" or not to "E," vitamin E's role in health and disease is the question, *JAMA* 292: 671–673, 2004.

125. Yu, S. G., Thomas, A. M., Gapor, A., Tan, B., Qureshi, N., Qureshi, A. A., Dose-response impact of various tocotrienols on serum lipid parameters in 5-week-old female chickens, *Lipids* 41: 453–461, 2006.

13 Fruits, Nuts, and Brain Aging

Nutritional Interventions Targeting Age-Related Neuronal and Behavioral Deficits

James A. Joseph, Barbara Shukitt-Hale, and Lauren M. Willis

Agricultural Research Service, U.S. Department of
Agriculture Human Nutrition Research Center on Aging
at Tufts University, Boston, Massachusetts, USA

CONTENTS

13.1 INTRODUCTION

For many years there has been a great deal of ongoing research from numerous labs showing declines in both motor and cognitive functions in aging even in the absence of neurodegenerative disease. The alterations in motor function may include decreases in balance, muscle strength, and coordination (Bartus, Flicker et al., 1983; Joseph, Hunt et al., 1992), while memory deficits are seen on cognitive tasks that require the use of spatial learning and memory (Ingram, Spangler et al., 1994; Shukitt-Hale, Mouzakis et al., 1998). The deficits have been observed in both animals (Bartus, 1990; Ingram, Spangler et al., 1994; Shukitt-Hale, Mouzakis et al., 1998) and humans (Muir, 1997; West, 1996). Age-related deficits in motor performance may result from deficits in the striatal dopamine (DA) system (Joseph, Hunt et al., 1992) or cerebellum (Bickford, 1993; Bickford, Heron et al., 1992).

Alterations in cognition appear to occur primarily in secondary memory systems that reflect the storage of newly acquired information (Bartus, Dean et al., 1982; Joseph, Hunt et al., 1992). It is thought that the hippocampus mediates "place" learning and that the prefrontal cortex is critical to acquiring the rules that govern performance in procedural knowledge. The dorsomedial striatum appears to mediate response and cue learning (Devan, Goad et al., 1996; McDonald and White, 1994; Oliveira, Bueno et al., 1997; Zyzak, Otto et al., 1995). Unfortunately, although the behavioral changes have been well studied and elucidated, specification of the mechanisms involved had been elusive. One avenue of exploration that has received a great deal of attention in recent years involves elucidating the role of oxidative stress (OS) and inflammation. OS results from an overwhelming of the cellular defenses by reactive oxygen species (ROS) resulting in the accumulation of free radicals and ROS and reactive nitrogen species (Andersen, 2004). These free radicals have been shown to be cytotoxic, inducing damage to DNA, lipids, and proteins, leading to compromised cellular function (Ames, Shigena et al., 1993; Andersen, 2004). Additionally, the ROS could also result from increased inflammatory processes. The studies reviewed below have provided clear evidence that both of these insults are intimately involved in the loss of behavioral and neuronal function in aging (Hauss-Wegrzyniak, Vannucchi et al., 2000; Shukitt-Hale, 1999; Shukitt-Hale, Casadesus et al., 2000).

Importantly, it is known that plants, including food plants (fruits and nuts), synthesize a vast array of secondary chemical compounds that, while not involved in their primary metabolism, are important in serving a variety of ecological functions that enhance the plants' survivability. Interestingly, these antioxidant/anti-inflammatory compounds may be responsible for the multitude of beneficial effects that have been reported for fruits on an array of behavioral and neuronal benefits that are discussed below.

13.2   OXIDATIVE STRESS/INFLAMMATORY INTERACTIONS

13.2.1   OXIDATIVE STRESS

If the increases in ROS in the aging brain are greater than the endogenous defenses, it could be particularly devastating to neuronal function. This is particularly important since studies have found indications of increased OS in brain aging, including: reductions in redox-active iron (Gilissen, Jacobs et al., 1999; Savory, Rao et al., 1999), increases in Bcl-2 (Sadoul, 1998), membrane lipid peroxidation (Yu, 1994), and increases in cellular hydrogen peroxide (Cavazzoni, Barogi et al., 1999). Additionally, there appear to be significant lipofuscin accumulation (Gilissen, Jacobs et al., 1999) and alterations in membrane lipids (Denisova, Erat et al., 1998). Recent studies have also suggested the involvement of lipid rafts with OS sensitivity (Shen, Lin et al., 2004). Importantly, the consequences of these increases in OS can ultimately induce changes in gene expression (Annunziato, Pannaccione et al., 2002; Dalton, Shertzer et al., 1999; Davies, 2000; Hughes and Reynolds, 2005; Perez-Campo, Lopez-Torres et al., 1998; Waring, 2005), further increasing vulnerability to OS as seen in aging (Halliwell, 2001; Rego and Oliveira, 2003) and age-related neurodegenerative diseases [e.g., Alzheimer's disease (Lovell, Ehmann et al., 1995; Marcus, Thomas et al., 1998; Smith, Carney et al., 1991) and Parkinson's disease (Dexter, Holley et al., 1994; Spencer, Jenner et al., 1998)].

Additional factors that may contribute to OS vulnerability in aging may involve microvasculature changes in vulnerable motor and memory control areas such as the hippocampus, as well as increases in oxidized proteins and lipids (Floyd and Hensley, 2002), and in the membrane microenvironment and structure (Joseph, Denisova et al., 1998; Joseph, Shukitt-Hale et al., 2001). There also may be critical declines in endogenous antioxidant protection, ratio of oxidized to total glutathione (Olanow, 1992) and glutamine synthetase (Carney, Smith et al., 1994). Taken together, these findings indicate that there are increases in OS in aging, that the central nervous system (CNS) may be particularly vulnerable to these increases (see Joseph, Denisova et al., 1998a and Joseph, Denisova et al., 1998b for review) and that the efficacy of antioxidants may be reduced in aging.

Finally, decrements in calcium homeostasis, alterations in cellular signaling cascades and decreases in calcium buffering (i.e., increases in intracellular calcium) have been shown to be significantly reduced in senescence (Joseph, Shukitt-Hale et al., 1999; Landfield and Eldridge, 1994; Toescu and Verkhratsky, 2004). The consequences of such long-lasting increases in cytosolic calcium may involve cell death induced by several mechanisms (e.g., xanthine oxidase activation; Cheng, Wixom et al., 1994), with subsequent pro-oxidant generation and loss of functional capacity of the cell. Similarly, amyloid beta (Aβ) peptides (which are oxidative stressors *in vitro*) have been shown to negatively promote calcium influx and disrupt calcium homeostasis in a manner similar to that seen in Alzheimer's disease (Foster and Kumar, 2002; Hartmann, Vormstein et al., 1996) and aging (Joseph, Shukitt-Hale et al., 1999; Toescu and Verkhratsky, 2004). Recent results from

Brewer, Lim et al. (2005) also suggest that hippocampal neurons obtained from aged animals have a greater sensitivity to Aβ application than those derived from middle-aged or embryonic animals, and that the effects of the Aβ are preceded by increases in OS.

However, the other "partner" in this *pas de deux* of brain aging is inflammation. OS may contribute to these age-related diseases by inducing the expression of pro-inflammatory cytokines through activation of the OS signal nuclear factor kappa B (Durany, Munch et al., 1999; Munch, Schinzel et al., 1998), among others (discussed in more detail below). These signals up-regulate the inflammatory responses, leading to a further increase in ROS (Lane, 2003), which results in a continuous increase in OS and inflammation, and thus vulnerability to further stressors.

13.2.2 INFLAMMATION

In addition to OS, CNS inflammatory events may have an important role in affecting neuronal and behavioral deficits in aging (Bodles and Barger, 2004). They emerge from several sources, such as glia and microglia, cytokines, and cyclooxygenases (COX-2). Studies indicate that activated glial cells even increase in the normal aging brain, resulting in increased immunoreactivity in markers for both microglia and astrocytes (Rozovsky, Finch et al., 1998; Sheng, Mrak et al., 1998; Sloane, Hollander et al., 1999) as compared with young animals. Additionally, increased glial fibrillary acid protein expression is observed by middle age (Rozovsky, Finch et al., 1998), followed by further increases in the elderly even without the influence of a defined stressor (McGeer and McGeer, 1995). This is an important finding when one considers that glial cells mediate the endogenous immune system in the CNS (Kreutzberg, 1996). Their activation is the hallmark of inflammatory processes in the brain (Orr, Rowe et al., 2002), and the mediators produced by the activated microglia can produce a myriad of inflammatory molecules that include cytokines, growth factors, and complement proteins (Chen, Frederickson et al., 1996; Darley-Usmar, Wiseman et al., 1995; McGeer and McGeer, 1995). These proinflammatory mediators can initiate the production of additional signals [e.g., interleukin-1, interleukin-6, and tumor necrosis factor-α (Luterman, Haroutunian et al., 2000; Tarkowski, Liljeroth et al., 2003)], further activating microglia in a positive-feedback loop to continue and enhance the cascade (Floyd, 1999).

It is also important to note that stress signals such as cytokines, and cyclooxygenases, among others, may act as generators of additional ROS that are associated with decrements in neuronal function or glial neuronal interactions (Rosenman, Shrikant et al., 1995; Schipper, 1996; Steffen, Breier et al., 1996; Stella, Estelles et al., 1997; Woodroofe, 1995) and perhaps, ultimately, the deficits in behavior that have been observed in aging.

In support of these findings, rodent studies have suggested that young animals exposed to OS show neuronal and behavioral changes similar to those seen in aged animals. Thus, young animals irradiated with particles of high energy and charge show behavioral deficits paralleling those observed in aging (Joseph, Erat et al., 1998; Joseph, Shukitt-Hale et al., 2000; Shukitt-Hale, Casadesus et al., 2000). Particles of high energy and charge (specifically, 600 MeV or 1 GeV of ^{56}Fe) also disrupt the functioning of DA-mediated behaviors, such as motor behavior (Joseph, Hunt et al., 1992), spatial learning, and memory (Rabin, Joseph et al., 1998), and amphetamine-induced conditioned taste aversion (Bickford, Shukitt-Hale et al., 1999). Given these considerations, the question as to the use of interventions such as those involving nutrition to increase endogenous antioxidant/anti-inflammatory protection arises.

13.3 NUTRITIONAL INTERVENTIONS

13.3.1 BERRY POLYPHENOLS AS NEUROPROTECTIVE AGENTS

One emerging therapeutic option to address a multiplicity of neurodegenerative processes may be the use of nutritional interventions (Joseph, Shukitt-Hale et al., 2005). Anecdotal evidence that dietary intake of fruits and vegetables improves brain function in humans has recently gained support in epidemiological studies of dietary antioxidants such as berries, fruits, and red wine (Hardy, Hardy et al., 2003). Recent animal studies of dietary antioxidants have corroborated these observations and have shown that nutritional compounds can have profound effects on physiological processes such as cardiovascular health (Stoclet, Chataigneau et al., 2004), cancer prevention (Dulak, 2005), and neurological performance (Ramassamy, 2006; Stoclet, Chataigneau et al., 2004; see also Bastianetto, Krantic et al., 2008).

A plethora of research suggests that polyphenolic compounds contained in colorful fruits and vegetables exhibit potent antioxidant and anti-inflammatory activities that can reduce the age-related sensitivity to OS or inflammation. Research from this laboratory found that crude blueberry (BB) or strawberry (SB) extracts significantly attenuated age-related motor and cognitive deficits in rodents, as well as radiation-induced cognitive-behavioral decrements in young rats, similar to those seen in old rats (Joseph, Shukitt-Hale et al., 1999). The rodents in all diet groups, but not the control group, showed improved working memory (short-term memory) performance in the Morris water maze (MWM), demonstrated as one-trial learning following the 10-min retention interval. A later study suggested that, in addition to MWM performance, BB supplementation was also effective in reversing cognitive declines in object recognition (Goyarzu, Malin et al., 2004). Furthermore, the beneficial effects of blueberries were seen even when superimposed on an already well-balanced, healthy rodent diet ("chow") that was more representative of a balanced human diet.

In the case of motor performance, BB supplementation improved performance on tests of motor function that assessed balance and coordination (e.g., rod walking and the accelerating rotarod) (Joseph, Shukitt-Hale et al., 1999). However, in subsequent experiments, we have shown that the cognitive/neuronal variables are sensitive to a greater number of fruits than those seen with respect to motor behavior. Thus far, only BB, cranberry (Shukitt-Hale, Galli et al., 2005), SB (Shukitt-Hale, Cheng et al., 2006), Concord grape juice (Shukitt-Hale, Carey et al., 2006) and blackberry have been effective in reversing the motor behavioral deficits. Rats on the BB diet have generally shown the greatest increases in motor performance, as well as increases in carbachol-stimulated GTPase activity and oxotremorine-enhanced DA release (both markers of muscarinic receptor sensitivity).

Although studies into the neuroprotective properties of fruits and vegetables initially focused on the antioxidant potential of nutritional compounds, further studies have demonstrated that nutritional antioxidants can actually exert a multiplicity of effects, including modulating cell signaling and inflammatory processes. These actions have been attributed to the fact that berries contain high levels of polyphenols (Joseph, Shukitt-Hale et al., 2007). The anthocyanins, a subset of polyphenols, are particularly abundant in brightly colored fruits, such as berries and grapes (Prior, Cao et al., 1998). Anthocyanins are potent antioxidants, able to donate hydrogen in order to stabilize free radicals (Kong, Chia et al., 2003) and lipid peroxidation mediators (Narayan, Naidu et al., 1999), preventing the spread of oxidative damage in the cell. Berries and red wine are both rich sources of anthocyanins and flavonoid glycosides, which impart the foods with red, violet, blue, and purple colors (Kong, Chia et al., 2003; Prior, 2003). Polyphenols, including anthocyanins, in berry fruits have been shown to exhibit antioxidative activity (Fraga, 2007), to modulate neuronal signaling cascades (Joseph, Shukitt-Hale et al., 2007; Weinreb, Mandel et al., 2004)), and to decrease inflammation (Yoon and Baek, 2005), all of which contribute to motor and cognitive functions during aging. As an additional example, resveratrol (see Basianetto et al., 2007), the red wine polyphenol, has been shown to prevent cognitive impairment on the passive avoidance paradigms, the elevated plus maze, and closed field activity tests (Sharma and Gupta, 2002). Additionally, resveratrol administration will improve maze performance and locomotor activity in an animal model of Huntington's disease (Kumar, Padi et al., 2006) and prevent behavioral decrements seen with chronic ethanol consumption (Assuncao, Santos-Marques et al., 2007). Polyphenols from grape seeds and juice have also been shown to mediate cognitive-behavioral performance, improving memory performance in aged rats (Balu, Sangeetha et al., 2005) and in estrogen-depleted spontaneously hypertensive rats (Peng, Clark et al., 2005).

In the case of motor function, berry supplementation has also been shown to improve performance on tests of motor function that assess balance and coordination. In the accelerating rotarod and rod walking tests of motor function, BB supplementation was able to improve the performance of aged animals, whereas animals supplemented with spinach did not exhibit behavioral improvements on these tests (Joseph, Shukitt-Hale et al., 1999). So far, only dietary supplementations with blueberries (Joseph, Shukitt-Hale et al., 1999; Lau, Shukitt-Hale et al., 2005), cranberries (Shukitt-Hale, Galli et al., 2005), SBs (Shukitt-Hale, Cheng et al., 2006), Concord grape juice (Shukitt-Hale, Carey et al., 2006), and blackberries (Shukitt-Hale et al., in preparation) have been shown to reverse age-related motor behavioral deficits.

The aforementioned studies have clearly demonstrated the efficacy of berry fruits and Concord grapes in impacting cognitive and motor functions. The putative mechanisms behind these improvements are diverse, and could include direct antioxidant effects and actions on neuronal cell signaling cascades and inflammatory processes. Recent cell studies (Joseph, Fisher et al., 2006; Joseph, Shukitt-Hale et al., 2007) have suggested that BB pretreatment prior to the application of an oxidative stressor such as DA can protect the cells against decrements in calcium buffering and viability, and this protection is associated with increases in such protective signals as in the presence of oxidative stressors, such as DA BB pretreatment-enhanced mitogen-activated protein kinase, but reduced putative stress signals, such as cyclic AMP response element binding protein (Joseph, Fisher et al., 2006). A subsequent study indicated that lipopolysaccharide treatment enhanced markers of inflammatory signaling (nuclear factor kappa B COX-2) in BV-2 mouse microglial cells, but if the cells were treated with BB extract prior to lipopolysaccharide, the increases in inflammatory signals were reduced (Lau, Bielinski et al., 2007).

13.3.2 POLYUNSATURATED FATTY ACIDS AND COGNITION: ANIMAL STUDIES

As has been shown above, berry fruit consumption may produce significant effects on behavior and neuronal function. Equally important may be the inclusion of monounsaturated fatty acids and polyunsaturated fatty acids (PUFAs) in diet. The majority of studies have focused on the n-, and n-6 PUFAs found in fish and nuts (Bourre, 2004). While fish contain primarily the long-chain fatty acids eicosapentaenoic acid (EPA) and docosahexaenoic acid (DHA), nuts such as walnuts contain the monounsaturated fatty acid oleic acid (C18:1) and the n-6 and n-3 polyunsaturated long-chain fatty acids linoleic acid (n-6 C18:2, LA) and alpha-linolenic acid (n-3 C18:3, ALA) (Crews, Hough et al., 2005). LA is the precursor to arachidonic acid (AA), and ALA is metabolized to either EPA or docosapentaenoic acid. Although numerous studies have demonstrated that consumption of n-3-deficient diets will impair cognitive functioning (McCann and Ames, 2005), few intervention studies have demonstrated a positive effect of fish oil, DHA or EPA on cognitive abilities in aged animals (Barcelo-Coblijn, Hogyes et al., 2003; Carrie, Guesnet et al., 2000).

Aged rats provided a diet supplemented with fish oil did not exhibit improved MWM performance (Barcelo-Coblijn, Hogyes et al., 2003; Carrie, Guesnet et al., 2000). However, rats given diets supplemented with ALA and LA throughout their lifetime exhibited an increased life span and improved brightness-discrimination learning ability during senescence (Yamamoto, Okaniwa et al., 1991). ALA and LA were also shown to modulate behavior following dietary intervention: young rats administered ALA and LA at a ratio of approximately 1:4 exhibited improved MWM performance (Yehuda, Rabinovitz et al., 1998). Dietary ALA and LA supplementation has also been shown to impact learning and memory in the senescence-accelerated mouse (SAMP8), a model of cognitive deterioration that exhibits age-related deficits on behavioral tests (Butterfield and Poon, 2005). Dietary supplementation with oils rich in ALA or LA resulted in improved performance on the Sidman avoidance test and in light/dark discrimination learning in these mice (Umezawa, Ohta et al., 1995). Interestingly, the most effective ALA/LA ratio in improving cognitive performance was found to be 1:4, and walnuts contain ALA and LA at this ratio (Venkatachalam and Sathe, 2006). Additionally, recent cognitive and behavioral studies showed that dietary supplementation of aged rats with walnuts (at 6% of the diet; approximately equivalent to 1 ounce per day for humans) improved motor behavior and MWM performance (Willis, Shukitt-Hale et al., 2008). It is interesting to note that brain uptake of the two main fatty acids in walnuts, ALA and LA, is quite low in the rat although the metabolites of ALA and LA, specifically DHA, EPA, and AA, do become incorporated into neural tissue (Lin and Salem, 2007). Additionally, it could be that given the additional components of walnuts, such as polyphenols, vitamin EE, and melatonin, there could be synergistic effects with the fatty acids found in the walnuts. The mechanism by which ALA and LA supplementation improves behavior and cognition in aged animals remains to be elucidated but may lie in the ability of dietary PUFAs to modulate neuronal membrane properties and stress signaling cascades.

In conclusion, foods high in polyphenolic compounds, such as berries, and perhaps walnuts represent an intriguing alternative or adjunctive therapy for the prevention and treatment of age-related brain dysfunction. Berry fruits have been shown to be effective in improving motor and cognitive performance in aged animals. Moreover, the much lauded antioxidant properties of the berry fruits may pale in comparison with their effects on OS and inflammatory signaling cascades. Given the importance of these oxidative stressors on ever-increasing aging populations, as sensitivity to OS and inflammation continue to increase in an aging population, the nutritional intervention can play an increasing role in maintaining cognitive, motor, and neuronal functions and possibly reverse decrements in these parameters in senescence.

13.4 SUMMARY

By the year 2050, 30% of the total population of the United States will be older than 65 years. As the aged population expands, the economic burden of care and treatment of those with age-related health disorders also increase, necessitating the immediate implementation of therapeutics to prevent or even reverse age-related health disorders. One such potential therapeutic option is the use of nutrition with berry fruit and fatty acids derived from walnuts or fish. Research has recently shown that consumption of the aforementioned substances can dramatically impact the aging brain, possibly leading to improved cognition and motor abilities. It has been postulated that these behavioral and neuronal declines are the result of an increasing vulnerability to oxidative and inflammatory insults, thus creating a "fertile environment" for the subsequent development of age-related neurodegeneration. However, fruits and vegetables may have direct effects on OS and inflammation in aging, and data also indicate that polyphenolic compounds from berries and fatty acids from nuts and fish may have a plethora of additional effects involving enhanced signaling and neurogenesis, leading to an improvement in motor and cognitive functions. The present chapter has provided some background on OS and inflammatory stress in age-related neuronal degeneration and reviewed some recent advances in the effects of berry grape or walnut supplementation on motor, cognitive, and signaling functions in the aged brain and in cell models.

REFERENCES

Ames, B. M., M. K. Shigena, et al. (1993). Oxidants, antioxidants and anticarcinogens: Oxygen radicals and degenerative disease. *Proc Natl Acad Sci U S A* 90: 7915–7922.

Andersen, J. K. (2004). Oxidative stress in neurodegeneration: Cause or consequence? *Nat Med* 10 Suppl: S18–S25.

Annunziato, L., A. Pannaccione, et al. (2002). Modulation of ion channels by reactive oxygen and nitrogen species: A pathophysiological role in brain aging? *Neurobiol Aging* 23(5): 819–834.

Assuncao, M., M. J. Santos-Marques, et al. (2007). Red wine antioxidants protect hippocampal neurons against ethanol-induced damage: A biochemical, morphological and behavioral study. *Neuroscience* 146(4): 1581–1592.

Balu, M., P. Sangeetha, et al. (2005). Age-related oxidative protein damages in central nervous system of rats: Modulatory role of grape seed extract. *Int J Dev Neurosci* 23(6): 501–507.

Barcelo-Coblijn, G., E. Hogyes, et al. (2003). Modification by docosahexaenoic acid of age-induced altera-tions in gene expression and molecular composition of rat brain phospholipids. *Proc Natl Acad Sci U S A* 100(20): 11321–11326.

Bartus, R. T. (1990). Drugs to treat age-related neurodegenerative problems. The final frontier of medical science? *J Am Geriatr Soc* 38(6): 680–695.

Bartus, R. T., R. L. Dean, et al. (1982). The cholinergic hypothesis of geriatric memory dysfunction. *Science* 217: 408–417.

Bartus, R. T., C. Flicker, et al. (1983). Logical principles for the development of animal models of age-related memory impairments. *Aging*: 263–299.

Bastianetto, S., S. Krantic, et al. (2008). Polyphenols as potential inhibitors of amyloid aggregation and toxic-ity: Possible significance to Alzheimer's disease. *Mini Rev Med Chem* 8(5): 429–435.

Bickford, P. (1993). Motor learning deficits in aged rats are correlated with loss of cerebellar noradrenergic function. *Brain Res* 620(1): 133–138.

Bickford, P., C. Heron, et al. (1992). Impaired acquisition of novel locomotor tasks in aged and norepinephrine-depleted F344 rats. *Neurobiol Aging* 13(4): 475–481.

Bickford, P. C., B. Shukitt-Hale, et al. (1999). Effects of aging on cerebellar noradrenergic function and motor learning: Nutritional interventions. *Mech Ageing Dev* 111(2–3): 141–154.

Bodles, A. M., and S. W. Barger (2004). Cytokines and the aging brain—What we don't know might help us. *Trends Neurosci* 27(10): 621–626.

Bourre, J. M. (2004). Roles of unsaturated fatty acids (especially omega-3 fatty acids) in the brain at various ages and during ageing. *J Nutr Health Aging* 8(3): 163–174.

Brewer, G. J., A. Lim, et al. (2005). Age-related calcium changes, oxyradical damage, caspase activation and nuclear condensation in hippocampal neurons in response to glutamate and beta-amyloid. *Exp Gerontol* 40(5): 426–437.

Butterfield, D. A., and H. F. Poon (2005). The senescence-accelerated prone mouse (SAMP8): A model of age-related cognitive decline with relevance to alterations of the gene expression and protein abnormalities in Alzheimer's disease. *Exp Gerontol* 40(10): 774–783.

Carney, J. M., C. D. Smith, et al. (1994). Aging- and oxygen-induced modifications in brain biochemistry and behavior. *Ann N Y Acad Sci*: 44–453.

Carrie, I., P. Guesnet, et al. (2000). Diets containing long-chain n-3 polyunsaturated fatty acids affect behaviour differently during development than ageing in mice. *Br J Nutr* 83(4): 439–447.

Cavazzoni, M., S. Barogi, et al. (1999). The effect of aging and an oxidative stress on peroxide levels and the mitochondrial membrane potential in isolated rat hepatocytes. *FEBS Lett* 449(1): 53–56.

Chen, S., R. C. Frederickson, et al. (1996). Neuroglial-mediated immunoinflammatory responses in Alzheimer's disease: Complement activation and therapeutic approaches. *Neurobiol Aging* 17(5): 781–787.

Cheng, Y., P. Wixom, et al. (1994). Effects of extracellular ATP on Fe^{2+}-induced cytotoxicity in PC-12 cells. *J Neurochem* 66: 895–902.

Crews, C., P. Hough, et al. (2005). Study of the main constituents of some authentic walnut oils. *J Agric Food Chem* 53(12): 4853–4860.

Dalton, T. P., H. G. Shertzer, et al. (1999). Regulation of gene expression by reactive oxygen. *Annu Rev Pharmacol Toxicol* 39: 67–101.

Darley-Usmar, V., H. Wiseman, et al. (1995). Nitric oxide and oxygen radicals: A question of balance. *FEBS Lett* 369(2–3): 131–135.

Davies, K. J. (2000). Oxidative stress, antioxidant defenses, and damage removal, repair, and replacement systems. *IUBMB Life* 50(4–5): 279–289.

Denisova, N. A., S. A. Erat, et al. (1998). Differential effect of aging on cholesterol modulation of carbachol-stimulated low-K(m) GTPase in striatal synaptosomes. *Exp Gerontol* 33(3): 249–265.

Devan, B. D., E. H. Goad, et al. (1996). Dissociation of hippocampal and striatal contributions to spatial naviga-tion in the water maze. *Neurobiol Learn Mem* 66: 305–323.

Dexter, D. T., A. E. Holley, et al. (1994). Increased levels of lipid hydroperoxides in the parkinsonian substantia nigra: An HPLC and ESR study. *Mov Disord* 9(1): 92–97.

Dulak, J. (2005). Nutraceuticals as anti-angiogenic agents: Hopes and reality. *J Physiol Pharmacol* 56 Suppl 1: 51–67.

Durany, N., G. Munch, et al. (1999). Investigations on oxidative stress and therapeutical implications in demen-tia. *Eur Arch Psychiatry Clin Neurosci* 249 Suppl 3: 68–73.

Floyd, R. A. (1999). Neuroinflammatory processes are important in neurodegenerative diseases: An hypothesis to explain the increased formation of reactive oxygen and nitrogen species as major factors involved in neurodegenerative disease development. *Free Radic Biol Med* 26(9–10): 1346–1355.

Floyd, R. A., and K. Hensley (2002). Oxidative stress in brain aging. Implications for therapeutics of neurodegenerative diseases. *Neurobiol Aging* 23(5): 795–807.

Foster, T. C., and A. Kumar (2002). Calcium dysregulation in the aging brain. *Neuroscientist* 8(4): 297–301.

Fraga, C. G. (2007). Plant polyphenols: How to translate their *in vitro* antioxidant actions to *in vivo* conditions. *IUBMB Life* 59(4–5): 308–315.

Gilissen, E. P., R. E. Jacobs, et al. (1999). Magnetic resonance microscopy of iron in the basal forebrain cholinergic structures of the aged mouse lemur. *J Neurol Sci* 168(1): 21–27.

Goyarzu, P., D. H. Malin, et al. (2004). Blueberry supplemented diet: Effects on object recognition memory and nuclear factor-kappa B levels in aged rats. *Nutr Neurosci* 7(2): 75–83.

Halliwell, B. (2001). Role of free radicals in the neurodegenerative diseases: Therapeutic implications for antioxidant treatment. *Drugs Aging* 18(9): 685–716.

Hardy, G., I. Hardy, et al. (2003). Nutraceuticals—A pharmaceutical viewpoint: Part II. *Curr Opin Clin Nutr Metab Care* 6(6): 661–671.

Hartmann, A., M. Vormstein, et al. (1996). [An absent correlation between antioxidant blood concentrations and the remission response of preoperatively treated breast carcinomas]. *Strahlentherapie und Onkologie* 172(8): 434–438.

Hauss-Wegrzyniak, B., M. G. Vannucchi, et al. (2000). Behavioral and ultrastructural changes induced by chronic neuroinflammation in young rats. *Brain Res* 859(1): 157–166.

Hughes, K. A., and R. M. Reynolds (2005). Evolutionary and mechanistic theories of aging. *Annu Rev Entomol* 50: 421–445.

Ingram, D. K., E. L. Spangler, et al. (1994). New pharmacological strategies for cognitive enhancement using a rat model of age-related memory impairment. *Ann N Y Acad Sci* 717: 16–32.

Joseph, J., B. Shukitt-Hale, et al. (2001). Copernicus revisited: Amyloid beta in Alzheimer's disease. *Neurobiol Aging* 22(1): 131–146.

Joseph, J. A., N. Denisova, et al. (1998a). Age-related neurodegeneration and oxidative stress: Putative nutritional intervention. *Neurol Clin* 16(3): 747–755.

Joseph, J. A., N. Denisova, et al. (1998b). Membrane and receptor modifications of oxidative stress vulnerability in aging. Nutritional considerations. *Ann N Y Acad Sci* 854: 268–276.

Joseph, J. A., S. Erat, et al. (1998). CNS effects of heavy particle irradiation in space: Behavioral implications. *Adv Space Res* 22(2): 209–216.

Joseph, J. A., D. R. Fisher, et al. (2006). Blueberry extract alters oxidative stress-mediated signaling in COS-7 cells transfected with selectively vulnerable muscarinic receptor subtypes. *J Alzheimers Dis* 9(1): 35–42.

Joseph, J. A., W. A. Hunt, et al. (1992). Possible 'accelerated striatal aging' induced by [56]Fe heavy-particle irradiation: Implications for manned space flights. *Radiat Res* 130(1): 88–93.

Joseph, J. A., B. Shukitt-Hale, et al. (1999). Reversals of age-related declines in neuronal signal transduction, cognitive, and motor behavioral deficits with blueberry, spinach, or strawberry dietary supplementation. *J Neurosci* 19(18): 8114–8121.

Joseph, J. A., B. Shukitt-Hale, et al. (2000). CNS-induced deficits of heavy particle irradiation in space: The aging connection. *Adv Space Res* 25(10): 2057–2064.

Joseph, J. A., B. Shukitt-Hale, et al. (2005). Oxidative stress and inflammation in brain aging: Nutritional considerations. *Neurochem Res* 30(6–7): 927–935.

Joseph, J. A., B. Shukitt-Hale, et al. (2007). Fruit polyphenols and their effects on neuronal signaling and behavior in senescence. *Ann N Y Acad Sci* 1100: 470–485.

Kong, J. M., L. S. Chia, et al. (2003). Analysis and biological activities of anthocyanins. *Phytochemistry* 64(5): 923–933.

Kreutzberg, G. W. (1996). Microglia: A sensor for pathological events in the CNS. *Trends Neurosci* 19(8): 312–318.

Kumar, P., S. S. Padi, et al. (2006). Effect of resveratrol on 3-nitropropionic acid-induced biochemical and behavioural changes: Possible neuroprotective mechanisms. *Behav Pharmacol* 17(5–6): 485–492.

Landfield, P. W., and J. C. Eldridge (1994). The glucocorticoid hypothesis of age-related hippocampal neurodegeneration: Role of dysregulated intraneuronal Ca^{2+}. *Ann N Y Acad Sci* 746: 308–321.

Lane, N. (2003). A unifying view of ageing and disease: The double-agent theory. *J Theor Biol* 225(4): 531–540.

Lau, F. C., D. F. Bielinski, et al. (2007). Inhibitory effects of blueberry extract on the production of inflammatory mediators in lipopolysaccharide-activated BV2 microglia. *J Neurosci Res* 85(5): 1010–1017.

Lau, F. C., B. Shukitt-Hale, et al. (2005). The beneficial effects of fruit polyphenols on brain aging. *Neurobiol Aging* 26 Suppl 1: 128–132.

Lin, Y. H., and N. Salem, Jr. (2007). Whole body distribution of deuterated linoleic and {alpha}-linolenic acids and their metabolites in the rat. *J Lipid Res* 48(12): 2709–2724.

Lovell, M. A., W. D. Ehmann, et al. (1995). Elevated thiobarbituric acid-reactive substances and antioxidant enzyme activity in the brain in Alzheimer's disease. *Neurology* 45(8): 1594–1601.

Luterman, J. D., V. Haroutunian, et al. (2000). Cytokine gene expression as a function of the clinical progression of Alzheimer disease dementia. *Arch Neurol* 57(8): 1153–1160.

Marcus, D. L., C. Thomas, et al. (1998). Increased peroxidation and reduced antioxidant enzyme activity in Alzheimer's disease. *Exp Neurol* 150(1): 40–44.

McCann, J. C., and B. N. Ames (2005). Is docosahexaenoic acid, an n-3 long-chain polyunsaturated fatty acid, required for development of normal brain function? An overview of evidence from cognitive and behavioral tests in humans and animals. *Am J Clin Nutr* 82(2): 281–295.

McDonald, R. J., and N. M. White (1994). Parallel information processing in the water maze: Evidence for independent memory systems involving dorsal striatum and hippocampus. *Behav Neural Biol* 61: 260–270.

McGeer, P. L., and E. G. McGeer (1995). The inflammatory response system of the brain: Implications for therapy of Alzheimer and other neurodegenerative diseases. *Brain Res Rev* 21: 195–218.

Muir, J. L. (1997). Acetylcholine, aging, and Alzheimer's disease. *Pharmacol Biochem Behav* 56: 687–696.

Munch, G., R. Schinzel, et al. (1998). Alzheimer's disease—Synergistic effects of glucose deficit, oxidative stress and advanced glycation endproducts. *J Neural Transm* 105(4–5): 439–461.

Narayan, M. S., K. A. Naidu, et al. (1999). Antioxidant effect of anthocyanin on enzymatic and non-enzymatic lipid peroxidation. *Prostaglandins Leukot Essent Fatty Acids* 60(1): 1–4.

Olanow, C. W. (1992). An introduction to the free radical hypothesis in Parkinson's disease. *Ann Neurol* 32 Suppl: S2–S9.

Oliveira, M. G. M., O. F. A. Bueno, et al. (1997). Strategies used by hippocampal, and caudate-putamen-lesioned rats in a learning task. *Neurobiol Learn Mem* 68: 32–41.

Orr, C. F., D. B. Rowe, et al. (2002). An inflammatory review of Parkinson's disease. *Prog Neurobiol* 68(5): 325–340.

Peng, N., J. T. Clark, et al. (2005). Antihypertensive and cognitive effects of grape polyphenols in estrogen-depleted, female, spontaneously hypertensive rats. *Am J Physiol Regul Integr Comp Physiol* 289(3): R771–R775.

Perez-Campo, R., M. Lopez-Torres, et al. (1998). The rate of free radical production as a determinant of the rate of aging: Evidence from the comparative approach. *J Comp Physiol [B]* 168(3): 149–158.

Prior, R. L. (2003). Fruits and vegetables in the prevention of cellular oxidative damage. *Am J Clin Nutr* 78 Suppl 3: 570S–578S.

Prior, R. L., G. Cao, et al. (1998). Antioxidant capacity as influenced by total phenolic and anthocyanin content, maturity and variety of *Vaccinium* species. *J Agric Food Chem* 46: 2586–2593.

Rabin, B. M., J. A. Joseph, et al. (1998). Effects of exposure to different types of radiation on behaviors mediated by peripheral or central systems. *Adv Space Res* 22(2): 217–225.

Ramassamy, C. (2006). Emerging role of polyphenolic compounds in the treatment of neurodegenerative diseases: A review of their intracellular targets. *Eur J Pharmacol* 545(1): 51–64.

Rego, A. C., and C. R. Oliveira (2003). Mitochondrial dysfunction and reactive oxygen species in excitotoxicity and apoptosis: Implications for the pathogenesis of neurodegenerative diseases. *Neurochem Res* 28(10): 1563–1574.

Rosenman, S., P. Shrikant, et al. (1995). Cytokine-induced expression of vascular cell adhesion molecule-1 (VCAM-1) by astrocytes and astrocytoma cell lines. *J Immunol* 154(4): 1888–1899.

Rozovsky, I., C. E. Finch, et al. (1998). Age-related activation of microglia and astrocytes: *In vitro* studies show persistent phenotypes of aging, increased proliferation, and resistance to down-regulation. *Neurobiol Aging* 19(1): 97–103.

Sadoul, R. (1998). Bcl-2 family members in the development and degenerative pathologies of the nervous system. *Cell Death Differ* 5(10): 805–815.

Savory, J., J. K. Rao, et al. (1999). Age-related hippocampal changes in Bcl-2:Bax ratio, oxidative stress, redox-active iron and apoptosis associated with aluminum-induced neurodegeneration: Increased susceptibility with aging. *Neurotoxicology* 20(5): 805–817.

Schipper, H. (1996). Astrocytes, brain aging, and neurodegeneration. *Neurob Aging* 17(3): 467–480.

Sharma, M., and Y. K. Gupta (2002). Chronic treatment with trans resveratrol prevents intracerebroventricular streptozotocin induced cognitive impairment and oxidative stress in rats. *Life Sci* 71(21): 2489–2498.

Shen, H. M., Y. Lin, et al. (2004). Essential roles of receptor-interacting protein and TRAF2 in oxidative stress-induced cell death. *Mol Cell Biol* 24(13): 5914–5922.

Sheng, J. G., R. E. Mrak, et al. (1998). Enlarged and phagocytic, but not primed, interleukin-1 alpha-immunoreactive microglia increase with age in normal human brain. *Acta Neuropathol (Berl)* 95(3): 229–234.

Shukitt-Hale, B. (1999). The effects of aging and oxidative stress on psychomotor and cognitive behavior. *Age* 22: 9–17.

Shukitt-Hale, B., A. Carey, et al. (2006). The effects of Concord grape juice on cognitive and motor deficits in aging. *Nutrition* 22(3): 295–302.

Shukitt-Hale, B., G. Casadesus, et al. (2000). Spatial learning and memory deficits induced by exposure to iron-56-particle radiation. *Radiat Res* 154(1): 28–33.

Shukitt-Hale, B., V. Cheng, et al. (2006). Differential brain regional specificity to blueberry and strawberry polyphenols in improved motor and cognitive function in aged rats. *Soc Neurosci Absrtr* 32: 81.

Shukitt-Hale, B., R. Galli, et al. (2005). Dietary supplementation with fruit polyphenolics ameliorates age-related deficits in behavior and neuronal markers of inflammation and oxidative stress. *Age* 27: 49–57.

Shukitt-Hale, B., G. Mouzakis, et al. (1998). Psychomotor and spatial memory performance in aging male Fischer 344 rats. *Exp Gerontol* 33(6): 615–624.

Sloane, J. A., W. Hollander, et al. (1999). Increased microglial activation and protein nitration in white matter of the aging monkey. *Neurobiol Aging* 20(4): 395–405.

Smith, C. D., J. M. Carney, et al. (1991). Excess brain protein oxidation and enzyme dysfunction in normal aging and in Alzheimer disease. *Proc Natl Acad Sci U S A* 88: 10540–10543.

Spencer, J. P., P. Jenner, et al. (1998). Conjugates of catecholamines with cysteine and GSH in Parkinson's disease: Possible mechanisms of formation involving reactive oxygen species. *J Neurochem* 71(5): 2112–2122.

Steffen, B., G. Breier, et al. (1996). ICAM-1, VCAM-1, and MAdCAM-1 are expressed on choroid plexus epithelium but not endothelium and mediate binding of lymphocytes *in vitro*. *Am J Pathol* 148(6): 1819–1838.

Stella, N., A. Estelles, et al. (1997). Interleukin-1 enhances the ATP-evoked release of arachidonic acid from mouse astrocytes. *J Neurosci* 17(9): 2939–2946.

Stoclet, J. C., T. Chataigneau, et al. (2004). Vascular protection by dietary polyphenols. *Eur J Pharmacol* 500(1–3): 299–313.

Tarkowski, E., A. M. Liljeroth, et al. (2003). Cerebral pattern of pro- and anti-inflammatory cytokines in dementias. *Brain Res Bull* 61(3): 255–260.

Toescu, E. C., and A. Verkhratsky (2004). Ca^{2+} and mitochondria as substrates for deficits in synaptic plasticity in normal brain ageing. *J Cell Mol Med* 8(2): 181–190.

Umezawa, M., A. Ohta, et al. (1995). Dietary alpha-linolenate/linoleate balance influences learning and memory in the senescence-accelerated mouse (SAM). *Brain Res* 669(2): 225–233.

Venkatachalam, M., and S. K. Sathe (2006). Chemical composition of selected edible nut seeds. *J Agric Food Chem* 54(13): 4705–4714.

Waring, P. (2005). Redox active calcium ion channels and cell death. *Arch Biochem Biophys* 434(1): 33–42.

Weinreb, O., S. Mandel, et al. (2004). Neurological mechanisms of green tea polyphenols in Alzheimer's and Parkinson's diseases. *J Nutr Biochem* 15(9): 506–516.

West, R. L. (1996). An application of pre-frontal cortex function theory to cognitive aging. *Psych Bull* 120: 272–292.

Willis, L. M., B. Shukitt-Hale, et al. (2008). Dose-dependent effects of walnuts on motor and cognitive function in aged rats. *Br J Nutr* Sep 9:1–5.

Woodroofe, M. N. (1995). Cytokine production in the central nervous system. *Neurology* 45 6 Suppl 6: S6–S10.

Yamamoto, N., Y. Okaniwa, et al. (1991). Effects of high-linoleate and a high alpha-linolenate diet on the learning ability of aged rats. Evidence against an autoxidation-related lipid peroxide theory of aging. *J Gerontol* 46(1): 17–22.

Yehuda, S., S. Rabinovitz, et al. (1998). Modulation of learning and neuronal membrane composition in the rat by essential fatty acid preparation: Time-course analysis. *Neurochem Res* 23(5): 627–634.

Yoon, J. H., and S. J. Baek (2005). Molecular targets of dietary polyphenols with anti-inflammatory properties. *Yonsei Med J* 46(5): 585–596.

Yu, B. P. (1994). Cellular defenses against damage from reactive oxygen species [published erratum appears in *Physiol Rev* 1995 Jan;75(1):preceding 1]. *Physiol Rev* 74(1): 139–162.

Zyzak, D. R., T. Otto, et al. (1995). Cognitive decline associated with normal aging in rats: A neuropsychological approach. *Learn Mem* 2(1): 1–16.

14 Modulation of Multiple Pathways Involved in the Maintenance of Neuronal Function by Fisetin

Pamela Maher

The Salk Institute for Biological Studies, La Jolla, California, USA

CONTENTS

14.1 INTRODUCTION

It is becoming increasingly clear that neurological disorders are multifactorial, involving disruptions in multiple cellular systems. Thus, while each disease has its own initiating mechanisms and pathologies, certain common pathways appear to be involved in most, if not all, neurological disorders described to date. These include alterations in redox homeostasis, gene transcription, protein modification and processing, neurotrophic factor signaling, mitochondrial function, and the immune response. Therefore, it is unlikely that hitting a single target will result in significant benefits to patients with a neurological disorder.[1,2] However, current drug research efforts are almost exclusively focused on single protein targets and the identification of small molecules that can modulate these targets with high affinity.[3] Thus, for the treatment of neurological disorders, it may be necessary to use combinations of drugs directed against different targets. However, this approach is subject to a number of potential problems, including pharmacokinetic and bioavailability challenges, which in central nervous system (CNS) disorders are exacerbated by the difficulty of getting multiple compounds across the blood-brain barrier and the potential for drug-drug interactions. An alternative approach is to identify small molecules that have multiple biological activities that are relevant to neurological disorders.

Flavonoids are polyphenolic compounds that are widely distributed in fruits and vegetables and therefore regularly consumed in the human diet.[4–6] Flavonoids were historically characterized on the basis of their antioxidant and free radical scavenging effects. However, more recent studies have shown that flavonoids have a wide range of activities that could make them particularly effective as neuroprotective agents for the treatment of a variety of neurological disorders. In this chapter, I describe the multiple physiological benefits of one flavonoid, fisetin, and indicate how these activities might be useful for the treatment of a variety of neurological disorders.

14.2 FISETIN CAN MAINTAIN INTRACELLULAR GLUTATHIONE LEVELS AND PROMOTE THE SURVIVAL OF NERVE CELLS IN RESPONSE TO MULTIPLE TOXIC INSULTS

Over the last few years, we have characterized the molecular and cellular events underlying nerve cell death induced by oxidative stress using the mouse hippocampal cell line HT22 and primary cortical neurons.[7] This pathway of programmed cell death, termed oxidative glutamate toxicity or oxytosis,[7] is initiated by the addition of glutamate to the extracellular medium. Glutamate inhibits the uptake of cystine, which is required for glutathione (GSH) synthesis, causing the depletion of GSH in neurons. Subsequently, this decrease in cellular GSH results in the production of reactive oxygen species (ROS) by mitochondria. ROS accumulation gives rise to Ca^{2+} influx from the extracellular medium, which leads to a form of cell death with characteristics of both apoptosis and necrosis. In addition to the elucidation of the steps involved in oxytosis, HT22 cells have been used to screen potentially therapeutic drugs for the treatment of clinical conditions involving oxidative stress.[7]

Using this system, we found that fisetin was very effective at preventing oxidative glutamate toxicity and that at least part of its effectiveness was due to its ability to maintain GSH levels in cells.[8] Of approximately 30 flavonoids tested in this study, only two, fisetin and quercetin, were able to maintain GSH levels in the presence of oxidative stress, indicating that this is not a common property of flavonoids. More recently, we have extended these studies to several additional toxicity paradigms and provided further evidence for the importance of GSH maintenance in the action of fisetin.

Stroke is the leading cause of adult disability and the third leading cause of death in the United States. Worldwide, approximately 6 million people died of stroke in 2005, with a projected increase over the next decade of 12%.[9] Ischemic stroke occurs when the normal blood supply to the brain is disrupted, usually due to artery blockage by a blood clot, thereby depriving the brain of oxygen and metabolic substrates and hindering the removal of waste products.[10] The nerve cell damage caused by cerebral ischemia results in functional impairment and/or death. Despite the significant advances that have been made in understanding the pathophysiology of cerebral ischemia on the cellular and molecular levels, only one drug, recombinant tissue-type plasminogen activator, is approved by the U.S. Food and Drug Administration for use in patients with ischemic stroke.[11] Unfortunately, the utilization of recombinant tissue-type plasminogen activator is limited by its short time window of efficacy and its potential to cause intracerebral hemorrhage.[12–14] Thus, there is a critical need for additional safe and effective treatments for stroke.

In an attempt to meet this need, we developed an *in vitro* screen for testing compounds that might have therapeutic value for the treatment of stroke.[15] This screen utilizes the toxin iodoacetic acid (IAA), a well-known irreversible inhibitor of the glycolytic enzyme glyceraldehyde 3-phosphate dehydrogenase in combination with the HT22 nerve cell line. The changes seen following IAA treatment of nerve cells are very similar to changes that have been seen in animal models of ischemic stroke and include alterations in membrane potential, breakdown of phospholipids, loss of ATP and GSH, and an increase in ROS. Using this screen, we tested a variety of flavonoids and found that fisetin was highly effective in protecting the HT22 cells from IAA toxicity. Fisetin not only prevented cell death but also stood out among the flavonoids tested as being able to maintain both GSH and ATP levels in the presence of IAA. Further studies in the rabbit small clot embolism model of stroke demonstrated that fisetin could significantly improve the behavioral outcome when administered 5 min after the initiation of an embolic stroke.[15] Consistent with our animal data, fisetin was also shown to reduce ischemic damage and infarct volume in the permanent focal middle cerebral artery occlusion model of stroke in rats.[16]

Another cell death paradigm in which we have recently tested fisetin is that of peroxynitrite toxicity. Peroxynitrite-mediated toxicity has been implicated in many neurological disorders. In the brain, focal traumatic brain injury,[17,18] ischemic stroke,[19] and numerous neurodegenerative disorders, including Alzheimer's disease (AD), Parkinson's disease (PD), multiple sclerosis (MS), and amyotrophic lateral sclerosis (ALS),[20] have all been linked to the increased production of peroxynitrite. Although peroxynitrite itself is not a free radical, it is a uniquely damaging molecule because it can initiate strong oxidation reactions through decomposition into a hydroxyl radical and nitrogen dioxide.[21] It can also form a highly reactive nitroderivative in the presence of transition metals[22,23] and interact directly with protein and non-protein thiol groups, leading to the depletion of cellular antioxidant defenses, including GSH.[24,25]

Using primary cultures of cortical neurons in combination with the peroxynitrite generator SIN-1, we found an ~50% decrease in both intracellular GSH levels and cellular viability that could be prevented by treatment with 10 μM fisetin (Figure 14.1).[26] Similar results with fisetin

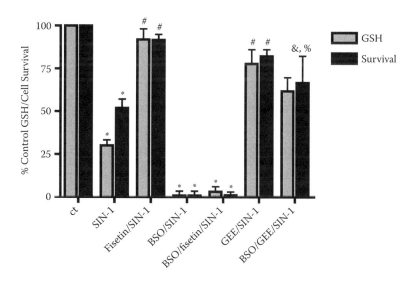

FIGURE 14.1 Effects of fisetin on SIN-1-mediated decreases in GSH and viability in rat primary cortical neurons. GSH levels and survival of rat primary cortical neurons (7 days in culture) treated with the peroxynitrite donor SIN-1 for 6 hr and/or the neuroprotective flavonoid fisetin (10 μM) and/or the exogenous GSH source GSH monoethyl ester (5 mM) in the absence or presence of the GSH synthesis inhibitor BSO (1 mM). Cell viability was measured by the MTT assay. GSH levels were measured using monochlorobimane. Similar results were obtained in three independent experiments. *Significantly different ($p < 0.05$) from control. #Significantly different from SIN-1 alone. &Significantly different from BSO alone. %Significantly different from fisetin + SIN-1 treatment. After Ref. 26.

were obtained when authentic peroxynitrite was used as the toxic insult. The protection by fisetin, as well as its ability to maintain GSH levels, was inhibited by treatment with buthionine sulfoximine (BSO), an inhibitor of glutamate cysteine ligase (GCL),[27] the rate-limiting enzyme for GSH biosynthesis. In contrast, BSO had no effect on the ability of GSH monoethyl ester, a cell-permeable form of GSH, to maintain GSH levels and protect the neurons from peroxynitrite toxicity, indicating that BSO was not blocking the neuroprotective effect of fisetin through a toxic effect unrelated to GSH.

How does fisetin maintain GSH levels? In general, intracellular GSH levels are regulated by a complex series of mechanisms that include substrate availability and transport, rates of synthesis and regeneration, GSH utilization and GSH efflux to extracellular compartments.[28] Because glutamate and glycine occur at relatively high intracellular concentrations, cysteine is limiting for GSH synthesis in many types of cells, including nerve cells. In the extracellular environment, cysteine is readily oxidized to form cystine, so for most cell types, cystine transport mechanisms are essential to provide them with the cysteine needed for GSH synthesis. Cystine uptake in many types of cells is mediated by system x_c^-, a Na^+-independent cystine/glutamate antiporter.[29] System x_c^- is a member of the disulfide-linked heteromeric amino acid transporter family and consists of a light chain (xCT) that confers substrate specificity and a heavy chain (4F2hc) that is shared among a number of amino acid transporters. The results with BSO suggested that fisetin increases GSH levels by increasing cystine import and/or enhancing GCL activity.

Interestingly, both xCT[30] and GCL[31] are regulated at the transcriptional level by the transcription factor NF-E2-related factor 2 (Nrf2). Nrf2 binds to the antioxidant response element (ARE; also EpRE) within the promoter region of various genes, thereby regulating the inducible production of a variety of proteins involved in the protection of cells from oxidative stress and in the maintenance of redox homeostasis.[32–35] Treatment of cells with an Nrf2 inducer results in the accumulation and translocation of Nrf2 to the nucleus, where it heterodimerizes with Maf family proteins to induce gene transcription. We have found that fisetin can increase the nuclear levels of Nrf2 in a variety of nerve cells, including HT22 cells,[36] retinal ganglion cells,[37] and primary cortical neurons (Figure 14.2). In primary cortical neurons treated with peroxynitrite,[26] Nrf2 levels decreased significantly, but treatment with fisetin was able to prevent this decrease. The peroxynitrite-mediated decreases in Nrf2 levels correlated with a decrease in the levels of both subunits of GCL, and this was also prevented by

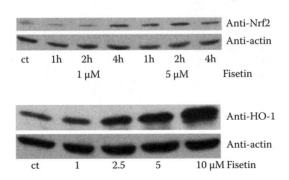

FIGURE 14.2 Fisetin induces the expression of Nrf2, the ARE-specific transcription factor, and HO-1, an Nrf-2-dependent protein. Rat primary cortical neurons (7 days in culture) were untreated (ct) or treated with 1 or 5 μM fisetin for 1, 2, or 4 hr (Nrf2) or with 1–10 μM fisetin for 24 hr (HO-1). Nuclei were prepared (Nrf2) and equal amounts of protein were analyzed by SDS-PAGE and immunoblotting with Nrf2 antibodies or cell lysates were prepared and equal amounts of protein were analyzed by SDS-PAGE and immunoblotting with HO-1 antibodies. Immunoblotting with anti-actin is shown as a loading control. Similar results were obtained in three independent experiments.

fisetin treatment. These findings are in agreement with earlier studies which showed that fisetin could increase the expression of various ARE-dependent genes in multiple non-neuronal cell lines, including MCF-7 breast carcinoma cells,[38] Cos-1 cells,[39] and hepatoma cells,[40] as well as rat C6 glioma cells.[41] The precise mechanisms whereby fisetin increases Nrf2 levels are still under investigation.

14.3 FISETIN CAN ACTIVATE NEUROTROPHIC FACTOR SIGNALING PATHWAYS

Neurotrophic factors play critical roles in promoting the differentiation, survival, and functional maintenance of nerve cells. Changes in the levels of neurotrophic factors and/or their receptors are implicated in the pathophysiology of a variety of neurodegenerative diseases, including AD, PD, ALS, and Huntington's disease.[42–45] Given these results, neurotrophic factors appear to be a logical choice for the treatment of these diseases. However, because neurotrophic factors are proteins, their clinical use has been limited by difficulties in delivery to the brain and unsuitable pharmacokinetics.[44] An alternative approach is to identify small molecules that can substitute for neurotrophic factors.

As described above, our initial studies with fisetin showed that it could promote neuronal survival. To test the idea that fisetin could also promote neuronal differentiation, we looked at its ability to induce neurite outgrowth in PC12 cells, a well-studied model of neuronal differentiation.[46] In response to classical neurotrophic factors such as nerve growth factor (NGF), PC12 cells undergo a series of physiological changes culminating in a differentiated phenotype characterized by the extension of long processes called neurites. These changes are the result of the activation of a coordinated series of signaling pathways, including the Ras-ERK cascade and the PI3-kinase-Akt cascade. Several years ago, we showed that fisetin was able to induce neurite outgrowth in PC12 cells with an EC_{50} of ~5 µM.[47] Fisetin induced the differentiation of 70–80% of the cells, an efficacy that was similar to that of NGF (Figure 14.3). Fisetin was also found to induce the activation of the

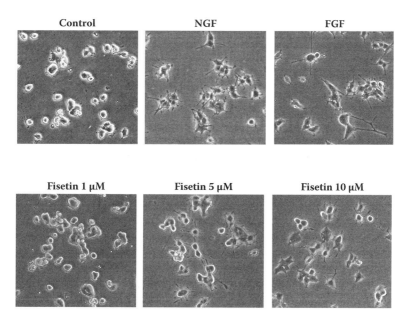

FIGURE 14.3 Fisetin induces PC12 cell differentiation. PC12 cells in N2 medium were untreated or treated with 1, 5, or 10 µM fisetin, FGF-2 (25 ng/ml), or NGF (50 ng/ml). Twenty-four hours later, the cells were examined and photographed by phase-contrast microscopy. Similar results were obtained in five independent experiments. After Ref. 47.

Ras-ERK cascade in the PC12 cells with a time course that was delayed compared with the induction of this kinase cascade by NGF. However, the induction of differentiation by fisetin was dependent on the Ras-ERK cascade because inhibitors of several steps in the cascade blocked differentiation. In contrast, inhibitors of other signaling cascades implicated in neuronal differentiation had no effect on the ability of fisetin to induce differentiation of the PC12 cells.

14.4 FISETIN CAN ENHANCE COGNITIVE FUNCTION

The transcription factor cAMP-response element-binding (CREB) protein interacts with the CRE in the promoter region of genes that encode proteins involved in the regulation of nerve cell function, including learning and memory.[48,49] A number of studies in a wide range of animal species have shown that CREB plays a key role in the formation of long-term memory (LTM).[50] CREB is a constitutively nuclear protein whose activity is regulated by phosphorylation of both subunits of the homodimer.[51] Phosphorylation promotes the interaction of CREB with the transcriptional co-activator CREB binding protein or its homolog p300, which stimulates the transcriptional activity of CREB. Several kinases can phosphorylate CREB on Ser133 and positively regulate its transcriptional activity, including protein kinase A, calmodulin-dependent kinase IV, MNK1, MNK2, and MSK1. The latter three kinases are all substrates of the MAP kinase ERK, and so CREB activity can be regulated by the ERK signaling pathway.

These observations, coupled with our studies on fisetin in the PC12 cells, led us to test the idea that fisetin could activate CREB and enhance memory. Using rat hippocampal slices, we demonstrated that 1 µM fisetin could induce rapid phosphorylation of CREB and that this phosphorylation was dependent on ERK activation since inhibitors of ERK activation also blocked CREB phosphorylation (Figure 14.4).[52]

Given these results and the known associations between CREB and learning and memory, it was next asked whether fisetin could affect long-term potentiation (LTP) in the hippocampal slices. LTP is an *in vitro* assay that is considered to be a good model of how memory is formed at the cellular level.[53] Fisetin had no effect on basal synaptic responses in the CA1 area of rat hippocampal slices.[52] However, it induced LTP in slices exposed to a weak tetanic stimulus (15 pulses at 100 Hz) that by itself failed to induce LTP. The facilitation of LTP by fisetin was dose dependent, with a maximal effect seen at 1 µM, and it persisted for at least 60 min. Importantly, the facilitation of LTP by fisetin was blocked by two inhibitors of ERK activation, PD98059 and U0126. Together, these data strongly support the hypothesis that ERK-dependent CREB activation by fisetin is responsible for the facilitation of LTP by fisetin. Further support for this hypothesis comes from studies with the phosphodiesterase 4 inhibitor rolipram, which enhances CREB phosphorylation by preventing the breakdown of cAMP. Rolipram also had no effect on basal synaptic responses in rat hippocampal slices but facilitated LTP induced by a weak tetanic stimulus in a manner very similar to fisetin.[54] However, in contrast to rolipram, fisetin did not increase cAMP levels in the hippocampal slices.[52]

To determine if the biochemical and electrophysiological effects of fisetin seen in hippocampal slices could be translated into alterations in LTM in animals, we tested fisetin in mice using the object recognition task.[55] A major reason for choosing this specific memory assay was that it had been to be proven very effective in measuring CREB-dependent functions. In this test, during the training period, mice are presented with two identical objects that they explore for a fixed period. The mice are presented 1 day later with two objects, one of which was presented previously during the training and is thus familiar to the mice and the other is new to them, to test for LTM. The better the mice remember the familiar object, the more time they will spend exploring the novel object. To test the effects of fisetin in this memory task, we administered it orally to the mice before the start of the training period.[52] Rolipram, administered by injection, was used as a positive control. As shown in Figure 14.5, fisetin significantly increased the time the mice spent exploring the novel object, indicating a significant effect on LTM consistent with our cell- and tissue-based results.

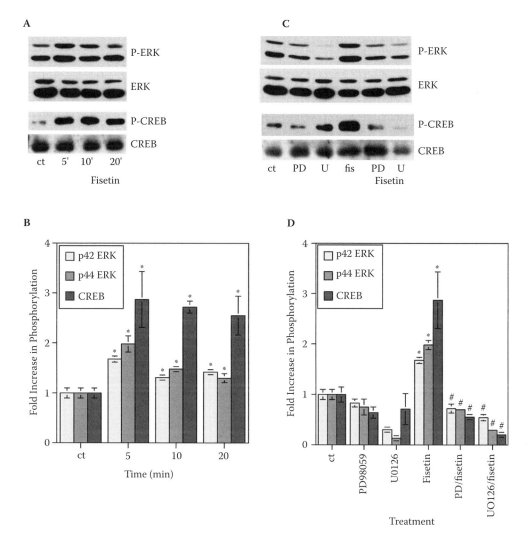

FIGURE 14.4 Fisetin activates ERK1 (p44), ERK2 (p42), and CREB in rat hippocampal slices. (A) Hippocampal slices in artificial cerebral spinal fluid were treated with 1 μM fisetin for 5–20 min, and then equal amounts of protein were analyzed by SDS-PAGE and immunoblotting with antibodies to phospho-ERK and phospho-CREB along with antibodies to the unphosphorylated forms of the proteins demonstrating no change in overall protein levels. Similar results were obtained in two independent experiments. (B) The average phosphoprotein signal from the blots, quantified by densitometry and normalized to total protein, is plotted with the S.D. *Significantly different from control (p < 0.005). (C) Hippocampal slices were pretreated for 30 min with the MEK inhibitors PD98059 (PD) (50 μM) and U0126 (U) (10 μM) before the addition of 1 μM fisetin for 5 min. Samples were analyzed as in (A). (D) The average phosphoprotein signal from the blots, quantified by densitometry and normalized to total protein, is plotted with the S.D. *Significantly different from control (p < 0.0005). #Significantly different from fisetin alone (p < 0.0001). Similar results were obtained in two independent experiments. From Ref. 52. © 2006 by the National Academy of Sciences of the USA.

14.5   FISETIN CAN REGULATE PROTEIN HOMEOSTASIS (PROTEOSTASIS)

In order to gain additional insight into the mechanisms underlying the neurotrophic activity of fisetin, we developed a novel assay for identifying and characterizing small molecules that can prevent the death of neurons following loss of neurotrophic support.[56] The transfer of cortical

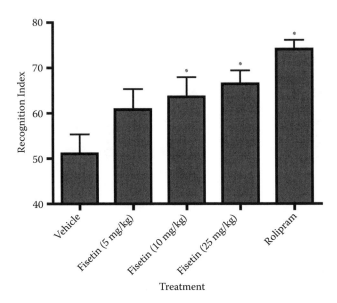

FIGURE 14.5 Fisetin enhances LTM in mice. The effect of different oral doses of fisetin on object recognition over a 10-min test period. Rolipram, injected intraperitoneally at 0.1 mg/kg, served as a positive control. Data represent the mean ± S.E.M. of 10 mice per treatment group. Data were analyzed by one-way ANOVA followed by post hoc comparisons with Fisher's test. *Significantly different from vehicle control (p < 0.02). Similar results were obtained in two independent, blinded experiments done by Psychogenics. From Ref. 52. © 2006 by the National Academy of Sciences of the USA.

neurons from embryonic rats or mice to cell culture causes an abrupt and dramatic loss of trophic support that can be compensated for by plating the cells at high density.[57] However, if primary cultures are instead plated at low density in the absence of serum and the presence of the minimal N2 supplement, they die within 24 hr. This observation was used as the basis for our assay for further testing of the hypothesis that small molecules such as fisetin can substitute for neurotrophic factors. These low-density cultures can be rescued by the addition of fetal calf serum, the growth conditioned medium of high-density cultures, or a defined combination of exogenous protein growth factors. We found that they could also be rescued by fisetin with a maximal effective dose of 2 μM, which provided a level of rescue similar to that seen with 10% fetal calf serum.

In further studies, we explored the mechanisms underlying the fisetin-mediated neuroprotection seen in this assay.[56] We first tested the roles of both the Ras-ERK cascade and GSH maintenance in the survival-promoting effects of fisetin on low-density primary cultures but surprisingly found that neither of these mechanisms was involved. Instead, we found that modest activation (Figure 14.6) of the ubiquitin-proteasome pathway by fisetin appeared to play a critical role in its survival-promoting effects on primary neurons since the proteasome inhibitor lactacystin blocked fisetin-induced cell survival in this assay.

It is not clear at this time how fisetin increases proteasome activity in the primary cortical neurons. Many of the protein components of the proteasome are transcriptionally regulated by Nrf2,[58] and Nrf2 inducers can increase proteasome activity in several cell types.[59–61] However, preliminary data suggest that the regulation of proteasome activity by fisetin is likely to be more complex.

The ubiquitin-proteasome pathway mediates the majority of the proteolysis seen in the cytoplasm and nucleus of mammalian cells. As such, it plays an important role in the regulation of a variety of physiological and pathophysiological processes.[62,63] Of specific interest, while proteasome inhibitors can stimulate neurite outgrowth in nerve cell lines,[64,65] several recent studies have demonstrated that proteasome activity is required for axon initiation, elongation, and maintenance

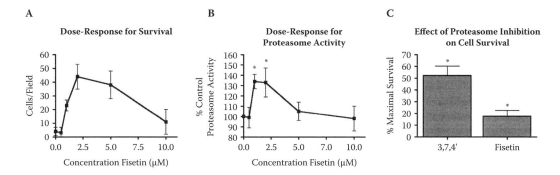

FIGURE 14.6 Role of proteasome activity in the survival promoting effects of fisetin. (A) Primary cortical neurons were plated at 5×10^5 cells/ml in 35-mm dishes alone or in the presence of increasing concentrations of fisetin. Cell survival was measured after 24 hr by counting the number of cells with neurites greater than two cell body diameters in length. Results are the average of five independent experiments with the S.D. (B) Primary cortical neurons were plated at 1×10^6 cells/ml in B27 medium and grown for 2 days. The cells were then left untreated or treated with increasing concentrations of fisetin. After 6 hr, the cells were harvested and assayed for proteasome activity as described previously.[56] (C) Primary cortical neurons were plated at 5×10^5 cells/ml in 35-mm dishes alone or in the presence of 10 μM 3,7,4′trihydroxyflavone or 2 μM fisetin. Some dishes were pre-treated for 30 min with the proteasome inhibitor lactacystin (0.5 μM) before the addition of the flavonoids. Cell survival was measured after 24 hr as described in (A). The results are presented as the level of survival relative to the flavonoid alone and are the average of three independent experiments with the S.D. After Ref. 56.

in primary post-mitotic neurons.[66–68] A requirement for axon maintenance for nerve cell survival in our neurotrophic factor withdrawal assay is entirely consistent with the pattern of cell death that appears to involve the loss of processes prior to the demise of the cell body.

A number of recent studies have shown that proteasome activity is decreased in a variety of neurological disorders, including AD, PD, ALS, and stroke, as well as during normal aging.[69–71] This decrease in proteasome activity is thought to play a critical role in the accumulation of abnormal and oxidized proteins. Thus, enhancing proteasome activity could prove beneficial for treating neurological disorders. Although the increase in proteasome activity brought about by fisetin is modest, it is similar to the increases seen with several other compounds.[60,61] Furthermore, dramatic increases in activity may not be compatible with the maintenance of normal cell function.

Consistent with the ability of fisetin to alter protein stability is a recent report showing that fisetin can inhibit beta-amyloid peptide fibril formation in a cell-free assay system.[72] Furthermore, the authors showed that fisetin could prevent extracellular beta-amyloid peptide toxicity in the HT22 cells. Since beta-amyloid peptide is thought to play a key role in the nerve cell loss that is the hallmark of AD, these results suggest that fisetin may be able to reduce the burden of beta-amyloid peptide through multiple mechanisms.

14.6 ANTI-INFLAMMATORY EFFECTS OF FISETIN

Over the past few years, several studies have demonstrated that fisetin can modulate inflammatory responses. Using the mouse macrophage cell line RAW264.7, Wang et al.[73] looked at effects of fisetin and other flavonoids on macrophage activation by bacterial lipopolysaccharide (LPS). A hallmark of immune cell activation by LPS is an increase in nitric oxide production measured as nitrite accumulation in the culture medium. Fisetin showed a dose-dependent inhibition of LPS-induced nitrite accumulation reaching a maximum of 71% inhibition at 10 μM. Fisetin also significantly reduced the production of prostaglandin E2 (PGE2) by the LPS-stimulated macrophages. The increase in nitric oxide production by LPS-stimulated macrophages is due to an increase in iNOS expression, which, in turn, is regulated by the transcription factor NF-κB. Fisetin was shown to reduce the increase in both iNOS mRNA expression and NF-κB activation by LPS at the same doses that inhibited nitric oxide production. It is likely that the inhibition of LPS-stimulated increases in PGE2 production by fisetin was also due to the effects of fisetin on NF-κB activation since cyclooxygenase 2, which regulates the production of PGE2, is also regulated by NF-κB.

In the human lung adenocarcinoma cell line H1299, higher doses of fisetin (25–50 μM) were shown to reduce NF-κB activation induced by LPS and a variety of other immunomodulatory compounds, including tumor necrosis factor alpha (TNF-α).[74] Further studies with TNF-α demonstrated that fisetin reduced both NF-κB-dependent reporter gene expression and protein expression through a mechanism that involved the inhibition of TNF-α-induced activation of IKKs, the upstream kinases in the NF-κB activation pathway. In contrast to the study with macrophages,[73] no other flavonoid tested had this effect. The precise mechanisms underlying fisetin-mediated inhibition of IKK activation remain to be determined but may involve its antioxidant activity since mild oxidative stress has been shown to activate IKKs.[75,76]

Macrophages play an important role in MS, a neuroinflammatory disease whose hallmark is demyelination. In this disease, monocyte-derived macrophages accumulate in the CNS and produce toxic factors that contribute to the demyelination process. In addition, macrophages phagocytose myelin, leading to further activation and production of toxic factors. Fisetin, along with several other flavonols, was found to dose dependently inhibit myelin phagocytosis by the RAW264.7 macrophage cell line.[77]

Microglia are the resident immune cell population of the CNS, comprising 10–15% of the total cell population.[78–80] They play important protective roles in the CNS, such as removing pathogens and promoting tissue regeneration after injury. Microglia are also implicated in the pathogenesis of a variety of acute and chronic neurological disorders, including stroke, AD, PD, HIV-associated dementia, and MS. Activated microglia can produce a wide array of pro-inflammatory and cytotoxic factors, including cytokines, ROS, excitatory neurotransmitters, and eicosanoids [see fisetin and lipoxygenases (LOXs) below], that may work in concert to promote neurodegeneration. A very recent study demonstrated that fisetin could reduce LPS-induced microglial activation and neurotoxicity.[81] With the use of the BV-2 microglial cell line, it was shown that fisetin was very effective at blocking LPS-induced nitric oxide production measured as accumulation of nitrite in the culture medium with an EC_{50} of ~7 μM. The same dose of fisetin also reduced LPS-induced increases in extracellular PGE2 levels and increases in the expression of the cyclooxygenase 2, iNOS, and interleukin-1β genes. Similar to the results obtained with other cells treated with LPS as described above, the effects of fisetin on microglial activation appeared to be mediated by inhibition of LPS-stimulated NF-κB activation. Similar data were obtained with primary microglial cells isolated from the cerebral cortices of 1-day-old mice. Importantly, the authors also showed neuroprotective effects of fisetin in a nerve/microglia co-culture system. In this assay, neuroblastoma cells were co-cultured with LPS-activated microglia with or without pretreatment with fisetin. In the absence of fisetin, the LPS-activated microglia reduced the viability of the neuroblastoma cells by ~50%. However, following pretreatment of the microglia with ~7 μM fisetin, the viability of the neuroblastoma cells was reduced by ~10% only.

Together, these results support the idea that fisetin has anti-inflammatory activity in the CNS that can target both resident and infiltrating immune cells and so might be effective in a variety of conditions involving the dysregulation of the immune system in the brain.

14.7 OTHER ACTIVITIES OF FISETIN THAT COULD BE RELEVANT TO ITS ABILITY TO MAINTAIN NEURONAL FUNCTION

14.7.1 IRON CHELATION/INHIBITION OF LIPID PEROXIDATION

Although it is not entirely clear how relevant the antioxidant activity of flavonoids as measured in test tube assays is to their effects *in vivo*, fisetin is a relatively good antioxidant with a Trolox equivalent antioxidant capacity value of ~3.[8,36] Furthermore, fisetin was also shown to be very good at inhibiting lipid peroxidation and chelating iron.[82] All these properties could contribute to the beneficial effects of fisetin on CNS cells.

14.7.2 INHIBITION OF PARP ACTIVITY

PARP-1 [poly(ADP-ribose) polymerase-1] is a nuclear enzyme that functions as a sensor of DNA damage.[83] It catalyzes the formation of poly(ADP-ribose) polymers from NAD^+. Overactivation of PARP-1 causes high levels of poly(ADP-ribose) polymer formation, resulting in the rapid depletion of cellular NAD^+ and ATP and, eventually, cell death. PARP-1 gene deletion and/or PARP inhibitors have been shown to prevent cell death in a variety of models for neurological disorders, including excitotoxicity, oxidative stress, and the MPTP model of PD.[83] PARP-1 is also thought to play a role in promoting inflammatory responses, probably through its ability to co-activate the transcription factor NF-κB. In a recent study,[84] it was shown that fisetin is an effective PARP-1 inhibitor both in the test tube and in cells, albeit at relatively high concentrations. Whether or not this activity contributes to the neuroprotective actions of fisetin remains to be determined.

14.7.3 INHIBITION OF LIPOXYGENASE ACTIVITY

LOXs metabolize 20-carbon unsaturated fatty acids, such as arachidonic acid, which are produced from membrane phospholipids by the action of phospholipases, to eicosanoids, including hydroxyperoxyeicosatetraenoic acids, hydroxyeicosatetraenoic acids, and leukotienes.[85] The LOXs are dioxygenases that incorporate molecular oxygen into specific positions of arachidonic acid and can be distinguished on the basis of the site of oxygen insertion. 5-LOX, 12-LOX, and 15-LOX are all expressed in the brain. There is good evidence that both 5-LOX and 12-LOX contribute to the nerve cell loss seen in several neurological disorders. For example, we[86] and others[87–89] have shown that 12-LOX activation plays an important role in the nerve cell death induced by GSH depletion. 12-LOX was also shown to be increased in AD,[90] and 12-LOX inhibition protected cortical neurons from beta-amyloid peptide-induced death.[91] More recently, we found that the 12-LOX inhibitor baicalein could significantly improve behavioral outcome in the rabbit small clot embolism model when administered both 5 and 60 min after the initiation of an embolic stroke.[92] Using a double transgenic mouse model generated by crossing 5-LOX deficient mice with the Tg2576 mouse model of AD, Firuzi et al.[93] recently showed that loss of 5-LOX reduced the beta-amyloid peptide burden in the brains of the mice by 64–80%, suggesting that inhibition of 5-LOX might have benefits in AD.

In earlier studies, it was shown that fisetin is an effective inhibitor of 12-LOX activity in the fish gill,[94] 5-LOX activity in stimulated peritoneal leukocytes,[95] and 15-LOX activity in rabbit reticulocytes.[96] As shown in Figure 14.7, we have confirmed and extended these studies. Fisetin inhibits 12-LOX from human platelets with an IC_{50} of 0.14 μM, 5-LOX from human peripheral blood mononuclear leukocytes with an IC_{50} of 0.585 μM, and 15-LOX from rabbit reticulocytes with an IC_{50} of 0.341 μM. In all cases, the IC_{50} values for fisetin are as good as or better than those for known LOX inhibitors.

Together, these results suggest that an important activity of fisetin in the brain is the inhibition of LOX activity. This action may contribute to both the neuroprotective and anti-inflammatory effects of fisetin and thereby promote the functional maintenance of the CNS.

14.7.4 INHIBITION OF CYCLIN-DEPENDENT KINASES

Cyclin-dependent kinases (CDKs) regulate a variety of activities, including cell division, apoptosis, transcription, differentiation, and neuronal function. CDK1, CDK4, CDK5, and CDK6 have all been implicated in nerve cell death initiated by a range of insults, and inhibition of these kinases can provide neuroprotection.[97] Fisetin has been shown to inhibit several of these CDKs, including

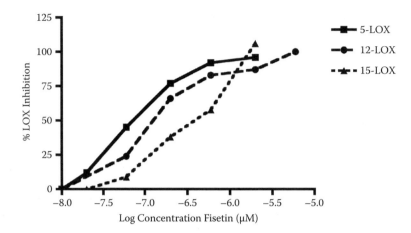

FIGURE 14.7 Fisetin inhibits LOX activity. Peripheral blood mononuclear leukocytes (5-LOX), sonicated human platelets (12-LOX), or rabbit reticulocytes (15-LOX) were preincubated with the indicated doses of fisetin prior to the addition of a LOX substrate (arachidonic acid for 5-LOX and 12-LOX and linoleic acid for 15-LOX) for 10–15 min. 5-LOX activity was determined from EIA quantitation of the 5-LOX product leukotriene B4,[111] 12-LOX activity was determined by spectrophotometric quantitation of the 12-LOX product 12-hydroxyeicosatetraenoic acid,[112] and 15-LOX activity was determined by spectrophotometric determination of the 15-LOX product 13-HPODE.[113] NDGA was used as a positive control for 5-LOX inhibition (IC_{50} = 0.076), baicalein was used as a positive control for 12-LOX inhibition (IC_{50} = 0.741), and PD146176 was used as a positive control for 15-LOX inhibition (IC_{50} = 1.5 μM).

CDK1, CDK5, and CDK6, with low micromolar affinity.[98] This activity might also contribute to the neuroprotective effects of fisetin.

14.7.5 INHIBITION OF NADPH OXIDASE ACTIVITY

NADPH oxidases are widespread in mammalian cells and are an important source of ROS that can mediate both physiological and pathological processes in the CNS.[99] NADPH oxidases utilize molecular oxygen to produce superoxide. The superoxide can be rapidly converted to hydrogen peroxide by superoxide dismutases. Alternatively, the superoxide can react with nitric oxide to produce the toxic molecule peroxynitrite. Furthermore, the production of ROS by NADPH oxidases in microglial cells is thought to contribute to the inflammatory process. A recent study by Steffen et al.[100] demonstrated that certain mono-*O*-methylated flavonoids were potent inhibitors of NADPH oxidases. Thus, while fisetin had no effect on NADPH oxidase activity, an *in vitro* methylated form of fisetin was a potent inhibitor with an IC_{50} of 8.2 μM. Interestingly, fisetin is methylated in human liver,[101] and so *in vivo* fisetin might be able to act as an NADPH oxidase inhibitor via its metabolic conversion to a methylated form.

14.7.6 ACTIVATION OF SIRTUINS

Sirtuins or Sir2-family proteins are NAD^+-dependent histone/protein deacetylases. They have been implicated in a variety of cellular processes, including DNA repair, maintenance of genomic integrity, transcriptional repression, and promotion of longevity in yeast, worms, and flies.[102] Furthermore, there is evidence that Sir2 activation can provide neuroprotection in Huntington's disease [103] and AD.[104] Many of the activities of Sir2 are associated with its histone deacetylase activity. Several years ago, fisetin, along with several other flavonols and the stilbene resveratrol, was shown to activate SIRT1, a human Sir2 family protein.[105] However, the results with resveratrol have since been

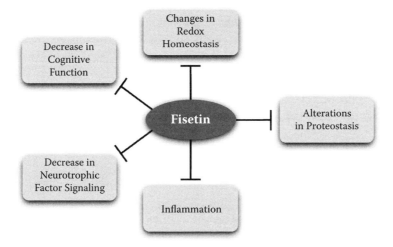

FIGURE 14.8 Fisetin affects multiple pathways implicated in neurological disorders. As discussed in this chapter, fisetin can increase neuronal function and survival through maintenance of redox homeostasis, activating neurotrophic factor signaling pathways, regulating protein homeostasis (proteostasis[114]), and inhibiting inflammatory responses. Fisetin can also enhance cognitive function. Therefore, it has the potential to act as a multifactorial drug for the treatment of a variety of neurological disorders.

called into question since its activating effect on SIRT1 appeared to be specific to the synthetic SIRT1 substrate used in the original assay.[102,106,107] Presumably, this concern also applies to the results with fisetin, suggesting that fisetin may not actually possess this activity.

14.8 METABOLISM OF FISETIN

Flavonoids are known to be extensively metabolized following oral consumption, resulting in glucuronidated, sulfated, and methylated metabolites.[108] Although the metabolism of fisetin has not been studied *in vivo*, it has been shown to be methylated by human liver extracts[101] but not extensively metabolized by rat liver microsomes.[109] In addition, in contrast to a number of other flavonoids, fisetin is readily taken up by cells.[110]

There is an ongoing debate about whether flavonoids such as fisetin can reach levels in the brain that are sufficient to affect neuronal function. However, following a single intraperitoneal injection, fisetin was detected in the brains of rats, and this correlated with a significant reduction in cerebral damage in a stroke model.[16] Similarly, as described above, we have seen significant protection in a rabbit stroke model following intravenous injection of fisetin,[15] and we found that oral administration of fisetin could enhance learning and memory in mice.[52] Thus, fisetin appears to be able to affect neuronal function *in vivo*, although whether this is a direct effect remains to be determined.

14.9 SUMMARY

Fisetin has positive effects on a number of common pathways associated with a variety of neurological disorders (Figure 14.8). While the precise relationships among the pleiotropic effects of fisetin on nerve cells remain to be determined, this combination of actions suggests that fisetin has the potential to maintain neuronal function even in the presence of diverse neurological insults. Therefore, it might have therapeutic value as a multifactorial drug for the treatment of a variety of neurological disorders.

ACKNOWLEDGMENTS

This work was supported by the U.S. Public Health Service through grants NS28121 and AG025337 and the Michael J. Fox Foundation.

REFERENCES

1. Cavalli, A., Bolognesi, M. L., Minarini, A., Rosini, M., Tumiatiit, V., Recanatini, M., and Melchiorre, C. Multi-target-directed ligands to combat neurodegenerative diseases. *J. Med. Chem.* 2008, 51, 347–372.
2. Hellerstein, M. K. A critique of the molecular target-based drug discovery paradigm based on principles of metabolic control: Advantages of pathway-based discovery. *Met. Engineer* 2008, 10, 1–9.
3. Pangalos, M. N., Schechter, L. E., and Hurko, O. Drug development for CNS disorders: Strategies for balancing risk and reducing attrition. *Nature Rev. Drug Discov.* 2007, 6, 521–532.
4. Middleton, E., Kandaswami, C., and Theoharides, T. C. The effects of flavonoids on mammalian cells: Implications for inflammation, heart disease and cancer. *Pharmacol. Rev.* 2000, 52, 673–751.
5. Heim, K. E., Tagliaferro, A. R., and Bobilya, D. J. Flavonoid antioxidants: Chemistry, metabolism and structure-activity relationships. *J. Nutr. Biochem.* 2002, 13, 572–584.
6. Ross, J. A., and Kasum, C. M. Dietary flavonoids: Bioavailability, metabolic effects, and safety. *Annu. Rev. Nutr.* 2002, 22, 19–34.
7. Tan, S., Schubert, D., Maher, and P. Oxytosis: A novel form of programmed cell death. *Curr. Top. Med. Chem.* 2001, 1, 497–506.
8. Ishige, K., Schubert, D., and Sagara, Y. Flavonoids protect neuronal cells from oxidative stress by three distinct mechanisms. *Free Radic. Biol. Med.* 2001, 30, 433–446.
9. Ingall, T. Stroke — Incidence, mortality and risk. *J. Insur. Med.* 2004, 36, 143–152.
10. Lapchak, P. A., and Araujo, D. M. Advances in ischemic stroke treatment: Neuroprotective and combination therapies. *Expert Opin. Emerg. Drugs* 2007, 12, 97–112.
11. Green, A. R., and Shuaib, A. Therapeutic strategies for the treatment of stroke. *Drug Disc. Today* 2006, 11, 681–693.
12. Lyden, P. D., and Zivin, J. A. Hemorrhagic transformation after cerebral ischemia: Mechanisms and incidence. *Cerebrovasc. Brain Metab. Rev.* 1993, 5, 1–16.
13. Lapchak, P. A. Development of thrombolytic therapy for stroke: A perspective. *Expert Opin. Investig. Drugs* 2002, 11, 1623–1632.
14. Lapchak, P. A. Hemorrhagic transformation following ischemic stroke: Significance, causes and relationship to therapy and treatment. *Curr. Neurol. Neurosci. Rep.* 2002, 2, 38–43.
15. Maher, P., Salgado, K. F., Zivin, J. A., and Lapchak, P. A. A novel approach to screening for new neuroprotective compounds for the treatment of stroke. *Brain Res.* 2007, 1173, 117–125.
16. Rivera, F., Urbanavicius, J., Gervaz, E., Morquio, A., and Dajas, F. Some aspects of the in vivo neuroprotective capacity of flavonoids: Bioavailability and structure-activity relationship. *Neurotox. Res.* 2004, 6, 543–553.
17. Singh, I. N., Sullivan, P. G., Deng, Y., Mbye, L. H., and Hall, E. D. Time course of post-traumatic mitochondrial oxidative damage and dysfunction in a mouse model of focal traumatic brain injury: Implications for neuroprotective therapy. *J. Cereb. Blood Flow Metab.* 2006, 26, 1407–1418.
18. Xiong, Y. and Hall, E. D. Pharmacological evidence for a role of peroxynitrite in the pathophysiology of spinal cord injury. *Exp. Neurol.* 2009, 216, 105–114.
19. Eliasson, M. J. L., Huang, Z., Ferrante, R. J., Sasamata, M., Molliver, M. E., Snyder, S. H., and Moskowitz, M. A. Neuronal nitric oxide synthase activation and peroxynitrite formation in ischemic stroke linked to neuronal damage. *J. Neurosci.* 1999, 19, 5910–5918.
20. Torreilles, F., Salman-Tabcheh, S., Guerin, M., and Torreilles, J. Neurodegenerative disorders: The role of peroxynitrite. *Brain Res. Brain Res. Rev.* 1999, 30, 153–163.
21. Beckman, J. S., Beckman, T. W., Chen, J., Marshall, P. A., and Freeman, B. A. Apparent hydroxyl radical production by peroxynitrite: Implications for endothelial injury from nitric oxide and superoxide. *Proc. Natl. Acad. Sci. U S A* 1990, 87, 1620–1624.
22. Beckman, J. S., Ischiropoulos, H., Zhu, L., van der Woerd, M., Smith, C., Chen, J., Harrison, J., Martin, J. C., and Tsai, M. Kinetics of superoxide dismutase- and iron-catalyzed nitration of phenolics by peroxynitrite. *Arch. Biochem. Biophys.* 1992, 298, 438–445.

23. Ischiropoulos, H., Zhu, L., Chen, J., Tsai, M., Martin, J. C., Smith, C. D., and Beckman, J. S. Peroxynitrite-mediated tyrosine nitration catalyzed by superoxide dismutase. *Arch. Biochem. Biophys.* 1992, 298, 431–437.

24. Radi, R., Beckman, J. S., Bush, K. M., and Freeman, B. A. Peroxynitrite oxidation of sulfhydryls. The cytotoxic potential of superoxide and nitric oxide. *J. Biol. Chem.* 1991, 266, 4244–4250.

25. Szabo, C., Ischiropoulos, H. and Radi, R. Peroxynitrite: biochemistry, pathophysiology and development of therapeutics. *Nature Rev. Drug Discov.* 2007, 6, 662–678.

26. Burdo, J., Schubert, D., and Maher, P. Glutathione production is regulated via distinct pathways in stressed and non-stressed cortical cultures. *Brain Res.* 2008, 1189, 12–22.

27. Griffith, O. W., and Meister, A. Potent and specific inhibition of glutathione synthesis by buthionine sulfoximine (*S-n*-butyl homocysteine sulfoximine). *J. Biol. Chem.* 1979, 254, 7558–7560.

28. Meister, A., and Anderson, M. E. Glutathione. *Ann. Rev. Biochem.* 1983, 52, 711–760.

29. Sato, H., Tamba, M., Ishii, T., and Bannai, S. Cloning and expression of a plasma membrane cystine/glutamate exchange transporter composed of two distinct proteins. *J. Biol. Chem.* 1999, 274, 11455–11458.

30. Ishii, T., Itoh, K., Takahashi, S., Sato, H., Yanagawa, T., Yasutake, K., Bannai, S., and Yamamoto, M. Transcription factor Nrf2 coordinately regulates a group of oxidative stress-inducible genes in macrophages. *J. Biol. Chem.* 2000, 275, 16023–16029.

31. Chan, J. Y., and Kwong, M. Impaired expression of glutathione synthetic enzyme genes in mice with targeted deletion of the Nrf2 basic-leucine zipper protein. *Biochim. Biophys. Acta* 2000, 1517, 19–26.

32. Hayes, J. D., Ellis, E. M., Neal, G. E., Harrison, D. J., and Manson, M. M. Cellular response to cancer chemopreventive agents: Contribution of the antioxidant responsive element to the adaptive response to oxidative and chemical stress. *Biochem. Soc. Symp.* 1999, 64, 141–168.

33. Nguyen, T., Sherratt, P. J., and Pickett, C. B. Regulatory mechanisms controlling gene expression mediated by the antioxidant response element. *Annu. Rev. Pharmacol. Toxicol.* 2003, 43, 233–260.

34. Chen, C., and Kong, A.-N. T. Dietary chemopreventive compounds and ARE/EpRE signaling. *Free Rad. Biol. Med.* 2004, 36, 1505–1516.

35. Kensler, T. W., Wakabayashi, N., and Biswal, S. Cell survival responses to environmental stresses via the Keap1-Nrf2-ARE pathway. *Annu. Rev. Pharmacol. Toxicol.* 2007, 47, 89–116.

36. Maher, P. A comparison of the neurotrophic activities of the flavonoid fisetin and some of its derivatives. *Free Rad. Res.* 2006, 40, 1105–1111.

37. Maher, P., and Hanneken, A. Flavonoids protect retinal ganglion cells from oxidative stress-induced death. *Invest. Ophthalmol. Vis. Sci.* 2005, 46, 4796–4803.

38. Valerio, L. G., Kepa, J. K., Pickwell, G. V., and Quattrochi, L. C. Induction of human NAD(P)H:quinone oxidoreductase (NQO1) gene expression by the flavonol quercetin. *Toxicol. Lett.* 2001, 119, 49–57.

39. Myhrstad, M. C. W., Carlsen, H., Nordstrom, O., Blomhoff, R., and Moskaug, J. O. Flavonoids increase the intracellular glutathione level by transactivation of the γ-glutamylcysteine synthetase catalytical subunit promoter. *Free Rad. Biol. Med.* 2002, 32, 386–393.

40. Hou, D.-X., Fukuda, M., Johnson, J. A., Miyamori, K., Ushikai, M., and Fujii, M. Fisetin induces transcription of NADPH:quinone oxidoreductase gene through an antioxidant responsive element-involved activation. *Int. J. Oncol.* 2001, 18, 1175–1179.

41. Chen, T. J., Jeng, J. Y., Lin, C. W., Wu, C. Y., and Chen, Y. C. Quercetin inhibition of ROS-dependent and -independent apoptosis in rat glioma C6 cells. *Toxicol.* 2006, 223, 113–126.

42. Price, R. D., Milne, S. A., Sharkey, J., and Matsuoka, N. Advances in small molecules promoting neurotrophic function. *Pharmacol. Ther.* 2007, 115, 292–306.

43. Chen, S., Le, W. Neuroprotective therapy in Parkinson disease. *Amer. J. Ther.* 2006, 13, 445–457.

44. Levy, Y. S., Gilgun-Sherki, Y., and Melamed, E., Offen, D. Therapeutic potential of neurotrophic factors in neurodegenerative diseases. *BioDrugs* 2005, 19, 97–127.

45. Zuccato, C., and Cattaneo, E. Role of brain-derived neurotrophic factor in Huntington's disease. *Prog. Neurobiol.* 2007, 81, 294–330.

46. Keegan, K., and Halegoua, S. Signal transduction pathways in neuronal differentiation. *Curr. Opin. Neurobiol.* 1993, 3, 14–19.

47. Sagara, Y., Vahnnasy, J., and Maher, P. Induction of PC12 cell differentiation by flavonoids is dependent upon extracellular signal-regulated kinase activation. *J. Neurochem.* 2004, 90, 1144–1155.

48. Mayr, B., and Montminy, M. Transcriptional regulation by the phosphorylation-dependent factor CREB. *Nature Rev. Mol. Cell Biol.* 2001, 2, 599–609.

49. Carlezon, W. A., Duman, R. S., and Nestler, E. J. The many faces of CREB. *Trends Neurosci.* 2005, 28, 436–445.

50. Tully, T., Bourtchouladze, R., Scott, R., and Tallman, J. Targeting the CREB pathway for memory enhancers. *Nat. Rev. Drug Discov.* 2003, 2, 267–277.

51. Johannessen, M., Delghandi, M. P., and Moens, U. What turns CREB on? *Cell. Signal.* 2004, 16, 1211–1227.

52. Maher, P., Akaishi, T., and Abe, K. The flavonoid fisetin promotes ERK-dependent long-term potentiation and enhances memory. *Proc. Natl. Acad. Sci. U S A* 2006, 103, 16568–16573.

53. Bliss, T. V. P., and Collingridge, G. L. A synaptic model of memory: Long-term potentiation in the hippocampus. *Nature* 1993, 361, 31–39.

54. Barad, M., Bourtchouladze, R., Winder, D. G., Golan, H., and Kandel, E. Rolipram, a type IV-specific phosphodiesterase inhibitor, facilitates the establishment of long-lasting long-term potentiation and improves memory. *Proc. Natl. Acad. Sci. U S A* 1998, 95, 15020–15025.

55. Bevins, R. A., and Besheer, J. Object recognition in rats and mice: A non-trial non-matching-to-sample learning task to study "recognition memory." *Nature Protocols* 2006, 1, 1306–1311.

56. Maher, P. The flavonoid fisetin promotes nerve cell survival from trophic factor withdrawal by enhancement of proteasome activity. *Arch. Biochem. Biophys.* 2008, 476(2), 139–144.

57. Abe, K., Takayanagi, M., and Saito, H. Effects of recombinant human basic fibroblast growth factor and its modified protein CS23 on survival of primary cultured neurons from various regions of fetal rat brain. *Japan J. Pharmacol.* 1990, 53, 221–227.

58. Kwak, M.-K., Wakabayashi, N., Itoh, K., Motohashi, H., Yamamoto, M., and Kensler, T. W. Modulation of gene expression by cancer chemopreventative dithiolethiones through the Keap1-Nrf2 pathway. *J. Biol. Chem.* 2003, 278, 8135–8145.

59. Kwak, M.-K., Wakabayashi, N., Greenlaw, J. L., Yamamoto, M., and Kensler, T. W. Antioxidants enhance mammalian proteasome expression through the Keap1-Nrf2 signaling pathway. *Mol. Cell. Biol.* 2003, 23, 8786–9794.

60. Kwak, M.-K., and Kensler, T. W. Induction of 26S proteasome subunit PSMB5 by the bifunctional inducer 3-methylcholanthrene through the Nrf2-ARE, but not the AhR/Arnt-XRE, pathway. *Biochem. Biophys. Res. Commun.* 2006, 345, 1350–1357.

61. Kwak, M.-K., Huang, B., Chang, H., Kim, J.-A., and Kensler, T. W. Tissue specific increase of the catalytic subunits of the 26S proteasome by indirect antioxidant dithiolethione in mice: Enhanced activity for degradation of abnormal protein. *Life Sci.* 2007, 80, 2411–2420.

62. Ciechanover, A. Proteolysis: From the lysosome to ubiquitin and the proteasome. *Nature Rev. Mol. Cell Biol.* 2005, 6, 79–86.

63. Kisselev, A. F., and Goldberg, A. L. Proteasome inhibitors: From research tools to drug candidates. *Chem. Biol.* 2001, 8, 739–758.

64. Giasson, B. I., Bruening, W., Durham, H. D., and Mushynski, W. E. Activation of stress-activated protein kinases correlates with neurite outgrowth induced by protease inhibition in PC12 cells. *J. Neurochem.* 1999, 72, 1081–1087.

65. Obin, M., Mesco, E., Gong, X., Haas, A. L., Joseph, J., and Taylor, A. Neurite outgrowth in PC12 cells. *J. Biol. Chem.* 1999, 274, 11789–11795.

66. Laser, H., Mack, T. G. A., Wagner, D., and Coleman, M. P. Proteasome inhibition arrests neurite outgrowth and causes "dying-back" degeneration in primary culture. *J. Neurosci. Res.* 2003, 74, 906–916.

67. Klimaschewski, L., Hausott, B., Ingorokva, S., and Pfaller, K. Constitutively expressed catalytic proteasomal subunits are up-regulated during neuronal differentiation and required for axon initiation, elongation and maintenance. *J. Neurochem.* 2006, 96, 1708–1717.

68. Yi, J. J., and Ehlers, M. D. Emerging roles for ubiquitin and protein degradation in neuronal function. *Pharmacol. Rev.* 2007, 59, 14–39.

69. Chung, K. K. K., Dawson, V. L., and Dawson, T. M. The role of the ubiquitin-proteasomal pathway in Parkinson's disease and other neurodegenerative disorders. *Trends Neurosci.* 2001, 24, S7–S14.

70. Ciechanover, A., and Brundin, P. The ubiquitin proteasome system in neurodegenerative diseases: Sometimes the chicken, sometimes the egg. *Neuron* 2003, 40, 427–446.

71. Betarbet, R., Sherer, T. B., and Greenamyre, J. T. Ubiquitin-proteasome system and Parkinson's disease. *Exper. Neurol.* 2005, 191, S17–S27.

72. Kim, H., Park, B.-S., Lee, K.-G., Choi, C. Y., Jang, S. S., Kim, Y.-H., and Lee, S.-E. Effects of naturally occurring compounds on fibril formation and oxidative stress of β-amyloid. *J. Agric. Food. Chem.* 2005, 53, 8537–8541.

73. Wang, L., Tu, Y.-C., Lian, T.-W., Hung, J.-T., Yen, J.-H., and Wu, M.-J. Distinctive antioxidant and anti-inflammatory effects of flavonols. *J. Agric. Food Chem.* 2006, 54, 9798–9804.

74. Sung, B., Pandey, M. K., and Aggarwal, B. B. Fisetin, an inhibitor of cyclin-dependent kinase 6, down-regulates nuclear factor-kB-regulated cell proliferation, antiapoptotic and metastatic gene products through the suppression of TAK-1 and receptor-interacting protein-regulated IkBa kinase activation. *Mol. Pharmacol.* 2007, 71, 1703–1714.

75. Yamamoto, Y., and Gaynor, R. B. IκB kinases: Key regulators of the NF-κB pathway. *Trends Biochem. Sci.* 2004, 29, 72–79.

76. Bubici, C., Papa, S., Dean, K., and Franzoso, G. Mutual cross-talk between reactive oxygen species and nuclear factor-kappa B: Molecular basis and biological significance. *Oncogene* 2006, 25, 6731–6748.

77. Hendriks, J. J. A., de Vries, H. E., van der Pol, S. M. A., van den Berg, T. K., van Tol, E. A. F., and Dijkstra, C. D. Flavonoids inhibit myelin phagocytosis by macrophages, a structure-activity relationship study. *Biochem. Pharmacol.* 2003, 65, 877–885.

78. Rock, R. B., and Peterson, P. K. Microglia as a pharmacological target in infectious and inflammatory diseases on the brain. *J. Neuroimmune Pharmacol.* 2006, 1, 117–126.

79. Garden, G. A., and Moller, T. Microglia biology in health and disease. *J. Neuroimmune Pharmacol.* 2006, 1, 127–137.

80. Dringen, R. Oxidative and antioxidative potential of brain microglial cells. *Antioxid. Redox Signal.* 2005, 7, 1223–1233.

81. Zheng, L. T., Ock, J., Kwon, B.-M., and Suk, K. Suppressive effects of flavonoid fisetin on lipopolysaccharide-induced microglial activation and neurotoxicity. *Int. Immunopharmacol.* 2008, 8, 484–494.

82. van Acker, S. A. B. E., van den Berg, D.-J., Tromp, M. N. J. L., Griffioen, D. H., van Bennekom, W. P., van der Vijgh, W. J. F., and Bast, A. Structural aspects of antioxidant activity of flavonoids. *Free Radic. Biol. Med.* 1996, 20, 331–342.

83. Kauppinen, T. M. Multiple roles for poly(ADP-ribose)polymerase-1 in neurological disease. *Neurochem. Int.* 2007, 50, 954–958.

84. Geraets, L., Moonen, H. J. J., Brauers, K., Wouters, E. F. M., Bast, A., and Hageman, G. J. Dietary flavones and flavonoles are inhibitors of poly(ADP-ribose)polymerase-1 in pulmonary epithelial cells. *J. Nutr.* 2007, 137, 2190–2195.

85. Phillis, J. W., Horrocks, L. A., and Farooqui, A. A. Cyclooxygenases, lipoxygenases, and epoxygenases in CNS: Their role and involvement in neurological disorders. *Brain Res. Rev.* 2006, 52, 201–243.

86. Li, Y., Maher, P., and Schubert, D. A role of 12-lipoxygenase in nerve cell death caused by glutathione depletion. *Neuron* 1997, 19, 453–463.

87. Canals, S., Casarejos, M. J., de Bernardo, S., Rodriguez-Martin, E., and Mena, M. A. Nitric oxide triggers the toxicity due to glutathione depletion in midbrain cultures through 12-lipoxygenase. *J. Biol. Chem.* 2003, 278, 21542–21549.

88. Kramer, B. C., Yabut, J. A., Cheong, J., Jnobaptiste, R., Robakis, T., and Olanow, C. W., Mytilineou, C. Toxicity of glutathione depletion in mesencephalic cultures: A role for arachidonic acid and its lipoxygenase metabolites. *Eur. J. Neurosci.* 2004, 19, 280–286.

89. Wang, H., Li, J., Follett, P. L., Zhang, Y., Cotanche, D. A., Jensen, F. E., Volpe, J. J., and Rosenberg, P. A. 12-Lipoxygenase plays a key role in cell death caused by glutathione depletion and arachidonic acid in rat oligodendrocytes. *Eur. J. Neurosci.* 2004, 20, 2049–2058.

90. Pratico, D., Zhukareva, V., Yao, Y., Uryu, K., Funk, C. D., Lawson, J. A., Trojanowski, J. Q., and Lee, V. M.-Y. 12/15-Lipoxygenase is increased in Alzheimer's disease. *Am. J. Pathol.* 2004, 164, 1655–1662.

91. Lebeau, A., Terro, F., Rostene, W., and Pelaprat, D. Blockade of 12-lipoxygenase expression protects cortical neurons from apoptosis induced by beta-amyloid peptide. *Cell Death Differ.* 2004, 11, 875–884.

92. Lapchak, P. A., Maher, P., Schubert, D., and Zivin, J. A. Baicalein, an antioxidant 12/15-lipoxygenase inhibitor improves clinical rating scores following multiple infarct embolic strokes. *Neuroscience* 2007, 150, 585–591.

93. Firuzi, O., Zhuo, J., Chinnici, C. M., Wisniewski, T., and Pratico, D. 5-Lipoxygenase gene disruption reduces amyloid-beta pathology in a mouse model of Alzheimer's disease. *FASEB J.* 2008, 22, 1169–1178.

94. Hsieh, R. J., German, J. B., and Kinsella, J. E. Relative inhibitory potencies of flavonoids on 12-lipoxygenase of fish gill. *Lipids* 1988, 23, 322–326.

95. Laughton, M. J., Evans, P. J., and Moroney, M. A., Hoult, J. R. S., Halliwell, B. Inhibition of mammalian 5-lipoxygenase and cyclo-oxygenase by flavonoids and phenolic dietary additives. *Biochem. Pharmacol.* 1991, 42, 1673–1681.

96. Sadik, C. D., Sies, H., and Schewe, T. Inhibition of 15-lipoxygenases by flavonoids: Structure-activity relations and mode of action. *Biochem. Pharmacol.* 2003, 65, 773–781.

97. Knockaert, M., Greengard, P., and Meijer, L. Pharmacological inhibitors of cyclin-dependent kinases. *Trends Pharmacol. Sci.* 2002, 23, 417–425.

98. Lu, H., Chang, D. J., Baratte, B., Meijer, L., and Schulze-Gahmen, U. Crystal structure of a human cyclin-dependent kinase 6 complex with a flavonol inhibitor, fisetin. *J. Med. Chem.* 2005, 48, 737–743.

99. Sun, G. Y., Horrocks, L. A., and Farooqui, A. A. The roles of NADPH oxidase and phospholipases A2 in oxidative and inflammatory responses in neurodegenerative diseases. *J. Neurochem.* 2007, 103, 1–16.

100. Steffen, Y., Gruber, C., Schewe, T., and Sies, H. Mono-*O*-methylated flavonols and other flavonoids as inhibitors of endothelial NADPH oxidase. *Arch. Biochem. Biophys.* 2008, 469, 209–219.

101. De Santi, C., Pietrabissa, A., Mosca, F., and Pacifici, G. M. Methylation of quercetin and fisetin, flavonoids widely distributed in edible vegetables, fruits and wine, by human liver. *Int. J. Clin. Pharmacol. Ther.* 2002, 40, 207–212.

102. Grubisha, O., Smith, B. C., and Denu, J. M. Small molecule regulation of Sir2 protein deacetylases. *FEBS J.* 2005, 272, 4607–4616.

103. Anekonda, T. S., Reddy, P. H. Neuronal protection by sirtuins in Alzheimer's disease. *J. Neurochem.* 2006, 96, 305–313.

104. Kim, D., Nguyen, M. D., Dobbin, M. M., Fischer, A., Sanabenesi, F., Rodgers, J. T., Delalle, I., Baur, J. A., Sui, G., Armour, S. M., Puigserver, P., Sinclair, D. A., and Tsai, L. H. SIRT1 deacetylase protects against neurodegeneration in models for Alzheimer's disease and amyotrophic lateral sclerosis. *EMBO J.* 2007, 26, 3169–3179.

105. Howitz, K. T., Bitterman, K. J., Cohen, H. Y., Lamming, D. W., Lavu, S., Wood, J. G., Zipkin, R. E., Chung, P., Kisielewski, A., Zhang, L.-L., Scherer, B., and Sinclair, D. A. Small molecule activators of sirtuins extend *Saccharomyces cerevisiae* lifespan. *Nature* 2003, 425, 191–196.

106. Kaeberlein, M., McDonagh, T., Heltweg, B., Hixon, J., Westman, E. A., Caldwell, S. D., Napper, A., Curtis, R., DiStefano, P. S., Fields, S., Bedalov, A., and Kennedy, B. K. Substrate-specific activation of sirtuins by resveratrol. *J. Biol. Chem.* 2005, 280, 17038–17045.

107. Garber, K. A mid-life crisis for aging theory. *Nature Biotechnol.* 2008, 26, 371–374.

108. Manach, C., Williamson, G., Morand, C., Scalbert, A., and Remesy, C. Bioavailability and bioefficacy of polyphenols in humans: I. Review of 97 bioavailability studies. *Am. J. Clin. Nutr.* 2005, 81, 230S–242S.

109. Nielsen, S. E., Breinholt, V., Justesen, U., Cornett, C., and Dragsted, L. O. *In vitro* biotransformation of flavonoids by rat liver microsomes. *Xenobiotica* 1998, 28, 389–401.

110. Watjen, W., Michels, G., Steffan, B., Niering, P., Chovolou, Y., Kampkotter, A., Tan-Thi, Q.-H., Proksch, P., and Kahl, R. Low concentrations of flavonoids are protective in rat H4IIE cells whereas high concentrations cause DNA damage and apoptosis. *J. Nutr.* 2005, 135, 525–531.

111. Carter, G. W., Young, P. R., Albert, D. H., Bouska, J., Dyer, R., Bell, R. L., Summers, J. B., and Brooks, D. W. 5-Lipoxygenase inhibitory activity of zileuton. *J. Pharmacol. Exp. Ther.* 1991, 256, 929–937.

112. Sekiya, K., and Okuda, H. Selective inhibition of platelet lipoxygenase by baicalein. *Biochem. Biophys. Res. Commun.* 1982, 105, 1090–1095.

113. Auerbach, B. J., Kiely, J. S., and Conrnicell, J. A. A spectrophotometric microtiter-based assay for the detection of hydroperoxy derivatives of linoleic acid. *Anal. Biochem.* 1992, 201, 375–380.

114. Balch, W. E., Morimoto, R. I., Dillin, A., and Kelly, J. W. Adapting proteostasis for disease intervention. *Science* 2008, 319, 916–919.

15 Dietary Flavonoids as Neuroprotective Agents

Jeremy P. E. Spencer, David Vauzour, Katerina Vafeiadou, and Ana Rodriguez Mateos

Molecular Nutrition Group, School of Chemistry, Food and Pharmacy, University of Reading, Reading, UK

CONTENTS

15.1 INTRODUCTION

Representing one of the most important lifestyle factors, diet can strongly influence the incidence and onset of cardiovascular disease and neurodegenerative disorders. Various phytochemical constituents of foods and beverages, in particular a class of compounds called flavonoids, have been avidly investigated in recent years. They have been proposed to exert a multiplicity of neuroprotective actions within the brain, including a potential to protect neurons against injury induced by neurotoxins [1], an ability to suppress neuroinflammation [2], and a potential to promote memory, learning, and cognitive functions [3]. These effects appear to be underpinned by two common processes. First, they are capable of interactions with critical protein and lipid kinase signaling cascades in the brain, leading to an inhibition of apoptosis triggered by neurotoxic species and to a promotion of neuronal survival and synaptic plasticity. Second, they induce beneficial effects on the vascular system, leading to changes in cerebrovascular blood flow capable of causing angiogenesis and neurogenesis and changes in neuronal morphology. This chapter highlights the neuroprotective mechanisms of flavonoids through their ability to interact with neuronal signaling pathways, their potential to inhibit neuroinflammation, and their impact on the vascular system.

15.2   SYNTHESIS, SOURCE, AND STRUCTURE OF FLAVONOIDS

Flavonoids are synthesized in plants from the reaction of a chalcone precursor with three molecules of malonyl-CoA. Under the action of the enzymes chalcone synthase and chalcone flavanone isomerase, the chalcone precursor is isomerized into a flavanone [4,5]. Hydroxylation in position 3 of the C-ring allows the differentiation of flavanonols from flavanones since they share a similar structure based on the 2,3-dihydro-2-phenylchromen-4-one skeleton. From these central intermediates, the pathway diverges into several side branches, each resulting in a different class of flavonoids (Figure 15.1). Due to their origin in plants, flavonoids are major constituents of fruit, vegetables, and plant-derived beverages, such as wine, tea, cocoa, and fruit juices. Most commonly, flavonoids share a common structure consisting of two aromatic rings (A and B), which are bound together by three carbon atoms, forming an oxygenated heterocycle (ring C) (Figure 15.1). Based on variations in the saturation of the basic flavan ring system, their alkylation and/or glycosylation, and the hydroxylation pattern of the molecules, flavonoids may be divided into seven subclasses: flavonols, flavones, flavanones, flavanonols, flavanols, anthocyanidins, and isoflavones (Figure 15.1). Flavanols are found both as monomers and as oligomers referred to as condensed tannins or proanthocyanidins. These compounds differ in their nature based on their constitutive units (e.g., catechins and epicatechin), their sequence, and the positions of interflavanic linkages (C4–C6 or C4–C8 in the B-type series, with additional C2–O–C7 or C2–O–C5 bonds in A-type structures) [6].

The flavanols, sometimes referred to as flavan-3-ols, are found predominantly in green and black teas, red wine, and chocolate. Variations in their structures lie in the hydroxylation pattern of the B ring and the presence of gallic acid in position 3. The lack of a double bond at positions 2–3 and the presence of a 3-hydroxyl group on the C ring create two centers of asymmetry. Typical dietary flavanols include catechin, epicatechin, epigallocatechin (EGC), and (–)-EGC-3-gallate (EGCG) (Figure 15.1). Flavanols exist also as oligomers or polymers, referred to as condensed tannins or proanthocyanidins, which are found in high concentration in cocoa, tea, red wine, and fruits such as apples, grapes, and strawberries. These differ in nature based on their constitutive units (e.g., catechins and epicatechin), their sequence, and the position of interflavanic linkages. The sources of anthocyanidins, such as pelargonidin, cyanidin, and malvidin, include red wine and berry fruits such as blueberries, blackberries, cherries, and strawberries. These compounds exist as glycosides in plants, are water soluble, and appear red or blue according to pH level. Individual anthocyanidins arise from the variation in number and arrangement of the hydroxyl and methoxy groups around the three rings (Figure 15.1). Flavones such as apigenin and luteolin are found in parsley, chives, artichoke, and celery. Hydroxylation in position 3 of the flavone structure gives rise to the 3-hydroxyflavones also known as the flavonols (e.g., kaempferol, quercetin), which are found in onions, leeks, and broccoli (Figure 15.1). The diversity of these compounds stems from the varying positions of phenolic –OH groups around the three rings. Dietary flavanones include naringenin, hesperetin, and taxifolin, and are found predominantly in citrus fruit and tomatoes. Hydroxylation of flavanones in position 3 of the C-ring gives rise to the flavanonols (Figure 15.1). Finally, isoflavones such as daidzein and genistein are a subclass of the flavonoid family found in soy and soy products. They have a large structural variability, and more than 600 isoflavones have been identified to date and are classified according to the oxidation level of the central pyran ring (Figure 15.1).

15.3   ABSORPTION, METABOLISM, AND DISTRIBUTION OF FLAVONOIDS

Although flavonoids have been identified as powerful antioxidants *in vitro* [7–9], their ability to act as effective antioxidants *in vivo* will depend on the extent of their biotransformation and conjugation during absorption from the gastrointestinal (GI) tract, in the liver and, finally, in cells. Following oral ingestion and during absorption, flavonoids are extensively metabolized, resulting in significant alteration of their redox potentials. For example, most flavonoid glycosides and aglycones present in plant-derived foods are extensively conjugated and metabolized during absorption (reviewed by

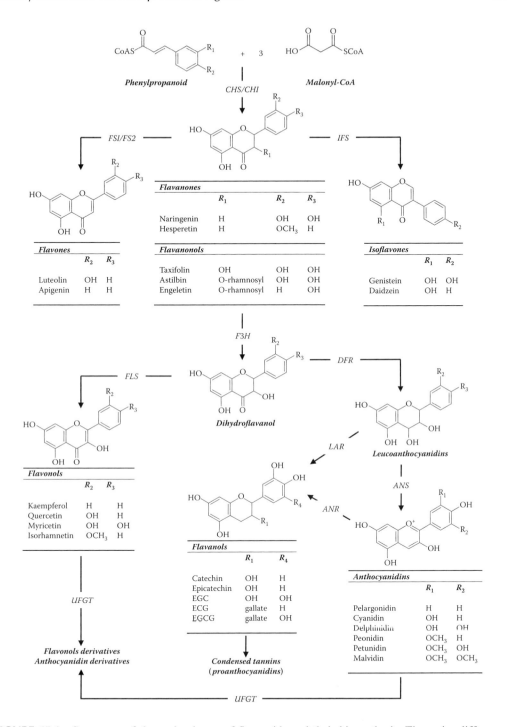

FIGURE 15.1 Structures of the main classes of flavonoids and their biosynthesis. The major differences between the individual groups reside in the hydroxylation pattern of the ring structure, the degree of saturation of the C-ring, and the substitution in the 3-position. All flavonoids are derived from chalcone precursors that are derived from phenylpropanoid and three malonyl- CoAs and biosynthesised by chalcone synthase. Various enzymes act to bring about the formation of the various flavonoid classes: chalcone isomerase (CHI), flavone synthase (FSI/FS2), isoflavone synthase (IFS), flavanone 3-hydroxylase (F3H), dihydroflavonol reductase (DFR), anthocyanidin synthase (ANS), leucoanthocyanidin reductase (LAR), anthocyanidin reductase (ANR), UDP glucose-flavonoid 3-O-glucosyl transferase (UFGT), and flavonol synthase (FLS).

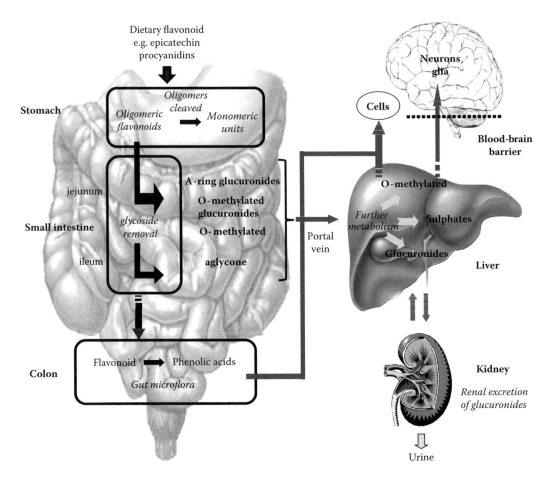

FIGURE 15.2 **(See color insert following page 234.)** Summary of the formation of GI tract and hepatic metabolites and conjugates of polyphenols in humans. Cleavage of oligomeric flavonoids such as procyanidins may occur in the stomach in an environment of low pH. All classes of flavonoids undergo extensive metabolism in the jejunum and ileum of the small intestine, and resulting metabolites enter the portal vein and undergo further metabolism in the liver. Colonic microflora degrades flavonoids into smaller phenolic acids that may also be absorbed. The fate of flavonoids is renal excretion, although some may enter cells and tissues.

Spencer et al. [10,11]). In the upper GI tract, dietary polyphenols are substrates for a number of enzymes, such as phase I enzymes (hydrolyzing and oxidizing) and phase II enzymes (conjugating and detoxifying). During the transfer across the small intestine, and again in the liver, they are transformed into glucuronides, sulfates, and *O*-methylated forms [10,12,13]. Further transformations occur in the colon, where the enzymes of the gut microflora act to metabolize flavonoids to simple phenolic acids, which are absorbed and further metabolized in the liver [14] (Figure 15.2). In addition, flavonoids may undergo at least three types of intracellular metabolism: (1) oxidative metabolism, (2) P450-related metabolism, and (3) conjugation with thiols, particularly GSH [15]. Circulating metabolites of flavonoids, such as glucuronides, sulfates, and conjugated *O*-methylated forms, or intracellular metabolites, such as flavonoid-GSH adducts, have greatly reduced antioxidant potential [16]. Indeed, studies have indicated that although such conjugates and metabolites may participate directly in plasma antioxidant reactions and in scavenging reactive oxygen species (ROS) and reactive nitrogen species in the circulation, their effectiveness is reduced relative to their parent aglycones [17–21].

15.4 EFFECTS OF FLAVONOIDS ON THE BRAIN

15.4.1 Do Flavonoids Access the Brain?

In order for flavonoids to access the brain, they must first cross the blood-brain barrier (BBB), which controls entry of xenobiotics into the brain [22] (Figure 15.2). Flavanones such as hesperetin, naringenin, and their *in vivo* metabolites, along with some dietary anthocyanins, cyanidin-3-rutinoside, and pelargonidin-3-glucoside, have been shown to traverse the BBB in relevant *in vitro* and *in situ* models [23]. Their degree of BBB penetration is dependent on compound lipophilicity [24], meaning that less polar *O*-methylated metabolites may be capable of greater brain uptake than the more polar flavonoid glucuronides. However, evidence exists to suggest that certain drug glucuronides may cross the BBB [25] and exert pharmacological effects [26,27], suggesting that there may be a specific uptake mechanism for glucuronides *in vivo*. Their brain entry may also depend on their interactions with specific efflux transporters expressed in the BBB, such as P-glycoprotein [28], which appears to be responsible for the differences between naringenin and quercetin flux into the brain in situ [23]. In animals, flavanones have been found to enter the brain following their intravenous administration [29], while EGCG [30], epicatechin [31], and anthocyanins [32,33] are found in the brain after their oral administration. Furthermore, several anthocyanins have been identified in different regions of the rat [34] and pig brains [35] of blueberry-fed animals, with 11 intact anthocyanins found in the cortex and cerebellum. These results indicate that flavonoids transverse the BBB and are able to localize in the brain, suggesting that they are candidates for direct neuroprotective and neuromodulatory actions.

15.4.2 Protection against Neuronal Injury Induced by Neurotoxins

Neurodegeneration in Parkinson's, Alzheimer's, and other neurodegenerative diseases appears to be triggered by multifactorial events, including neuroinflammation, glutamatergic excitotoxicity, increases in iron, and/or depletion of endogenous antioxidants [36–38]. There is a growing body of evidence to suggest that flavonoids may be able to counteract the neuronal injury underlying these disorders [1,39,40]. For example, a *Ginkgo biloba* extract has been shown to protect hippocampal neurons from nitric oxide- and beta-amyloid-induced neurotoxicity [41], and studies have demonstrated that the consumption of green tea may have beneficial effect in reducing the risk of Parkinson's disease [42]. In agreement with the latter study, tea extracts and EGCG have also been shown to attenuate 6-hydroxydopamine-induced toxicity [43], to protect against hippocampal injury during transient global ischemia [44], and to prevent nigral damage induced by MPTP [45].

The death of nigral neurons in Parkinson's disease is thought to involve the formation of the endogenous neurotoxin 5-*S*-cysteinyl-dopamine (5 *S*-Cys-DA) [46,47] (Figure 15.3). Recent investigations have shown that 5-*S*-cysteinyl-catecholamine conjugates possess strong neurotoxicity and initiate a sustained increase in intracellular ROS in neurons leading to DNA oxidation, caspase-3 activation, and delayed neuronal death [48,49] (Figure 15.3). Such adducts may be generated by reactive species [50] and have been observed to be elevated in the human substantia nigra of patients who died of Parkinson's disease [46], suggesting that such species may be potential endogenous nigral toxins. However, 5-*S*-cysteinyl-dopamine-induced neuronal injury is counteracted by nanomolar concentrations of various flavonoids, including pelargonidin, quercetin, hesperetin, caffeic acid, the 4′-*O*-Me derivatives of catechin, and epicatechin [50]. Furthermore, in the presence of the flavanol (+)-catechin, tyrosinase-induced formation of 5-*S*-Cys-DA was inhibited by a mechanism linked to the capacity of catechin to undergo tyrosinase-induced oxidation to yield cysteinyl-catechin adducts [51]. In contrast, the inhibition afforded by flavanones such as hesperetin was not accompanied by the formation of cysteinyl-hesperetin adducts, indicating that its inhibitory effects are mediated by direct interaction with tyrosinase [51].

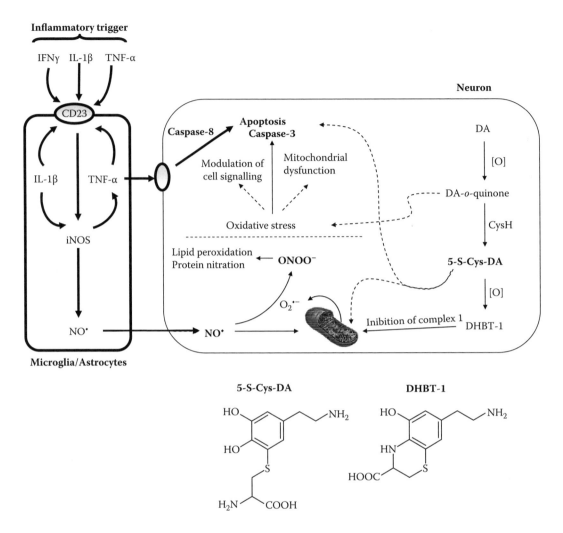

FIGURE 15.3 Involvement of neuroinflammation, endogenous neurotoxins, and oxidative stress in dopaminergic neurodegeneration. Structures of the 5-*S*-Cys-DA and the dihydrobenzothiazine-1 (DHBT-1) are shown.

ROS and reactive nitrogen species have also been proposed to play a role in the pathology of many neurodegenerative diseases [37] (Figure 15.3). There is abundant evidence that flavonoids are effective in blocking this oxidant-induced neuronal injury, although their potential to do so is thought not to rely on direct radical or oxidant scavenging activity [16,39,52]. Instead, they are believed to act by modulating a number of protein kinase and lipid kinase signaling cascades, such as the phosphatidylinositol 3-kinase (PI3K)/Akt, tyrosine kinase, protein kinase C (PKC), and mitogen-activated protein kinase (MAPK) signaling pathways [39,53]. Inhibitory or stimulatory actions at these pathways are likely to profoundly affect neuronal function by altering the phosphorylation state of target molecules, leading to changes in caspase activity, and/or by gene expression [53]. For example, flavonoids have been observed to block oxidative-induced neuronal damage by preventing the activation of caspase-3, providing evidence in support of their potent antiapoptotic action [16,54,55]. The flavanols epicatechin and 3′-*O*-methyl-epicatechin also protect neurons against oxidative damage via a mechanism involving the suppression of c-jun amino-terminal kinase (JNK) and downstream partners c-jun and pro-caspase-3 [54]. Flavanones such as hesperetin and its metabolite 5-nitro-hesperetin have been observed to inhibit oxidant-induced neuronal apoptosis via a mechanism involving the activation/phosphorylation of signaling proteins important in the prosurvival pathways [56]. Similarly, the flavone baicalein has been shown to significantly inhibit 6-hydroxydopamine-induced JNK activation and neuronal cell death and quercetin may suppress JNK activity and apoptosis induced by hydrogen peroxide [57,58], 4-hydroxy-2-nonenal [59], and tumor necrosis factor-alpha (TNF-α) [60].

15.4.3 Inhibition of Neuroinflammation

Neuroinflammatory processes in the CNS are believed to play a crucial role in the development of neurodegenerative diseases such as Alzheimer's disease and Parkinson's disease [61] and neuronal injury associated with stroke [62]. Glial cells (microglia and astrocytes) activation leads to the production of cytokines and other inflammatory mediators that may contribute to the apoptotic cell death of neurons observed in many neurodegenerative diseases. In particular, increases in cytokine production [interleukin-1β (IL-1β); TNF-α] [63], inducible nitric oxide synthase (iNOS) and nitric oxide (NO$^\bullet$), and increased NADPH oxidase activation [64] all contribute to glial-induced neuronal death (Figure 15.3). These events are controlled by MAPK signaling, which mediates both the transcriptional and posttranscriptional regulation of iNOS and cytokines in activated microglia and astrocytes [65–67]. While ibuprofen, a nonsteroidal anti-inflammatory drug, has been shown to delay the onset of neurodegenerative disorders, such as Parkinson's disease [68], most existing drug therapies for neurodegenerative disorders have failed to prevent the underlying degeneration of neurons. Consequently, there is a desire to develop alternative strategies capable of preventing the progressive neuronal loss resulting from neuroinflammation.

Flavonoid-rich blueberry extracts have been observed to inhibit NO$^\bullet$, IL-1β, and TNF-α production in activated microglia cells [69,70], while the flavonol quercetin [71], the flavones wogonin and bacalein [72], the flavanols catechin and EGCG [73], and the isoflavone genistein [74] have all been shown to attenuate microglial and/or astrocyte-mediated neuroinflammation via mechanisms that include inhibition of (1) iNOS and cyclooxygenase-2 expression, (2) NO$^\bullet$ production, (3) cytokine release, and (4) NADPH oxidase activation and subsequent ROS generation in astrocytes and microglia. Flavonoids may exert these effects via direct modulation of protein and lipid kinase signaling pathways [1,39,53], for example via the inhibition of MAPK signaling cascades, such as p38 or ERK1/2 (extracellular signal-related kinase 1/2), which regulate both iNOS and TNF-α expression in activated glial cells [66]. In this respect, fisetin inhibits p38 MAPK phosphorylation in LPS-stimulated BV-2 microglial cells [75] and the flavone luteolin inhibits IL-6 production in activated microglia via inhibition of the JNK signaling pathway [23,45]. The effects of flavonoids on these kinases may influence downstream pro-inflammatory transcription factors important in iNOS transcription. One of these, nuclear factor-kappa B, responds to p38 signaling and is involved in iNOS induction [67], suggesting that there is interplay between signaling pathways, transcription factors, and cytokine production in determining the neuroinflammatory response in the CNS. In this respect, flavonoids have also been shown to prevent transcription factor activation, with the flavonol quercetin able to suppress nuclear factor-kappa B, signal transducer and activator of transcription-1, and activating protein-1 activation in LPS- and IFN-γ-activated microglial cells [71].

15.4.4 Flavonoid-Induced Improvements in Memory, Learning, and Cognitive Performance

There is growing interest in the potential of phytochemicals to improve memory, learning, and general cognitive ability [1,3]. A recent prospective study which aimed to examine flavonoid intake in relation to cognitive function and decline has provided strong evidence that dietary flavonoid intake is associated with better cognitive evolution (i.e., the preservation of cognitive performance with ageing) [76]. In this PAQUID study (Personnes Agées QUID), 1640 subjects (aged 65 years or older) free from dementia at baseline and with reliable dietary assessment data were examined for their cognitive performance (Mini-Mental State Examination, Benton's Visual Retention Test, "Isaacs" Set Test) four times over a 10-year period. After adjustment for age, sex, and educational level, flavonoid intake was found to be associated with significantly better cognitive performance at baseline and with a significantly better evolution of the performance over time. In particular, subjects included in the two highest quartiles of flavonoid intake had better cognitive evolution than subjects in the lowest quartile, and after 10 years, subjects with the lowest flavonoid intake had lost on average 2.1 points on the Mini-Mental State Examination, whereas subjects with the highest quartile had lost 1.2 points. Such data provide a strong indication that regular flavonoid consumption may have a positive effect on neurocognitive performance as we age.

There has been much interest in the neurocognitive effects of soy isoflavones (Figure 15.1), primarily in postmenopausal women [77–80]. Isoflavone supplementation has been observed to have a favorable effect on cognitive function [81], particularly verbal memory, in postmenopausal women [82], and 6- and 12-week supplementations were observed to have a positive effect on frontal lobe function [83]. Furthermore, animal studies have also indicated that isoflavones are capable of improving cognitive function [84–86]. However, there is still uncertainty regarding their effects as some large intervention trials have reported that isoflavone supplementation does not lead to cognitive improvements [87]. The rationale behind the potential of isoflavones to exert positive effects on cognitive function is believed to lie primarily in their potential to mimic the actions and functions of estrogens in the brain [78]. For example, postmenopausal women who undertake estrogen-replacement therapy have a significantly lower risk for the onset of Alzheimer's disease [88]. They may also be effective by affecting the synthesis of acetylcholine and neurotrophic factors such as brain-derived neurotrophic factor and nerve growth factor in the hippocampus and frontal cortex [89,90].

There is also extensive evidence to suggest that berries, in particular blueberries, are effective in reversing age-related deficits in motor function and spatial working memory [91–96]. In addition to spatial memory, blueberry supplementation has been shown to improve "object recognition memory" [97] and "inhibitory fear conditioning learning" [95,96]. Blueberry intake appears to have a pronounced effect on short-term memory [95] and has also been shown to improve long-term reference memory following 8 weeks of supplementation. [93]. Tests using a radial arm maze have supported these findings and have provided further evidence for the efficacy of blueberries [94]. Indeed, these have shown that improvements in spatial memory may emerge within 3 weeks, the equivalent of about 3 years in humans. Beneficial effects of flavonoid-rich foods and beverages on psychomotor activity in older animals have also been reported [92,98]. In addition to those with berries, animal studies with tea [99] and pomegranate juice [100], or pure flavonols such as quercetin, rutin [101], or fisetin [102] have provided further evidence that dietary flavonoids are beneficial in reversing the course of neuronal and behavioral aging.

The flavonoid-rich plant extract *G. biloba* has also been shown to induce positive effects on memory, learning, and concentration [103–105]. *G. biloba* has a prominent effect on brain activity and short-term memory in animals and humans suffering from cognitive impairment [106–108] and promotes spatial learning in aged rodents [108–111]. Furthermore, *G. biloba* promotes inhibitory avoidance conditioning in rats with high-dose intake, leading to short-term, but not long-term, passive avoidance learning in senescent mice [110,112]. However, the pharmacological mechanisms by which *G. biloba* promotes cognitive effects are unclear, with its ability to elicit a reduction in levels of ROS [113,114], increase cerebral blood flow (CBF) [115], modulate brain fluidity [110], interact with the muscarinic cholinergic system [116], and protect the striatal dopaminergic system [117] all being suggested as possible mechanisms of brain action.

15.4.5 CEREBROVASCULAR EFFECTS OF FLAVONOIDS

Dementia is a serious degenerative disease affecting predominantly elderly people, with the two most common forms of this illness being Alzheimer's disease and vascular dementia. The factors

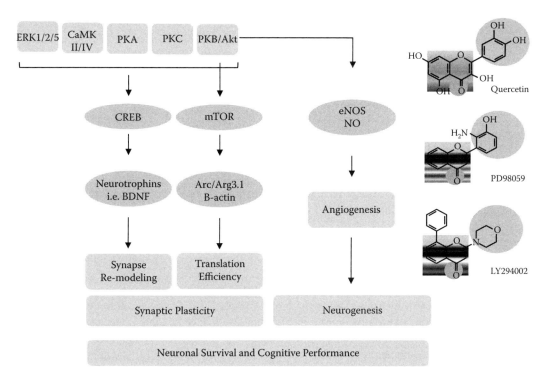

FIGURE 15.4 Signaling pathways underlying neuronal survival and cognitive performance. Flavonoids activate the ERK-CREB pathway and the PI3K-mTOR cascade, leading to changes in synaptic plasticity. They are also capable of influencing neurogenesis through the activation of PI3K-Akt-eNOS. The structures of MEK inhibitor PD98059 and the PI3K inhibitor LY294002, which have close structural homology to that of flavonoids, are also shown.

affecting dementia are age, hypertension, arteriosclerosis, diabetes mellitus, smoking, atrial fibrillation, and having the ApoE4 genotype [118]. There is evidence to suggest that flavonoids may be capable of preventing many forms of cerebrovascular disease, including those associated with stroke and dementia [119,120]. There is powerful evidence for the beneficial effects of flavonoids on peripheral [121] and cerebrovascular [122,123] blood flow, and these vascular effects are potentially significant as increased cerebrovascular function is known to facilitate adult neurogenesis in the hippocampus [124] (Figure 15.4). Indeed, new hippocampal cells are clustered near blood vessels, proliferate in response to vascular growth factors, and may influence memory [125]. As well as new neuronal growth, increases in neuronal spine density and morphology are considered vital for learning and memory [126]. Changes in spine density, morphology, and motility have been shown to occur with paradigms that induce synaptic and altered sensory experiences and lead to alterations in synaptic connectivity and strength between neuronal partners, affecting the efficacy of synaptic communication. These events are mediated at the cellular and molecular levels and are strongly correlated with memory and learning.

Efficient CBF is also vital for optimal brain function, with several studies indicating that there is a decrease in CBF in patients with dementia [127,128]. Brain imaging techniques, such as functional magnetic resonance imaging and transcranial Doppler ultrasound, have shown that there is a correlation between CBF and cognitive function in humans. For example, CBF velocity is significantly lower in patients with Alzheimer's disease and low CBF is also associated with incipient markers of dementia. In contrast, nondemented subjects with higher CBF were less likely to develop dementia. Flavonoids have been shown to exert a positive effect on CBF in humans [123,129]. After consumption of a flavanol-rich cocoa drink, the blood oxygen level-dependent functional magnetic resonance imaging showed an increase in blood flow in certain regions of the brain, along with a modification of the blood oxygen level-dependent response to task switching. Furthermore, arterial spin-labeling sequence magnetic resonance imaging [130] also indicated that cocoa flavanols increase CBF up to a maximum of 2 hours after ingestion of the flavanol-rich drink. In support of these findings, an increase in CBF through the middle cerebral artery has been reported after the consumption of flavanol-rich cocoa using transcranial Doppler ultrasound [129].

15.5 MECHANISMS OF ACTION OF FLAVONOIDS

15.5.1 ANTIOXIDANT EFFECTS

Historically, the beneficial effects of flavonoids were believed to be due to their antioxidant capacity. As a consequence, their ability to act as hydrogen-donating species was used to explain their protective effects against oxidative stress-associated diseases [131]. For example, flavonoids have been shown to act as classical electron (or hydrogen)-donating antioxidants [132], as scavengers of reactive species [133,134], and as inhibitors of low-density lipoprotein [135] and DNA oxidation [136]. Furthermore, flavonoids have also been shown to have the capacity to increase plasma total antioxidant capacity [137] and to decrease specific markers of oxidative stress, such as F2-isoprostanes [138] and lymphocyte 8-hydroxy-2'-deoxyguanosine levels [139]. However, although the brain is potentially susceptible to oxidative injury, due to high levels of oxidizable lipids and low levels of antioxidant protection, the bioavailability of flavonoids in the brain [31] suggests that their antioxidant activity would be negligible relative to that of ascorbic acid and α-tocopherol, which are present at much higher concentrations [140]. Ultimately, a precise understanding of whether or not flavonoids do influence oxidative stress in the brain will be wholly dependent on the physiological relevance of the markers of oxidative stress measured [141]. However, it appears that the effects of flavonoids may also be mediated by their interactions with specific neuronal intracellular signaling cascades, such as the MAPK signaling pathway and the PI3K/Akt signaling cascade [1,39,53,142].

15.5.2 INTERACTION OF FLAVONOIDS WITH THE CELL SIGNALING CASCADES

Flavonoids have been shown to exert many cellular effects through selective actions on components of a number of protein kinase and lipid kinase signaling cascades, such as the Akt/PKB tyrosine kinase, PKC, PI3K, and members of the MAPK family, such as the ERK, JNK, and p38 kinase [39,53] (Figure 15.4). Inhibitory or stimulatory actions at these pathways are likely to profoundly affect neuronal function by altering the phosphorylation state of target molecules and/or by modulating gene expression. In the mature brain, postmitotic neurones utilize MAPK and PI3K cascades in the regulation of key functions such as synaptic plasticity and memory formation and the MAPK pathway is essential for the control of neuronal proliferation, death, and survival [143]. In general, activation of the ERK pathway is regulated by growth factors and is associated with cell survival, while activation of JNK and p38 is stimulated by various cellular stresses and is involved in cell death [144].

Oxidative stress-induced activation of caspase-3 in neurons has been observed to be blocked by flavonoids, providing compelling evidence in support of a potent antiapoptotic action of flavonoids in these cells [52,54,55]. The flavanols epicatechin and 3'-O-methyl-epicatechin have been shown to protect neurons against oxidative damage via a mechanism involving the suppression of JNK and downstream partners c-jun and pro-caspase-3 [52,54]. There are a number of potential sites where flavonoids may interact with the JNK pathway. For instance, flavonoid-mediated inhibition of oxidative stress-induced apoptosis may occur by preventing the activation of JNK by influencing one of the many upstream MAPKKK-activating proteins that transduce signals to JNK. For example, their ability to inhibit JNK activation may proceed via flavonoid-induced modulation of the ASK1 phosphorylation state and its association with 14-3-3 protein, which is essential for suppression of cellular apoptosis [56]. Other potential mechanisms include an ability to preserve Ca^{2+} homeostasis, thereby preventing Ca^{2+}-dependent activation of JNK [54] or an attenuation of the proapoptotic signaling cascade lying downstream of JNK.

The effects of flavonoid-rich foods on neurocognitive function have been linked to the ability of flavonoids to interact with the cellular and molecular architecture responsible for memory and learning [1,3], including those involved in long-term potentiation and synaptic plasticity [39] (Figure 15.4). These effects are likely to lead to the enhanced neuronal connection and communication and thus a greater capacity for memory acquisition, storage, and retrieval [3]. For example, the flavanol (−)-epicatechin, especially in combination with exercise, has been observed to enhance the retention of rat spatial memory by a mechanism involving increased angiogenesis and neuronal spine density in the dentate gyrus of the hippocampus and an up-regulation of genes associated with learning in the hippocampus [145]. Fisetin, a flavonoid found in strawberries, has been shown to improve long-term potentiation and to enhance object recognition in mice by a mechanism dependent on the activation of ERK and CREB [102]. Similarly, the flavanol (−)-epicatechin induces both ERK1/2 and CREB activation in cortical neurons and subsequently increases CREB-regulated gene expression [146], while nanomolar concentrations of quercetin are effective in enhancing CREB activation [147]. Blueberry-induced improvements in memory have been shown to be mediated by increases in the phosphorylation state of ERK1/2, rather than in that of calcium calmodulin kinase (CaMKII and CaMKIV) or PKA [94]. Other flavonoids have also been found to influence the ERK pathway, with the citrus flavanone hesperetin capable of activating ERK1/2 signaling in cortical neurons [56] and flavanols such as ECGC restoring both PKC and ERK1/2 activities in 6-hydroxy dopamine toxicity and serum-deprived neurons [43,148].

It is likely that the actions of flavonoids on the ERK pathway result from their effects on upstream kinases, such as MEK1 and MEK2, and potentially on membrane receptors. This appears likely as flavonoids have close structural homology to specific inhibitors of ERK signaling, such as PD98059 (2′-amino-3′-methoxyflavone) (Figure 15.4). PD98059 is a flavone that has been shown to act *in vivo* as a highly selective noncompetitive inhibitor of MEK1 activation and the MAPK cascade [149–152]. PD98059 acts via its ability to bind to the inactive forms of MEK, thus preventing its activation by upstream activators such as c-Raf [151]. This raises the possibility that flavonoids, and their metabolites, may also act on this pathway in a similar manner. In support of this, the flavonol quercetin and, to a lesser extent, its *O*-methylated metabolites have been shown to induce neuronal apoptosis via a mechanism involving the inhibition of ERK, rather than by induction of proapoptotic signaling through JNK [147]. Flavonoids also have close structural homology with pharmacological inhibitors of the PI3K pathway, such as the LY294002, an inhibitor that was modeled on the structure of quercetin [153] (Figure 15.4). LY294002 and quercetin fit into the ATP binding pocket of the enzyme, and it appears that the number and substitution of hydroxyl groups on the B ring and the degree of unsaturation of the C2–C3 bond determine this particular bioactivity. Similar to the PI3 inhibitor LY294002, quercetin and other flavonoids have been shown to inhibit PI3K activity [147], with inhibition directed at the ATP binding site of the kinase [153].

15.6 SUMMARY

The neuroprotective actions of dietary flavonoids involve a number of effects within the brain, including a potential to protect neurons against injury induced by neurotoxins, an ability to suppress neuroinflammation, and a potential to promote memory, learning, and cognitive functions. This multiplicity of effects appears to be underpinned by two processes. First, they interact with important neuronal signaling cascades in the brain, leading to an inhibition of apoptosis triggered by neurotoxic species and to a promotion of neuronal survival and differentiation. These include selective actions on a number of protein kinase and lipid kinase signaling cascades, most notably the PI3K/Akt and MAPK pathways, which regulate prosurvival transcription factors and gene expression. It appears that the concentrations of flavonoids encountered in the brain may be sufficiently high to exert such pharmacological activity on receptors, kinases, and transcription factors. Second, they are known to induce beneficial effects on the peripheral and cerebral vascular system that lead to changes in cerebrovascular blood flow. Such changes are likely to induce angiogenesis, new nerve cell growth in the hippocampus, and changes in neuronal morphology, all processes known to be important in maintaining optimal neuronal function and neurocognitive performance.

Consumption of flavonoid-rich foods, such as berries and cocoa, throughout life holds a potential to limit neurodegeneration and prevent or reverse age-dependent deteriorations in cognitive performance. However, at present, the precise temporal nature of the effects of flavonoids on these events is unclear. For example, it is presently unclear as to when one needs to begin consuming flavonoids in order to obtain maximum benefits. It is also unclear which flavonoids are most effective in inducing these changes. However, due to the intense interest in the development of drugs capable of enhancing brain function, flavonoids may represent important precursor molecules in the quest to develop a new generation of brain-enhancing drugs.

ACKNOWLEDGMENTS

The authors were sponsored by the Biotechnology and Biological Sciences Research Council (grant nos. BB/C518222/1, BB/F008953/1, and BB/G005702/1) and the Medical Research Council (grant ref. G0400278).

REFERENCES

1. Spencer, J.P.E., Flavonoids: Modulators of brain function? *Br J Nutr* 99 E Suppl 1, ES60, 2008.
2. Vafeiadou, K. et al., Neuroinflammation and its modulation by flavonoids, *Endocr Metab Immune Disord Drug Targets* 7, 211, 2007.
3. Spencer, J.P.E., Food for thought: The role of dietary flavonoids in enhancing human memory, learning and neuro-cognitive performance, *Proc Nutr Soc* 67, 238, 2008.
4. Dixon, R.A., and Steele, C.L., Flavonoids and isoflavonoids—A gold mine for metabolic engineering, *Trends Plant Sci* 4, 394, 1999.
5. Winkel-Shirley, B., Flavonoid biosynthesis. A colorful model for genetics, biochemistry, cell biology, and biotechnology, *Plant Physiol* 126, 485, 2001.
6. Cheynier, V., Polyphenols in foods are more complex than often thought, *Am J Clin Nutr* 81, 223S, 2005.
7. Rice-Evans, C.A. et al., Structure-antioxidant activity relationships of flavonoids and phenolic acids, *Free Radic Biol Med* 20, 933, 1996.
8. Rice-Evans, C., Flavonoid antioxidants, *Curr Med Chem* 8, 797, 2001.
9. Rice-Evans, C., Plant polyphenols: Free radical scavengers or chain-breaking antioxidants? *Biochem Soc Symp* 61, 103, 1995.
10. Spencer, J.P.E. et al., Bioavailability of flavan-3-ols and procyanidins: Gastrointestinal tract influences and their relevance to bioactive forms *in vivo*, *Antioxid Redox Signal* 3, 1023, 2001.
11. Spencer, J.P.E. et al., Biomarkers of the intake of dietary polyphenols: Strengths, limitations and application in nutrition research, *Br J Nutr* 1, 12, 2007.
12. Spencer, J.P.E. et al., The small intestine can both absorb and glucuronidate luminal flavonoids, *FEBS Lett* 458, 224, 1999.
13. Spencer, J.P.E., Metabolism of tea flavonoids in the gastrointestinal tract, *J Nutr* 133, 3255S, 2003.
14. Scheline R.R. (1999). Metabolism of Oxygen Heterocyclic Compounds, in *CRC Handbook of Mammalian Metabolism of Plant Compounds,* pp. 243–295, CRC Press, Inc., Boca Raton, FL.
15. Spencer, J.P.E. et al., Intracellular metabolism and bioactivity of quercetin and its *in vivo* metabolites, *Biochem J* 372, 173, 2003.
16. Spencer, J.P.E. et al., Contrasting influences of glucuronidation and O-methylation of epicatechin on hydrogen peroxide-induced cell death in neurons and fibroblasts, *Free Radic Biol Med* 31, 1139, 2001.
17. Miyake, Y. et al., Identification and antioxidant activity of flavonoid metabolites in plasma and urine of eriocitrin-treated rats, *J Agric Food Chem* 48, 3217, 2000.

18. Terao, J. et al., Protection by quercetin and quercetin 3-*O*-beta-d-glucuronide of peroxynitrite-induced antioxidant consumption in human plasma low-density lipoprotein, *Free Radic Res* 35, 925, 2001.
19. Shirai, M. et al., Inhibitory effect of a quercetin metabolite, quercetin 3-*O*-beta-d-glucuronide, on lipid peroxidation in liposomal membranes, *J Agric Food Chem* 49, 5602, 2001.
20. Yamamoto, N. et al., Inhibitory effect of quercetin metabolites and their related derivatives on copper ion-induced lipid peroxidation in human low-density lipoprotein, *Arch Biochem Biophys* 372, 347, 1999.
21. da Silva, E.L. et al., Quercetin metabolites inhibit copper ion-induced lipid peroxidation in rat plasma, *FEBS Lett* 430, 405, 1998.
22. Abbott, N.J., Astrocyte-endothelial interactions and blood-brain barrier permeability, *J Anat* 200, 629, 2002.
23. Youdim, K.A. et al., Flavonoid permeability across an in situ model of the blood-brain barrier, *Free Radic Biol Med* 36, 592, 2004.
24. Youdim, K.A. et al., Interaction between flavonoids and the blood-brain barrier: *In vitro* studies, *J Neurochem* 85, 180, 2003.
25. Aasmundstad, T.A. et al., Distribution of morphine 6-glucuronide and morphine across the blood-brain barrier in awake, freely moving rats investigated by *in vivo* microdialysis sampling, *J Pharmacol Exp Ther* 275, 435, 1995.
26. Sperker, B. et al., The role of beta-glucuronidase in drug disposition and drug targeting in humans, *Clin Pharmacokinet* 33, 18, 1997.
27. Kroemer, H.K., and Klotz, U., Glucuronidation of drugs. A re-evaluation of the pharmacological significance of the conjugates and modulating factors, *Clin Pharmacokinet* 23, 292, 1992.
28. Lin, J.H., and Yamazaki, M., Role of P-glycoprotein in pharmacokinetics: Clinical implications, *Clin Pharmacokinet* 42, 59, 2003.
29. Peng, H.W. et al., Determination of naringenin and its glucuronide conjugate in rat plasma and brain tissue by high-performance liquid chromatography, *J Chromatogr B Biomed Sci Appl* 714, 369, 1998.
30. Suganuma, M. et al., Wide distribution of [H-3](−)-epigallocatechin gallate, a cancer preventive tea polyphenol, in mouse tissue, *Carcinogenesis* 19, 1459, 1998.
31. Abd El Mohsen, M.M. et al., Uptake and metabolism of epicatechin and its access to the brain after oral ingestion, *Free Radic Biol Med* 33, 1693, 2002.
32. El Mohsen, M.A. et al., Absorption, tissue distribution and excretion of pelargonidin and its metabolites following oral administration to rats, *Br J Nutr* 95, 51, 2006.
33. Talavera, S. et al., Anthocyanin metabolism in rats and their distribution to digestive area, kidney, and brain, *J Agric Food Chem* 53, 3902, 2005.
34. Passamonti, S. et al., Fast access of some grape pigments to the brain, *J Agric Food Chem* 53, 2005.
35. Kalt, W. et al., Identification of anthocyanins in the liver, eye, and brain of blueberry-fed pigs, *J Agric Food Chem* 56, 705, 2008.
36. Barzilai, A., and Melamed, E., Molecular mechanisms of selective dopaminergic neuronal death in Parkinson's disease, *Trends Mol Med* 9, 126, 2003.
37. Jellinger, K.A., Cell death mechanisms in neurodegeneration, *J Cell Mol Med* 5, 1, 2001.
38. Spires, T.L., and Hannan, A.J., Nature, nurture and neurology: Gene-environment interactions in neurodegenerative disease. FEBS Anniversary Prize Lecture delivered on 27 June 2004 at the 29th FEBS Congress in Warsaw, *FEBS J* 272, 2347, 2005.
39. Spencer, J.P.E., The interactions of flavonoids within neuronal signalling pathways, *Gen Nutr* 2, 257, 2007.
40. Mandel, S., and Youdim, M.B., Catechin polyphenols: Neurodegeneration and neuroprotection in neurodegenerative diseases, *Free Radic Biol Med* 37, 304, 2004.
41. Luo, Y. et al., Inhibition of amyloid-beta aggregation and caspase-3 activation by the *Ginkgo biloba* extract EGb761, *Proc Natl Acad Sci U S A* 99, 12197, 2002.
42. Checkoway, H. et al., Parkinson's disease risks associated with cigarette smoking, alcohol consumption, and caffeine intake, *Am J Epidemiol* 155, 732, 2002.
43. Levites, Y. et al., Attenuation of 6-hydroxydopamine (6-OHDA)-induced nuclear factor-kappaB (NF-kappaB) activation and cell death by tea extracts in neuronal cultures, *Biochem Pharmacol* 63, 21, 2002.
44. Lee, S. et al., Protective effects of the green tea polyphenol (−)-epigallocatechin gallate against hippocampal neuronal damage after transient global ischemia in gerbils, *Neurosci Lett* 287, 191, 2000.
45. Levites, Y. et al., Green tea polyphenol (-)-epigallocatechin-3-gallate prevents *N*-methyl-4-phenyl-1,2,3,6-tetrahydropyridine-induced dopaminergic neurodegeneration, *J Neurochem* 78, 1073, 2001.

46. Spencer, J.P.E. et al., Conjugates of catecholamines with cysteine and GSH in Parkinson's disease: Possible mechanisms of formation involving reactive oxygen species, *J Neurochem* 71, 2112, 1998.

47. Spencer, J.P.E. et al., Superoxide-dependent depletion of reduced glutathione by l-DOPA and dopamine. Relevance to Parkinson's disease, *NeuroReport* 6, 1480, 1995.

48. Spencer, J.P.E. et al., 5-*S*-cysteinyl-conjugates of catecholamines induce cell damage, extensive DNA base modification and increases in caspase-3 activity in neurons, *J Neurochem* 81, 122, 2002.

49. Hastings, T.G., Enzymatic oxidation of dopamine: The role of prostaglandin H synthase, *J Neurochem* 64, 919, 1995.

50. Vauzour, D. et al., Peroxynitrite induced formation of the neurotoxins 5-*S*-cysteinyl-dopamine and DHBT-1: Implications for Parkinson's disease and protection by polyphenols, *Arch Biochem Biophys* 476, 145, 2008.

51. Vauzour, D. et al., Inhibition of the formation of the neurotoxin 5-*S*-cysteinyl-dopamine by polyphenols, *Biochem Biophys Res Commun* 362, 340, 2007.

52. Spencer, J.P.E. et al., Epicatechin and its *in vivo* metabolite, 3'-*O*-methyl epicatechin, protect human fibroblasts from oxidative-stress-induced cell death involving caspase-3 activation, *Biochem J* 354, 493, 2001.

53. Williams, R.J. et al., Flavonoids: Antioxidants or signalling molecules? *Free Radic Biol Med* 36, 838, 2004.

54. Schroeter, H. et al., Flavonoids protect neurons from oxidized low-density-lipoprotein-induced apoptosis involving c-Jun N-terminal kinase (JNK), c-Jun and caspase-3, *Biochem J* 358, 547, 2001.

55. Schroeter, H. et al., Phenolic antioxidants attenuate neuronal cell death following uptake of oxidized low-density lipoprotein, *Free Radic Biol Med* 29, 1222, 2000.

56. Vauzour, D. et al., Activation of pro-survival Akt and ERK1/2 signalling pathways underlie the anti-apoptotic effects of flavanones in cortical neurons, *J Neurochem* 103, 1355, 2007.

57. Wang, L. et al., Inhibition of c-Jun N-terminal kinase ameliorates apoptosis induced by hydrogen peroxide in the kidney tubule epithelial cells (NRK-52E), *Nephron* 91, 142, 2002.

58. Ishikawa, Y., and Kitamura, M., Anti-apoptotic effect of quercetin: Intervention in the JNK- and ERK-mediated apoptotic pathways, *Kidney Int* 58, 1078, 2000.

59. Uchida, K. et al., Activation of stress signaling pathways by the end product of lipid peroxidation. 4-hydroxy-2-nonenal is a potential inducer of intracellular peroxide production, *J Biol Chem* 274, 2234, 1999.

60. Kobuchi, H. et al., Quercetin inhibits inducible ICAM-1 expression in human endothelial cells through the JNK pathway, *Am J Physiol* 277, C403, 1999.

61. Hirsch, E.C. et al., Neuroinflammatory processes in Parkinson's disease, *Parkinsonism Relat Disord* 11 Suppl 1, S9, 2005.

62. Zheng, Z. et al., Stroke: Molecular mechanisms and potential targets for treatment, *Curr Mol Med* 3, 361, 2003.

63. Kozuka, N. et al., Lipopolysaccharide and proinflammatory cytokines require different astrocyte states to induce nitric oxide production, *J Neurosci Res* 82, 717, 2005.

64. Bal-Price, A. et al., Stimulation of the NADPH oxidase in activated rat microglia removes nitric oxide but induces peroxynitrite production, *J Neurochem* 80, 73, 2002.

65. Marcus, J.S. et al., Cytokine-stimulated inducible nitric oxide synthase expression in astroglia: Role of Erk mitogen-activated protein kinase and NF-kappaB, *Glia* 41, 152, 2003.

66. Bhat, N.R. et al., Extracellular signal-regulated kinase and p38 subgroups of mitogen-activated protein kinases regulate inducible nitric oxide synthase and tumor necrosis factor-alpha gene expression in endotoxin-stimulated primary glial cultures, *J Neurosci* 18, 1633, 1998.

67. Bhat, N.R. et al., p38 MAPK-mediated transcriptional activation of inducible nitric-oxide synthase in glial cells. Roles of nuclear factors, nuclear factor kappa B, cAMP response element-binding protein, CCAAT/enhancer-binding protein-beta, and activating transcription factor-2, *J Biol Chem* 277, 29584, 2002.

68. Casper, D. et al., Ibuprofen protects dopaminergic neurons against glutamate toxicity *in vitro*, *Neurosci Lett* 289, 201, 2000.

69. Lau, F.C. et al., Nutritional intervention in brain aging: Reducing the effects of inflammation and oxidative stress, *Subcell Biochem* 42, 299, 2007.

70. Lau, F.C. et al., Inhibitory effects of blueberry extract on the production of inflammatory mediators in lipopolysaccharide-activated BV2 microglia, *J Neurosci Res* 85, 1010, 2007.

71. Chen, J.C. et al., Inhibition of iNOS gene expression by quercetin is mediated by the inhibition of IkappaB kinase, nuclear factor-kappa B and STAT1, and depends on heme oxygenase-1 induction in mouse BV-2 microglia, *Eur J Pharmacol* 521, 9, 2005.

72. Lee, H. et al., Flavonoid wogonin from medicinal herb is neuroprotective by inhibiting inflammatory activation of microglia, *FASEB J* 17, 1943, 2003.

73. Li, R. et al., (–)-Epigallocatechin gallate inhibits lipopolysaccharide-induced microglial activation and protects against inflammation-mediated dopaminergic neuronal injury, *J Neurosci Res* 78, 723, 2004.

74. Wang, X. et al., Genistein protects dopaminergic neurons by inhibiting microglial activation, *NeuroReport* 16, 267, 2005.

75. Zheng, L.T. et al., Suppressive effects of flavonoid fisetin on lipopolysaccharide-induced microglial activation and neurotoxicity, *Int Immunopharmacol* 8, 484, 2008.

76. Letenneur, L. et al., Flavonoid intake and cognitive decline over a 10-year period, *Am J Epidemiol* 165, 1364, 2007.

77. Lee, Y.B. et al., Soy isoflavones and cognitive function, *J Nutr Biochem* 16, 641, 2005.

78. Birge, S.J., Is there a role for estrogen replacement therapy in the prevention and treatment of dementia? *J Am Geriatr Soc* 44, 865, 1996.

79. White, L.R. et al., Brain aging and midlife tofu consumption, *J Am Coll Nutr* 19, 242, 2000.

80. File, S.E. et al., Eating soya improves human memory, *Psychopharmacology (Berl)* 157, 430, 2001.

81. Casini, M.L. et al., Psychological assessment of the effects of treatment with phytoestrogens on postmenopausal women: A randomized, double-blind, crossover, placebo-controlled study, *Fertil Steril* 85, 972, 2006.

82. Kritz-Silverstein, D. et al., Isoflavones and cognitive function in older women: The SOy and Postmenopausal Health In Aging (SOPHIA) Study, *Menopause* 10, 196, 2003.

83. File, S.E. et al., Cognitive improvement after 6 weeks of soy supplements in postmenopausal women is limited to frontal lobe function, *Menopause* 12, 193, 2005.

84. Pan, Y. et al., Soy phytoestrogens improve radial arm maze performance in ovariectomized retired breeder rats and do not attenuate benefits of 17beta-estradiol treatment, *Menopause* 7, 230, 2000.

85. Lund, T.D. et al., Visual spatial memory is enhanced in female rats (but inhibited in males) by dietary soy phytoestrogens, *BMC Neurosci* 2, 20, 2001.

86. Lee, Y.B. et al., Soy isoflavones improve spatial delayed matching-to-place performance and reduce cholinergic neuron loss in elderly male rats, *J Nutr* 134, 1827, 2004.

87. Fournier, L.R. et al., The effects of soy milk and isoflavone supplements on cognitive performance in healthy, postmenopausal women, *J Nutr Health Aging* 11, 155, 2007.

88. Henderson, V.W., Estrogen-containing hormone therapy and Alzheimer's disease risk: Understanding discrepant inferences from observational and experimental research, *Neuroscience* 138, 1031, 2006.

89. Pan, Y. et al., Evidence for up-regulation of brain-derived neurotrophic factor mRNA by soy phytoestrogens in the frontal cortex of retired breeder female rats, *Neurosci Lett* 261, 17, 1999.

90. Pan, Y. et al., Effect of estradiol and soy phytoestrogens on choline acetyltransferase and nerve growth factor mRNAs in the frontal cortex and hippocampus of female rats, *Proc Soc Exp Biol Med* 221, 118, 1999.

91. Joseph, J.A. et al., Long-term dietary strawberry, spinach, or vitamin E supplementation retards the onset of age-related neuronal signal-transduction and cognitive behavioral deficits, *J Neurosci* 18, 8047, 1998.

92. Joseph, J.A. et al., Reversals of age-related declines in neuronal signal transduction, cognitive, and motor behavioral deficits with blueberry, spinach, or strawberry dietary supplementation, *J Neurosci* 19, 8114, 1999.

93. Casadesus, G. et al., Modulation of hippocampal plasticity and cognitive behavior by short-term blueberry supplementation in aged rats, *Nutr Neurosci*, 7, 309, 2004.

94. Williams, C.M. et al., Blueberry-induced changes in spatial working memory correlate with changes in hippocampal CREB phosphorylation and brain-derived neurotrophic factor (BDNF) levels, *Free Radic Biol Med* 45, 295, 2008.

95. Ramirez, M.R. et al., Effect of lyophilised *Vaccinium* berries on memory, anxiety and locomotion in adult rats, *Pharmacol Res* 52, 457, 2005.

96. Barros, D. et al., Behavioral and genoprotective effects of *Vaccinium* berries intake in mice, *Pharmacol Biochem Behav* 84, 229, 2006.

97. Goyarzu, P. et al., Blueberry supplemented diet: Effects on object recognition memory and nuclear factor-kappa B levels in aged rats, *Nutr Neurosci* 7, 75, 2004.

98. Shukitt-Hale, B. et al., Effects of Concord grape juice on cognitive and motor deficits in aging, *Nutrition* 22, 295, 2006.

99. Chan, Y.C. et al., Favorable effects of tea on reducing the cognitive deficits and brain morphological changes in senescence-accelerated mice, *J Nutr Sci Vitaminol (Tokyo)* 52, 266, 2006.

100. Hartman, R.E. et al., Pomegranate juice decreases amyloid load and improves behavior in a mouse model of Alzheimer's disease, *Neurobiol Dis* 24, 506, 2006.

101. Pu, F. et al., Neuroprotective effects of quercetin and rutin on spatial memory impairment in an 8-arm radial maze task and neuronal death induced by repeated cerebral ischemia in rats, *J Pharmacol Sci* 104, 329, 2007.

102. Maher, P. et al., Flavonoid fisetin promotes ERK-dependent long-term potentiation and enhances memory, *Proc Natl Acad Sci U S A* 103, 16568, 2006.

103. Cohen-Salmon, C. et al., Effects of *Ginkgo biloba* extract (EGb 761) on learning and possible actions on aging, *J Physiol Paris* 91, 291, 1997.

104. Clostre, F., *Gingko biloba* extract (EGb 761). State of knowledgement in the dawn of the year 2000, *Ann Pharm Fr* 57, 1S8, 1999.

105. Diamond, B.J. et al., *Ginkgo biloba* extract: Mechanisms and clinical indications, *Arch Phys Med Rehab* 81, 668, 2000.

106. Itil, T.M. et al., The pharmacological effects of *Gingko biloba*, a plant extract, on the brain of dementia patients in comparison with tacrine, *Psychopharmacology* 34, 391, 1998.

107. Shif, O. et al., Effects of *Ginkgo biloba* administered after spatial learning on water maze and radial arm maze performance in young adult rats, *Pharmacol Biochem Behav* 84, 17, 2006.

108. Winter, J.C., The effects of an extract of *Ginkgo biloba*, EGb 761, on cognitive behavior and longevity in the rat, *Physiol Behav* 63, 425, 1998.

109. Hoffman, J.R. et al., *Ginkgo biloba* promotes short-term retention of spatial memory in rats, *Pharmacol Biochem Behav* 77, 533, 2004.

110. Stoll, S. et al., *Ginkgo biloba* extract (EGb 761) independently improves changes in passive avoidance learning and brain membrane fluidity in the aging mouse, *Pharmacopsychiatry* 29, 144, 1996.

111. Wang, Y. et al., The *in vivo* synaptic plasticity mechanism of EGb 761-induced enhancement of spatial learning and memory in aged rats, *Br J Pharmacol* 148, 147, 2006.

112. Topic, B. et al., Enhanced conditioned inhibitory avoidance by a combined extract of *Zingiber officinale* and *Ginkgo biloba*, *Phytother Res* 16, 312, 2002.

113. Oyama, Y. et al., *Gingko biloba* extract protects brain neurons against oxidative stress induced by hydrogen peroxide, *Brain Res* 712, 349, 1996.

114. Oyama, Y. et al., Myricetin and quercetin, the flavonoid constituents of *Ginkgo biloba* extract, greatly reduce oxidative metabolism in both resting and Ca(2+)-loaded brain neurons, *Brain Res* 635, 125, 1994.

115. Gajewski, A., *Gingko biloba* and memory for a maze, *Psychol Rep* 84, 481, 1999.

116. Chopin, P., and Briley, M., Effects of four non-cholinergic cognitive enhancers in comparison with tacrine and galanthamine on scopolamine-induced amnesia in rats, *Psychopharmacology* 106, 26, 1992.

117. Ramassamy, C. et al., Prevention by a *Ginkgo biloba* extract (GBE 761) of the dopaminergic neurotoxicity of MPTP, *J Pharm Pharmacol* 42, 785, 1990.

118. Breteler, M.M., Vascular risk factors for Alzheimer's disease: An epidemiologic perspective, *Neurobiol Aging* 21, 153, 2000.

119. Commenges, D. et al., Intake of flavonoids and risk of dementia, *Eur J Epidemiol* 16, 357, 2000.

120. Dai, Q. et al., Fruit and vegetable juices and Alzheimer's disease: The Kame Project, *Am J Med* 119, 751, 2006.

121. Schroeter, H. et al., (-)-Epicatechin mediates beneficial effects of flavanol-rich cocoa on vascular function in humans, *Proc Natl Acad Sci U S A* 103, 1024, 2006.

122. Dinges, D.F., Cocoa flavanols, cerebral blood flow, cognition, and health: Going forward, *J Cardiovasc Pharmacol* 47 Suppl 2, S221, 2006.

123. Francis, S.T. et al., The effect of flavanol-rich cocoa on the fMRI response to a cognitive task in healthy young people, *J Cardiovasc Pharmacol* 47 Suppl 2, S215, 2006.

124. Gage, F.H., Mammalian neural stem cells, *Science* 287, 1433, 2000.

125. Palmer, T.D. et al., Vascular niche for adult hippocampal neurogenesis, *J Comp Neurol* 425, 479, 2000.

126. Harris, K.M., and Kater, S.B., Dendritic spines: Cellular specializations imparting both stability and flexibility to synaptic function, *Annu Rev Neurosci* 17, 341, 1994.

127. Nagahama, Y. et al., Cerebral correlates of the progression rate of the cognitive decline in probable Alzheimer's disease, *Eur Neurol* 50, 1, 2003.

128. Ruitenberg, A. et al., Cerebral hypoperfusion and clinical onset of dementia: The Rotterdam Study, *Ann Neurol* 57, 789, 2005.

129. Fisher, N.D. et al., Cocoa flavanols and brain perfusion, *J Cardiovasc Pharmacol* 47 Suppl 2, S210, 2006.

130. Wang, Z. et al., Assessment of functional development in normal infant brain using arterial spin labeled perfusion MRI, *Neuroimage* 39, 973, 2008.

131. Rice-Evans, C.A. et al., The relative antioxidant activities of plant-derived polyphenolic flavonoids, *Free Radic Res* 22, 375, 1995.

132. Hertog, M.G., and Hollman, P.C., Potential health effects of the dietary flavonol quercetin, *Eur J Clin Nutr* 50, 63, 1996.

133. Garcia-Alonso, M. et al., Electron spin resonance spectroscopy studies on the free radical scavenging activity of wine anthocyanins and pyranoanthocyanins, *Mol Nutr Food Res* 49, 1112, 2005.

134. Halliwell, B. et al., Health promotion by flavonoids, tocopherols, tocotrienols, and other phenols: Direct or indirect effects? Antioxidant or not? *Am J Clin Nutr* 81, 268S, 2005.

135. Yamamoto, N. et al., Inhibitory effect of quercetin metabolites and their related derivatives on copper ion-induced lipid peroxidation in human low-density lipoprotein [Full text delivery], *Arch Biochem Biophys* 372, 347, 1999.

136. Duthie, S.J., and Dobson, V.L., Dietary flavonoids protect human colonocyte DNA from oxidative attack *in vitro*, *Eur J Nutr* 38, 28, 1999.

137. Rietveld, A., and Wiseman, S., Antioxidant effects of tea: Evidence from human clinical trials, *J Nutr* 133, 3285S, 2003.

138. O'Reilly, J.D. et al., Consumption of flavonoids in onions and black tea: Lack of effect on F2-isoprostanes and autoantibodies to oxidized LDL in healthy humans, *Am J Clin Nutr* 73, 1040, 2001.

139. Hodgson, J.M. et al., Regular ingestion of tea does not inhibit *in vivo* lipid peroxidation in humans, *J Nutr* 132, 55, 2002.

140. Halliwell, B. et al., The gastrointestinal tract: A major site of antioxidant action? *Free Radic Res* 33, 819, 2000.

141. Halliwell, B. et al., Establishing biomarkers of oxidative stress: The measurement of hydrogen peroxide in human urine, *Curr Med Chem* 11, 1085, 2004.

142. Kong, A.N. et al., Signal transduction events elicited by natural products: Role of MAPK and caspase pathways in homeostatic response and induction of apoptosis, *Arch Pharm Res* 23, 1, 2000.

143. Chang, L., and Karin, M., Mammalian MAP kinase signalling cascades, *Nature* 410, 37, 2001.

144. Torii, S. et al., Regulatory mechanisms and function of ERK MAP kinases, *J Biochem (Tokyo)* 136, 557, 2004.

145. van Praag H. et al., Plant-derived flavanol (−)epicatechin enhances angiogenesis and retention of spatial memory in mice, *J Neurosci* 27, 5869, 2007.

146. Schroeter, H. et al., (−)-Epicatechin stimulates ERK-dependent cyclic AMP response element activity and upregulates GLUR2 in cortical neurons, *J Neurochem* 101, 1596, 2007.

147. Spencer, J.P.E. et al., Modulation of pro-survival Akt/protein kinase B and ERK1/2 signaling cascades by quercetin and its *in vivo* metabolites underlie their action on neuronal viability, *J Biol Chem* 278, 34783, 2003.

148. Reznichenko, L. et al., Green tea polyphenol (−)-epigallocatechin-3-gallate induces neurorescue of long-term serum-deprived PC12 cells and promotes neurite outgrowth, *J Neurochem* 93, 1157, 2005.

149. Dudley, D.T. et al., A synthetic inhibitor of the mitogen-activated protein kinase cascade, *Proc Natl Acad Sci U S A* 92, 7686, 1995.

150. Pang, I. et al., Inhibition of MAP kinase kinase blocks the differentiation of PC-12 cells induced by nerve growth factor, *J Biol Chem* 270, 13585, 1995.

151. Alessi, D.R. et al., PD 098059 is a specific inhibitor of the activation of mitogen-activated protein kinase kinase *in vitro* and *in vivo*, *J Biol Chem* 270, 27489, 1995.

152. Lazar, D.F. et al., Mitogen-activated protein kinase kinase inhibition does not block the stimulation of glucose utilization by insulin, *J Biol Chem* 270, 20801, 1995.

153. Vlahos, C.J. et al., A specific inhibitor of phosphatidylinositol 3-kinase, 2-(4-morpholinyl)-8-phenyl-4*H*-1-benzopyran-4-one (LY294002), *J Biol Chem* 269, 5241, 1994.

16 Actions of Bioactive Phytochemicals in Cell Function and Alzheimer's Disease Pathology

Richard E. Hartman

Department of Psychology, Loma Linda
University, Loma Linda, California, USA

CONTENTS

16.1 INTRODUCTION

Phytochemicals are broadly defined as compounds produced by plants and include the phenols, terpenes, and organosulfurs. Many of these chemicals have pigment, odorant, and/or irritant properties that may help the plant's biochemical defenses against metabolic byproducts [e.g., reactive oxygen species (ROS), protein misfolding] and environmental insults (e.g., pathogens, insects, ultraviolet radiation). Consumption of plants or their phytochemicals can confer some of these beneficial properties and modulate a number of biological pathways, including inflammatory, enzymatic, neurotransmitter systems, apoptosis, and neurogenesis.[1]

Several dietary, medicinal, and isolated phytochemicals have been shown to modulate various aspects of Alzheimer's disease (AD), the most common neurodegenerative disorder of aging and cause of dementia. This chapter presents an overview of AD etiology followed by a survey of phytochemicals that may affect its neurophysiological and/or neuropathological sequelae.

16.2   AMYLOID β–RELATED NEUROTOXICITY IN ALZHEIMER'S DISEASE BRAIN

AD currently affects approximately 10% of the population over the age of 65 years and 50% of the population over the age of 85 years, but the incidence is expected to rise as the population ages. The first, and most prominent, symptom is loss of memory for recent events, followed by a progressive decline in general cognition (e.g., memory, language, executive functions) and motor abilities. Neuropathological evidence of AD includes the accumulation of protein deposits ("plaques") between and surrounding the brain's neurons and neurofibrillary tangles inside the neurons. In addition to the characteristic plaques and tangles found during postmortem examination of AD brains, evidence of mitochondrial dysfunction, inflammation, astrogliosis, microglial activation, synaptic loss, neuronal damage, and apoptosis is also observed.

The plaques are primarily composed of amyloid-β (Aβ) peptides[2] that are enzymatically snipped from the much larger amyloid precursor protein (APP) by the γ-secretases and β-secretases. Other proteins (e.g., apolipoproteins) and non-proteins (e.g., ROS, hemes, metals) are also found within the plaques, which generally begin to accumulate in the medial temporal lobes years before the first behavioral symptoms emerge. Brain structures in this region include the entorhinal cortex and hippocampal formation, which play important roles in learning and memory. With age, the plaques gradually spread throughout the cortical and subcortical areas.[3]

Neurons within the plaques often have abnormally twisted axons and dendrites resulting from the neurofibrillary tangles. These tangled cytoskeletal microtubules destabilize the structure of long neuronal processes and disrupt intracellular transport mechanisms. The tau protein normally plays a role in stabilizing cytoskeletal microtubules within neurons, but the Aβ-related build-up of abnormal tau in the cell destabilizes the microtubules, which eventually leads to the cell's demise. Neurotransmitter systems, especially acetylcholine (ACh) and glutamate, become dysfunctional as the neurons that produce these chemicals atrophy and die.[4] Thus, the age-related accumulation of Aβ in the brain is associated with neuronal dysfunction that ultimately leads to the behavioral symptoms associated with AD.

Aβ's putative toxicity probably involves several inter-related mechanisms. Aβ causes intracellular tau disruption and neuronal death in hippocampal cell cultures.[5] It also induces hypersensitivity to excitotoxic damage by glutamate[6] and oxidative stress, which occurs when the build-up of potentially harmful ROS cannot be effectively controlled. For example, Aβ-heme peroxidase complexes form within plaques that can cause inflammation, release of ROS, and damage to muscarinic ACh receptors, and these effects are prevented by antioxidant compounds.[7–11] The plaques eventually disrupt synaptic structures, and synaptic loss within plaques provides a better predictor of cognitive dysfunction than the amount of plaque deposition.[12–15] This observation suggests that certain individuals may be more or less susceptible to the effects of Aβ deposition. The deleterious effect of Aβ on synaptic function is also demonstrated by experiments that assess long-term potentiation (LTP), which is a neuronal model of learning and memory.[16–20] Finally, numerous studies have demonstrated that the age-related accumulation of brain Aβ is associated with progressive cognitive impairments in transgenic mouse models of AD. *In vivo* imaging shows that Aβ plaques can form quite rapidly (over the course of 24 hours) in the brains of these mice and that signs of neurodegeneration around the plaques are seen shortly thereafter.[21]

16.2.1   Accelerating Amyloid-β Deposition Increases the Risk for Alzheimer's Disease

Thus, the gradual accumulation of potentially toxic Aβ in the brain is associated with progressive oxidative stress and various downstream events that cause structural damage to neurons. This process eventually leads to functional deficits, cognitive and behavioral impairments, and death. Pathophysiological conditions that accelerate Aβ accumulation in the brain increase the risk of developing AD. For example, Down syndrome is characterized by the overproduction of APP in the brain, which leads to elevated Aβ production and deposition and dementia by around 50 years of age.[22,23]

Furthermore, several inheritable mutations in the genes for APP or constituents of γ-secretase lead to elevated APP and Aβ production. These genes are associated with early onset of AD,[24,25] and their identification has spawned the development of several lines of APP transgenic mice. These transgenic mouse models of AD generally express relatively high brain levels of human APP and develop age-related neuropathology and cognitive deficits coincident with the accumulation of Aβ aggregates and deposits.[26–34]

Mounting evidence also suggests that brain inflammation and oxidative stress resulting from traumatic brain injury, stroke, or even chronic low-level insult (e.g., hypoxia due to breathing problems[35] or high-cholesterol diet[36–42]) can induce accumulation of APP and Aβ in the brain and elevate the risk of developing AD. Indeed, oxidative stress, a common component of all types of brain injury, is sufficient to induce Aβ accumulation,[43] initiating a vicious circle of progressive oxidative load in the brain.

16.2.2   Inhibiting Amyloid-β Deposition Decreases the Risk for Alzheimer's Disease

Further evidence for the support of Aβ accumulation as a causative factor in AD comes from experiments using APP transgenic mouse models of AD. Many studies have described systemic treatments that lower levels of Aβ in the brains of these mice (e.g., monoclonal anti-Aβ antibodies, dietary manipulations) and prevent or even reverse neuropathology and behavioral deficits.[16,34,44–51] Interestingly, reducing the brain's oxidative load can improve cognitive function in APP transgenic mice without reducing Aβ levels,[52,53] suggesting that Aβ contributes to oxidative overload in the brain that gradually impacts the function of cortical circuits involved in learning and memory.

16.2.3   Summary: Alzheimer's Disease Etiology

Thus, AD is associated with the abnormal build-up of Aβ in the brain, which induces events that lead to even greater Aβ accumulation. The idea that this process creates a vicious circle of neurodegenerative decline is known as the amyloid cascade hypothesis of AD.[54–56] Current pharmacological approaches for treating AD include stabilizing glutamatergic activity by blocking NMDA glutamate channels (e.g., memantine) and inhibiting acetylcholinesterase (AChE), an enzyme that breaks down ACh and has been shown to promote the aggregation of Aβ (e.g., galantamine, tacrine, donepezil, rivastigmine). However, mounting epidemiological and experimental evidence suggests that diet and other sources of bioactive phytochemicals can also decrease the risk of developing AD.[57–59]

16.3   BIOACTIVE PHYTOCHEMICALS AND ALZHEIMER'S DISEASE

There is a growing body of literature demonstrating that phytochemical compounds can affect various aspects of AD via antioxidant and other pathways. Antioxidant compounds found in plants include minerals (e.g., selenium, zinc), vitamins (e.g., ascorbic acid, α-tocopherol), and other organic compounds (e.g., phenols, terpenes, organosulfurs), all of which can neutralize ROS by giving up electrons to oxygen ions, peroxides, and free radicals. However, evidence suggests that these compounds may also work by a variety of other mechanisms. Indeed, many currently available pharmaceuticals have roots in traditional herbal medicines. For example, galantamine is a pharmaceutical

AChE inhibitor derived from daffodils, and the anti-inflammatory blood thinner aspirin is derived from salicylic acid, a polyphenol found in willow bark. Both of these drugs are currently used to control AD. The following subsection provides a survey of the epidemiological and experimental evidence for the effects of other plants and isolated phytochemicals on AD processes.

16.3.1 EPIDEMIOLOGICAL EVIDENCE OF PHYTOCHEMICAL EFFECTS ON ALZHEIMER'S DISEASE

Regular consumption of a variety of fruits and vegetables may decrease the risk for or slow the progression of AD. For example, a large study of elderly Japanese Americans found that drinking fruit and vegetable juices was associated with a lower risk for AD,[60] and a group of studies on elderly French subjects showed that daily consumption of phenol-rich fruits and vegetables significantly decreased the risk of developing AD with age.[61–63]

Epidemiological evidence that isolated dietary phytochemicals can affect AD remains elusive. For example, one study reported that dietary tocopherols (isoforms of vitamin E), vitamin C, β-carotene, and tea were not correlated with the risk of developing AD.[60] However, another study found a lower incidence of AD with high intake of food-based α-tocopherols and γ-tocopherols[64] and a large study from the Netherlands associated high intake of dietary (but not supplemental) vitamins E and C with a lower risk for AD. In that study, the effects were stronger for tobacco smokers, who also benefited from dietary β-carotene and polyphenols.[65] Finally, a recent study found that supplemental vitamin E and/or C did not reduce the risk of developing AD over 5 years of follow-up.[66] Thus, the evidence suggests that acquiring phytovitamins through a varied diet may provide more protection against AD than the use of supplemental vitamins.

Some epidemiological evidence suggests that dietary phospholipids, such as the omega-3 fatty acids [e.g., α-linolenic acid, eicosapentaenoic acid, and docosahexaenoic acid (DHA)], which are found in flax, nuts, algae, and the oil of fish that eat algae, may protect against developing AD.[62,67–69] Other sources of bioactive phytochemicals include colorful, flavorful, and aromatic spices, which can contain high concentrations of phenols, terpenes, and organosulfurs. For example, a study of elderly Asians showed that those whose diets included curry performed significantly better on neuropsychological tests of cognitive performance.[70] Curry is a mixture of spices including the bright yellow turmeric and curcumin, its associated polyphenol.

Additionally, moderate wine consumption by the elderly is associated with reduced risk for AD,[71] but this effect could be due to grape polyphenols, such as resveratrol and ethanol (which is a derivative of phytosugars). Interestingly, smoking tobacco may also protect against Aβ deposition. In a postmortem examination of AD brains, there were significantly lower levels of Aβ in smokers' entorhinal cortex, which plays a role in learning and memory.[72] However, another study showed no protective effect of smoking on the risk of developing symptoms of dementia.[73]

Thus, epidemiological evidence suggests that eating a wide variety of fruits and vegetables may provide several bioactive phytochemical compounds that may work collectively to lower the risk for AD. Relatively few experimental clinical trials have been published assessing the effects of plants or phytochemical compounds on AD in humans. However, a number of experimental preclinical studies using APP transgenic and/or *in vitro* models have provided evidence that various aspects of AD can be manipulated by plants and their phytochemicals. The following subsection surveys the experimental literature describing the potential effects of bioactive phytochemicals on AD.

16.3.2 EXPERIMENTAL EVIDENCE OF PHYTOCHEMICAL EFFECTS ON ALZHEIMER'S DISEASE

16.3.2.1 Ginkgo Biloba and Its Phytochemicals

The leaves of the ginkgo biloba tree have been used to improve cognition for centuries. In part because of this history, ginkgo biloba is one of the most studied plants in terms of its effects on AD. Its biologically active compounds include polyphenols (e.g., kaempferol, quercetin) and terpenes (ginkgolides and bilobalides). Ginkgo biloba is often studied experimentally using a commercially available extract (EGb761) that is standardized to contain 24% polyphenols and 6% terpenes.

Several clinical trials of EGb761 have suggested that daily treatment for 3–6 months may provide mild cognitive benefit over placebo in elderly demented patients.[74–78] Several other studies have compared the clinical effects of the extract with clinically used AChE inhibitors. One study found that ginkgo biloba extract was as effective as the pharmaceutical AChE inhibitor donepezil.[78] A meta-analysis of placebo-controlled studies of demented patients that continued for at least 6 months determined that the reported efficacy of EGb761 was similar to that of four pharmaceutical AChE inhibitors (tacrine, donepezil, rivastigmine, and metrifonate). Tacrine was the only one of the five treatments that was associated with a high dropout rate due to negative side effects.[79] Another study demonstrated that the extract produced electroencephalogram changes in elderly demented patients similar to those produced by tacrine.[80]

However, other studies have found no significant cognitive effect of ginkgo biloba in elderly demented[81–83] or non-demented[84] subjects. The difficulty of demonstrating a strong clinical effect in humans is not surprising, considering that the currently available pharmaceutical therapies provide only modest and short-term clinical benefits. However, a number of animal and *in vitro* studies have provided evidence that the phenols and terpenes in ginkgo biloba can modulate aspects of AD pathology.

For example, ginkgo biloba extract was shown to eliminate cognitive deficits in APP transgenic mice without reducing levels of brain Aβ.[85] Beneficial effects on synaptic function may underlie ginkgo biloba's putative cognitive effects, as an extract enriched with terpene ginkgolides and bilobalides (70% vs. 6% in EGb761) prevented Aβ-induced neurotoxicity and inhibition of LTP (a neuronal model of learning and memory) in rodent hippocampal neurons. Ginkgolide J, the major terpenoid component of the extract, provided similar protection as the whole extract.[86]

EGb761 has antioxidant properties and can modulate a variety of cellular signaling pathways.[87–91] For example, the extract protected cultured rat hippocampal neurons from Aβ- and protein kinase C-induced neurotoxicity and oxidative stress via effects on nitric oxide. The isolated polyphenolic components of the extract produced a similar but less potent effect, whereas the isolated terpene component was ineffective at preventing neurotoxicity.[88,89] This pattern of results suggests that the combined phytochemicals have a synergistic antioxidant effect. Similarly, ginkgo biloba extract significantly reduced ROS-induced apoptosis in mice[90] and prevented oxidative stress and reversed neuritic dystrophy associated with Aβ plaques in transgenic APP mice.[92] EGb761 and its polyphenolic components also attenuated the build-up of ROS in cultured APP-producing neurons and transgenic *Caenorhabditis elegans* nematodes that express human Aβ.[93,94]

Besides its antioxidant properties, ginkgo biloba extract inhibits Aβ aggregation, caspase-3 activity, and apoptosis in cultured APP transgenic neurons[95] and increases α-secretase processing of APP in rat hippocampal slices. Cleavage of APP by α-secretase (rather than γ- and β-secretases) not only prevents the production of Aβ but also yields the potentially neuroprotective peptide sAPPα.[96] Cholesterol may provide another pathway for the effects of ginkgo biloba on AD, as treating aged rats with EGb761 lowered circulating levels of cholesterol and brain levels of APP and Aβ. *In vitro* experiments suggested that the inhibition of Aβ production was associated with enhanced clearance of intracellular cholesterol.[97]

Thus, there is some evidence that ginkgo biloba may provide benefits to elderly demented individuals similar to those of the current clinically approved pharmaceutical AChE inhibitors. Because ginkgo biloba is generally not a dietary plant and because phytochemical concentrations in the whole leaf are probably too low to provide acute benefits at usable doses, any significant beneficial effects of ginkgo biloba will likely come from concentrated extracts and pharmaceutical derivatives rather than from incorporation of whole ginkgo biloba leaf into the diet.

16.3.2.2 Pomegranate and Its Phytochemicals

Pomegranates have been used as food and medicine for centuries. Chemical assays show that they contain very high concentrations of bioactive polyphenols, including phenolic acid tannins such as the punicalagins that hydrolyze (break down in water) to smaller phenols such as ellagic acid and gallic acid.[85,98–100] Several human and animal studies of pomegranate juice have demonstrated a variety of biological effects, including antioxidation.[101–107] Additionally, one report suggests that it inhibits the activation of oxidation-sensitive genes in response to cellular stress, and another demonstrated modulation of endothelial nitric oxide synthase expression. In addition to the antioxidant properties of the whole fruit juice, other studies have reported anti-apoptotic, anticancer, antibacterial, and cardiovascular effects.[101–109]

Animal experiments in which mice were fed diluted pomegranate juice through their water demonstrate the neuroprotective effects of these phytochemicals. In these studies, dietary supplementation with pomegranate juice provided the mice with an average daily dose of polyphenols roughly equivalent in human terms to 250–500 ml of full-strength pomegranate juice. The juice improved maze performance and significantly reduced levels of Aβ in the hippocampus of APP transgenic mice.[16] Additionally, when fed to pregnant mice, it protected their neonatal offspring from hypoxic-ischemic brain injury.[110]

Ellagic acid extracted from pomegranate husks was shown to inhibit β-secretase activity,[111] suggesting that it may prevent the formation of Aβ from APP *in vivo*. Additionally, ellagic acid suppresses the pro-inflammatory nuclear transcription factor-κB (NF-κB) activation pathway[112] and modulates other cell-signaling pathways.[113] However, *in vitro* experiments assessing its antioxidant and anti-apoptotic properties suggest that the whole juice may provide synergistic benefit over isolated phytochemical components.[106] Interestingly, one study suggests that the sugar fraction of pomegranate juice, which consists of conjugated sucroses, fructoses, and glucoses, may have significant antioxidant properties independent from those of the phenolic compounds.[104] Thus, dietary pomegranate juice may provide significant behavioral and neuropathological protection against AD and a host of other age-related diseases.

16.3.2.3   Turmeric and Curcumin

A diet that includes high amounts of the spice mixture curry is associated with improved cognitive performance in the elderly.[70] Curcumin is a phenolic acid and yellow pigment found in the curry spice turmeric (a member of the ginger family), which has some interesting AD-related properties. Dietary curcumin prevented learning deficits, oxidative stress, synaptic damage, and cortical microgliosis (while increasing plaque-associated microgliosis) in a rat model of AD that uses intracerebroventricular infusion of Aβ.[114] Curcumin also lowered levels of oxidized proteins and plaque burden in APP transgenic mice.[115–117] Its antioxidant properties were further demonstrated by its reduction of heme-Aβ peroxidase damage to muscarinic ACh receptors.[7]

In addition to its effect on Aβ plaque burden *in vivo*, curcumin also prevented and reversed Aβ aggregation *in vitro*. Interestingly, dietary or injected curcumin binds to amyloid fibrils in the brain similarly to histological stains such as thioflavin S and Congo red and can be viewed under fluorescent light to visualize Aβ plaques. Other mechanisms by which curcumin could act on AD processes include inhibition of pro-inflammatory NF-κB activity[112] and modulation of other cell-signaling pathways.[113]

16.3.2.4   Garlic and Its Phytochemicals

Garlic has a number of aromatic sulfur-containing phytochemicals, including *S*-allyl cysteine (SAC) and di-allyl-disulfide, known collectively as organosulfurs. Adding aged garlic extract, SAC, or di-allyl-disulfide to the diets of APP transgenic mice prevented cognitive deficits and lowered brain levels of inflammation, Aβ plaques, and abnormal tau. The whole extract was more effective than the isolated components, suggesting a synergistic effect.[118,119] Isolated SAC also inhibited and reversed Aβ aggregation *in vitro* by binding to Aβ and altering its conformation,[120] suggesting another mechanism by which garlic may reduce Aβ deposition. Finally, garlic extract and its organosulfur components have inhibited Aβ-induced generation of ROS, pro-inflammatory NF-κB, caspase-3 activation, DNA fragmentation, and apoptosis.[112,121]

These observations suggest that garlic-based organosulfurs may act on AD processes by several pathways.

16.3.2.5   Omega-3 Fatty Acids

As mentioned in the epidemiology section, dietary phospholipids such as the omega-3 fatty acids (found mainly in flax, nuts, algae, and fish oil) may decrease the risk of developing AD. Phospholipids increase the fluidity of neuronal membranes and promote α-secretase processing of APP, which not only prevents the formation of the Aβ peptide but also yields sAPPα, a neuroprotective peptide.[122] Dietary omega-3 essential fatty acids have been shown to reduce learning deficits, Aβ plaques, and synaptic neuropathology, while increasing cerebral blood volume in APP transgenic mice.[123–126] One well-studied omega-3 fatty acid, DHA, comprises around 15% of the brain's total fatty acids and 30%–40% of its gray matter.[68] DHA protects against an Aβ-induced model of neurotoxicity in the rat brain,[127,128] and has anti-inflammatory and anticancer properties.[68] Therefore, diets high in flax and other seeds or nuts with high omega-3 content may provide benefit to the brain and reduce the risk for AD by several potential mechanisms.

16.3.2.6   Phytovitamins

Some epidemiological studies have found a protective effect of dietary antioxidant tocopherols (vitamin E isoforms) against the risk of developing AD, and studies of transgenic APP mice support those data. One study found that vitamin E prevented oxidative stress associated with Aβ plaques and reversed neuritic dystrophy associated with amyloid plaques in APP transgenic mice,[92] while another study found that chronic dietary administration of vitamin E to young, but not to old, APP transgenic mice reduced Aβ deposition.[129] Furthermore, dietary administration of vitamin E to APP transgenic mice before and after repetitive traumatic brain injury ameliorated behavioral impairments, oxidative stress, and injury-accelerated Aβ formation.[130] Therefore, human and animal data suggest that dietary tocopherols, which have well-known antioxidant properties, may serve to protect the brain from oxidative stress that could eventually lead to increased Aβ deposition and AD.

Another study looked at the effects of a diet deficient in folic acid (an isoform of vitamin B$_9$ found in many fruits and vegetables) on neuropathology in APP transgenic mice. Although Aβ levels were not affected, significant neurodegeneration within the hippocampus was noted in the brains of folic acid-deprived mice.[131] Another *in vitro* study found that folic acid deprivation increased expression of the genes involved in encoding the γ- and β-secretases along with increased levels of Aβ.[132] Together with other data showing the neuroprotective effects of folic acid on the developing nervous system and the data on dietary tocopherols, several pieces of evidence suggest that various phytovitamins may play a protective role against neurodegenerative processes such as AD.

16.3.2.7   Tea and Its Phytochemicals

Although one epidemiological study showed that tea consumption was not correlated with the risk for AD, other lines of evidence suggest that consumption of tea may protect against oxidative stress[133] and tea has been used as a medicinal tonic for centuries. Tea leaves contain several bioactive phytochemicals, including polyphenol catechins and tannins (up to 25% of the mass of a tea leaf), and psychoactive xanthines, such as caffeine. Several experimental studies have shown that some of its compounds may have protective effects against AD.

For example, an extract of tea catechins reduced cognitive deficits, inflammation, and oxidative stress in rats subjected to intermittent oxygen deprivation as a model of obstructive sleep apnea, which produces brain damage that is associated with an increased risk for AD.[35,134] Isolated compounds in tea have been studied in more depth. Epigallocatechin-3-gallate (EGCG), a polyphenol catechin found in tea, improved cognitive performance in APP transgenic mice,[135] reduced production and deposition of brain Aβ, and increased levels of α-secretase processing and the subsequent neuroprotective peptide sAPPα.[135,136] The effect was observed whether EGCG was injected or administered via the drinking water. EGCG has also been shown to prevent fibrillogenesis of both

Aβ and α-synuclein by directly binding to the peptides and preventing aggregation[137] and to prevent abnormal protein folding in Huntington's disease models.[138] Tannic acid, a phenolic tannin responsible for the astringency in tea's taste, has also been shown to inhibit Aβ aggregation *in vitro*.[139] Thus, the polyphenols isolated from tea may protect against AD by a variety of mechanisms.

16.3.2.8 Caffeine

This psychoactive xanthine probably explains the worldwide popularity of tea and coffee (another plant with relatively high concentrations of the polyphenol caffeic acid[140,141] and antioxidant properties[142–145]). Caffeine functions as an insecticide in the plant and as a psychostimulant in mammals, primarily via competitive inhibition of adenosine receptors. Because of its stimulant effect, which is generally devoid of the euphoric properties of other phytostimulants, such as cocaine and ephedrine, caffeine has been used for centuries as a general enhancer of cognition. Caffeine added to the drinking water of APP transgenic mice at a dose equivalent to roughly five cups of coffee per day prevented the learning deficits observed in the noncaffeinated transgenic mice. The cognitive benefit was associated with inhibition of β-secretase and decreased Aβ deposition, suggesting that drinking caffeine decreased the production of Aβ. Caffeine also decreased Aβ production in neuronal cultures[146,147] and prevented cholesterol-induced disruption of the blood-brain barrier in rabbits, which develop age-related brain Aβ deposition similar to that seen in human AD patients and APP transgenic mice.[148] Interestingly, "caffeinol," a mixture of caffeine and ethanol, has been shown to have a potent synergistic neuroprotective effect in rodent stroke models.[149] Therefore, caffeine may improve general cognition via its mild stimulant effects and can prevent AD neuropathology by inhibiting Aβ production and protecting the brain from insult.

16.3.2.9 Phytocannabinoids

Cannabis is another plant with a long history of medicinal and recreational use. It contains a wide variety of phytochemicals that bind with CB_1 and CB_2 cannabinoid receptors. Collectively known as the phytocannabinoids, these terpene compounds include tetrahydrocannabinol (THC), cannabidiol, and cannabinol. CB_1 receptors are expressed mainly in the brain and are thought to be responsible for the well-known psychoactive effects of cannabis, whereas CB_2 receptors are expressed mainly in the periphery and thought to play a role in inflammatory processes. Aging is associated with a gradual loss of cannabinoid receptor binding in the brain and may even be further reduced in the hippocampus and caudate of AD brains.[150] Other studies have found that Aβ plaques in AD and Down syndrome brains express high levels of CB_2 receptors but that CB_1 receptor expression in plaques is unchanged or even reduced.[151–153] The increased expression of CB_2 receptors in brain Aβ plaques suggests a role for inflammatory mediation by the endogenous cannabinoids (e.g., anandamide),[154] and this notion is supported by the observation that both THC and anandamide can inhibit production of the free radical nitric oxide.[155,156] Indeed, the cannabinoids are antioxidant, anti-inflammatory, and neuroprotective against excitotoxicity *in vitro* and against acute brain damage *in vivo*.[157,158]

THC strongly inhibits AChE activity and prevents AChE-induced Aβ fibrillogenesis by binding directly to a site on AChE that mediates Aβ aggregation.[159] Interestingly, the cannabinoids can also stimulate neurogenesis within the adult hippocampus.[160] Other experimental evidence for a role of the cannabinoids in AD comes from studies of several psychoactive and nonpsychoactive synthetic cannabinoids. Independent of their antioxidant and/or psychoactive properties, these have been shown to prevent cognitive impairment, neurodegeneration, and microglial activation in rats subjected to models of Aβ-induced neurotoxicity[152,161] and chronic brain inflammation[162] and to block Aβ-induced microglial activation and neurotoxicity *in vitro*.[152] Thus, several studies suggest a number of mechanisms by which the phytocannabinoids may affect AD processes, but there are currently no reported epidemiological data on the incidence of AD among long-term cannabis users.

16.3.2.10   Nicotine

Nicotine is a psychoactive alkaloid that functions as an insecticide in the tobacco plant and has stimulant properties in mammals, primarily due to its nicotinic ACh receptor agonism. Similar to caffeine, nicotine has a long history as a general cognitive enhancer due to its stimulant qualities. As noted earlier, tobacco smokers may have reduced levels of Aβ in the entorhinal cortex,[72] although another study found no protective effect of tobacco smoking on risk of developing AD.[73] Other lines of evidence are provided by experiments showing that nicotine decreased accumulation of Aβ in the cortex and hippocampus of APP transgenic mice.[163,164] Possible mechanisms may include activity at the nicotinic ACh receptor, which results in decreased levels of Aβ and nitric oxide, and inhibition of NF-κB and apoptosis.

In addition to the potential cognitive enhancing, anti-inflammatory, and anti-Aβ effects attributed to nicotine, nornicotine is a long-lived psychoactive nicotine metabolite and minor constituent of tobacco that can inhibit Aβ aggregation by forming permanent covalent bonds with Aβ and glucose via glycation.[165] This study provides another explanation as to the putative neuroprotective effects of nicotine.

16.3.2.11   Huperzine A

Huperzine A is a terpene alkaloid and AChE inhibitor derived from a traditional medicinal plant called Chinese club moss.[166] In some clinical studies, treatment with huperzine A for 8 weeks was associated with mild cognitive improvement over placebo.[167,168] As compared with tacrine (a pharmaceutical AChE inhibitor), huperzine A protected cultured neurons from Aβ-induced neurotoxicity and oxidative stress[169] and caspase-3-induced apoptosis [170] to an equal degree.

However, its neuroprotective properties may be independent of its AChE inhibitory properties, since its enantiomer [(molecular mirror image; (+)-huperzine A] is 50-fold less potent at inhibiting AChE but equally potent at protecting cultured neurons from Aβ-induced toxicity.[171] Evidence for other possible mechanisms of huperzine A includes protection of cultured cells exposed to oxygen glucose deprivation,[169] non-competitive antagonism of the NMDA glutamate receptor (similar to memantine),[172] and increased sAPPα release from cultured human neurons that overexpress APP. The elevated sAPPα was suppressed by muscarinic ACh receptor antagonists and protein kinase C inhibitors. Thus, huperzine may join galantamine in the world of plant-derived AChE inhibitory pharmaceuticals. Indeed, huprine X is a hybrid AChE inhibitor synthesized by combining parts of huperzine A with tacrine, which inhibits AChE to a much larger extent than either huperzine A or tacrine alone.[173–177]

16.3.2.12   Other Phytochemicals

Experimental evidence suggests that a number of other foods and isolated phytochemicals have potential anti-AD properties. For example, the dark blue pigments in blueberries are polyphenols with potent antioxidant properties, and dietary blueberries improved cognitive performance in APP transgenic mice without decreasing Aβ plaque levels.[52] Furthermore, a blueberry-enriched diet fed to aged rats elevated levels of blueberry polyphenols in the brain and improved cognition.[178] Various raw fruit extracts have prevented *in vitro* Aβ-induced calcium flux deficits[179] and neurotoxicity induced by oxidative stress.[53]

Experiments with other isolated phytochemicals also suggest possible roles in AD-related processes. For example, the common dietary polyphenol luteolin, found in the leaves of many plants, was found to reduce levels of Aβ in the brains of APP transgenic mice. Luteolin also demonstrated *in vitro* inhibition of the enzyme glycogen synthase kinase 3, which plays a role in the cleavage of APP to Aβ.[180] Resveratrol, a polyphenol stilbenoid found in grapes and nuts, was shown to increase clearance and decrease levels of Aβ *in vitro* by intracellular proteasome-facilitated degradation of Aβ[181] and can modulate a number of other cell-signaling pathways.[113] Its biochemical effects may explain the epidemiological evidence for a decreased risk for AD among elderly subjects who consume moderate amounts of wine. The phenolic acids rosmarinic acid (from rosemary) and nor-dihydroguaiaretic acid (from creosote) prevented and reversed Aβ aggregation *in vitro*,[117,182] and myricetin and quercetin (polyphenols that are found in a variety of fruits, vegetables, and spices) reduced heme-Aβ damage to muscarinic ACh receptors.[7–9]

Finally, phytochemicals may improve AD symptoms via effects on the cellular processes of learning and memory. For example, fisetin, an isolated polyphenol found in strawberries and other fruits and vegetables, enhanced long-term memory in normal mice and induced cAMP response element binding phosphorylation and enhanced LTP in rat hippocampal slices.[183] Fisetin's effect on cAMP response element binding (which plays an important role in learning and memory mechanisms) and its functional effects of enhanced LTP and cognition suggest that it may improve synaptic plasticity. Additionally, caffeine and nicotine have long histories of anecdotal and experimental use as general cognitive enhancers, most likely due to their psychostimulant effects. Finally, cannabis also has a history of use for cognitive and creative enhancement but has acute deleterious effects on short-term memory processes. Thus, the cellular effects of phytocannabinoids on learning and memory processes may suggest targets for cognitive enhancement via selective cannabinoid antagonists.

16.3.3 SUMMARY

Several epidemiological and experimental studies show that dietary and isolated phytochemicals may have beneficial effects on cell function and AD-related pathology. In particular, phenols, terpenes, and organosulfurs all have well-characterized antioxidant properties and have demonstrated a variety of other mechanisms by which they may affect AD processes. Other phytochemically mediated pathways related to AD include modulation of the enzymatic processes that produce Aβ from APP, inhibition of Aβ aggregation via direct binding, increased intracellular clearance of Aβ and cholesterol, modulation of the glutamatergic and cholinergic neurotransmitter systems, anti-inflammatory effects via modulation of NF-κB and microglia, anti-apoptotic effects via inhibition of caspase-3, protection of the blood-brain barrier, and even stimulation of neurogenesis.

Low-level accumulation of Aβ in the brain occurs over the lifetime of individuals, and a variety of acute or chronic brain insults can accelerate this process and increase the risk of developing AD. Because age- and insult-related neuropathologies include the build-up of oxidative Aβ and because this gradually increases the oxidative load on the brain, the Aβ deposition process may self-propagate. Traumatic brain injury often occurs in young patients, planting early seeds for future AD neuropathology, and elderly individuals have an increased risk for stroke, which adds to the overall oxidative stress in the brain and subsequent Aβ cascade. Whereas accelerating brain Aβ accumulation can increase the risk of developing AD, reducing Aβ accumulation may decrease the risk. Currently, pharmacological strategies for controlling AD progression include modulation of the glutamatergic system by blocking NMDA receptor channels and preventing the degradation of ACh by inhibition of AChE. Unfortunately, these treatment strategies provide mild to moderate benefits at most.

However, a lifetime of consuming high levels of bioactive phytochemicals and low levels of cholesterol may help attenuate Aβ-related neuropathology associated with age and/or insult. Therefore, an AD-protective diet could reduce the susceptibility to the progression of AD neuropathology and symptoms. Because AD is a progressive disease of the elderly, delaying the onset by as little as a few years would significantly decrease its incidence. Given that Aβ accumulation seems to cause oxidative stress and inflammation, leading to the accumulation of even more Aβ, keeping oxidative stress in check with an AD-protective diet may reduce the slow but steady accumulation of Aβ in the brain with aging.[184,185] Interestingly, several studies suggest that the wide variety of phytochemicals and their isoforms found in whole plant preparations may provide synergistic benefit over isolated phytochemical compounds.[88,106,186,187]

In summary, epidemiological and experimental lines of evidence suggest that diets consisting of a wide variety of brightly colored and spicy foods should provide a broad degree of chronic

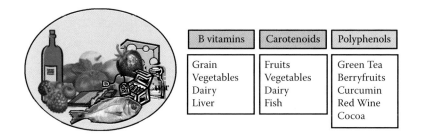

B vitamins	Carotenoids	Polyphenols
Grain Vegetables Dairy Liver	Fruits Vegetables Dairy Fish	Green Tea Berryfruits Curcumin Red Wine Cocoa

COLOR FIGURE 4.1   Many foods contain compounds that have been shown to have neuroprotective effects. Maintaining a healthy lifestyle is one way to preserve a healthy brain.

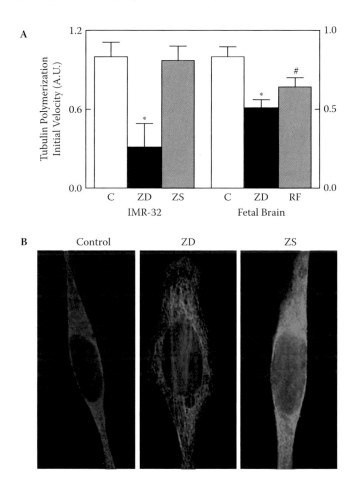

COLOR FIGURE 11.2   Zinc deficiency affects microtubule structure and polymerization in human neuroblastoma IMR-32 cells and in fetal brain supernatants. (A) For the cell study, IMR-32 cells were incubated for 24 h in control (C), zinc-deficient (ZD), or zinc-supplemented (ZS) media. For the animal study, Sprague-Dawley rats were fed from day 0 until day 19 of gestation a control diet (25 μg Zn/g) given ad libitum (C), a zinc-deficient diet containing 0.5 μg Zn/g (ZD), or the control diet at a restricted intake equal to that consumed by the ZD group (RF). In vitro tubulin polymerization kinetics in cells or in fetal brain supernatants were assessed as previously described [21]. The slope in the linear portion of the curves was calculated. Results are shown as the mean ± SEM. (*) and (#) indicate significantly lower compared with the C and ZS groups (*) or with the C group (#) ($p < 0.05$, one-way analysis of variance test). (B) Differentiated IMR-32 cells were incubated for 24 h in control (C), zinc deficient (ZD), or zinc-supplemented (ZS) media. β-tubulin was immunodetected by using a specific anti-β-tubulin antibody, followed by an FITC-labeled secondary antibody. Images of cells incubated in control (C), zinc-deficient (ZD), or zinc-supplemented (ZS) media are shown. This figure is an adaptation of results previously published [20,21,27].

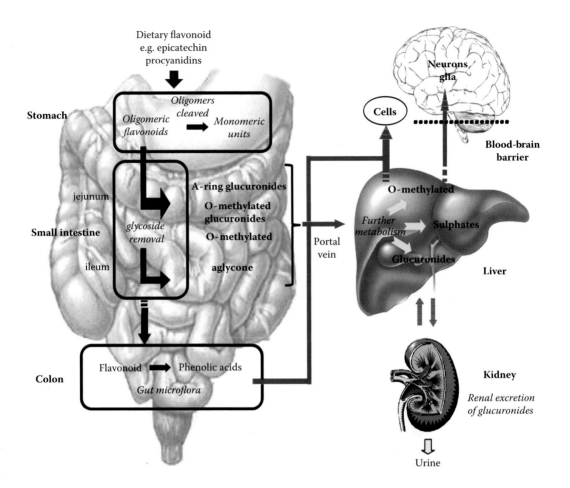

COLOR FIGURE 15.2 Summary of the formation of GI tract and hepatic metabolites and conjugates of polyphenols in humans. Cleavage of oligomeric flavonoids such as procyanidins may occur in the stomach in an environment of low pH. All classes of flavonoids undergo extensive metabolism in the jejunum and ileum of the small intestine, and resulting metabolites enter the portal vein and undergo further metabolism in the liver. Colonic microflora degrades flavonoids into smaller phenolic acids that may also be absorbed. The fate of flavonoids is renal excretion, although some may enter cells and tissues.

Spice		Active Phytochemical

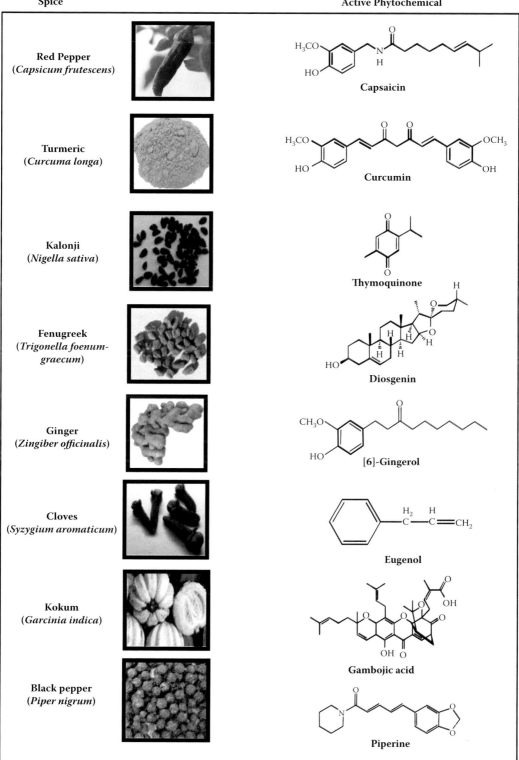

COLOR FIGURE 20.1   Structure and sources of various spice-derived phytochemicals.

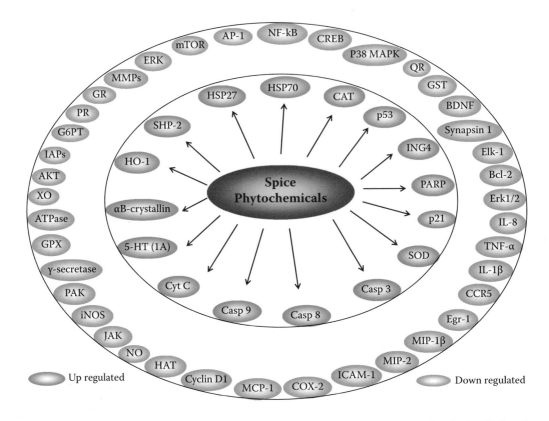

COLOR FIGURE 20.2 Molecular targets modulated by various spice-derived phytochemicals. AP-1, activator protein-1; BDNF, brain-derived neurotrophic factor; CAT, catalase; Casp, caspase; CCR5, chemokine (C-C motif) receptor 5; COX-2, cyclooxygenase-2; CREB, cyclic adenosine monophosphate (cAMP) response element-binding protein; Cyt C, cytochrome *c*; Egr-1, early growth response factor 1; GPX, glutathione peroxidase; GST, glutathione-*S*-transferase; HAT, histone acetyl transferase; (HO)-1, heme oxygenase HSP, heat shock proteins; ICAM-1, intracellular adhesion molecule 1; IAPs, inhibitor of apoptosis; iNOS, inducible nitric oxide synthase; MAPKs, mitogen-activated protein kinases; MCP-1, macrophage chemotactic protein-1; MCP, monocyte chemoattractant protein 1; MIP-2, macrophage inflammatory protein; MMPs, matrix-metallo proteases; mTOR, mammalian target of ripamycin; NF-κB, nuclear factor kappa B; PR, progesterone receptor; QR, quinone reductase; SOD, superoxide dismutase; STAT, signal transducers and activator of transcription protein; TNF, tumor necrosis factor; VEGF, vascular endothelial growth factor; XO, xanthine oxidase.

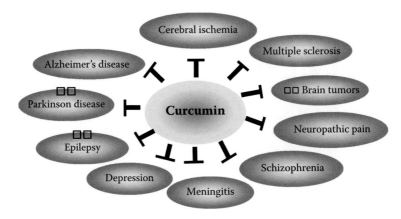

COLOR FIGURE 20.3 Potential role of curcumin in the treatment of various neurodegenerative diseases.

background protection against the deleterious effects of aging, including Aβ deposition and its associated consequences. In addition to the antioxidant effects of these compounds, a host of effects on other biochemical pathways that may provide additional lifetime benefits have been identified.

REFERENCES

1. Ramassamy, C. Emerging role of polyphenolic compounds in the treatment of neurodegenerative diseases: A review of their intracellular targets. *Eur J Pharmacol,* 545, 51–64, 2006.
2. Glenner, G. G., and Wong, C. W. Alzheimer's disease: Initial report of the purification and characterization of a novel cerebrovascular amyloid protein. *Biochem Biophys Res Commun,* 120, 885–890, 1984.
3. Mann, D. M. The pathogenesis and progression of the pathological changes of Alzheimer's disease. *Ann Med,* 21, 133–136, 1989.
4. Lanari, A., et al. Neurotransmitter deficits in behavioural and psychological symptoms of Alzheimer's disease. *Mech Ageing Dev,* 127, 158–165, 2006.
5. Stein, T. D., et al. Neutralization of transthyretin reverses the neuroprotective effects of secreted amyloid precursor protein (APP) in APPSW mice resulting in tau phosphorylation and loss of hippocampal neurons: Support for the amyloid hypothesis. *J Neurosci,* 24, 7707–7717, 2004.
6. Wenk, G. L. Neuropathologic changes in Alzheimer's disease: Potential targets for treatment. *J Clin Psychiatry,* 67 Suppl 3, 3–7; quiz 23, 2006.
7. Atamna, H., and Boyle, K. Amyloid-beta peptide binds with heme to form a peroxidase: Relationship to the cytopathologies of Alzheimer's disease. *Proc Natl Acad Sci U S A,* 103, 3381–3386, 2006.
8. Atamna, H., and Frey, W. H. 2nd. A role for heme in Alzheimer's disease: Heme binds amyloid beta and has altered metabolism. *Proc Natl Acad Sci U S A,* 101, 11153–11158, 2004.
9. Fawcett, J. R., et al. Inactivation of the human brain muscarinic acetylcholine receptor by oxidative damage catalyzed by a low molecular weight endogenous inhibitor from Alzheimer's brain is prevented by pyrophosphate analogs, bioflavonoids and other antioxidants. *Brain Res,* 950, 10–20, 2002.
10. Reddy, P. H. Amyloid precursor protein-mediated free radicals and oxidative damage: Implications for the development and progression of Alzheimer's disease. *J Neurochem,* 96, 1–13, 2006.
11. Reddy, P. H., et al. Gene expression profiles of transcripts in amyloid precursor protein transgenic mice: Up-regulation of mitochondrial metabolism and apoptotic genes is an early cellular change in Alzheimer's disease. *Hum Mol Genet,* 13, 1225–1240, 2004.
12. Dong, H., et al. Spatial relationship between synapse loss and beta-amyloid deposition in Tg2576 mice. *J Comp Neurol,* 500, 311–321, 2007.
13. Lacor, P. N., et al. Abeta oligomer-induced aberrations in synapse composition, shape, and density provide a molecular basis for loss of connectivity in Alzheimer's disease. *J Neurosci,* 27, 796–807, 2007.
14. Love, S., et al. Premorbid effects of APOE on synaptic proteins in human temporal neocortex. *Neurobiol Aging,* 27, 797–803, 2006.
15. Scheff, S. W., et al. Synaptic alterations in CA1 in mild Alzheimer disease and mild cognitive impairment. *Neurology,* 68, 1501–1508, 2007.
16. Hartman, R. E., et al. Pomegranate juice decreases amyloid load and improves behavior in a mouse model of Alzheimer's disease. *Neurobiol Dis,* 24, 506–515, 2006.
17. Rowan, M. J., et al. Synaptic plasticity in animal models of early Alzheimer's disease. *Philos Trans R Soc Lond B Biol Sci,* 358, 821–828, 2003.
18. Rowan, M. J., et al. Mechanisms of the inhibitory effects of amyloid beta-protein on synaptic plasticity. *Exp Gerontol,* 39, 1661–1667, 2004.
19. Rowan, M. J., et al. Synaptic plasticity disruption by amyloid beta protein: Modulation by potential Alzheimer's disease modifying therapies. *Biochem Soc Trans,* 33, 563–567, 2005.
20. Bisel, B. E., Henkins, K. M., and Parfitt, K. D. Alzheimer amyloid beta-peptide A-beta25–35 blocks adenylate cyclase-mediated forms of hippocampal long-term potentiation. *Ann N Y Acad Sci,* 1097, 58–63, 2007.
21. Meyer-Luehmann, M., et al. Rapid appearance and local toxicity of amyloid-beta plaques in a mouse model of Alzheimer's disease. *Nature,* 451, 720–724, 2008.
22. Head, E., and Lott, I. T. Down syndrome and beta-amyloid deposition. *Curr Opin Neurol,* 17, 95–100, 2004.
23. Mehta, P. D., et al. Increased amyloid beta protein levels in children and adolescents with Down syndrome. *J Neurol Sci,* 254, 22–27, 2007.
24. Selkoe, D. J. The origins of Alzheimer disease: A is for amyloid. *JAMA,* 283, 1615–1617, 2000.

25. Selkoe, D. J. Alzheimer's disease results from the cerebral accumulation and cytotoxicity of amyloid beta-protein. *J Alzheimers Dis,* 3, 75–80, 2001.

26. Arendash, G. W., et al. Progressive, age-related behavioral impairments in transgenic mice carrying both mutant amyloid precursor protein and presenilin-1 transgenes. *Brain Res,* 891, 42–53, 2001.

27. Dodart, J. C., et al. Does my mouse have Alzheimer's disease? *Genes Brain Behav,* 1, 142–155, 2002.

28. Gordon, M. N., et al. Correlation between cognitive deficits and Abeta deposits in transgenic APP+PS1 mice. *Neurobiol Aging,* 22, 377–385, 2001.

29. Irizarry, M. C., et al. APPSw transgenic mice develop age-related A beta deposits and neurophil abnormalities, but no neuronal loss in CA1. *J Neuropathol Exp Neurol,* 56, 965–973, 1997.

30. Kawarabayashi, T., et al. Age-dependent changes in brain, CSF, and plasma amyloid (beta) protein in the Tg2576 transgenic mouse model of Alzheimer's disease. *J Neurosci,* 21, 372–381, 2001.

31. King, D. L., and Arendash, G. W. Maintained synaptophysin immunoreactivity in Tg2576 transgenic mice during aging: Correlations with cognitive impairment. *Brain Res,* 926, 58–68, 2002.

32. King, D. L., and Arendash, G. W. Behavioral characterization of the Tg2576 transgenic model of Alzheimer's disease through 19 months. *Physiol Behav,* 75, 627–642, 2002.

33. King, D. L., et al. Progressive and gender-dependent cognitive impairment in the APP(SW) transgenic mouse model for Alzheimer's disease. *Behav Brain Res,* 103, 145–162, 1999.

34. Hartman, R. E., et al. Treatment with an amyloid-beta antibody ameliorates plaque load, learning deficits, and hippocampal long-term potentiation in a mouse model of Alzheimer's disease. *J Neurosci,* 25, 6213–6220, 2005.

35. Erkinjuntti, T., et al. Snoring and dementia. *Age Ageing,* 16, 305–310, 1987.

36. Yanagisawa, K. Cholesterol and pathological processes in Alzheimer's disease. *J Neurosci Res,* 70, 361–366, 2002.

37. Woodruff-Pak, D. S., Agelan, A., and Del Valle, L. A rabbit model of Alzheimer's disease: Valid at neuropathological, cognitive, and therapeutic levels. *J Alzheimers Dis,* 11, 371–383, 2007.

38. Wu, C. W., et al. Brain region-dependent increases in beta-amyloid and apolipoprotein E levels in hypercholesterolemic rabbits. *J Neural Transm,* 110, 641–649, 2003.

39. Sparks, D. L., et al. Induction of Alzheimer-like beta-amyloid immunoreactivity in the brains of rabbits with dietary cholesterol. *Exp Neurol,* 126, 88–94, 1994.

40. Shie, F. S., et al. Diet-induced hypercholesterolemia enhances brain A beta accumulation in transgenic mice. *NeuroReport,* 13, 455–459, 2002.

41. Levin-Allerhand, J. A., Lominska, C. E., and Smith, J. D. Increased amyloid-β levels in APPSWE transgenic mice treated chronically with a physiological high-fat high-cholesterol diet. *J Nutr Health Aging,* 6, 315–319, 2002.

42. Refolo, L. M., et al. Hypercholesterolemia accelerates the Alzheimer's amyloid pathology in a transgenic mouse model. *Neurobiol Dis,* 7, 321–331, 2000.

43. Mazur-Kolecka, B., Dickson, D., and Frackowiak, J. Induction of vascular amyloidosis-beta by oxidative stress depends on APOE genotype. *Neurobiol Aging,* 27, 804–814, 2006.

44. Allinson, T. M., et al. The role of ADAM10 and ADAM17 in the ectodomain shedding of angiotensin converting enzyme and the amyloid precursor protein. *Eur J Biochem,* 271, 2539–2547, 2004.

45. Dodart, J. C., et al. Immunization reverses memory deficits without reducing brain Abeta burden in Alzheimer's disease model. *Nat Neurosci,* 5, 452–457, 2002.

46. Kotilinek, L. A., et al. Reversible memory loss in a mouse transgenic model of Alzheimer's disease. *J Neurosci,* 22, 6331–6335, 2002.

47. Wilcock, D. M., et al. Amyloid-beta vaccination, but not nitro-nonsteroidal anti-inflammatory drug treatment, increases vascular amyloid and microhemorrhage while both reduce parenchymal amyloid. *Neuroscience,* 144, 950–960, 2007.

48. Oddo, S., et al. Abeta immunotherapy leads to clearance of early, but not late, hyperphosphorylated tau aggregates via the proteasome. *Neuron,* 43, 321–332, 2004.

49. Patel, N. V., et al. Caloric restriction attenuates Abeta-deposition in Alzheimer transgenic models. *Neurobiol Aging,* 26, 995–1000, 2005.

50. Love, R. Calorie restriction may be neuroprotective in AD and PD. *Lancet Neurol,* 4, 84, 2005.

51. Wang, J., et al. Caloric restriction attenuates beta-amyloid neuropathology in a mouse model of Alzheimer's disease. *FASEB J,* 19, 659–661, 2005.

52. Joseph, J. A., et al. Blueberry supplementation enhances signaling and prevents behavioral deficits in an Alzheimer disease model. *Nutr Neurosci,* 6, 153–162, 2003.

53. Park, L., et al. Nox2-derived radicals contribute to neurovascular and behavioral dysfunction in mice overexpressing the amyloid precursor protein. *Proc Natl Acad Sci U S A,* 105, 1347–1352, 2008.

54. Hardy, J. An "anatomical cascade hypothesis" for Alzheimer's disease. *Trends Neurosci,* 15, 200–201, 1992.

55. Hardy, J., and Allsop, D. Amyloid deposition as the central event in the aetiology of Alzheimer's disease. *Trends Pharmacol Sci,* 12, 383–388, 1991.

56. Hardy, J. A., and Higgins, G. A. Alzheimer's disease: The amyloid cascade hypothesis. *Science,* 256, 184–185, 1992.

57. Mattson, M. P. Existing data suggest that Alzheimer's disease is preventable. *Ann N Y Acad Sci,* 924, 153–159, 2000.

58. Morris, M. C., et al. Dietary niacin and the risk of incident Alzheimer's disease and of cognitive decline. *J Neurol Neurosurg Psychiatry,* 75, 1093–1099, 2004.

59. Pope, S. K., Shue, V. M., and Beck, C. Will a healthy lifestyle help prevent Alzheimer's disease? *Annu Rev Public Health,* 24, 111–132, 2003.

60. Dai, Q., et al. Fruit and vegetable juices and Alzheimer's disease: The Kame Project. *Am J Med,* 119, 751–759, 2006.

61. Commenges, D., et al. Intake of flavonoids and risk of dementia. *Eur J Epidemiol,* 16, 357–363, 2000.

62. Barberger-Gateau, P., et al. Dietary patterns and risk of dementia: The Three-City cohort study. *Neurology,* 69, 1921–1930, 2007.

63. Letenneur, L., et al. Flavonoid intake and cognitive decline over a 10-year period. *Am J Epidemiol,* 165, 1364–1371, 2007.

64. Morris, M. C., et al. Relation of the tocopherol forms to incident Alzheimer disease and to cognitive change. *Am J Clin Nutr,* 81, 508–514, 2005.

65. Engelhart, M. J., et al. Dietary intake of antioxidants and risk of Alzheimer disease. *JAMA,* 287, 3223–3229, 2002.

66. Gray, S. L., et al. Antioxidant vitamin supplement use and risk of dementia or Alzheimer's disease in older adults. *J Am Geriatr Soc,* 56, 291–295, 2008.

67. Grant, W. B. Dietary links to Alzheimer's disease: 1999 update. *J Alzheimers Dis,* 1, 197–201, 1999.

68. Horrocks, L. A., and Yeo, Y. K. Health benefits of docosahexaenoic acid (DHA). *Pharmacol Res,* 40, 211–225, 1999.

69. Peers, R. J. Alzheimer's disease and omega-3 fatty acids: Hypothesis. *Med J Aust,* 153, 563–564, 1990.

70. Ng, T. P., et al. Curry consumption and cognitive function in the elderly. *Am J Epidemiol,* 164, 898–906, 2006.

71. Orgogozo, J. M., et al. Wine consumption and dementia in the elderly: A prospective community study in the Bordeaux area. *Rev Neurol (Paris),* 153, 185–192, 1997.

72. Court, J. A., et al. Attenuation of Abeta deposition in the entorhinal cortex of normal elderly individuals associated with tobacco smoking. *Neuropathol Appl Neurobiol,* 31, 522–535, 2005.

73. Letenneur, L., Larrieu, S., and Barberger-Gateau, P. Alcohol and tobacco consumption as risk factors of dementia: A review of epidemiological studies. *Biomed Pharmacother,* 58, 95–99, 2004.

74. Kanowski, S., et al. Proof of efficacy of the ginkgo biloba special extract EGb 761 in outpatients suffering from mild to moderate primary degenerative dementia of the Alzheimer type or multi-infarct dementia. *Pharmacopsychiatry,* 29, 47–56, 1996.

75. Le Bars, P. L., Kieser, M., and Itil, K. Z. A 26-week analysis of a double-blind, placebo-controlled trial of the ginkgo biloba extract EGb 761 in dementia. *Dement Geriatr Cogn Disord,* 11, 230–237, 2000.

76. Maurer, K., et al. Clinical efficacy of *Ginkgo biloba* special extract EGb 761 in dementia of the Alzheimer type. *J Psychiatr Res,* 31, 645–655, 1997.

77. Oken, B. S., Storzbach, D. M., and Kaye, J. A. The efficacy of *Ginkgo biloba* on cognitive function in Alzheimer disease. *Arch Neurol,* 55, 1409–1415, 1998.

78. Mazza, M., et al. *Ginkgo biloba* and donepezil: A comparison in the treatment of Alzheimer's dementia in a randomized placebo-controlled double-blind study. *Eur J Neurol,* 13, 981–985, 2006.

79. Wettstein, A. Cholinesterase inhibitors and *Gingko* extracts—Are they comparable in the treatment of dementia? Comparison of published placebo-controlled efficacy studies of at least six months' duration. *Phytomedicine,* 6, 393–401, 2000.

80. Itil, T. M., et al. The pharmacological effects of ginkgo biloba, a plant extract, on the brain of dementia patients in comparison with tacrine. *Psychopharmacol Bull,* 34, 391–397, 1998.

81. Schneider, L. S., et al. A randomized, double-blind, placebo-controlled trial of two doses of *Ginkgo biloba* extract in dementia of the Alzheimer's type. *Curr Alzheimer Res,* 2, 541–551, 2005.

82. van Dongen, M. C., et al. The efficacy of ginkgo for elderly people with dementia and age-associated memory impairment: New results of a randomized clinical trial. *J Am Geriatr Soc,* 48, 1183–1194, 2000.

83. van Dongen, M., et al. Ginkgo for elderly people with dementia and age-associated memory impairment: A randomized clinical trial. *J Clin Epidemiol,* 56, 367–376, 2003.

84. Solomon, P. R., et al. Ginkgo for memory enhancement: A randomized controlled trial. *JAMA,* 288, 835–840, 2002.

85. Stackman, R. W., et al. Prevention of age-related spatial memory deficits in a transgenic mouse model of Alzheimer's disease by chronic *Ginkgo biloba* treatment. *Exp Neurol,* 184, 510–520, 2003.

86. Vitolo, O., et al. Protection against beta-amyloid induced abnormal synaptic function and cell death by Ginkgolide J. *Neurobiol Aging,* 30, 257–265, 2007.

87. DeFeudis, F. V., and Drieu, K. *Ginkgo biloba* extract (EGb 761) and CNS functions: Basic studies and clinical applications. *Curr Drug Targets,* 1, 25–58, 2000.

88. Bastianetto, S., et al. The *Ginkgo biloba* extract (EGb 761) protects hippocampal neurons against cell death induced by beta-amyloid. *Eur J Neurosci,* 12, 1882–1890, 2000.

89. Bastianetto, S., Zheng, W. H., and Quirion, R. The *Ginkgo biloba* extract (EGb 761) protects and rescues hippocampal cells against nitric oxide-induced toxicity: Involvement of its flavonoid constituents and protein kinase C. *J Neurochem,* 74, 2268–2277, 2000.

90. Schindowski, K., et al. Age-related increase of oxidative stress-induced apoptosis in mice prevention by *Ginkgo biloba* extract (EGb761). *J Neural Transm,* 108, 969–978, 2001.

91. Zimmermann, M., et al. Ginkgo biloba extract: From molecular mechanisms to the treatment of Alzheimer's disease. *Cell Mol Biol (Noisy-le-grand),* 48, 613–623, 2002.

92. Garcia-Alloza, M., et al. Plaque-derived oxidative stress mediates distorted neurite trajectories in the Alzheimer mouse model. *J Neuropathol Exp Neurol,* 65, 1082–1089, 2006.

93. Smith, J. V., and Luo, Y. Elevation of oxidative free radicals in Alzheimer's disease models can be attenuated by *Ginkgo biloba* extract EGb 761. *J Alzheimers Dis,* 5, 287–300, 2003.

94. Luo, Y. Alzheimer's disease, the nematode *Caenorhabditis elegans,* and ginkgo biloba leaf extract. *Life Sci,* 78, 2066–2072, 2006.

95. Luo, Y., et al. Inhibition of amyloid-beta aggregation and caspase-3 activation by the *Ginkgo biloba* extract EGb761. *Proc Natl Acad Sci U S A,* 99, 12197–12202, 2002.

96. Colciaghi, F., et al. Amyloid precursor protein metabolism is regulated toward alpha-secretase pathway by *Ginkgo biloba* extracts. *Neurobiol Dis,* 16, 454–460, 2004.

97. Yao, Z. X., et al. *Ginkgo biloba* extract (EGb 761) inhibits beta-amyloid production by lowering free cholesterol levels. *J Nutr Biochem,* 15, 749–756, 2004.

98. Kelawala, N. S., and Ananthanarayan, L. Antioxidant activity of selected foodstuffs. *Int J Food Sci Nutr,* 55, 511–516, 2004.

99. Wang, R. F., et al. Bioactive compounds from the seeds of *Punica granatum* (pomegranate). *J Nat Prod,* 67, 2096–2098, 2004.

100. Seeram, N. P., et al. Comparison of antioxidant potency of commonly consumed polyphenol-rich beverages in the United States. *J Agric Food Chem,* 56, 1415–1422, 2008.

101. Aviram, M., et al. Pomegranate juice consumption reduces oxidative stress, atherogenic modifications to LDL, and platelet aggregation: Studies in humans and in atherosclerotic apolipoprotein E-deficient mice. *Am J Clin Nutr,* 71, 1062–1076, 2000.

102. Kaplan, M., et al. Pomegranate juice supplementation to atherosclerotic mice reduces macrophage lipid peroxidation, cellular cholesterol accumulation and development of atherosclerosis. *J Nutr,* 131, 2082–2089, 2001.

103. Aviram, M., et al. Pomegranate juice consumption for 3 years by patients with carotid artery stenosis reduces common carotid intima-media thickness, blood pressure and LDL oxidation. *Clin Nutr,* 23, 423–433, 2004.

104. Rozenberg, O., Howell, A., and Aviram, M. Pomegranate juice sugar fraction reduces macrophage oxidative state, whereas white grape juice sugar fraction increases it. *Atherosclerosis,* 188, 68–76, 2005.

105. de Nigris, F., et al. Beneficial effects of pomegranate juice on oxidation-sensitive genes and endothelial nitric oxide synthase activity at sites of perturbed shear stress. *Proc Natl Acad Sci U S A,* 102, 4896–4901, 2005.

106. Seeram, N. P., et al. *In vitro* antiproliferative, apoptotic and antioxidant activities of punicalagin, ellagic acid and a total pomegranate tannin extract are enhanced in combination with other polyphenols as found in pomegranate juice. *J Nutr Biochem,* 16, 360–367, 2005.

107. Rosenblat, M., et al. Pomegranate byproduct administration to apolipoprotein E-deficient mice attenuates atherosclerosis development as a result of decreased macrophage oxidative stress and reduced cellular uptake of oxidized low-density lipoprotein. *J Agric Food Chem,* 54, 1928–1935, 2006.

108. Braga, L. C., et al. Pomegranate extract inhibits *Staphylococcus aureus* growth and subsequent enterotoxin production. *J Ethnopharmacol,* 96, 335–339, 2005.

109. Braga, L. C., et al. Synergic interaction between pomegranate extract and antibiotics against *Staphylococcus aureus. Can J Microbiol,* 51, 541–547, 2005.

110. Loren, D. J., et al. Maternal dietary supplementation with pomegranate juice is neuroprotective in an animal model of neonatal hypoxic-ischemic brain injury. *Pediatr Res,* 57, 858–864, 2005.

111. Kwak, H. M., et al. beta-Secretase (BACE1) inhibitors from pomegranate (*Punica granatum*) husk. *Arch Pharm Res,* 28, 1328–1332, 2005.

112. Aggarwal, B. B., and Shishodia, S. Suppression of the nuclear factor-kappaB activation pathway by spice-derived phytochemicals: Reasoning for seasoning. *Ann N Y Acad Sci,* 1030, 434–441, 2004.

113. Aggarwal, B. B., and Shishodia, S. Molecular targets of dietary agents for prevention and therapy of cancer. *Biochem Pharmacol,* 71, 1397–1421, 2006.

114. Frautschy, S. A., et al. Phenolic anti-inflammatory antioxidant reversal of Abeta-induced cognitive deficits and neuropathology. *Neurobiol Aging,* 22, 993–1005, 2001.

115. Lim, G. P., et al. The curry spice curcumin reduces oxidative damage and amyloid pathology in an Alzheimer transgenic mouse. *J Neurosci,* 21, 8370–8377, 2001.

116. Yang, F., et al. Curcumin inhibits formation of amyloid beta oligomers and fibrils, binds plaques, and reduces amyloid *in vivo. J Biol Chem,* 280, 5892–5901, 2005.

117. Ono, K., et al. Curcumin has potent anti-amyloidogenic effects for Alzheimer's beta-amyloid fibrils *in vitro. J Neurosci Res,* 75, 742–750, 2004.

118. Chauhan, N. B. Effect of aged garlic extract on APP processing and tau phosphorylation in Alzheimer's transgenic model Tg2576. *J Ethnopharmacol,* 108, 385–394, 2006.

119. Chauhan, N. B., and Sandoval, J. Amelioration of early cognitive deficits by aged garlic extract in Alzheimer's transgenic mice. *Phytother Res,* 21, 629–640, 2007.

120. Gupta, V. B., and Rao, K. S. Anti-amyloidogenic activity of *S*-allyl-l-cysteine and its activity to destabilize Alzheimer's beta-amyloid fibrils *in vitro. Neurosci Lett,* 429, 75–80, 2007.

121. Peng, Q., Buz'Zard, A. R., and Lau, B. H. Neuroprotective effect of garlic compounds in amyloid-beta peptide-induced apoptosis *in vitro. Med Sci Monit,* 8, BR328– BR337, 2002.

122. Kojro, E., et al. Low cholesterol stimulates the nonamyloidogenic pathway by its effect on the alpha-secretase ADAM 10. *Proc Natl Acad Sci U S A,* 98, 5815–5820, 2001.

123. Calon, F., et al. Docosahexaenoic acid protects from dendritic pathology in an Alzheimer's disease mouse model. *Neuron,* 43, 633–645, 2004.

124. Cole, G. M., and Frautschy, S. A. Docosahexaenoic acid protects from amyloid and dendritic pathology in an Alzheimer's disease mouse model. *Nutr Health,* 18, 249–259, 2006.

125. Lim, G. P., et al. A diet enriched with the omega-3 fatty acid docosahexaenoic acid reduces amyloid burden in an aged Alzheimer mouse model. *J Neurosci,* 25, 3032–3040, 2005.

126. Hooijmans, C. R., et al. Changes in cerebral blood volume and amyloid pathology in aged Alzheimer APP/PS1 mice on a docosahexaenoic acid (DHA) diet or cholesterol enriched typical Western diet (TWD). *Neurobiol Dis,* 28, 16–29, 2007.

127. Hashimoto, M., et al. Docosahexaenoic acid-induced amelioration on impairment of memory learning in amyloid beta-infused rats relates to the decreases of amyloid beta and cholesterol levels in detergent-insoluble membrane fractions. *Biochim Biophys Acta,* 1738, 91–98, 2005.

128. Hashimoto, M., et al. Chronic administration of docosahexaenoic acid ameliorates the impairment of spatial cognition learning ability in amyloid beta-infused rats. *J Nutr,* 135, 549–555, 2005.

129. Sung, S., et al. Early vitamin E supplementation in young but not aged mice reduces Abeta levels and amyloid deposition in a transgenic model of Alzheimer's disease. *FASEB J,* 18, 323–325, 2004.

130. Conte, V., et al. Vitamin E reduces amyloidosis and improves cognitive function in Tg2576 mice following repetitive concussive brain injury. *J Neurochem,* 90, 758–764, 2004.

131. Kruman, I. I., et al. Folic acid deficiency and homocysteine impair DNA repair in hippocampal neurons and sensitize them to amyloid toxicity in experimental models of Alzheimer's disease. *J Neurosci,* 22, 1752–1762, 2002.

132. Fuso, A., et al. S-adenosylmethionine/homocysteine cycle alterations modify DNA methylation status with consequent deregulation of PS1 and BACE and beta-amyloid production. *Mol Cell Neurosci,* 28, 195–204, 2005.

133. Weinreb, O., et al. Neurological mechanisms of green tea polyphenols in Alzheimer's and Parkinson's diseases. *J Nutr Biochem,* 15, 506–516, 2004.

134. Burckhardt, I. C., et al. Green tea catechin polyphenols attenuate behavioral and oxidative responses to intermittent hypoxia. *Am J Respir Crit Care Med,* 177, 1135–1141, 2008.

135. Rezai-Zadeh, K., et al. Green tea epigallocatechin-3-gallate (EGCG) reduces beta-amyloid mediated cognitive impairment and modulates tau pathology in Alzheimer transgenic mice. *Brain Res,* 2008.

136. Rezai-Zadeh, K., et al. Green tea epigallocatechin-3-gallate (EGCG) modulates amyloid precursor protein cleavage and reduces cerebral amyloidosis in Alzheimer transgenic mice. *J Neurosci,* 25, 8807–8814, 2005.

137. Ehrnhoefer, D. E., et al. EGCG redirects amyloidogenic polypeptides into unstructured, off-pathway oligomers. *Nat Struct Mol Biol,* 15, 558–566, 2008.

138. Ehrnhoefer, D. E., et al. Green tea (-)-epigallocatechin-gallate modulates early events in huntingtin mis-folding and reduces toxicity in Huntington's disease models. *Hum Mol Genet,* 15, 2743–2751, 2006.

139. Ono, K., et al. Anti-amyloidogenic activity of tannic acid and its activity to destabilize Alzheimer's beta-amyloid fibrils *in vitro. Biochim Biophys Acta,* 1690, 193–202, 2004.

140. Nardini, M., et al. Absorption of phenolic acids in humans after coffee consumption. *J Agric Food Chem,* 50, 5735–5741, 2002.

141. Mattila, P., Hellstrom, J., and Torronen, R. Phenolic acids in berries, fruits, and beverages. *J Agric Food Chem,* 54, 7193–7199, 2006.

142. Natella, F., et al. Coffee drinking influences plasma antioxidant capacity in humans. *J Agric Food Chem,* 50, 6211–6216, 2002.

143. Yen, W. J., et al. Antioxidant properties of roasted coffee residues. *J Agric Food Chem,* 53, 2658–2663, 2005.

144. Mursu, J., et al. The effects of coffee consumption on lipid peroxidation and plasma total homocysteine concentrations: A clinical trial. *Free Radic Biol Med,* 38, 527–534, 2005.

145. Natella, F., et al. Coffee drinking induces incorporation of phenolic acids into LDL and increases the resistance of LDL to *ex vivo* oxidation in humans. *Am J Clin Nutr,* 86, 604–609, 2007.

146. Arendash, G. W., et al. Caffeine protects Alzheimer's mice against cognitive impairment and reduces brain beta-amyloid production. *Neuroscience,* 142, 941–952, 2006.

147. Leighty, R. E., et al. Use of artificial neural networks to determine cognitive impairment and therapeutic effectiveness in Alzheimer's transgenic mice. *J Neurosci Methods,* 167, 358–366, 2008.

148. Chen, X., et al. Caffeine blocks disruption of blood brain barrier in a rabbit model of Alzheimer's disease. *J Neuroinflammation,* 5, 12, 2008.

149. Aronowski, J., et al. Ethanol plus caffeine (caffeinol) for treatment of ischemic stroke: Preclinical experience. *Stroke,* 34, 1246–1251, 2003.

150. Westlake, T. M., et al. Cannabinoid receptor binding and messenger RNA expression in human brain: An *in vitro* receptor autoradiography and *in situ* hybridization histochemistry study of normal aged and Alzheimer's brains. *Neuroscience,* 63, 637–652, 1994.

151. Benito, C., et al. Cannabinoid CB$_2$ receptors and fatty acid amide hydrolase are selectively overexpressed in neuritic plaque-associated glia in Alzheimer's disease brains. *J Neurosci,* 23, 11136–11141, 2003.

152. Ramirez, B. G., et al. Prevention of Alzheimer's disease pathology by cannabinoids: Neuroprotection mediated by blockade of microglial activation. *J Neurosci,* 25, 1904–1913, 2005.

153. Nunez, E., et al. Glial expression of cannabinoid CB(2) receptors and fatty acid amide hydrolase are beta amyloid-linked events in Down's syndrome. *Neuroscience,* 151, 104–110, 2008.

154. Benito, C., et al. Cannabinoid CB$_2$ receptors in human brain inflammation. *Br J Pharmacol,* 153, 277–285, 2008.

155. Coffey, R. G., et al. Inhibition of macrophage nitric oxide production by tetrahydrocannabinol *in vivo* and *in vitro. Int J Immunopharmacol,* 18, 749–752, 1996.

156. Molina-Holgado, F., Lledo, A., and Guaza, C. Anandamide suppresses nitric oxide and TNF-alpha responses to Theiler's virus or endotoxin in astrocytes. *NeuroReport,* 8, 1929–1933, 1997.

157. de Lago, E., and Fernandez-Ruiz, J. Cannabinoids and neuroprotection in motor-related disorders. *CNS Neurol Disord Drug Targets,* 6, 377–387, 2007.

158. Mechoulam, R., et al. Cannabidiol—Recent advances. *Chem Biodivers,* 4, 1678–1692, 2007.

159. Eubanks, L. M., et al. A molecular link between the active component of marijuana and Alzheimer's disease pathology. *Mol Pharm,* 3, 773–777, 2006.

160. Wolf, S. A., and Ullrich, O. Endocannabinoids and the brain immune system: New neurones at the horizon? *J Neuroendocrinol,* 20 Suppl 1, 15–19, 2008.

161. Milton, N. G. Phosphorylated amyloid-beta: The toxic intermediate in Alzheimer's disease neurodegeneration. *Subcell Biochem,* 38, 381–402, 2005.

162. Marchalant, Y., Rosi, S., and Wenk, G. L. Anti-inflammatory property of the cannabinoid agonist WIN-55212-2 in a rodent model of chronic brain inflammation. *Neuroscience,* 144, 1516–1522, 2007.

163. Nordberg, A., et al. Chronic nicotine treatment reduces beta-amyloidosis in the brain of a mouse model of Alzheimer's disease (APPsw). *J Neurochem,* 81, 655–658, 2002.

164. Liu, Q., et al. Dissecting the signaling pathway of nicotine-mediated neuroprotection in a mouse Alzheimer disease model. *FASEB J,* 21, 61–73, 2007.

165. Dickerson, T. J., and Janda, K. D. Glycation of the amyloid beta-protein by a nicotine metabolite: A fortuitous chemical dynamic between smoking and Alzheimer's disease. *Proc Natl Acad Sci U S A,* 100, 8182–8187, 2003.

166. Peng, Y., et al. Huperzine A regulates amyloid precursor protein processing via protein kinase C and mitogen-activated protein kinase pathways in neuroblastoma SK-N-SH cells over-expressing wild type human amyloid precursor protein 695. *Neuroscience,* 150, 386–395, 2007.

167. Xu, S. S., et al. Efficacy of tablet huperzine-A on memory, cognition, and behavior in Alzheimer's disease. *Zhongguo Yao Li Xue Bao,* 16, 391–395, 1995.

168. Xu, S. S., et al. Huperzine-A in capsules and tablets for treating patients with Alzheimer disease. *Zhongguo Yao Li Xue Bao,* 20, 486–490, 1999.

169. Xiao, X. Q., Wang, R., and Tang, X. C. Huperzine A and tacrine attenuate beta-amyloid peptide-induced oxidative injury. *J Neurosci Res,* 61, 564–569, 2000.

170. Xiao, X. Q., Zhang, H. Y., and Tang, X. C. Huperzine A attenuates amyloid beta-peptide fragment 25–35-induced apoptosis in rat cortical neurons via inhibiting reactive oxygen species formation and caspase-3 activation. *J Neurosci Res,* 67, 30–36, 2002.

171. Zhang, H. Y., et al. Stereoselectivities of enantiomers of huperzine A in protection against beta-amyloid(25–35)-induced injury in PC12 and NG108-15 cells and cholinesterase inhibition in mice. *Neurosci Lett,* 317, 143–146, 2002.

172. Zhang, J. M., and Hu, G. Y. Huperzine A, a nootropic alkaloid, inhibits *N*-methyl-d-aspartate-induced current in rat dissociated hippocampal neurons. *Neuroscience,* 105, 663–669, 2001.

173. Badia, A., et al. Synthesis and evaluation of tacrine-huperzine A hybrids as acetylcholinesterase inhibitors of potential interest for the treatment of Alzheimer's disease. *Bioorg Med Chem,* 6, 427–440, 1998.

174. Camps, P., et al. Huprine X is a novel high-affinity inhibitor of acetylcholinesterase that is of interest for treatment of Alzheimer's disease. *Mol Pharmacol,* 57, 409–417, 2000.

175. Camps, P., et al. New tacrine-huperzine A hybrids (huprines): Highly potent tight-binding acetylcholinesterase inhibitors of interest for the treatment of Alzheimer's disease. *J Med Chem,* 43, 4657–4666, 2000.

176. Carlier, P. R., et al. Potent, easily synthesized huperzine A-tacrine hybrid acetylcholinesterase inhibitors. *Bioorg Med Chem Lett,* 9, 2335–2338, 1999.

177. Dvir, H., et al. 3D structure of *Torpedo californica* acetylcholinesterase complexed with huprine X at 2.1 Å resolution: Kinetic and molecular dynamic correlates. *Biochemistry,* 41, 2970–2981, 2002.

178. Andres-Lacueva, C., et al. Anthocyanins in aged blueberry-fed rats are found centrally and may enhance memory. *Nutr Neurosci,* 8, 111–120, 2005.

179. Joseph, J. A., Fisher, D. R., and Carey, A. N. Fruit extracts antagonize Abeta- or DA-induced deficits in Ca^{2+} flux in M1-transfected COS-7 cells. *J Alzheimers Dis,* 6, 403–11; discussion 443–449, 2004.

180. Rezai-Zadeh, K., et al. Flavonoid-mediated presenilin-1 phosphorylation reduces Alzheimer's disease beta-amyloid production. *J Cell Mol Med,* 2008.

181. Marambaud, P., Zhao, H., and Davies, P. Resveratrol promotes clearance of Alzheimer's disease amyloid-beta peptides. *J Biol Chem,* 280, 37377–37382, 2005.

182. Ono, K., et al. Nordihydroguaiaretic acid potently breaks down pre-formed Alzheimer's beta-amyloid fibrils *in vitro*. *J Neurochem,* 81, 434–440, 2002.

183. Maher, P., Akaishi, T., and Abe, K. Flavonoid fisetin promotes ERK-dependent long-term potentiation and enhances memory. *Proc Natl Acad Sci U S A,* 103, 16568–16573, 2006.

184. Kostrzewa, R. M., and Segura-Aguilar, J. Novel mechanisms and approaches in the study of neurodegeneration and neuroprotection. A review. *Neurotox Res,* 5, 375–383, 2003.

185. Polidori, M. C. Antioxidant micronutrients in the prevention of age-related diseases. *J Postgrad Med,* 49, 229–235, 2003.

186. Seeram, N. P., et al. Total cranberry extract versus its phytochemical constituents: Antiproliferative and synergistic effects against human tumor cell lines. *J Agric Food Chem,* 52, 2512–2517, 2004.

187. Lansky, E. P., et al. Pomegranate (*Punica granatum*) pure chemicals show possible synergistic inhibition of human PC-3 prostate cancer cell invasion across Matrigel. *Invest New Drugs,* 23, 121–122, 2005.

17 Does *Ginkgo biloba* Extract Exert an Effect on Alzheimer's Disease Progression?

Yves Christen

Institut Ipsen, Boulogne-Billancourt, France

CONTENTS

17.1 INTRODUCTION

Research into Alzheimer's disease (AD) has seen spectacular developments over the last few years. Most of the molecular events involved in the onset of the disease have been identified. The so-called amyloid cascade theory provides a unified picture that appears convincing despite some uncertainties and disputed areas. It provides a direct path to therapeutic projects to target the cause of the pathological process, as we can look at developing drugs that are able to act on one or another of the stages in the cascade [amyloid-β protein (Aβ) formation, oligomerization, aggregation, clearance, etc.]. In addition to experimental data from research into the molecular mechanisms associated with AD, epidemiological data have also spawned several therapeutic avenues, some of which have involved well-known drugs such as estrogens, anti-inflammatories, statins, anti-hypertensives, and antioxidants, among others.

Therefore, in the early 2000s, there was understandably a hope that we would soon have a sizeable therapeutic arsenal at our disposal. Sadly, clinical trials have not as yet borne out the expectations raised by the epidemiological surveys or the data obtained in the laboratory: trials on estrogens and cyclooxygenase 2 inhibitors have had to be stopped prematurely due to serious side effects, as have attempts at vaccinating with Aβ or clioquinol (a metal chelator). Statins have not as yet proved useful in this area, and cholinesterase inhibitors, which are used for the symptomatic treatment of AD, have proved ineffective in preventing the disease or in improving mild cognitive impairment (MCI), which is often an early form of it. Alzhemed, Flurizan, and many other compounds that have been tested in a clinical setting have given only negative results.

In the context of this somewhat disappointing picture on the treatment and prevention fronts, the situation regarding *Ginkgo biloba* extract stands out as meriting attention, both in terms of what we already know about it and because at the present time it is one of the rare clinical research avenues that have not reached a dead end.

17.2   *GINKGO BILOBA* AND *GINKGO BILOBA* EXTRACT

The *G. biloba* tree is phylogenetically unique today, an originality expressed by its remarkable biochemical composition. Several molecules (ginkgolides and bilobalide) exist only in this tree (1). It also contains flavonoids, which belong to a vast family of compounds of plant origin, but they appear in an unusual form — not the standard components such as quercetin and kaempferol—but instead combined with sugars (flavonol-*O*-glycosides). In Chinese traditional medicine, an infusion of boiled leaves mixed with other preparations is used to treat heart and lung diseases. However, the uses of *G. biloba* extracts in Western medicine owe little to these traditional applications (1). The *G. biloba* extract EGb 761 is standardized to contain 24% flavonoids and 6% terpene lactones (ginkgolides, 3.1%; bilobalide, 2.9%) and <5 ppm of ginkolic acid. The vast majority of laboratory and clinical studies concern this particular extract. EGb 761 is a drug marketed first in France in 1975 and then in other countries. EGb 761 has been approved by government agencies as an ethical drug and is reimbursed by the government-run health insurance programs in France, Germany, Spain, Belgium, Switzerland, Portugal, Russia, Romania, Austria, and Turkey, as well as Argentina, Brazil, and China, among others. In the United States, on the other hand, it is regarded as a food supplement. EGb 761 is indicated in diverse circulatory, cerebral, and neurosensory deficits. Several studies demonstrate its interest in age-associated cognitive disorders, AD in particular. The drug classification developed by the French Caisse Nationale d'Assurance Maladie, in common with the World Health Organization classification, the German certification, and the Belgian authorities list EGb 761, together with cholinesterase inhibitors, as a treatment for dementia.

Several other *G. biloba* extracts have been marketed in various countries, notably the United States. According to information from manufacturers, the composition of some of these extracts is close to that of EGb 761 (24% flavonoids and 6% terpene), although it should be borne in mind that it is practically impossible to manufacture two identical plant extracts from the same plant unless exactly the same extraction process is used. This makes it difficult to know the actual effects of these other extracts, as well as their potential toxicity.

17.3 THE EFFECTS OF *GINKGO BILOBA* EXTRACT

17.3.1 A BRIEF HISTORY

Research into EGb 761's modes of action has developed over the last decades as advances have been made in pharmacological and biological knowledge (1). When it was first marketed in France in 1975, EGb 761 was known primarily for its vascular effect. Several sets of experiments had demonstrated its different actions according to different contexts (on arteries, veins, and capillaries). At that time, its beneficial effect on the brain and sensory organs was explained by this vascular action. Subsequently, EGb 761's neural effect came to light, largely thanks to neurobiological research into AD.

17.3.2 FREE RADICAL SCAVENGER EFFECT

EGb 761 and, to a lesser extent, other *G. biloba* extracts have been shown in various *in vitro* and in vivo models to be powerful free radical scavengers (2). This effect is linked not only to the flavonoids contained in the extract (flavonoids are well-known antioxidants) but also to its terpenes. Several of these models have demonstrated a protective action on neurons, in cases of cerebral ischemia, for example (3). The free radical scavenging effect of EGb 761 has even been shown in humans in open-heart surgery (4) and in healthy subjects (5). Oxidative processes seem to be involved in AD and even in MCI. Aβ can produce reactive oxygen species, and free radicals participate in neuronal death. Oxidative stress could be both a cause and a consequence of Aβ damage. Antioxidants are therefore good candidates as therapeutic agents, and notably with regard to AD prevention. Smith and Luo (6) found that the elevation of oxidative free radicals in two AD models (neuroblastoma cell line expressing AD-associated double mutation and transgenic *Caenorhabditis elegans* expressing human Aβ) is attenuated by EGb 761. Ramassamy et al. (7) found that EGb 761 protects the frontal cortices of AD patients against oxidative damage. Using multiphoton microscopy, Garcia-Alloza et al. (8) found that EGb 761 reduces the oxidative stress resulting from the senile plaques in the brain of APPswe/PS1d9 transgenic mice.

17.3.3 EFFECT ON GENE EXPRESSION

Pioneering studies by Mizuno and Packer (9) have shown an effect of EGb 761 on down-regulation of the transcription factor AP-1. More exhaustive work using high-density oligonucleotide microarrays has shown changes in the expression of several hundreds of genes. In vivo essays confirm this fact, including at the level of the brain (10). This is especially interesting because some people believe that flavonoids administered per os are not bioavailable. This study confirms not only the bioavailability of flavonoids and other components of EGb 761 but also their effects at the level of the brain. Many of the genes up-regulated by EGb 761 could exert a beneficial effect against brain aging and AD: the mitochondrial DNA-encoded cytochrome oxidase subunit III, the heme oxygenase 1 (HO-1), mitochondrial superoxide dismutase, NADH dehydrogenase subunit I, SIRT1, UCP-3 (uncoupling protein 3), and transthyretin (TTR), among others.

The activation of TTR is especially interesting. TTR is a tetrameric protein produced mainly by liver and in the choroids plexus of the brain and secreted into plasma and cerebrospinal fluid. It is responsible for thyroid hormone and retinol transport. However, TTR is also a major amyloid fibril protein involved in several forms of amyloidosis. TTR mutations are the cause of familial amyloid polyneuropathy. Many experimental data suggest a protective effect of TTR against AD and Aβ fibrillation. Stein and Johnson (11) found that the lack of neurodegeneration in transgenic mice overexpressing mutant amyloid precursor protein is associated with increased levels of TTR, and Buxbaum et al. (12) discovered that the in vivo overexpression of wild-type TTR transgene protects

AD mice from the behavioral and biochemical effects of Aβ toxicity. It appears that there is a physical interaction between TTR and Aβ. TTR may behave in a chaperone-like manner for molecular species of Aβ larger than monomers. Interestingly, TTR was the most highly up-regulated protein by dietary supplementation of EGb 761 in the rat hippocampus, with a 16-fold enhancement (10).

HO-1 is another gene of interest. It is considered a protective gene not only in AD but also in hypertension, atherosclerosis, and diabetes (13). The HO system is essential for iron homeostasis because it degrades iron into heme. Iron can be involved in reactive oxygen species formation and cause oxidative stress. Zhuang et al. found that EGb 761 increased HO-1 activity in a dose- and time-dependent manner and significantly reduced neuronal cell loss (14). They also found that the neuroprotective action in ischemic reperfusion injury model is dependent on its effect on HO-1 [this protective effect is abolished in HO-1 knockout mice; cf Ref. (3)]. These data are of interest not only for the treatment of ischemic brain injury but possibly also for AD because more and more lines of evidence suggest a relationship between vascular disorder and AD.

EGb 761 also activates SIRT1 (15), the target gene of resveratrol and caloric restriction that has been shown to be involved in longevity and possibly in AD (16).

17.3.4 ACTION ON THE AMYLOID CASCADE

Aβ, the main component of the senile plaques formed in AD, is synthesized from the precursor protein APP. The action of two particular enzymes, β- and γ- secretases, cleaves the Aβ sequence at the N- and C- terminal ends. Another enzyme, α-secretase, also cleaves APP within the Aβ sequence, preventing its formation. There are several types of Aβ, depending on the number of amino acids it is made up of. The two main types have 40 and 42 amino acids, respectively. Aβ42 aggregates more readily and is a more toxic form. While protein aggregation is the hallmark of neurodegenerative disorders, more recent research suggests that the most toxic forms are not the large aggregates but the soluble oligomers that can spread and damage synapses (17). The destructive action of Aβ appears to be the root cause of AD, which in turn leads to all the other alterations (including those resulting in the formation of tangles of hyperphosphorylated tau protein). From a therapeutic point of view, several possibilities present themselves: reducing Aβ formation (by inhibiting β- and γ-secretases, by activating α-secretase, or by indirect approaches, such as lowering cholesterol levels), inhibiting the formation or action of Aβ oligomers, inhibiting their toxic action, inhibiting fibrillation of Aβ, promoting Aβ degradation (by activating enzymes, such as neprilysin and insulin-degrading enzyme, or via the intermediary of apolipoprotein E, which Jiang et al. (18) have recently shown to promote proteolytic degradation of Aβ), changing its conformation to make it nonpathological (by acting on some of the chaperone proteins), and promoting its elimination (in the proteasome) or its clearance, among others. EGb 761 acts on several of these stages (19): it induces α-secretase activity (20); reduces blood cholesterol levels (21); inhibits Aβ oligomerization (22); prevents Aβ-induced cell death (23), in particular cell death caused by soluble oligomers (24,25); inhibits activation of caspases and hence apoptosis (26); decreases oligomerization and fibrillation of Aβ (22); and probably promotes Aβ degradation as it induces the release of apolipoprotein E (27).

17.3.5 ANTI-APOPTOSIS EFFECT

Regardless of their causes, cell death and synapse damage are decisive factors in AD pathology. EGb 761 protects neurons against cell death in several models, not only those involving the harmful action of Aβ (23). It affords protection against neuronal death due to apoptosis caused by several neurotoxic agents such as glutamate (28), nitric oxide (29), hydrogen peroxide (30,31), simvastastin (32), staurosporine (33), MPP+ and paraquat (34), and even HIV-1 (35). It not only protects against neuronal death but also increases the level of brain-derived neurotrophic factor (36), and it enhances hippocampal neurogenesis in both young (6 months) and old (22 months) transgenic mouse models of AD (TgAPP/PS1) (36). This action is mediated by activation of the cAMP response element binding protein (36,37) and could be of interest for the treatment of AD because this disease is related not only to neuronal death but also to a negative balance between neuronal death and impaired recovery [Rodriguez et al. (38) recently found an impaired neurogenesis in the dentate gyrus in the triple transgenic model of AD].

17.3.6 EFFECT ON THE BRAIN

EGb 761 protects not only neurons in various *in vitro* models but also the brain in vivo. This effect has been shown in cerebral ischemia models (3) and in the protection of the hippocampus (a structure that is vulnerable to the effects of aging and is involved in memory) in elderly mice (39), as well as in models of AD and HIV-associated neurological disease (35).

EGb 761 leads to a progressive reversal of the structural changes in dystrophic neurites associated with senile plaques (dense core plaques are associated with curvature of nearby neuritis) even by oral administration (8).

17.3.7 EFFECT ON MEMORY

EGb 761 protects the brain, but its effect is also seen in behavior. Several experiments on rodents have demonstrated a beneficial effect on memory in these animals, especially the elderly mice (40–43), and in Tg2576 AD model mice (44). Bilobalide, a component of EGb 761 (a GABA A receptor antagonist), diminishes cognitive impairment in Ts65Dn Down syndrome model mice (45). Along the same lines, it is remarkable that EGb 761 enhances LTP (an increase in synaptic efficacy that is a memory-associated pathway) exclusively in aged mice, not in the young ones (46).

Several studies on humans have demonstrated a beneficial effect on memory in subjects who are elderly (47,48) and suffering from dementia or MCI (49,50) or multiple sclerosis (51). In healthy subjects, the majority of studies have found a positive effect on cognitive abilities (52–54), with the exception of the study by Solomon et al. (55).

17.3.8 EFFECT ON ALZHEIMER'S DISEASE SYMPTOMS

EGb 761 has been tested as a symptomatic treatment for AD. Several clinical studies have demonstrated a beneficial effect on cognitive parameters assessed using ADAS-Cog (56,57) or such other scales as SKT and CGI (58–63). However, this effect was not found by Schneider et al. (64), in the context of a study in which no conclusion could be formed due to the fact that the placebo subjects showed no deterioration, McCarney et al. (65), and van Dongen et al. (66). It is worth noting, however, that the author of this negative study is also the co-author of the meta-analyses of the Cochrane collaboration database, which conclude that "there is promising evidence of improvement in cognition and function" under the influence of treatment with EGb 761 (67). The odds ratio for global improvement is exactly the same for EGb 761 (2.16) and for Donepezil (2.18) according to a Cochrane meta-analysis for this product (68).

According to the data of Schneider et al. (64), the effect of EGb 761 on cognitive functions could be particularly evident in AD patients with behavioral and psychological symptoms of dementia. But this is not confirmed by other recent trials.

17.4 CAN EXTRACT OF *GINKGO BILOBA* SLOW THE PROGRESSION OF ALZHEIMER'S DISEASE?

17.4.1 NEED FOR PREVENTION

Regardless of any symptomatic effects that EGb 761 may have, it would be much more useful to prove its effects on preventing AD or slowing its progression. The symptomatic drugs recognized by the U.S. Food and Drug Administration and other government agencies only have a slight effect on symptoms; they do not prevent the onset of dementia or prevent it from progressing. A preventive drug would be extremely useful, as it would significantly reduce the prevalence of the disease and hence the burden it represents in human and financial terms. From what we know of EGb 761's mode of action, this hope is a genuine possibility.

17.4.2 THEORETICAL ARGUMENTS SUPPORTING THE POSSIBILITY THAT GINKGO BILOBA EXTRACT HAS A PREVENTIVE EFFECT

We have many reasons to believe that antioxidants can prevent AD or have an effect on its progression, especially the transition from MCI to AD (69). The implication of reactive oxygen species in neuron aging, MCI, and AD seems to have been demonstrated at a theoretical level (69–71). At a practical level, several epidemiological surveys, although not all, have suggested a beneficial effect of antioxidant-rich foods such as fruit, vegetables, and wine (69). Finally, one clinical study concluded that vitamin E could slow the progression toward severe dementia (72).

17.4.3 EXPERIMENTAL ARGUMENTS IN SUPPORT OF THIS EFFECT

Experimental studies show that EGb 761 not only has an effect on behavior or neurochemical parameters. Contrary to symptomatic treatment of AD with drugs such as cholinesterase inhibitors, it acts on the pathogenic mechanisms (the amyloid cascade in particular) and protects the brain against several degenerative processes associated with aging and ischaemia or with neurodegenerative disorders. EGb 761 not only slows these degenerative processes or prevents them but also induces healing processes: it facilitates neurogenesis in the hippocampus (36) and reverses alterations linked with dystrophic neuritis around senile plaques (8).

17.4.4 RESULTS OF EPIDEMIOLOGICAL STUDIES IN HUMANS

17.4.4.1 EPIDOS Study

Two epidemiological studies have demonstrated a significant effect of EGb 761 on prevention of AD. They involved sizeable population samples, and it is worthwhile noting that they were set up independently of funding by any pharmaceutical laboratory that sells EGb 761 (although with help from other laboratories).

The first of these studies used the EPIDOS cohort from 1992 to 1999, involving 7598 French women in good health and older than 75 years, in order to evaluate the risk factors for fracture of the femur neck. The 1462 women of the Toulouse site were then regularly monitored, and nearly half of them ($n = 714$) underwent prospective longitudinal evaluation of any cognitive disorders and evolution toward dependency and/or dementia. After potential confounding factors were included in a multivariate analysis, women with dementia were found to have had less continuous exposure to vasodilators, including EGb 761 (odds ratio = 0.31 – CI = 95%, p = 0.018). If EGb 761 treatment is isolated from other vasodilators, the results are comparable (odds ratio = 0.38 – CI = 95%) but not statistically significant, probably due to insufficient population size (73). This demonstration of a preventive effect could appear as a surprise since there is a strong bias against the drug in this study: many women of the EPIDOS cohort were probably taking EGb 761 because they had memory problems, which could be a sign of the beginning of AD (such a bias does not exist in the case of epidemiological studies involving anti-inflammatory drugs or estrogens). This means that the preventive effect of EGb 761 could be very important.

17.4.4.2 PAQUID Study

A second study used the PAQUID cohort (74). The participants were 3534 subjects, aged 65 years or older, representative of Gironde and Dordogne (in the southwest of France). Of the sample, 225

(6.4%) consumed EGb 761 and 888 (25.1%) did other treatments for memory impairment at base-line. Similar to the EPIDOS study, there is in this one a strong bias against the drugs taken for the treatment of memory impairment (people who received these treatments were at higher risk for dementia than the group who took no treatment). One of the main interests of Dartigues et al. (74) on this PAQUID cohort was to make possible a comparison between the group of people who took EGb 761 and subjects who took other drugs from the same category (historically classified in France as vasodilators) because the bias is the same for these two categories. In this particular study (and contrary to the data of the EPIDOS study), the initial consumption of EGb 761 did not modify the risk of dementia but the consumption of other treatments for memory impairment was associated with a higher risk of dementia. The logical explanation is that this increase is probably related to the original bias of inclusion (at-risk people more numerous in the treatments groups), not to a negative effect of vasodilators. But the beneficial effect of EGb 761 compensated this negative bias. More important, Dartigues et al. (74) investigated the probability of survival without dementia. This is one of the most important criteria for such a study because dementia is associated to a higher risk of premature death. Consequently, a drug with a beneficial effect on survival will be associated with a higher risk of dementia just because it protects people against death. On the other hand, drugs with a toxic effect can be evaluated statistically protective against dementia just because they kill the more affected people. Remarkably, EGb 761 is associated with a higher probability of survival (relative risk = 0.76, 95% CI = 0.62–0.93), information that fits well with animal experiments: EGb 761 increases life span in the rat (42) and in *C. elegans* (75). In the PAQUID study, this criterion is the only one, with the well-known female bias, associated with a lower risk of mortality. All the other characteristics are neutral (education, MMS at the baseline, married or not, depressive symptoms) or negative (disability, diabetes, high number of medications). When dementia and mortality are taken together (survival without dementia), EGb 761 still appears beneficial (RR = 0.86, against 1.08 for the consumers of other treatments for memory impairment). Globally, the PAQUID study confirms the beneficial effect observed in the EPIDOS study even whether some differences exist (mainly the fact that other vasodilators were beneficial in the EPIDOS study; one possible explanation is that this particular study only included women).

17.4.5 INTERVENTION STUDIES

17.4.5.1 Oregon Study

The existence of a preventive effect cannot be proven by epidemiological observations alone. Large interventional studies involving many individuals who are randomly assigned (and thus without bias) to either a treatment group or the control group and followed for several years are required.

Three interventional studies have been undertaken: the Oregon study and the GEM study in United States and the GuidAge study in France. The smaller one, the Oregon study, is currently achieved. In this study, Dodge et al. (76) performed a randomized, placebo-controlled and double-blind 42-month pilot study with 118 cognitively intact [clinical dementia rating (CDR) = 0] subjects aged 85 years or older. Due to the small number of people, the beneficial effect of Ginkgo extract (240 mg daily) did not reach statistical significance (p = 0.06) in intention-to-treat analysis. However, in the secondary analysis, when the authors controlled the medication adherence level, the treated group had a lower risk of progression from a CDR of 0 to that of 0.5 (hazard ratio = 0.33, p = 0.02) and a smaller decline in memory scores (p = 0.04). Kaye et al. (77) also performed imaging studies on the same groups of people. Using MRI, they assessed the volumes of total brain, ventricular volume, white matter high signal, and hippocampus. There was a significantly (p = 0.02) lower rate of total brain volume in the treated group versus the placebo group. According to Kaye et al., their data "support the clinical efficacy data adjusted for mediation adherence" shown in their previous study and "suggest a 'disease-modifying' effect in a randomized controlled trial" (77).

17.4.5.2 GEM Study

Larger prevention trials with greater statistical power, such as GEM and GuidAge, are indeed needed to confirm these observations. The GEM study tested the effectiveness of 120 mg of EGb 761 twice daily versus placebo in lowering the incidence of AD and or dementia in normal elderly people or those with MCI. This randomized, double-blind and placebo-controlled clinical trial conducted in four academic sites in the United States involved a total of 3071 subjects age 75 years or older followed up for 8 years. Neuropsychological tests were performed every 6 months. After 8 years, more than 440 persons developed dementia. Mortality and subjects lost to follow-up were well below expected rates, and EGb 761 was well tolerated (78,79). Unfortunately, the data recently published are negative (80). *G. biloba* showed no benefit for reducing all-cause dementia (hazard ratio = 1.12) or dementia of the Alzheimer type. The only beneficial effect found was a reduction of the number of vascular dementia without Alzheimer's dementia (hazard ratio = 0.41, p = 0.05).

17.4.5.3 GuidAge Study

The GuidAge study also assesses the efficacy of EGb 761 (120 mg twice daily) in the prevention of AD. The participants are 2854 patients, aged 70 years or older, with a spontaneous memory complaint. They are followed up for 5 years. The results are expected in 2010.

17.4.5.4 Limits of Current Intervention Studies

Despite the large size of these studies, they may not provide conclusive results, principally because it has not been possible to examine different doses and because of the selection of "patients." Selecting normal and young subjects means extremely long and expansive studies because few people will become demented in 5 or even 10 years. Selecting people with MCI, memory complaint, or any pathological symptoms means that they already started AD and it could be too late for having a preventive effect. Furthermore, in so far as the normal diet includes efficient antioxidants, notably flavonoids, it will not be easy to distinguish the specific effects of EGb 761. Positive results will provide evidence of efficacy but will not indicate the most efficient dose or the age at which treatment should start.

17.5 SUMMARY

The search for treatments that could prevent AD or slow its progression is essential. However, it is a particularly arduous task given the clinical trials required for proof. Given the current state of knowledge, EGb 761 would seem to be an appropriate treatment of choice for at least two reasons: (a) from what we know about its pharmacological effects, it could be used as a drug to prevent and slow the progression of AD, and (b) the absence of any major side effects is a clear advantage, as a drug of this nature would need to be administered in the long term and in elderly patients who are relatively fragile. Unfortunately, the GEM study found no beneficial effect. We are now waiting for the data of the GuidAge study.

REFERENCES

1. DeFeudis, F. V., *Ginkgo biloba Extract (EGb 761): From Chemistry to the Clinic*, Ullstein Medical, Wiesbaden, 1998.
2. Christen, Y., *Ginkgo biloba* extract and Alzheimer's disease: Is the neuroprotection explained merely by antioxidant action? In: *Oxidative Stress and Age-Related Neurodegeneration* (Y. Luo and L. Packer, Eds.), Taylor & Francis, Boca Raton, 43–58, 2006.
3. Saleem, S., Zhuang, H., Biswal, S., Christen, Y., and Doré S., *Ginkgo biloba* neuroprotective action is dependent on heme oxygenase 1 in ischemic reperfusion brain injury. *Stroke*, 39, 3389–3396, 2008.

4. Pietri, S., Seguin, J. R., d'Arbigny, P., Drieu, K., and Culcasi, M., *Ginkgo biloba* extract (EGb 761) pretreatment limits free radical-induced oxidative stress in patients undergoing coronary bypass surgery. *Cardiovasc. Drugs Therapy*, 11, 121–131, 1997.

5. Kudolo, G. B., Wang, W., Dorsey, S., and Blodgett, J., Oral ingestion of *Ginkgo biloba* extract reduces thiobarbituric acid reacting (TBAR) substances in washed platelets of healthy subjects. *J. Herbal Pharmacother.*, 3, 1–15, 2003.

6. Smith, J. V., and Luo, Y., Elevation of oxidative free radicals in Alzheimer's disease models can be attenuated by *Ginkgo biloba* extract EGb 761. *J. Alzheimer Dis.*, 5, 287–300, 2003.

7. Ramassamy, C., Averill, D., Beffert, U., Bastianetto, S., Theroux, L., Lussier-Cacan, S., Cohn, J. S., Christen, Y., Davignon, J., Quirion, R., and Poirier, J., Oxidative damage and protection by antioxidants in the frontal cortex of Alzheimer's disease is related to the apolipoprotein E genotype. *Free Radic. Biol. Med.*, 27, 544–553, 1999.

8. Garcia-Alloza, M., Dodwell, S. A., Meyer-Luehmann, M., Hyman, B. T., and Backsai, B. J., Plaque-derived oxidative stress mediates distorted neurite trajectories in the Alzheimer mouse model. *J. Neuropathol. Exp. Neurol.*, 1082–1089, 2006.

9. Mizuno, M., and Packer, L., *Ginkgo biloba* extract EGb 761 is a suppressor of AP-1 transcription factor stimulated by PMA. *Biochem. Molec. Biol. Int.*, 39, 395–401, 1996.

10. Watanabe, C. M. H., Wolffram, S., Ader, P., Rimbach, G., Packer, L., Maguire, J. J., Schultz, P. G., and Gohil, K., The in vivo neuromodulatory effects of the herbal medicine ginkgo biloba. *Proc. Natl. Acad. Sci. U S A*, 98, 6577–6580, 2001.

11. Stein, T. D., and Johnson, J. A., Lack of neurodegeneration in transgenic mice overexpressing mutant amyloid precursor protein is associated with increased levels of transthyretin and the activation of cell survival pathways. *J. Neurosci.*, 22, 7380–7388, 2002.

12. Buxbaum, J. N., Ye, Z., Reixach, N., Friske, L., Levy, C., Das, P., Golde, T., Masliah, E., Roberts, A. R., and Bartfai, T., Transthyretin protects Alzheimer's mice from the behavioral and biochemical effects of A{beta} toxicity, *Proc. Natl. Acad. Sci. U S A*, 105, 2681–2686, 2008.

13. Schipper, H. M., Heme oxygenase-1: Role in brain aging and neurodegeneration. *Exp. Gerontol.*, 35, 821–830, 2000.

14. Zhuang, H., Pin, S., Christen, Y., and Dore, S. Induction of heme oxygenase 1 by *Ginkgo biloba* in neuronal cultures and potential implications in ischemia. *Cell. Mol. Biol.*, 48, 647–653, 2002.

15. Longpré, F., Garneau, P., Christen, Y., and Ramassamy, C., Protection by EGb 761 against beta-amyloid-induced neurotoxicity: Involvement of NF-kappaB, SIRT1 and MAPKs pathways and inhibition of amyloid fibril formation. *Free. Radic. Biol. Med.*, 41, 1781–1794, 2006.

16. Qin, W., Yang, T., Ho, L., Zhao, Z., Wang, J., Chen, L., Zhao, W., Thiyagarajan, M., MacGrogan, D., Rodgers, J. T., Puigserver, P., Sadoshima, J., Deng, H., Pedrini, S., Gandy S., Sauve, A. A., and Pasinetti, G. M., Neuronal SIRT1 activation as a novel mechanism underlying prevention of Alzheimer disease amyloid neuropathology by calorie restriction. *J. Biol. Chem.*, 281, 21745–21754, 2006.

17. Selkoe, D. J., Triller, A., and Christen, Y., Eds., *Synaptic Plasticity and the Mechanism of Alzheimer's Disease*, Springer, Heidelberg, 2008.

18. Jiang, Q., Lee, C. Y. D., Mandrekar, S., Wilkinson, B., Cramer, P., Zelcer, N., Mann, K., Lamb, B., Wilson, T. M., Collins, J. L., Richardson, J. C., Smith, J. D., Comery, T. A., Riddell, D., Holtzman, D. M., Tontonoz P., and Landreth, G. E., ApoE promotes the proteolytic degradation of Aβ. *Neuron*, 58, 681–693, 2008.

19. Ramassamy, C., Longpré, F., and Christen Y., *Ginkgo biloba* extract (EGb 761) in Alzheimer's disease: Is there any evidence? *Curr. Alzheimer Res.*, 4, 253–262, 2007

20. Colciaghi, F., Borroni, B., Zimmermann, M., Belkner, A., Longhi, C., Padovani, A., Cattabeni, F., Christen, Y., and Di Luca, M., Amyloid precursor protein metabolism is regulated toward alpha-secretase pathway by *Ginkgo biloba* extracts. *Neurobiol. Dis.*, 16, 454–460, 2004.

21. Yao, Z. X., Han, Z., Drieu, K., and Papadopoulos, V., *Ginkgo biloba* extract (EGb 761) inhibits beta-amyloid production by lowering free cholesterol levels. *J. Nutr. Biochem.*, 15, 749–756, 2004.

22. Wu, Y., Wu, Z., Butko, P., Christen, Y., Lambert, M., Klein, W., Link, C. D., and Luo, Y., Amyloid β-induced pathological behaviors are suppressed by *Ginkgo biloba* extract EGb 761 and ginkgolide in transgenic *Caenorhabditis elegans*. *J. Neurosci.*, 26, 13102–13113, 2006.

23. Bastianetto, S., Ramassamy, C., Doré, S., Christen, Y., Poirier, J., and Quirion, R., The *Ginkgo biloba* extract (EGb 761) protects hippocampal neurons against cell death induced by β-amyloid. *Eur. J. Neurosci.*, 12, 1882–1890, 2000.

24. Yao, Z. X., Drieu, K., and Papadopoulos, V., The *Ginkgo biloba* extract EGb 761 rescues the PC 12 neuronal cells from β-amyloid-induced cell death by inhibiting the formation of β-amyloid-derived diffusible neurotoxic ligands. *Brain Res.,* 889, 181–190, 2001.

25. Chromy, B. A., Nowak, R. J., Lambert, M. P., Viola, K. L., Chang, L., Velasco, P. T., Jones, B. W., Fernandez, S. J., Lacor, P. N., Horowitz, P., Finch, C. E., Krafft, G. A., and Klaein, W. L., Self-assembly of Aβ(1–42) into globular neurotoxins. *Biochemistry,* 42, 12749–12760, 2003.

26. Luo, Y., Smith, J. V., Paramasivan, V., Burdick, A., Curry, K. J., Bufford, J. P., Khan, I., Netzer, W. J., Xu, H., and Butko, P., Inhibition of amyloid-β aggregation and caspase-3 activation by the *Ginkgo biloba* extract EGb 761. *Proc. Natl. Acad. Sci. U S A,* 99, 2197–2202, 2002.

27. Ramassamy, C., Delisle, M. C., Schoofs, A., Christen, Y., and Poirier, J., Effects of *Ginkgo biloba* extract (EGb 761) on apolipoprotein E and apolipoprotein J release by astrocytes after lipoperoxidation and on the level of hippocampal apolipoprotein E after entorhinal cortex "lesioning." In: *Effects of Ginkgo biloba Extract (EGb 761) on Neuronal Plasticity* (Y. Christen, M.-T. Droy-Lefaix and J. F. Macias-Nunez, Eds.), Elsevier, Paris, 53–60, 1996.

28. Ahlemeyer, B., Mowes, A., and Krieglstein, J., Inhibition of serum deprivation and staurosporine-induced neuronal apoptosis by *Ginkgo biloba* extracts and some of its constituents. *Eur. J. Pharmacol.,* 367, 423–430, 1999.

29. Bastianetto S., Zheng, W. H. and. Quirion, R., The *Ginkgo biloba* extract (EGb 761) protects and rescues hippocampal cells against nitric oxide-induced toxicity: Involvement of its flavonoid constituents. *J. Neurochem.,* 74, 2268–2277, 2000.

30. Bastianetto, S., Ramassamy, C., Christen, Y., Poirier, J., and Quirion, R., *Ginkgo biloba* extract (EGb 761) prevents cell death induced by oxidative stress in hippocampal neuronal cell cultures. In: *Advances in Ginkgo biloba Extract Research. Ginkgo biloba Extract (EGb 761) Study: Lessons from Cell Biology* (L. Packer and Y. Christen, Eds.), Elsevier, Paris, 85–99, 1998.

31. Oyama, Y., Chikahisa, L., Ueha, T., Kanemaru, K., and Noda, K., *Ginkgo biloba* extract protects brain neurons against oxidative stress induced by hydrogen peroxide. *Brain Res.,* 712, 349–352, 1996.

32. Altiok, N., Ersoz, M., Karpuz, V., and Koyuturk, M., *Ginkgo biloba* extract regulates differentially the cell death induced by hydrogen peroxide and simvastatin. *Neurotoxicology,* 27, 158–163, 2006.

33. Massieu, L., Moran, J., and Christen, Y., Effect of *Ginkgo biloba* (EGb761) on staurosporine-induced neuronal death and caspase activity in cortical cultured neurons. *Brain Res.,* 1002, 76–85, 2004.

34. Gagné, B., Gélinas, S., Lagacé, B., Chiasson, K., Ramassamy, C., and Martinolli, M.-G., Effect of quercetin, kaempferol and *Ginkgo biloba* extracts on MPP⁺ and paraquat induced cytotoxicity in PC12 cells. *Soc. Neurosci.,* 28, no. 100, 1, 2002.

35. Zou, W., Kim, B. O., Zhou, B. Y., Liu, Y., Messing, A., and He, J. J., Protection of *Ginkgo biloba* extract EGb 761 against HIV-1 Tat neurotoxicity: Involvement of glial fibrillary acidic protein. *Am. J. Pathol.,* 171, 1923–1935, 2007.

36. Tchantchou, F., Xu, Y., Wu, Y., Christen, Y., and Luo, Y., EGb 761 enhances adult hippocampal neurogenesis and phosphorylation of CREB in transgenic mouse model of Alzheimer's disease. *FASEB J.,* 21, 2400–2408, 2007.

37. Xu, Y., Cui C., Pang, C., Christen, Y., and Luo Y., Restoration of impaired phosphorylation of cyclic AMP response element-binding protein (CREB) by EGb 761 and its constituents in Aβ-expressing neuroblastoma cells. *Eur. J. Neurosci.,* 26, 2931–2939, 2007.

38. Rodriguez, J. J., Jones, V. C., Tabuchi, M., Allan, S. M., Knight, E. M., LaFerla, F. M., Oddo, S., and Verkhratsky, A., Impaired adult neurogenesis in the dentate gyrus of a triple transgenic mouse model of Alzheimer's disease. *PLoS One,* 3, e2935, 2008; doi: 10.1371/journal.pone.0002935.

39. Barkats, M., Venault, P., Christen, Y., and Cohen-Salmon, C., Effects of long term treatment with EGb 761 on age-dependent structural changes in the hippocampi of three inbred mouse strains. *Life Sci.,* 56, 213–222, 1995.

40. Pardon, M. C., Barkats, M., Venault, P., Christen, Y., and Cohen-Salmon, C., Effect of long-term treatment with *Ginkgo biloba* extract (EGb 761) on age-dependent structural changes in the hippocampus and spatial memory performance of inbred mice. In: *Advances in Ginkgo biloba Extract Research. Effects of Ginkgo biloba Extract (EGb 761) on Neuronal Plasticity* (Y. Christen, M.-T. Droy-Lefaix and J. F. Macias-Nunez, Eds.), Elsevier, Paris, 21–34, 1996.

41. Winter, E., Effect of an extract of *Ginkgo biloba* on learning and memory in mice. *Pharmacol. Biochem. Behav.,* 38, 109–114, 1991.

42. Winter, J. C., The effects of an extract of *Ginkgo biloba*, EGb 761, on cognitive behavior and longevity in the rat. *Physiol. Behav.,* 63, 425–433, 1998.

43. Wirth, S., Stemmelin, J., Will, B., Christen, Y., and Di Scala, G., Facilitative effects of EGb 761 on olfactory recognition in young and aged rats. *Pharmacol. Biochem. Behav.,* 65, 321–326, 2000.

44. Stackman, R. W., Eckenstein, F., Frei, B., Kulhanek, D., Nowlin, J., and Quinn, J. F., Prevention of age-related spatial memory deficits in a transgenic mouse model of Alzheimer's disease by chronic *Ginkgo biloba* treatment. *Exp. Neurol.,* 184, 510–520, 2003.

45. Fernandez, F., Morishita, W., Zuniga, E., Nguyen, J., Blank, M., Malenka, R. C., and Garner, C. C., Pharmacotherapy for cognitive impairment in a mouse model of Down syndrome. *Nat. Neurosci.,* 10, 411–413, 2007.

46. Williams, B., Watanabe, C. M. H., Schultz, P. G., Rimbach, G., and Krucker, T., Age-related effects of *Ginkgo biloba* extract on synaptic plasticity and excitability. *Neurobiol. Aging*, 25, 955–962, 2004.

47. Israel, L., Dell'Accio, E., Martin, G., and Hugonot, R., Extrait de *Ginkgo biloba* et exercices d'entraînement de la mémoire. Evaluation comparative chez des personnes âgées ambulatoires. *Psychol. Méd.,* 19, 1431–1439, 1987.

48. Allain, H., Raoul, P., Lieury, A., LeCoz, F., Gandon, J. M., and d'Arbigny P., Effect of two doses of *Ginkgo biloba* extract (EGb 761) on the dual-coding test in elderly subjects. *Clin. Ther.,* 15, 549–558, 1993.

49. Rai, G. S., Shovlin, C., and Wesnes, K. A., A double-blind, placebo controlled study of *Ginkgo biloba* extract (tanakan) in elderly outpatients with mild to moderate memory impairment. *Curr. Med. Res. Opin.,* 12, 350–355, 1991.

50. Ercoli, L. M., Small, G. W., Silverman, D. H. S., Siddarth, P., Dorsey, D., Miller, K., Kaplan, A., Skura, S., Byrd, G., Huang, S. C., and Phelps, M. E., The effects of *Ginkgo biloba* on cognitive and cerebral metabolic function in age-associated memory impairment. *Soc. Neurosci. Nov.,* 9, 2003; abstract no. 127.11.

51. Kenney, C., Norman, M., Jacobson, M., Lampinen, S., Nguyen, N. P., and Corey-Bloom, J., A double-blind, placebo-controlled, modified crossover pilot study of the effects of *Ginkgo biloba* on cognitive and functional abilities in multiple sclerosis. *Neurology,* 58, A458, 2002.

52. Mix, J. A., and Crews, Jr., W. D., A double-blind, placebo-controlled, randomized trial of *Ginkgo biloba* extract EGb 761 in a sample of cognitively intact older adults: Neuropsychological findings. *Human Psychopharmacol.,* 17, 267–277, 2002.

53. Cieza, A., Maier, P., and Pöppel, E., Effects of *Ginkgo biloba* on mental functioning in healthy volunteers. *Arch. Med. Res.,* 24, 373–381, 2003.

54. Rigney, U., Kimber, S., and Hindmarch, I., The effects of acute doses of standardized *Ginkgo biloba* extract on memory and psychomotor performance in volunteers. *Phytother. Res.,* 13, 408–415, 1999.

55. Solomon, P. R., Adams, F., Silver, A., Zimmer, J., and De Veaux, R., *Ginkgo* for memory enhancement: A randomized controlled trial. *JAMA,* 288, 835–840, 2002.

56. Le Bars, P. L., Katz, M. M., Berman, N., Itil, T., Freedman, A. M., and Schalzberg, A. F., A placebo-controlled double-blind, randomized trial of an extract of *Ginkgo biloba* for dementia. *JAMA,* 278, 1327–1332, 1997.

57. Wettstein, A., Cholinesterase inhibitors and *Ginkgo* extracts — Are they comparable in the treatment of dementia? Comparison of published placebo-controlled efficacy studies of at least six months' duration. *Phytomedicine,* 6, 393–401, 2000.

58. Taillandier, J., Ammar, A., Rabourdin, J. P., Ribeyre, J. P., Pichon, J., Niddam, S., and Pierart, H., Traitement des troubles du vieillissement cerebral par l'extrait de *Ginkgo biloba*. Etude longitudinale multicentrique à double insu face au placebo. *Presse Med.,* 15, 1583–1587, 1986.

59. Hofferberth, B., The efficacy of EGb 761 in patients with senile dementia of the Alzheimer type. A double-blind, placebo-controlled study on different levels of investigation. *Human Psychopharmacol.,* 9, 215–222, 1994.

60. Kanowski, S., and Hoerr, R., *Ginkgo biloba* extract EGb 761 in dementia: Intent-to-treat analyses of a 24-week, multi-center, double-blind, placebo-controlled, randomized trial. *Pharmacopsychiatry,* 36, 297–303, 2003.

61. Maurer, K., Ihl, R., Dieks, T., and Frölich, L., Clinical efficacy of *Ginkgo biloba* special extract EGb 761 in dementia of the Alzheimer type. *J. Psychiatr. Res.,* 11, 645–655, 1997.

62. Mazza, M., Capuano, A., Bria, P., and Mazza, S., *Ginkgo biloba* and donepezil: A comparison in the treatment of Alzheimer's dementia in a randomized placebo-controlled double-blind study. *Eur. J. Neurol.,* 13, 981, 2006.

63. Ihl, R., Tribanek, M., and Napryeyenko, O., A 240 mg once-daily formulation of *Ginkgo biloba* extract EGb 761 is effective in both Alzheimer's disease and vascular dementia: Results from a randomized trial. *Alzheimers Dement.,* 4, T167, 2008.

64. Schneider, L. S., DeKosky, S. T., Farlow, M. R., Tariot, P. N., Hoerr, R., and Kieser, M., A random-ized, double-blind placebo-controlled trial of two doses of *Ginkgo biloba* extract in dementia of the Alzheimer's type. *Curr. Alzheimer Res.,* 5, 541–551, 2005.

65. McCarney, R., Fisher, P., Iliffe, S., van Haselen, R., Griffin, M., van der Meulen J., and Walker, J., *Ginkgo biloba* for mild to moderate dementia in a community setting: A pragmatic, randomised, parallel-group, double-blind, placebo-controlled trial. *Int. J. Geriatr. Psychiatry,* 2008; doi: 10.10.1002/gps.2055.

66. van Dongen M., van Rossum, E., Kessels, A., Sielhorst, H., and Knipschild, P., The efficacy of ginkgo elderly people with dementia and age-associated memory impairment: New results of a randomized clinical trial. *J. Am. Geriat. Soc.,* 56, 1154–1166, 2000.

67. Birks, J., Grimley Evans, J., and Van Dongen, M., *Ginkgo biloba* for cognitive impairment and dementia (Cochrane Review). In: *The Cochrane Library,* no. 4, Update Software, Oxford, 2002.

68. Birks, J. S., and Harvey, R., Donepezil for dementia due to Alzheimer's disease. *Cochrane Database Syst. Rev.,* (4), CD001190, 2000.

69. Ancelin, M.-L., Christen, Y., and Ritchie, K., Is antioxidant therapy a viable alternative for mild cognitive impairment? Examination of the evidence. *Dement. Geriatr. Cogn. Disord.,* 24, 1–19, 2007.

70. Christen, Y., Oxidative stress and Alzheimer's disease. *Am. J. Clin. Nutr.,* 71 (Suppl.), 621S–629S, 2000.

71. Luo, Y., and Packer, L., Eds. *Oxidative Stress and Age-Related Neurodegeneration.* Taylor & Francis, Boca Raton, 2006.

72. Sano, M., Ernesto, C., Thomas, R. G., Klauber, M. R., Schafer, K., Grundman, M., Woodbury, P., Growdon, J., Cotman, C. W., Pfeiffer, E., Schneider, L. S., and Thal, L. J., A controlled trial of selegiline, alpha-tocopherol, or both as treatment for Alzheimer's disease. The Alzheimer's Disease Cooperative Study. *N. Engl. J. Med.,* 336, 1216–1222, 1997.

73. Andrieu, S., Gillette, S., Amouyal, K., Nourhashemi, F., Reynish, W., Ousset, P. J., Albarède, J.-L., Vellas, B., and Grandjean, H., Association of Alzheimer's disease onset with ginkgo biloba and other symptomatic cognitive treatments in a population of women aged 75 years and older from the EPIDOS Study. *J. Gerontol. A Biol. Sci. Med. Sci.,* 58, 372–377, 2003.

74. Dartigues, J.-F., Carcaillon, L., Helmer, C., Lechevallier, N., Lafuma, A., and Khoshnood, B., Vasodilators and nootropics as predictors of dementia and mortality in the PAQUID cohort. *J. Am. Geriat. Soc.,* 55, 395–399, 2007.

75. Smith, J. V., and Luo, Y., *Ginkgo biloba* extract EGb 761 extends life span and attenuates H_2O_2 levels in a *Caenorhabditis elegans* model of Alzheimer's disease. In: *Oxidative Stress and Age-Related Neurodegeneration* (Y. Luo and L. Packer, Eds.), Taylor & Francis, Boca Raton, 301–326, 2006.

76. Dodge, H. H., Zitzelberger, T., Oken, B. S., Howieson, D., and Kaye, J., A randomized placebo-controlled trial of *Ginkgo biloba* for the prevention of cognitive decline. *Neurology,* 70, 1809–1817, 2008.

77. Kaye, J. A., Dodge, H., Zitzelberger, T., and Moore, M., MRI evidence for a disease modifying effect of *Ginkgo biloba* extract in a dementia prevention trial. *Alzheimers Dement.,* 4, T4, 2008.

78. DeKosky, S. T., Williamson, J., Fitzpatrick, A., Ives, D. G., Saxton, J., Lopez, O. L., Burke, G., Carlson, M., Fried, L., Kuller, L., Robbins, J., Tracy, R., Dunn, L. O., Kronmal, R., Nahin, R., and Furberg, C., The *Ginkgo* in Evaluation of Memory (GEM) study: Primary outcomes in a longitudinal dementia prevention trial. *Alzheimers Dement.,* 4, T165, 2008.

79. DeKosky, S. T., Fitzpatrick, A., Ives, D. G., Saxton, J., Williamson, J., Lopez, O. L., Burke, G., Fried, L., Kuller, L. H., Robbins, J., Tracy, R., Woolard, N., Dunn, L., Kronmal, R., Nahin, R., Furberg, C., GEMS Investigators, The Ginkgo Evaluation of Memory (GEM) study: Design and baseline data of a randomized trial of *Ginkgo biloba* extract in prevention of dementia. *Contemp. Clin. Trials,* 27, 238–253, 2006.

80. DeKosky, S. T., Williamson, J. D., Fitzpatrick, A., Kronmal, R. A., Ives, D. G., Saxton, J., Lopez, O. L., Burke, G., Carlson, M. C., Fried, L., Kuller, L. H., Robbins, J., Tracy, R., Woolard, N., Dunn, L., Snitz, B. E., Nahin, R., and Furberg, C., for the GEM Study Investigators, *Ginkgo biloba* for prevention of dementia. A randomized controlled trial. *JAMA,* 300, 2253–2262, 2008.

18 Green Tea Polyphenols Protect Neurons against Alzheimer's Disease and Parkinson's Disease

Baolu Zhao

State Key Laboratory of Brain and Cognitive Science,
Institute of Biophysics, Academia Sinica, Beijing, China

CONTENTS

18.1 INTRODUCTION

Throughout history, tea has been one of the most popular drinks all over the world. Epidemiological studies have shown that tea is beneficial to human health by protecting the body from cancer and other diseases. Most people thought that the beneficial effect of tea came from the vitamins and trace elements in it, which is not correct and complete. The most effective components in green tea are polyphenols, which are about 30% of the weight of dry tea; the four main components of polyphenols are EC [(–)-epicatechin], EGC [(–)-epigallocatechin], EGC [(–)-epicatechin gallate], and EGCG [(–)-epigallocatechin gallate] (Figure 18.1). The main polyphenol in black tea is theaflavin, the structure of which is shown in Figure 18.1. This chapter summarizes the health effects of green tea polyphenols (GTPs) and reviews their possible antioxidant mechanisms.

The maximum concentrations of catechins EGCG, EGC, and EC in the plasma of healthy people reach about 77.9±22.2, 223.4±35.2, and 124.03±7.86 ng/ml, respectively, and the half-lives are 3.4±0.3, 1.7±0.4, and 2.0±0.4 h, respectively, after green tea consumption (20 mg/kg) for 1.3–1.6 h.

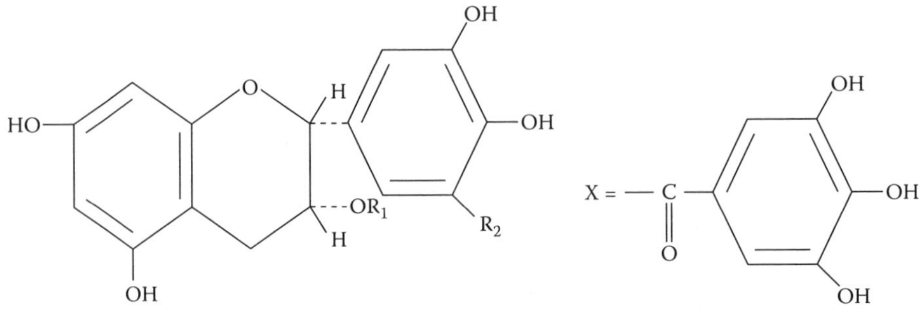

(EC): (−)-Epicatechin R_1 = H, R_2 = H;

(EGC): (−)-Epigallocatechin R_1 = H, R_2 = OH;

(ECG): (−)-Epicatechin gallate R_1 = X, R_2 = H;

(EGCG): (−)-Epigallocatechin gallate R_1 = X, R_2 = OH.

FIGURE 18.1 Chemical structure of green TC.

About 77% EGCG, 31% EGC, and 21% EC are still in their original structure after 1 h. 4′-O-methyl-EGC was detected in the plasma and urea, and it reached a maximum concentration (1.2–2.2 µg/ml) at about 1.7 h, and the half-life was about 4.4±1.1 h (1).

Radioactivity levels of about 30.7%, 40.6%, and 3.9% were detected in the stomach, small intestine, and colon, respectively, and just a little was found in other tissues such as the brain after ^{3}H-EGCG was fed to mice through the stomach for 1 h. However, after 6 h, the concentration increased rapidly and then slowly reached a maximum of about 0.22 µM in the brain at 24 h. If drank several times after the first consumption, the concentration of ^{3}H-EGCG in the brain will be about six times higher than that for just a single consumption (2). These results show that GTP can not only enter into plasma but also penetrate the blood-brain barrier, which is necessary for treating neurodegenerative diseases.

18.2 OXIDATIVE STRESS IN AD AND PD

Extra free radicals undoubtedly result in heavy injury to biological bodies. Free radicals peroxidize membrane lipids (1) and oxidize proteins (3), resulting in damage of plasma membrane and cross-linking of cytoskeletal proteins. In addition, free radicals damage RNA (4), nuclear (5), and DNA (6). In the brain, the high metabolic rate, the low concentrations of glutathione and antioxidant enzyme catalase, and the high proportion of polyunsaturated fatty acids make the brain tissue particularly susceptible to oxidative damage (7,8). Oxidative stress, an imbalance toward the pro-oxidant side of the pro-oxidant/antioxidant homeostasis, occurs in several brain neurodegenerative disorders. Among these neurodegenerative brain disorders are those in which protein aggregation is observed, including Alzheimer's disease (AD) and Parkinson's disease (PD).

18.2.1 OXIDATIVE STRESS AND AD

Accumulated data demonstrated that oxidative damage occurs in AD brain (6). Aβ peptide has been proven to produce hydrogen peroxide (H_2O_2) through metal ion reduction, with concomitant release of thiobarbituric acid-reactive substances (TBARSs), a process probably mediated by formation of hydroxyl radicals (9,10). The cytotoxicity of Aβ fibrils has also been implicated to be relative to an oxidative mechanism. Aβ fibril-induced H_2O_2 was detected by several laboratories (4,5). There are considerable lines of evidence consistent with the importance of oxidative stress in the pathology of AD [for recent reviews, see Refs. (6)–(8)]. Evidence supporting the notion of free radical oxidative stress in AD brain includes increased redox-active metal ions in AD brain; increased lipid peroxidation detected by decreased levels of polyunsaturated fatty acids and increased levels of the lipid peroxidation products, acrolein, TBARSs, and iso- and neuroprostanes; increased protein oxidation; increased oxidation of DNA and RNA; and decreased activity of oxidatively prone enzymes, such as glutamine synthetase.

Oxidation of proteins is normally caused by free radicals, and this process, from a chemical thermodynamics standpoint, is an exothermic event. Oxidative reactions of peptides are mediated mainly by the hydroxyl radical (OH). There are two possible oxidative pathways that can occur: (a) backbone oxidation and (b) side-chain oxidation. Backbone oxidation is initiated by carbon abstraction of hydrogen by the free radical, leading to the formation of a carbon-centered radical. In the presence of oxygen, this radical is converted to a peroxyl radical. This can lead to the formation of an alkoxyl radical and subsequent hydroxylation of the peptide backbone. The oxidation of amino acid side chains greatly depends on their structure. An important oxidative process with profound functional and structural consequences involves the irreversible nitration of tyrosine residues by peroxynitrite ($ONOO^-$) (9). The levels of protein oxidation in membrane systems can be indirectly monitored by use of electron paramagnetic resonance spin-labeling techniques (10). The changes in protein conformation due to protein oxidation are deduced from the relevant electron paramagnetic resonance parameter.

18.2.2   Oxidative Stress and PD

PD is a progressive neurodegenerative disorder; the hallmark of this disease is selective loss of dopaminergic neurons in the substantia nigra pars compacta (11). Recently, the death of dopaminergic neurons has been reported to occur by apoptosis (12–14). Oxidative stress has been widely believed to be an important pathogenetic mechanism of neuronal apoptosis in PD (15). Overproduction of reactive oxygen species (ROS) can lead to oxidative damage in the brain of PD, as shown by increased lipid peroxidation and DNA damage in the substantia nigra. Increased protein oxidation is also apparent in many areas of the brain, while the substantia nigra is particularly vulnerable (16). Under physiological conditions, 6-hydroxydopamine (6-OHDA) is rapidly and nonenzymatically oxidized by molecular oxygen to form hydrogen peroxide (H_2O_2) and the corresponding p-quinone (17). The former can react with iron (II) to form the reactive and damaging hydroxyl free radical. The latter then undergoes an intramolecular cyclization followed by a cascade of oxidative reactions, resulting in the formation of an insoluble polymeric pigment related to neuromelanin (18,19).

18.3   GREEN TEA POLYPHENOLS PROTECT NEURONS AGAINST PD AND AD

GTPs and their major constituents, such as EGCG, have diverse pharmacological activities, such as antimutagenic and anticarcinogenic effects. It is believed that these beneficial effects of GTPs are due to their potent antioxidative properties. In fact, it was demonstrated that GTPs serve as powerful antioxidants against free radicals, such as DPPH radicals (20), superoxide anion (21–23), lipid free radicals, and hydroxyl radicals (24,25). In the central nervous system, there is also some evidence showing that oral administration of GTPs and flavonoid-related compounds has preventive effects on iron-induced lipid peroxide accumulation and age-related accumulation of neurotoxic lipid peroxides in rat brain (26,27).

Tea catechins (TCs) are usually expected as scavengers of free radicals, but different components have different functions. Herein, we studied the effect of TCs on the PC12 and SY5Y cells exposed to 6-OHDA and found that TCs could protect PC12 cells against apoptosis caused by 6-OHDP.

We investigated the effects of exposure of PC12 and SY5Y cells to 6-OHDA alone and in association with pretreatment of TCs. Exposure of PC12 and SY5Y cells to 6-OHDA induced a concentration-dependent decrease in cell viability determined by MTT assay and apoptosis of PC12 and SY5Y cells observed by flow cytometry, fluorescence microscopy, and DNA fragmentation. TCs displayed significantly inhibitory effects against PC12 and SY5Y cell death. EGCG and ECG were more effective than TC, but EGC, EC, and (+)-C were less effective. 6-OHDA-induced apoptosis was greatly inhibited by GTPs at 200–400 μμM. From 50 to 400 μμM, the protective effects increased with the concentrations, and EGCG was better than GTPs at the same concentrations. From the data of flow cytometry, the apoptotic cells were inhibited greatly by 200–400 μμM of GTPs and EGCG (the inhibitory ratios of GTPs were 83.1% and 84.8%; those of EGCG, 88.3% and 90.3%). The nuclear changes characteristic of apoptosis disappeared, especially in the EGCG-protected PC12 and SY5Y cells. The DNA ladder also disappeared in the channels of 200–400 μM concentrations of GTPs and EGCG (28–30).

Nitric oxide (NO) and related pathways are thought to play an important role in the pathogenesis of PD. The above-mentioned *in vitro* experiments suggested that GTPs might protect dopamine neurons through inhibition of NO and ROS. Immunohistochemistry, terminal deoxynucleotidyl transferase-mediated dUTP nick-end labeling assay, electron spin resonance (ESR) spin trapping, enzyme-linked immunosorbent assay, and molecular biological methods were used to investigate the effects of GTP in a unilateral 6-OHDA-treated rat model of PD. GTP treatment dose-dependently protected dopaminergic neurons by preventing them from midbrain and striatal 6-OHDA-induced increase in (a) both ROS and NO levels, (b) lipid peroxidation, (c) nitrite/nitrate content, (d) inducible NO synthase, and (e) protein-bound 3-nitrotyrosine. Moreover, GTP treatment dose-dependently preserved the free radical-scavenging capability of both the midbrain and the striatum. These results support the *in vivo* protection of GTP against 6-OHDA and suggest that GTP treatment might represent a neuroprotective treatment of PD (31); the pathway is shown in Figure 18.2.

In an animal study, Levites et al. (32) demonstrated the neuroprotective property of green tea extract and ECG in an MPTP-treated mice model of PD. MPTP neurotoxin caused dopamine neuron loss in the substantia nigra concomitant with a depletion in striatal dopamine and tyrosine hydroxylase protein levels. Pretreatment with either green tea extract (0.5 and 1 mg/kg) or ECG (2 and 10 mg/kg) prevented these effects. In addition, the neurotoxin caused an elevation in the activities of striatal antioxidant enzymes SOD (240%) and catalase (165%), with both effects being prevented by ECG. ECG itself also increased the activities of both enzymes in the brain. The neuroprotective effects are not likely to be caused by inhibition of MPTP conversion to its active metabolite 1-methyl-4-phenylpyridinium by monoamine oxidase-B as both green tea and ECG are very poor inhibitors of this enzyme *in vitro* (770 mg/ml and 660 mM, respectively). The brain-penetrating property of polyphenols, as well as their antioxidant and iron-chelating properties, may make such compounds an important class of drugs to be developed for treatment of neurodegenerative diseases in which oxidative stress has been implicated (32).

Levites et al. (33) also demonstrated highly potent antioxidant-radical-scavenging activities of green tea and black tea extracts on brain mitochondrial membrane fraction against iron (2.5 μM)-induced lipid peroxidation. Both extracts (0.6–3 μM total polyphenols) were shown to attenuate the neurotoxic action of 6-OHDA-induced neuronal death. 6-OHDA at a concentration of 350 μM and that at 50 μM activated the iron-dependent inflammatory redox-sensitive NF-κB in PC12 and SH-SY5Y cells, respectively. Immunofluorescence and electromobility shift assays showed increased nuclear translocation and binding activity of NF-κB after exposure to 6-OHDA in SH-SY5Y cells, with a concomitant disappearance from the cytoplasm. Introduction of green tea extract (0.6; 3 μM total polyphenols) before 6-OHDA inhibited both NF-κB nuclear translocation and binding activity induced by this toxin in SH-SY5Y cells. Neuroprotection was attributed to the potent antioxidant and iron-chelating actions of the polyphenolic constituents of tea extracts, preventing nuclear translocation and activation of cell death promoting NF-κB (33).

Levites et al. (34) furthermore demonstrated that EGCG restored reduced activities of protein kinase C (PKC) and extracellular signal-regulated kinases (ERK1/2) caused by 6-OHDA toxicity.

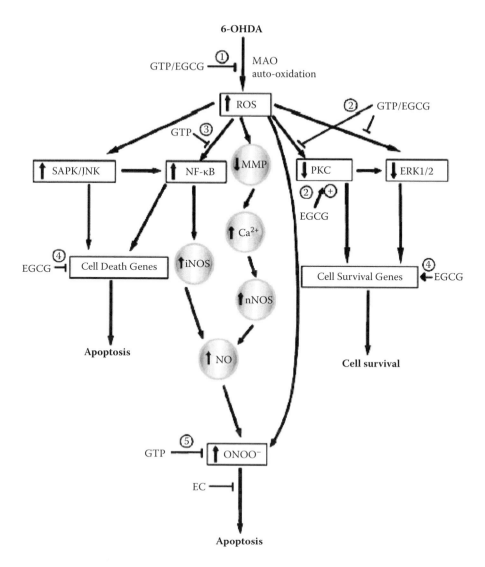

FIGURE 18.2 A hypothetical model diagram of the potential targets of GTP or EGCG is suggested in cell signaling pathways affected by 6-OHDA-induced oxidative stress: (a) direct inhibition of oxidized 6-OHDA and/or scavenging of ROS; (b) inhibiting the negative effect of 6-OHDA on both PKC and ERK1/2, and EGCG can direct phosphorylative activation of PKC; (c) attenuating NF-κB translocation to the nucleus and inhibiting its activation; (d) modulating the expression of cell death and cell cycle genes; and (e) modulating the intracellular NO level and inhibiting the generation of peroxynitrite.

However, the neuroprotective effect of EGCG on cell survival was abolished by pretreatment with PKC inhibitor GF 109203X (1 μM). Because EGCG increased phosphorylated PKC, Levites et al. suggested that PKC isoenzymes are involved in the neuroprotective action of EGCG against 6-OHDA. In addition, gene expression analysis revealed that EGCG prevented both the 6-OHDA-induced expression of several mRNAs, such as Bax, Bad, and Mdm2, and the decrease in Bcl-2, Bcl-w, and Bcl xL. These results suggest that the neuroprotective mechanism of EGCG against oxidative stress-induced cell death includes stimulation of PKC and modulation of cell survival/cell cycle genes (34).

GTPs are usually expected as a potent chemopreventive agent due to their ability of scavenging free radicals and chelating metal ions. However, not all the actions of GTPs are necessarily beneficial. We demonstrated that high concentrations of GTPs significantly enhanced the neurotoxicity by treatment of sodium nitroprusside (SNP), an NO donor. SNP induced apoptosis in human neuroblastoma SY5Y cells in a concentration and time-dependent manner as estimated by cell viability assessment, FACS scan analysis, and DNA fragmentation assay, whereas treatment with GTP alone had no effect on cell viability. Pretreatment with low-dose GTP (50 and 100 μM) had only a slightly deleterious effect in the presence of SNP, while high-dose GTP (200 and 500 μM) synergistically damaged the cells severely. Further research showed that co-incubation of GTP and SNP caused loss of mitochondrial membrane potential, depletion of intracellular GSH, and accumulation of ROS and exacerbated NO-induced neuronal apoptosis via a Bcl-2-sensitive pathway (35).

Recent studies suggest that green tea flavonoids may be used for the prevention and treatment of a variety of neurodegenerative diseases. Rezai-Zadeh et al. (36) reported that EGCG, the main polyphenolic constituent of green tea, reduces Aβ generation in both murine neuron-like cells (N2a) transfected with the human Swedish mutant amyloid precursor protein (APP) and in primary neurons derived from Swedish mutant APP-overexpressing mice (Tg APPsw line 2576). In concert with these observations, we found that EGCG markedly promotes cleavage of the α-C-terminal fragment of APP and elevates the N-terminal APP cleavage product, soluble APP-α. These cleavage events are associated with elevated α-secretase activity and enhanced hydrolysis of tumor necrosis factor α-converting enzyme, a primary candidate of α-secretase. As a validation of these findings *in vivo*, we treated Tg APPsw transgenic mice overproducing Aβ with EGCG and found decreased Aβ levels and plaques associated with promotion of the nonamyloidogenic α-secretase proteolytic pathway. These data raise the possibility that EGCG dietary supplementation may provide effective prophylaxis for AD (36).

Recently, Obregon et al. (37) showed that GTP EGCG exerts a beneficial role on reducing brain Aβ levels, resulting in mitigation of cerebral amyloidosis in a mouse model of AD. EGCG seems to accomplish this by modulating APP processing, resulting in enhanced cleavage of the α-COOH-terminal fragment of APP and corresponding elevation of the NH2-terminal APP product, soluble APP-α. These beneficial effects were associated with increased α-secretase cleavage activity but no significant alteration in β- or α-secretase activities. To gain insight into the molecular mechanism whereby EGCG modulates APP processing, we evaluated the involvement of three candidate α-secretase enzymes, ADAM (a disintegrin and metalloprotease)-9, ADAM-10, and ADAM-17, in EGCG-induced nonamyloidogenic APP metabolism. Results show that EGCG treatment of N2a cells stably transfected with Swedish mutant human APP (SweAPP N2a cells) leads to markedly elevated active (mature form, ~60 kDa) ADAM-10 protein. Elevation of active ADAM-10 correlates with increased α-COOH-terminal fragment cleavage and elevated soluble APP-α. To specifically test the contribution of ADAM-10 to nonamyloidogenic APP metabolism, we employed small interfering RNA knockdown of ADAM-9, ADAM-10, or ADAM-17 mRNA. Results show that ADAM-10 (but not ADAM-9 or ADAM-17) is critical for EGCG-mediated α-secretase cleavage activity. In summary, ADAM-10 activation is necessary for EGCG promotion of nonamyloidogenic (α-secretase cleavage) APP processing. Thus, ADAM-10 represents an important pharmacotherapeutic target for the treatment of cerebral amyloidosis in AD (37).

Considering the multi-etiological character of AD, the current pharmacological approaches using drugs oriented toward a single molecular target possess limited ability to modify the course of the disease and thus offer a partial benefit to patients. In line with this concept, novel strategies include the use of a cocktail of several drugs and/or the development of a single molecule possessing two or more active neuroprotective-neurorescue moieties that simultaneously manipulate multiple targets involved in AD pathology. A consistent observation in AD is a dysregulation of metal ion (Fe^{2+}, Cu^{2+}, and Zn^{2+}) homeostasis and consequential induction of oxidative stress, associated with β-amyloid aggregation and neurite plaque formation. In particular, iron has been demonstrated to modulate the AD amyloid precursor holoprotein expression by a pathway similar to that of ferritin L- and H-mRNA translation through iron-responsive elements in their 5′-untranslated regions. Mandel et al. (38) discussed two scenarios concerning multiple therapy targets in AD sharing in common the implementation of iron chelation activity: (a) novel multimodal brain-permeable iron-chelating drugs, possessing neuroprotective-neurorescue, and APP-processing regulatory activities, and (b) natural plant polyphenols (flavonoids), such as green tea EGCG and curcumin, reported to have access to the brain and to possess multifunctional activities, such as metal chelation-radical scavenging, anti-inflammation, and neuroprotection (38).

18.4   ANTIOXIDANT EFFECT OF GTP

All above-mentioned biological functions of tea polyphenols connect with their antioxidant properties. Therefore, a lot of studies about antioxidant properties appeared in the literature. We have also studied scavenging effects of GTP on oxygen free radicals in different systems and their mechanisms.

18.4.1   Scavenging Effect of GTP on Oxygen Free Radicals

We trapped O_2^- and .OH generated from photoradiation of riboflavine/EDTA, Fenton reaction, and polymorphonuclear leukocytes with DMPO (5,5-dimethyl-1-pyroline-1-oxide) and measured by ESR. The scavenging effects of TCs on the oxygen free radicals were studied and compared with those of rosemary, curcumin, and vitamins C and E. It was found that TCs had the strongest scavenging effects on oxygen free radicals produced from cell systems. Unlike other antioxidants, they practically eliminated all active oxygen free radicals (21).

18.4.2   The Synergic Effect of GTP on Free Radicals

It is thought that there are four components of the GTP and that each component has a different biological function, while all may act synergistically to scavenge toxic oxygen free radicals. We studied the synergic scavenging effects of these four components of polyphenol on O_2^- and found that the combination with the ratio EGCG/ECG/EC/EGC=5:2:2:1, which is equivalent to that naturally found in tea, had the strongest effect. The effect was stronger than other ratios that even contained greater amounts of EGCG. This suggests that GTP successfully evolved in order to survive in a toxic environment (22).

18.4.3   GTP Inhibition of Lipid Peroxidation and Scavenging
of Lipid Radicals in Synaptosomes

Using Fe^{2+}/Fe^{3+} to stimulate lipid peroxidation of synaptosomes, we studied the GTP inhibition of lipid peroxidation and found that EGCG, ECG, EGC, and EC showed different inhibitory effects on TBARSs formed in the system. Preincubation of synaptosomes with one of them before exposure to Fe^{2+}/Fe^{3+} treatment resulted in a decrease in the amount of peroxidation, and it was dose dependent. It was found that the IC_{50} levels of EGCG, ECG, EGC, and EC were 0.1, 0.19, 0.24, and 0.35 mM, respectively (39).

Lipid peroxidation plays an important role in the free radical-generating process. We used 4-POBN to trap the lipid free radical when synaptosomes were incubated with Fe^{2+}/Fe^{3+}. EGCG, ECG, EGC, and EC were then added to this system in order to measure their ability to scavenge lipid-free radicals. The results showed that the scavenging rates were increased with the increase of the concentrations of added polyphenols. The IC_{50} levels of ECG, EGCG, EC, and EGC were 7.31, 14.9, 22.14, and 59.28 mM, respectively (22).

Another aspect of the relationship between the structures and activities of antioxidant that we looked at was the chemical structures, including steric structures, and the free radical-scavenging activities of TCs and their corresponding epimers. In particular, we examined the steric hindrance and the stability of the catechins in relation to the abilities to scavenge both small and big free radicals. Scavenging effects were evaluated not only on superoxide anions, singlet oxygen, and the free radicals from AAPH using ESR spin-trapping techniques but also on the DPPH radical using direct ESR measurement (24).

18.4.4 Inhibitory Effect of GTP on the Oxidant Activity of Peroxynitrite

Peroxynitrite is a very strong oxidant, and it can oxidate most of the cell membrane component. Using spin-trapping agent tNB (2-mehtyl-2-nitrosopropane), we trapped a free radical from a reaction of DMSO (dimethyl sulfoxide) with peroxynitrite. It was identified as the $.CH_3$ free radical spin adduct of tNB by the analysis of its spectrum, and it was found that the yield of $tNB-CH_3$ radical adduct was dose dependent on the concentrations of peroxynitrite and DMSO. This is a useful index for studying the oxidation property of peroxynitrite. The scavenging effects of GTP on $.CH_3$ free radicals generated from the oxidation of DMSO by peroxynitrite were studied and compared with the scavenging effects of quercertin and vitamin C. Although they could all effectively inhibit the oxidation of peroxynitrite in this system, the inhibitory effect of GTP was the strongest, while that of vitamin C was the weakest (40). Reaction with peroxynitrite was found to increase the 8-oxodeoxyguanosine levels in calf thymus DNA by 35- to 38-fold. This oxidation of deoxyguanosine and the peroxynitrite-mediated nitration of tyrosine to 3-nitrotyrosine were significantly inhibited by ascorbic, glutathione, and EGCG. In order to inhibit 50% of the oxidation, we required 1.1, 7.6, and 0.25 mM ascorbate, glutathione, and EGCG, respectively. In order to inhibit 50% of the nitration, we needed the respective concentrations of 1.4, 4.6, and 0.11 mM. Thus, EGCG was significantly more efficient in inhibiting both reactions. Reaction of EGCG with peroxynitrite alone resulted in the formation of a number of products. Ultraviolet spectra of two of these suggest that the tea polyphenol and/or its oxidation products are nitrated by peroxynitrite (41). One experiment also showed that catechins were able to effectively inhibit peroxynitrite-mediated tyrosine nitration *in vitro* (42).

18.4.5 Inhibition Effects of Green Tea Extract on Low-Density Lipoprotein Oxidation

Five groups of 20 female New Zealand white rabbits were fed with a restricted amount of high-fat (30%) semipurified diet supplemented with cholesterol (0.15%,w/w) for 21 weeks. The vitamin E content of the control diet was 40 mg/kg diet. The animals received green tea, black tea, vitamin E (200 mg/kg diet), or β-carotene (20 mg/kg) in their drinking supply. The serum cholesterol concentrations (in the order of 18–23 mM) were not significantly different between the groups, but vitamin E was substantially increased by 3-fold within 8 weeks in plasma and low-density lipoprotein ($p < 0.01$). Compared with controls, in vitamin E-supplemented animals, vitamin E was slightly increased (1.2-fold) in rabbits fed with green tea and black tea ($p < 0.05$). Green tea consumption reduced aortic lesion formation by 31%, while black tea, vitamin E, and β-carotene had no effect (43–45).

18.5 SUMMARY

This chapter has reviewed studies on the preventive effects of GTPs on neurodegenerative diseases and especially summarized the results about the protective effects of GTPs on neurons against cell apoptosis caused by 6-OHDA in PD models and Aβ-induced apoptosis of hippocampal neuronal cells in AD models *in vitro* and *in vivo*. The signal pathways of GTPs against AD and PD and their antioxidant properties have also been analyzed in this chapter.

ACKNOWLEDGMENTS

This work was supported by grants from the National Natural Science Foundation of China (29935080, 973, and 2006CB500706) and the E-Institutes of Shanghai Municipal Education Commission (Project No. E-04010).

REFERENCES

1. Lee S.-R., Suh S.-I., Kim S.-P. (2000). Protective effects of the green tea polyphenol (2)-epigallocatechin gallate against hippocampal neuronal damage after transient global ischemia in gerbils. *Neurosci. Lett.* **287**: 191–194.

2. Suganuma M., Okabe S., Oniyama M., Tada Y., Ito H., Fujiki H. (1998). Wide distribution of ^{3}H-(−)-epigallocatechin gallate, a cancer preventive tea polyphenol, in mouse tissue. *Carcinogenesis* **19**: 1771–1776.

3. Butterfield D. A., Kanski J. (2001). Brain protein oxidation in age-related neurodegenerative disorders that are associated with aggregated proteins. *Mech. Ageing Dev.* **122**: 945–962.

4. Stadtman E. R. (1990). Metal ion-catalyzed oxidation of proteins: Biochemical mechanism and biological consequences. *Free Radic. Biol. Med.* **9**: 315–325.

5. Nunomura A., Perry G., Pappolla M. A., Wade R., Hirai K., Chiba S., Smith M. A. (1999). RNA oxidation is a prominent feature of vulnerable neurons in Alzheimer's disease. *J. Neurosci.* **19**: 1959–1964.

6. Gabbita S. P., Lovell M. A., Markesbery W. R. (1998). Increased nuclear DNA oxidation in the brain in Alzheimer's disease. *J. Neurochem.* **71**: 2034–2040.

7. Mecocci P. L., MacGarvey U., Beal M. F. (1994). Oxidative damage to mitochondrial DNA is increased in Alzheimer's disease. *Ann. Neurol.* **36**: 747–750.

8. Smith M. A., Perry G., Richey P. L., Sayre L. M., Anderson V. E., Beal M. F., Kowall N. (1996). Oxidative damage in Alzheimer's disease. *Nature* **382:** 120–121.

9. Smith M. A., Hirai K., Hsiao K., Pappolla M. A., Harris P. L., Siedlak S. L., Tabaton M., Perry G. (1998). Amyloid-β deposition in Alzheimer transgenic mice is associated with oxidative stress. *J. Neurochem.* **70**: 2212–2215.

10. Huang X., Atwood C. S., Hartshorn M. A., Multhaup G., Goldstein L. E., Scarpa R. C., Cuajungco M. P., Gray D. N., Lim J., Moir R. D., Tanzi R. E., Bush A. I. (1999). The amyloid-β-peptide of Alzheimer's disease directly produces hydrogen peroxide through metal ion reduction. *Biochemistry* **38**: 7609–7616.

11. Huang X., Cuajungco M. P., Atwood C. S., Hartshorn M. A., Tyndall J. D., Hanson G. R., Stokes K. C., Leopold M., Multhaup G., Goldstein L. E., Scarpa R. C., Saunders A. J., Lim J., Moir R. D., Glabe C., Bowden E. F., Masters C. L., Fairlie D. P., Tanzi R. E., Bush A. I. (1999). Cu(II) potentiation of Alzheimer Aβ neurotoxicity. Correlation with cell-free hydrogen peroxide production and metal reduction. *J. Biol. Chem.* **74**: 37111–37116.

12. Behl C., Davis J. B., Lesley R., Schubert D. (1994). Hydrogen peroxide mediates amyloid β protein toxicity. *Cell* **77**: 817–827.

13. Markesbery W. R., Carney J. M. (1999). Oxidative alterations in Alzheimer's disease. *Brain Path.* **9**: 133–146.

14. Markesbery W. R. (1997). Oxidative stress hypothesis in Alzheimer's disease. *Free Radic. Biol. Med.* **23**: 134–147.

15. Christen Y. (2000). Oxidative stress and Alzheimer disease. *Am. J. Clin. Nutr.* **71**: 621S–629S.

16. Smith M. A. (2000). Oxidative stress in Alzheimer's disease. *Biochim. Biophys. Acta* **1502**: 139–144.

17. Varadarajan S. (2000). Review: Alzheimer's amyloid-peptide-associated free radical oxidative stress and neurotoxicity. *J. Struct. Biol.* **130**: 184–208.

18. Beckman J. S. (1996). Oxidative damage and tyrosine nitration from peroxynitrite. *Chem. Res. Toxicol.* **9**: 836–844

19. Xin W.-J., Zhao B.-L., Zhang J.-Z. (1984). Studies on the property of sulfhydryl binding site on the lung normal cell and cancer cell membrane of Chinese hamster with maleimide spin labels. *Sci. Sinica* B **28**: 1008–1014.

20. Nanjo F., Goto K., Seto R., Suzuki M., Sakai M., Hara Y. (1996). Scavenging effect of tea catechins and their derivatives on 1,1-diphenyl-2-picrylhdrazyl radical. *Free Radic. Biol. Med.* **21**: 895–902.

21. Zhao B.-L., Li X.-J., He R.-G., Cheng S.-J., Xin W.-J. (1989). Scavenging effect of extracts of green tea and natural antioxidants on active oxygen radicals. *Cell Biophys.* **14**: 175–181.

22. Guo Q., Zhao B.-L., Li M.-F., Shen S.-R., Xin W.-J. (1996). Studies on protective mechanisms of four components of green tea polyphenols (GTP) against lipid peroxidation in synaptosomes. *Biochem. Biophys. Acta* **1304**: 210–222.

23. Guo Q., Zhao B.-L., Hou J.-W., Xin W.-J. (1999). ESR study on the structure-antioxidant activity relationship of tea catechins and their epimers. *Biochim. Biophys. Acta* **1427**: 13–23.

24. Zhao B.-L., Guo Q., Xin W.-J. (2001). Free radical scavenging by green tea polyphenols. *Methods Enzymol.* **335**: 217–231.

25. Nie G.-J., Wei T.-T., Zhao B.-L. (2001). Polyphenol protection of DNA against damage. *Methods Enzymol.* **335**: 232–244.

26. Inanami O., Watanabe Y., Syuto B., Nakano M., Tsuji M., Kuwabara M. (1998). Oral administration of (-)catechin protects against ischemia-reperfusion-induced neuronal death in the gerbil. *Free Radic. Res.* **29**: 359–365.

27. Yoneda T., Hiramatsu M., Skamoto N., Togasaki K., Komatsu M., Yamaguchi K. (1995). Antioxidant effects of "β catechin." *Biochem. Mol. Biol. Intern.* **35**: 995–1008.

28. Nie G. J., Jin C.-F., Zhao B.-L. (2002). Distinct effects of tea catechins on 6-hydroxydopamine-induced apoptosis in PC12 cells. *Arch. Biochem. Biophys.* **397**: 84–90.

29. Nie G. J., Cao Y. L., Zhao B. L. (2002). Protective effects of green tea polyphenols and their major component, (−)-epigallocatechin-3-gallate (EGCG), on 6-hydroxyldopamine-induced apoptosis in PC12 cells. *Redox Report* **7**: 170–177.

30. Guo S.-H., Bezard E., Zhao B.-L. (2005). Protective effect of green tea polyphenols on the SH-SY5Y cells against 6-OHDA induced apoptosis through ROS-NO pathway. *Free Radic. Biol. Med.* **39**: 682–695.

31. Guo S., Yan J., Yang T., Yang X., Bezard E., Zhao B. L. (2007). Protective effect of green tea polyphenols on rat model of Parkinson's disease caused by 6-OHDA through ROS-NO pathway. *Biol. Psychiatr.* **62**: 1353–1362.

32. Levites Y., Weinreb O., Maor G., Youdim M. B. H., Mandel S. (2001). Green tea polyphenol epigallocatechin-3-gallate prevents MPTP induced dopaminergic neurodegeneration. *J. Neurochem.* **78**: 1073–1082.

33. Levites Y., Youdima M. B. H., Mao G., Mandel S. (2002). Attenuation of 6-OHDA-induced nuclear factor-NF-κB activation and cell death by tea extracts in neuronal cultures. *Biochem. Pharm.* **63**: 21–29.

34. Levites Y., Amit T., Youdim M. B. H., Mandel S. (2002). Involvement of protein kinase C activation and cell survival/cell cycle genes in green tea polyphenol (−)-epigallocatechin 3-gallate neuroprotective action. *J. Biol. Chem.* **77**: 30574–30580.

35. Zhang Y., Zhao B. L. (2003). Green tea polyphenols enhance sodium nitroprusside induced neurotoxicity in human neuroblastoma SH-SY5Y cells. *J. Neur. Chem.* **86**: 1189–1200.

36. Rezai-Zadeh K., Shytle D., Sun N., Mori T., Hou H., Jeanniton D., Ehrhart J., Townsend K., Zeng J., Morgan D., Hardy J., Town T., Tan J. (2005). Green tea epigallocatechin-3-gallate (EGCG) modulates amyloid precursor protein cleavage and reduces cerebral amyloidosis in Alzheimer transgenic mice. *J. Neurosci.* **25**: 8807–8814.

37. Obregon D. F., Rezai-Zadeh K., Bai Y., Sun N., Hou H., Ehrhart J., Zeng J., Mori T., Arendash G. W., Shytle D., Town T., Tan J. (2006). ADAM10 activation is required for green tea (−)-epigallocatechin-3-gallate-induced α-secretase cleavage of amyloid precursor protein. *J. Biol. Chem.* **281**: 16419–16427.

38. Mandel S., Amit T., Bar-Am O., Youdim M. B. H. (2007). Iron dysregulation in AD: Multimodal brain permeable iron chelating drugs, possessing neuroprotective-neurorescue and amyloid precursor protein-processing regulatory activities as therapeutic agents. *Prog. Neurobiol.* **82**: 348–360.

39. Shen S.-R., Yang X.-Q., Yang F.-J., Zhao B.-L., Xin W.-J. (1993). Synergistic antioxidant effect of green tea polyphenols. *Tea Sci.* **13**: 141–146.

40. Zhao B.-L., Wang J.-C., Hou J.-W., Xin W.-J. (1996). Scavenging effect of green tea polyphenols on ONOO⁻. *Sci. Bult. Sinica* **41**: 925–927.

41. Zhao B.-L., Wang J.-C., Xin,W.-J. (1996). Scavenging effect of green tea polyphenols, on peroxynitrites. *Chinese Sci. Bull.* **41**: 923–927.

42. Fiala E. S., Sodum R. S., Bhattacharya M., Li H. (1996). Epigallocatechin gallate, a polyphenolic tea antioxidant, inhibits peroxynitrite-mediated formation of 8-oxodeoxyguanosine and 3-nitrotyrosine. *Experientia* **52**: 922–926.

43. Pannala A., Rice-Evans C. A., Halliwell B., Surinder S. (1997). Inhibition of peroxynitrite-mediated tyrosine nitration by catechin polyphenols. *Biochem. Biophys. Res. Commun.* **232**: 164–168.

44. Tijiburg L. B., Wiseman S. A., Meijer G. W., Weststrate J. A. (1997). Effects of green tea, black tea and dietary lipophilic antioxidants on LDL oxidizability and atherosclerosis in hypercholesterolaemic rabbits. *Atherosclerosis* **135**: 37–47.

45. Luo M., Kannar K., Wahlqvist M. L., O'Brien R. C. (1997). Inhibition of LDL oxidation by green tea extract. *Lancet* **349**: 360–361.

19 Transport of Flavonoids into the Brain

Paul E. Milbury

Gerald Human Nutrition Center on Aging at Tufts
University, Boston, Massachusetts, USA

CONTENTS

19.1 INTRODUCTION

"Plant secondary metabolites" is the generic term used to describe approximately 30,000 phytochemicals produced by plants as hormones, as attractants (such as for color or scent attractants for pollinators or seed dispersers), and for protection against pests. Flavonoids are the predominant polyphenolic secondary metabolites present in edible plants. Although flavonoids have been extensively studied for more than 100 years as pigments and phytoalexins, nutritional biochemists, with typical hubris, have believed that these compounds were irrelevant to animals because they were usually found in very small amounts in plasma and there was no identified physiological function for these compounds in animals. They appeared to be "non-essential," especially since no apparent flavonoid deficit symptom had been observed. In recent decades, with the classification of flavonoids as dietary antioxidants, studies have identified a growing number of *in vivo* flavonoid protective functions that now extend far beyond radical quenching.[1] Many of these functions had been previously predicted from earlier *in vitro* studies.[2]

The last decade has seen that increased focus was placed on studies of the bioavailability of these compounds in order to better understand the said functions in vivo. Unfortunately, much of the early bioavailability (and function) data were generated using the aglycone forms of flavonoids, with attention paid to glycosylated and metabolite forms only recently. The vast majority of flavonoids exist in plants as *O*- or *C*- glycosylated, acylated, or prenylated and sometimes sulfated derivatives, which give rise to more than 6000 forms of these compounds.[3] To compound matters, while neuroprotective effects have been observed for many flavonoids *in vitro*,[4] neuroprotection in animal studies has either not been forthcoming or been only achieved using particular delivery mechanisms.[5]

19.2 FLAVONOIDS, DRUG DELIVERY, AND METABOLISM

In 1989, a serendipitous observation that grapefruit juice had an adverse impact on delivery of the drug felodipine[6] led to the understanding that some flavonoids could significantly impact intestinal metabolism of CYP3A substrates and influence the activity of CYP3A enzymes.[7] The drug industry has, as a matter of course, extensively studied the uptake and delivery of their drugs, and there has therefore been a clear understanding of the importance of both influx and efflux transporters in drug uptake and clearance. Concern for possible flavonoid interference in drug transport led researchers to investigate the ability of flavonoids to inhibit transporters and thereby potentially inhibit drug bioavailability. Flavonoid inhibitors of a number of transporters were indeed quickly identified, and they included biochanin, morin, phloretin, and silymarin[8] and tea catechins.[9] These compounds were found to inhibit P-glycoprotein (Pgp)-mediated transport. Other flavonoid- and phenolic-rich extracts of herbals have subsequently been demonstrated to inhibit other transporters, including human organic anion-transporting polypeptide 2B1 (OATP2B1)[10] and multidrug resistance-associated protein 1 (MRP1), MRP4, and MRP5,[11] which suggested to researchers that flavonoids were likely to interfere with both the uptake and the efflux of drugs at the enterocyte in the intestine.

Unfortunately, this focus on drug flavonoid-transporter interaction yielded proportionately less data on the transport flavonoids themselves than on flavonoid interference in drug transport. Because of the importance of tea flavonoids, such as epicatechin, in reducing the risk for some cancers and an accumulating body of evidence suggesting that dietary flavonoids are health beneficial by other means, such as reducing mortality from vascular diseases, experimentation began with *in vitro* models to understand flavonoid transport itself.[12] However, to date, relatively little is known about the actual *in vivo* transport kinetics of flavonoids. By 2003, it had become generally clear that the potency of flavonoids in exhibiting their demonstrated *in vitro* protective effects would depend heavily on their bioavailability and lipophilicity.[13] Concern that the potential of flavonoids having direct bioactivity in the brain may be severely restricted by the blood-brain barrier (BBB) was raised.[14]

19.3 FLAVONOID BIOAVAILABILITY

The classic pharmacological definition of *bioavailability* refers to the proportion of an orally administered compound that appears in plasma over time (measured as area under the curve) relative to the amount present in plasma when it is administered intravenously. A more modern concept of pharmacokinetics has evolved as a result of studies of the processes of drug action through the various processes of liberation, absorption, distribution, metabolism, and excretion and is often represented by the acronym LADME. Under this scheme, the bioavailability is defined as L, liberation of the compound from the dietary matrix (food or supplement); A, absorption of the compound from the gastrointestinal lumen into the blood circulation; D, distribution of the compound into body tissues; M, metabolic fate of the compound, either in tissues or by gastrointestinal microflora; and E, elimination from the body.

Flavonoids in general have relatively low oral bioavailability primarily due to their limited ability to transit from the gastrointestinal lumen into the blood circulation.[15,16] Compound solubility has a major influence at all stages of bioavailability and especially so in the first two phases of the LADME scheme. Flavonoid aglycones are generally not very soluble in water and are only moderately soluble in organics such as ethanol, methanol, and acetonitrile. As cited previously, flavonoids are present in plant foods predominantly as glycosidic conjugates, which, compared with their aglycones, are relatively more water soluble. Unfortunately, extensive glycosidic conjugation also negatively impacts bioavailability of dietary flavonoids as it does with many polyphenolic compounds.[1] It has become clear that with more than 6000 forms of flavonoid compounds and the influence that the various sugars have on bioavailability and absorption,[17] the amount of flavonoids ultimately in circulation in plasma will be highly dependent on their original form within the food matrix and on any polymorphism in cleavage, and conjugation enzymes may exist in any individual subject consuming those complex mixtures.

The most common forms of flavonoid glycosidic conjugates include linkage to one or more galactose, glucose, arabinose, or rhamnose sugar units. Flavonoid glycoside conjugates within the food matrix survive the stomach and are passed to the small intestine,[18] where the sugar moieties are removed by enzymes such as lactase-phlorizin hydrolase,[19,20] a broad-specific β-glucosidase enzyme,[21,22] or by intestinal epithelial cell (enterocyte) β-glucosidases.[23] Hydrolysis can occur in the oral cavity as soon as flavonoids are consumed;[24] however, while this may account for flavonoid bioactivity against oral cancers,[25] oral absorption is far from the primary mode of systemic flavonoid bioavailability. Quercetin aglycones, but not their glycosides, have been shown to be absorbed directly in the stomach;[26] however, the bulk of flavonoid absorption occurs lower in the gastrointestinal tract. Although studied for many years, the mechanism of flavonoid absorption by intestinal epithelial cells remained controversial and unclear.[15] This was undoubtedly due in part to the existence and influence of numerous flavonoid conjugate structures and the tendency to assess bioavailability in early years with analytical methods incorporating hydrolysis to aglycones for quantification.

Some, but not all, intact flavonoid glycosides were demonstrated to be absorbed by the intestinal epithelial cell via the sodium-dependent glucose transporter 1 (SLGT1).[27–29]

There are also *in vitro* and *ex vivo* lines of evidence for the existence of a transporter other than sodium-dependent glucose transporter 1 that may function to permit the brush boarder passage of anthocyanins,[30] and this transporter appears to be shared with other flavonols, such as quercetin-3-glucoside, perhaps due to their similar structures. Indeed, the transporter substrate preference for quercetin-3-glucoside suggests that the presence of a variety of flavonoids within the food matrix may significantly impact the absorption of particular individual flavonoids. The sugar conjugates of flavonoids dramatically affect their absorption as shown by the example of the quercetin-3-rutinoside, which is not hydrolyzed and absorbed in the small intestine[31] but which can be hydrolyzed by gut microflora and absorbed in the colon.[32]

Once present as the aglycone within the enterocyte, flavonoids are either passed directly to the circulation or rapidly conjugated with glucuronide, sulfate, or methyl moieties by UGT, UDP-glucuronosyltransferase, sulfotransferase, or catechol-*O*-methyltransferase.[33] Either prior to initial hydrolysis or after reconjugation, flavonoid conjugates are subject to return to the intestinal lumen via the actions of MRP2.[34] This fact may account for the low absorption levels of some flavonoids. Glucuronidated or sulfated flavonoids are also easily transferred to the plasma by MRP3. Once in circulation, flavonoids, both aglycones and conjugated products generated during transit through the enterocytes, are exposed to hepatic phase I and phase II metabolism.[35]

At this point, it is clear that most dietary flavonoid glycosides are very poorly absorbed; however, hydrolysis to their aglycone forms occurs easily and efficiently by both epithelial cells and gut microflora and absorbed at various locations throughout the gastrointestinal tract. It is also clear that the majority, if not all, of aglycones that are taken up are quickly conjugated. This conjugation results in their return to the lumen of the gastrointestinal tract or their systemic elimination via the kidney and hence accounts for the observed low bioavailability of flavonoids. Pharmacokinetic data for all the numerous individual flavonoid glycosides are far from complete. The discussion to this point has pertained to all the phases of LADME except D — distribution of the compound into body tissues and in particular to the brain.

19.4 BIOAVAILABILITY TO THE BRAIN

The study of tissue distribution of flavonoids is in its early stages, with only a few compounds from each of the classes of flavonoids studied to date. In many regards, this aspect of bioavailability has been neglected in the rush to define mechanistic rationalization for observed flavonoid health-beneficial effects. Most of the mechanistic studies to date have been conducted *in vitro* using cultured cells. Without knowledge of the potential physiological concentration of flavonoids achievable at the target tissues or knowledge of the metabolism of flavonoids by target tissues, such *in vitro* studies can only suggest the potential mechanisms of action of flavonoids. Many studies have investigated the effects of aglycone forms of flavonoids administered directly to cells in culture. As intimated earlier, epithelial cells of oral, intestinal, and colonic tissues may well benefit from direct contact with flavonoid aglycones, glycosides, and metabolites, as well as their glucuronidation, sulfation, and methylation products. However, in order to have physiological effects in other tissues of the body, flavonoids must transit barriers beyond the gastrointestinal lining. These barriers are formed by the vascular endothelium, a class of epithelial cells lining the intima of blood vessels. The vascular endothelium regulates a wide variety of functions ranging from maintenance of vascular smooth muscle tone, host-defense reactions, and angiogenesis to maintenance of interstitial tissue fluid homeostasis and control of macromolecule passage into interstitial tissues.

The boundary between the blood and the central nervous system represents a special case with vascular endothelium unlike that protecting other interstitial tissues. To be precise, there are two fluid barriers protecting the central nervous system — (1) the BBB formed by the brain capillary endothelial cells and (2) the blood-cerebrospinal fluid barrier formed by the choroid plexus. For this discussion, we concentrate on the BBB.

The vascular system permeates every tissue of the body, but it is clear that blood vessels also vary tremendously with regard to their properties of structure and function to provide specific nutrient and waste transport properties.[36] Even among the vasculature of the central nervous system, a significant degree of difference can be discerned with regard to permeability between the vessels of the various regions. For example, blood-borne macromolecules can penetrate the areas adjacent the lateral septal nucleus, the medial portion of the hippocampus, and the dorsal portion of the thalamus with relative ease;[37] however, the microvascular bed of cerebral tissue is a highly regulated barrier to not only macromolecules but also small organic drugs and ions. These microvessels are the capillaries responsible for the BBB.

The capillary endothelial cells that form the BBB differ significantly from other vascular endothelial cells in three major respects. The capillary endothelial cells that form the BBB are not fenestrated, are connected to adjacent cells by high-resistance tight junctions, and show much lower levels of pinocytosis.[38] These characteristics contribute to reducing unregulated diffusion of molecules across the BBB.[39] As a consequence of the physical barrier presented by the BBB capillary endothelial cells, the permeability of most noncharged compounds could be inversely correlated to size and directly correlated to lipophilicity. A number of compounds that vary from this general observation, such as the hydrophilic amino acids and glucose that are highly permeable and many lipophilic xenobiotics that are not permeable, suggest that the BBB also acts as a selective biochemical barrier. Specific transporter proteins are found on the plasma membranes of both luminal (blood) and abluminal (brain) sides of the BBB endothelial cells. As is the case of all endothelia, the BBB functions to actively deliver nutrients to and remove waste byproducts from the brain, and it does so via these specific transport systems similar to transporters found in the liver. Among these transporters identified in the brain microvasculature are multidrug resistance protein 1 (MDR1a/b, MDR1), MRP1, MRP2, MRP4, OATP2, and OATP3 (see Figure 19.1).

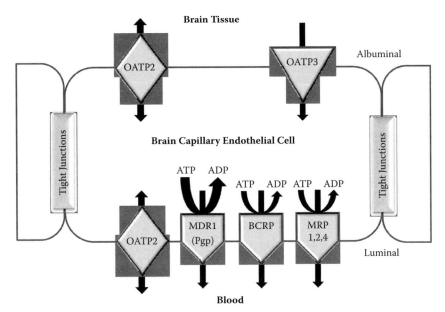

FIGURE 19.1 Multispecific transporters identified in brain capillary endothelial cells. Substrates are directionally transported as indicated by arrows. OATP2 has been identified on both the blood and tissue sides of the endothelial barrier, whereas OATP3 has been detected only on the capillary blood side. ATP-dependent transporters MRP1, MRP2, MRP4, BCRP, and MDR1 (or Pgp) are expressed on the luminal or blood side of the barrier membrane.

The microvasculature endothelium also has the full complement of phase II detoxification enzymes. These transporters and enzymes function together to remove xenobiotic compounds from tissues[40] and are likely to function similarly in the brain to remove most exogenous polyphenolic compounds, including flavonoids and their metabolites.

MDR1 is a member of the ATP-binding cassette (ABC) superfamily of transporters that impart multidrug resistance in mammalian cells. The protein is also known as traffic ATPase and Pgp or ABCB1 and is a plasma membrane efflux transporter requiring energy derived from ATP hydrolysis to pump substrates out of the cells.[41] The MRPs also belong to the family of ABC transporters. An inventory of all known and putative human ABC transporters can be found on the Web site of M. Müller (University of Wageningen, The Netherlands),[42] and their structure and function have been reviewed.[43] A striking characteristic of MDRs and MRPs is their polarized expression on the luminal side of endothelial cells in the BBB.[44,45] This physiological placement is indicative of their efflux function. The ABC efflux transporters have an extremely wide spectrum of substrates including numerous drugs, steroids, and cytokines. However, these varied compounds share the traits of being hydrophobic, generally either positively or neutrally charged, and planar in structure.[46] These are characteristics shared by most flavonoids, and flavonoids have been demonstrated to interact with ABC transporters.[47,48]

ABCG2, another protein that is variously named breast cancer resistance protein (BCRP), MXR, and ABCP, is yet another described ABC transporter that was originally identified by its ability to confer drug resistance independent of MRP1 and Pgp. BCRP has been located mainly in the luminal surface of the endothelial cells of the BBB.[49] As a cautionary note, it is important to understand that species differences can profoundly affect the results of various transporter models. For example, the expression of BCRP in mice brains is minimal and perhaps inconsequential to xenobiotic efflux,[50] while its expression in pigs and in humans is significant and it is thought to play a role in toxin removal.[51]

Unlike the efflux transporters just described, OATPs are bidirectional in their transport of xenobiotics. The OATPs are a group of membrane transporters classified within the solute carrier family 21A and exhibit a wide spectrum of amphipathic transport substrates. Several very similar OATPs (OATP1, OATP2, and OATP3) were isolated and structurally characterized in rats and found to function as a multispecific and sodium ion-independent transporter for anionic and neutral organic compounds, including conjugated metabolites, bile acids, and xenobiotics.[52–54] Although first isolated from a liver cDNA library, human OATP was found to be predominantly active in the brain and was found to be similar to, but not clearly a functional counterpart of, the rat OATP proteins. As more similar protein transporters emerged, an attempt was made to organize the nomenclature. OATP was renamed OATP-A, and subsequent similar protein transporters received the designations OATP-B, OATP-C, OATP-D, and OATP-E.[55] By the time 36 known OATPs have been cloned, 11 were identified from human tissues.[56] Within this superfamily of OATP proteins, OATP1B1 (also known as LST-1, OATP-C, OATP-2, and *SLC21A6*) is perhaps the most studied human OATP protein.[57] As research into these transporters continued, it was discovered that several other previously noted sodium ion-independent transporters, such as prostaglandin transporters found in mice, rats, and humans, were also members of the OATP family.[55]

Because of the potential of flavonoid interference with drug delivery, a great deal of research was devoted to the study of the interaction of flavonoids with BBB transporters. However, while compounds competing for transporter interaction can result in interruption of transport of one of the compounds, it does not necessarily follow that both compounds could be transported.

Decades of *in vitro* studies have shown that flavonoids can interact with proteins and inhibit or induce a significant variety of mammalian enzyme systems.[58] These "interactions" are often mediated by binding of proteins to receptors, leading to signal transduction pathway activation. Recent interest has increased, driven by the interest in resveratrol's life extension qualities, in the theory of xenohormesis, which has led to an intriguing speculation that, similar to the loss of the ability to synthesize vitamin C or certain amino acids among some animal lineages, perhaps animal phyla as a whole "have lost an ancestral capacity to synthesize flavonoid-like 'physiological regulators'" and yet they may have retained the ability to respond to them when encountered in the diet.

A few plant polyphenol binding sites have been characterized on mammalian proteins, such as the affinity of quercetin and resveratrol for the nucleotide binding sites of protein kinases. There are also other examples in which polyphenols do not compete with an enzyme's nucleotide substrates but instead bind elsewhere on the proteins.[59,60] While there binding domains do not appear to be conserved polyphenol interaction domains, they are protein locations where polyphenols are bound in hydrophobic pockets by van der Waals contacts and H-bonds involving the hydroxyl groups. These conditions exist on many proteins, and such binding may or may not effect or interfere with enzyme, receptor, or transporter activation.

Despite an increase in research in flavonoid brain bioavailability in the last 4 years, the extent to which various dietary flavonoids and flavonoid metabolites cross the BBB in humans is not known.[61] For a nutrient to be beneficial, it must be bioavailable (i.e., it must be readily absorbed from the digestive tract, enter the bloodstream, and reach the tissues or organs where it is bioactive). An excellent example of the difficulties posed with regard to flavonoid bioavailability and action is presented by one of the currently "popular" flavonoids of public interest. Scientists believe that *in vitro* and animal study results suggest that resveratrol has the potential of being an extraordinarily effective chemopreventive and anticarcinogenic agent. However, although resveratrol is well absorbed (at least 70%) by the gut, its tissue bioavailability is negligible due to its immediate metabolism to sulfates and glucuronides.[62–65] This resveratrol metabolism primarily occurs in the gut and, although some resveratrols may enter the bloodstream, liver metabolism conjugates the remaining resveratrol within 30 minutes. As a consequence, unmetabolized resveratrol is undetectable in blood.

Because of these facts, some researchers have suggested that there is a question as to whether the therapeutic evidence for resveratrol's benefits is due to resveratrol or its metabolites.[66] Evidence from nutritional physiology is yet inconclusive. This controversy remains to be resolved; however, for other flavonoids, the answer to the question "Are they present in target tissues?" is slowly forthcoming.

19.5   EVIDENCE FOR FLAVONOID PRESENCE IN BRAIN TISSUE AFTER FEEDING

A number of investigators have sought to directly measure in vivo levels of flavonoids in brain tissue after feeding. These are critical experiments in determining the plausibility of theories regarding the mechanism of action of bioactives. Epicatechin, as an aglycone and in its conjugated epicatechin glucuronide and *O*-methylated glucuronide forms, has been detected using liquid chromatography with mass spectroscopy detection in the brain of rats receiving epicatechin daily (100 mg/kg body weight) via gavage.[67] These investigators suggest, however, that the glucuronides they observed were formed in the brain rather than transported across the BBB. When the radiolabeled flavonoid [^{3}H](–)-epigallocatechin gallate was fed to mice, radioactivity that was increased by 6.8-fold upon second gavaging was found in the brain,[68] suggesting that repeated dosing significantly increases the levels of this flavonoid in tissues. Although the researchers utilized microautoradiography to confirm that the administered [^{3}H](–)-epigallocatechin gallate was incorporated into the cells of tissues, such data were not presented for the brain but rather only for the colon and lung. This study might suggest that flavonoids, or metabolites derived from them, may be present in, and interact with, many organs in the body following oral intake. The use of labeled flavonoids in the determination of tissue accumulation in vivo is a powerful tool and should be exploited in the future. Unfortunately, there are few flavonoids that have been radiolabeled. Another problem avoided by the use of isotope-labeled flavonoids pertains to the fact that, unless special care is taken to eliminate blood or ensure flavonoids are not present in blood at the time of brain harvesting, analytical results can be compromised. For this reason, data should be interpreted carefully unless contamination from blood remaining in the microvasculature is ensured.

Table 19.1 presents data accumulated from feeding or gavage studies representing all six classes of flavonoids. In many cases, doses are pharmacological, but a few studies present findings from dietarily achievable doses. Whether a flavonoid is designed to be used as a pharmaceutical or whether it is derived from the diet, levels in target tissues will need to be determined from direct

TABLE 19.1
Flavonoid Feeding Studies in which Flavonoids Were Directly Measured in the Brain

Model	Reference	Flavonoid Administered	Dose	Duration	Administration	Level Identified in the Brain
Flavanols						
Mice	1	CTN986, a highly water-soluble favonol triglycoside	0.16, 0.4, and 1.0 mg/kg	Acute	I.V.	LC/MS/MS CTN986 detected in brain.
Pig	2	Quercetin	Long-term feeding, daily intake of 50 mg/kg; last meal 60–90 minutes prior to collection	4 weeks	80% of voluntary feed intake, twice daily	HPLC with fluorescence detection. Aglycones plus hydrolyzed conjugates and corrected for residual blood. Quercetin 0.02 nmol/g wet tissue, isorhamnetin 0.001, and tamarixetin n.d. Brain contained only deconjugated quercetin.
Rat	3	Quercetin	~800 mg/(kg body wt · day) in week 1 to ~500 mg/(kg body wt · day) during week 11	11 weeks	Ad libitum	Measured by HPLC with coulometric array detector and correction for residual blood. Of all tissues, the lowest concentrations were found in the brain.
Pig	3 (second experiment)	Quercetin	500 mg/(kg body wt · day), divided into three meals a day	3 days	Three meals per day	0.1% quercetin diet: quercetin<LOD; isorhamnetin, 0.19 nmol/g tissue; tamarixetin, 0.14.1% quercetin diet: quercetin, 0.065 nmol/g tissue; isorhamnetin, 0.46; tamarixetin, 0.17; quercetin, 0.07 nmol/g tissue; isorhamnetin<LOD; tamarixetin<LOD.

(continued)

TABLE 19.1 (CONTINUED)
Flavonoid Feeding Studies in which Flavonoids Were Directly Measured in the Brain

Model	Reference	Flavonoid Administered	Dose	Duration	Administration	Level Identified in the Brain
Flavanols						
Rat	4	St. John's wort extract and isoquercitrin	SJW, 1600 mg/kg; isoquercitrin, 100 mg/kg	8 days	Oral gavage	CNS levels for quercetin (340 ng/g) and isorhamnetin/tamarixetin (50 ng/g) at 4 hours. Repeated doses: maximal cumulation for quercetin (367 ng/g) occurred after 5 days, while isorhamnetin/tamarixetin (640 ng/g) did not reach its maximal cumulation level within the 8-day test period.
Rat	5	Quercetin	Lecithin/quercetin preparation (30 mg/kg)	Acute	Single intraperitoneal administration	Quercetin was detected in brain tissue 30 min and 1 hour after administration.
Flavonone						
Rat	6	Naringenin	20 mg/kg	Acute	I.V.	Measured by HPLC with ultraviolet detection. Naringin detection limit, 0.4 µg/g. Naringenin and its glucuronide conjugate: 2.1160.4 and 1.6560.19 µg/g, respectively, at 10 minutes post-administration.
Rat	7	[³H]naringenin and unlabeled naringenin		2 and 18 hours post-gavage	Gastric gavage	Total identified metabolites detected after 18 hours were only 1%–5% of the levels detected after 2 hours. However, the brain retained 27%, at 2 hours relative to total detected metabolites.
Rat	8	Naringin	30 mg/kg	In vivo microdialysis	I.V.	Microbore HPLC with electrochemical detection. Naringin was undetectable unless co-administered with cyclosporin A (20 mg/kg).
In vitro ECV304/C6 co-culture and in situ, Rat	9	Naringenin and quercetin				Uptake rates [K(in)] into right cerebral hemisphere were 0.145 and 0.019 ml/min/g for naringenin and quercetin, respectively. Quercetin K(in) was comparable with colchicines.

Isoflavone

Species	#	Flavonoid	Dose	Duration	Administration	Results/Methods
Rat	10	Puerarin, a natural isoflavone-C-glucoside, isolated from puerariae radix (the root of kudzu)	50 mg/kg of body weight	12 days	Orally administered in suspension by gavage to the rats four times at intervals of 72 hours	Brain and blood collected 24 hours after final administration. Brain was perfused with ice-cold normal saline. LC-MS/MS with multiple reaction monitoring analysis detected the presence of puerarin and its metabolites in brain tissue.
Rat	11	Genistein ([1-^{14}C] genistein (2.07 GBq mmol[21])	4 mg of genistein per kilogram of BW (1.833 MBq [^{14}C]genistein)	Acute	After an overnight fast, gavage, genistein suspension in corn oil	Radioactive content was quantitated by liquid scintillation counting. Brain frontal lobe genistein. Male and female: 2 hours - 69, 28 ng eq/g tissue 7 hours - 22, 17 24 hours - 10, 4
Rat	12	Genistein	Soy- and alfalfa-free feed fortified with genistein aglycone at various levels equivalent to 0, 18.5, 370, or 1852 mmol/kg	Long-term Postnatal, 21 and 140 days	Ad libitum; a single female rat dosed via gavage with genistein showed brain genistein at 0.27 pmol/mg 2 hours after dosing	Low genistein levels measured in brain tissue using LCES/MS were confirmed by repeating these analyses using MS/MS. In diets fortified with 0, 5, 100, or 500 mg/g of genistein total and (aglycone) - <0.02 (<0.02), <0.02 (<0.02), 0.32 (<0.02), 0.67 (0.04) pmol/mg, respectively

Flavan-3-ol (Catechins)

Species	#	Flavonoid	Dose	Duration	Administration	Results/Methods
Rat	13	Epicatechin	100 mg/kg BW/ day	1, 5, or 10 days (decapitation exactly 2 hours following the final oral supplementation)	Via gavage	After perfusing with ice-cold phosphate-buffered saline, cerebellum. After HPLC with UV detection — LC/MS to confirm identities. Metabolites correspond to approximately 200 nM in the final brain extract. Calculates to 0.4 nmol/g, but authors state the levels found were too low for accurate quantification.
Rat	14	Epigallocatechin-3-gallate	10 mg/kg	Acute	I.V. or oral (100 mg/ kg)	Liquid chromatography technique coupled with tandem mass spectrometry (LC-MS/ MS). The brain distribution result indicated that EGCG may potentially penetrate through the BBB at a lower rate.

(continued)

TABLE 19.1 (CONTINUED)
Flavonoid Feeding Studies in which Flavonoids Were Directly Measured in the Brain

Model	Reference	Flavonoid Administered	Dose	Duration	Administration	Level Identified in the Brain
Flavan-3-ol (Catechins)						
Human	15	Flavan-3-ols	Ingestion of green tea	Acute	Oral	By LC/MS/MS, flavan-3-ol methyl, glucuronide, and sulfate metabolites appeared in the bloodstream but did not pass through the blood-cerebrospinal fluid barrier.
Rat	16	Grape seed extract (GSE) monomers, oligomers, and polymers of flavan-3-ols	GSE (1 g/kg body weight) dissolved in water (1 g of GSE per 6 ml)	Acute	Gavage	Measured by HPLC with diode array detection and tandem mass spectrometry. No metabolite was detected in brain extracts.
Rat	17	Epigallocatechin-3-gallate	500 mg/kg BW	Acute	Oral	HPLC with chemiluminescence detection levels reached 0.5 nmol/g in the brain.
Mouse	18	[^{3}H] epigallocatechin-3-gallate	200 μl of 0.05% solution w/3.7 MBq EGCG	Acute; collection at 6, 12, and 24 hours	Gavage	Microautoradiography and oxidizer: 0.32% of total radioactivity was incorporated into brain after 24 hours. Levels at 2, 6, and 24 hours were 1.7, 9, and 22.5×10^3 dpm/100 mg tissue, respectively.
Flavone						
Rat	19	Tangeretin, a methoxyflavone from citrus fruits	10 mg/kg/day	28 days	Chronic oral administration	Tangeretin detected in hypothalamus, striatum, and hippocampus at 3.88, 2.36, and 2.00 ng/mg, respectively.
Atlantic killifish	20	5,7-dimethoxyflavone (5,7-DMF) and unmethylated analogue chrysin	5μM 5,7-DMF, or 5 μM chrysin	8 hours	In seawater	HPLC with PDA detection. Accumulation (20- to 100-fold) in all tissues examined, with the highest accumulation in the liver and brain, whereas chrysin was barely detectable in any tissues except the liver. In brain, 5,7-DMF = 307 μM and chrysin = 2 μM.
Rat	21	Baicalein	60 mg/kg	Collection by microdialysis probes	I.V.	HPLC with electrochemical detection (various brain regions - 3 to 4 μg/g tissue). CSF = 0.21 μg/ml and plasma = 9.81 μg/ml.

Anthocyanins

Animal	Ref	Substance	Dose	Duration	Administration	Results
Pig	22	Anthocyanins from blueberries	0, 1, 2, or 4% w/w blueberries (*Vaccinium corymbosum* L. "Jersey")	4 weeks	Ad libitum	Pigs fasted for 18–21 hours prior to euthanasia. No anthocyanins detected in plasma or urine. LC-MS/MS demonstrated 11 intact anthocyanins in eye, cortex, and cerebellum. Anthocyanins detected in whole brain.
Rat	23	Anthocyanins as purified grape	17.8 μmol dose/kg BW	Acute	Via gavage	
Rat	24	Anthocyanins from blueberries	2% blueberry diet	70 days	Ad libitum	Several anthocyanins found in cerebellum, cortex, hippocampus, and striatum but not in controls.
Rat	25	Anthocyanins from blueberry extract	Blueberry extract providing 14.8 mmol anthocyanin per kg diet	15 days	25 g diet/rat/day	Measured by HPLC with PDA detector with identification by HPLC-ESI-MS/MS analysis. In the brain, 0.21 nmol Cy 3-glc/g tissue and 0.25 nmol Cy 3-glc equiv/g tissue.
Rat	26	Anthocyanins and ellagitannins from raspberry juice	2.77 ml of raspberry juice	Single acute; tissues collected at 0, 1, 2, 3, 4, 6, 12, and 24 hours after administration	Via gavage	Measured by HPLC-PDA-MS/MS. Brains were perfused in situ with chilled 0.15 M KCl and then removed. Anthocyanins were not detected in the brain.
Rat	25	Blackberry extract	318 μmol/day, 15 g of blackberry extract per kg diet (i.e., 14.8 mmol anthocyanins per kg diet)	15 days	25 g diet/rat/day ad libitum	Low concentrations of anthocyanins (0.25 nmol Cy 3-glc equiv/g tissue) were detected in brains.
Rat	23	Anthocyanin mixture extracted from *Vitis vinifera* grapes	3.8 μmol anthocyanins (2 mg), 8 mg/kg of body weight of anthocyanins	10 min	Via gavage	Using HPLC with PDA-MS. Some unmetabolized anthocyanins were detected in the brains of some rats (average total anthocyanins = 192 ng/g).

References

[1] Guo, J., et al., Pharmacokinetics and tissue distribution of a water-soluble flavonol triglycoside, CTN986, in mice. *Planta Medica*, 74, 228, 2008.

[2] Bieger, J., et al., Tissue distribution of quercetin in pigs after long-term dietary supplementation. *J. Nutr.*, 138, 1417, 2008.

(continued)

TABLE 19.1 (CONTINUED)
Flavonoid Feeding Studies in which Flavonoids Were Directly Measured in the Brain

3 de Boer, V. C., et al., Tissue distribution of quercetin in rats and pigs. *J. Nutr.*, 135, 1718, 2005.

4 Paulke, A., et al., St. John's wort flavonoids and their metabolites show antidepressant activity and accumulate in brain after multiple oral doses. *Pharmazie*, 63, 296, 2008.

5 Dajas, F., et al., Cell culture protection and in vivo neuroprotective capacity of flavonoids. *Neurotoxicity Res.*, 5, 425, 2003.

6 Peng, H. W., et al., Determination of naringenin and its glucuronide conjugate in rat plasma and brain tissue by high-performance liquid chromatography. *J. Chromatogr. B Biomed. Sci. Appl.*, 714, 369, 1998.

7 El Mohsen, M. A., et al., The differential tissue distribution of the citrus flavanone naringenin following gastric instillation [erratum appears in *Free Radic. Res.*, 39, 221, 2005]. *Free Radic. Res.*, 38, 1329, 2004.

8 Tsai, T. H., Determination of naringin in rat blood, brain, liver, and bile using microdialysis and its interaction with cyclosporin A, a P-glycoprotein modulator. *J. Agric. Food Chem.*, 50, 6669, 2002.

9 Youdim, K. A., et al., Flavonoid permeability across an in situ model of the blood-brain barrier [erratum appears in *Free Radic. Biol. Med.*, 36, 1342, 2004]. *Free Radic. Biol. Med.*, 36, 592, 2004.

10 Prasain, J. K., et al., Identification of puerarin and its metabolites in rats by liquid chromatography-tandem mass spectrometry. *J. Agric. Food Chem.*, 52, 3708, 2004.

11 Coldham, N. G., and Sauer, M. J., Pharmacokinetics of [¹⁴C]genistein in the rat: Gender-related differences, potential mechanisms of biological action, and implications for human health. *Toxicol. Appl. Pharm.*, 164, 206, 2000.

12 Chang, H. C., et al., Mass spectrometric determination of genistein tissue distribution in diet-exposed Sprague-Dawley rats. *J. Nutr.*,130, 1963, 2000.

13 Abd El Mohsen, M. M., et al., Uptake and metabolism of epicatechin and its access to the brain after oral ingestion. *Free Radic. Biol. Med.*, 33, 1693, 2002.

14 Lin, L. C., et al., Pharmacokinetics of (−)-epigallocatechin-3-gallate in conscious and freely moving rats and its brain regional distribution. *J. Agric. Food Chem.*, 55, 1517, 2007.

15 Zini, A., et al., Do flavan-3-ols from green tea reach the human brain? *Nutr. Neurosci.*, 9, 57, 2006.

16 Tsang, C., et al., The absorption, metabolism and excretion of flavan-3-ols and procyanidins following the ingestion of a grape seed extract by rats [erratum appears in *Br. J. Nutr.*, 95, 847, 2006]. *Br. J. Nutr.*, 94, 170, 2005.

17 Nakagawa, K., and Miyazawa, T., Absorption and distribution of tea catechin, (−)-epigallocatechin-3-gallate, in the rat. *J. Nutr. Sci. Vitaminol.*, 43, 679, 1997.

18 Suganuma, M., et al., Wide distribution of [³H](−)-epigallocatechin gallate, a cancer preventive tea polyphenol, in mouse tissue. *Carcinogenesis*, 19, 1771, 1998.

19 Datla, K. P., et al., Tissue distribution and neuroprotective effects of citrus flavonoid tangeretin in a rat model of Parkinson's disease. *NeuroReport*, 12, 3871, 2001.

20 Tsuji, P. A., Winn, R. N., and Walle, T., Accumulation and metabolism of the anticancer flavonoid 5,7-dimethoxyflavone compared to its unmethylated analog chrysin in the Atlantic killifish. *Chem. Biol. Interact.*, 164, 85, 2006.

21 Tsai, T. H., et al., The effects of the cyclosporin A, a P-glycoprotein inhibitor, on the pharmacokinetics of baicalein in the rat: A microdialysis study. *Br. J. Pharm.*, 137, 1314, 2002.

22 Kalt, W., et al., Identification of anthocyanins in the liver, eye, and brain of blueberry-fed pigs. *J. Agric. Food Chem.*, 56, 705, 2008.

23 Passimonti, S., et al., Fast access of some grape pigments to the brain. *J. Agric. Food Chem.*, 53, 7029, 2005.

24 Andres-Lacueva, C., et al., Anthocyanins in aged blueberry-fed rats are found centrally and may enhance memory. *Nutr. Neurosci.*, 8, 111, 2005.

25 Talavéra, S., et al. Anthocyanin metabolism in rats and their distribution to digestive area, kidney and brain. *J. Agric. Food Chem.*, 53, 3902, 2005.

26 Borges, G., et al., The bioavailability of raspberry anthocyanins and ellagitannins in rats. *Mol. Nutr. Food Res.*, 51, 714, 2007.

measurements in order to validate proposed mechanisms of action. Given the number and diversity of dietary flavonoids available in diets rich in berries, fruits, and vegetables, there is considerable work to be done. Fortunately, analytical instruments and techniques are improving yearly. Inspection of data from direct measurements shows clear differences between flavonoids from particular classes with regard to their ability to permeate the BBB. Future studies directed more on flavonoids rather than on drugs may offer more precise data concerning the actions of efflux transporters on particular flavonoids. Such "bioavailability and neuropharmacokinetic data" will permit correlation with effects on cell signaling pathways, neurogenesis, and neurophysiology and prevention of neurodegenerative disorders.

REFERENCES

1. Williamson, G., and Manach, C., Bioavailability and bioefficacy of polyphenols in humans: II. Review of 93 intervention studies. *Am. J. Clin. Nutr.*, 81, 243S, 2005.
2. Middleton, E., Jr., Kandaswami, C., and Theoharides, T. C., The effects of plant flavonoids on mammalian cells: Implications for inflammation, heart disease, and cancer. *Pharmacol. Rev.*, 52, 673, 2000.
3. Harborne, J. B., and Williams, C. A., Advances in flavonoid research since 1992. *Phytochemistry*, 55, 481, 2000.
4. Dajas, F., et al., Neuroprotection by flavonoids. *Braz. J. Med. Biol. Res.*, 36, 1613, 2003.
5. Dajas, F., et al., Cell culture protection and in vivo neuroprotective capacity of flavonoids. *Neurotoxicity Res.*, 5, 425, 2003.
6. Bailey, D. G., et al., Ethanol enhances the hemodynamic effects of felodipine. *Clin. Invest. Med.*, 12, 357, 1989.
7. Saito, M., et al., Undesirable effects of citrus juice on the pharmacokinetics of drugs: Focus on recent studies. *Drug Saf.*, 28, 677, 2005.
8. Zhang, S., and Morris, M. E., Effects of the flavonoids biochanin A, morin, phloretin, and silymarin on P-glycoprotein-mediated transport. *J. Pharmacol. Exp. Ther.*, 304, 1258, 2003.
9. Kitagawa, S., Nabekura, T., and Kamiyama, S., Inhibition of P-glycoprotein function by tea catechins in KB-C2 cells. *J. Pharm. Pharmacol.*, 56, 1001, 2004.
10. Fuchikami, H., et al., Effects of herbal extracts on the function of human organic anion-transporting polypeptide OATP-B. *Drug Metab. Dispos.*, 34, 577, 2006.
11. Wu, C.-P., et al., Modulatory effects of plant phenols on human multidrug resistance proteins 1, 4, and 5 (ABCC1, 4 and 5). *FEBS J.*, 272, 4725, 2005.
12. Vaidyanathan, J. B., and Walle, T., Transport and metabolism of the tea flavonoid (−)-epicatechin by the human intestinal cell line Caco-2. *Pharm Res.*, 18, 1420, 2001.
13. Mandel, S., et al., Green tea polyphenol (−)-epigallocatechin-3-gallate protects rat PC12 cells from apoptosis induced by serum withdrawal independent of P13-Akt pathway. *Neurotoxicity Res.*, 5, 419, 2003.
14. Youdim, K. A., et al., Interaction between flavonoids and the blood-brain barrier: *In vitro* studies. *J. Neurochem.*, 85, 180, 2003.
15. Walle, T., Absorption and metabolism of flavonoids. *Free Radic. Biol. Med.*, 36, 829, 2004.
16. Manach, C., and Donovan, J. L., Pharmacokinetics and metabolism of dietary flavonoids in humans. *Free Radic. Res.*, 38, 771, 2004.
17. Hollman, P. C., et al., The sugar moiety is a major determinant of the absorption of dietary flavonoid glycosides in man. *Free Radic. Res.*, 31, 569, 1999.
18. Rios, L., et al., Cocoa proanthocyanidins are stable during gastric transit in humans. *Am. J. Clin. Nutr.*, 76, 1106, 2002.
19. Sesink, A. L., et al., Intestinal uptake of quercetin-3-glucoside in rats involves hydrolysis by lactase phlorizin hydrolase. *J. Nutr.*, 133, 773, 2003.
20. Day, A. J., et al., Dietary flavonoid and isoflavone glycosides are hydrolysed by the lactase site of lactase phlorizin hydrolase. *FEBS Lett.*, 468, 166, 2000.
21. Ioku, K., et al., Glucosidase activity in the rat small intestine toward quercetin monoglucosides. *Biosci. Biotechnol. Biochem.*, 62, 1428, 1998.
22. Day, A. J., et al., Deglycosylation of flavonoid and isoflavonoid glycosides by human small intestine and liver β-glucosidase activity. *FEBS Lett.*, 436, 71, 1998.
23. Nemeth, K., et al., Deglycosylation by small intestinal epithelial cell β-glucosidases is a critical step in the absorption and metabolism of dietary flavonoid glycosides in humans. *Eur. J. Nutr.*, 42, 29, 2003.

24. Walle, T., et al., Flavonoid glucosides are hydrolyzed and thus activated in the oral cavity in humans. *J. Nutr.*, 135, 48, 2005.
25. Browning, A. M., Walle, U. K., and Walle, T., Flavonoid glycosides inhibit oral cancer cell proliferation — Role of cellular uptake and hydrolysis to the aglycones. *J. Pharm. Pharmacol.*, 57, 1037, 2005.
26. Crespy, V., et al., Quercetin, but not its glycosides, is absorbed from the rat stomach. *J. Agric. Food. Chem.*, 50, 618, 2002.
27. Hollman, P. C. H., et al., Absorption of dietary quercetin glycosides and quercetin in healthy ileostomy volunteers. *Am. J. Clin. Nutr.*, 62, 1276, 1995.
28. Walgren, R. A., et al., Cellular uptake of dietary flavonoid quercetin 4'-β-glucoside by sodium-dependent glucose transporter SGLT1. *J. Pharmacol. Exp. Ther.*, 294, 837, 2000.
29. Wolffram, S., Blöck, M., and Ader, P., Quercetin-3-glucoside is transported by the glucose carrier SGLT1 across the brush border membrane of rat small intestine. *J. Nutr.*, 132, 630, 2002.
30. Walton, M. C., et al., The flavonol quercetin-3-glucoside inhibits cyanidin-3-glucoside absorption *in vitro*. *J. Agric. Food Chem.*, 54, 4913, 2006.
31. Crespy, V., et al., Part of quercetin absorbed in the small intestine is conjugated and further secreted in the intestinal lumen. *Am. J. Physiol.*, 277, G120, 1999.
32. Karakaya, S., Bioavailability of phenolic compounds. *Crit. Rev. Food Sci. Nutr.*, 44, 453, 2004.
33. Lu, H., Meng, X., and Yang, C. S., Enzymology of methylation of tea catechins and inhibition of catechol-*O*-methyltransferase by (2)-epigallocatechin gallate. *Drug Metab. Disp.*, 31, 572, 2003.
34. Walgren, R. A., et al., Efflux of dietary flavonoid quercetin 4'-β-glucoside across human intestinal Caco-2 cell monolayers by apical multidrug resistance-associated protein-2. *J. Pharmacol. Exp. Ther.*, 294, 830, 2000.
35. Van der Woude, H., et al., Identification of 14 quercetin phase II mono- and mixed conjugates and their formation by rat and human phase II *in vitro* model systems. *Chem. Res. Toxicol.*, 17, 1520, 2004.
36. Aird, W. C., Phenotypic heterogeneity of the endothelium: I. Structure, function, and mechanisms. *Circ. Res.*, 100, 158, 2007.
37. Ueno, M., et al., Blood-brain barrier permeability in the periventricular areas of the normal mouse brain. *Acta Neuropathol.*, 99, 385, 2000.
38. Pollay, M., and Roberts, P. A., Blood-brain barrier: A definition of normal and altered function. *Neurosurgery*, 6, 675, 1980.
39. Pardridge, W. M., Drug and gene targeting to the brain with molecular Trojan horses. *Nat. Rev. Drug Disc.*, 1, 131, 2002.
40. Schinkel, A. H., The physiological function of drug-transporting P-glycoproteins. *Semin. Cancer Biol.*, 8, 161, 1997.
41. Litman, T., et al., From MDR to MXR: New understanding of multidrug resistance systems, their properties and clinical significance. *Cell. Mol. Life Sci.*, 58, 931, 2001.
42. Müller, M., 48 human ATP-binding cassette transporters. http://www.nutrigene.4t.com/humanabc.htm, 2001.
43. Borst, P., and Oude Elferink, R., Mammalian ABC transporters in health and disease. *Annu. Rev. Biochem.*, 71, 537, 2002.
44. Thiebaut, F., et al., Cellular localization of the multidrug-resistance gene product P-glycoprotein in normal human tissues. *Proc. Natl. Acad. Sci. U S A*, 84, 7735, 1987.
45. Cordon-Cardo, C., et al., Multidrug-resistance gene (P-glycoprotein) is expressed by endothelial cells at blood-brain barrier sites. *Proc. Natl. Acad. Sci. U S A*, 86, 695, 1989.
46. Kusuhara, H., Suzuki, H., and Sugiyama, Y., The role of P-glycoprotein and canalicular multispecific organic anion transporter in the hepatobiliary excretion of drugs. *J. Pharm. Sci.*, 87, 1025, 1998.
47. Di Pietro, A., et al., Modulation by flavonoids of cell multidrug resistance mediated by P-glycoprotein and related ABC transporters. *Cell. Mol. Life Sci.*, 59, 307, 2002.
48. Boumendjel, A., et al., Recent advances in the discovery of flavonoids and analogs with high-affinity binding to P-glycoprotein responsible for cancer cell multidrug resistance. *Med. Res. Rev.*, 22, 512, 2002.
49. Cisternino, S., et al., Expression, up-regulation, and transport activity of the multidrug-resistance protein ABCG2 at the mouse blood-brain barrier. *Cancer Res.*, 64, 3296, 2004.
50. Lee, Y. J., et al., Investigation of efflux transport of dehydroepiandrosterone sulfate and mitoxantrone at the mouse blood-brain barrier: A minor role of breast cancer resistance protein. *J. Pharmacol. Exp. Ther.*, 312, 44, 2005.
51. Cooray, H. C., et al., Localisation of breast cancer resistance protein in microvessel endothelium of human brain. *NeuroReport*, 13, 2059, 2002.

52. Jacquemin, E., et al., Expression cloning of a rat liver Na(+)-independent organic anion transporter. *Proc. Natl. Acad. Sci. U S A*, 91, 133, 1994.

53. Noe, I. B., et al., Isolation of a multispecific organic anion and cardiac glycoside transporter from rat brain. *Proc. Natl. Acad. Sci. U S A*, 94, 10346, 1997.

54. Abe, T., et al., Molecular characterization and tissue distribution of a new organic anion transporter subtype (oatp3) that transports thyroid hormones and taurocholate and comparison with oatp2. *J. Biol. Chem.*, 273, 22395, 1998.

55. Tamai, I., et al., Molecular identification and characterization of novel members of the human organic anion transporter (OATP) family. *Biochem. Biophys. Res. Commun.*, 273, 251, 2000.

56. Hagenbuch, B., and Meier, P. J., Organic anion transporting polypeptides of the OATP/SLC21 family: Phylogenetic classification as OATP/SLCO superfamily, new nomenclature and molecular/functional properties. *Pflugers Arch.*, 447, 653, 2004.

57. Wang, X., Wolkoff, A. W., and Morris, M. E., Flavonoids as a novel class of human organic anion-transporting polypeptide OATP1B1 (OATP-C) modulators. *Drug Metab. Dispos.*, 33, 1666, 2005.

58. Middleton, E. Jr., Kandaswami, C., and Theoharides, T. C., The effects of plant flavonoids on mammalian cells: Implications for inflammation, heart disease, and cancer. *Pharmacol Rev.*, 52, 673, 2000.

59. Gledhill, J. R., et al., Mechanism of inhibition of bovine F1-ATPase by resveratrol and related polyphenols. *Proc. Natl. Acad. Sci. U S A*, 104, 13632, 2007.

60. Howitz, K. T., et al., Small molecule activators of sirtuins extend *Saccharomyces cerevisiae* lifespan. *Nature*, 425, 191, 2003.

61. Youdim, K. A., et al., Flavonoid permeability across an in situ model of the blood-brain barrier. *Free Radic. Biol. Med.*, 36, 592, 2004.

62. Gescher, A. J., and Steward, W. P., Relationship between mechanisms, bioavailability, and preclinical chemopreventive efficacy of resveratrol: A conundrum. *Cancer Epidemiol. Biomark. Prev.*, 12, 953, 2003.

63. Wenzel, E., et al., Bioactivity and metabolism of trans-resveratrol orally administered to Wistar rats. *Mol. Nutr. Food Res.*, 49, 482, 2005.

64. Wenzel, E., and Somoza, V., Metabolism and bioavailability of trans-resveratrol. *Mol. Nutr. Food Res.*, 49, 472, 2005.

65. Walle, T., et al., High absorption but very low bioavailability of oral resveratrol in humans. *Drug Metab. Dispos.*, 32, 1377, 2004.

66. Yu, C. W., et al., Human, rat, and mouse metabolism of resveratrol. *Pharm. Res.*, 19, 1907, 2002.

67. Abd El Mohsen, M. M., et al., Uptake and metabolism of epicatechin and its access to the brain after oral ingestion. *Free Radic. Biol. Med.*, 33, 1693, 2002.

68. Suganuma, M., Wide distribution of [^{3}H](−)-epigallocatechin gallate, a cancer preventive tea polyphenol, in mouse tissue. *Carcinogenesis*, 19, 1771, 1998.

20 Prevention and Treatment of Neurodegenerative Diseases by Spice-Derived Phytochemicals

Bharat B. Aggarwal[1], Kuzhuvelil B. Harikumar, and Sanjit Dey

Cytokine Research Laboratory, Department of
Experimental Therapeutics, University of Texas M. D.
Anderson Cancer Center, Houston, Texas, USA

CONTENTS

20.1 INTRODUCTION

Why many chronic illnesses are more common among individuals from certain countries is not fully understood. Extensive observational and experimental research carried out within the last half-century has indicated that lifestyle plays a major role in chronic disease [1]. Interaction of lifestyle factors with genetic factors leads to the disease process. Food is perhaps one of the most important lifestyle components that could modulate the course of various diseases. Hippocrates proclaimed almost 25 centuries ago, "Let food be thy medicine, and medicine by thy food." This is no different from the common saying, "You are what you eat," or recommendations from the National Institutes of Health to increase intake of fruits and vegetables for their preventive properties. The components in fruits and vegetables that prevent disease and the mechanisms by which it is done are increasingly becoming evident. One food substance consumed primarily on the Indian subcontinent is a variety of spices. These spices are used in Indian cooking as preservatives and for taste and appearance, but increasingly appear to have medicinal value as well. More than 45 different spices have been identified in Indian cuisine. These include turmeric, red chili, fenugreek, cinnamon, cumin, coriander, nutmeg, cloves, asafetida, fennel, and ginger (Figure 20.1). Several of these spices have been described in Ayurveda for the treatment of various inflammatory diseases.

In the past century, various phytochemicals have been identified in spices that exhibit antioxidant and anti-inflammatory effects. Since inflammation plays a major role in most chronic illnesses [2–3], these spices would tend to have benefit for those who consume them. Observational studies have shown reduced incidence of these chronic diseases in spice-consuming countries, but now experimental evidence supports the notion that the reduction is due to dietary factors, namely, spices. The focus of the current review is to describe the potential of spice-derived phytochemicals for prevention and therapy of neurodegenerative and other diseases.

20.2   PROTECTION FROM NEUROTOXICITY

We have summarized the effect of different spice-derived phytochemicals on several degenerative diseases in Table 20.1 and the various molecular targets modulated by these phytochemicals is depicted in Figure 20.2. Most of the information available about spice-derived phytochemicals for neurological diseases is on curcumin, a phytochemical derived from the "golden spice," turmeric, popularly known as curry powder (Figure 20.3). Numerous studies have indicated that curcumin exhibits antioxidant effects in the brain. Oral administration of curcumin to rats has been shown to significantly decrease the lipid peroxidation (LPO) in the brain induced by carbon tetrachloride, paraquat, and cyclophosphamide [4]. Curcumin was also found to significantly lower the serum cholesterol levels in these animals. Rajakumar and Rao [5] examined the effect of dehydrozingerone and curcumin on LPO in rat brain homogenates. Both test compounds inhibited the formation of conjugated dienes and spontaneous LPO. These compounds also inhibited LPO induced by ferrous ions, ferric-ascorbate, and ferric-ADP-ascorbate. In all these cases, curcumin was more active than dehydrozingerone and dl-alpha-tocopherol. In another study, the same group compared the effect of demethoxycurcumin (DMC), bisdemethoxycurcumin (BDMC), and acetylcurcumin with curcumin on iron-stimulated LPO in rat brain homogenates [6]. All compounds were found to be equally active, and more potent than alpha-tocopherol. These results showed that the methoxy and phenolic groups contribute little to the activity. Spectral studies showed that all compounds could interact with iron. Thus, the inhibition of iron-catalyzed LPO by curcuminoids may involve iron chelation.

In another study, the neuroprotective effect of curcumin was examined using ethanol as a model for brain injury [7]. Oral administration of curcumin to rats was found to significantly inhibit the peroxidation of brain lipids and enhance glutathione (GSH) in ethanol-intoxicated rats, suggesting it had a protective role against ethanol-induced brain injury.

Excessive nitric oxide (NO) production in the brain has been correlated with neurotoxicity [8]. NO production from the neuroglial cells surrounding neurons contributes significantly to the pathogenesis of neurodegenerative diseases; suppression of NO production in these cells may be beneficial in treating many disorders. The effect of spice-derived phytochemicals on the release of NO from lipopolysaccharide (LPS)/interferon gamma (IFN-γ)-stimulated C6 astrocyte cells has been investigated [9]. Results indicated that curcumin, caffeic acid, and apigenein inhibited NO production.

Curcumin is a potent stimulator of stress-induced expression of Hsp27, alpha B crystallin, and Hsp70. When C6 rat glioma cells were exposed to arsenite, $CdCl_2$, or heat in the presence of curcumin, induction of the synthesis of all three proteins was markedly stimulated [10]. Curcumin prolonged the stress-induced activation of the heat shock element (HSE)-binding activity of heat shock transcription factor in cultured cells. Induction of Hsp27, alpha B crystalline, and Hsp70 in the liver and adrenal glands of heat-stressed rats was also enhanced by prior injection of curcumin.

Another mechanism by which curcumin could exhibit neuroprotective effect is through suppression of activator protein-1 (AP-1). For instance, in rat cerebellar granule cells, glutamate-induced apoptosis was found to be suppressed by curcumin through inhibition of AP-1 [11]. Suppression of

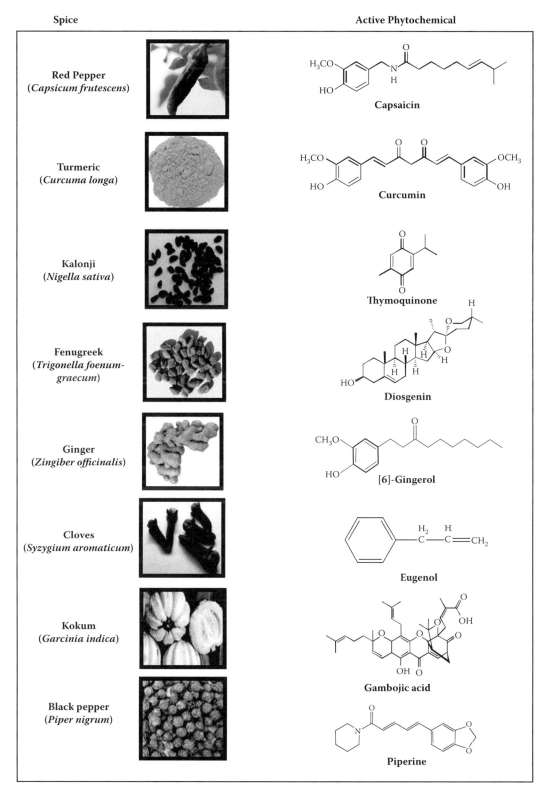

Spice | Active Phytochemical

Red Pepper (*Capsicum frutescens*) — **Capsaicin**

Turmeric (*Curcuma longa*) — **Curcumin**

Kalonji (*Nigella sativa*) — **Thymoquinone**

Fenugreek (*Trigonella foenum-graecum*) — **Diosgenin**

Ginger (*Zingiber officinalis*) — **[6]-Gingerol**

Cloves (*Syzygium aromaticum*) — **Eugenol**

Kokum (*Garcinia indica*) — **Gambojic acid**

Black pepper (*Piper nigrum*) — **Piperine**

FIGURE 20.1 **(See color insert following page 234.)** Structure and sources of various spice-derived phytochemicals.

TABLE 20.1

Effect of Spice-Derived Phytochemicals on Neurodegenerative Diseases

Disease	Dose	Effects	References
Alzheimer's disease			
In vitro		Curcumin protected nerve cells and EC from Abeta-induced cytotoxicity	[101]
In vitro	0.1–1.0 μM	Curcumin inhibited Abeta fibril formation	[36]
In vitro		Curcumin interacted with Cu and Fe	[101]
In vitro	0.1–10.0 μM	Curcumin inhibited the peroxidase activity of Abeta-heme complex	[103]
In vitro	0.1 μM	Curcumin enhanced uptake of Abeta by macrophages from AD	[104]
In vitro	0.1 μM	Curcumin enhanced Abeta uptake by increasing expression of MGAT3 and TLRs	[105]
In vitro		Curcumin (a p300 inhibitor) blocked IL-1beta-induced proMMP-9 expression	[106]
In vitro		Antioxidant and metal-binding properties of spice ingredients prevent or treat neurodegenerative diseases such as AD	[15]
Tg mice	160, 5000 ppm, diet	Curcumin reduces oxidative damage and oxidative pathology	[108]
Tg mice	500 ppm, diet, 50 μM, i.v.	Curcumin inhibits Abeta oligomers and fibrils, binds plaques, and reduces amyloid[a]	[35]
Mice	7.5 mg/kg/day i.v.	Curcumin cleared and reduced existing plaques; reversed changes in dystrophic dendrites, abnormal curvature, and dystrophy size	[109]
Tg Mice		Curcumin suppressed PAK translocation in aged Tg2576 transgenic mice	[40]
Patients	Capsule or diet	Curcumin has no side effects, increased the level of Abeta in serum	[110]
Huntington's disease			
In vitro	100 μM	Gambogic acid inhibited the *in vitro* aggregation of amino-terminal mutant Huntington fragment	[111]
Multiple sclerosis			
In vitro		Curcumin modulated IFN-β and IL-12 signaling	[112]
Mice	50, 100 μg, i.p.	Curcumin decreased EAE and IL-12 production, inhibited IL-12-induced JAK2 and TYK2 phosphorylation	[113]
Mice	2.5 mg/mL, d.w.	Curcumin delayed recovery from EAE	[114]
Rat	1 mg/kg	Thymoquinone inhibited experimental autoimmune encephalomyelitis, inhibits NF-κB	[29]
Mice	0.5 mg/kg	Capsaicin inhibited EAE[b]	[115]
Rats	1 mg/kg, i.p.	Capsaicin inhibited EAE	[116]
Parkinson's disease			
In vitro	50 μM	Curcumin reversed peroxynitrate mediated inhibition brain mitochondria complex I	[47]
Rats	50 mg/kg, p.o.	Curcumin attenuated the loss of dopaminergic neurons, protected rats from 6-OHDA Induced Parkinson disease	[52]
	0.1, 1 μmol/kg	Eugenol protected mice from 6-OHDA induced Parkinson's disease	[54]

TABLE 20.1 (CONTINUED)
Effect of Spice-Derived Phytochemicals on Neurodegenerative Diseases

Disease	Dose	Effects	References
Epilepsy			
Mice	3, 30 mg/kg, i.p.	Curcumin decreased the severity of epilepsy, attenuated kainate-induced histone modifications	[117]
Rats	50 mg/kg, i.p.	Curcumin protected from KA-induced neuronal damage, reduced the level of NO, decreased the expression of c-jun, COX-2, BDNF, and iNOS[c]	[60,61]
In vitro	2–100 μmol	Eugenol suppressed epileptiform field potentials and spreading depression in rat neocortical and hippocampal tissues	[83]
Cerebral injury			
Rats	20, 100, 200 mg/kg, i.p.	Curcumin protected rat brain against I/R injury through modulation of XO, O2$^-$, MDA, GPx, SOD, and LDH	[63]
Rats	100, 300 mg/kg, i.p.	Curcumin protected rat brain from cerebral ischemia; modulated the activity of GPx and SOD	[67]
Mongolian gerbils	30 mg/kg, i.p., or 2 g/kg, diet	Curcumin protects I/R-induced neuronal cell death and glial activation; decreased LPO	
		Curcumin prevented mitochondrial dysfunction and the apoptosis; curcumin levels increased in plasma and brain within 1 h	[64]
Rats	200 mg/kg, i.p.	Curcumin reduced the neuronal damage, decreased the level of LPO, increased the level of GSH and activities of SOD and CAT	[65]
Rats	100 mg/kg, p.o.	Curcumin protected infarct in MCAO models and as antioxidant effect	[123]
Rats	500 mg/kg, i.p.	Curcumin delayed neuronal death, increased antioxidant system and levels of peroxynitrite[d]	[124]
Rats	1, 2 mg/kg, i.v.	Curcumin prevented blood-brain barrier damage, improved neurological deficit, decreased mortality and the level of iNOS	[66]
Rats	2 mg/kg	Gambogic acid inhibited kainic acid-triggered neuronal cell death and decreased infarct volume in the transient MCAO model of stroke[e]	[118]
Rats	1 mg/kg	Thymoquinone inhibited experimental autoimmune encephalomyelitis; inhibited NF-κB	[29]
Rats	5 mg/kg	Thymoquinone protected hippocampus from forebrain ischemia, increased the level of GSH, SOD	[69]
Rats	2.5, 5,10 mg/kg	Thymoquinone inhibited LPO, protected rat brain against I/R injury and CAT	[68]
Rats	50 mg/kg	Thymoquinone prevented hippocampal neurodegeneration	[24]
Brain cancer			
In vitro	5, 10, 20 μM	Curcumin inhibited MMP gene expression in astroglioma cells	[37]
In vitro	5–20 μM	Curcumin induced growth arrest in sub G1 phase, sensitized glioma cells to TRAIL-induced apoptosis	[88]
In vitro	1–4 μg/ml	Curcumin inhibited HATs, induced apoptosis	[89]
In vitro	10–25 μM	Curcumin inhibited the growth and overexpression of NF-κB, AP-1, Akt, JNK, DNA repair enzymes and suppressed chemoresistance	[94]
In vitro	4, 10 μM	Curcumin arrested the growth of glioma cells in G2/M transition, increased the expression of ING4, p53, p21	[95]
In vitro	25, 50 μM	Curcumin induced apoptosis via mitochondrial pathway, activated calpain, and down-regulated NF-κB, and IAPs	[92]

(continued)

TABLE 20.1 (CONTINUED)
Effect of Spice-Derived Phytochemicals on Neurodegenerative Diseases

Disease	Dose	Effects	References
In vitro	100–250 μM	Capsaicin induced apoptosis in human glioblastoma cells	[25]
In vitro	50 μmol	Capsaicin-induced apoptosis of glioma cells is mediated by TRPV1 vanilloid receptor; activated p38MAPK	[125]
In vitro	0.5 μM	Gambogic acid bound to TrkA, prevented glutamate-induced neuronal cell death, induced neurite outgrowth in PC12 cells	[118]
In vitro	1–2 μM	Gambogic acid inhibited growth and induced apoptosis in glioma cells[e]	[119]
Rats	0.5–2 mg/kg, i.v	Gambogic acid inhibited the growth of orthotopic glioma, induced apoptosis	[119]
Depression			
In vitro	0–40 μM	Curcumin protected nerve cells via BDNF/TrkB signaling pathway	[126]
In vitro	12.5–100 μM	Piperine inhibited the growth of cultured neurons from embryonic rat brain	[80]
Mice	1–20 mg	Piperine showed antidepressant activity, modulated serotonergic system	[77]
Mice	2.5, 5, 10 mg/kg	Piperine protected mice from CMS, up-regulated BDNF	[78]
Mice	10, 30, 100 mg/kg	Eugenol showed antidepressant-like activity and induced expression of metallothionein-III in the hippocampus	[81]
Prion disease			
Mice	50, 500 mg/kg p.o.	Curcumin at low dose, the animal survived from infection	[120]
Meningitis			
Mice		Inhibited the symptoms of meningitis	[100]
Neuropathic pain			
Mice	15, 30, 60 mg/kg, i.p.	Curcumin attenuated thermal hyperalgesia in a diabetic mouse model of neuropathic pain	[62]
Rats	60 mg/kg, p.o	Curcumin suppressed diabetic neuropathic pain through inhibition of NO and TNF	[121]
Rats	40 mg/kg	Eugenol alleviated neuropathic pain	[122]

[1] Mice were given 500 ppm curcumin in diet and on day of perfusion, 50 μM curcumin was given intravenously.

[2] Used arvanil, a synthetic capsaicin-anandamide hybrid.

[3] Used manganese complex of curcumin.

[4] Curcuma oil was used.

[5] Gambojic amide is used.

AAAs, abdominal aortic aneurysms; AB, aortic banding; A-beta, amyloid beta; AD, Alzheimer's disease; AAGE, advanced glycation; BDNF, brain-derived neurotphic factor; CMS, chronic mild stress; CRP, C-reactive protein; Cu, copper; DSS, dextran sulfate sodium; EAE, experimental autoimmune encephalomyelitis; ET-1, endothelin-1; Fe, Ferrous; GST-PI, glutathione-S-transferase; HASMC, human airway smooth muscle cells; i.p., intraperitoneal; I/R, ischemia-reperfusion; i.v., intravenous; IDV, idinavir; IFN, interferon; IL, interleukin; LPO, lipid peroxidation; MCAO, middle cerebral artery occlusion; MCP-1, macrophage chemotactic protein-1; MGAT3, beta-1,4-mannosyl-glycoprotein 4-beta-N-acetylglucosaminyltransferase; MMIF, macrophage migration inhibitory factor; MPO, myeloperoxidae; MVEC, microvascular endothelial cells; NF-κB, nuclear factor kappa B; NOS, nitric oxide synthase; p.o., orally; RA, rheumatoid arthritis; RANTES, regulated upon activation, normal T cell expressed and secreted; TBARS, thiobarbituric acid reacting substances; TLR, Toll-like receptors; TNBS, 2,4,6-trinitrobenzene sulfonic acid; TNF, tumor necrosis factor; VEGF, vascular endothelial growth factor.

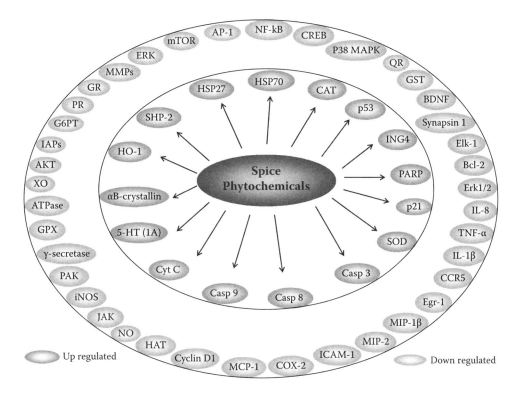

FIGURE 20.2 (See color insert following page 234.) Molecular targets modulated by various spice-derived phytochemicals. AP-1, activator protein-1; BDNF, brain-derived neurotrophic factor; CAT, catalase; Casp, caspase; CCR5, chemokine (C-C motif) receptor 5; COX-2, cyclooxygenase-2; CREB, cyclic adenosine monophosphate (cAMP) response element-binding protein; Cyt C, cytochrome c; Egr-1, early growth response factor 1; GPX, glutathione peroxidase; GST, glutathione-S-transferase; HAT, histone acetyl transferase; (HO)-1, heme oxygenase HSP, heat shock proteins; ICAM-1, intracellular adhesion molecule 1; IAPs, inhibitor of apoptosis; iNOS, inducible nitric oxide synthase; MAPKs, mitogen-activated protein kinases; MCP-1, macrophage chemotactic protein-1; MCP, monocyte chemoattractant protein 1; MIP-2, macrophage inflammatory protein; MMPs, matrixmetallo proteases; mTOR, mammalian target of ripamycin; NF-κB, nuclear factor kappa B; PR, progesterone receptor; QR, quinone reductase; SOD, superoxide dismutase; STAT, signal transducers and activator of transcription protein; TNF, tumor necrosis factor; VEGF, vascular endothelial growth factor; XO, xanthine oxidase.

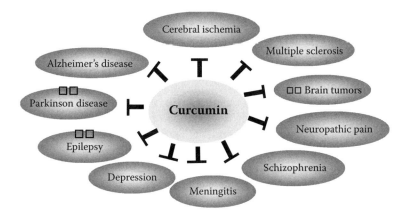

FIGURE 20.3 (See color insert following page 234.) Potential role of curcumin in the treatment of various neurodegenerative diseases.

the Janus kinase/signal transducers and activator of transcription (JAK-STAT) pathway has been described as another mechanism for the neuroprotective effects of curcumin [12]. Treatment of microglial cells with curcumin led to an increase in phosphorylation and association with JAK1/2 of SHP-2, which inhibited the initiation of JAK-STAT inflammatory signaling in activated microglia. This led to suppression of the ganglioside-, LPS-, or IFN-γ-stimulated induction of cyclooxygenase-2, intracellular adhesion molecule-1, monocyte chemoattractant protein 1, and inducible NO synthase. These data suggested that curcumin suppresses JAK-STAT signaling via activation of SHP-2, thus attenuating the inflammatory response of brain microglial cells.

Evidence suggests that increases in dopamine (DA) in striata may factor into neurodegenerative processes during acute ischemia, hypoxia, and excitotoxicity. Molecular events involved in DA toxicity were examined in a rat model of intrastriatal DA injection [13]. Intrastriatal injections of DA results in apoptotic cell death. Injections of DA produced a strong and prolonged AP-1 activity containing c-*fos*, c-*jun*, and phosphorylated c-jun protein. DA injections also stimulated the activity of nuclear factor–kappa B (NF-κB). Injection of curcumin decreased the DA-induced apoptosis. Moreover, preinjection of the antioxidant GSH significantly inhibited both DA-induced activation of AP-1 and NF-κB and subsequent apoptosis. Thus, DA-induced apoptosis *in vivo* is possibly inhibited by curcumin through suppression of the activation of AP-1 and NF-κB.

There are numerous studies to suggest that spice-derived phytochemicals protect the brain from various toxic agents. Recently, curcumin has been implicated as a neuroprotectant in the treatment of various neurological disorders. 3-Nitropropionic acid (3-NP) is a well-known fungic toxin causing neurotoxicity. Systemic administration of 3-NP causes motor and cognitive deficits that are associated with excessive free radical generation. Kumar et al. [14] investigated the possible neuroprotective mechanisms of curcumin in attenuating 3-NP-induced neurotoxicity. The effect of curcumin was examined in 3-NP-induced cognitive impairment and oxidative stress in rats. Intraperitoneal (i.p.) administration of 3-NP caused loss in body weight, declining motor function, poor memory retention, and changes in oxidative stress (LPO and reduced GSH and nitrite levels) parameters in the brain. Long-term treatment with curcumin once daily before 3-NP administration reduced the 3-NP-induced motor and cognitive impairment. Biochemical analysis revealed that curcumin administration significantly attenuated 3-NP-induced oxidative stress in the brains of rats. It also significantly restored the decreased succinate dehydrogenase activity. The results of this study clearly indicated that curcumin (through its antioxidant activity) offered neuroprotection against 3-NP-induced behavioral and biochemical alterations.

Dairam and coworkers [15] investigated the neuroprotective effects of curcuminoids against lead-induced neurotoxicity. They found that lead significantly increased LPO and reduced the viability of primary hippocampal neurons in culture. This lead-induced toxicity was significantly curtailed by the coincubation of neurons with curcuminoids. Rats were trained in a water maze and thereafter dosed for 5 days with lead alone or in conjunction with curcumin, DMC, or BDMC. Animals treated with curcumin or DMC (but not BDMC) had more GSH and less oxidized protein in the hippocampus than those treated with lead alone. These animals also had faster escape latencies when compared to lead-treated animals, indicating that curcumin- and DMC-treated animals retained their spatial reference memory. The findings of this study indicate that curcumin is capable of playing a major role in preventing and treating heavy metal-induced neurotoxicity.

Jung et al. [16] have investigated whether curcumin can protect primary rat microglial cells from the brain mononuclear phagocytes causing neuronal damage through NO production. Curcumin decreased NO production in LPS-stimulated microglial cells, suppressed both mRNA and protein levels of inducible nitric oxide synthase (iNOS), and altered phosphorylation of all mitogen-activated protein kinases (MAPKs), NF-κB, and AP-1. Analysis of the features of specific MAPK inhibitors revealed that a series of signaling cascades—including JNK, p38, and NF-κB—play a critical role in curcumin-mediated NO inhibition in microglial cells. Thus, curcumin holds promise for the prevention and treatment of both NO- and microglial cell-mediated neurodegenerative disorders.

In the central nervous system, heme oxygenase (HO)-1 has been reported to operate as a fundamental defensive mechanism for neurons exposed to an oxidant challenge. Treatment of astrocytes with curcumin up-regulated expression of HO-1 protein at both the cytoplasmic and nuclear levels [17]. Astrocytes exposed to curcumin exhibited a significant expression of two-phase II detoxification enzymes namely quinone reductase and GST. Moreover, elevated expression of HO-1 mRNA and protein were detected in cultured hippocampal neurons exposed to curcumin. Higher concentrations of curcumin caused a substantial cytotoxic effect, with no change in HO-1 protein expression. Interestingly, preincubation with curcumin resulted in an enhanced cellular resistance to glucose oxidase-mediated oxidative damage; this cytoprotective effect was considerably attenuated by zinc protoporphyrin IX, an inhibitor of heme oxygenase activity. These findings give additional support to the possible use of curcumin as a dietary preventive agent against oxidative stress-related diseases.

The pervasive action of oxidative stress on neuronal function and plasticity after traumatic brain injury (TBI) is becoming increasingly recognized. Whether curcumin can counteract the oxidative damage to the injured brain has been investigated [18]. The effect of curcumin on the synaptic plasticity and cognition in the injured brain was also examined. The analysis was focused on the brain-derived neurotrophic factor (BDNF) system based on its action on synaptic plasticity and cognition by modulating synapsin I and cyclic adenosine monophosphate response element-binding protein (CREB). Because a high-fat diet has been shown to exacerbate the effects of TBI on synaptic plasticity and cognitive function, rats were exposed to a regular diet or a diet high in saturated fat with or without curcumin for 4 weeks, and then given a mild fluid percussion injury. Supplementation of curcumin in the diet dramatically reduced oxidative damage and normalized levels of BDNF, synapsin I, and CREB that had been altered by TBI. Furthermore, curcumin supplementation counteracted the cognitive impairment caused by TBI. Curcumin can prevent oxidative stress-induced injury to the brain by acting through the BDNF system to affect synaptic plasticity and cognition.

Zhu et al. [19] investigated the effect of curcumin on *tert*-butyl hydroperoxide (*t*-BHP)-induced oxidative damage in rat cortical neurons. They showed that exposure of the neurons to *t*-BHP for 60 min caused loss of mitochondrial membrane potential (MMP), cytochrome *c* release from the mitochondria, subsequent activation of caspase-3 and cleavage of PARP, and cell apoptosis. After removal of *t*-BHP and further treatment with curcumin, curcumin abrogated the loss of MMP and release of cytochrome *c*, blocked activation of caspase 3, and altered the expression of members of the Bcl-2 family. It prevented decrease of cellular GSH and decreased intracellular ROS generation, and eventually attenuated *t*-BHP-induced apoptosis in cortical neurons. Thus, curcumin attenuates oxidative damage to cortical neurons.

The potential neuroprotective effects curcumin exhibits against lead-induced neurotoxicity in rats has also been investigated [20]. Exposure of rats to lead for 45 days caused an increase in LPO and a decrease in reduced GSH, superoxide dismutase (SOD), and catalase levels in the cerebellum, corpus striatum, hippocampus, and frontal cortex as compared with controls. Cotreatment with curcumin and lead for 45 days caused a significant decrease in LPO with concomitant decrease in lead levels in all brain regions and reversed the changes in GSH levels, SOD, and catalase activities. Therefore, curcumin prevents lead-induced neurotoxicity.

Daniel et al. [21] showed that through metal binding, curcumin protects against lead- and cadmium-induced LPO in rat brain homogenates and against lead-induced tissue damage in rat brains. Coronal brain sections of rats injected with lead acetate in the presence and absence of curcumin were compared microscopically to determine the extent of lead-induced damage to the cells in the hippocampal CA1 and CA3 regions. Electrochemical, ultraviolet spectrophotometric, and infrared spectroscopic analyses showed that the chelation of lead and cadmium by curcumin was the mechanism of neuroprotection against such heavy metal insult to the brain. There was interaction between curcumin and both cadmium and lead, with the possible formation of a complex between the metals and this ligand, suggesting that curcumin could be used therapeutically to chelate these toxic metals, potentially reducing their neurotoxicity and damage to tissue. Luo et al. [22] showed that curcumin exhibits protective effects on infection (pertussis bacilli)-induced brain edema in rats. This effect was associated with antioxidant properties, inhibition of the activity of cytokines, and the induction of HSP70 expression.

Vajragupta and colleagues [23] showed that manganese complexes of curcumin and its derivatives exhibited neuroprotective activity.

Neuropharmacological tests in mice supported the idea that SOD-mimicking complexes were able to penetrate the brain. They also had a role in the modulation of brain neurotransmitters under aberrant conditions. The complexes significantly reduced the impairment of learning and memory caused by transient ischemic/reperfusion injury and showed significant neurocognitive protection, whereas manganese acetate and curcumin had no effect at similar doses. In addition, treatment with the manganese complex of diacetylcurcumin (AcylCpCpx) and curcumin significantly attenuated 1-methyl-4-phenyl-1,2,3,6-tetrahydropyridine (MPTP)-induced striatal DA depletion in mice, and increased the density of dopaminergic neurons when compared with MPTP-treated mice. These results support the important role of manganese in importing SOD activity and consequently, the enhancement of radical scavenging activity. AcylCpCpx and CpCpx seem to be the most promising neuroprotective agents for vascular dementia.

Whether thymoquinone (TQ) can prevent neurodegeneration in the hippocampus after chronic toluene exposure in rats has also been examined [24]. Kanter [24] showed that chronic toluene exposure caused severe degenerative changes, shrunken cytoplasma, slightly dilated cisternae of endoplasmic reticulum, markedly swollen mitochondria with degenerated cristae, and nuclear membrane breakdown with chromatin disorganization in hippocampal neurons. The distorted nerve cells were mainly absent in the TQ-treated rats, suggesting that TQ therapy causes morphologic improvement of neurodegeneration in the hippocampus after chronic toluene exposure.

Diosgenin derived from fenugreek was also found to affect the brain. Lee et al. [25] showed that diosgenin could induce the release of growth hormones in rats.

As noted, the pharmacokinetic properties of curcumin in mice have also been examined [26]. After i.p. administration of curcumin (0.1 g/kg) in a murine model, about 2.25 µg/mL of curcumin appeared in the plasma in the first 15 min. One hour after administration, the levels of curcumin in the intestines, spleen, liver, and kidneys were 177, 26, 27, and 8 µg/g, respectively. Only traces (0.41 µg/g) were observed in the brain at 1 h.

20.3 MULTIPLE SCLEROSIS

Experimental autoimmune encephalomyelitis (EAE) is a demyelinating disease of the central nervous system widely accepted to be an animal model for multiple sclerosis. Inflammation plays a major role in neuropathological processes associated with neutrophilic infiltrates, such as EAE and traumatic injury of the brain. Whether curcumin may influence inflammation in the CNS through the modulation of the CXC chemokine, macrophage inflammatory protein (MIP)-2, has been investigated [27]. Astrocytes prepared from the neonatal brains of mice were stimulated with LPS in the presence or absence of various amounts of curcumin. The latter inhibited the LPS-induced induction of MIP-2 gene expression, the production of MIP-2 protein, and the transcription of MIP-2 promoter activity. Thus, curcumin potently inhibits MIP-2 production at the level of gene transcription and offers further support for its potential use in the treatment of inflammatory conditions of the CNS.

Mohamed et al. [28,29] showed that TQ inhibits the activation of NF-κB in the brain and spinal cord of rats with EAE. They showed that the treatment of rats with TQ prevents myelin basic protein-induced EAE. TQ inhibited the perivascular cuffing and infiltration of mononuclear cells in the brain and spinal cord, increased levels of red blood cells and GSH, and inhibited the activation of NF-κB in the brain and spinal cord, consistent with the clinical signs of the disease.

20.4 ALZHEIMER'S DISEASE

As the average life span in developed countries increases, so does the incidence of Alzheimer's disease (AD) or senile dementia. The disease is named after the German physician Alois Alzheimer, who described its symptoms and pathology about 100 years ago, when he was treating an elderly lady who was progressively losing memory and speech, displaying hallucinations and delusions, and finally becoming totally helpless. Autopsy of her brain showed strands of thick, insoluble deposits, which presumably interfered with normal brain function by disturbing the electrical activity. Currently, there are 5.2 million people (one out of every eight people aged 65 years or older) living in the United States with this neurological disorder. Individuals with AD live an average of 8 years, costing $148 billion annually (with average lifetime costs of $174,000 per patient). Every 72 seconds somebody in the United States will develop AD; 50% of nursing home residents currently suffer from AD or related disorders. Ten million baby boomers are expected to develop AD in their lifetimes (with the incidence expected to double or triple by 2050). Some of symptoms of AD include the erosion of memory, changes in personality, and the slow but inevitable loss of brain cells. It is not clear what causes the plaques associated with AD, but a variety of factors seem to be involved. Mutations or variants of the gene for the blood protein Apolipoprotein E are associated with the disorder, as are a variety of lifestyle factors, such as diet and drugs. The mere process of aging is a factor, hence the name "senile dementia." Smoking, drinking, impaired nutrition, lack of exercise, and excessive exposure to the sun could be contributory factors.

India is one of the developing countries where AD is less prevalent, even when adjusted for age [30]. Although AD affects about 3.1% of Americans between 70 and 79 years old on average, only 0.7% of the people in India of that age are affected. Why India has the lowest rate and the United States the highest is not fully understood. Oxidative damage and inflammation have been implicated in most age-related neurodegenerative diseases, including AD. Amyloid beta peptide (Abeta), a proteolytic fragment of the amyloid precursor protein, is a major component of the plaques found in the brains of AD patients. These plaques are thought to cause the loss of cholinergic neurons observed in the basal forebrain of AD patients. Curcumin has been shown to affect AD through numerous mechanisms. For instance, up-regulation of 75-kDa neurotrophin receptor (p75NTR), a nonselective neurotrophin receptor belonging to the death receptor family, has been reported in neurons affected by AD. The expression of p75NTR has been found to correlate with beta-amyloid sensitivity in vivo and *in vitro*, suggesting a possible role for p75NTR as an Abeta receptor.

Human neuroblastoma cell lines were used to investigate the involvement of p75NTR in Abeta-induced cell death [31]. Abeta peptides bound to p75NTR, resulting in activation of NF-κB, a major mediator of inflammation. Blocking the interaction of Abeta with p75NTR using nerve growth factor or inhibition of NF-κB activation by curcumin abolished Abeta-induced apoptotic cell death. These results suggest that p75NTR might be a death receptor for Abeta, and thus a possible therapy for AD. Kim et al. [32] also showed that curcumin protected neurons from Abeta (25–35)-induced apoptosis. They suggested that the hydroxy group at the *para*-position in curcumin is critical for the expression of biological activity. In addition to curcumin, shogoals from ginger (*Zingiber officinale*) were found to protect human neuroblastoma and normal human umbilical vein endothelial cells from Abeta (25–35) insult [32].

Abeta1–40 has been shown to activate nuclear transcription factor early growth response-1 (Egr-1), which results in the increased expression of cytokines and chemokines in monocytes. Whether curcumin suppressed Egr-1 activation and the concomitant expression of chemokines was investigated [33]. Curcumin inhibited the activation of Egr-1 DNA-binding activity, abrogated the Abeta-induced expression of cytokines [tumor necrosis factor (TNF)-alpha and interleukin (IL)-1beta] and chemokines (MIP-1beta, MCP-1, and IL-8) in monocytes, and suppressed MAP kinase activation and the phosphorylation of ERK-1/2 and its downstream target, Elk-1. Curcumin inhibited Abeta1–40-induced expression of CCR5, but not of CCR2b. This inhibition involved abrogation of Egr-1 DNA binding in the CCR5 promoter by curcumin. Finally, curcumin inhibited chemotaxis of THP-1 monocytes in response to chemoattractants. The inhibition of Egr-1 is one of the mechanisms by which curcumin could ameliorate the inflammation and progression of AD.

Whether curcumin can protect against Abeta-induced damage in rats was examined [34]. Lipoprotein carrier-mediated, intracerebroventricular infusion of Abeta peptides induced oxidative damage, synaptophysin loss, a microglial response, and widespread Abeta deposits. Dietary curcumin (2000 ppm), but not ibuprofen, suppressed oxidative damage (isoprostane levels) and synaptophysin loss. Both ibuprofen and curcumin reduced microgliosis in cortical layers, but curcumin increased microglial labeling within and adjacent to Abeta-ir deposits. In a second group of middle-aged female Sprague-Dawley rats, 500-ppm dietary curcumin prevented Abeta-infusion-induced spatial memory deficits in the Morris water maze and postsynaptic density-95 loss and reduced Abeta deposits. Because of its low side-effect profile and long history of safe use, curcumin may find clinical application for AD prevention.

Yang et al. [35] have investigated whether curcumin could bind and prevent the aggregation of Abeta in AD models. Under aggregating conditions *in vitro*, curcumin inhibited aggregation as well as disaggregating fibrillar Abeta40. Curcumin was a better Abeta40 aggregation inhibitor than ibuprofen and naproxen, and prevented Abeta42 oligomer formation and toxicity between 0.1 and 1 μM; curcumin also decreased Abeta fibril formation. The effects of curcumin did not depend on Abeta sequence but on fibril-related conformation. AD and Tg2576 murine brain sections incubated with curcumin revealed preferential labeling of amyloid plaques. In vivo studies showed that curcumin injected peripherally into aged Tg2576 mice crossed the blood-brain barrier (BBB) and bound the plaques. When fed to aged Tg2576 mice with advanced amyloid accumulation, curcumin labeled plaques and reduced amyloid levels and plaque burdens. Hence, curcumin directly binds small Abeta to block aggregation and fibril formation *in vitro* and in vivo. Low-dose curcumin effectively disaggregates Abeta as well as prevents fibril and oligomer formation, supporting the rationale for curcumin use in clinical trials on the prevention or treatment of AD. That curcumin can inhibit the formation of Abeta fibrils (fAbeta), as well as destabilize preformed fAbeta, has also been shown by other research groups [36–38]. Ferulic acid was found to be less active than curcumin.

Curcumin is structurally similar to Congo red, and has been demonstrated to bind Abeta amyloid and prevent further oligomerization of Abeta monomers onto growing amyloid beta-sheets. Reasoning that the oligomerization kinetics and mechanisms of amyloid formation are similar in Parkinson's disease (PD) and AD, the effect of curcumin on alpha-synuclein (AS) protein aggregation was examined [39]. The *in vitro* model of AS aggregation was developed by treating purified AS protein (wild-type) with Fe^{3+} (Fenton reaction). It was observed that the addition of curcumin inhibited aggregation and increased AS solubility. Curcumin inhibits AS oligomerization into higher molecular weight aggregates and, therefore, should be further explored as a potential therapeutic compound for AD and related disorders. Another potential mechanism by which curcumin could mediate its effects is through modulation of p21-activated kinase (PAK). The PAK family of kinases are known to regulate actin filaments and the morphogenesis of dendritic spines through Rho family GTPases Rac and Cdc42. Active PAK has been shown to be markedly reduced in AD cytosol, accompanied by downstream loss of the spine actin-regulatory protein drebrin. Abeta oligomer was implicated in PAK defects. PAK was found to be aberrantly activated and translocated from cytosol to membrane in AD brains and in 22-month-old Tg2576 transgenic AD mice [40]. Curcumin comparatively suppressed PAK translocation in aged Tg2576 transgenic AD mice and in Abeta42 oligomer-treated cultured hippocampal neurons.

Ryu et al. [38] evaluated radiolabeled curcumin as a potential probe for Abeta plaque imaging. Partition coefficient measurement and biodistribution in normal mice demonstrated that [18F]8 has a suitable lipophilicity and reasonable initial brain uptake. Metabolism studies also indicated that [18F]8 is metabolically stable in the brain and is a suitable radioligand for Abeta plaque imaging [38].

Another potential mechanism by which curcumin could exhibit activity against AD is through inhibition of gamma-secretase [41]. To minimize the metal chelation properties of curcumin, Narlawar et al. [41] synthesized curcumin-derived oxazoles and pyrazoles. The reduced rotational freedom and the absence of stereoisomers were thought to enhance the inhibition of gamma-secretase. Accordingly, the replacement of 1,3-dicarbonyl moieties by isosteric heterocycles turned curcumin analogue oxazoles and pyrazoles into potent gamma-secretase inhibitors. They were potent inhibitors of gamma-secretase and displayed activity in the low micromolar range. Another pyrazole derivative of curcumin, called CNB-001, was synthesized and found to exhibit far superior activity in neuroprotection [42] when examined in cell culture assays for trophic factor withdrawal, oxidative stress, excitotoxicity, and glucose starvation, as well as toxicity from both intracellular and extracellular amyloids.

In another study, the anti-amyloidogenic effects of dietary curcumin and its more stable metabolite, tetrahydrocurcumin (THC), were examined either when administered chronically to aged Tg2576 APPsw mice or acutely to LPS-injected wild-type mice [43]. Despite dramatically higher drug plasma levels after THC compared to curcumin gavage, the resulting brain levels of parent compounds were similar, correlating with the reduction in LPS-stimulated iNOS, nitrotyrosine, F2 isoprostanes, and carbonyls. In both the acute (LPS) as well as chronic inflammation (Tg2576) models, THC and curcumin similarly reduced IL-1β. Despite these similarities, only curcumin was effective in reducing amyloid plaque burden, insoluble Abeta, and carbonyls. THC had no impact on plaques or insoluble Abeta, but reduced both tris-buffered saline (TBS)-soluble Abeta and pJNK. Curcumin (but not THC) prevented Abeta aggregation. The THC metabolite was detected in the brain and plasma of mice chronically fed the parent compound. These data indicate that the dienone bridge present in curcumin, but not in THC, is necessary to reduce plaque deposition and protein oxidation in an Alzheimer's model. Nevertheless, THC did reduce neuroinflammation and the effects of soluble Abeta, which may be attributable to limiting JNK-mediated transcription. Thus, curcumin and THC display a very different activity profile against AD.

20.5 SPONGIFORM ENCEPHALOPATHIES

Conversion of the native, predominantly alpha-helical conformation of prion protein (PrP) into the beta-stranded conformation is characteristic of the transmissible spongiform encephalopathies such as Creutzfeld-Jakob disease [44]. Curcumin, an extended planar molecule, inhibits *in vitro* conversion of PrP and formation of protease resistant PrP in neuroblastoma cells [45]. Curcumin recognizes the converted beta forms of PrP (oligomers and fibrils), but not the native form. Curcumin binds to the prion fibrils in the left-handed chiral arrangement, labels the plaques in the brains of variant Creutzfeld-Jakob disease cases, and stains the same structures as antibodies against PrP. In contrast to thioflavin, T curcumin also binds to the alpha-helical intermediate of PrP present at an acidic pH at a stoichiometry of 1:1. Congo red competes with curcumin for binding to the alpha-intermediate, as well as to the beta form, of PrP. However, it is toxic and also binds to the native form of PrP. Thus, the partially unfolded structural intermediate of the PrP can be targeted by curcumin.

20.6 PARKINSON'S DISEASE

Although the cause of dopaminergic cell death in PD remains unknown, the role of oxidative stress has been strongly implicated. Because of their ability to combat oxidative stress, spice-derived phenolic compounds continue to be considered as potential agents for long-term use in PD. Oxidative stress has been implicated in the degeneration of dopaminergic neurons in the substantia nigra (SN) of PD patients. An important biochemical feature of presymptomatic PD is a significant depletion of the thiol antioxidant GSH in these neurons, resulting in oxidative stress, mitochondrial dysfunction, and ultimately, cell death [46].

Treatment of dopaminergic murine neuronal cells with curcumin was found to restore depletion of GSH levels, protect against protein oxidation, and preserve the mitochondrial complex I activity that is normally impaired due to GSH loss [47]. Using systems biology and dynamic modeling, researchers examined the mechanism of curcumin action in a model of mitochondrial dysfunction linked to GSH metabolism that corroborates the major findings. Thus, curcumin may also have therapeutic potential for neurodegenerative diseases involving GSH depletion-mediated oxidative stress.

The prevalence of PD is higher in men than in women [48]. Although the reason for this gender difference is not clear, female steroid hormones or their receptors may be involved in the pathogenesis of PD. It has been demonstrated that ligand-activated estrogen receptor beta suppressed dopaminergic neuronal death in an *in vitro* PD model that used 1-methyl-4-phenylpyridinium ions (MPP(+)) [49]. They showed that (similar to estrogen) MPP(+) treatment caused the up-regulation of JNK and dopaminergic neuronal death, which was blocked by curcumin.

Selective damage to mitochondrial complex I within the dopaminergic neurons of the SN is the central event of PD. Peroxynitrite is one of the free radicals most likely mediating complex I damage [50]. Peroxynitrite inhibits brain complex I mainly by 3-nitrotyrosine and nitrosothiol formation, but how these modifications alter the structure-function relation of complex I is unclear. Curcumin pretreatment protected brain mitochondria against peroxynitrite *in vitro* by direct detoxification and prevention of 3-nitrotyrosine formation, and in vivo by elevation of total cellular GSH levels [47]. These results suggest a potential therapeutic role for curcumin against nitrosative stress in neurological disorders.

It has been shown that curcumin acts as a powerful free radical scavenger in vivo in the brain, and interferes with oxidative stress caused by the parkinsonian neurotoxin, MPTP [51]. MPTP treatment caused a significant depletion in GSH and increased the specific activity of SOD, catalase, and PLO in both the striatum and midbrain on the third and seventh days. Curcumin treatment blocked these changes. This provides direct evidence for the involvement of curcumin in neuroprotection against oxidative stress.

Whether curcumin can be neuroprotective in the 6-OHDA model of PD was examined [52]. Unilateral infusion of 6-OHDA into the medial forebrain bundle produced a significant loss of tyrosine hydroxylase (TH)-positive cells in the SN as well as decreased DA content in the striata of the vehicle-treated animals. Rats pretreated with curcumin showed a clear protection of TH-positive cells in the SN and DA in the striata. The ability of curcumin to exhibit neuroprotection in the 6-OHDA model of PD may be related to its antioxidant capabilities and its ability to penetrate into the brain.

Eugenol (derived from cloves) has also been examined as a means to prevent the progression of Parkinson's disease. Experimental results indicate that demethyldiisoeugenol is a potentially effective antioxidant and can protect rat brain homogenates and LDL against oxidation [53]. Kabuto et al. [54] showed that eugenol prevents 6-hydroxydopamine-induced DA depression and PLO inductivity in mouse striatum.

20.7   AGE-ASSOCIATED NEURODEGENERATION

Various age-dependent degenerative changes have been reported in the brain. Bala and coworkers [55] investigated the influence of chronically administered curcumin on normal aging-related parameters: PLO, lipofuscin concentration, intraneuronal lipofuscin accumulation, and activities of the enzymes SOD, GPx, Na(+), K(+), and -ATPase in different brain regions (cerebral cortex, hippocampus, cerebellum, and medulla) of 6- and 24-month-old rats [55]. Under normal aging, LPO and lipofuscin concentration were found to increase with age; however, the activities of SOD, GPx, Na(+), K(+), and -ATPase decreased with age. Chronic curcumin treatment of both 6- and 24-month-old rats resulted in significant decreases in lipid peroxides and lipofuscin content in brain regions. The activities of SOD, GPx, Na(+), K(+), and -ATPase, however, showed significant increases in various brain regions. Thus, curcumin exhibits antioxidative, antilipofusinogenesic, and antiaging effects in the brain.

20.8   SCHIZOPHRENIA

Tardive dyskinesia (TD) is a motor disorder of the orofacial region resulting from chronic neuroleptic treatment. The high incidence and irreversibility of this hyperkinetic disorder has been considered a major clinical issue in the treatment of schizophrenia. The molecular mechanism related to the pathophysiology of TD is not completely understood. Various animal studies have demonstrated an enhanced oxidative stress and increased glutamatergic transmission, as well as inhibition in the glutamate uptake after the chronic administration of haloperidol. The effect of curcumin on haloperidol-induced TD has been investigated by using different behavioral (orofacial dyskinetic movements, stereotypy, locomotor activity, percent retention), biochemical—lipid peroxidation, reduced GSH levels, antioxidant enzyme levels (SOD and catalase)—and neurochemical (neurotransmitter level) parameters [56]. Chronic administration of haloperidol significantly increased vacuous chewing movements, tongue protrusions, and facial jerking in rats, which was inhibited by curcumin. Chronic administration of haloperidol also resulted in increased DA receptor sensitivity, as evidenced by increased locomotor activity and stereotypy and decreased retention time on elevated plus maze paradigm. Pretreatment with curcumin reversed these behavioral changes.

Haloperidol also induced oxidative damage in all major regions of the brain, which was attenuated by curcumin, especially in the subcortical region containing striatum. Chronic administration of haloperidol caused a decrease in turnover of DA, serotonin, and norepinephrine in both cortical and subcortical regions, which was again dose-dependently reversed by treatment with curcumin. The findings of this study suggested the involvement of free radicals in the development of neuroleptic-induced TD and pointed to curcumin as a possible therapeutic option to treat this hyperkinetic movement disorder.

20.9 EPILEPSY

Animal models of epilepsy have revealed the basic molecular and cellular mechanisms of epileptogenesis. Generalized limbic seizures and subsequent status epilepticus can be induced by pilocarpine, the muscarinic acetylcholine receptor agonist, or kainate, the glutamate receptor agonist. There has been increasing evidence that chromatin remodeling might play a critical role in gene regulation even in nondividing cells, such as neurons. One form of chromatin remodeling is histone aminoterminal modification that can generate synergistic or antagonistic affinities for the interactions of transcriptional factors, in turn, causing changes in gene activity [57]. Two widely studied histone modification processes are histone acetylation and phosphorylation. Although histone hyperacetylation indicates an increase in gene activity, hypoacetylation marks gene repression. Both states are controlled by a dynamic interplay of histone acetyltransferase (HAT) and histone deacetylase. The up-regulation of acetylation and phosphorylation of histones, coupled with status epilepticus after kainate administration, has been reported. Both c-*fos* and c-*jun* mRNA have been sequentially induced in response to kainate, in different hippocampal subpopulations starting from the dentate gyrus and spreading to the cornus ammonis regions, which are correlated with the spatiotemporal distribution of histone H4 hyperacetylation. Both histone modifications are associated with the c-*fos* gene promoter after kainate stimulation, while only histone acetylation is associated with the c-*jun* gene. Pretreatment with curcumin, which has a HAT inhibitory activity specific to CBP/p300, attenuates histone modifications, immediate early gene (IEG) expression, and also the severity of status epilepticus after kainate treatment [58]. Histone modifications may have a crucial role in the development of epilepsy induced by kainate.

Kainic acid (KA)-induced oxidative stress is associated with hippocampal cell death. Recent studies suggest that curcumin may provide protection from KA-induced oxidative stress. Shin et al. [59] investigated the effect of curcumin treatment on hippocampal reactive astrocytes in mice with KA-induced seizures. Eighteen hours after curcumin treatment, mice were treated with KA (30 mg/kg, i.p.), and then sacrificed after a further 48 h. Histological evaluation revealed cell death in the hippocampus treated with KA alone, but not in mice treated with curcumin. In addition, curcumin treatment reduced the KA-induced immunoreactivity of caspase-3. Similarly, immunoreactivity analyses indicated that KA caused up-regulation of hippocampal GFAP, eNOS, and HO-1 levels, all of which were reduced in animals that received the curcumin treatment. Thus, curcumin is a potent inhibitor of reactive astrocyte expression and prevents hippocampal cell death.

Sumanont and colleagues [60] examined the in vivo effect of manganese complexes of curcumin (C1) and AcylCpCpx (C2) on NO levels enhanced by KA and l-arginine (l-Arg) in the hippocampus of rats. Injection of KA and l-Arg significantly increased the concentration of NO, whereas C1 and C2 (50 mg/kg, i.p.) significantly reversed the effects. Following KA-induced seizures, severe neuronal cell damage was observed in the CA1 and CA3 subfields of hippocampal regions 3 days after KA administration. Pretreatment with C1 and C2 (50 mg/kg, i.p.) significantly attenuated KA-induced neuronal cell death in both the CA1 and CA3 regions of rat hippocampus. C1 and C2 had more potent neuroprotective effects than their parent compounds, curcumin and AcylCpCpx. These are effective neuroprotective agents in the treatment of acute brain pathologies associated with NO-induced neurotoxicity and oxidative stress-induced neuronal damage, such as epilepsy, stroke, and TBI [60,61].

20.10   NEUROPATHIC PAIN

The major role of inflammation in neuropathic pain has been established. Diabetic neuropathic pain, an important microvascular complication in diabetes mellitus, is recognized as one of the most difficult types of pain to treat. A lack of understanding of its etiology, inadequate pain relief, and development of tolerance to classical antinociceptives warrant the investigation of newer treatment agents. Whether cucumin exhibits the antinociceptive effects through modulation of TNF-alpha and NO release in streptozotocin-induced murine diabetes has been examined [62]. Four weeks after a single i.p. injection of streptozotocin (200 mg/kg), mice were tested in the tail immersion and hot plate assays. Diabetic mice exhibited significant hyperalgesia along with increased plasma glucose and decreased body weights as compared with control mice. Chronic treatment with curcumin (15, 30, and 60 mg/kg body weight; p.o.) for 4 weeks starting from the fourth week of streptozotocin injection) significantly attenuated thermal hyperalgesia and hot plate latencies. Curcumin also inhibited TNF-alpha and NO release. These results indicate an antinociceptive activity of curcumin, possibly through its inhibitory action on NO and TNF-alpha release, and point toward its potential to attenuate diabetic neuropathic pain.

20.11   CEREBRAL ISCHEMIA

Oxidative stress has been implicated in the pathogenesis of postischemic cerebral injury. Curcumin has been examined for its neuroprotective effects in experimental models of cerebral ischemia/ reperfusion (I/R) in rat forebrains [63]. The effects of a single i.p. dose of the drug (50, 100, or 200 mg/kg) administered 0.5 h after the onset of ischemia were examined by assessing oxidative stress-related biochemical parameters in the rat forebrain. At the highest dose level (200 mg/kg), curcumin decreased the I/R-induced xanthine oxidase activity, superoxide anion production, malondialdehyde level, and GSH peroxidase, SOD, and lactate dehydrogenase activities. Thus, curcumin was found to protect the rat forebrain against I/R insult.

Wang et al. [64] investigated the mechanisms underlying the neuroprotective effects of curcumin. Global cerebral ischemia was induced in Mongolian gerbils by transient occlusion of the common carotid arteries. Histochemical analysis indicated extensive neuronal death and increased reactive astrocytes and microglial cells in the hippocampal CA1 area at 4 days after I/R. These ischemic changes were preceded by a rapid increase in PLO and followed by a decrease in MMP, increased cytochrome *c* release, and caspase-3 activation and apoptosis. Administration of curcumin by i.p. injections (30 mg/kg b.w.) or by supplementation to the AIN76 diet (2 g/kg diet) for 2 months significantly attenuated ischemia-induced neuronal death as well as glial activation. Curcumin administration also decreased PLO, mitochondrial dysfunction, and the apoptotic indices. Biochemical changes resulting from curcumin also correlated well with its ability to ameliorate the changes in locomotor activity induced by I/R. Bioavailability studies indicated a rapid increase in curcumin in plasma and brain within 1 h after treatment. Thus, curcumin exhibits neuroprotective effects against I/R-induced neuronal damage.

The effects of curcumin have been examined on the death of hippocampal CA1 neurons following transient forebrain ischemia in rats [65]. Treatment with curcumin (200 mg/kg/day, i.p.) at three different times (immediately, 3 h, and 24 h after ischemia) significantly reduced evidence of neuronal damage 7 days after ischemia. Also, treatment of ischemic rats with curcumin decreased the elevated levels of MDA and increased GSH contents, catalase, and SOD activities to normal levels. *In vitro*, curcumin was as potent an antioxidant as butylated hydroxytoluene. Therefore, curcumin attenuates forebrain ischemia-induced neuronal injury and oxidative stress in hippocampal tissue in humans when treated immediately or even when treatment is delayed up to 24 h.

The mechanism by which curcumin exhibits protective effects against cerebral I/R injury is not fully understood. Disruption of the BBB occurs after stroke. Protection of the BBB has become an important target of stroke interventions in experimental therapeutics. Whether curcumin prevents cerebral I/R injury by protecting BBB integrity has been investigated [66]. A single injection of curcumin (1 and 2 mg/kg, i.v.) 30 min after focal cerebral I/R in rats significantly diminished infarct volume, improved neurological deficit, decreased mortality, reduced water content in the brain, and reduced the extravasation of Evans blue dye in the ipsilateral hemisphere. In cultured astrocytes, curcumin significantly inhibited iNOS expression and NO_x (nitrites/nitrates content) production induced by LPS/TNF. Furthermore, curcumin prevented ONOO(−) donor SIN-1-induced cerebral capillary endothelial cell damage. Thus, curcumin ameliorates cerebral I/R injury by preventing ONOO(−)-mediated BBB damage.

Curcumin was also found to have neuroprotective effects in focal cerebral ischemia induced by middle cerebral artery occlusion (MCAO) in rats [67]. Administration of curcumin (100 and 300 mg/kg, i.p) 30 min after MCAO produced reduction in infarct volumes. Ischemia-induced cerebral edema was reduced in a dose-dependent manner. Curcumin inhibited the increase in LPO after MCAO in the ipsilateral and contralateral hemispheres of the brain, prevented I/R injury, mediated the decrease in GSH peroxide activity, and reduced peroxynitrite formation, and hence, the extent of tyrosine nitration in the cytosolic proteins. These results suggested the neuroprotective potential of curcumin in cerebral ischemia and that this potential is mediated through curcumin's antioxidant activity.

The effect of *Nigella sativa* oil (NSO) and TQ on transient global cerebral ischemia-induced oxidative injury has been examined [68]. Transient global cerebral ischemia induced a significant increase in the level of thiobarbituric acid reactive substances. Pretreatment with TQ and NSO induced a significant decrease in MDA levels as compared with the ischemic group. These results suggest that TQ and NSO may have protective effects on the LPO process during IR in the rat hippocampus. Whether TQ exhibits neuroprotective effects against transient forebrain ischemia-induced neuronal damage in the rat hippocampus has been investigated [69]. Pretreatment with TQ attenuated forebrain ischemia-induced neuronal damage as manifested by significantly decreasing the number of dead hippocampal neuronal cells, decreasing the elevated levels of malondialdehyde, increasing GSH, catalase, and SOD activities, and inhibiting *in vitro* nonenzymatic PLO in hippocampal homogenates induced by iron ascorbate. Thus, TQ is effective in protecting rats against transient forebrain ischemia-induced damage in the rat hippocampus.

20.12 DEPRESSION

Curcuma longa is a major constituent of Xiaoyao-san, a traditional Chinese medicine used to effectively manage stress- and depression-related disorders in China. The effects of curcumin on depressive behaviors in mice, using two animal models of depression, have been investigated [70]. Curcumin treatment at 5 and 10 mg/kg (p.o.) significantly reduced the duration of immobility in both the tail suspension and forced swimming tests. The doses that affected the immobile response did not affect locomotor activity. In addition, neurochemical assays showed that curcumin produced a marked increase in serotonin and noradrenaline levels at 10 mg/kg in both the frontal cortex and hippocampus. DA levels were also increased in the frontal cortex and the striatum. Moreover, curcumin was found to inhibit monoamine oxidase (MAO) activity in the mouse brain. These findings suggested that the antidepressant-like effects of curcumin might involve the central monoaminergic neurotransmitter systems.

In another study, Xu et al. [71] examined the effect of curcumin on stress-induced damage to hippocampal neurons that may contribute to the pathophysiology of depression. Their results suggested that curcumin administration (10 and 20 mg/kg, p.o.) increased hippocampal neurogenesis in chronically stressed rats, similar to treatment (10 mg/kg, i.p.) with the classic antidepressant imipramine (sold as Antideprin, Deprenil, Deprimin, Deprinol, Depsonil, Dynaprin, Eupramin, Imipramil, Irmin, Janimine, Melipramin, Surplix, and Tofranil). They also showed that these new cells mature and become neurons. In addition, curcumin significantly prevented the stress-induced decrease in 5-HT(1A) mRNA and BDNF protein levels in hippocampal subfields; these two molecules are involved in hippocampal neurogenesis. These results raise the possibility that increases in cell proliferation and neuronal populations may be a mechanism by which curcumin treatment overcomes stress-induced behavioral abnormalities and hippocampal neuronal damage. Moreover, curcumin treatment, via up-regulation of 5-HT(1A) receptors and BDNF, may reverse or protect hippocampal neurons from further damage in response to chronic stress, which may underlie the therapeutic actions of curcumin.

That curcumin may also alleviate stress-induced depressive-like behaviors and hypothalamic-pituitary-adrenal (HPA) axis dysfunction in rats was examined [72]. They found that subjecting animals to a chronic stress protocol for 20 days resulted in performance deficits in the shuttle-box task and several physiological effects, such as an abnormal adrenal gland weight/body weight ratio and increased thickness of the adrenal cortex, as well as elevated serum corticosterone levels and reduced glucocorticoid receptor mRNA expression. These changes were reversed by chronic curcumin administration (5 or 10 mg/kg, p.o.). Chronic stress also induced a down-regulation of BDNF protein levels and reduced the ratio of phosphorylated CREB (pCREB) to CREB levels (pCREB/CREB) in the hippocampus and frontal cortex of stressed rats. Furthermore, these stress-induced decreases in BDNF and pCREB/CREB were also blocked by chronic curcumin administration (5 or 10 mg/kg, p.o.). These results provide compelling evidence that the behavioral effects of curcumin in chronically stressed animals (and by extension, humans) may be related to curcumin's modulating effects on the HPA axis and neurotrophin factor expressions.

Whether curcumin would have an influence on depressive behaviors, was examined in the forced swimming test and bilateral olfactory bulbectomy (OB) models of depression in rats [71]. Chronic treatment with curcumin (14 days) reduced the immobility time in the forced swimming test and reversed OB-induced behavioral abnormalities, such as hyperactivity in the open field as well as deficits in step-down passive avoidance. In addition, OB-induced low levels of serotonin (5-HT), noradrenaline (NA), high 5-hydroxyindoleacetic acid (5-HIAA), and 4-dihydroxyphenylacetic acid in the hippocampus were observed—and were completely reversed by curcumin administration. Slight decreases in 5-HT, NA, and DA levels were found in the frontal cortex of OB rats, which curcumin treatment reversed. These results suggest that these antidepressant effects may be mediated by actions in the central monoaminergic neurotransmitter systems.

Numerous lines of evidence suggest piperine exhibits antidepressive effects as well. For instance, piperine was found to inhibit the activity of both MAO A and B from rat brain mitochondria [73]. Piperine was very effective in stimulating 5HT synthesis in the brain caused by release of 3H-5HT from an *in vitro* synaptosomal preparation [74,75]. Thus, piperine affects the central serotonergic system. Convulsions of E1 mice were completely suppressed by piperine injected intraperitoneally [76]. The 5-HT level was significantly higher in the cerebral cortex of piperine-treated mice than in control mice and the DA level in the piperine-treated mice was markedly higher in the hypothalamus, while the norepinephrine levels were lower in other parts of the brain.

Li et al. [77] investigated the antidepressant effects of piperine in mice exposed to a chronic mild stress (CMS) procedure. Repeated administration of piperine for 14 days reversed the CMS-induced changes in sucrose consumption, plasma corticosterone levels, and open field activity. Furthermore, the decreased proliferation of hippocampal progenitor cells was ameliorated and the level of BDNF in the hippocampus of CMS-stressed mice was up-regulated by piperine treatment. In addition, piperine (6.25–25 µM) or fluoxetine (1 µM) protected primary cultured hippocampal neurons from the lesions induced by 10-µM corticosterone (CORT). Piperine reversed the CORT-induced reduction of BDNF mRNA expression in cultured hippocampal neurons. Thus, up-regulation of the progenitor cell proliferation of the hippocampus and cytoprotective activity might be the mechanisms involved in the antidepressant effects of piperine, which is closely related to the elevation of hippocampal BDNF levels.

The mechanism of antidepressant effects of piperine and its derivative, antiepilepsirine (AES), were investigated in two depressive models: forced swimming test and tail suspension test [78]. They found that after 2 weeks of chronic administration, piperine and AES significantly reduced the duration of immobility in both the forced swimming test and tail suspension test, without accompanying changes in locomotor activity in the open-field test. But at the dose of 80 mg/kg, the antidepressant activity of both piperine and AES returned to the control level in the tail suspension test and forced swimming test. In the monoamine assay, chronic AES administration significantly elevated the DA level in the striatum, hypothalamus, and hippocampus, and also increased the serotonin level in the hypothalamus and hippocampus. In contrast, chronic treatment with piperine only enhanced the serotonin level in the hypothalamus and hippocampus, but did not influence the DA level. Moreover, both piperine and AES showed no effects on levels of noradrenaline in these brain regions. The MAO activity assay also indicated that piperine and AES showed a minor MAO inhibitory activity. Thus, the antidepressant effects of piperine and AES might depend on the augmentation of neurotransmitter synthesis or the reduction of neurotransmitter reuptake. Antidepressant properties of piperine were thought to be mediated via the regulation of serotonergic systems, whereas the mechanisms of antidepressant action of AES might be due to its dual regulation of both serotonergic and dopaminergic systems.

Mujumdar et al. [79] found that piperine poteniates phentobarbitone-induced hypnosis in rats, possibly through inhibition of liver microsomal enzyme systems.

Unchern et al. [80] showed that piperine induced marked injuries on cultured hippocampal neurons, whereas it induced only a marginal toxicity on cultured astrocytes. The membrane-permeable free radical scavengers, d-alpha-tocopherol and trolox, protected neurons from piperine toxicity. A lipoxygenase inhibitor, nordihydroguaiaretic acid, attenuated piperine toxicity. Thus, piperine induces selective neurotoxicity *in vitro*.

Eugenol, may have clinical relevance in depression, epilepsy, and headache. That eugenol has an antidepressant activity comparable to that of imipramine was shown by Irie et al. [81] using a forced swim test and tail suspension test in mice. Both eugenol and imipramine induced BDNF in the hippocampus with and without induction of metallothionein-III (MT-III), respectively. It may be possible that MT-III expression is involved in the exhibition of antidepressant activity by eugenol, but not by imipramine. In another study, eugenol was found to protect neuronal cells from excitotoxic and oxidative injury in primary cortical cultures [82]. These results suggest that eugenol may play a protective role against ischemic injury by modulating both NMDA receptor and superoxide radicals.

When investigated for its neurophysiologic properties, eugenol was found to suppress epileptiform field potentials and inhibit worsening of depression, likely via inhibition of synaptic plasticity [83]. The results indicate the potential of eugenol in the treatment of epilepsy and cephalic pain. Ardjmand et al. [84] showed that eugenol depresses synaptic transmission but does not prevent the induction of long-term potentiation in the CA1 region of rat hippocampal slices.

20.13 BRAIN TUMORS

Numerous *in vitro* and in vivo studies have shown that curcumin exhibits antitumor activity against various types of brain tumors [37,85–98]. Gliomas are the most common and lethal primary tumors of the CNS. Malignant gliomas are a debilitating class of brain tumors that are resistant to radiation and chemotherapeutic drugs, contributing to the poor prognosis associated with these tumors. Overexpression of transcription factors such as NF-κB and AP-1 contributes to glioma survival, radioresistance, and chemoresistance. Curcumin, which may inhibit these pathways, was therefore investigated for its potential therapeutic role in gliomas [94]. Curcumin reduced cell survival in a p53- and caspase-independent manner, an effect correlated with the inhibition of AP-1 and NF-κB signaling pathways via prevention of constitutive JNK and Akt activation. Curcumin sensitized glioma cells to cisplatin, etoposide, camptothecin, doxorubicin, and radiation. These effects correlated with reduced expression of bcl-2 and IAP family members, as well as DNA repair enzymes (MGMT, DNA-PK, Ku70, Ku80, and ERCC-1). These findings supported a role for curcumin as an adjunct to traditional chemotherapy and radiation in the treatment of brain cancer.

Another study reported that *in vitro* growth of glioma cells (U251) was significantly inhibited by treatment with curcumin and that a low dose of curcumin induced G2/M cell cycle arrest [95]. The high dose of curcumin not only enhanced G2/M cell cycle arrest, but also the S phase of cell cycle arrest. However, no obvious pre-G1 peak was observed. Curcumin significantly up-regulated the expression of p53, p21, and ING4. The results demonstrate that curcumin exerts inhibitory action on glioma cell growth and proliferation via induction of cell cycle arrest (instead of induction of apoptosis in a p53-dependent manner) and that ING4 is possibly involved in the signaling pathways.

Karmakar et al. [92] reported that curcumin decreased the viability of glioma (U87MG) cells, revealed the morphological and biochemical features of apoptosis, induced the activation of caspase-8 and the cleavage of Bid to tBid, increased the Bax/Bcl-2 ratio, and released cytochrome *c* from the mitochondria, followed by the activation of caspase-9 and caspase-3 for apoptosis [92]. Curcumin also increased cytosolic levels of Smac/Diablo to suppress proteins inhibiting apoptosis and down-regulate NF-kB, thus favoring apoptosis. Increased activities of calpain and caspase-3 cleaved 270 kDa alpha-spectrin at specific sites, generating 145 kDa spectrin breakdown products (SBDP) and 120 kDa SBDP, respectively, leading to apoptosis in U87MG cells [92].

DNA alkylating agents, including temozolomide (TMZ) and 1,3-bis[2-chloroethyl]-1-nitroso-urea (BCNU), are the most common form of chemotherapy in the treatment of gliomas. Despite their frequent use, the therapeutic efficacy of these agents is limited by the development of resistance. Previous studies suggest that the mechanism of this resistance is complex and involves multiple DNA repair pathways. To better define the pathways contributing to the mechanisms underlying glioma resistance, Chen et al. [11] examined the contribution of the Fanconi anemia (FA) DNA repair pathway. The FA-deficient cells were more sensitive to TMZ/BCNU relative to their corrected, isogenic counterparts. Inhibition of FA pathway activation by curcumin caused increased sensitivity to TMZ/BCNU in the U87 glioma cell line. Therefore, cellular resistance to DNA alkylating agents mediated through the FA repair pathway can be overcome by curcumin.

G6P translocase (G6PT) is thought to play a crucial role in transducing intracellular signaling events in brain tumor–derived cancer cells. The contribution of G6PT to the control of brain tumor-derived U87 glioma cell survival using siRNA- and curcumin-mediated suppression of G6PT was examined [90]. siRNA suppressed up to 91% of G6PT gene expression, which led to necrosis and late apoptosis. In a manner similar to siRNA, curcumin inhibited G6PT gene expression by more than 90% and triggered U87 glioma cell death. Overexpression of recombinant G6PT rescued the cells from curcumin-induced cell death. Thus, targeting G6PT expression provides another novel mechanistic rationale for the use of curcumin in the development of new anticancer strategies.

Acetylation of histones and nonhistone proteins is an important posttranslational modification involved in the regulation of gene expression in mammalian cells. Dysfunction of HAT is associated with the manifestation of several diseases. Whether curcumin can induce histone hypoacetylation in brain cancer cells leading to apoptosis was examined [89]. In addition, curcumin induces recontrolling of neural stem cell fates. It induces effective neurogenesis, synaptogenesis, and migration of neural progenitor cells *in vitro* in brain-derived adult neural stem cells. The neurogenic effect of curcumin was also confirmed in vivo. Curcumin actively suppressed differentiation in astrocytes, while promoting differentiation into the neurons associated with decreases in histone H3 and H4 acetylation. Histone hypoacetylation plays an important role in determining stem cell fate through controlling the simultaneous expression of many genes. Thus, curcumin could mediate the effects of histone against cancer and neurogenesis through inhibition of HAT.

Prolactinomas are the most prevalent functional pituitary adenomas. Dopamine (D_2) receptor (D_2R) agonists, such as bromocriptine, are the first line of therapy. However, drug intolerance/resistance to D_2R agonists exists. Apart from D_2R agonists, there is no established medical therapy for prolactinomas, therefore identifying novel therapeutics is crucial. Miller et al. [99] examined whether curcumin can inhibit the proliferation of pituitary tumor cells. Using rat lactotroph cell lines, they showed that curcumin inhibited the proliferation of pituitary tumor cells. Inhibitory effects of curcumin persisted even upon removal of curcumin, and curcumin also blocked the colony formation ability of pituitary tumor cells. The growth inhibitory effect of curcumin was accompanied by decreased expression of cyclin D3 and ser 780 phosphorylation of Rb. In addition, curcumin also induced apoptosis and suppressed intracellular levels and release of both prolactin and growth hormone. Low concentrations of curcumin enhanced the growth inhibitory effect of bromocriptine on cell proliferation. Thus, curcumin inhibits pituitary tumor cell proliferation, induces apoptosis, and decreases hormone production and release, suggesting its role in the management of prolactinomas.

Whether curcumin sensitizes malignant glioma cell lines to TNF-related apoptosis-inducing ligand (TRAIL)-induced apoptosis has been examined [88]. Curcumin at subtoxic concentrations sensitized glioma cells to TRAIL-induced cytotoxicity, enhanced accumulation of hypodiploid U87MG cells in the sub-G1 cell cycle phase, induced the cleavage of procaspase-3, -8, and -9, and released cytochrome c from mitochondria. These data indicate that curcumin sensitizes glioma cells to TRAIL-induced apoptosis through the activation of both extrinsic (receptor-mediated) and intrinsic (chemical-induced) pathways of apoptosis.

The abnormal expression of matrix metalloproteinases (MMPs) plays an important role in the invasion of malignant gliomas into the surrounding normal brain tissue. That curcumin exhibits inhibitory activity against expression of various MMP genes in human astroglioma cells has been examined [37]. Curcumin inhibited the phorbol 12-myristate 13-acetate (PMA)-induced mRNA expression of MMP-1, -3, -9, and -14; repressed the activity of AP-1, an upstream modulator of MMP-1, -3, and -9 gene expression; and suppressed the PMA-induced MAP kinase activities involved in modulating the MMP. Curcumin was also found to significantly repress the *in vitro* invasion of glioma cells. In another study, the effect of curcumin on PMA-induced MMP-9 expression in human astroglioma cell lines was examined [86]. Curcumin significantly inhibited the MMP-9 enzymatic activity and protein expression induced by PMA, correlating with decreased MMP-9 mRNA levels, the suppression of MMP-9 promoter activity, and inhibition of NF-κB and AP-1. It also repressed the phosphorylation of ERK, JNK, and p38 MAP kinase, which were dependent on the protein kinase C pathway. Therefore, the broad spectrum inhibition of MMP gene expression by curcumin might provide a novel therapeutic strategy for treating gliomas.

Other mechanisms by which curcumin exhibits antitumor effects involve suppressing the expression of cyclin D1 and inducing expression of the cyclin-dependent kinase inhibitor p21. Choi et al. [98] showed that transcription of the *p21* gene is activated by Egr-1 independently of p53 in response to curcumin treatment in U87-MG human glioblastoma cells. Egr-1 expression is induced by curcumin through ERK and JNK, but not the p38 MAPK pathways. Transient expression of Egr-1 enhanced curcumin-induced p21 promoter activity, whereas suppression of Egr-1 expression by small interfering RNA abrogated the ability of curcumin to induce p21 promoter activity. In addition, stable knockdown of Egr-1 expression in U87-MG cells suppressed curcumin-induced p21 expression. Thus, the antiproliferative effects of curcumin in glioma are mediated in part through the transcriptional activation of p21.

We examined the anticancer efficacy and mechanisms of curcumin in U87-MG and U373-MG malignant glioma cells [93]. Curcumin induced G(2)/M arrest and nonapoptotic autophagic cell death in both cell types. It inhibited the Akt/mTOR/p70S6K pathway and activated the ERK1/2 pathway, resulting in induction of autophagy. Activation of the Akt pathway inhibited curcumin-induced autophagy and cytotoxicity, whereas inhibition of the ERK1/2 pathway inhibited curcumin-induced autophagy and induced apoptosis, thus resulting in enhanced cytotoxicity. These results imply that the effect of autophagy on cell death may be pathway-specific. In the subcutaneous xenograft model of U87-MG cells, curcumin inhibited tumor growth significantly and induced autophagy. These results suggest that curcumin has high anticancer efficacy *in vitro* and in vivo through various mechanisms that warrant further investigation toward possible clinical application in patients with malignant glioma.

20.14 MENINGITIS

Angiostrongylus cantonensis can invade the central nervous system, leading to human eosinophilic meningitis or eosinophilic meningoencephalitis. The administration of curcumin has been reported to relieve the symptoms of meningitis [100]. The potential efficacy of curcumin in *A. cantonensis*-induced eosinophilic meningitis in BALB/c mice was examined. Assay indicators for the therapeutic effect included the larvicidal effect, eosinophil counts, and MMP-9 activity in angiostrongyliasis. Eosinophils were mildly reduced in treatment groups compared with untreated infected mice. However, there were no significant differences in larvicidal effects or MMP-9 activity. This study suggested that anti-inflammatory treatment with curcumin alone has low efficacy, and the treatment does not interfere with MMP-9 expression and is not useful for larvicidal effects. The possible reasons include low levels of curcumin crossing the BBB and that those larvae that survive stimulate MMP-9 production, which promotes BBB damage, with leukocytes then crossing the BBB to cause meningitis.

20.15 SUMMARY

The evidence presented above suggests that spice-derived phytochemicals have a significant potential against various neurodegenerative diseases. Both observational and experimental evidence supports this thesis. More systematic clinical trials, however, are needed to prove the potential of these spices. Although spices are highly safe and inexpensive, such trials in humans are difficult and most likely will be highly expensive because of the long-term monitoring (up to 20–30 years) that could be required. In the meantime, however, an increased daily consumption of dietary spices can only provide benefits with no harm done.

ACKNOWLEDGMENT

We would like to thank Maude E. Veech for carefully proofreading the manuscript and providing valuable comments. Dr. Aggarwal is a Ransom Horne, Jr., Professor of Cancer Research. This work was supported by a grant from the Clayton Foundation for Research (to B. B. A.) and National Institutes of Health core grant CA16672. We would also like to thank the Department of Biotechnology, Government of India for the award of Overseas Research Associateship to S.D.

REFERENCES

1. Kolonel, L.N., Altshuler, D., and Henderson, B.E., The multiethnic cohort study: Exploring genes, lifestyle and cancer risk, *Nat Rev Cancer* 4, 519, 2004.
2. Aggarwal, B.B., et al., Inflammation and cancer: How hot is the link?, *Biochem Pharmacol* 72, 1605, 2006.
3. Galimberti, D., Fenoglio, C., and Scarpini, E., Inflammation in nerodegenerative disorders: Friend or foe?, *Curr Aging Sci* 1, 30, 2008.
4. Soudamini, K.K., et al., Inhibition of lipid peroxidation and cholesterol levels in mice by curcumin, *Indian J Physiol Pharmacol* 36, 239, 1992.
5. Rajakumar, D.V., and Rao, M.N., Antioxidant properties of dehydrozingerone and curcumin in rat brain homogenates, *Mol Cell Biochem* 140, 73, 1994.
6. Sreejayan, N., and Rao, M.N., Curcuminoids as potent inhibitors of lipid peroxidation, *J Pharm Pharmacol* 46, 1013, 1994.
7. Rajakrishnan, V., et al., Neuroprotective role of curcumin from *Curcuma longa* on ethanol-induced brain damage, *Phytother Res* 13, 571, 1999.

8. Calabrese, V., et al., Nitric oxide in the central nervous system: Neuroprotection versus neurotoxicity, *Nat Rev Neurosci* 8, 766, 2007.

9. Soliman, K.F., and Mazzio, E.A., *In vitro* attenuation of nitric oxide production in C6 astrocyte cell culture by various dietary compounds, *Proc Soc Exp Biol Med* 218, 390, 1998.

10. Kato, K., et al., Stimulation of the stress-induced expression of stress proteins by curcumin in cultured cells and in rat tissues in vivo, *Cell Stress Chaperones* 3, 152, 1998.

11. Chen, C.C., Taniguchi, T., and D'Andrea, A., The Fanconi anemia (FA) pathway confers glioma resistance to DNA alkylating agents, *J Mol Med* 85, 497, 2007.

12. Kim, H.Y., et al., Curcumin suppresses Janus kinase-STAT inflammatory signaling through activation of Src homology 2 domain-containing tyrosine phosphatase 2 in brain microglia, *J Immunol* 171, 6072, 2003.

13. Luo, Y., et al., Intrastriatal dopamine injection induces apoptosis through oxidation-involved activation of transcription factors AP-1 and NF-kappaB in rats, *Mol Pharmacol*, 56, 254, 1999.

14. Kumar, P., et al., Possible neuroprotective mechanisms of curcumin in attenuating 3-nitropropionic acid-induced neurotoxicity, *Methods Find Exp Clin Pharmacol* 29, 19, 2007.

15. Dairam, A., et al., Curcuminoids, curcumin, and demethoxycurcumin reduce lead-induced memory deficits in male Wistar rats, *J Agric Food Chem* 55, 1039, 2007.

16. Jung, K.K., et al., Inhibitory effect of curcumin on nitric oxide production from lipopolysaccharide-activated primary microglia, *Life Sci* 79, 2022, 2006.

17. Scapagnini, G., et al., Curcumin activates defensive genes and protects neurons against oxidative stress, *Antioxid Redox Signal* 8, 395, 2006.

18. Wu, A., Ying, Z., and Gomez-Pinilla, F., Dietary curcumin counteracts the outcome of traumatic brain injury on oxidative stress, synaptic plasticity, and cognition, *Exp Neurol* 197, 309, 2006.

19. Zhu, Y.G., et al., Curcumin protects mitochondria from oxidative damage and attenuates apoptosis in cortical neurons, *Acta Pharmacol Sin* 25, 1606, 2004.

20. Shukla, P.K., et al., Protective effect of curcumin against lead neurotoxicity in rat, *Hum Exp Toxicol* 22, 653, 2003.

21. Daniel, S., et al., Through metal binding, curcumin protects against lead- and cadmium-induced lipid peroxidation in rat brain homogenates and against lead-induced tissue damage in rat brain, *J Inorg Biochem* 98, 266, 2004.

22. Luo, F., et al., Protective effect and mechanism of pretreatment with curcumin on infectious brain edema in rats, *Zhonghua Er Ke Za Zhi* 41, 940, 2003.

23. Vajragupta, O., et al., Manganese complexes of curcumin and its derivatives: Evaluation for the radical scavenging ability and neuroprotective activity, *Free Radic Biol Med* 35, 1632, 2003.

24. Kanter, M., *Nigella sativa* and derived thymoquinone prevents hippocampal neurodegeneration after chronic toluene exposure in rats, *Neurochem Res* 33, 579, 2008.

25. Lee, H.Y., et al., Induction of growth hormone release by dioscin from *Dioscorea batatas* DECNE, *J Biochem Mol Biol* 40, 1016, 2007.

26. Pan, M.H., Huang, T.M., and Lin, J.K., Biotransformation of curcumin through reduction and glucuronidation in mice, *Drug Metab Dispos* 27, 486, 1999.

27. Tomita, M., et al., Astrocyte production of the chemokine macrophage inflammatory protein-2 is inhibited by the spice principle curcumin at the level of gene transcription, *J Neuroinflammation* 2, 8, 2005.

28. Mohamed, A., et al., Improvement of experimental allergic encephalomyelitis (EAE) by thymoquinone; an oxidative stress inhibitor, *Biomed Sci Instrum* 39, 440, 2003.

29. Mohamed, A., et al., Thymoquinone inhibits the activation of NF-kappaB in the brain and spinal cord of experimental autoimmune encephalomyelitis, *Biomed Sci Instrum* 41, 388, 2005.

30. Thakur, M.K., Alzheimer's disease—A challenge in the new millennium, *Current Sci* 79, 29, 2000.

31. Kuner, P., Schubenel, R., and Hertel, C., Beta-amyloid binds to p57NTR and activates NFkappaB in human neuroblastoma cells, *J Neurosci Res* 54, 798, 1998.

32. Kim, D.S., Kim, D.S., and Oppel, M.N., Shogaols from *Zingiber officinale* protect IMR32 human neuroblastoma and normal human umbilical vein endothelial cells from beta-amyloid(25–35) insult, *Planta Med* 68, 375, 2002.

33. Giri, R.K., Rajagopal, V., and Kalra, V.K., Curcumin, the active constituent of turmeric, inhibits amyloid peptide-induced cytochemokine gene expression and CCR5-mediated chemotaxis of THP-1 monocytes by modulating early growth response-1 transcription factor, *J Neurochem* 91, 1199, 2004.

34. Frautschy, S.A., et al., Phenolic anti-inflammatory antioxidant reversal of Abeta-induced cognitive deficits and neuropathology, *Neurobiol Aging* 22, 993, 2001.

35. Yang, F., et al., Curcumin inhibits formation of amyloid beta oligomers and fibrils, binds plaques, and reduces amyloid in vivo, *J Biol Chem* 280, 5892, 2005.

36. Ono, K., et al., Curcumin has potent anti-amyloidogenic effects for Alzheimer's beta-amyloid fibrils *in vitro*, *J Neurosci Res* 75, 742, 2004.

37. Kim, S.Y., Jung, S.H., and Kim, H.S., Curcumin is a potent broad spectrum inhibitor of matrix metalloproteinase gene expression in human astroglioma cells, *Biochem Biophys Res Commun* 337, 510, 2005.

38. Ryu, E.K., et al., Curcumin and dehydrozingerone derivatives: Synthesis, radiolabeling, and evaluation for beta-amyloid plaque imaging, *J Med Chem* 49, 6111, 2006.

39. Pandey, N., et al., Curcumin inhibits aggregation of alpha-synuclein, *Acta Neuropathol* 115, 479, 2008.

40. Ma, Q.L., et al., p21-activated kinase-aberrant activation and translocation in Alzheimer disease pathogenesis, *J Biol Chem* 283, 14132, 2008.

41. Narlawar, R., et al., Curcumin derivatives inhibit or modulate beta-amyloid precursor protein metabolism, *Neurodegener Dis* 4, 88, 2007.

42. Liu, Y., et al., A broadly neuroprotective derivative of curcumin, *J Neurochem* 105, 1336, 2008.

43. Begum, A.N., et al., Curcumin structure-function, bioavailablity and efficacy in models of neuroinflammation and Alzheimer's, *J Pharmacol Exp Ther* 326, 196–208, 2008.

44. Wisniewski, T., and Sigurdsson, E.M., Therapeutic approaches for prion and Alzheimer's diseases, *FEBS J* 274, 3784, 2007.

45. Hafner-Bratkovic, I., et al., Curcumin binds to the alpha-helical intermediate and to the amyloid form of prion protein—a new mechanism for the inhibition of PrP(Sc) accumulation, *J Neurochem* 104, 1553, 2008.

46. Munch, G., et al., Advanced glycation end products in neurodegeneration: More than early markers of oxidative stress?, *Ann Neurol* 44, S85, 1998.

47. Mythri, R.B., et al., Mitochondrial complex I inhibition in Parkinson's disease: How can curcumin protect mitochondria?, *Antioxid Redox Signal* 9, 399, 2007.

48. Shulman, L.M., Gender differences in Parkinson's disease, *Gend Med* 4, 8, 2007.

49. Sawada, H., et al., Estradiol protects dopaminergic neurons in a MPP^+ Parkinson's disease model, *Neuropharmacology* 42, 1056, 2002.

50. Ebadi, M., and Sharma, S.K., Peroxynitrite and mitochondrial dysfunction in the pathogenesis of Parkinson's disease, *Antioxid Redox Signal* 5, 319, 2003.

51. Rajeswari, A., Curcumin protects mouse brain from oxidative stress caused by 1-methyl-4-phenyl-1,2,3,6-tetrahydropyridine, *Eur Rev Med Pharmacol Sci* 10, 157, 2006.

52. Zbarsky, V., et al., Neuroprotective properties of the natural phenolic antioxidants curcumin and naringenin but not quercetin and fisetin in a 6-OHDA model of Parkinson's disease, *Free Radic Res* 39, 1119, 2005.

53. Ko, F.N., et al., Antioxidant properties of demethyldiisoeugenol, *Biochim Biophys Acta* 1258, 145, 1995.

54. Kabuto, H., Tada, M., and Kohno, M., Eugenol [2-methoxy-4-(2-propenyl)phenol] prevents 6-hydroxy-dopamine-induced dopamine depression and lipid peroxidation inductivity in mouse striatum, *Biol Pharm Bull* 30, 423, 2007.

55. Bala, K., Tripathy, B.C., and Sharma, D., Neuroprotective and anti-ageing effects of curcumin in aged rat brain regions, *Biogerontology* 7, 81, 2006.

56. Bishnoi, M., Chopra, K., and Kulkarni, S.K., Protective effect of Curcumin, the active principle of turmeric (*Curcuma longa*) in haloperidol-induced orofacial dyskinesia and associated behavioural, biochemical and neurochemical changes in rat brain, *Pharmacol Biochem Behav* 88, 511, 2008.

57. Crosio, C., et al., Chromatin remodeling and neuronal response: Multiple signaling pathways induce specific histone H3 modifications and early gene expression in hippocampal neurons, *J Cell Sci* 116, 4905, 2003.

58. Taniura, H., Sng, J.C., and Yoneda, Y., Histone modifications in status epilepticus induced by kainate, *Histol Histopathol*, 21, 785, 2006.

59. Shin, H.J., et al., Curcumin attenuates the kainic acid-induced hippocampal cell death in the mice, *Neurosci Lett* 416, 49, 2007.

60. Sumanont, Y., et al., Prevention of kainic acid-induced changes in nitric oxide level and neuronal cell damage in the rat hippocampus by manganese complexes of curcumin and diacetylcurcumin, *Life Sci* 78, 1884, 2006.

61. Sumanont, Y., et al., Effects of manganese complexes of curcumin and diacetylcurcumin on kainic acid-induced neurotoxic responses in the rat hippocampus, *Biol Pharm Bull* 30, 1732, 2007.

62. Sharma, S., et al., Curcumin attenuates thermal hyperalgesia in a diabetic mouse model of neuropathic pain, *Eur J Pharmacol* 536, 256, 2006.
63. Ghoneim, A.I., et al., Protective effects of curcumin against ischaemia/reperfusion insult in rat forebrain, *Pharmacol Res* 46, 273, 2002.
64. Wang, Q., et al., Neuroprotective mechanisms of curcumin against cerebral ischemia-induced neuronal apoptosis and behavioral deficits, *J Neurosci Res* 82, 138, 2005.
65. Al-Omar, F.A., et al., Immediate and delayed treatments with curcumin prevents forebrain ischemia-induced neuronal damage and oxidative insult in the rat hippocampus, *Neurochem Res* 31, 611, 2006.
66. Jiang, J., et al., Neuroprotective effect of curcumin on focal cerebral ischemic rats by preventing blood-brain barrier damage, *Eur J Pharmacol* 561, 54, 2007.
67. Thiyagarajan, M., and Sharma, S.S., Neuroprotective effect of curcumin in middle cerebral artery occlusion induced focal cerebral ischemia in rats, *Life Sci* 74, 969, 2004.
68. Hosseinzadeh, H., et al., Effect of thymoquinone and *Nigella sativa* seeds oil on lipid peroxidation level during global cerebral ischemia-reperfusion injury in rat hippocampus, *Phytomedicine* 14, 621, 2007.
69. Al-Majed, A.A., Al-Omar, F.A., and Nagi, M.N., Neuroprotective effects of thymoquinone against transient forebrain ischemia in the rat hippocampus, *Eur J Pharmacol* 543, 40, 2006.
70. Xu, Y., et al., Antidepressant effects of curcumin in the forced swim test and olfactory bulbectomy models of depression in rats, *Pharmacol Biochem Behav* 82, 200, 2005.
71. Xu, Y., et al., Curcumin reverses impaired hippocampal neurogenesis and increases serotonin receptor 1A mRNA and brain-derived neurotrophic factor expression in chronically stressed rats, *Brain Res* 1162, 9, 2007.
72. Xu, Y., et al., Curcumin reverses the effects of chronic stress on behavior, the HPA axis, BDNF expression and phosphorylation of CREB, *Brain Res* 1122, 56, 2006.
73. Kong, L.D., Cheng, C.H., and Tan, R.X., Inhibition of MAO A and B by some plant-derived alkaloids, phenols and anthraquinones, *J Ethnopharmacol* 91, 351, 2004.
74. Liu, G.Q., et al., Stimulation of serotonin synthesis in rat brain after antiepilepsirine, an antiepileptic piperine derivative, *Biochem Pharmacol* 33, 3883, 1984.
75. Liu, G.Q., and Garattini, S., Stimulation of serotonin synthesis in rat brain after administration of antiepilepsirine, an antiepileptic piperine derivative, *Chin Med J (Engl)* 99, 411, 1986.
76. Mori, A., Kabuto, H., and Pei, Y.Q., Effects of piperine on convulsions and on brain serotonin and catecholamine levels in E1 mice, *Neurochem Res* 10, 1269, 1985.
77. Li, S., et al., Antidepressant like effects of piperine in chronic mild stress treated mice and its possible mechanisms, *Life Sci* 80, 1373, 2007.
78. Li, S., et al., Antidepressant-like effects of piperine and its derivative, antiepilepsirine, *J Asian Nat Prod Res* 9, 421, 2007.
79. Mujumdar, A.M., et al., Effect of piperine on pentobarbitone induced hypnosis in rats, *Indian J Exp Biol* 28, 486, 1990.
80. Unchern, S., et al., Piperine, a pungent alkaloid, is cytotoxic to cultured neurons from the embryonic rat brain, *Biol Pharm Bull* 17, 403, 1994.
81. Irie, Y., et al., Eugenol exhibits antidepressant-like activity in mice and induces expression of metallothionein-III in the hippocampus, *Brain Res* 1011, 243, 2004.
82. Wie, M.B., et al., Eugenol protects neuronal cells from excitotoxic and oxidative injury in primary cortical cultures, *Neurosci Lett* 225, 93, 1997.
83. Muller, M., et al., Effect of eugenol on spreading depression and epileptiform discharges in rat neocortical and hippocampal tissues, *Neuroscience* 140, 743, 2006.
84. Ardjmand, A., et al., Eugenol depresses synaptic transmission but does not prevent the induction of long-term potentiation in the CA1 region of rat hippocampal slices, *Phytomedicine* 13, 146, 2006.
85. Ambegaokar, S.S., et al., Curcumin inhibits dose-dependently and time-dependently neuroglial cell proliferation and growth, *Neuro Endocrinol Lett* 24, 469, 2003.
86. Woo, M.S., et al., Curcumin suppresses phorbol ester-induced matrix metalloproteinase-9 expression by inhibiting the PKC to MAPK signaling pathways in human astroglioma cells, *Biochem Biophys Res Commun* 335, 1017, 2005.
87. Nagai, S., et al., Inhibition of cellular proliferation and induction of apoptosis by curcumin in human malignant astrocytoma cell lines, *J Neurooncol* 74, 105, 2005.
88. Gao, X., et al., Curcumin differentially sensitizes malignant glioma cells to TRAIL/Apo2L-mediated apoptosis through activation of procaspases and release of cytochrome c from mitochondria, *J Exp Ther Oncol* 5, 39, 2005.

89. Kang, S.K., Cha, S.H., and Jeon, H.G., Curcumin-induced histone hypoacetylation enhances caspase-3-dependent glioma cell death and neurogenesis of neural progenitor cells, *Stem Cells Dev* 15, 165, 2006.

90. Belkaid, A., et al., Silencing of the human microsomal glucose-6-phosphate translocase induces glioma cell death: Potential new anticancer target for curcumin, *FEBS Lett* 580, 3746, 2006.

91. Karmakar, S., et al., Curcumin activated both receptor-mediated and mitochondria-mediated proteolytic pathways for apoptosis in human glioblastoma T98G cells, *Neurosci Lett* 407, 53, 2006.

92. Karmakar, S., Banik, N.L., and Ray, S.K., Curcumin suppressed anti-apoptotic signals and activated cysteine proteases for apoptosis in human malignant glioblastoma U87MG cells, *Neurochem Res* 32, 2103, 2007.

93. Aoki, H., et al., Evidence that curcumin suppresses the growth of malignant gliomas *in vitro* and in vivo through induction of autophagy: Role of Akt and extracellular signal-regulated kinase signaling pathways, *Mol Pharmacol* 72, 29, 2007.

94. Dhandapani, K.M., Mahesh, V.B., and Brann, D.W., Curcumin suppresses growth and chemoresistance of human glioblastoma cells via AP-1 and NFkappaB transcription factors, *J Neurochem* 102, 522, 2007.

95. Liu, E., et al., Curcumin induces G2/M cell cycle arrest in a p53-dependent manner and upregulates ING4 expression in human glioma, *J Neurooncol* 85, 263, 2007.

96. Shinojima, N., et al., Roles of the Akt/mTOR/p70S6K and ERK1/2 signaling pathways in curcumin-induced autophagy, *Autophagy* 3, 635, 2007.

97. Panchal, H.D., et al., Early anti-oxidative and anti-proliferative curcumin effects on neuroglioma cells suggest therapeutic targets, *Neurochem Res* 33, 1701–1710, 2008.

98. Choi, B.H., et al., p21 Waf1/Cip1 expression by curcumin in U-87MG human glioma cells: Role of early growth response-1 expression, *Cancer Res* 68, 1369, 2008.

99. Miller, M., et al., Curcumin (diferuloylmethane) inhibits cell proliferation, induces apoptosis and decreases hormone levels and secretion, in pituitary tumor cells, *Endocrinology* 149, 4158–4167, 2008.

100. Shih, P.C., et al., Efficacy of curcumin therapy against *Angiostrongylus cantonensis*-induced eosinophilic meningitis, *J Helminthol* 81, 1, 2007.

101. Kim, D.S., Park, S.Y., and Kim, J.K., Curcuminoids from *Curcuma longa* L. (Zingiberaceae) that protect PC12 rat pheochromocytoma and normal human umbilical vein endothelial cells from betaA(1–42) insult, *Neurosci Lett* 303, 57, 2001.

102. Baum, L., and Ng, A., Curcumin interaction with copper and iron suggests one possible mechanism of action in Alzheimer's disease animal models, *J Alzheimers Dis* 6, 367, 2004.

103. Atamna, H., and Boyle, K., Amyloid-beta peptide binds with heme to form a peroxidase: Relationship to the cytopathologies of Alzheimer's disease, *Proc Natl Acad Sci U S A* 103, 3381, 2006.

104. Zhang, L., et al., Curcuminoids enhance amyloid-beta uptake by macrophages of Alzheimer's disease patients, *J Alzheimers Dis* 10, 1, 2006.

105. Fiala, M., et al., Innate immunity and transcription of MGAT-III and Toll-like receptors in Alzheimer's disease patients are improved by bisdemethoxycurcumin, *Proc Natl Acad Sci U S A* 104, 12849, 2007.

106. Wu, C.Y., et al., IL-1beta induces proMMP-9 expression via c-Src-dependent PDGFR/PI3K/Akt/p300 cascade in rat brain astrocytes, *J Neurochem* 105, 1499, 2008.

107. Dairam, A., et al., Antioxidant and iron-binding properties of curcumin capsaicin, and *S*-allylcysteine reduce oxidative stress in rat brain homogenate, *J Agric Food Chem*, 56, 3350, 2008.

108. Lim, G.P., et al., The curry spice curcumin reduces oxidative damage and amyloid pathology in an Alzheimer transgenic mouse, *J Neurosci* 21, 8370, 2001.

109. Garcia-Alloza, M., et al., Curcumin labels amyloid pathology in vivo, disrupts existing plaques, and partially restores distorted neurites in an Alzheimer mouse model, *J Neurochem* 102, 1095, 2007.

110. Baum, L., et al., Six-month randomized, placebo-controlled, double-blind, pilot clinical trial of curcumin in patients with Alzheimer disease, *J Clin Psychopharmacol* 28, 110, 2008.

111. Wang, J., et al., Reversal of a full-length mutant huntingtin neuronal cell phenotype by chemical inhibitors of polyglutamine-mediated aggregation, *BMC Neurosci* 6, 1, 2005.

112. Fahey, A.J., Adrian Robins, R., and Constantinescu, C.S., Curcumin modulation of IFN-beta and IL-12 signalling and cytokine induction in human T cells, *J Cell Mol Med* 11, 1129, 2007.

113. Natarajan, C., and Bright, J.J., Curcumin inhibits experimental allergic encephalomyelitis by blocking IL-12 signaling through Janus kinase-STAT pathway in T lymphocytes, *J Immunol* 168, 6506, 2002.

114. Verbeek, R., van Tol, E.A., and van Noort, J.M., Oral flavonoids delay recovery from experimental autoimmune encephalomyelitis in SJL mice, *Biochem Pharmacol* 70, 220, 2005.

115. Malfitano, A.M., et al., Arvanil inhibits T lymphocyte activation and ameliorates autoimmune encephalomyelitis, *J Neuroimmunol* 171, 110, 2006.

116. Cabranes, A., et al., Decreased endocannabinoid levels in the brain and beneficial effects of agents activating cannabinoid and/or vanilloid receptors in a rat model of multiple sclerosis, *Neurobiol Dis* 20, 207, 2005.
117. Sng, J.C., Taniura, H., and Yoneda, Y., Histone modifications in kainate-induced status epilepticus, *Eur J Neurosci* 23, 1269, 2006.
118. Jang, S.W., et al., Gambogic amide, a selective agonist for TrkA receptor that possesses robust neurotrophic activity, prevents neuronal cell death, *Proc Natl Acad Sci U S A* 104, 16329, 2007.
119. Qiang, L., et al., Inhibition of glioblastoma growth and angiogenesis by gambogic acid: An *in vitro* and in vivo study, *Biochem Pharmacol* 75, 1083, 2008.
120. Riemer, C., et al., Evaluation of drugs for treatment of prion infections of the central nervous system, *J Gen Virol* 89, 594, 2008.
121. Sharma, S., Chopra, K., and Kulkarni, S.K., Effect of insulin and its combination with resveratrol or curcumin in attenuation of diabetic neuropathic pain: Participation of nitric oxide and TNF-alpha, *Phytother Res* 21, 278, 2007.
122. Guenette, S.A., et al., Pharmacokinetics of eugenol and its effects on thermal hypersensitivity in rats, *Eur J Pharmacol* 562, 60, 2007.
123. Shukla, P.K., et al., Anti-ischemic effect of curcumin in rat brain, *Neurochem Res* 33, 1036, 2008.
124. Rathore, P., et al., Curcuma oil: Reduces early accumulation of oxidative product and is anti-apoptogenic in transient focal ischemia in rat brain, *Neurochem Res* 33, 1672–1682, 2008.
125. Amantini, C., et al., Capsaicin-induced apoptosis of glioma cells is mediated by TRPV1 vanilloid receptor and requires p38 MAPK activation, *J Neurochem* 102, 977, 2007.
126. Wang, R., et al., Curcumin protects against glutamate excitotoxicity in rat cerebral cortical neurons by increasing brain-derived neurotrophic factor level and activating TrkB, *Brain Res* 1210, 84, 2008.

21 Neurohormetic Properties of the Phytochemical Resveratrol

Andrea Lisa Holme[1] and Shazib Pervaiz[2,3,4,5]

[1]National University Medical Institutes, Yong Loo Lin School of Medicine, National University of Singapore, Singapore

[2]Department of Physiology, Yong Loo Lin School of Medicine, [3]Graduate School for Integrative Sciences and Engineering, National University of Singapore, [4]Duke-NUS Graduate Medical School, [5]Singapore-MIT Alliance, Singapore

CONTENTS

21.1 CELL DEATH AND INJURY IN NEUROLOGICAL DISORDERS

Mammalian brains are protected by a four-tier system comprised of (1) an outer protective layer derived from the skull, (2) protection from the skull by the envelope membrane system, meninges (the dura, arachnoid, and the pia mater), (3) a structure of connective tissue separating the skull from the brain, which contains the arachnoid, a subarachnoid space that contains the protective buffering cerebrospinal fluid, and (4) the blood-brain barrier that protects the brain from blood-based toxins. At the cellular level, the brain is composed of neurons that transfer and retain information, and the identification of novel neurons such as mirror neurons suggests that the brain is by no means fully characterized (1). The "work horses" are the multifunctional glia cells, which are present at a ratio of 10:1 (glia/neurons) (2). The glia cells provide a support network to neurons, while the multifunctional astrocytes (a subtype of glia cells) control neuronal signals using chemical mediators (3). Microglia cells are the "macrophages of the brain" and arise from myeloid progenitor cells. A host of neurotransmitters regulate signaling by glia cells, such as noradrenaline (norepinephrine), acetylcholine, dopamine, and glutamate (4).

Many neurological disorders correlate with an imbalance between the rate of free radical production and their clearance via the cellular antioxidant defense systems, which gives rise to reductive-oxidative (redox) stress (5). This is exemplified in polyglutamine (polyQ) diseases, such as Huntington's disease (HD), spinomuscular bulbar atrophy/Kennedy disease, and spinocerebellar ataxia types 1, 2, 3, 6, 7, and 17, where the mutant protein has a gain of function mutation giving rise to polyglutamine (polyQ) (6,7). The disease etiology is a consequence of polyQ aggregation and its cleavage during apoptosis releases a neurotoxic N-terminal repeat (8). Another protein aggregation-based neurodegenerative disease is Parkinson's disease (PD), which is characterized by a progressive loss of dopaminergic neurons with the presence of intracytoplasmic inclusions termed Lewy bodies, which stain positive for α-synuclein and ubiquitin (9,10). While the underlying mechanisms still remain cloudy, compartmentalization studies suggest a link between axon degeneration and neural cell death (11). Brody et al., in an elegant study using a model of brain injury, demonstrated that secretion of the β-amyloid peptide played a protective function and its secretion correlated with improvement in the neurological status (12). However, β-amyloid can be toxic due to the formation of β-amyloid plaques, neurofibrillary tangles, and dystrophic neuritis, pathognomonic of the progressive neurodegenerative disease known as Alzheimer's disease (AD).

One of the most common forms of brain injury results from ischemia/hypoxia of the brain as seen in *stroke*. Ischemic brain damage triggers a cascade of events, such as loss of ATP and redox imbalance, activation of Ca^{2+} channels, and the release of glutamate. Glutamate activates nerve cell receptors, especially NMDA receptors, which in turn activate enzymes that digest cellular material via the catabolic process of autophagy. Ion channel abnormalities have also been implicated in epileptic disorders, in addition to head trauma, tumors, and central nervous system (CNS) infections, e.g., cytomegalovirus, which targets astrocytes and produces inflammatory cytokines such as interferon γ (13).

Inflammation also plays a critical role in the pathology of the nervous system, and the best example of that is the demyelinating autoimmune disease multiple sclerosis (MS) (14). MS is characterized by T-cell infiltration, macrophage activation, presence of autoantibodies, as well as tissue-damaging oxidative stress (15,16). It appears that mitochondrial injury is especially damaging to axons due to the severe depletion of intracellular ATP (17). Similar events have been reported with other autoimmune neurological disorders, such as transverse myelitis, Guillain-Barré syndrome, and acute disseminated encephalomyelitis. An underlying instigator in these diseases appears to be the high levels of the proinflammatory cytokine, interleukin (IL-6), which targets oligodendrocytes and causes demyelination (18). This is also seen in spinal muscle atrophy, where the cell autonomous degeneration of the lower motor neurons is initiated by loss-of-function mutations in the survival motor neuron 1 (*smn1*) gene that encodes the SMN protein, which is involved in processing mRNA (19).

By far, the most important brain pathology related to a defect or deficiency in cell fate signaling is brain neoplasia, primary as well as metastatic. Glial tumors account for more than half of all brain tumors and require growth factors such as epidermal growth factor, fibroblast growth factor, platelet-derived growth factor, ciliary neurotrophic factor, leukemia inhibitory factor receptor, and IL-6. Involvement of the sonic hedgehog pathway as well as phosphatidylinositol-3-kinase (PI3K)/ Akt activation have been implicated as survival signals in brain tumors such as medulloblastomas and glioblastomas (20–22). In addition, loss of the protease caspase 8 was shown to promote metastasis in neuroblastoma cells by regulating integrin expression (23).

21.2   RESVERATROL AS A BIOACTIVE AND NEUROPROTECTIVE COMPOUND

The "French paradox" refers to the observation that despite the high content of fat in the diet, regular consumption of wine by the French appears to prevent the development of cardiovascular disease (24). Indeed, this epidemiological observation has been given biological merit by the identification of the stilbene, resveratrol (RSV), as the effective component. In addition to RSV's beneficial effects on the cardiovascular and immune systems, there is now a growing body of evidence strongly supporting a neuroprotective effect in a variety of neurological disease models such as AD, PD, HD, amyotrophic lateral sclerosis, epilepsy, cancer, and nerve injury (25,26).

RSV was isolated in 1940 and identified as part of the plants' defense mechanism against pathological stressors (27,28). It was largely overlooked until the pioneering work by Jang et al.

demonstrating the cancer chemopreventive activity of RSV in a murine model of chemical carcino-genesis (29). This was followed by evidence linking its chemopreventive potential to its ability to trigger apoptotic cell death in human tumor cells (30). More recent discoveries have highlighted its antioxidant potential by demonstrating its ability to prevent oxidative modification of proteins (28,31) as well as retard the process of ageing by functioning as a caloric restriction mimic and a modulator of biogenesis (32–34). Furthermore, RSV is also an immune modulator as shown by its ability to regulate the aryl hydrocarbon receptor in Treg cells and T17 helper cells (35). Despite these and a plethora of other effects attributed to RSV, it is remarkable that the cellular responses to RSV, such as regulation of gene expression and antioxidant and metal chelator properties, are mostly dependent on its concentration (36). This is partly due to the relatively low bioavailability and formation of conjugates that may not be as effective as the parent molecule itself; oral or intra-venous RSV in animals is metabolized into glucuronic acid and sulfate conjugates (37–39).

21.3 CELL DEATH SIGNALING IN THE NERVOUS SYSTEM AND ITS MODULATION BY RSV

When appraising neurological cell death, survival, and proliferation/regeneration, we draw upon what is known from other cell types (40). However, in many instances, the site/cell of defect and degenera-tion is still unclear owing to the difficulty in resolving the spatiotemporal sequence of events (41,42). This is coupled with the fact that there are novel neural-specific mechanisms as well as undiscovered factors, resulting in an incomplete understanding of observed end point responses. This can be illus-trated within the context of neurogenesis and stem cell biology. During development, stem cells die in an orderly process to shape a normal brain. Bieberich et al. ascertained that these events are partially controlled by the lipid ceramide and prostate apoptosis response-4 (PAR-4) signals; however, the sig-naling cascade(s) evoked are still being unraveled (43). It is also becoming apparent that "living stress" can affect the brain's function and behavior in both adults and neonates (44). A number of experiments show that exposure to chronic stress increases glucocorticoids, which are not only responsible for the "flight or fight" phenomena but could also elicit severe neurotoxicity (45,46). Furthermore, the redox status of neural progenitor cells dictates their differentiation potential. This redox status appears to be regulated by Sirutin-1 (a NAD+ histone deacetylase), and by dint of its ability to activate SIRT-1, RSV mimics the effect of differentiation-inducing slight prooxidant milieu (47).

21.3.1 APOPTOTIC SIGNAL TRANSDUCTION AND REGULATION BY RSV

Although with the tremendous advancement in molecular biology and state-of-the-art techniques the intracellular networks and their interconnections during apoptosis are well characterized in many biological systems, the basis for these observations is largely attributed to the work of Wylie and Kerr who described the two distinct forms of cell death, namely apoptosis and necrosis (acci-dental), based upon morphological and histological hallmarks (48). Generally, by the very nature of the process, necrosis is a somewhat "explosive" form of death resulting from massive cell swelling and early rupture of the plasma membrane and tends to affect whole tissues. Apoptosis, on the other hand, could be referred to as an "implosive" phenomenon leading to changes at a single cell level, such as reduction in cell volume (shrinkage), membrane blebbing, and involution. Since the pioneer-ing work of Wylie and Kerr, a large body of work has added to our knowledge of the mechanisms involved not only in apoptosis commitment and execution but also in corpse disposal.

The most extensively studied apoptotic pathway involves activation of death receptors of the tumor necrosis factor (TNF) family, such as Fas (CD95/Apo1), TNF-R, and death receptor 5. Clustering of these receptors brought about by binding of the specific ligand, activates the assem-bly of the death-initiating signaling complex, which requires protein-protein interaction through specific domains termed death domains (DD) and death effector domains (DED). For example, in the case of Fas (CD95/Apo1), binding of the FasL (CD95L/Apo1L) triggers oligomerization of Fas (CD95/Apo1) containing a short cytoplasmic tail that lacks independent ability to signal but contains the DD. This allows for recruitment of the adaptor protein FADD (Fas-associated protein with death domain)/MORT-1 via a DD-DD interaction. FADD in turn engages the upstream cysteine protease procaspase 8 via DED-DED interaction. This assembly brings about activation of caspase 8, which can either directly trigger a cascade of caspase activation (type I) or activate the mitochondrial death amplification loop (type II), leading to the disassembly of the cell. A total of 14 caspases (caspase 1–14) have been identified in the mammalian system, which function at the death commitment phase (such as caspases 8, 10, and 2), amplification phase (such as caspases 9, 12, and 6), or the execution phase (such as caspase 3, 7, and 6). These proteases exist as inactive proenzymes and are activated upon proteolytic processing, either by upstream caspases or other proteases, such as the calcium (Ca^{2+})-dependent calpains.

Although neurons express these death receptors, they are generally insensitive to receptor-mediated death, suggesting they have a novel role besides death (49). There are data to suggest that death receptors can (1) induce death of nontransformed cells, (2) have immunoregulatory functions, and (3) exhibit a unique expression pattern in the CNS (49). This is exemplified by the TNF-related apoptosis-inducing ligand (TRAIL), which is not expressed within the human brain, while TRAIL receptors are found differently distributed on neurons, oligodendrocytes, and astrocytes. If, under pathologic circumstances, the CNS is inflamed, immune cells such as macrophages and T cells upregulate TRAIL and use this death ligand against tumor cells as well as against neurons and oligodendrocytes within the inflamed CNS (50). In parallel, TRAIL has an immunoregulatory impact on the activation and proliferation of encephalitogenic T cells outside the brain (50).

It is well established that RSV triggers activation of death signaling in tumor cells via death receptor up-regulation or clustering (36). Although RSV's death-inducing ability is not well studied in the nervous system, Fulda et al. showed that RSV could sensitize neuronal cancers to TRAIL-induced apoptosis via p21-mediated cell cycle arrest and down-regulation of the apoptosis inhibitory protein, survivin (51). Along these lines, RSV has also been shown to induce differentiation and a caspase 3-dependent apoptosis of human medulloblastoma cells (52). Furthermore, RSV treatment of U87 glioblastoma cells in the presence or absence of X-radiation resulted in cell cycle arrest (53). RSV is also able to induce an S-phase arrest in an SH-SY5Y neuroblastoma cell line that protected cells from paclitaxel cytotoxicity (54). In the same cell line, RSV was shown to activate or inhibit ERK-1/2 activation, depending upon its concentration (55).

21.3.2 MITOCHONDRIA IN THE NEUROPROTECTIVE EFFECT OF RSV

The neuroprotective properties of RSV are closely related to its ability to scavenge free radicals and modulate reactive oxygen species (ROS) generation from the mitochondrial electron transport chain. Neuronal cells do not easily switch to glycolysis when oxidative phosphorylation becomes limited (56). In fact, glucose is not the only fuel used in the nervous system as lactate plays an important role as an intermediary in numerous metabolic processes and perhaps a mediator of redox state among various compartments both within and between cells (57–62). A cell-to-cell lactate shuttle exists between neurons and astrocytes, and it appears that early oxidative metabolism in neurons is sustained by late activation of the astrocyte-neuron lactate shuttle (63).

RSV is able to control glycolysis and the pentose-phosphate pathway through the regulation of nitric oxide (NO) (64). A number of neurological pathologies involve an increase in iNOS as well as in the activity of the mitochondrial enzyme MnSOD (65–67). NO regulates the activity of cytochrome c oxidase and can down-regulate energy production and thus activate activated protein kinase (AMPK) and the glucose transporters to regulate the balance between glycolysis and the pentose-phosphate pathway (64,68,69). RSV can directly inhibit the mitochondria F1F0ATPase, which catalyses the terminal step in oxidative phosphorylation (70). RSV is reported to noncompetitively inhibit rat brain and liver F1F0ATPase/ATP synthase (IC_{50} of 12–28 μM) (70). Others have shown that RSV inhibits the F1F0ATPase in the heart and liver with an IC_{50} of 13–15 µM, but lower doses (picomolar-nanomolar) of RSV stimulated the F1F0ATPase activity in liver by 10% but not in the heart (71). RSV affects the mitochondria respiration rate (0.4 mg of protein/ml) in rat brains (72). This study also showed that RSV inhibited the mitochondrial respiratory chain through complexes I–III and competes with coenzyme Q. This allows RSV to provide protection from (a) lipid peroxidation of brain synaptosomes induced by the Fenton reaction and (b) scavenge the superoxide anion generated from rat forebrain mitochondria in a concentration-dependent manner. This protects against brain ischemia and interestingly works well even when RSV is administered immediately after the insult (73). This protective property is also shown in another study where RSV works as an antioxidant modulator of complex III and has a membrane stabilizing effect (74).

Efficient death execution involves a cross talk between intracellular organelles, such as the nucleus, endoplasmic reticulum, lysosomes, and the mitochondria. Most of the evidence strongly suggests a convergence of signals at the mitochondria to trigger a change in the mitochondrial permeability. For instance, endoplasmic reticulum stress induced relocalization of Ca^{2+} and activates µ-calpain, which cleaves caspase 12 and Bcl-xL, and the cleaved proapoptotic peptides induce mitochondrial permeability transition (MPT). Similarly, stimuli such as ROS trigger the release of lysosomal enzymes (cathepsins and others) that result in recruitment of the mitochondrial death pathway via induction of MPT. The mitochondrial death pathway is triggered downstream of early caspase activation and obligatory for nonreceptor-induced apoptosis, such as exposure to DNA damaging agents, withdrawal of growth factors, ROS, and irradiation. The connection between upstream caspase activation and engagement of the mitochondrial amplification pathway is the proapoptotic members of the Bcl-2 family, such as Bax, Bak, Bid, Bik, Bcl-XS, and Bim. This is typical for type II pathways, where the upstream activation of caspase 8 is not strong enough to engage downstream caspases, but instead results in cleavage-induced activation of Bid, which in turns acts as a signal for mitochondrial recruitment and oligomerization of proapoptotic proteins Bax and/or Bak. This conformational modification of Bax/Bak and localization to the mitochondria induces a change in the permeability of the mitochondrial outer membrane, thus facilitating the release of mitochondrial intermembranous proteins, such as cytochrome C, apoptosis inducing factor (AIF), and Smac/Diablo. The release of cytochrome C serves as a rate-limiting step in the execution of type II signal, as evidenced by the inhibitory activity of the antiapoptotic proteins members of the Bcl-2 family (Bcl-2, Bcl-xL, Mcl-1, Bcl-w) that are localized to the mitochondrial membranes and neutralize the effects of their proapoptotic counterparts. Cytosolic cytochrome C is an essential factor required for the assembly of the apoptosome, a complex formed with apoptosis protease activating factor 1 (Apaf-1), procaspase 9, and dATP. This assembly is critical for the activation of caspase 9, which in turn activates the downstream executioner caspase 3, resulting in efficient apoptosis. The critical role of the apoptosome in type II systems is supported by the block in apoptotic signaling in the absence of cytosolic cytochrome C and upon down-regulation or defective intracellular localization of Apaf-1. AIF, on the other hand, has been shown to translocate to the nucleus and amplify death signaling via activation of DNAse in a caspase-independent manner; however, whether the release of mitochondrial AIF occurs strictly in the absence of upstream caspase activation is still controversial; however, it is a major player in glutamate excitotoxicity (75,76). A third important protein released from the mitochondria in response to apoptotic stimuli is Smac/Diablo, which functions in an indirect way to facilitate downstream death signaling. This is accomplished by the opposing effect of Smac/Diablo on the inhibitory activity of the family of proteins termed IAPs (inhibitor of apoptosis proteins; NAIP, X-chromosomal IAP, human IAP1 and IAP2, survivin), which is highly conserved in many biological systems (77,78). These proteins abut death signaling by targeting specific caspases to the proteasome degradation pathway.

The mitochondria of HD patients are severally compromised (6,79–84). Using a yeast system expressing GFP-labeled polyQ peptides, it was shown that cellular respiration was reduced due to a change in mitochondrial respiratory chain complex II+III while the ROS levels increased. RSV is reported to be able to partially prevent this via its effect on the mitochondrial respiratory chain (85). Furthermore, an RSV derivative, 4-(methoxybenzylidene)-(3-methoxyphenyl) amine (MBMPA), elicited a more potent neuroprotective effect than RSV upon ischemia maintaining ATP levels (86). A study by Parker et al. using both *Caenorhabditis elegans* and mutant polyglutamine-specific cell death in neuronal cells derived from HdhQ111 knock-in mice showed that RSV can specifically rescue neurons from HD in a forkhead-dependent manner (87). RSV's protective function via SIRTs activation is not confined to HD as demonstrated in a Parkinson study where the inhibition of SIRT-2 rescued α-synuclein toxicity and modified the cellular inclusion morphology (88). RSV is also reported to prevent the toxicity of dopamine. SH-SY5Y cells undergo apoptosis upon exposure to dopamine (300 and 500 μM for 24 hours), and this could be prevented by pretreatment with RSV (5 μM) for 1 hour (89). Chao et al. demonstrated the protective abilities of RSV against 6-hydroxydopamine in rat renal cortical slices (90). Gursoy et al. showed that RSV (100 μM) as well as other phenolic compounds can inhibit the release of dopamine triggered by the loss of ATP during anoxia, however the ATP levels were not restored (91). RSV also inhibited cell death associated with kainic acid-induced excitotoxicity (92–94). Furthermore, RSV is known to increase glutamate uptake and glutamine synthetase activity (via Nrf2) and inhibit excitatory synaptic transmission in rat hippocampus (95). Similarly, RSV (50 μM) inhibited H_2O_2-induced oxidative stress in astrocytes via increasing the GSH content, glutamate uptake, and S100B secretion (96–98). Finally, Zhang et al. showed that RSV could blunt glutamine toxicity induced by oxygen/glucose deprivation in pyramidal neurons and in acute rat hippocampal slices by maintaining the neuronal membrane potential (99).

21.3.3 TRANSCRIPTIONAL REGULATION OF CELL DEATH BY RSV

A host of other proteins, including transcription factors like p53 and NF-κB, and stress-activated proteins (ERK, JNK, p38) are also interlinked to the death execution pathway. The tumor suppressor p53 deserves a special mention due to its ability to fuel the apoptotic signal through mechanisms that involve its transcriptional activity or by a direct effect independent of target gene(s) transcription. DNA damage-induced activation/phosphorylation of p53 triggers its localization to the nucleus and activates transcription of proapoptotic proteins Bax, Noxa, PUMA, Apaf-1, and CD95 (Fas/Apo1) and repression of antiapoptotic protein Bcl-2. Alternatively, p53 has been shown to translocate to the mitochondria and stimulate the mitochondrial generation of ROS via an increase in proline oxidase activity. Mitochondrial ROS generation could facilitate the egress of intermembranous proteins via oxidative modification of membrane lipids, such as cardiolipin, or amplify death signal through a direct effect on caspase activation. Indeed, more recent evidence indicates that caspase 3-dependent cleavage of a component of the mitochondrial electron transport chain complex I (p75) stimulates ROS production in the mitochondria upon apoptotic signaling. Although the involvement of intracellular ROS has been demonstrated in a variety of apoptosis models, evidence is accumulating that depending upon the level of ROS and the cell type, the effect of ROS on cell survival and death pathways could be highly varied. A second, more direct mechanism through which p53 affects apoptosis is by sequestering the antiapoptotic protein Bcl-xL via a direct protein-protein interaction to facilitate oligomerization of Bax at the mitochondria.

Immature neurons are tropic-dependent and sensitive to cell death, as opposed to mature neurons, which are no longer dependent on these factors and are refractory to cell death (100). This insensitivity is partially contributed to the Bcl-2 family "apoptotic brake." The protective ability of Bcl-2 in the brain has been demonstrated using transgenic mice expressing Bcl-2 in neurons and exposed to insults (101–104). RSV can up-regulate Bcl-2 through gene regulation in a number of systems, thus conferring a protection ability (105–107). However, RSV's antiproliferative effects against medulloblastoma cells (UW228-2 and UW228-3) have been reported via inhibition of STAT-3 signaling and activation of cell death (108); RSV treatment resulted in a decrease in STAT-3-activated genes such as survivin, cyclin D1, Cox-2, and c-Myc. In addition, the antitumor activity of RSV has also been documented in in vivo implanted gliomas, with evidence of retarded tumor growth rat and increase in animal survival (109). Jiang et al. demonstrated that RSV's cytotoxic effects were mediated via activation of caspases 1, 3, and 9, cell cycle arrest in G_0/G_1, cytosolic translocation of cytochrome C, and reciprocal mitochondrial translocation of Bax in human glioma U251 cells (110). Similar effects have been reported with active analogues of RSV. Chen et al. showed that 3,4,5,4'-tetrahydroxystilbene (MR-4; 0.5 µM) inhibited proliferation of transformed cells and was more potent against cancer cells while having little effect on normal cells (111). MR-4 causes apoptotic cell death with a rapid perinuclear clustering of mitochondria, thus suggesting mitochondria as a target of this small molecule compound. It should be pointed out that perinuclear redistribution of mitochondria precedes mitochondrial ROS production and cell death (112,113).

21.4   FREE RADICALS, LIPID PEROXIDES, EICOSANOIDS, AND RSV

Free radicals are chemicals composed of an oxygen, sulfur, nitrogen, or carbon center. These species are involved not only in the destructive processes underpinning the pathological conditions but also considered as secondary messengers regulating essential physiological processes in the cell (114,115). ROS are considered the most destructive due to the production of the highly reactive superoxide and hydroxyl radicals. Redox stress arises due to the loss or inability of the natural antioxidant defense systems to deal with the free radicals produced (116–120).

Many neurological disorders, such as PD, ALS, and HD, are characterized by an abnormal accumulation of free radicals resulting in the release of the excitotoxin glutamate (121). Investigations using *in vivo* knock-out models of the antioxidant enzymes superoxide dismutase (SOD), SOD-1 and SOD-2, suggest that the loss of these enzymes have many of the characteristics of neurodegenerative diseases (122). Therefore, compounds able to regulate the expression of such proteins are of special interest. In this regard, RSV could be promising as it can up-regulate Bcl 2 as well as increase the activity of SOD-2 (MnSOD) in the brain by 14-fold (123). Another redox protein that RSV modulates is the iron-protoporphyrin IX heme of heme oxidase (HO): HO-1 is stress-induced, whereas HO-2 is constitutively expressed (124). RSV has been shown to activate HO via NF-κB activation at concentrations of 1–10 µM; however, the inverse is true at concentrations above 20 µM, which inhibits IKBα phosphorylation, thereby inhibiting HO-1 induction. In a separate study using pheochromocytoma cells (PC12), RSV was shown to increase HO-1 expression via activation of the NF-E2-related factor 2 (Nrf2) and the transient activation of AKT and ERK-1/2 (124). It has also been demonstrated that under hypoxia and oxidative stress, iron released from the heme group is readily susceptible to metal radical reactions such as the Fenton reaction (125). Interestingly, biometals are known to be involved in dementia, and RSV functions as a metal chelator and inhibitor of metal-dependent peroxidation in neurons (126–130).

RSV also confers a protective influence by acting as a preconditioner, i.e., to produce resistance to an insult. This ability has been demonstrated in cardioprotection as seen in the modulation of a "redox-thiol" switch (131). Similar protective effect is observed in the cerebral ischemia model and involves the preactivation of SIRT-1 and peroxisome proliferator-activated receptor coactivator-1 α (PGC-1α) (132,133). This neuroprotective effect RSV also involves suppression of matrix metalloprotease 9 (MMP-9), which is involved in the various phases of stroke (89,134). Kawasaki et al. established that MMP-2 and MMP-9 work in parallel to transmit pain signals mediating chronic pain in spinal nerve ligation (135). Both MMPs induce IL-1β cleavage to its active form, which subsequently activates spinal microglia, known mediators of spinal changes involved in neuropathic pain. As mitogen-activated protein kinases (MAPKs) are involved in MMP-9 signaling in microglia (p38MAPK) and astrocytes (ERK), the inhibitory effect of RSV on MAPK pathways, observed in many other systems, could be of particular relevance to the nervous system as well (136–140).

Lipids act as structural components of cellular membranes, secondary messengers, and insulators, e.g., myelin. Evidence suggests that alterations in the lipid composition could result in neuronal death, as illustrated by the spongiform appearance in brains upon loss of the lipid PI(3,5)P2 from suppression of the *Vac14* or *FIG4* gene (141). These enzymes regulate the production of this lipid, and loss or low levels results in the loss of neurons. Another degenerative mechanism is derived from oxygen-dependent degradation of lipids during peroxidation, when lipids give up electrons to free radicals (142). Such damage induces changes in the trans- and peripheral membrane proteins, lipid symmetry, and protein folding and has been associated with pathological states. RSV is reported to inhibit lipid peroxidation via prevention of the generation of ROS as well as the regulation of the cyclooxygenase (COX) and lipoxygenase (LOX) pathways, which are reported to actively contribute to the inflammation seen in neurological diseases (143). A study by Lin et al. showed that RSV could induce p53-dependent apoptosis in glioma cells through the plasma integrin $\alpha V\beta3$ that also contains the thyroid hormone (T4) receptor (144). The apoptotic RSV signal involves the activation of PKC and ERK-1/2, nuclear translocation of COX-2, and p53 activation. It appears that if T4 is activated, the nuclear COX-2 accumulation is prevented, and there is instead an accumulation of Bcl-2, which inhibits apoptosis in glioma cells.

Eicosanoids are synthesized via COX from the oxygenation of either omega-3 (ω-3) or omega-6 (ω-6) essential fatty acids. There are four groups of eicosanoids, namely, prostaglandins (PG), prostacyclins, thromboxanes, and leukotrienes (LT). They can act as signaling molecules or regulate gene transcription; e.g., leukotriene B4 (LTB4) is a ligand for peroxisome proliferator-activated receptors α (PPAR-α) (145). They can also produce highly reactive and damaging peroxides. COX has two isoforms, constitutive (COX-1) and inducible (COX-2). In the CNS, microglial activation is involved in PG-induced inflammation (146). Inhibition of COX-1 and COX-2 activities has been proven to reduce brain injury after ischemia excitotoxicity and 1-methyl-4-phenyl-1,2,3,6-tetrahydropyridine(MPTP)-induced neurodegeneration. RSV is reported to be a noncompetitive inhibitor of the cyclooxygenase activity of COX-1 as well as the hydroperoxidase activity of COX-1 and COX-2 (147–149). Indeed, the ability of RSV to inhibit the generation of inflammatory stimuli contributes to its ability to inhibit the COX peroxidase reaction as well as inhibit the transcription of the terminal synthase responsible for PG-2 synthesis and the production of 8-iso-PGF2α (150). RSV also reduces COX-2 expression in mouse BV-2 microglial cells through inhibition of NF-κB activation (151).

As opposed to COX, the role of LOX in neuropathology is not as well documented. However there is evidence to suggest that at least in AD the 12-LOX metabolite, 12(S)-HETE (12(S)-hydroxy-(5Z, 8 Z, 10E, 14 Z)-eicosatetraenoic acid), promotes c-Jun-dependent apoptosis (152). The activation of caspase 8 upon treatment of cortical neurons with β-amyloid has been suggested to occur through a c-Jun-mediated induction of CD95L (153). The use of RSV as an inhibitor of the COX and LOX pathway has been suggested in many disease models including neurological disorders (154).

Arachidonic acid (ArA) is an eicosanoids precursor and is cleaved from its phospholipid molecule by phospholipase A2 and metabolized by LOX or COX. ArA can enhance glutamate release as well as depolarization-evoked Ca^{2+} accumulation and stimulate sphingomyelinase to produce ceramide (155,156). Ceramide induces apoptosis by inhibiting the mitochondrial electron transport chain, mitochondrial permeabilization, and caspase 3 activation. It has been shown that in an ischemia model using hippocampal CA1 neurons in rat slice preparations, superfusion with oxygen- and glucose-deprived medium produces a rapid depolarization by the ArA cascade via phospholipase A-2 (PLA-2) and the free radicals produced contribute to the irreversible depolarization produced by *in vitro* ischemia. This can be reversed by a number of agents, including RSV (157). The ROS formed by ArA metabolism by COX/LOX generates lipid peroxides and the cytotoxic byproducts malondialdehyde, 4-hydroxynonenal (HNE), and acrolein, which covalently bind to cellular proteins and alter their function (158,159). Of note, RSV has been shown to significantly block the toxic effects of HNE (160,161), thereby providing insights into the protective effects of this phytochemical in oxidized lipid-induced toxicity.

21.5 THE NEUTROTROPIC SIGNAL AXIS OF INSULIN AND ESTROGEN AND ITS MODULATION BY RSV

The health of the nervous system is influenced by diet and lifestyle, and unsurprisingly neurogenesis and neuroprotection are regulated by hormones (162). The steroid hormones, estrogen and testosterone, are two such protective agents (163). Indeed, the age-dependent decline in these steroids can somewhat explain the impact of ageing on brain function, much of which is due to the loss of the ability to blunt oxidative stress. An "estrogen protection" has been proposed for a multitude of disorders, and there is a distinct sex difference—stroke and PD are low in premenopausal women, while it is higher in men (164,165). Estrogen's neuroprotective action can be genomic via the estrogen receptor α/β (ER-α/β) or nongenomic via regulation of cell signaling pathways (166). The genomic pathway requires the activation of the receptor transcription factors, ER-α and ER-β, and they can compensate for each other. ERs are highly expressed in the forebrain, prooptic area, hypothalamus, and amygdala and can be found in astrocytes, glial cells, and peripheral nervous system (167). The two forms can also be expressed in the same cell. Once estrogen is bound to its receptors, the receptors dimerize to interact with DNA in the presence of coactivators and transcription factors such as AP-1 (c-fos, c-jun) and regulate gene expression of proteins such as Bcl-2, Bcl-XL, choline acetyltransferase, oxytocin, somatostatin, insulin-like growth factor 1 (IGF-1) as well as neurofilamentous proteins such as Tau and GAP43. Neuroprotection can also be conferred by estradiol and IGF-I, which have synergistic effects with AKT (168). Other nonclassical estrogen effects are related to ERs' ability to interact with cell signaling proteins (169). It can induce adenylyl cyclase activity within minutes, increase Ca^{2+} concentration, stimulate phospholipase C (PLC), activate endothelial nitric oxide synthase (eNOS), rapidly induce ERK-1/2 signaling, activate src, and directly interact with PI3K (ER-α). In neuronal cells, a synergistic action between nuclear and cytoplasmic ERs occurs (170). Steroids and hormones can directly regulate the action of neural ion transmission channels (166). It is hypothesized that components of the intracellular signaling pathways are clustered in lipid rafts with caveolin and ER (ER-α interacts with caveolin 1), thereby interacting with this structure and regulating the intracellular signaling (171). Estrogen also regulates the excitotoxic glutamate by directly binding to receptive receptors or modulating its release (172,173). AD appears to be related to estrogen's ability to up-regulate acetylcholine via acetylcholine transferase activity, and there is a correlative up-regulation of the ERs (174,175). Estrogen can rescue ischemia neurons from death by up-regulation of Bcl-2 and inhibition of caspases (176). Another study has shown that Bcl-2 overexpression can protect from aggregated proteins related to prion and AD (177). RSV is a phytoestrogen by virtue of its structure, which is similar to that of diethylstilbestrol and can bind to ER (27). Upon binding to ERs, it can activate as well as act as a superagonist when combined with estradiol. RSV (0.1–1.0 μM) can stimulate ^{14}C-catecholamine synthesis from [^{14}C] tyrosine, which was associated with the activation of tyrosine hydroxylase by virtue of its estrogenic activity (128). However, RSV at ≥1.0 μM inhibits catecholamine secretion.

Another major survival promoting signal is from the insulin family, such as proinsulin and IGF-1. These factors work by inducing differentiation and proliferation and by decreasing apoptosis. The signaling mechanisms that are involved in this are via the IGF-1, which activates insulin receptor substrate 1 (IRS-1) and PI3K/AKT systems, which in turn feed into the forkhead box (FOXO) transcription factors and SIRTs. In the presence of growth factors and insulin, FOXO factors are held in check, and the loss of these results in their activation and the induction of cell cycle arrest, stress resistance, and apoptosis. Interestingly, it appears that IGF-1 decreases with age, and infusion of IGF-1 in rats can reduce the ageing processes (178). ER-α (in an estrogen-dependent manner) can physically interact with IGF-I receptor and modulate the PI3K/AKT and decrease glycogen synthase kinase 3β (GSK-3β) to modify downstream pathways (168). RSV is reported to be able to physically inhibit protein aggregation and prevent the activity of microtubule-associated protein Tau, possibly by regulating GSK-β's activity. Tau hyperphosphorylation disrupts its normal function of regulating axonal transport and leads to the accumulation of neurofibrillary tangles and toxic species of soluble Tau. Furthermore, degradation of hyperphosphorylated Tau by the proteasome is inhibited by the actions of β-amyloid. β-Amyloid degradation can be enhanced by the activation of the plasmin protease cascade or increased lipidation of apolipoprotein E (ApoE) (179,180). Interestingly, RSV has been reported to break down β-amyloid through a proteasome subunit β5-dependent mechanism (181). The mechanism has yet to be determined in any detail but was reported not to be due to an increase in transcription or an increase in total proteasome activity, but a loss of intracellular cleavage of β-amyloid was noted when treated with siRNA for subunit β5.

21.6   CALORIC RESTRICTION, AGING, AND RSV

While it is still debatable if caloric restriction can extend the life span of an organism, it appears that consuming a diet low in calories improves muscle function as well as protects against neurological disorders. It should be noted that caloric restriction does not refer to nutritional starvation, but the term was coined to refer to the observations made in 1934 by Clive McCay and Mary Crowell wherein laboratory rats fed a severely reduced calorie diet, while maintaining vital nutrient levels, resulted in life spans of up to twice as long as otherwise expected (182). These findings have since been verified in human trials, and there is convincing evidence of health beneficial effects, particularly neuromuscular system, of diet high in nutritional value but low in caloric content (183). There are several mechanisms by which calorie restriction can work, such as (a) mitohormesis, which proposes that low-intensity stress is good, as it preconditions the cells, thus activating antiaging mechanisms such as increased mitochondrial biogenesis, (b) reducing insulin signaling/sensitivity allows for retention of animals' leanness, (c) SIRTuins (SIRTs) or "silent information regulators" are reported to effect life span and quality of life, (d) increase in levels of dihydroepiandrostanedione (DHEA) upon calorie restriction, and (f) significant decreases in ROS production as well as the activation of the redox-sensitive transcription factors NF-κB, Nrf-1, and HIF-1 (184–190).

RSV's effects on life extension are not fully understood, but they appear to mimic several of the biochemical effects of calorie restriction. These include metabolic shutdown and activation of SIRT-1, FOXO, PGC-1α (involved in preconditioning), AMPK/mTOR (energy depletion activates AMPK, which in turn activates CREB/p300), and SIRT-dependent inactivation of p53 and NF-κB. Interestingly, PGC-1α is reported to bypass the conventional HIF-1α pathway during hypoxia and activates VEGF via an estrogen-dependent pathway using the estrogen-related receptor α (191). A role for SIRTs has also been reported in autophagy via deacetylation of autophagic Atg proteins; inhibition of autophagy with 3-methyladenine decreases SIRT-1 expression (192). This study also showed that SIRT−/− cells shared remarkable phenotypic similarities with the Atg-5−/− mice, which is a function of severe energy depletion as supplementation of pyruvate to the mother before birth prevents perinatal death in these mice. Furthermore, SIRT-1 can ameliorate insulin resistance by silencing expression of protein tyrosine phosphatase 1B, a major negative regulator of insulin's action (193). SIRT-1−/− animals also exhibit reduced oxidative stress in SIRT-1 knock-out mice, but this appears to be a double-edged sword, as these mice are sensitized to stress such as calorie restriction (194). The loss of resistance to oxidative stress/damage appears to be due to the loss of SIRT-1, which allows activation of IGF-I/IRS-2/Ras/ERK1/2 signaling. It is likely that RSV's ability to activate SIRT and AMPK will lead to regulation of energy homeostasis via AMPK, which contributes to neuroprotection. AMPK is believed to regulate glucose metabolism via SIRT, and it is reported that in L6 myotubes, RSV (100 μM) stimulates glucose uptake via the Glut4 transporter (195). Dasgupta and Milbrandt reported similar observations in neurons stimulated with RSV. This resulted in inhibition of Neuro2a proliferation and the promotion of neuronal differentiation, mitochondrial biogenesis, and activation of AMPK in LKB1-dependent fashion, while in CNS neurons, Ca^{2+}/calmodulin-dependent protein kinase β appears to be the mediator (196). However, the precise mechanism is still far from clear, as similar effects were also reported in SIRT−/− cells within 2 hours of treatment. In a mixed neuron/glia culture, RSV was shown to induce SIRT-1 activity that in turn inhibits NF-κB signaling and protects against β-amyloid toxicity (180,197). In addition, a significant decrease in iNOS induction and cathespin B is observed, which might account for the protection against neurodegenerative apoptosis associated with AD.

21.7   SUMMARY

The relationship between life span and disease, especially neurological diseases, appears to be underlined by a detrimental total ROS production, which causes tissue atrophy, dysfunction, and mutations. RSV's multifunctional ability to regulate disease and life depends on the cellular phenotype, suggesting that it and its analogues are of potential benefit. Longevity genes regulate the pathways of IGF-1, TOR, and mitochondria, and the dampening of the ageing pathways can be seen with calorie restriction or mutations. RSV is a stilbene, which is multifunctional in cellular regulation. RSV has had a significant impact in understanding the ageing processes/diseases, and translation to humans appears promising. The development of compounds such as trimethoxy-RSV for AD is an example of this. It is interesting to note that in neurological studies, RSV's well-documented killing effect seen in cancer models is absent even at high concentrations. Efforts are needed to understand these differences and establish the role of RSV and it metabolites as neuroprotective agents.

REFERENCES

1. Iacoboni M, Dapretto M. The mirror neuron system and the consequences of its dysfunction. Nat Rev Neurosci 2006;7(12):942–51.
2. Biber K, Neumann H, Inoue K, Boddeke HW. Neuronal "on" and "off" signals control microglia. Trends Neurosci 2007;30(11):596–602.
3. Newman EA. New roles for astrocytes: Regulation of synaptic transmission. Trends Neurosci 2003;26(10):536–42.
4. Ibanez CF. Message in a bottle: Long-range retrograde signaling in the nervous system. Trends Cell Biol 2007;17(11):519–28.
5. Martin LJ. Transgenic mice with human mutant genes causing Parkinson's disease and amyotrophic lateral sclerosis provide common insight into mechanisms of motor neuron selective vulnerability to degeneration. Rev Neurosci 2007;18(2):115–36.
6. Imarisio S, Carmichael J, Korolchuk V, Chen CW, Saiki S, Rose C, Krishna G, Davies JE, Ttofi E, Underwood BR, et al. Huntington's disease: From pathology and genetics to potential therapies. Biochem J 2008;412(2):191–209.
7. Shimura H, Hattori N, Kubo S, Mizuno Y, Asakawa S, Minoshima S, Shimizu N, Iwai K, Chiba T, Tanaka K, et al. Familial Parkinson disease gene product, parkin, is a ubiquitin-protein ligase. Nat Genet 2000;25(3):302–5.
8. Walker FO. Huntington's disease. Lancet 2007;369(9557):218–28.
9. Hattori N, Mizuno Y. Pathogenetic mechanisms of parkin in Parkinson's disease. Lancet 2004;364(9435):722–4.
10. Schapira AH. Mitochondria in the aetiology and pathogenesis of Parkinson's disease. Lancet Neurol 2008;7(1):97–109.
11. Doering LC. Probing modifications of the neuronal cytoskeleton. Mol Neurobiol 1993;7(3–4):265–91.
12. Brody DL, Magnoni S, Schwetye KE, Spinner ML, Esparza TJ, Stocchetti N, Zipfel GJ, Holtzman DM. Amyloid-beta dynamics correlate with neurological status in the injured human brain. Science 2008;321(5893):1221–4.
13. van den Pol AN, Robek MD, Ghosh PK, Ozduman K, Bandi P, Whim MD, Wollmann G. Cytomegalovirus induces interferon-stimulated gene expression and is attenuated by interferon in the developing brain. J Virol 2007;81(1):332–48.
14. Popovich PG, Longbrake EE. Can the immune system be harnessed to repair the CNS? Nat Rev Neurosci 2008;9(6):481–93.
15. Zozulya AL, Wiendl H. The role of regulatory T cells in multiple sclerosis. Nat Clin Pract Neurol 2008;4(7):384–98.
16. Oksenberg JR, Baranzini SE, Sawcer S, Hauser SL. The genetics of multiple sclerosis: SNPs to pathways to pathogenesis. Nat Rev Genet 2008;9(7):516–26.
17. Waxman SG. Mechanisms of disease: Sodium channels and neuroprotection in multiple sclerosis—Current status. Nat Clin Pract Neurol 2008;4(3):159–69.
18. Krishnan C, Kaplin AI, Pardo CA, Kerr DA, Keswani SC. Demyelinating disorders: Update on transverse myelitis. Curr Neurol Neurosci Rep 2006;6(3):236–43.
19. Lunn MR, Wang CH. Spinal muscular atrophy. Lancet 2008;371(9630):2120–33.
20. Schuller U, Heine VM, Mao J, Kho AT, Dillon AK, Han YG, Huillard E, Sun T, Ligon AH, Qian Y, et al. Acquisition of granule neuron precursor identity is a critical determinant of progenitor cell competence to form Shh-induced medulloblastoma. Cancer Cell 2008;14(2):123–34.
21. Huang PH, Cavenee WK, Furnari FB, White FM. Uncovering therapeutic targets for glioblastoma: A systems biology approach. Cell Cycle 2007;6(22):2750–4.

22. Furnari FB, Fenton T, Bachoo RM, Mukasa A, Stommel JM, Stegh A, Hahn WC, Ligon KL, Louis DN, Brennan C, et al. Malignant astrocytic glioma: Genetics, biology, and paths to treatment. Genes Dev 2007;21(21):2683–710.

23. Stupack DG, Teitz T, Potter MD, Mikolon D, Houghton PJ, Kidd VJ, Lahti JM, Cheresh DA. Potentiation of neuroblastoma metastasis by loss of caspase-8. Nature 2006;439(7072):95–9.

24. Nanji AA, French SW. Alcoholic beverages and coronary heart disease. Atherosclerosis 1986;60(2):197–8.

25. Calabrese V, Cornelius C, Mancuso C, Pennisi G, Calafato S, Bellia F, Bates TE, Giuffrida Stella AM, Schapira T, Dinkova Kostova AT, et al.. Cellular stress response: A novel target for chemoprevention and nutritional neuroprotection in aging, neurodegenerative disorders and longevity. Neurochem Res 2008;33(12):2444–71.

26. Raval AP, Lin HW, Dave KR, Defazio RA, Della Morte D, Kim EJ, Perez-Pinzon MA. Resveratrol and ischemic preconditioning in the brain. Curr Med Chem 2008;15(15):1545–51.

27. Pervaiz S. Resveratrol—From the bottle to the bedside? Leuk Lymphoma 2001;40(5–6):491–8.

28. Pervaiz S. Resveratrol: From grapevines to mammalian biology. FASEB J 2003;17(14):1975–85.

29. Jang M, Cai L, Udeani GO, Slowing KV, Thomas CF, Beecher CW, Fong HH, Farnsworth NR, Kinghorn AD, Mehta RG, et al. Cancer chemopreventive activity of resveratrol, a natural product derived from grapes. Science 1997;275(5297):218–20.

30. Clement MV, Hirpara JL, Chawdhury SH, Pervaiz S. Chemopreventive agent resveratrol, a natural product derived from grapes, triggers CD95 signaling-dependent apoptosis in human tumor cells. Blood 1998;92(3):996–1002.

31. Harikumar KB, Aggarwal BB. Resveratrol: A multitargeted agent for age-associated chronic diseases. Cell Cycle 2008;7(8):1020–35.

32. Guarente L. Sirtuins in aging and disease. Cold Spring Harb Symp Quant Biol 2007;72:483–8.

33. Lopez-Lluch G, Irusta PM, Navas P, de Cabo R. Mitochondrial biogenesis and healthy aging. Exp Gerontol 2008;43(9):813–9.

34. Lagouge M, Argmann C, Gerhart-Hines Z, Meziane H, Lerin C, Daussin F, Messadeq N, Milne J, Lambert P, Elliott P, et al. Resveratrol improves mitochondrial function and protects against metabolic disease by activating SIRT1 and PGC-1alpha. Cell 2006;127(6):1109–22.

35. Quintana FJ, Basso AS, Iglesias AH, Korn T, Farez MF, Bettelli E, Caccamo M, Oukka M, Weiner HL. Control of T(reg) and T(H)17 cell differentiation by the aryl hydrocarbon receptor. Nature 2008;453(7191):65–71.

36. Pervaiz S. Chemotherapeutic potential of the chemopreventive phytoalexin resveratrol. Drug Resist Update 2004;7(6):333–44.

37. Marier JF, Vachon P, Gritsas A, Zhang J, Moreau JP, Ducharme MP. Metabolism and disposition of resveratrol in rats: Extent of absorption, glucuronidation, and enterohepatic recirculation evidenced by a linked-rat model. J Pharmacol Exp Ther 2002;302(1):369–73.

38. Boocock DJ, Faust GE, Patel KR, Schinas AM, Brown VA, Ducharme MP, Booth TD, Crowell JA, Perloff M, Gescher AJ, et al. Phase I dose escalation pharmacokinetic study in healthy volunteers of resveratrol, a potential cancer chemopreventive agent. Cancer Epidemiol Biomarkers Prev 2007;16(6):1246–52.

39. Abd El-Mohsen M, Bayele H, Kuhnle G, Gibson G, Debnam E, Kaila Srai S, Rice-Evans C, Spencer JP. Distribution of [3H]*trans*-resveratrol in rat tissues following oral administration. Br J Nutr 2006;96(1):62–70.

40. Ekshyyan O, Aw TY. Apoptosis: A key in neurodegenerative disorders. Curr Neurovasc Res 2004;1(4):355–71.

41. Conforti L, Adalbert R, Coleman MP. Neuronal death: Where does the end begin? Trends Neurosci 2007;30(4):159–66.

42. Clarke G, Collins RA, Leavitt BR, Andrews DF, Hayden MR, Lumsden CJ, McInnes RR. A one-hit model of cell death in inherited neuronal degenerations. Nature 2000;406(6792):195–9.

43. Wang G, Krishnamurthy K, Chiang YW, Dasgupta S, Bieberich E. Regulation of neural progenitor cell motility by ceramide and potential implications for mouse brain development. J Neurochem 2008;106(2):718–33.

44. Mirescu C, Peters JD, Gould E. Early life experience alters response of adult neurogenesis to stress. Nat Neurosci 2004;7(8):841–6.

45. Sandi C. Stress, cognitive impairment and cell adhesion molecules. Nat Rev Neurosci 2004;5(12):917–30.

46. Kaufer D, Ogle WO, Pincus ZS, Clark KL, Nicholas AC, Dinkel KM, Dumas TC, Ferguson D, Lee AL, Winters MA, et al. Restructuring the neuronal stress response with anti-glucocorticoid gene delivery. Nat Neurosci 2004;7(9):947–53.

47. Prozorovski T, Schulze-Topphoff U, Glumm R, Baumgart J, Schroter F, Ninnemann O, Siegert E, Bendix I, Brustle O, Nitsch R, et al. Sirt1 contributes critically to the redox-dependent fate of neural progenitors. Nat Cell Biol 2008;10(4):385–94.

48. Kerr JF, Wyllie AH, Currie AR. Apoptosis: A basic biological phenomenon with wide-ranging implications in tissue kinetics. Br J Cancer 1972;26(4):239–57.

49. Reich A, Spering C, Schulz JB. Death receptor Fas (CD95) signaling in the central nervous system: Tuning neuroplasticity? Trends Neurosci 2008;31(9):478–86.

50. Aktas O, Schulze-Topphoff U, Zipp F. The role of TRAIL/TRAIL receptors in central nervous system pathology. Front Biosci 2007;12:2912–21.

51. Fulda S, Debatin KM. Sensitization for tumor necrosis factor-related apoptosis-inducing ligand-induced apoptosis by the chemopreventive agent resveratrol. Cancer Res 2004;64(1):337–46.

52. Wang Q, Li H, Wang XW, Wu DC, Chen XY, Liu J. Resveratrol promotes differentiation and induces Fas-independent apoptosis of human medulloblastoma cells. Neurosci Lett 2003;351(2):83–6.

53. Leone S, Fiore M, Lauro MG, Pino S, Cornetta T, Cozzi R. Resveratrol and X rays affect gap junction intercellular communications in human glioblastoma cells. Mol Carcinog 2008;47(8):587–98.

54. Rigolio R, Miloso M, Nicolini G, Villa D, Scuteri A, Simone M, Tredici G. Resveratrol interference with the cell cycle protects human neuroblastoma SH-SY5Y cell from paclitaxel-induced apoptosis. Neurochem Int 2005;46(3):205–11.

55. Miloso M, Bertelli AA, Nicolini G, Tredici G. Resveratrol-induced activation of the mitogen-activated protein kinases, ERK1 and ERK2, in human neuroblastoma SH-SY5Y cells. Neurosci Lett 1999; 264(1–3):141–4.

56. Chang DT, Reynolds IJ. Mitochondrial trafficking and morphology in healthy and injured neurons. Prog Neurobiol 2006;80(5):241–68.

57. Galeffi F, Foster KA, Sadgrove MP, Beaver CJ, Turner DA. Lactate uptake contributes to the NAD(P)H biphasic response and tissue oxygen response during synaptic stimulation in area CA1 of rat hippocampal slices. J Neurochem 2007;103(6):2449–61.

58. Gilbert E, Tang JM, Ludvig N, Bergold PJ. Elevated lactate suppresses neuronal firing in vivo and inhibits glucose metabolism in hippocampal slice cultures. Brain Res 2006;1117(1):213–23.

59. Abe T, Takahashi S, Suzuki N. Oxidative metabolism in cultured rat astroglia: Effects of reducing the glucose concentration in the culture medium and of d-aspartate or potassium stimulation. J Cereb Blood Flow Metab 2006;26(2):153–60.

60. Bouzier-Sore AK, Voisin P, Canioni P, Magistretti PJ, Pellerin L. Lactate is a preferential oxidative energy substrate over glucose for neurons in culture. J Cereb Blood Flow Metab 2003;23(11):1298–306.

61. Chen CJ, Liao SL, Kuo JS. Gliotoxic action of glutamate on cultured astrocytes. J Neurochem 2000;75(4):1557–65.

62. Peng L, Zhang X, Hertz L. High extracellular potassium concentrations stimulate oxidative metabolism in a glutamatergic neuronal culture and glycolysis in cultured astrocytes but have no stimulatory effect in a GABAergic neuronal culture. Brain Res 1994;663(1):168–72.

63. Gladden LB. Lactate metabolism: A new paradigm for the third millennium. J Physiol 2004;558(Pt 1):5–30.

64. Bolanos JP, Herrero-Mendez A, Fernandez-Fernandez S, Almeida A. Linking glycolysis with oxidative stress in neural cells: A regulatory role for nitric oxide. Biochem Soc Trans 2007;35(Pt 5):1224–7.

65. Bayir H, Kagan VE, Clark RS, Janesko-Feldman K, Rafikov R, Huang Z, Zhang X, Vagni V, Billiar TR, Kochanek PM. Neuronal NOS-mediated nitration and inactivation of manganese superoxide dismutase in brain after experimental and human brain injury. J Neurochem 2007;101(1):168–81.

66. Ste-Marie L, Hazell AS, Bemeur C, Butterworth R, Montgomery J. Immunohistochemical detection of inducible nitric oxide synthase, nitrotyrosine and manganese superoxide dismutase following hyperglycemic focal cerebral ischemia. Brain Res 2001;918(1–2):10–9.

67. Kifle Y, Monnier J, Chesrown SE, Raizada MK, Nick HS. Regulation of the manganese superoxide dismutase and inducible nitric oxide synthase gene in rat neuronal and glial cells. J Neurochem 1996;66(5):2128–35.

68. Almeida A, Cidad P, Delgado-Esteban M, Fernandez E, Garcia-Nogales P, Bolanos JP. Inhibition of mitochondrial respiration by nitric oxide: Its role in glucose metabolism and neuroprotection. J Neurosci Res 2005;79(1–2):166–71.

69. Misiti F, Meucci E, Zuppi C, Vincenzoni F, Giardina B, Castagnola M, Messana I. O(2)-dependent stimulation of the pentose phosphate pathway by *S*-nitrosocysteine in human erythrocytes. Biochem Biophys Res Commun 2002;294(4):829–34.

70. Zheng J, Ramirez VD. Inhibition of mitochondrial proton F0F1-ATPase/ATP synthase by polyphenolic phytochemicals. Br J Pharmacol 2000;130(5):1115–23.

71. Kipp JL, Ramirez VD. Effect of estradiol, diethylstilbestrol, and resveratrol on F0F1-ATPase activity from mitochondrial preparations of rat heart, liver, and brain. Endocrine 2001;15(2):165–75.

72. Zini R, Morin C, Bertelli A, Bertelli AA, Tillement JP. Effects of resveratrol on the rat brain respiratory chain. Drugs Exp Clin Res 1999;25(2–3):87–97.

73. Morin C, Zini R, Albengres E, Bertelli AA, Bertelli A, Tillement JP. Evidence for resveratrol-induced preservation of brain mitochondria functions after hypoxia-reoxygenation. Drugs Exp Clin Res 2003;29(5–6):227–33.

74. Zini R, Morin C, Bertelli A, Bertelli AA, Tillement JP. Resveratrol-induced limitation of dysfunction of mitochondria isolated from rat brain in an anoxia-reoxygenation model. Life Sci 2002;71(26):3091–108.

75. Camins A, Pallas M, Silvestre JS. Apoptotic mechanisms involved in neurodegenerative diseases: Experimental and therapeutic approaches. Methods Find Exp Clin Pharmacol 2008;30(1):43–65.

76. Matute C, Alberdi E, Ibarretxe G, Sanchez-Gomez MV. Excitotoxicity in glial cells. Eur J Pharmacol 2002;447(2–3):239–46.

77. Winsauer G, Resch U, Hofer-Warbinek R, Schichl YM, de Martin R. XIAP regulates bi-phasic NF-kappaB induction involving physical interaction and ubiquitination of MEKK2. Cell Signal 2008;20(11):2107–12.

78. Schile AJ, Garcia-Fernandez M, Steller H. Regulation of apoptosis by XIAP ubiquitin-ligase activity. Genes Dev 2008;22(16):2256–66.

79. Zhang H, Das S, Li QZ, Dragatsis I, Repa J, Zeitlin S, Hajnoczky G, Bezprozvanny I. Elucidating a normal function of huntingtin by functional and microarray analysis of huntingtin-null mouse embryonic fibroblasts. BMC Neurosci 2008;9:38.

80. Pandey M, Varghese M, Sindhu KM, Sreetama S, Navneet AK, Mohanakumar KP, Usha R. Mitochondrial NAD+-linked State 3 respiration and complex-I activity are compromised in the cerebral cortex of 3-nitropropionic acid-induced rat model of Huntington's disease. J Neurochem 2008;104(2):420–34.

81. Rasouri S, Lagouge M, Auwerx J. [SIRT1/PGC-1: A neuroprotective axis?]. Med Sci (Paris) 2007;23(10):840–4.

82. Majumder P, Raychaudhuri S, Chattopadhyay B, Bhattacharyya NP. Increased caspase-2, calpain activations and decreased mitochondrial complex II activity in cells expressing exogenous huntingtin exon 1 containing CAG repeat in the pathogenic range. Cell Mol Neurobiol 2007;27(8):1127–45.

83. Fernandes HB, Baimbridge KG, Church J, Hayden MR, Raymond LA. Mitochondrial sensitivity and altered calcium handling underlie enhanced NMDA-induced apoptosis in YAC128 model of Huntington's disease. J Neurosci 2007;27(50):13614–23.

84. Banoei MM, Houshmand M, Panahi MS, Shariati P, Rostami M, Manshadi MD, Majidizadeh T. Huntington's disease and mitochondrial DNA deletions: Event or regular mechanism for mutant huntingtin protein and CAG repeats expansion? Cell Mol Neurobiol 2007;27(7):867–75.

85. Solans A, Zambrano A, Rodriguez M, Barrientos A. Cytotoxicity of a mutant huntingtin fragment in yeast involves early alterations in mitochondrial OXPHOS complexes II and III. Hum Mol Genet 2006;15(20):3063–81.

86. Choi SY, Kim S, Son D, Lee P, Lee J, Lee S, Kim DS, Park Y, Kim SY. Protective effect of (4-methoxybenzylidene)-(3-methoxynophenyl)amine against neuronal cell death induced by oxygen and glucose deprivation in rat organotypic hippocampal slice culture. Biol Pharm Bull 2007;30(1):189–92.

87. Parker JA, Arango M, Abderrahmane S, Lambert E, Tourette C, Catoire H, Neri C. Resveratrol rescues mutant polyglutamine cytotoxicity in nematode and mammalian neurons. Nat Genet 2005;37(4):349–50.

88. Outeiro TF, Kontopoulos E, Altmann SM, Kufareva I, Strathearn KE, Amore AM, Volk CB, Maxwell MM, Rochet JC, McLean PJ, et al. Sirtuin 2 inhibitors rescue alpha-synuclein-mediated toxicity in models of Parkinson's disease. Science 2007;317(5837):516–9.

89. Lee MK, Kang SJ, Poncz M, Song KJ, Park KS. Resveratrol protects SH-SY5Y neuroblastoma cells from apoptosis induced by dopamine. Exp Mol Med 2007;39(3):376–84.

90. Chao J, Yu MS, Ho YS, Wang M, Chang RC. Dietary oxyresveratrol prevents parkinsonian mimetic 6-hydroxydopamine neurotoxicity. Free Radic Biol Med 2008;45(7):1019–26.

91. Gursoy M, Buyukuysal RL. Resveratrol protects rat striatal slices against anoxia-induced dopamine release. Neurochem Res 2008;33(9):1838–44.

92. Wang Q, Yu S, Simonyi A, Rottinghaus G, Sun GY, Sun AY. Resveratrol protects against neurotoxicity induced by kainic acid. Neurochem Res 2004;29(11):2105–12.

93. Gupta YK, Briyal S, Chaudhary G. Protective effect of *trans*-resveratrol against kainic acid-induced seizures and oxidative stress in rats. Pharmacol Biochem Behav 2002;71(1–2):245–9.

94. Virgili M, Contestabile A. Partial neuroprotection of in vivo excitotoxic brain damage by chronic administration of the red wine antioxidant agent, *trans*-resveratrol in rats. Neurosci Lett 2000;281(2–3):123–6.

95. Kode A, Rajendrasozhan S, Caito S, Yang SR, Megson IL, Rahman I. Resveratrol induces glutathione synthesis by activation of Nrf2 and protects against cigarette smoke-mediated oxidative stress in human lung epithelial cells. Am J Physiol Lung Cell Mol Physiol 2008;294(3):L478–88.

96. Vieira de Almeida LM, Pineiro CC, Leite MC, Brolese G, Leal RB, Gottfried C, Goncalves CA. Protective effects of resveratrol on hydrogen peroxide induced toxicity in primary cortical astrocyte cultures. Neurochem Res 2008;33(1):8–15.

97. de Almeida LM, Pineiro CC, Leite MC, Brolese G, Tramontina F, Feoli AM, Gottfried C, Goncalves CA. Resveratrol increases glutamate uptake, glutathione content, and S100B secretion in cortical astrocyte cultures. Cell Mol Neurobiol 2007;27(5):661–8.

98. dos Santos AQ, Nardin P, Funchal C, de Almeida LM, Jacques-Silva MC, Wofchuk ST, Goncalves CA, Gottfried C. Resveratrol increases glutamate uptake and glutamine synthetase activity in C6 glioma cells. Arch Biochem Biophys 2006;453(2):161–7.

99. Zhang H, Schools GP, Lei T, Wang W, Kimelberg HK, Zhou M. Resveratrol attenuates early pyramidal neuron excitability impairment and death in acute rat hippocampal slices caused by oxygen-glucose deprivation. Exp Neurol 2008;212(1):44–52.

100. Reich A, Spering C, Schulz JB. Death receptor Fas (CD95) signaling in the central nervous system: Tuning neuroplasticity? Trends Neurosci 2008;31(9):478–86.

101. Mostafapour SP, Del Puerto NM, Rubel EW. bcl-2 Overexpression eliminates deprivation-induced cell death of brainstem auditory neurons. J Neurosci 2002;22(11):4670–4.

102. Hansson O, Petersen A, Leist M, Nicotera P, Castilho RF, Brundin P. Transgenic mice expressing a Huntington's disease mutation are resistant to quinolinic acid–induced striatal excitotoxicity. Proc Natl Acad Sci U S A 1999;96(15):8727–32.

103. Offen D, Beart PM, Cheung NS, Pascoe CJ, Hochman A, Gorodin S, Melamed E, Bernard R, Bernard O. Transgenic mice expressing human Bcl-2 in their neurons are resistant to 6-hydroxydopamine and 1-methyl-4-phenyl-1,2,3,6-tetrahydropyridine neurotoxicity. Proc Natl Acad Sci U S A 1998;95(10):5789–94.

104. Farlie PG, Dringen R, Rees SM, Kannourakis G, Bernard O. bcl-2 transgene expression can protect neurons against developmental and induced cell death. Proc Natl Acad Sci U S A 1995;92(10):4397–401.

105. Jha RK, Yong MQ, Chen SH. The protective effect of resveratrol on the intestinal mucosal barrier in rats with severe acute pancreatitis. Med Sci Monit 2008;14(1):BR14–19.

106. Hu Y, Rahlfs S, Mersch-Sundermann V, Becker K. Resveratrol modulates mRNA transcripts of genes related to redox metabolism and cell proliferation in non-small-cell lung carcinoma cells. Biol Chem 2007;388(2):207–19.

107. Jang JH, Surh YJ. Protective effect of resveratrol on beta-amyloid-induced oxidative PC12 cell death. Free Radic Biol Med 2003;34(8):1100–10.

108. Yu LJ, Wu ML, Li H, Chen XY, Wang Q, Sun Y, Kong QY, Liu J. Inhibition of STAT3 expression and signaling in resveratrol-differentiated medulloblastoma cells. Neoplasia 2008;10(7):736–44.

109. Tseng SH, Lin SM, Chen JC, Su YH, Huang HY, Chen CK, Lin PY, Chen Y. Resveratrol suppresses the angiogenesis and tumor growth of gliomas in rats. Clin Cancer Res 2004;10(6):2190–202.

110. Jiang H, Zhang L, Kuo J, Kuo K, Gautam SC, Groc L, Rodriguez AI, Koubi D, Hunter TJ, Corcoran GB, et al. Resveratrol-induced apoptotic death in human U251 glioma cells. Mol Cancer Ther 2005;4(4).554–61.

111. Gosslau A, Chen M, Ho CT, Chen KY. A methoxy derivative of resveratrol analogue selectively induced activation of the mitochondrial apoptotic pathway in transformed fibroblasts. Br J Cancer 2005;92(3):513–21.

112. De Vos K, Goossens V, Boone E, Vercammen D, Vancompernolle K, Vandenabeele P, Haegeman G, Fiers W, Grooten J. The 55-kDa tumor necrosis factor receptor induces clustering of mitochondria through its membrane-proximal region. J Biol Chem 1998;273(16):9673–80.

113. Dewitt DA, Hurd JA, Fox N, Townsend BE, Griffioen KJ, Ghribi O, Savory J. Peri-nuclear clustering of mitochondria is triggered during aluminum maltolate induced apoptosis. J Alzheimers Dis 2006;9(2):195–205.

114. Pervaiz S, Clement MV. Superoxide anion: Oncogenic reactive oxygen species? Int J Biochem Cell Biol 2007;39(7–8):1297–304.

115. Pervaiz S, Clement MV. A permissive apoptotic environment: Function of a decrease in intracellular superoxide anion and cytosolic acidification. Biochem Biophys Res Commun 2002;290(4):1145–50.

116. Jacob C, Giles GI, Giles NM, Sies H. Sulfur and selenium: The role of oxidation state in protein structure and function. Angew Chem Int Ed Engl 2003;42(39):4742–58.

117. Ying W. NAD+/NADH and NADP+/NADPH in cellular functions and cell death: Regulation and biological consequences. Antioxid Redox Signal 2008;10(2):179–206.

118. Veal EA, Day AM, Morgan BA. Hydrogen peroxide sensing and signaling. Mol Cell 2007;26(1):1–14.

119. Kakkar P, Singh BK. Mitochondria: A hub of redox activities and cellular distress control. Mol Cell Biochem 2007;305(1–2):235–53.

120. Townsend DM. *S*-Glutathionylation: Indicator of cell stress and regulator of the unfolded protein response. Mol Interv 2007;7(6):313–24.

121. Mates JM, Segura JA, Alonso FJ, Marquez J. Pathways from glutamine to apoptosis. Front Biosci 2006;11:3164–80.

122. Voehringer DW, Meyn RE. Redox aspects of Bcl-2 function. Antioxid Redox Signal 2000;2(3):537–50.

123. Robb EL, Winkelmolen L, Visanji N, Brotchie J, Stuart JA. Dietary resveratrol administration increases MnSOD expression and activity in mouse brain. Biochem Biophys Res Commun 2008;372(1):254–9.

124. Chen CY, Jang JH, Li MH, Surh YJ. Resveratrol upregulates heme oxygenase–1 expression via activation of NF-E2–related factor 2 in PC12 cells. Biochem Biophys Res Commun 2005;331(4):993–1000.

125. Gericke GS. Reactive oxygen species and related haem pathway components as possible epigenetic modifiers in neurobehavioural pathology. Med Hypotheses 2006;66(1):92–9.

126. Gonzalez A, Pariente JA, Salido GM. Ethanol stimulates ROS generation by mitochondria through Ca2+ mobilization and increases GFAP content in rat hippocampal astrocytes. Brain Res 2007;1178:28–37.

127. Martinez R, Quintana K, Navarro R, Martin C, Hernandez ML, Aurrekoetxea I, Ruiz-Sanz JI, Lacort M, Ruiz-Larrea MB. Pro-oxidant and antioxidant potential of catecholestrogens against ferrylmyoglobin-induced oxidative stress. Biochim Biophys Acta 2002;1583(2):167–75.

128. Frankel D, Schipper HM. Cysteamine pretreatment of the astroglial substratum (mitochondrial iron sequestration) enhances PC12 cell vulnerability to oxidative injury. Exp Neurol 1999;160(2):376–85.

129. Fremont L, Belguendouz L, Delpal S. Antioxidant activity of resveratrol and alcohol-free wine polyphenols related to LDL oxidation and polyunsaturated fatty acids. Life Sci 1999;64(26):2511–21.

130. Belguendouz L, Fremont L, Linard A. Resveratrol inhibits metal ion-dependent and independent peroxidation of porcine low-density lipoproteins. Biochem Pharmacol 1997;53(9):1347–55.

131. Das S, Khan N, Mukherjee S, Bagchi D, Gurusamy N, Swartz H, Das DK. Redox regulation of resveratrol-mediated switching of death signal into survival signal. Free Radic Biol Med 2008;44(1):82–90.

132. Das S, Fraga CG, Das DK. Cardioprotective effect of resveratrol via HO-1 expression involves p38 map kinase and PI-3-kinase signaling, but does not involve NFkappaB. Free Radic Res 2006;40(10):1066–75.

133. Das S, Tosaki A, Bagchi D, Maulik N, Das DK. Resveratrol-mediated activation of cAMP response element-binding protein through adenosine A3 receptor by Akt-dependent and -independent pathways. J Pharmacol Exp Ther 2005;314(2):762–9.

134. Gao D, Zhang X, Jiang X, Peng Y, Huang W, Cheng G, Song L. Resveratrol reduces the elevated level of MMP-9 induced by cerebral ischemia-reperfusion in mice. Life Sci 2006;78(22):2564–70.

135. Kawasaki Y, Xu ZZ, Wang X, Park JY, Zhuang ZY, Tan PH, Gao YJ, Roy K, Corfas G, Lo EH, et al. Distinct roles of matrix metalloproteases in the early- and late-phase development of neuropathic pain. Nat Med 2008;14(3):331–6.

136. Shin CY, Lee WJ, Choi JW, Choi MS, Park GH, Yoo BK, Han SY, Ryu JR, Choi EY, Ko KH. Role of p38 MAPK on the down-regulation of matrix metalloproteinase-9 expression in rat astrocytes. Arch Pharm Res 2007;30(5):624–33.

137. Wu CY, Hsieh HL, Jou MJ, Yang CM. Involvement of p42/p44 MAPK, p38 MAPK, JNK and nuclear factor–kappa B in interleukin-1beta-induced matrix metalloproteinase-9 expression in rat brain astrocytes. J Neurochem 2004;90(6):1477–88.

138. Hsieh HL, Yen MH, Jou MJ, Yang CM. Intracellular signalings underlying bradykinin-induced matrix metalloproteinase–9 expression in rat brain astrocyte-1. Cell Signal 2004;16(10):1163–76.

139. Woo MS, Park JS, Choi IY, Kim WK, Kim HS. Inhibition of MMP-3 or -9 suppresses lipopolysaccharide-induced expression of proinflammatory cytokines and iNOS in microglia. J Neurochem 2008;106(2):770–80.

140. Aggarwal BB, Bhardwaj A, Aggarwal RS, Seeram NP, Shishodia S, Takada Y. Role of resveratrol in prevention and therapy of cancer: Preclinical and clinical studies. Anticancer Res 2004;24(5A):2783–840.

141. Zhang Y, Zolov SN, Chow CY, Slutsky SG, Richardson SC, Piper RC, Yang B, Nau JJ, Westrick RJ, Morrison SJ, et al. Loss of Vac14, a regulator of the signaling lipid phosphatidylinositol 3,5-bisphosphate, results in neurodegeneration in mice. Proc Natl Acad Sci U S A 2007;104(44):17518–23.
142. Montine TJ, Neely MD, Quinn JF, Beal MF, Markesbery WR, Roberts LJ, Morrow JD. Lipid peroxidation in aging brain and Alzheimer's disease. Free Radic Biol Med 2002;33(5):620–6.
143. Sun AY, Simonyi A, Sun GY. The "French paradox" and beyond: Neuroprotective effects of polyphenols. Free Radic Biol Med 2002;32(4):314–8.
144. Lin HY, Tang HY, Keating T, Wu YH, Shih A, Hammond D, Sun M, Hercbergs A, Davis FB, Davis PJ. Resveratrol is pro-apoptotic and thyroid hormone is anti-apoptotic in glioma cells: Both actions are integrin and ERK mediated. Carcinogenesis 2008;29(1):62–9.
145. Devchand PR, Keller H, Peters JM, Vazquez M, Gonzalez FJ, Wahli W. The PPARalpha-leukotriene B4 pathway to inflammation control. Nature 1996;384(6604):39–43.
146. Tzeng SF, Hsiao HY, Mak OT. Prostaglandins and cyclooxygenases in glial cells during brain inflammation. Curr Drug Targets Inflamm Allergy 2005;4(3):335–40.
147. Zykova TA, Zhu F, Zhai X, Ma WY, Ermakova SP, Lee KW, Bode AM, Dong Z. Resveratrol directly targets COX-2 to inhibit carcinogenesis. Mol Carcinog 2008.
148. Handler N, Brunhofer G, Studenik C, Leisser K, Jaeger W, Parth S, Erker T. "Bridged" stilbene derivatives as selective cyclooxygenase-1 inhibitors. Bioorg Med Chem 2007;15(18):6109–18.
149. Jang M, Pezzuto JM. Effects of resveratrol on 12-O-tetradecanoylphorbol-13-acetate-induced oxidative events and gene expression in mouse skin. Cancer Lett 1998;134(1):81–9.
150. Candelario-Jalil E, de Oliveira AC, Graf S, Bhatia HS, Hull M, Munoz E, Fiebich BL. Resveratrol potently reduces prostaglandin E2 production and free radical formation in lipopolysaccharide-activated primary rat microglia. J Neuroinflammation 2007;4:25.
151. Heynekamp JJ, Weber WM, Hunsaker LA, Gonzales AM, Orlando RA, Deck LM, Jagt DL. Substituted $trans$-stilbenes, including analogues of the natural product resveratrol, inhibit the human tumor necrosis factor alpha–induced activation of transcription factor nuclear factor kappaB. J Med Chem 2006;49(24):7182–9.
152. Lebeau A, Terro F, Rostene W, Pelaprat D. Blockade of 12-lipoxygenase expression protects cortical neurons from apoptosis induced by beta-amyloid peptide. Cell Death Differ 2004;11(8):875–84.
153. Morishima Y, Gotoh Y, Zieg J, Barrett T, Takano H, Flavell R, Davis RJ, Shirasaki Y, Greenberg ME. Beta-amyloid induces neuronal apoptosis via a mechanism that involves the c-Jun N-terminal kinase pathway and the induction of Fas ligand. J Neurosci 2001;21(19):7551–60.
154. Bastianetto S, Zheng WH, Quirion R. Neuroprotective abilities of resveratrol and other red wine constituents against nitric oxide-related toxicity in cultured hippocampal neurons. Br J Pharmacol 2000;131(4):711–20.
155. Won SJ, Kim DY, Gwag BJ. Cellular and molecular pathways of ischemic neuronal death. J Biochem Mol Biol 2002;35(1):67–86.
156. Lipton P. Ischemic cell death in brain neurons. Physiol Rev 1999;79(4):1431–568.
157. Tanaka E, Niiyama S, Sato S, Yamada A, Higashi H. Arachidonic acid metabolites contribute to the irreversible depolarization induced by $in vitro$ ischemia. J Neurophysiol 2003;90(5):3213–23.
158. Schneider C, Porter NA, Brash AR. Routes to 4-hydroxynonenal: Fundamental issues in the mechanisms of lipid peroxidation. J Biol Chem 2008;283(23):15539–43.
159. Muralikrishna Adibhatla R, Hatcher JF. Phospholipase A2, reactive oxygen species, and lipid peroxidation in cerebral ischemia. Free Radic Biol Med 2006;40(3):376–87.
160. Kutuk O, Basaga H. Apoptosis signalling by 4-hydroxynonenal: A role for JNK-c-Jun/AP-1 pathway. Redox Rep 2007;12(1):30–4.
161. Kutuk O, Poli G, Basaga H. Resveratrol protects against 4-hydroxynonenal-induced apoptosis by blocking JNK and c-JUN/AP-1 signaling. Toxicol Sci 2006;90(1):120–32.
162. Gomez-Pinilla F. Brain foods: The effects of nutrients on brain function. Nat Rev Neurosci 2008;9(7):568–78.
163. McCarthy MM. Estradiol and the developing brain. Physiol Rev 2008;88(1):91–124.
164. Wooten GF, Currie LJ, Bovbjerg VE, Lee JK, Patrie J. Are men at greater risk for Parkinson's disease than women? J Neurol Neurosurg Psychiatry 2004;75(4):637–9.
165. Macrae IM, Carswell HV. Oestrogen and stroke: The potential for harm as well as benefit. Biochem Soc Trans 2006;34(Pt 6):1362–5.
166. Vasudevan N, Pfaff DW. Non-genomic actions of estrogens and their interaction with genomic actions in the brain. Front Neuroendocrinol 2008;29(2):238–57.

167. Pozzi S, Benedusi V, Maggi A, Vegeto E. Estrogen action in neuroprotection and brain inflammation. Ann NY Acad Sci 2006;1089:302–23.

168. Mendez P, Wandosell F, Garcia-Segura LM. Cross-talk between estrogen receptors and insulin-like growth factor-I receptor in the brain: Cellular and molecular mechanisms. Front Neuroendocrinol 2006;27(4):391–403.

169. Beyer C, Ivanova T, Karolczak M, Kuppers E. Cell type-specificity of nonclassical estrogen signaling in the developing midbrain. J Steroid Biochem Mol Biol 2002;81(4–5):319–25.

170. Nadal A, Diaz M, Valverde MA. The estrogen trinity: Membrane, cytosolic, and nuclear effects. News Physiol Sci 2001;16:251–5.

171. Luoma JI, Boulware MI, Mermelstein PG. Caveolin proteins and estrogen signaling in the brain. Mol Cell Endocrinol 2008;290(1–2):8–13.

172. Chan CR, Hsu JT, Chang IT, Young YC, Lin CM, Ying C. The effects of glutamate can be attenuated by estradiol via estrogen receptor dependent pathway in rat adrenal pheochromocytoma cells. Endocrine 2007;31(1):44–51.

173. Sato K, Matsuki N, Ohno Y, Nakazawa K. Estrogens inhibit l-glutamate uptake activity of astrocytes via membrane estrogen receptor alpha. J Neurochem 2003;86(6):1498–505.

174. Shen ZX. Brain cholinesterases: III. Future perspectives of AD research and clinical practice. Med Hypotheses 2004;63(2):298–307.

175. Shen ZX. Acetylcholinesterase provides deeper insights into Alzheimer's disease. Med Hypotheses 1994;43(1):21–30.

176. Amantea D, Russo R, Bagetta G, Corasaniti MT. From clinical evidence to molecular mechanisms underlying neuroprotection afforded by estrogens. Pharmacol Res 2005;52(2):119–32.

177. Ferreiro E, Eufrasio A, Pereira C, Oliveira CR, Rego AC. Bcl-2 overexpression protects against amyloid-beta and prion toxicity in GT1–7 neural cells. J Alzheimers Dis 2007;12(3):223–8.

178. Lynch CD, Lyons D, Khan A, Bennett SA, Sonntag WE. Insulin-like growth factor-1 selectively increases glucose utilization in brains of aged animals. Endocrinology 2001;142(1):506–9.

179. Jacobsen JS, Comery TA, Martone RL, Elokdah H, Crandall DL, Oganesian A, Aschmies S, Kirksey Y, Gonzales C, Xu J, et al. Enhanced clearance of Abeta in brain by sustaining the plasmin proteolysis cascade. Proc Natl Acad Sci U S A 2008;105(25):8754–9.

180. Jiang Q, Lee CY, Mandrekar S, Wilkinson B, Cramer P, Zelcer N, Mann K, Lamb B, Willson TM, Collins JL, et al. ApoE promotes the proteolytic degradation of Abeta. Neuron 2008;58(5):681–93.

181. Marambaud P, Zhao H, Davies P. Resveratrol promotes clearance of Alzheimer's disease amyloid-beta peptides. J Biol Chem 2005;280(45):37377–82.

182. Fernandes G. Progress in nutritional immunology. Immunol Res 2008;40(3):244–61.

183. Malik VS, Hu FB. Popular weight-loss diets: From evidence to practice. Nat Clin Pract Cardiovasc Med 2007;4(1):34–41.

184. Schulz TJ, Zarse K, Voigt A, Urban N, Birringer M, Ristow M. Glucose restriction extends *Caenorhabditis elegans* life span by inducing mitochondrial respiration and increasing oxidative stress. Cell Metab 2007;6(4):280–93.

185. Ralser M, Benjamin IJ. Reductive stress on life span extension in *C. elegans*. BMC Res Notes 2008;1:19.

186. Hunt-Newbury R, Viveiros R, Johnsen R, Mah A, Anastas D, Fang L, Halfnight E, Lee D, Lin J, Lorch A, et al. High-throughput in vivo analysis of gene expression in *Caenorhabditis elegans*. PLoS Biol 2007;5(9):e237.

187. Zarse K, Schulz TJ, Birringer M, Ristow M. Impaired respiration is positively correlated with decreased life span in *Caenorhabditis elegans* models of Friedreich Ataxia. FASEB J 2007;21(4):1271–5.

188. Picard F, Kurtev M, Chung N, Topark-Ngarm A, Senawong T, Machado De Oliveira R, Leid M, McBurney MW, Guarente L. Sirt1 promotes fat mobilization in white adipocytes by repressing PPAR-gamma. Nature 2004;429(6993):771–6.

189. Roth GS, Lane MA, Ingram DK. Caloric restriction mimetics: The next phase. Ann NY Acad Sci 2005;1057:365–71.

190. Kassi E, Papavassiliou AG. Could glucose be a proaging factor? J Cell Mol Med 2008.

191. Arany Z, Foo SY, Ma Y, Ruas JL, Bommi-Reddy A, Girnun G, Cooper M, Laznik D, Chinsomboon J, Rangwala SM, et al. HIF-independent regulation of VEGF and angiogenesis by the transcriptional coactivator PGC-1alpha. Nature 2008;451(7181):1008–12.

192. Lee IH, Cao L, Mostoslavsky R, Lombard DB, Liu J, Bruns NE, Tsokos M, Alt FW, Finkel T. A role for the NAD-dependent deacetylase Sirt1 in the regulation of autophagy. Proc Natl Acad Sci U S A 2008;105(9):3374–9.

193. Sun C, Zhang F, Ge X, Yan T, Chen X, Shi X, Zhai Q. SIRT1 improves insulin sensitivity under insulin-resistant conditions by repressing PTP1B. Cell Metab 2007;6(4):307–19.
194. Jiang WJ. Sirtuins: Novel targets for metabolic disease in drug development. Biochem Biophys Res Commun 2008;373(3):341–4.
195. Breen DM, Sanli T, Giacca A, Tsiani E. Stimulation of muscle cell glucose uptake by resveratrol through sirtuins and AMPK. Biochem Biophys Res Commun 2008;374(1):117–22.
196. Dasgupta B, Milbrandt J. Resveratrol stimulates AMP kinase activity in neurons. Proc Natl Acad Sci U S A 2007;104(17):7217–22.
197. Chen J, Zhou Y, Mueller-Steiner S, Chen LF, Kwon H, Yi S, Mucke L, Gan L. SIRT1 protects against microglia-dependent amyloid-beta toxicity through inhibiting NF-kappaB signaling. J Biol Chem 2005;280(48):40364–74.

22   Sirtuin[1] and Resveratrol

Antoni Camins[1,2], Carme Pelegrí[2,3], Jordi Vilaplana[2,3],
Rosa Cristòfol[4], Coral Sanfeliu[4], and Mercè Pallàs[1,2]

[1]Unitat de Farmacologia i Farmacognòsia i Institut de
Biomedicina (IBUB), Facultat de Farmàcia, Universitat de
Barcelona, Nucli Universitari de Pedralbes, Barcelona, Spain

[2]Centro de Investigación de Biomedicina en Red de
Enfermedades Neurodegenerativas (CIBERNED),
Instituto de Salud Carlos III, Madrid, Spain

[3]Departament de Fisiologia, Facultat de Farmàcia, Universitat de
Barcelona, Nucli Universitari de Pedralbes, Barcelona, Spain

[4]Department d'Isquèmia Cerebral i Neurodegeneració,
Institut d'Investigacions Biomèdiques de Barcelona,
CSIC-IDIBAPS, Barcelona, Spain

CONTENTS

22.1   INTRODUCTION

Protein acetylation is involved in the regulation of several cellular functions such as protein-protein interactions, protein stability, and DNA recognition by proteins. For instance, the acetylation of histone proteins alters gene transcription [1]. Thus, removal of acetyl groups from lysine residues results in compaction of chromatin and, hence, repression of gene transcription. The process of acetylation and deacetylation regulated by proteins with acetyltransferase activity is important for cellular processes. These proteins are usually known as histone acetyltransferases [2]. Nowadays, there is a growing interest for histone deacetylases (HDACs) because of their potential clinical applications [3]. HDACs have been divided into four groups. Class I and class II HDACs are similar to the yeast Rpd3p and Hda1p proteins. Class III HDACs are similar to the yeast transcriptional repressor Sir2p and are referred to as sirtuins. Class I and class II HDACs are characterized by

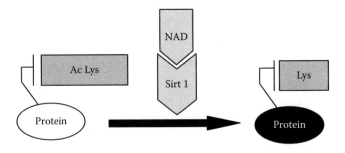

FIGURE 22.1 Enzymatic activity of SIRT1, using NAD$^+$ as a cofactor, producing a deacetylated substrate and nicotinamide.

their sensitivity to inhibition by trichostatin A, while class III HDACs are dependent on nicotin-amide adenine dinucleotide (NAD). Class IV HDACs include the deacetylase HDACII. The focus of this chapter is on sirtuins (class III HDAC), which are a conserved family of NAD$^+$-dependent deacetylases and named after the founding member, *Saccharomyces cerevisiae* silent information regulator 2 (Sir2) protein [4]. Its products function in a complex as transcriptional repressors or silencers, acting largely through histone deacetylation, at the telomeres, mating-type loci, and the rDNA gene loci [5] . The *SIR2* gene, required for silencing of rDNA loci, is evolutionarily conserved from prokaryotes to humans. Sir2 is a NAD-dependent class III protein deacetylase [6,7], with ADP-ribosyltransferase activity *in vitro*. Analysis of SIRT1 enzymatic activity has revealed that it functions differently from previously described HDACs. Studies using purified SIRT1 revealed that for every acetyl lysine group removed, one molecule of NAD is cleaved, and nicotinamide and *O*-acetyl-ADP-ribose are produced (Figure 22.1). Therefore, SIRT1 appears to possess two enzymatic activities: the deacetylation of a target protein and the metabolism of NAD [8,9]. These two activities suggest that SIRT1 could act as a metabolic or oxidative sensor, regulating cellular machinery based on such information.

The need for NAD$^+$ in the deacetylase activity of SIRT1 has led to the suggestion that enzymatic activity could be regulated by the concentration of NAD$^+$, the ratio NAD$^+$/NADH, or by the intra-cellular concentration of nicotinamide [10].

Studies performed in yeast showed that Sir2 deletion leads to histone hyperacetylation. Subsequent studies were focused on this enzyme as a key mediator of *S. cerevisiae*'s replicative life span [10–12].

With reference to the mammalian Sir2 gene family (or sirtuins), seven homologues (SIRT1-7) have been characterized until now, among them the nuclear SIRT1, which is the closest homologue to Sir2, based on amino acid identity, and the best understood in terms of cellular activity and func-tions. Among the nonhistone cellular substrates of SIRT1, tumor suppressor p53, the transcription factor nuclear factor–kappa B (NF-κB), and the FOXO family of transcription factors have been identified. All of them are involved in the transcriptional control of key genes in cell proliferation and cell survival. Moreover, SIRT1 also deacetylates the nuclear receptor peroxisome-proliferator activated receptor-γ (PPARγ) and its transcriptional coactivator, PPARγ coactivator-α (PGC-α), which regulates a wide range of metabolic activities in muscle, adipose tissues, and liver, linked to hepatic nuclear factor1 (HNF-4α). SIRT1 substrates, therefore, have apparent functions that can link nutrient availability and energy metabolism to adaptive changes in transcriptional profiles that affect cell survival in multiple systems.

Sirtuins are currently an object of interest in various fields of aging medicine, ranging from oncology to gerontology, because of their role as a longevity factor in multiple model organisms. The interest in SIRT1 has also intensified over the past 2 years with further discoveries of its role in cancer, metabolic diseases, and neurodegenerative disorders. The main focus of this review shall be SIRT1's role in neuronal aging and neurodegenerative diseases, and its possible modulation by a natural activator, resveratrol.

22.2 SIRT1 AND NEUROPROTECTION

Several important roles of SIRT1 have been described in the central nervous system (CNS), mainly in neuronal development and neuroprotection. It is known that there are high levels of SIRT1 expression in the heart, brain, spinal cord, and dorsal root ganglia [13]. Previous studies demonstrated that high SIRT1 levels in the embryonic brain suggest that it might have a role in neuronal and/or brain development. This notion is in agreement with some of the phenotypes associated with SIRT1 knockout mice, in which postnatal survival is infrequent and which have developmental defects [14,15].

As in other mammalian cells, SIRT1 promotes survival and stress tolerance in CNS neurons. However, data in this regard are scarce, because earlier studies have not used neuronal cells. In the adult rat brain, SIRT1 can be found in the hippocampus, cerebellum, and cerebral cortex.

Interestingly, SIRT1 expression is regulated by oxidative stress because the antioxidant vitamin E has been shown to reduce both the oxidative damage and the reduction of SIRT1 caused by a high fat and sugar diet, with the restoration of SIRT1 levels [16]. This study suggests that SIRT1 levels in the brain are affected by oxidative stress and energy homeostasis. A role for SIRT1 in the protection of cardiac myocytes against ischemia-induced apoptosis has been well documented [17,18]. A recent interesting study using organotypic hippocampal slice culture as an *in vitro* model of cerebral ischemia showed that pretreatment using resveratrol, an activator of SIRT1, mimics ischemic preconditioning via SIRT1 [19]. When SIRT1 is inactivated by sirtinol after ischemic preconditioning or resveratrol pretreatment, neuroprotection is abolished. This study demonstrated a neuroprotective role of SIRT1 in ischemic injury, which could be elicited by a small molecule such as resveratrol, and it is therefore of substantial clinical interest. In contrast, another earlier report had shown that the sirtuin inhibitor nicotinamide enhances neuronal cell survival in acute anoxic injury [20], although an involvement of SIRT1 in this case was not clearly demonstrated.

22.3 RESVERATROL

Resveratrol (3,5,4'-trihydroxystilbene) was first isolated from the roots of white hellebore (*Veratrum grandiflorum* O. Loes) in 1940 and later, in 1963, from the roots of *Polygonum cuspidatum*, a plant used in traditional Chinese and Japanese medicine. Initially characterized as a phytoalexin, resveratrol, a polyphenol present in black grapes and its derivatives, attracted little interest until 1992, when it was postulated to account for some of the cardioprotective effects of red wine. Since then, dozens of reports have shown that resveratrol can prevent or slow down the progression of a wide variety of illnesses, including cancer, cardiovascular disease, and ischemic injuries, as well as enhance stress resistance and extend the life span of various organisms from yeast to vertebrates [21,22]. Recent reports indicate that resveratrol treatment alone has a range of beneficial effects in mice, but does not increase the longevity of *ad libitum*-fed animals when started midlife in contrast to high-fat diet-fed-mice [22–24]. Chapter 24 discusses resveratrol properties in greater detail.

The mechanism by which resveratrol exerts such a range of beneficial effects across species and disease models is not yet clear [25], although at the beginning it was proposed that the antioxidant properties of this drug may explain the majority of its beneficial effects. Attempts to show its favorable effects *in vitro* have met with almost universal success, and have led to the identification of multiple direct targets for this compound. However, results from pharmacokinetic studies indicate that circulating resveratrol is rapidly metabolized, and cast doubt on the physiological relevance of the high concentrations typically used for *in vitro* experiments [26,27]. Further experiments are needed to show whether resveratrol or its metabolites accumulate sufficiently in tissues to recapitulate *in vitro* observations, or whether alternative higher-affinity targets, such as quinone reductase 2, have the key roles in its protective effects [28,29]. *In vivo* results have, therefore, become increasingly important in the attempts to understand how effective resveratrol is in the treatment of different diseases. It is also unclear what conclusion should be drawn from the studies described so far. The benefits of resveratrol, as we noted above, can be explained by its antioxidant properties or better if this substance acts through a specific genetic pathway that has evolved to increase disease and stress resistance. With regard to the latter proposal, there is already ample evidence for the existence of health-promoting pathways that are activated by caloric restriction. It has been known since the 1930s that a severe lowering of caloric intake dramatically slows down the rate of aging in mammals and delays the onset of numerous diseases of aging, including cancer, cardiovascular disease, diabetes, and neurodegeneration. The hypothesis that resveratrol might use the same pathways activated by caloric restriction in mammals is attractive because it appears to do so in lower organisms; however, proving this hypothesis will require a better understanding of these processes.

In reference to antioxidant action of resveratrol, it is widely accepted that resveratrol exerts antioxidant effects, but it is not yet clear if this is primarily a direct scavenging effect or the result of the activation of pathways that up-regulate cells' natural antioxidant defenses. Reactive oxygen species (ROS) have been shown to have a role in the initiation and progression of cancer by directly damaging DNA and other macromolecules. In addition to its possible modulation of antioxidant enzymes involved in the phase II response, resveratrol has an intrinsic antioxidant capacity that could be related to its chemopreventive effects. *In vivo*, resveratrol has been shown to increase plasma antioxidant capacity and decrease lipid peroxidation; however, it is difficult to assess whether these effects are direct or the result of up-regulation of endogenous antioxidant enzymes. In addition, clinical trials of antioxidant molecules have yielded disappointing results, suggesting that phytochemicals could possess other properties that are more relevant to cancer prevention. Oxidation of low-density lipoprotein (LDL) particles is strongly associated with the risk of coronary heart disease and myocardial infarction. Resveratrol prevents LDL oxidation *in vitro* by chelating copper, as well as by directly scavenging free radicals (although other components of red wine are superior free radical scavengers) [30]. Treatment of normal rats with resveratrol does not affect lipid peroxidation, as reflected by the presence of thiobarbituric acid-reactive substances [31]. However, resveratrol can be detected in LDL particles from humans after consumption of red wine, which is rich in this compound, and the pure compound prevents increases in lipid peroxidation induced by tumors or ultraviolet irradiation [32,33], in addition to blocking gentamicin-induced nephrotoxicity [34]. In stroke-prone, spontaneously hypertensive rats, resveratrol significantly reduces markers of oxidative stress such as glycated albumin in serum, and 8-hydroxyguanosine in urine [35]. Furthermore, in guinea pigs, resveratrol induces the activities of QR1 and catalase in cardiac tissue, and decreases the concentration of ROS generated by menadione [36]. These results indicate that resveratrol can suppress pathological increases in the peroxidation of lipids and other macromolecules *in vivo*, but whether the mechanism is direct, indirect, or both is not yet clear.

Another mechanism by which resveratrol could combat tumor formation is induction of cell cycle arrest and apoptosis. The antiproliferative and proapoptotic effects of resveratrol in tumor cell lines have been extensively documented *in vitro* and are supported by down-regulation of cell cycle proteins and increases in apoptosis in tumor models *in vivo*. Although resveratrol has been found to target leukemic cells preferentially *in vitro* in some studies, the specificity of these effects remains unclear because other researchers have found that resveratrol inhibits growth and induces apoptosis in normal hematopoietic cells at similar doses. Some level of specificity could arise from the apparent increased susceptibility of cycling cells to the effects of resveratrol [37]. A more precise mechanism by which resveratrol could act is sensitization of tumor cells to other inducers of apoptosis. Resveratrol has been shown to sensitize several tumor lines, but not normal human fibroblasts, to TRAIL (tumor necrosis factor-related apoptosis-inducing ligand)-induced apoptosis. It remains to be seen whether the proapoptotic effects of resveratrol *in vivo* are related to these *in vitro* observations, or secondary to other effects, such as inhibition of angiogenesis.

The last protective mechanism related with resveratrol is its role as activator of SIRT1. Resveratrol increases the affinity of SIRT1 for its acetylated substrates, possibly inducing a conformational

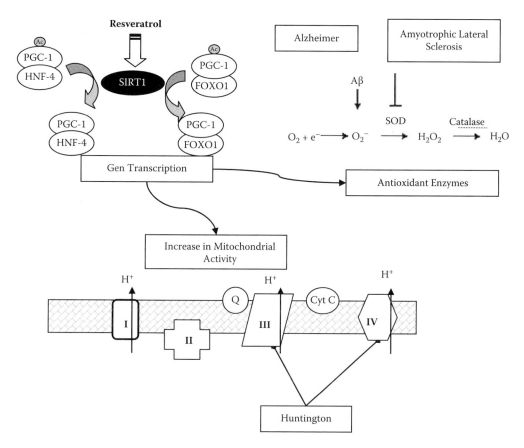

FIGURE 22.2   Scheme illustrating the effect of resveratrol on mitochondrial biogenesis and its influence in the antioxidant enzymes gene transcription through the SIRT1/PGC-1α/FOXO pathway. Main points of dysfunction for some neurodegenerative diseases are also indicated.

change of SIRT1 [38]. The rat brain has receptors for polyphenols such as resveratrol. This indicates that this substance and its derivatives can pass the blood-brain barrier, and several studies suggest that it may have a protective effect in some neurodegenerative processes, as we will describe below in greater extent. Hence, the axis SIRT1/PGC-1α activated by resveratrol (Figure 22.2) is a signaling pathway involved in several cellular contexts and each of the actors involved may promote a separate slowdown in the neurodegenerative process [39]. This neuroprotective action is very likely because the central factor of this signal, PGC-1α, promotes mitochondrial activity, while neurodegenerative diseases are linked to mitochondrial failures. It is strongly suggested that the activation of the axis SIRT1/PGC-1α by resveratrol could be a key feature of the mechanisms of neuroprotection by this polyphenol and could lead to new therapeutic prospects (Figure 22.2).

22.4   RESVERATROL AND HUNTINGTON'S DISEASE

Using the neurotoxin 3-nitropropionic acid, a mitochondrial complex II inhibitor, and a well-established experimental model of Huntington's disease, it has been reported that the beneficial effects of resveratrol against this neurotoxin might be attributed to its antioxidant activity [40]. However, several findings have converged on the notion that SIRT1's neuroprotective effect could be extended to degenerating neurons. Parker and coworkers [41] showed that resveratrol, acting through Sir2 and SIRT1 activation, respectively, protected *Caenorhabditis elegans* and mouse neurons against the cytotoxicity of the mutant polyglutamine protein huntingtin. Huntingtin is the product of the gene mutated in the hereditary neurodegenerative disorder Huntington's disease, whose expansion of a polyglutamine stretch resulted in a mutant polypeptide that could form cytotoxic aggregates in neurons [42]. Although *C. elegans* has no huntingtin orthologue, overexpression of a huntingtin fragment in touch receptor neurons resulted in a gain-of-function mechanosensory defect that could model the disease. Both resveratrol and an increased sir-2.1 gene dosage alleviated the worm neuronal dysfunction in a DAF16-dependent manner. Furthermore, resveratrol decreased cell death associated with neurons cultured from a mutant huntingtin (109Q) knocking mice, in a manner that is reversible by two SIRT1 inhibitors, sirtinol and nicotinamide [43].

22.5 RESVERATROL AND ALZHEIMER'S DISEASE

A link between SIRT1 and Alzheimer's disease (AD) is also becoming increasingly evident [44,45]. The amyloid hypothesis [46] depicts that extracellular plaques consist of aggregated beta-amyloid (Aβ) peptide generated from proteolytic cleavages of the amyloid precursor protein (APP) as the etiological agent of AD pathology [47]. Both intracellular and extracellular soluble oligomeric forms of Aβ could, in fact, initiate synaptic malfunctions and the onset of AD symptoms [48,49]. NF-κB signaling in microglia is known to be critically involved in neuronal death induced by Aβ peptides [50]. Chen and collaborators [51] showed that stimulation of microglia with Aβ increased acetylation of RelA/p65 subunit of NF-κB at lysine 310. Overexpression of SIRT1 and resveratrol treatment markedly reduced NF-κB signaling stimulated by Aβ and had strong neuroprotective effects. This result connects the known role of SIRT1 in modulating NF-κB activity [52] with AD. It should be kept in mind that, for AD, as with other neurodegenerative diseases, the beneficial effect of resveratrol is multifaceted. Its immediate effect is more likely associated with its activity as an antioxidant [53,54], but at a more extended time frame, its activation of SIRT1 and modulation of NF-κB signaling may result in other beneficial effects, such as anti-inflammation.

Another possible link between SIRT1 and AD came from the potential benefits of caloric restriction (CR) on AD symptoms and progression. It is well known in the epidemiology of neurodegenerative diseases that the incidence of sporadic Parkinson's disease (PD) and AD are both correlated with multiple genetic factors, diet, and social behavior [55]. High caloric diets are associated with the risk of AD, and CR has been proposed to protect against both PD and AD [56]. Firmer evidence for this idea was obtained when Patel [57] showed that short-term CR substantially decreased the accumulation of Aβ plaques in two AD-prone APP/presenilin transgenic mice lines, and also decreased gliosis marked by astrocytic activation. In another study, Wang and colleagues [58] also showed that a CR dietary regimen prevents Aβ peptide generation and neuritic plaque deposition in the brain of another mouse model of AD (Tg2576 mice). In this latter study, the authors suggested that CR resulted in the promotion of APP processing via the nonamyloidogenic α-secretase-mediated pathway. They observed a larger-than-twofold increase in the concentration of brain sAPPα (a product of α-secretase cleavage) and a statistically significant 30% increase in ADAM10 (a putative α-secretase) levels in CR animals compared to controls. There also appeared to be a moderate increase in the levels of the insulin degrading enzyme, which has been associated to brain amyloid clearance [59]. In another recent report, the same group showed that CR resulted in reduced contents of Aβ in the temporal cortex of squirrel monkeys, in a manner that was inversely correlated with SIRT1 protein concentrations in the same brain region [60]. It is not particularly clear in the above reports whether CR's effects in attenuating amyloid production were mediated through SIRT1 activation. Recent evidence suggests that this may indeed be the case, and may actually involve a novel signaling crosstalk [61].

22.6 RESVERATROL AND PARKINSON'S DISEASE

PD is a neurodegenerative disease that is also characterized at the clinical level by bradykinesia, tremor, and rigidity, and at the cellular level by a loss of dopamine neurons of the gray matter and the frequent presence of intraneuronal inclusions named Lewy bodies, mainly composed of fibrillar α-synuclein [62]. Like AD, the familial form of PD concerns only a small proportion of patients (10%). The majority of them are suffering from a sporadic form and, if the genetic causes are fairly well identified, the reasons for the emergence of the sporadic contrary are still unclear. The involvement of mitochondrial dysfunction in the PD has been established for more than two decades when it was discovered that the administration of 1-methyl-4-phenyl-1,2,3,4-tetrahydropyridine causes the emergence of parkinsonism in laboratory animals and also in humans, through its active metabolite ion MPP^+, which inhibits the complex I of the chain of mitochondrial electron transport. It is well known that complex I is the major source of production of free radicals; the assumption is that the alteration of its functions could, beyond the declining production of adenosine triphosphate, give rise to increased oxidative stress, explaining the emergence of the disease. However, different authors working on different PD models conclude that SIRT1 activation does not play a major role in the protective effect of resveratrol against MPP^+ cytotoxicity, because sirtuin inhibitors such as nicotinamide and sirtinol did not counteract neuroprotection by resveratrol [63]. Instead, all works point to propose that antioxidative actions are responsible for neuroprotection by resveratrol against MPP^+ [63,64]. However, it has been recently described that genetic inhibition of SIRT2 via small interfering RNA rescued α-synuclein toxicity [65]. Furthermore, the inhibitors of this enzyme protected against dopaminergic cell death both *in vitro* and in a *Drosophila* model of PD, and found that inhibition of SIRT2 rescued α-synuclein toxicity and modified inclusion morphology in a cellular model of PD [65]. However, increased SIRT1 expression or activity delays the toxic effects induced by α-synuclein, the protein that forms insoluble aggregates in several age-onset pathologies including PD. Resveratrol could be an interesting candidate for potential application in the treatment of PD only by its antioxidant properties [63].

22.7   RESVERATROL IN STROKE AND BRAIN DAMAGE

Numerous studies have raised the possibility that resveratrol might be useful in protecting against brain damage following cerebral ischemia. Laboratory animals given intraperitoneal injections of resveratrol showed less motor impairment and significantly smaller infarct volume jointly with decreased delayed neuronal cell death and glial cell activation after ischemia. Similar effects were observed in wild-type, but not peroxisome proliferators-activated receptor-$\alpha^{-/-}$, mice [66]. Resveratrol administered intraperitoneally also prevented seizures induced by $FeCl_3$, kainic acid [67], or pentylenetetrazole [68], and partially restored cognition in rats receiving streptozotocin intracerebroventricularly [69]. These results indicate that resveratrol is capable of penetrating the blood-brain barrier and exerts strong neuroprotective effects, even at low doses, after stroke or on neurotoxin-injured brain. Moreover, more studies are necessary to determine if these neuroprotective effects are mediated through the stimulation of SIRT1 or by its antioxidant properties [21].

22.8   RESVERATROL AND AGING

As mentioned, in yeast, worms, and flies, extra copies of genes encoding sirtuins are associated with extended life span [11,70,71]. Inbred knockout mice that lack SIRT1 show developmental defects, have a low survival rate, and have a significantly shorter life span compared to wild-type mice, although out-breeding seems to improve the phenotype significantly [15]. It has been postulated that the main function of sirtuin proteins might be to promote survival and stress resistance in times of adversity [72]. An evolutionary advantage arising from the ability to modify life span in response to environmental conditions could have allowed these enzymes to be conserved as species evolved, and to take on new functions in response to new stresses and demands on the organism. This could explain why the same family of enzymes has dramatic effects on life span in different organisms with seemingly dissimilar causes of aging [73]. Caloric restriction and intermittent fasting are implicated in most of the theories for successful brain aging [74]. The data from lower organisms have provoked intense research into the function of sirtuin proteins in mammalian systems. An *in vitro* screen for activators of SIRT1 identified resveratrol as the most potent of 18 inducers of deacetylase activity [12]. Subsequent work has shown that resveratrol extends the life span of *S. cerevisiae*, *C. elegans*, and *Drosophila melanogaster*, but only if the gene that encodes Sir2 is present in these organisms. More recently, resveratrol was shown to extend the maximum life span of a species of short-lived fish by up to 59%, concomitant with the maintenance of learning and motor function with age and a dramatic decrease in aggregated proteins in elderly fish brains [75]; however, the extent to which this effect is Sir2-dependent, if at all, is not known. Moreover, resveratrol consistently recapitulates the protective effects of SIRT1 overexpression in cell culture, and Sir2/SIRT1 have been shown as essential mediators of effects on adipogenesis, NF-κB acetylation, protection from mutant huntingtin protein, and life span extension in lower organisms [11,41,70,76]. The question of whether enhanced SIRT1 activity and/or resveratrol treatment will increase mammalian life span looms large in the aging-research community.

22.9 HORMESIS: AXIS BETWEEN SIRT1 AND RESVERATROL?

"Hormesis" describes the phenomenon in which a mild stress (e.g., irradiation, heat, or toxins) can induce a protective response against subsequent stresses [77]. This hormetic response is credited for the paradoxical result that mildly stressed animals outlive their unstressed counterparts, which also possibly applies to humans. It has been suggested that caloric restriction then activation of SIRT1 might act as a mild stress to induce a hormetic response [78], which could account for enhanced stress tolerance and longevity in caloric-restricted mice, as well as the otherwise counterintuitive finding that such animals are better able to resist starvation [79]. In yeast, at least, this contention is strongly supported by the observation that both caloric restriction and mild stresses induce expression of Pnc1, an upstream activator of sirtuin proteins that is necessary and sufficient for life span extension [80]. The "xenohormesis hypothesis" postulates that sensing stress responses, such as resveratrol accumulation, in a food source might be sufficient to induce a hormetic response in animals eating that food. It can be imagined that throughout evolution, such stress markers in the surrounding vegetation would have served as strong predictors of a coming famine or direct stress to the animal. Reacting to these molecules would allow the hormetic response to begin ahead of any direct damage or energy deficit, and, more importantly, would not stake the life of the animal on the hope that the initial stress would be mild and/or protective. If the xenohormesis hypothesis is correct, then stressed plants might form an abundant reservoir for medicinal compounds that trigger conserved protective responses in humans [81]. The relatively low amounts of resveratrol in foods belie the possibility that there are numerous potential xenohormetic compounds in a stressed plant that could act additively or even synergistically. Indeed, another potentially xenohormetic compound, quercetin, behaves similarly to resveratrol in many assays and also inhibits sulfation of resveratrol, which predicts a greater-than-additive effect.

22.10 SUMMARY

In the past decade, sirtuin biology has traveled a long way from their original description as yeast NAD+-dependent class III HDACs that control yeast life span. In mammals, seven orthologues of Sir2 have been identified, SIRT1 to SIRT7, and the exact biological function of most of these sirtuins still remains only partially characterized. Of particular interest is the fact that SIRT1 not only deacetylates histones to mediate gene silencing, but is also able to interact with and deacetylate some well-known transcriptional regulators, thereby modulating specifically various biological processes. Hence, modulating the expression of SIRT1 or its activity, by using sirtuin activating compounds such as resveratrol, will have pleiotropic effects. SIRT1 activation reduces fat accumulation and adipocyte differentiation through repression of the activity of the adipogenic nuclear receptor PPARγ. SIRT1 also promotes mitochondrial function and energy expenditure and consequently protects mice from diet-induced obesity, through deacetylation and subsequent activation of PGC-1α in the skeletal muscle and in the brown adipose tissue. The SIRT1/PGC-1α interaction is also important in the liver, where SIRT1 activation upon fasting induces gluconeogenesis and prevents against hepatosteatosis. In addition, SIRT1 significantly enhances insulin secretion in the pancreatic β cells. In combination, these studies illustrate that SIRT1 is a major modulator of metabolism. SIRT1 activation also seems to be endowed with neuroprotective activities, as suggested from the study of models of Huntington disease or AD (Figure 22.2). Furthermore, other sirtuins might play important roles in some diseases, as illustrated by SIRT2, which is down-regulated in human gliomas and could be involved in cancer treatment. Obviously, more studies, in animal models and humans, are still needed to define the exact role of sirtuins in the pathophysiology of human diseases. It can, however, be predicted that therapeutic interventions aiming at activating or blocking sirtuins, depending on the context, will one day become helpful in the treatment of human diseases.

This chapter discusses the effects of SIRT1 modulation by resveratrol that have been observed *in vivo* and possible evolutionary explanations, as they relate to the development of human therapeutics, based on either resveratrol itself or new, more potent compounds that mimic its effects.

ACKNOWLEDGMENTS

We thank the Language Assessment Service of the University of Barcelona for revising the manuscript. This study was supported by several grants: SAF-2006-13092, BFU-2006-13092 from Ministerio de Educación y Ciencia, PI 041300 from Instituto de Salud Carlos III, 2005/SGR00893 from Generalitat de Catalunya, and from Fundació La Marató de TV3 (0310).

REFERENCES

1. Lee, K.K., and Workman, J.L. Histone acetyltransferase complexes: One size doesn't fit all. *Nat. Rev. Mol. Cell Biol.* 8, 284, 2007.
2. An, W. Histone acetylation and methylation: Combinatorial players for transcriptional regulation. *Subcell. Biochem.* 41, 351, 2007.
3. Yang, X.J., and Seto, E. HATs and HDACs: From structure, function and regulation to novel strategies for therapy and prevention. *Oncogene* 26, 5310, 2007.
4. Brachmann, C.B., Sherman, J.M., Devine, S.E. et al. The *SIR2* gene family, conserved from bacteria to humans, functions in silencing, cell cycle progression, and chromosome stability. *Genes Dev.* 9, 2888, 1995.
5. Blander, G., and Guarente, L. The Sir2 family of protein deacetylases. *Annu. Rev. Biochem.* 73, 417, 2004.
6. Denu, J.M. Linking chromatin function with metabolic networks: Sir2 family of NAD(+)-dependent deacetylases. *Trends Biochem. Sci.* 28, 41, 2003.
7. Imai, S., Armstrong, C.M., Kaeberlein, M. et al. Transcriptional silencing and longevity protein Sir2 is an NAD-dependent histone deacetylase. *Nature* 403, 795, 2000.
8. Frye, R.A. Characterization of five human cDNAs with homology to the yeast *SIR2* gene: Sir2-like proteins (sirtuins) metabolize NAD and may have protein ADP-ribosyltransferase activity. *Biochem. Biophys. Res. Commun.* 260, 273, 1999.
9. Michan, S., and Sinclair, D. Sirtuins in mammals: Insights into their biological function. *Biochem. J.* 404, 1, 2007.
10. Lin, S.J., Defossez, P.A., and Guarente, L. Requirement of NAD and SIR2 for life-span extension by calorie restriction in *Saccharomyces cerevisiae*. *Science* 289, 2126, 2000.
11. Kaeberlein, M., McVey, M., and Guarente, L. The SIR2/3/4 complex and SIR2 alone promote longevity in *Saccharomyces cerevisiae* by two different mechanisms 1. *Genes Dev.* 13, 2570, 1999.
12. Howitz, K.T., Bitterman, K.J., Cohen, H.Y. et al. Small molecule activators of sirtuins extend *Saccharomyces cerevisiae* lifespan. *Nature* 425, 191, 2003.
13. Sakamoto, J., Miura, T., Shimamoto, K. et al. Predominant expression of Sir2alpha, an NAD-dependent histone deacetylase, in the embryonic mouse heart and brain. *FEBS Lett.* 556, 281, 2004.

14. Cheng, H.L., Mostoslavsky, R., Saito, S. et al. Developmental defects and p53 hyperacetylation in Sir2 homolog (SIRT1)-deficient mice. *Proc. Natl. Acad. Sci. U. S. A.* 100, 10794, 2003.

15. McBurney, M.W., Yang, X., Jardine, K. et al. The mammalian SIR2alpha protein has a role in embryogenesis and gametogenesis. *Mol. Cell Biol.* 23, 38, 2003.

16. Wu, A., Ying, Z., and Gomez-Pinilla, F. Oxidative stress modulates Sir2alpha in rat hippocampus and cerebral cortex. *Eur. J. Neurosci.* 23, 2573, 2006.

17. Alcendor, R.R., Kirshenbaum, L.A., Imai, S. et al. Silent information regulator 2alpha, a longevity factor and class III histone deacetylase, is an essential endogenous apoptosis inhibitor in cardiac myocytes. *Circ. Res.* 95, 971, 2004.

18. Pillai, J.B., Isbatan, A., Imai, S. et al. Poly(ADP-ribose) polymerase-1-dependent cardiac myocyte cell death during heart failure is mediated by NAD^+ depletion and reduced Sir2alpha deacetylase activity. *J. Biol. Chem.* 280, 43121, 2005.

19. Raval, A.P., Dave, K.R., and Perez-Pinzon, M.A. Resveratrol mimics ischemic preconditioning in the brain. *J. Cereb. Blood Flow Metab.* 26, 1141, 2006.

20. Chong, Z.Z., Lin, S.H., Li, F. et al. The sirtuin inhibitor nicotinamide enhances neuronal cell survival during acute anoxic injury through AKT, BAD, PARP, and mitochondrial associated "anti-apoptotic" pathways. *Curr. Neurovasc. Res.* 2, 271, 2005.

21. Baur, J.A. and Sinclair, D.A. Therapeutic potential of resveratrol: The *in vivo* evidence. *Nat. Rev. Drug Discov.* 5, 493, 2006.

22. Baur, J.A., Pearson, K.J., Price, N.L. et al. Resveratrol improves health and survival of mice on a high-calorie diet. *Nature* 444, 337, 2006.

23. Pearson, K.J., Baur, J.A., Lewis, K.N. et al. Resveratrol delays age-related deterioration and mimics transcriptional aspects of dietary restriction without extending life span. *Cell Metab.* 8, 157–168, 2008.

24. Barger, J.L., Kayo, T., Vann, J.M. et al. A low dose of dietary resveratrol partially mimics caloric restriction and retards aging parameters in mice. *PLoS. ONE.* 3, e2264, 2008.

25. Fremont, L. Biological effects of resveratrol. *Life Sci.* 66, 663, 2000.

26. Walle, T., Hsieh, F., DeLegge, M.H. et al. High absorption but very low bioavailability of oral resveratrol in humans. *Drug Metab. Dispos.* 32, 1377, 2004.

27. Asensi, M., Medina, I., Ortega, A. et al. Inhibition of cancer growth by resveratrol is related to its low bioavailability. *Free Radic. Biol. Med.* 33, 387, 2002.

28. Hsieh, T.C., Wang, Z., Hamby, C.V. et al. Inhibition of melanoma cell proliferation by resveratrol is correlated with upregulation of quinone reductase 2 and p53. *Biochem. Biophys. Res. Commun.* 334, 223, 2005.

29. Buryanovskyy, L., Fu, Y., Boyd, M. et al. Crystal structure of quinone reductase 2 in complex with resveratrol. *Biochemistry* 43, 11417, 2004.

30. Holvoet, P. Oxidized LDL and coronary heart disease. *Acta Cardiol.* 59, 479, 2004.

31. Turrens, J.F., Lariccia, J., and Nair, M.G. Resveratrol has no effect on lipoprotein profile and does not prevent peroxidation of serum lipids in normal rats. *Free Radic. Res.* 27, 557, 1997.

32. Renaud, S., and de, L.M. Wine, alcohol, platelets, and the French paradox for coronary heart disease. *Lancet* 339, 1523, 1992.

33. Afaq, F., Adhami, V.M., and Ahmad, N. Prevention of short-term ultraviolet B radiation-mediated damages by resveratrol in SKH-1 hairless mice. *Toxicol. Appl. Pharmacol.* 186, 28, 2003.

34. Morales, A.I., Buitrago, J.M., Santiago, J.M. et al. Protective effect of *trans*-resveratrol on gentamicin-induced nephrotoxicity. *Antioxid. Redox. Signal.* 4, 893, 2002.

35. Mizutani, K., Ikeda, K., Kawai, Y. et al. Protective effect of resveratrol on oxidative damage in male and female stroke-prone spontaneously hypertensive rats. *Clin. Exp. Pharmacol. Physiol.* 28, 55, 2001.

36. Floreani, M., Napoli, E., Quintieri, L. et al. Oral administration of *trans*-resveratrol to guinea pigs increases cardiac DT-diaphorase and catalase activities, and protects isolated atria from menadione toxicity. *Life Sci.* 72, 2741, 2003.

37. Ferry-Dumazet, H., Garnier, O., Mamani-Matsuda, M. et al. Resveratrol inhibits the growth and induces the apoptosis of both normal and leukemic hematopoietic cells. *Carcinogenesis* 23, 1327, 2002.

38. Kaeberlein, M., McDonagh, T., Heltweg, B. et al. Substrate-specific activation of sirtuins by resveratrol. *J. Biol. Chem.* 280, 17038, 2005.

39. Nemoto, S., Fergusson, M.M., and Finkel, T. SIRT1 functionally interacts with the metabolic regulator and transcriptional coactivator PGC-1{alpha}. *J. Biol. Chem.* 280, 16456, 2005.

40. Kumar, P., Padi, S.S., Naidu, P.S. et al. Effect of resveratrol on 3-nitropropionic acid-induced biochemical and behavioural changes: Possible neuroprotective mechanisms. *Behav. Pharmacol.* 17, 485, 2006.

41. Parker, J.A., Arango, M., Abderrahmane, S. et al. Resveratrol rescues mutant polyglutamine cytotoxicity in nematode and mammalian neurons. *Nat. Genet.* 37, 349, 2005.

42. Borrell-Pages, M., Zala, D., Humbert, S. et al. Huntington's disease: From huntingtin function and dysfunction to therapeutic strategies. *Cell Mol. Life Sci.* 63, 2642, 2006.

43. Tang, B.L., and Chua, C.E.L. SIRT1 and neuronal diseases. *Mol. Aspects Med.* 29, 187, 2008.

44. Anekonda, T.S. Resveratrol—a boon for treating Alzheimer's disease? *Brain Res. Rev.* 52, 316, 2006.

45. Anekonda, T.S., and Reddy, P.H. Can herbs provide a new generation of drugs for treating Alzheimer's disease? *Brain Res. Brain Res. Rev.* 50, 361, 2005.

46. Hardy, J., and Selkoe, D.J. The amyloid hypothesis of Alzheimer's disease: Progress and problems on the road to therapeutics. *Science* 297, 353, 2002.

47. Pallas, M., and Camins, A. Molecular and biochemical features in Alzheimer's disease. *Curr. Pharm. Des* 12, 4389, 2006.

48. Wirths, O., Multhaup, G., and Bayer, T.A. A modified beta-amyloid hypothesis: Intraneuronal accumulation of the beta-amyloid peptide—the first step of a fatal cascade. *J. Neurochem.* 91, 513, 2004.

49. Cuello, A.C. Intracellular and extracellular Abeta, a tale of two neuropathologies. *Brain Pathol.* 15, 66, 2005.

50. Valerio, A., Boroni, F., Benarese, M. et al. NF-kappaB pathway: A target for preventing beta-amyloid (Abeta)-induced neuronal damage and Abeta42 production. *Eur. J. Neurosci.* 23, 1711, 2006.

51. Chen, J., Zhou, Y., Mueller-Steiner, S. et al. SIRT1 protects against microglia-dependent amyloid-beta toxicity through inhibiting NF-kappaB signaling. *J. Biol. Chem.* 280, 40364, 2005.

52. Yeung, F., Hoberg, J.E., Ramsey, C.S. et al. Modulation of NF-kappaB-dependent transcription and cell survival by the SIRT1 deacetylase. *EMBO J.* 23, 2369, 2004.

53. Frank, B., and Gupta, S. A review of antioxidants and Alzheimer's disease. *Ann. Clin. Psychiatry* 17, 269, 2005.

54. Pervaiz, S. Resveratrol: From grapevines to mammalian biology. *FASEB J.* 17, 1975, 2003.

55. Mattson, M.P., Duan, W., Chan, S.L. et al. Neuroprotective and neurorestorative signal transduction mechanisms in brain aging: Modification by genes, diet and behavior. *Neurobiol. Aging* 23, 695, 2002.

56. Mattson, M.P. Will caloric restriction and folate protect against AD and PD? *Neurology* 60, 690, 2003.

57. Patel, N.V., Gordon, M.N., Connor, K.E. et al. Caloric restriction attenuates Abeta-deposition in Alzheimer transgenic models. *Neurobiol. Aging* 26, 995, 2005.

58. Wang, R., Wang, B., He, W. et al. Wild-type presenilin 1 protects against Alzheimer disease mutation-induced amyloid pathology. *J. Biol. Chem.* 281, 15330, 2006.

59. Farris, W., Mansourian, S., Leissring, M.A. et al. Partial loss-of-function mutations in insulin-degrading enzyme that induce diabetes also impair degradation of amyloid beta-protein. *Am. J. Pathol.* 164, 1425, 2004.

60. Qin, W., Chachich, M., Lane, M. et al. Calorie restriction attenuates Alzheimer's disease type brain amyloidosis in squirrel monkeys (*Saimiri sciureus*). *J. Alzheimers. Dis.* 10, 417, 2006.

61. Qin, W., Yang, T., Ho, L. et al. Neuronal SIRT1 activation as a novel mechanism underlying the prevention of Alzheimer disease amyloid neuropathology by calorie restriction. *J. Biol. Chem.* 281, 21745, 2006.

62. Chen, L., and Feany, M.B. [alpha]-Synuclein phosphorylation controls neurotoxicity and inclusion formation in a Drosophila model of Parkinson disease. *Nat. Neurosci.* 8, 657, 2005.

63. Okawara, M., Katsuki, H., Kurimoto, E. et al. Resveratrol protects dopaminergic neurons in midbrain slice culture from multiple insults. *Biochem. Pharmacol.* 73, 550, 2007.

64. Alvira, D., Tajes, M., Verdaguer, E. et al. Inhibition of the cdk5/p25 fragment formation may explain the antiapoptotic effects of melatonin in an experimental model of Parkinson's disease. *J. Pineal Res.* 40, 251, 2006.

65. Outeiro, T.F., Kontopoulos, E., Altmann, S.M. et al. Sirtuin 2 inhibitors rescue alpha-synuclein-mediated toxicity in models of Parkinson's disease. *Science* 317, 516, 2007.

66. Inoue, H., Jiang, X.F., Katayama, T. et al. Brain protection by resveratrol and fenofibrate against stroke requires peroxisome proliferator-activated receptor [alpha] in mice. *Neuroscience Letters* 352, 203, 2003.

67. Gupta, Y.K., Briyal, S., and Chaudhary, G. Protective effect of *trans*-resveratrol against kainic acid-induced seizures and oxidative stress in rats. *Pharmacol. Biochem. Behav.* 71, 245, 2002.

68. Gupta, Y.K., Chaudhary, G., and Srivastava, A.K. Protective effect of resveratrol against pentylenetetrazole-induced seizures and its modulation by an adenosinergic system. *Pharmacology* 65, 170, 2002.

69. Gupta, Y.K., Chaudhary, G., Sinha, K. et al. Protective effect of resveratrol against intracortical FeCl$_3$-induced model of posttraumatic seizures in rats. *Methods Find. Exp. Clin. Pharmacol.* 23, 241, 2001.

70. Tissenbaum, H.A., and Guarente, L. Increased dosage of a sir-2 gene extends lifespan in *Caenorhabditis elegans*. *Nature* 410, 227, 2001.

71. Rogina, B., and Helfand, S.L. Sir2 mediates longevity in the fly through a pathway related to calorie restriction. *Proc. Natl. Acad. Sci. U. S. A.* 101, 15998, 2004.

72. Guarente, L., and Picard, F. Calorie restriction—the SIR2 connection. *Cell* 120, 473, 2005.

73. Koubova, J., and Guarente, L. How does calorie restriction work? *Genes Dev.* 17, 313, 2003.

74. Martin, B., Mattson, M.P., and Maudsley, S. Caloric restriction and intermittent fasting: Two potential diets for successful brain aging. *Ageing Res. Rev.* 5, 332, 2006.

75. Valenzano, D.R., Terzibasi, E., Genade, T. et al. Resveratrol prolongs lifespan and retards the onset of age-related markers in a short-lived vertebrate. *Curr. Biol.* 16, 296, 2006.

76. Picard, F., and Guarente, L. Molecular links between aging and adipose tissue. *Int. J. Obes.(Lond)* 29 Suppl 1, S36, 2005.

77. Mattson, M.P. Dietary factors, hormesis and health. *Ageing Res. Rev.* 7, 43–48, 2007.

78. Lamming, D.W., Wood, J.G., and Sinclair, D.A. Small molecules that regulate lifespan: Evidence for xenohormesis. *Mol. Microbiol.* 53, 1003, 2004.

79. Hipkiss, A.R. Dietary restriction, glycolysis, hormesis and ageing. *Biogerontology* 8, 221, 2007.

80. Leakey, J.E., Cunny, H.C., Bazare, J., Jr. et al. Effects of aging and caloric restriction on hepatic drug metabolizing enzymes in the Fischer 344 rat. I: The cytochrome P-450 dependent monooxygenase system. *Mech. Ageing Dev.* 48, 145, 1989.

81. Howitz, K.T., and Sinclair, D.A. Xenohormesis: Sensing the chemical cues of other species. *Cell* 133, 387, 2008.

23 Acetyl-l-Carnitine and Ferulic Acid Action in Aging and Neurodegenerative Diseases

Renã A. Sowell,[1,2] *Christopher D. Aluise,*[1,2]
and D. Allan Butterfield[1,2,3]

[1]Department of Chemistry, [2]Sanders-Brown Center on Aging, and [3]Center of Membrane Sciences, University of Kentucky, Lexington, Kentucky, USA

CONTENTS

23.1 INTRODUCTION

In 1956, Harman proposed the "free radical theory of aging," which suggested that free radicals and/or reactive oxygen species (ROS) contributed to the loss of molecular and cellular function in organisms over time (Harman 1956). This loss of function in aging is due in part to ROS exposure causing an imbalance in cellular homeostasis because the organism is not able to sufficiently scavenge free radicals. Harman's theory has more recently been accepted as the "*oxidative stress theory of aging*" (Muller et al. 2007) and has been linked to disorders including cancer, atherosclerosis, stroke, and diabetes (Mariani et al. 2005). Oxidative stress also has been implicated in several neurodegenerative disorders such as Alzheimer's disease (AD), Parkinson's disease (PD), and amyotrophic lateral sclerosis (ALS) (Markesbery 1997). Because oxidative stress can affect cellular processes such as metabolism, structural integrity, inflammation, and apoptosis (Terman et al. 2006), its link with many human diseases is not surprising. As a result, many researchers have investigated the molecular mechanisms underlying oxidative stress in various model systems of aging and age-related disorders (Humphries et al. 2006; Muller et al. 2007) in order to find clues that lead to the development of therapeutic molecules. While the development/synthesis of novel compounds that are able to combat oxidative stress is necessary, there is much interest in the application of naturally occurring compounds that may also offer beneficial protection.

Under physiological conditions, cellular ROS levels are regulated by antioxidant molecules and antioxidant enzymes, such as catalase (CAT), superoxide dismutase (SOD), and glutathione peroxidase (GPx). An overabundance of ROS renders cellular materials vulnerable to oxidative modifications that adversely can affect the structure and function of proteins, lipids, carbohydrates, and nucleic acids and can eventually lead to cell death (Smith et al. 1994; Mark et al. 1997; Markesbery 1997; Lovell et al. 2001; Butterfield 2002; Butterfield et al. 2002a). Although oxidative stress can occur in any cell, neuronal oxidative stress is particularly detrimental in part because brain cells do not divide or regenerate. Moreover, the brain is highly susceptible to oxidative insult because it consumes a vast amount of oxygen, has high levels of polyunsaturated fatty acids and redox-active transition metal ions, and contains low levels of antioxidant enzymes (Markesbery 1997; Butterfield et al. 2001; Butterfield et al. 2002a; Butterfield et al. 2002b; Floyd et al. 2002). Therefore, antioxidant compounds that are able to cross the blood-brain barrier (BBB) and increase endogenous brain antioxidant levels are desirable because they may provide neuroprotection against oxidative damage in aging and neurodegenerative disorders. Examples of such compounds include ferulic acid (FA), FA derivatives, and acetyl-l-carnitine (LAC), all of which are naturally occurring molecules. This review will focus on literature reports that have investigated the use of FA and its derivatives and LAC as means to suppress or halt the oxidative damage present in aging and age-related neurodegenerative conditions.

23.2   FA IN AGING AND NEURODEGENERATIVE DISEASES

FA [3-(4-hydroxy-3-methoxyphenyl)-2-propenoic acid], a derivative of cinnamic acid, is found in many edible plants, vegetables, and fruits (Graf 1992). Therapeutically, significant attention has been given to FA because it is a naturally occurring antioxidant molecule. The *trans* isomer predominates in plants, where FA is typically esterified and covalently bound to sugars, glycoproteins, polyamines, and other cell wall components (Graf 1992). This molecule is synthesized from phenylalanine and tyrosine via the shikimate pathway, in which both amino acids are converted to *p*-coumaric acid, which is then hydroxylated to caffeic acid. Caffeic acid is then methylated to form FA, with methionine serving as the methyl donor. The structure of FA contains a phenol whose hydroxyl group is *para*-substituted to propenoic acid (Figure 23.1); the antioxidant capacity of FA lies in its ability to trap free radicals in an extensive pi-conjugated network. Radicals, such as hydroxyl radical (·OH), encounter FA and immediately abstract an H atom from the

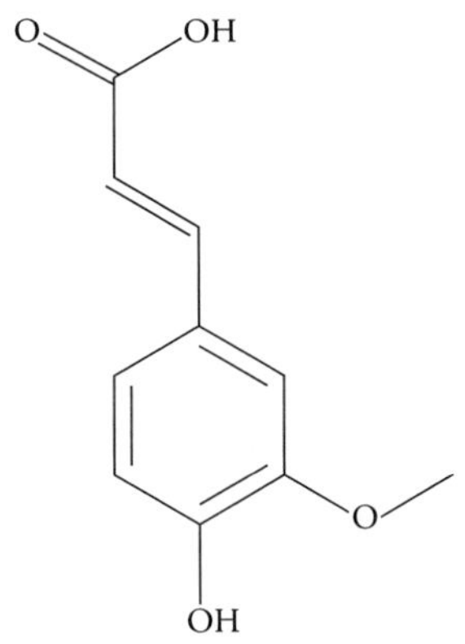

FIGURE 23.1   Structure of FA.

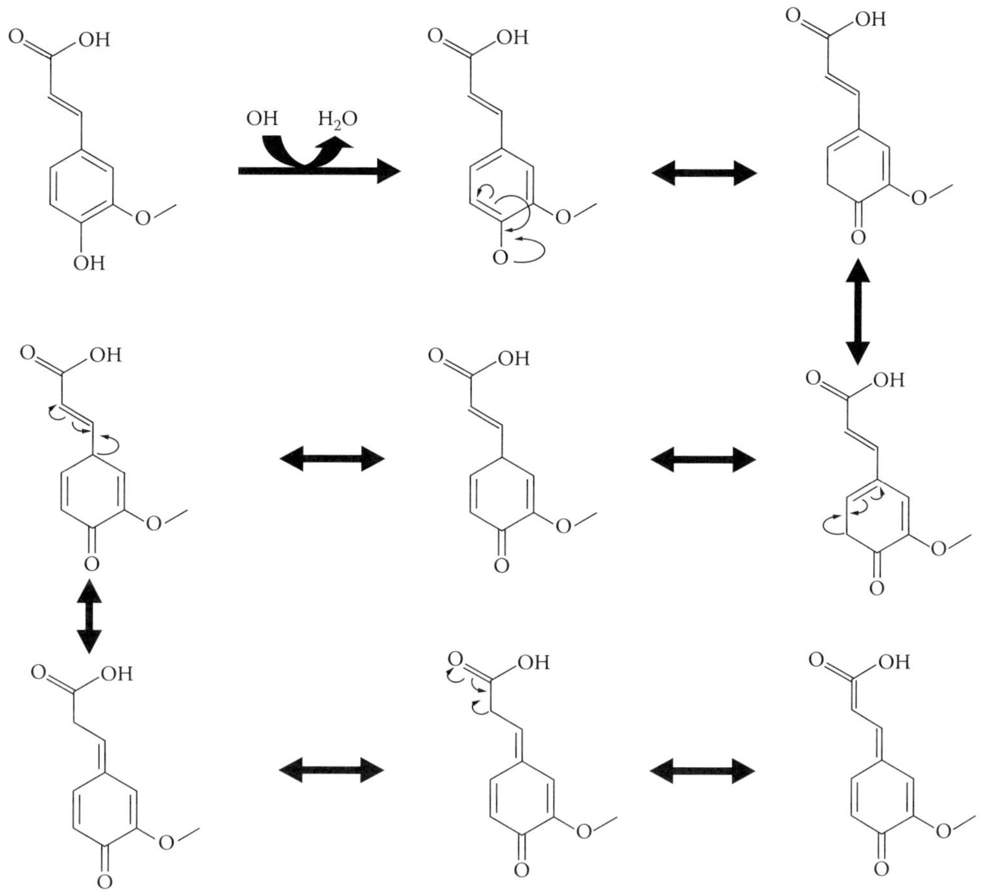

FIGURE 23.2 Diagram of ROS scavenging by FA.

phenolic –OH to make a phenoxy radical; the remaining unpaired electron on the oxygen atom is capable of resonating throughout the multiple conjugated sites of FA (Figure 23.2). Because of its radical scavenging properties, FA is sometimes referred to as a "chain-breaking" antioxidant. Toxic events such as oxidation of lipids, DNA, RNA, and proteins are mediated by free radical reactions that damage cellular components and, in turn, produce more free radical species that propagate until the "chain" is broken. FA intercepts ROS and other reactive radicals, thereby halting insidious modifications and preventing other downstream consequences.

It has been suggested that the neurotherapeutic capabilities of FA are partially hampered by the negative charge that exists on the carboxylic oxygen atom at physiological pH, as well as relatively low lipophilicity (Scapagnini et al. 2004). The presence of the BBB, an epithelial cell layer connected by tight junctions, prevents the passage of most peripherally localized materials into the brain space. Exogenous small molecules, lipophilics, and molecules with transport systems are the select species allowed to enter the brain. Therefore, derivatives of FA such as ferulic acid ethyl ester (FAEE) may be more neuroprotective because the ethyl ester moiety decreases the overall polarity of the molecule and neutralizes the negative charge, thus increasing its permeability at the BBB. FAEE is also reported to be a superior scavenger of ·OH and superoxide ($O_2{\cdot}^-$) relative to FA (Scapagnini et al. 2004). However, *in vivo* studies in which FA was peripherally administered to mice report both behavioral and neurobiochemical improvements when mice were subjected to insults, such as intracerebroventricular injection of amyloid beta (Aβ) (Yan et al. 2001; Cho et al. 2005) or buthionine sulfoximine (BSO) (Mamiya et al. 2008). Studies of this nature indicate potential neurological gains from FA that are similar to FA derivatives. Therefore, the applications of both FA and its derivatives for providing potential protection from detriments associated with aging and neurodegeneration are discussed below.

23.2.1 Cellular Protection from Oxidative Damage by FA and FA Derivatives

The capability of FA to combat ROS has been well characterized (Scott et al. 1993; Kanski et al. 2002; Ogiwara et al. 2002; Hsieh et al. 2005; Joshi et al. 2006). FA has been shown to scavenge ·OH, peroxynitrite, hypochlorous acid, 1,1-diphenyl-2-picrylhydrazyl, and oxidized low-density lipoprotein (Scott et al. 1993; Kanski et al. 2002; Hsieh et al. 2005). FA can also scavenge nitric oxide (NO) and $O_2{\cdot}^-$ (Ogiwara et al. 2002). Treatment of synaptosomes with Fe^{2+} and H_2O_2 (reagents for Fenton hydroxyl production) and 2,2-azobis(2-amidino-propane) dihydrochloride (AAPH) (a source for peroxyl and alkoxyl radicals) resulted in reduced free radicals and oxidative stress products when mice were pretreated *in vivo* with FAEE (Joshi et al. 2006). $FeSO_4$ administration to neurons caused increased levels of thiobarbituric acid substances, a measure of lipid peroxidation, which were attenuated with FA (Zhang et al. 2003). Intracerebroventricular injection of BSO, an inhibitor of glutathione (GSH) synthesis, enhanced protein carbonyl levels in mice brains, and this elevation was prevented by subcutaneous administration of FA (Mamiya et al. 2008). Also, FA suppressed malondialdehyde levels, another marker of lipid peroxidation, in rat brain homogenates (Sharma 1976). Several studies investigating the effects of radiation-mediated oxidative stress report increased protection of lipids, antioxidants, and DNA by FA (Hsieh et al. 2005; Maurya et al. 2005, 2006; Prasad et al. 2006; Srinivasan et al. 2006).

Although oxidative stress is not specific to a single neurological condition, many studies investigating the effects of FA on neurodegenerative disease pathogenesis focus on AD. AD is characterized behaviorally by cognitive dysfunction and memory impairment and pathologically by brain-localized senile plaques and neurofibrillary tangles. Amyloid β peptide (Aβ) is the major component of senile plaques. Aβ associates into oligomers/fibrils and causes oxidative stress by direct oxidation of biomolecules and by producing large amounts of ROS/reactive nitrogen species (RNS), possibly due to the methionine residue at position 35 (Varadarajan et al. 2001). *In vitro*, FA in a dose-dependent manner inhibited Aβ fibril formation and also destabilized preformed fibrils (Ono et al. 2005). FAEE scavenged Aβ(1–42)-generated ROS and protected against Aβ(1–42)-induced oxidative damage in rat primary neuronal cell culture (Sultana et al. 2005). FAEE injected intraperitoneally into rodents also protected synaptosomes from oxidative stress induced by Aβ(1–42) (Perluigi et al. 2006). The benefits of FA as an AD therapeutic also extend to regulation of acetylcholine levels, which are markedly decreased in AD brain (Davies et al. 1976; Coyle et al. 1983; Wevers et al. 2000). Orally administered FA significantly elevated choline acetlytransferase activity, an acetylcholine synthesizing enzyme, in trimethyltin-induced cognitively deficient mice, in addition to improving memory (Kim et al. 2007). It has also been reported that decreased cortical acetylcholine levels were attenuated by FA in Aβ(1–42)-treated mice (Yan et al. 2001).

Besides scavenging free radicals, FA and FAEE have a hormesis effect by promoting increased levels of antioxidant defense molecules. FAEE induces the expression of heme-oxygenase 1 (HO-1) at both protein and mRNA levels (Scapagnini et al. 2004; Sultana et al. 2005; Joshi et al. 2006; Perluigi et al. 2006). HO-1 is an inducible enzyme that catalyzes the breakdown of heme to carbon monoxide (CO) and biliverdin. Biliverdin is reduced to bilirubin, which at low levels is an excellent antioxidant (Poon et al. 2004). The protective properties of HO-1 are manifested in the antioxidant and anti-inflammatory capabilities of these products (Stocker et al. 1987; Clark et al. 2000; Poon et al. 2004). The mechanism by which FAEE up-regulates HO-1 is unclear, but is hypothesized to be through the Nrf-2 pathway (Scapagnini et al. 2004). In addition to HO-1, FAEE also up-regulates heat shock proteins (HSPs), namely HSP72 (Sultana et al. 2005; Perluigi et al. 2006) and HSP70

(Joshi et al. 2006). HSP72, an inducible member of the HSP70 family, is a chaperone protein used to refold denatured proteins. HSP72 has been shown to protect against stroke and ischemia (Lee et al. 2001; Kelly et al. 2002; Zheng et al. 2008). Also, heat shock proteins have been reported to protect central nervous system (CNS) cells against apoptosis and necrosis *in vitro* (Poon et al. 2004). In hepatocytes subjected to gamma radiation, FA significantly increased antioxidant levels of GSH, vitamins A, E, and C, ceruloplasmin, and uric acid (Srinivasan et al. 2006). FA also increased diminished activities of antioxidant enzymes such as SOD, GPx, and CAT (Srinivasan et al. 2006); similar findings were reported in diabetic rats treated with FA (Balasubashini et al. 2004). The return to baseline antioxidant levels may be due to ROS scavenging by FA, so as to preserve the levels of these molecules for normal cellular function (Srinivasan et al. 2006). Likewise, the generated ROS may damage antioxidant enzymes, thus lowering activity levels. FA restored decreased activities of SOD, GPx, and CAT, again possibly due to direct oxidant scavenging (Srinivasan et al. 2006).

23.2.2 FA Offers Protection from Inflammation

As previously described, the beneficial effects of FA and its derivatives are derived from an excellent ability to combat cellular oxidative stress. However, FA also has anti-inflammatory capabilities (Chawla et al. 1987; Ozaki 1992; Fernandez et al. 1998; Akihisa et al. 2000), partly due to the fact that inflammation is associated with large amounts of ROS and in part due to the suppression/up-regulation of key species in this cascade. Inflammation is a bodily defense mechanism generated in response to a pathogen; inflammation is characterized by activation of immune cells, cytokines, chemokines, and acute phase molecules that aid in the removal of the insult. The anti-inflammatory ability of FA appears to be fourfold: (1) scavenging preformed ROS that can contribute to inflammation; (2) scavenging ROS produced by inflammation; (3) suppression of key proteins/molecules contributing to inflammation; and (4) enhancing proteins/molecules that are inhibitory to inflammation. Neuroinflammation is implicated in several neurodegenerative conditions, namely PD, AD, ALS, Creutzfelt-Jakob disease, multiple sclerosis (MS), and Pick's disease (Klegeris et al. 2007).

Glial cells, such as microglia and astrocytes, largely mediate inflammation in the CNS. Upon activation, these cells produce cytokines, chemokines, and other proinflammatory agents for action against toxic antigens; however, they also produce large amounts of ROS/RNS, thereby contributing to oxidative stress and its consequences. Reactive agents such as lipopolysaccharide (LPS) and Aβ can trigger glial cell activation, which leads to the release of species such as NO and H_2O_2, capable of killing neighboring cells (Block et al. 2007). Factors released from dying cells can in turn activate more microglia, creating a toxic cycle if not properly regulated (Block et al. 2007). Activated microglia and astrocytes are implicated in conditions such as AD, PD, MS, and HIV/AIDS. Long-term administration of FA to mice centrally treated with Aβ(1–42) suppressed immunoreactivity of OX-42, a microglial active marker (Kim et al. 2004). Another study found microglial activation to be suppressed with FA, and more so with a nitric oxide releasing FA derivative (Wenk et al. 2004). Increases in glial fibrillary acidic protein (GFAP), a marker of glial cell activation, and proinflammatory cytokine interleukin 1β (IL-1β) by Aβ were successfully suppressed with FA, and also with sodium ferulate (SF) in a separate study (Yan et al. 2001; Jin et al. 2005). Along with endothelial nitric oxide synthase (eNOS) and 3-nitrotyrosine (3NT), astrocytic IL-1β immunoreactivity was also decreased in mice treated with FA and Aβ(1–42) compared with mice treated only with Aβ(1–42) (Cho et al. 2005). In addition to its inflammatory contributions, IL-1β has downstream effects related to apoptosis (see following section).

Inducible nitric oxide synthase (iNOS) is a protein directly involved in both oxidative/nitrosative stress and inflammation. iNOS produces NO from arginine and is only expressed in response to immunological stimuli (Bredt et al. 1994; Sessa 1994). Active microglia are capable of activating iNOS, thereby producing large amounts of RNS/ROS. Overproduction of NO can cause nitration of cellular materials, leading to dysfunction. Treatment of neurons with Aβ(1–42) caused an increase in iNOS levels, which was attenuated by FAEE administration (Sultana et al. 2005). Similar results were obtained using Fe^{2+}/H_2O_2 or AAPH as the oxidant species (Joshi et al. 2006). Accordingly, both of these reports show increased 3-NT levels after treatment with $A\beta(1-42)$, as well as Fe^{2+}/H_2O_2 and AAPH that are abrogated with FAEE administration. This suggests that prevention of nitrosative stress is possibly due to lowered iNOS expression (Sultana et al. 2005; Joshi et al. 2006), thereby supporting the duality of FAEE to act as both an antioxidant and anti-inflammatory molecule.

23.2.3 FA OFFERS PROTECTION AGAINST APOPTOSIS

Another downstream result of oxidative stress is cell death. Preemptive treatment with antioxidants such as FA to antagonize inflammation and/or oxidative stress is a way to minimize signs of apoptosis and synapse loss in degenerative conditions. As mentioned previously, IL-1β increases with subject exposure to Aβ. Interestingly, increases in IL-1 β levels by $A\beta(25-35)$ were accompanied by increases in the p38 mitogen-activating kinase (p38 MAPK) pathway (Jin et al. 2005). This led to increased caspase 3 levels and activity and to an increased Fas ligand expression, which is indicative of cell death; these trends were attenuated by SF treatment (Jin et al. 2005). SF has also been reported to abrogate $A\beta(1-40)$-induced apoptosis (Jin et al. 2006). Caspase 3 activation and activation of proapoptotic genes p53 and p21[waf1/cip1] decreased in neurons subject to Fe^{2+}-induced oxidative damage by FA treatment (Zhang et al. 2003). Mitochondrial release of cytochrome c is an initiating step in the apoptotic pathway, and this initiating activity is reportedly decreased when bound to and stabilized by FA (Yang et al. 2007). Similarly, FAEE decreased release of cytochrome c triggered by $A\beta(1-42)$, and restored the reduction in synaptosomal phospholipid asymmetry caused by this peptide, further supporting the notion that FA compounds are protective against Aβ-induced apoptosis (Mohmmad Abdul and Butterfield 2005). Excitotoxic events such as increased intracellular Ca^{2+} can also lead to apoptotic cell death. Yu et al. (2006) reported evidence of SF as a competitive N-methyl-d-aspartate (NMDA) receptor antagonist and hence neuroprotectant most likely by preventing Ca^{2+} influx ($[Ca^{2+}]_i$) and subsequent cell death.

23.3 LAC IN AGING AND NEURODEGENERATIVE DISEASES

LAC (γ-trimethyl-β-acetylbutyrobetaine) is the acetyl ester of the precursor molecule carnitine, a nonessential amino acid (Figure 23.3). Carnitine is synthesized from methionine and lysine amino acids in the brain, liver, and kidney. The acetyl moiety can be added to carnitine from the acetyl group on coenzyme A (CoA) with the aid of LAC transferase in order to form LAC. LAC is located

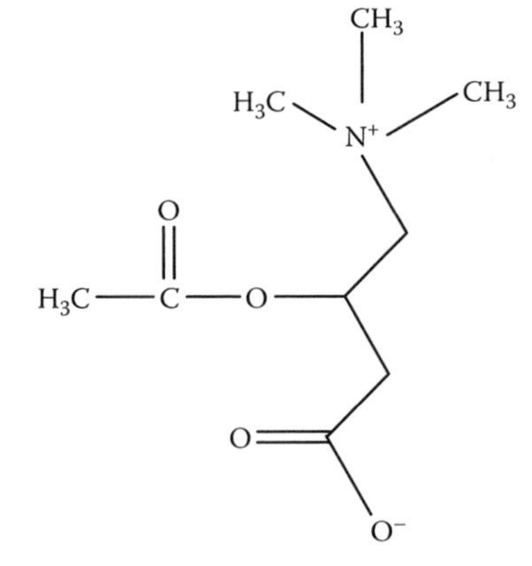

FIGURE 23.3 Structure of LAC.

in the inner mitochondrial matrix and is involved in β-oxidation/fatty acid metabolism, cellular energy production, neuronal maintenance and repair, regulation of mitochondrial enzymes, and buffering of toxic acyl-CoA metabolites (Calabrese et al. 2006b). In addition, LAC assists in the uptake of acetyl-CoA into the mitochondria, increases acetylcholine production, and promotes phospholipid synthesis of proteins and membranes (Calvani et al. 1999). The acetyl moiety also increases the permeability of carnitine allowing it to readily cross the BBB and transport fatty acids across mitochondrial membranes. Thus, LAC is a crucial molecule involved in the maintenance of normal mitochondrial function. Recent evidence suggests that, overall, LAC increases mitochondrial ATP production and provides protection against oxidative attack (Beal 2003; Mazzio et al. 2003).

While the molecular mechanisms of LAC treatment are discussed in this review, it is noteworthy to mention the other beneficial aspects that have been reported upon LAC administration. For example, LAC has been shown to induce synaptic long-term potentiation in aged animals (Ando et al. 2002) and to influence attention, learning, and memory in rodents (Spagnoli et al. 1991). In AD patients, LAC has been reported to improve memory function and cognition and/or to slow brain deterioration (Spagnoli et al. 1991; Nikitovic et al. 1998; Alam et al. 2003; Balogun et al. 2003; Poon et al. 2004). This is the case even in AD patients younger than 61 years old (Calabrese et al. 2000; Calabrese et al. 2002) and demonstrates the behavioral/symptomatic benefits of LAC administration in AD.

23.3.1 Restoration of Mitochondrial Function by LAC

Mitochondria are key to maintaining proper organism function and to organism survival. Approximately 90% of the cell's ATP is generated from the mitochondria, thus its term the "energy powerhouse" of the cell. However, the constant metabolism of oxygen by the mitochondria also generates ROS and thus can contribute to oxidative stress and cause cellular damage. This effect can be significantly enhanced upon impairments to mitochondrial functional integrity (Kidd 2005). In aging, AD, PD, Down's syndrome, MS, ALS, Huntington's disease, and other neurodegenerative disorders, one of the commonly suggested mechanisms of disease pathogenesis involves mitochondrial dysfunction (Kidd 2005); consequently, treatments that either maintain or improve mitochondrial function are necessary. LAC in combination with the mitochondrial molecule, α-lipoic acid, has been reported to reverse mitochondrial damage associated with aging (Ames et al. 2004). A double-blind placebo or LAC study in AD patients revealed that LAC improved metabolic measures determined by magnetic resonance spectroscopy, suggesting renewal of neuronal membranes and energy stores in the cortex (Pettegrew et al. 1995).

In the N-methyl-4-phenyl-1,2,3,6-tetrahyropyidine (MPTP)-induced PD mouse model, LAC provided protection against dopaminergic neuron loss in the substantia nigra (Bodis-Wollner et al. 1991). The potential of LAC as a neuroprotective agent against neurotoxic insults, such as MPTP, were also demonstrated in animal models of PD and AD, in which LAC controlled the mitochondrial acyl-CoA/CoA ratio, peroxisomal oxidation of fatty acids, and production of ketone bodies (Virmani et al. 2005). Furthermore, in MS patients, LAC treatment decreased nitrosative end products found in the cerebrospinal fluid and increased GSH levels (Calabrese et al. 2003). Damage to mitochondrial enzymes in the hypoxic brain region of stroke patients was partially attenuated by LAC treatment through decreases in hypoxic damage and enhancement of cell survival (Corbucci et al. 1992). Overall, substantial evidence exists for the use of LAC as a potential treatment against mitochondrial damage and dysfunction in aging and neurodegenerative disorders.

23.3.2 Cellular Protection from Oxidative Damage by LAC

As mentioned above for FA, naturally occurring antioxidants have the potential to provide significant protection against oxidative damage. LAC provides neuroprotection against oxidative attack by either mediating oxidative damage as an antioxidant molecule (in which the mechanisms are not completely understood) or by influencing the levels of other antioxidant and stress response-related molecules. Our laboratory recently reported that combined pretreatment of cortical neurons with LAC and α-lipoic acid reduces 4-hydroxy-nonenal (HNE) toxicity, protein oxidation, lipid peroxidation, and apoptosis (Abdul and Butterfield 2007). In addition, this combined antioxidant treatment increases cellular GSH, HSP levels (i.e., HO-1 and HSP72), decreases iNOS levels, and activates the PI3K, PKG, and ERK1/2 pathways, which are important prosurvival pathways for neuronal survival (Abdul and Butterfield 2007).

Similar results were also obtained utilizing a senescent rat model of aging. Calabrese et al. (2006a) demonstrated that senescent rats treated with LAC for four months caused an induction in HO-1, HSP70, SOD2, and GSH levels. Oxidatively modified levels of protein carbonyls and HNE-bound proteins (e.g., marker of lipid peroxidation) were also reduced in LAC-treated senescent rats and LAC was reported to prevent age-related mitochondrial respiratory chain complex expression (Calabrese et al. 2006a). In addition, astrocytes treated with LAC induced HO-1 in a dose- and time-dependent manner, which was correlated with an increase in HSP60 and the redox-sensitive transcription factor, Nrf2 (Calabrese et al. 2005). Pretreatment of astrocytes isolated from rat with LAC before lipopolysaccharide and interferon γ exposure (e.g., molecules that induce inflammation and nitrosative stress) prevented protein nitrosation, disturbances to mitochondrial respiratory chain complex activity, and antioxidant status (Calabrese et al. 2005). These reports provide additional evidence that LAC is beneficial for maintaining proper mitochondrial function and providing protection against oxidative damage. Furthermore, LAC appears to be influential in the expression of endogenous antioxidants (e.g., GSH, SOD-2) and in the so-called vitagenes (e.g., HO-1, HSP70), which are crucial for maintenance and repair functions in the brain (Calabrese et al. 2006b; Poon et al. 2004). Taken together, these pathways influenced by LAC antioxidant supplementation provide a potential means for delaying aging, reducing the risk of age-related disorders, and treating the symptoms associated with neurodegenerative disorders, such as AD.

Insights into specific proteins and brain regions that are protected by LAC treatment have been revealed by immunochemical and redox proteomics analysis of aged rats in our laboratory (Poon et al. 2006). LAC reduced age-associated protein oxidation in protein carbonyls of aged rats in cortex (CX), substantia nigra (SN), septum (SP), hippocampus (HP), and cerebellum (CB) regions of the brain (Poon et al. 2006). Significant increases in protein-bound HNE levels were only observed in CX, SP, and HP of aged rats and were attenuated with LAC administration. Aged rats were also observed to have elevated 3-NT-modified protein levels in SN and HP, which were reduced with LAC. Redox proteomics analysis of HP in aged rats relative to young rats revealed that significant increases to protein carbonylation levels of hemoglobin, cofilin 1, and β-actin proteins in aged rats were reduced after LAC treatment (only the reduction in cofilin 1 was statistically significant). LAC treatment in aged rats also restored the levels of mitochondrial aconitase, inositol monophosphatase, α-enolase, creatine kinase B chain, and tubulin α-1 chain to similar levels of young rats. In the CX brain region, LAC administration reduced the carbonyl levels associated with eight proteins, including heat shock cognate protein 70, β-actin, and peroxiredoxin 1. It should be noted that these changes were not statistically significant. However, LAC administration in aged rats restored the levels of F-actin capping protein β-subunit, Rab GDP dissociation inhibitor β, and ubiquitin to levels similar to young rats (Poon et al. 2006). The pathways associated with these altered proteins and oxidatively modified proteins in aged rats include antioxidant, mitochondria function, and plasticity. This provides further evidence and is consistent with the notion that LAC may be a suitable candidate for treatment in aging and age-related neurodegenerative diseases in order to reduce oxidative stress, mitochondrial decline, and improve learning and memory deficits that are a result of declines to neuronal plasticity.

23.3.3 Restoration of Cholinergic Functions by LAC

The structural similarity of LAC to acetylcholine potentially allows LAC to have cholinergic benefits such as stabilizing cholinergic neurotransmission (Virmani et al. 2004) and modifying acetylcholine production in the CNS (Carta et al. 1993). Cholinergic neuronal loss is associated with

AD (Davies et al. 1976; Coyle et al. 1983; Wevers et al. 2000), such that primary therapies for AD patients are based on cholinesterase inhibitors (e.g., Aricept®). The exact mechanism by which LAC provides protection in AD brains is not clearly understood; however, several reports provide insights about key pathways. For example, in aged rats and rat models of the aging CNS, LAC increases choline acetyltransferase (ChAT) activity and nerve growth factors (NGF) (Piovesan et al. 1994; Taglialatela et al. 1994). In streptozotocin-treated rats (e.g., a diabetic rodent model), LAC attenuated ChAT activity (Prickaerts et al. 1995). Reductions in ChAT activity, if not attenuated by interventions such as LAC treatment, can result in low levels of acetylcholine, which is important in maintaining normal neurotransmission (Ojika 1998).

23.3.4 LAC OFFERS PROTECTION AGAINST APOPTOSIS

Similar to FA, LAC also has beneficial aspects that are utilized in apoptotic pathways.

In a rat model of peripheral neuropathy, LAC prevents apoptosis induction, reduces cytosolic cytochrome c levels, and impairs caspase 3 protease activity, suggesting that LAC prevents regulated cell death (Di Cesare Mannelli et al. 2007). Scorziello et al. (1997) utilized a derivative of LAC, LAC amide (ST857), to pretreat cerebrellar granule cells that were exposed to $A\beta(25–35)$. ST857 improved cell survival, rescued cells from $A\beta$-induced neurotoxicity, and reduced modified glutamate-induced $[Ca^{2+}]_i$ levels (Scorziello et al. 1997). Increased intracellular Ca^{2+} levels can cause mitochondrial swelling, which results in caspase activation and ultimately leads to apoptotic neuronal cell death. LAC has been reported to improve neuronal survival in cerebellar granule cells possibly by increasing aspartate uptake (e.g., a marker of maturation of glutamatergic neurons) and maintaining functional NMDA receptors (Rampello et al. 1992).

23.3.5 NEUROTROPHIC ACTIONS PROVIDED BY LAC

LAC has also been reported to enhance NGF receptors (Kidd 2005). Growth factors generated in brain cells require similar levels of growth factor receptors in order to maintain a healthy state. Declines in NGF binding capacity have been reported in the HP and basal forebrain regions of aged rats (Angelucci et al. 1988). LAC-treated aged rats had twice the level of NGF binding capacity when compared with non-LAC-treated aged controls (Angelucci et al. 1988). Growth factors and their corresponding receptors have been of much interest as potential therapeutics against neurodegeneration (Kidd 2005); however, when administered peripherally, NGF does not cross the BBB. Thus, LAC because of its ability to cross the BBB may also be a potential therapeutic related to increasing NGF receptors.

23.4 SUMMARY

Investigations of FA and its derivatives and LAC have revealed their tremendous cellular protective benefits and have provided evidence in support of the oxidative stress theory of aging and its relation to age-related disorders. Much of the potential of FA to fight disease is manifested in its ability to combat oxidative stress and its downstream consequences, such as damage to cellular materials, inflammation, and cell death. LAC also fights oxidative stress and cell death in addition to mitochondrial damage, cholinergic dysfunction, and neurotrophic mediation. Both compounds are able to influence protein expression in antioxidant and stress-related pathways, which are key to maintaining redox balance in the cell and ultimately brain. FA and LAC administration appear to be potential therapeutic antioxidant interventions to treat conditions predicated by oxidative stress; for example, aging, age-related neuronal conditions like AD and PD, and other age-related conditions such as diabetes. Because FA and LAC are naturally occurring compounds, they can be easily incorporated into clinical trials in aging and neurodegenerative studies. As the search for the cure/treatment of human disease continues, FA, LAC, and other antioxidants deserve substantial attention as potential therapeutics and neuroprotectants.

ACKNOWLEDGMENTS

This work was supported in part by NIH grants to D.A.B. (AG-05119 and AG-10836).

REFERENCES

Abdul, H. M., and D. A. Butterfield, Involvement of PI3K/PKG/ERK1/2 signaling pathways in cortical neurons to trigger protection by cotreatment of acetyl-l-carnitine and alpha-lipoic acid against HNE-mediated oxidative stress and neurotoxicity: Implications for Alzheimer's disease. *Free Radic Biol Med*, 42(3), 371–84. 2007.

Akihisa, T., K. Yasukawa, M. Yamaura, et al., Triterpene alcohol and sterol ferulates from rice bran and their anti-inflammatory effects. *J Agric Food Chem*, 48(6), 2313–9. 2000.

Alam, J., and J. L. Cook, Transcriptional regulation of the heme oxygenase-1 gene via the stress response element pathway. *Curr Pharm Des*, 9(30), 2499–511. 2003.

Ames, B. N., and J. Liu, Delaying the mitochondrial decay of aging with acetylcarnitine. *Ann N Y Acad Sci*, 1033, 108–16. 2004.

Ando, S., S. Kobayashi, H. Waki, et al., Animal model of dementia induced by entorhinal synaptic damage and partial restoration of cognitive deficits by BDNF and carnitine. *J Neurosci Res*, 70(3), 519–27. 2002.

Angelucci, L., M. T. Ramacci, G. Taglialatela, et al., Nerve growth factor binding in aged rat central nervous system: Effect of acetyl-l-carnitine. *J Neurosci Res*, 20(4), 491–6. 1988.

Balasubashini, M. S., R. Rukkumani, P. Viswanathan, and V. P. Menon, Ferulic acid alleviates lipid peroxidation in diabetic rats. *Phytother Res*, 18(4), 310–4. 2004.

Balogun, E., M. Hoque, P. Gong, et al., Curcumin activates the haem oxygenase–1 gene via regulation of Nrf2 and the antioxidant-responsive element. *Biochem J*, 371(Pt 3), 887–95. 2003.

Beal, M. F., Bioenergetic approaches for neuroprotection in Parkinson's disease. *Ann Neurol*, 53 Suppl 3, S39–47; discussion S47–8. 2003.

Block, M. L., L. Zecca, and J. S. Hong, Microglia-mediated neurotoxicity: Uncovering the molecular mechanisms. *Nat Rev Neurosci*, 8(1), 57–69. 2007.

Bodis-Wollner, I., E. Chung, M. F. Ghilardi, et al., Acetyl-levo-carnitine protects against MPTP-induced parkinsonism in primates. *J Neural Transm Park Dis Dement Sect*, 3(1), 63–72. 1991.

Bredt, D. S., and S. H. Snyder, Nitric oxide: A physiologic messenger molecule. *Annu Rev Biochem*, 63, 175–95. 1994.

Butterfield, D. A., Amyloid beta-peptide (1–42)-induced oxidative stress and neurotoxicity: Implications for neurodegeneration in Alzheimer's disease brain. A review. *Free Radic Res*, 36(12), 1307–13. 2002.

Butterfield, D. A., A. Castegna, C. M. Lauderback, and J. Drake, Evidence that amyloid beta-peptide-induced lipid peroxidation and its sequelae in Alzheimer's disease brain contribute to neuronal death. *Neurobiol Aging*, 23(5), 655–64. 2002a.

Butterfield, D. A., J. Drake, C. Pocernich, and A. Castegna, Evidence of oxidative damage in Alzheimer's disease brain: Central role for amyloid beta-peptide. *Trends Mol Med*, 7(12), 548–54. 2001.

Butterfield, D. A., and C. M. Lauderback, Lipid peroxidation and protein oxidation in Alzheimer's disease brain: Potential causes and consequences involving amyloid beta-peptide-associated free radical oxidative stress. *Free Radic Biol Med*, 32(11), 1050–60. 2002b.

Calabrese, V., C. Colombrita, R. Sultana, et al., Redox modulation of heat shock protein expression by acetylcarnitine in aging brain: Relationship to antioxidant status and mitochondrial function. *Antioxid Redox Signal*, 8(3–4), 404–16. 2006a.

Calabrese, V., A. M. Giuffrida Stella, M. Calvani, and D. A. Butterfield, Acetylcarnitine and cellular stress response: Roles in nutritional redox homeostasis and regulation of longevity genes. *J Nutr Biochem*, 17(2), 73–88. 2006b.

Calabrese, V., A. Ravagna, C. Colombrita, et al., Acetylcarnitine induces heme oxygenase in rat astrocytes and protects against oxidative stress: Involvement of the transcription factor Nrf2. *J Neurosci Res*, 79(4), 509–21. 2005.

Calabrese, V., G. Scapagnini, A. Ravagna, et al., Disruption of thiol homeostasis and nitrosative stress in the cerebrospinal fluid of patients with active multiple sclerosis: Evidence for a protective role of acetylcarnitine. *Neurochem Res*, 28(9), 1321–8. 2003.

Calabrese, V., G. Scapagnini, A. Ravagna, A. M. Giuffrida Stella, and D. A. Butterfield, Molecular chaperones and their roles in neural cell differentiation. *Dev Neurosci*, 24(1), 1–13. 2002.

Calabrese, V., G. Testa, A. Ravagna, T. E. Bates, and A. M. Stella, HSP70 induction in the brain following ethanol administration in the rat: Regulation by glutathione redox state. *Biochem Biophys Res Commun*, 269(2), 397–400. 2000.

Calvani, M., and E. Arrigoni-Martelli, Attenuation by acetyl-l-carnitine of neurological damage and biochemical derangement following brain ischemia and reperfusion. *Int J Tissue React*, 21(1), 1–6. 1999.

Carta, A., M. Calvani, D. Bravi, and S. N. Bhuachalla, Acetyl-l-carnitine and Alzheimer's disease: Pharmacological considerations beyond the cholinergic sphere. *Ann N Y Acad Sci*, 695, 324–6. 1993.

Chawla, A. S., M. Singh, M. S. Murthy, M. Gupta, and H. Singh, Anti-inflammatory action of ferulic acid and its esters in carrageenan induced rat paw oedema model. *Indian J Exp Biol*, 25(3), 187–9. 1987.

Cho, J. Y., H. S. Kim, D. H. Kim, J. J. Yan, H. W. Suh, and D. K. Song, Inhibitory effects of long-term administration of ferulic acid on astrocyte activation induced by intracerebroventricular injection of beta-amyloid peptide (1–42) in mice. *Prog Neuropsychopharmacol Biol Psychiatry*, 29(6), 901–7. 2005.

Clark, J. E., R. Foresti, C. J. Green, and R. Motterlini, Dynamics of haem oxygenase–1 expression and bilirubin production in cellular protection against oxidative stress. *Biochem J*, 348 Pt 3, 615–9. 2000.

Corbucci, G. G., A. Menichetti, A. Cogliatti, et al., Metabolic aspects of acute cerebral hypoxia during extracorporeal circulation and their modification induced by acety l-carnitine treatment. *Int J Clin Pharmacol Res*, 12(2), 89–98. 1992.

Coyle, J. T., D. L. Price, and M. R. DeLong, Alzheimer's disease: A disorder of cortical cholinergic innervation. *Science*, 219(4589), 1184–90. 1983.

Davies, P., and A. J. Maloney, Selective loss of central cholinergic neurons in Alzheimer's disease. *Lancet*, 2(8000), 1403. 1976.

Di Cesare Mannelli, L., C. Ghelardini, M. Calvani, et al., Protective effect of acetyl-l-carnitine on the apoptotic pathway of peripheral neuropathy. *Eur J Neurosci*, 26(4), 820–7. 2007.

Fernandez, M. A., M. T. Saenz, and M. D. Garcia, Anti-inflammatory activity in rats and mice of phenolic acids isolated from *Scrophularia frutescens*. *J Pharm Pharmacol*, 50(10), 1183–6. 1998.

Floyd, R. A., and K. Hensley, Oxidative stress in brain aging. Implications for therapeutics of neurodegenerative diseases. *Neurobiol Aging*, 23(5), 795–807. 2002.

Graf, E., Antioxidant potential of ferulic acid. *Free Radic Biol Med*, 13(4), 435–48. 1992.

Harman, D., Aging: A theory based on free radical and radiation chemistry. *J Gerontol*, 11(3), 298–300. 1956.

Hsieh, C. L., G. C. Yen, and H. Y. Chen, Antioxidant activities of phenolic acids on ultraviolet radiation-induced erythrocyte and low density lipoprotein oxidation. *J Agric Food Chem*, 53(15), 6151–5. 2005.

Humphries, K. M., P. A. Szweda, and L. I. Szweda, Aging: A shift from redox regulation to oxidative damage. *Free Radic Res*, 40(12), 1239–43. 2006.

Jin, Y., Y. Fan, E. Z. Yan, Z. Liu, Z. H. Zong, and Z. M. Qi, Effects of sodium ferulate on amyloid-beta-induced MKK3/MKK6-p38 MAPK-Hsp27 signal pathway and apoptosis in rat hippocampus. *Acta Pharmacol Sin*, 27(10), 1309–16. 2006.

Jin, Y., E. Z. Yan, Y. Fan, Z. H. Zong, Z. M. Qi, and Z. Li, Sodium ferulate prevents amyloid-beta-induced neurotoxicity through suppression of p38 MAPK and upregulation of ERK-1/2 and Akt/protein kinase B in rat hippocampus. *Acta Pharmacol Sin*, 26(8), 943–51. 2005.

Joshi, G., M. Perluigi, R. Sultana, R. Agrippino, V. Calabrese, and D. A. Butterfield, *In vivo* protection of synaptosomes by ferulic acid ethyl ester (FAEE) from oxidative stress mediated by 2,2-azobis(2-amidino-propane)dihydrochloride (AAPH), or Fe(2+)/H(2)O(2): Insight into mechanisms of neuroprotection and relevance to oxidative stress-related neurodegenerative disorders. *Neurochem Int*, 48(4), 318–27. 2006.

Kanski, J., M. Aksenova, A. Stoyanova, and D. A. Butterfield, Ferulic acid antioxidant protection against hydroxyl and peroxyl radical oxidation in synaptosomal and neuronal cell culture systems *in vitro*: Structure-activity studies. *J Nutr Biochem*, 13(5), 273–281. 2002.

Kelly, S., Z. J. Zhang, H. Zhao, et al., Gene transfer of HSP72 protects cornu ammonis 1 region of the hippocampus neurons from global ischemia: Influence of Bcl-2. *Ann Neurol*, 52(2), 160–7. 2002.

Kidd, P. M., Neurodegeneration from mitochondrial insufficiency: Nutrients, stem cells, growth factors, and prospects for brain rebuilding using integrative management. *Altern Med Rev*, 10(4), 268–93. 2005.

Kim, H. S., J. Y. Cho, D. H. Kim, et al., Inhibitory effects of long-term administration of ferulic acid on microglial activation induced by intracerebroventricular injection of beta-amyloid peptide (1–42) in mice. *Biol Pharm Bull*, 27(1), 120–1. 2004.

Kim, M. J., S. J. Choi, S. T. Lim, et al., Ferulic acid supplementation prevents trimethyltin-induced cognitive deficits in mice. *Biosci Biotechnol Biochem*, 71(4), 1063–8. 2007.

Klegeris, A., E. G. McGeer, and P. L. McGeer, Therapeutic approaches to inflammation in neurodegenerative disease. *Curr Opin Neurol*, 20(3), 351–7. 2007.

Lee, J. E., M. A. Yenari, G. H. Sun, et al., Differential neuroprotection from human heat shock protein 70 over-expression in *in vitro* and *in vivo* models of ischemia and ischemia-like conditions. *Exp Neurol*, 170(1), 129–39. 2001.

Lovell, M. A., C. Xie, and W. R. Markesbery, Acrolein is increased in Alzheimer's disease brain and is toxic to primary hippocampal cultures. *Neurobiol Aging*, 22(2), 187–94. 2001.

Mamiya, T., M. Kise, and K. Morikawa, Ferulic acid attenuated cognitive deficits and increase in carbonyl proteins induced by buthionine-sulfoximine in mice. *Neurosci Lett*, 430(2), 115–8. 2008.

Mariani, E., M. C. Polidori, A. Cherubini, and P. Mecocci, Oxidative stress in brain aging, neurodegenerative and vascular diseases: An overview. *J Chromatogr B Analyt Technol Biomed Life Sci*, 827(1), 65–75. 2005.

Mark, R. J., M. A. Lovell, W. R. Markesbery, K. Uchida, and M. P. Mattson, A role for 4-hydroxynonenal, an aldehydic product of lipid peroxidation, in disruption of ion homeostasis and neuronal death induced by amyloid beta-peptide. *J Neurochem*, 68(1), 255–64. 1997.

Markesbery, W. R., Oxidative stress hypothesis in Alzheimer's disease. *Free Radic Biol Med*, 23(1), 134–47. 1997.

Maurya, D. K., and C. K. Nair, Preferential radioprotection to DNA of normal tissues by ferulic acid under *ex vivo* and *in vivo* conditions in tumor bearing mice. *Mol Cell Biochem*, 285(1–2), 181–90. 2006.

Maurya, D. K., V. P. Salvi, and C. K. Nair, Radiation protection of DNA by ferulic acid under *in vitro* and *in vivo* conditions. *Mol Cell Biochem*, 280(1–2), 209–17. 2005.

Mazzio, E., K. J. Yoon, and K. F. Soliman, Acetyl-l-carnitine cytoprotection against 1-methyl-4-phenylpyridinium toxicity in neuroblastoma cells. *Biochem Pharmacol*, 66(2), 297–306. 2003.

Mohmmad Abdul, H., and D. A. Butterfield, Protection against amyloid beta-peptide (1–42)-induced loss of phospholipid asymmetry in synaptosomal membranes by tricyclodecan-9-xanthogenate (D609) and ferulic acid ethyl ester: Implications for Alzheimer's disease. *Biochim Biophys Acta*, 1741(1–2), 140–8. 2005.

Muller, F. L., M. S. Lustgarten, Y. Jang, A. Richardson, and H. Van Remmen, Trends in oxidative aging theories. *Free Radic Biol Med*, 43(4), 477–503. 2007.

Nikitovic, D., A. Holmgren, and G. Spyrou, Inhibition of AP-1 DNA binding by nitric oxide involving conserved cysteine residues in Jun and Fos. *Biochem Biophys Res Commun*, 242(1), 109–12. 1998.

Ogiwara, T., K. Satoh, Y. Kadoma, et al., Radical scavenging activity and cytotoxicity of ferulic acid. *Anticancer Res*, 22(5), 2711–7. 2002.

Ojika, K., [Hippocampal cholinergic neurostimulating peptide]. *Seikagaku*, 70(9), 1175–80. 1998.

Ono, K., M. Hirohata, and M. Yamada, Ferulic acid destabilizes preformed beta-amyloid fibrils *in vitro*. *Biochem Biophys Res Commun*, 336(2), 444–9. 2005.

Ozaki, Y. Antiinflammatory effect of tetramethylpyrazine and ferulic acid. *Chem Pharm Bull (Tokyo)*, 40(4), 954–6. 1992.

Perluigi, M., G. Joshi, R. Sultana, et al., *In vivo* protective effects of ferulic acid ethyl ester against amyloid-beta peptide 1–42-induced oxidative stress. *J Neurosci Res*, 84(2), 418–26. 2006.

Pettegrew, J. W., W. E. Klunk, K. Panchalingam, J. N. Kanfer, and R. J. McClure, Clinical and neurochemical effects of acetyl-l-carnitine in Alzheimer's disease. *Neurobiol Aging*, 16(1), 1–4. 1995.

Piovesan, P., L. Pacifici, G. Taglialatela, M. T. Ramacci, and L. Angelucci, Acetyl-l-carnitine treatment increases choline acetyltransferase activity and NGF levels in the CNS of adult rats following total fimbria-fornix transection. *Brain Res*, 633(1–2), 77–82. 1994.

Poon, H. F., V. Calabrese, M. Calvani, and D. A. Butterfield, Proteomics analyses of specific protein oxidation and protein expression in aged rat brain and its modulation by l-acetylcarnitine: Insights into the mechanisms of action of this proposed therapeutic agent for CNS disorders associated with oxidative stress. *Antioxid Redox Signal*, 8(3–4), 381–94. 2006.

Poon, H. F., V. Calabrese, G. Scapagnini, and D. A. Butterfield, Free radicals: Key to brain aging and heme oxygenase as a cellular response to oxidative stress. *J Gerontol A Biol Sci Med Sci*, 59(5), 478–93. 2004.

Prasad, N. R., M. Srinivasan, K. V. Pugalendi, and V. P. Menon, Protective effect of ferulic acid on gamma-radiation-induced micronuclei, dicentric aberration and lipid peroxidation in human lymphocytes. *Mutat Res*, 603(2), 129–34. 2006.

Prickaerts, J., A. Blokland, W. Honig, F. Meng, and J. Jolles, Spatial discrimination learning and choline acetyltransferase activity in streptozotocin-treated rats: Effects of chronic treatment with acetyl-l-carnitine. *Brain Res*, 674(1), 142–6. 1995.

Rampello, L., G. Giammona, G. Aleppo, A. Favit, and L. Fiore, Trophic action of acetyl-l-carnitine in neuronal cultures. *Acta Neurol (Napoli)*, 14(1), 15–21. 1992.

Scapagnini, G., D. A. Butterfield, C. Colombrita, R. Sultana, A. Pascale, and V. Calabrese, Ethyl ferulate, a lipophilic polyphenol, induces HO-1 and protects rat neurons against oxidative stress. *Antioxid Redox Signal*, 6(5), 811–8. 2004.

Scorziello, A., O. Meucci, M. Calvani, and G. Schettini, Acetyl-l-carnitine arginine amide prevents beta 25–35-induced neurotoxicity in cerebellar granule cells. *Neurochem Res*, 22(3), 257–65. 1997.

Scott, B. C., J. Butler, B. Halliwell, and O. I. Aruoma, Evaluation of the antioxidant actions of ferulic acid and catechins. *Free Radic Res Commun*, 19(4), 241–53. 1993.

Sessa, W. C., The nitric oxide synthase family of proteins. *J Vasc Res*, 31(3), 131–43. 1994.

Sharma, O. P., Antioxidant activity of curcumin and related compounds. *Biochem Pharmacol*, 25(15), 1811–2. 1976.

Smith, M. A., S. Taneda, P. L. Richey, et al., Advanced Maillard reaction end products are associated with Alzheimer disease pathology. *Proc Natl Acad Sci U S A*, 91(12), 5710–4. 1994.

Spagnoli, A., U. Lucca, G. Menasce, et al., Long-term acetyl-l-carnitine treatment in Alzheimer's disease. *Neurology*, 41(11), 1726–32. 1991.

Srinivasan, M., A. R. Sudheer, K. R. Pillai, P. R. Kumar, P. R. Sudhakaran, and V. P. Menon, Influence of ferulic acid on gamma-radiation induced DNA damage, lipid peroxidation and antioxidant status in primary culture of isolated rat hepatocytes. *Toxicology*, 228(2–3), 249–58. 2006.

Stocker, R., Y. Yamamoto, A. F. McDonagh, A. N. Glazer, and B. N. Ames, Bilirubin is an antioxidant of possible physiological importance. *Science*, 235(4792), 1043–6. 1987.

Sultana, R., A. Ravagna, H. Mohmmad-Abdul, V. Calabrese, and D. A. Butterfield, Ferulic acid ethyl ester protects neurons against amyloid beta-peptide(1–42)-induced oxidative stress and neurotoxicity: Relationship to antioxidant activity. *J Neurochem*, 92(4), 749–58. 2005.

Taglialatela, G., D. Navarra, R. Cruciani, M. T. Ramacci, G. S. Alema, and L. Angelucci, Acetyl-l-carnitine treatment increases nerve growth factor levels and choline acetyltransferase activity in the central nervous system of aged rats. *Exp Gerontol*, 29(1), 55–66. 1994.

Terman, A., and U. T. Brunk, Oxidative stress, accumulation of biological "garbage," and aging. *Antioxid Redox Signal*, 8(1–2), 197–204. 2006.

Varadarajan, S., J. Kanski, M. Aksenova, C. Lauderback, and D. A. Butterfield, Different mechanisms of oxidative stress and neurotoxicity for Alzheimer's A beta(1–42) and A beta(25–35). *J Am Chem Soc*, 123(24), 5625–31. 2001.

Virmani, A., and Z. Binienda, Role of carnitine esters in brain neuropathology. *Mol Aspects Med*, 25(5–6), 533–49. 2004.

Virmani, A., F. Gaetani, and Z. Binienda, Effects of metabolic modifiers such as carnitines, coenzyme Q10, and PUFAs against different forms of neurotoxic insults: Metabolic inhibitors, MPTP, and methamphetamine. *Ann N Y Acad Sci*, 1053, 183–91. 2005.

Wenk, G. L., K. McGann-Gramling, B. Hauss-Wegrzyniak, et al., Attenuation of chronic neuroinflammation by a nitric oxide-releasing derivative of the antioxidant ferulic acid. *J Neurochem*, 89(2), 484–93. 2004.

Wevers, A., B. Witter, N. Moser, et al., Classical Alzheimer features and cholinergic dysfunction: Towards a unifying hypothesis? *Acta Neurol Scand Suppl*, 176, 42–8. 2000.

Yan, J. J., J. Y. Cho, H. S. Kim, et al., Protection against beta-amyloid peptide toxicity *in vivo* with long-term administration of ferulic acid. *Br J Pharmacol*, 133(1), 89–96. 2001.

Yang, F., B. R. Zhou, P. Zhang, Y. F. Zhao, J. Chen, and Y. Liang, Binding of ferulic acid to cytochrome c enhances stability of the protein at physiological pH and inhibits cytochrome c-induced apoptosis. *Chem Biol Interact*, 170(3), 231–43. 2007.

Yu, L., Y. Zhang, R. Ma, L. Bao, J. Fang, and T. Yu, Potent protection of ferulic acid against excitotoxic effects of maternal intragastric administration of monosodium glutamate at a late stage of pregnancy on developing mouse fetal brain. *Eur Neuropsychopharmacol*, 16(3), 170–7. 2006.

Zhang, Z., T. Wei, J. Hou, G. Li, S. Yu, and W. Xin, Iron-induced oxidative damage and apoptosis in cerebellar granule cells: Attenuation by tetramethylpyrazine and ferulic acid. *Eur J Pharmacol*, 467(1–3), 41–7. 2003.

Zheng, Z., J. Y. Kim, H. Ma, J. E. Lee, and M. A. Yenari, Anti-inflammatory effects of the 70 kDa heat shock protein in experimental stroke. *J Cereb Blood Flow Metab*, 28(1), 53–63. 2008.

24 Evidence Required for Causal Inferences about Effects of Micronutrient Deficiencies during Development on Brain Health

*DHA, Choline, Iron, and Vitamin D**

Joyce C. McCann and Bruce N. Ames

Children's Hospital Oakland Research Institute
(CHORI), Oakland, California, USA

CONTENTS

---

* This chapter is largely based on our four critical reviews [1–4]. It was only possible to include a few citations to the original studies due to limited space. Readers are referred to the four reviews for complete citations and in-depth discussion.

24.1 INTRODUCTION

A large body of research suggests that an inadequate dietary supply during development of any of a number of essential micronutrients may adversely affect brain function. Some studies also suggest positive effects of multivitamin and mineral supplementation during development on cognitive function. However, this enormous body of research has, in general, not been systematically evaluated to determine whether evidence is sufficient to support causal inferences. A causal relationship between micronutrient deficiencies and suboptimal brain function would have major public health implications, as large segments of the world (including the U.S.) population, particularly the poor, are known to be undernourished for a number of micronutrients. If such a relationship exists, a major effort to address micronutrient undernutrition as an adjunct to the various programs underway to improve dietary habits, particularly of the poor, will be well justified [5–7].

This chapter focuses on four micronutrients: docosahexaenoic acid (DHA), a long-chain polyunsaturated omega-3 fatty acid that rapidly accumulates in the brain during the brain growth spurt [1], which occurs in humans from the last trimester through 2 years of age; choline, a component of membrane phospholipids in the brain and precursor of the neurotransmitter acetylcholine [4]; iron, the most plentiful trace metal micronutrient in the brain [2]; vitamin D, which modulates the expression of a number of important genes in the brain [3]. We discuss the types of scientific evidence that must come together to make causal inferences, major conclusions of our four critical reviews [1–4], and public health implications of the findings. In these evaluations, particular, but not exclusive, attention was paid to micronutrient availability during the brain growth spurt—a critical period for brain development.

24.2 SCIENTIFIC EVIDENCE REQUIRED FOR CAUSAL INFERENCES

There are many different types of scientific evidence relevant to causal relationships between micronutrient availability and cognitive function. In humans, study designs that provide the strongest stand-alone evidence of causality are randomized controlled therapeutic treatment (or preventive) trials (RCTs) in which treatment with a specific micronutrient can be shown to result in improved cognitive performance compared with treatment with placebo. But even these studies, which are few, do not prove that effects of the micronutrient in question are due specifically to effects in the brain, as discussed below. Therefore, conclusions as to causal relationships must rely on a complex of different types of evidence, from human, animal, and *in vitro* mechanistic studies, none of which can stand alone in making an overall assessment, as we have discussed [8].

24.2.1 CAUSAL CRITERIA

In conducting these evaluations, we focused on five causal criteria, slightly modified from an early formulation.

24.2.1.1 A Plausible Biological Rationale

This criterion asks if there is a biological mechanism that can explain how inadequacy (or supplementation) of a micronutrient might alter cognitive or other brain function. For example, in the case of iron, rats deprived of iron have lower numbers of receptors in their brains for the neurotransmitter dopamine, functions of which are consistent with observed effects of iron deficiency on cognitive tasks that involve motor activity. DHA is highly concentrated in synaptic neural membranes and is known to be involved in critical signal transduction mechanisms in the brain. Choline is a known precursor of the neurotransmitter acetylcholine and is an important constituent of 25% of the total lipid weight of gray matter. Finally, the active form of vitamin D (calcitriol) is synthesized in the brain and has been shown to affect the expression of a number of genes involved in important brain functions, such as memory.

24.2.1.2 A Consistent Association

Epidemiological investigations of human populations examine associations between poor cognitive performance and micronutrient status and are a major source of evidence. This type of study can be the first clue that a causal relationship exists. However, for several reasons, this evidence does not demonstrate causality. One reason is that many epidemiological studies are cross-sectional, i.e., they determine micronutrient status and the outcome at one point in time. In such cases, it is impossible to know which came first, low micronutrient status or the outcome. For example, a cross-sectional study reported that people with Alzheimer's disease have low levels of vitamin D compared with healthy elders [9]. Since blood levels of vitamin D depend greatly on how much time is spent in the sun and how many foods containing vitamin D (e.g., fatty fish) are consumed, it is not possible to determine whether low vitamin D status is simply due to lack of outdoor exercise and a poor diet among individuals with Alzheimer's disease or to a causal relationship, i.e., the association between low vitamin D and Alzheimer's disease may have absolutely nothing to do with a causal relationship. In this example, outdoor exercise and eating a poor diet are what epidemiologists call "potential confounders." Whether an epidemiological study be cross-sectional or prospective (e.g., comparing micronutrient status in normal and Alzheimer's populations before the development of the disease), there are always potential confounders that must be accounted for, and to do this can be complex and uncertain. Even if an association remains after potential confounders have been adequately adjusted for, the result remains an association, and without additional evidence, it does not alone demonstrate causality.

24.2.1.3 Specificities of Cause and Effect

Confounders can generally be classified as violations of one of these two criteria, which require that the experiment allow one to specifically point to the micronutrient as the cause and to cognitive dysfunction as the effect. Examples of the violation of the specificity of cause criterion are the many observational studies that have compared the cognitive performance of infants or children who were either breast- or formula-fed. In almost all of these studies, breast-fed offspring perform better on a range of cognitive tests than formula-fed infants. Since breast milk, but not infant formula (until recently), contains DHA, this led to speculation that DHA in breast milk may have been responsible. However, breast milk contains many substances not present in formula, and other factors, such as maternal nurturing or home environment, could also be responsible for the effect. Thus, while the breast-feeding experiments are of great interest, they do not demonstrate that DHA is responsible for the observed effect.

Examples of violation of the specificity-of-effect criterion are the many studies comparing the cognitive performance of children with iron-deficiency anemia (ID+A) before and after treatment with iron. Most studies that examined children older than 2 years found that iron treatment improved cognitive performance. However, since a major characteristic of ID+A is fatigue, it cannot be ruled out that anemic children performed poorly because of chronic fatigue and that performance improved when iron treatment restored normal energy levels. Thus, it is difficult to conclude that improved performance was due specifically to effects of iron on the brain.

24.2.1.4 A Dose-Response Relationship

This criterion requires that the intensity of effect (e.g., cognitive performance or a biochemical measure of brain function) depend on the degree of micronutrient deficiency or supplementation. This is an important criterion, particularly in evaluating the potential significance for humans of experiments in animals, which are often conducted under extreme conditions. For example, one of the major functions of the active form of vitamin D, calcitriol, is to modulate gene expression. (An *in silico* study showed that some 900 human genes contain the vitamin D response element sequence, suggesting extraordinarily broad functionality of calcitriol [10].) In rodents, direct evidence of effects of calcitriol treatment on the expression of some 100 genes has been obtained.

However, these experiments were conducted using extremely high doses of calcitriol, which, in some cases, were injected directly into the brain. One study also reported effects on gene expression in offspring when dams were deprived of vitamin D during gestation [11], but these conditions were also extreme, as vitamin D was entirely removed from maternal diets and exposure to UV light was prevented beginning 6 weeks before mating. Without some additional evidence of effects at less extreme conditions, it is difficult to extrapolate these results to the actual variations in vitamin D status experienced by humans.

24.2.1.5   Ability to Experimentally Manipulate the Effect

This criterion was used to ask whether effects of micronutrient deficiency (or excess) on cognitive or other brain functions were reversible. For example, rat offspring whose dams were starved for α-linolenic acid (the essential fatty acid precursor of DHA) during gestation, and who were also diet-restricted for α-linolenic acid after birth, do not perform well on cognitive tests. However, if DHA is added back to the diet after birth, performance is normal. This ability to restore function with DHA lends credence to the interpretation that the effect of α-linolenic acid deficiency was due to inability of the dams and offspring to synthesize sufficient amounts of DHA.

Evidence of reversibility is also seen in rats after severe dietary restriction for iron, provided the period of iron restriction occurs after weaning, but not before. This result parallels iron treatment studies in anemic children. Poor cognitive function in anemic children older than 2 years is reversible in most cases after iron treatment, but it is not reversible in most cases in anemic children younger than 2 years. Without the additional evidence from the rodent studies, it would not be possible to decide whether effects of iron deficiency in very young children are so severe they are irreversible or whether the persistent poor performance of the young children examined was due to something other than iron deficiency. The parallel results obtained in rodent studies, where it is known with certainty that the only difference between control and treatment groups is iron availability, provide some additional supportive evidence for the interpretation that severe ID+A during the brain growth spurt may have irreversibly damaging effects on the brain.

24.3   THE FOUR MICRONUTRIENTS

This section briefly summarizes several major conclusions of our analyses of the four micronutrients [1–4,12,13].

24.3.1   DHA

DHA is a long-chain polyunsaturated omega-3 fatty acid (LCPUFA) that rapidly accumulates in the brain during the brain growth spurt and is especially concentrated in neuronal membranes. Although precise mechanisms are not defined, DHA appears to facilitate signal transduction in synaptic membranes and to be neuroprotective. It also binds to transcription factors, modifying the expression of critical genes involved in synaptic plasticity. DHA is mostly supplied to the developing brain through the placenta and breast milk. For some 20 years, a controversial question was whether brain development was adversely affected in infants who were bottle-fed using formula unsupplemented with DHA.*

Many studies have compared the cognitive/behavioral performance of children who were breast- or bottle-fed with either unsupplemented or DHA-supplemented formulas. Other studies in animals (mostly rodents) have been conducted under more severe conditions, achieved by restricting dams and offspring of the essential fatty acid precursor of DHA, α-linolenic acid. These three types of comparisons of

---

* LCPUFA-supplemented formulas always contain arachidonic acid as well as DHA because it is important to maintain the ratio of these two LCPUFAs.

cognitive performance (breast-fed versus unsupplemented formula-fed children, DHA-supplemented formula-fed versus unsupplemented formula-fed children, and offspring of dams depleted of α-linolenic acid versus controls) result in different degrees of depletion of brain DHA. For example, human autopsy studies indicate that breast-fed infants have 11–40% higher concentrations of DHA in their brains compared with infants bottle-fed with unsupplemented formula. In rodents, more severe dietary restriction protocols can be applied that can result in ≈85% lower brain concentrations of DHA in newborn pups [14], and autopsies of primate brains suggest that concentrations of brain DHA in infants bottle-fed with DHA-supplemented formula are also somewhat greater than when DHA is omitted [15].

Results of the three types of studies parallel the degree of depletion of brain DHA. In humans, almost all breast-feeding studies indicate that children who are breast-fed consistently score higher on a variety of performance tests compared with unsupplemented formula-fed children. Rodent studies also consistently reported that offspring of α-linolenic acid-restricted dams perform less well than controls. In contrast, RCTs that compared the performance of children fed DHA-supplemented or unsupplemented formulas did not consistently detect significant differences.

Taking into account the different degrees to which brain DHA is depleted in these experimental designs and the limited sensitivity of cognitive/behavioral tests, we suggested that the apparently conflicting results may not be inconsistent with what might be expected. Thus, weak or negative results in RCTs could reflect either brain plasticity (i.e., no adverse consequences of small reductions in brain DHA) or difficult to detect but possibly important weak effects.

24.3.2 CHOLINE

Choline is considered to be an essential micronutrient because, although it can be synthesized in the body, the rate of synthesis is not sufficient to maintain normal liver function. Choline is a precursor of the membrane phospholipids phosphatidylcholine and sphingomyelin, the neurotransmitter and developmental growth factor acetylcholine, and the methyl donor betaine. It is also required for metabolism of triglycerides and other fats and is a precursor for signaling lipids (platelet-activating factor and sphingosylphosphocholine). In the human and rodent brain, choline-containing phosphoglycerides represent more than 25% of the total lipid weight of gray matter and over 10% of the total lipid weight of myelin. In rodents, manipulation of choline availability to the dam during gestation results in a number of effects in the fetal brain, including changes in apoptosis, cell proliferation, migration of precursor cells, and several indicators of neuronal differentiation (for a recent review, see Sanders and Zeisel [16]).

All studies (34 at the time of our review) that examined effects of choline availability during development on cognitive performance and other brain functions in offspring have been in rodents, primarily rats. Experimental designs of these studies are quite similar. In most cases, the diet of dams was manipulated, often for the 6- to 7-day period in gestation during which cholinergic neurons are developing. Although some dietary deficiency studies were carried out, we concluded that the most consistent results were obtained by increasing the dietary supply of choline by approximately 2.5 times.

The most striking result consistently observed was enhanced cognitive performance of offspring, particularly in more complex tasks. Two of a number of replicated examples cited in our review [4] include enhanced performance in radial maze tests only when arm numbers were greater than eight and in interval timing tests when animals are required to estimate longer intervals.

These results are strengthened by evidence that choline supplementation during gestation also significantly modifies electrophysiological, biochemical, and morphological end points in the brain. We cite a number of results in our review [4] that were independently replicated and also studies that examined cognitive/behavioral performance and biochemical or other brain effects in the same animal groups, some reporting changes in both outcome measures. Replicated effects include changes in the electrophysiological responsiveness of cells in hippocampal slices and increased neuron diameter in offspring of supplemented dams.

24.3.3 IRON

The concentration of iron in the brain is far higher than all other trace metals except zinc. Iron is required by enzymes involved in specific brain functions including myelination and for the synthesis of the neurotransmitters serotonin and dopamine. Accumulation of iron by the human fetus begins early in pregnancy, increases dramatically in the third trimester, and continues after birth up to 30 to 50 years of age. Studies in rats indicate a similar pattern of accumulation of iron in the fetal and postnatal brain. Once in the brain, iron is sequestered, with very low turnover, in contrast to the rapid turnover of iron in plasma.

ID+A is a clinical condition easily detected by decreased blood levels of hemoglobin. Less severe degrees of iron deficiency (iron-deficiency without anemia [ID−A]) may also result in adverse effects but are less easily detectable.

Because of the widespread problem of ID+A in many parts of the world, a large number of observational studies and RCTs have been conducted in these populations to attempt to discern effects of iron deficiency on cognitive performance. Most animal studies are in rodents, and these have also emphasized ID+A.

In both humans and rodents, the consequences of ID+A for the brain appear to be most severe if iron deficiency occurs during the brain growth spurt. A consistent observation in human trials involving children is the clear association between ID+A and poor performance on cognitive/behavioral tests, regardless of age. A critical question motivating a great deal of research in the field is whether and how long effects of ID+A on cognitive/behavioral function persist after children are no longer iron-deficient. In general, poorer test performance of ID+A children tends to improve with iron treatment in children after the brain growth spurt (i.e., >2 years of age), but is more resistant to improvement in younger children. The resistance to improvement with iron treatment of impaired cognitive performance in children who were ID+A at an early age is supported by several long-term follow-up studies. As discussed in section 24.2.1.5, similar differences in ability to recover with iron treatment depending upon age are observed in rodents.

Evidence suggests the possible presence of cognitive or behavioral deficits in children who are ID−A. Given the significant public health consequences (see section 24.4) if there are effects of ID−A on cognition or behavior, it is surprising that so few studies have focused on this topic. We identified some 17 investigations in children or adolescents, most of which were conducted over 15 years ago and virtually no animal studies [2]. Most human studies were iron treatment trials that measured cognitive/behavioral performance before and after periods of iron treatment ranging from 1 week to 6 months; effects overall were very weak and were substantially negative. An important recent report in 9- to 12-month-old ID−A African American infants that focused on tests targeted at noncognitive social/emotional behavior suggests significant adverse effects of ID−A [17]. Almost all of the eight studies we examined involving children older than 2 years or adolescents reported weak but significant effects of ID−A.

24.3.4 VITAMIN D

While vitamin D is present in some foods (e.g., fatty fish, fortified milk), the major source for most people is the action of UV light from the sun on 7-dehydrocholesterol in the skin. The active form of vitamin D, calcitriol, is then formed after several enzymatic steps. An intermediate stable form of vitamin D is 25-hydroxycholecalciferol, which is measured in serum to indicate vitamin D status. The final activation step of calcitriol takes place in the kidney and in many other tissues throughout the body, including the brain. Calcitriol binds to the vitamin D receptor (VDR), which in turn binds to DNA to modulate expression of many genes in the body. Calcitriol also has nongenomic functions effected through binding to cell membrane receptors.

The classical hormonal function of calcitriol is to ensure healthy bones by controlling blood levels of calcium through regulating the expression of genes involved in its intestinal absorption, renal excretion, and movement in and out of bone. More recently, many other so-called noncalcemic functions of calcitriol have been identified, which include regulation of proliferative and apoptotic activity, immunomodulatory and prodifferentiation activity, and interaction with the renin-angiotensin system (involved in the regulation of blood pressure), insulin secretion, and neuroprotective functions.

Research into the possible functions of calcitriol in the brain is recent, and the relevant evidence base is not extensive. The experiments we analyzed employed three basic experimental designs. First, the supply of vitamin D or calcitriol was manipulated *in vitro* or *in vivo* by restricting exposure to UV light and limiting dietary intake, or through increasing supply by supplementing the cell culture medium or treating subjects orally or by injection. Second, the functionality of calcitriol was restricted by use of a mouse knock-out strain lacking a functional VDR. Third, experimental or population groups were examined for associations between blood concentrations of 25-hydroxy-cholecalciferol and behavioral or biochemical end points.

Our analysis indicated two apparently paradoxical findings. On the one hand, there was ample biological evidence to suggest an important role for vitamin D in brain development and function. For example, VDR and 1-α-hydroxylase (the terminal calcitriol-activating enzyme) are distributed throughout both the fetal and adult brain, and various calcitriol target genes have been identified in the brain that include genes involved in critical brain functions such as synaptogenesis (formation of synaptic connections), memory formation, and neurotransmission.

On the other hand, direct effects of severe vitamin D inadequacy on cognition/behavior appeared to be subtle. For example, the relatively few studies that examined linkages between vitamin D status and cognitive/behavioral function in humans, which for the most part were concerned with possible connections of vitamin D deficiency during development to schizophrenia, or in adults, to seasonal affective disorder (SAD), are negative, very weak, or, in the case of SAD, inconsistent. Results in rodents, while suggestive, are quite subtle and, as we discuss in our review [3], somewhat conflicting. This is surprising since both experimental designs used in the rodent studies should have resulted in severe loss of vitamin D functionality during development (i.e., knock-out mice lacked a functional VDR, and diet-restricted animals were shown to be severely deficient).

24.4 IMPLICATIONS FOR PUBLIC HEALTH

Determining whether micronutrient deficiencies in humans are causally related to deficits in cognitive/behavioral or other brain function has two major components. The first component is a determination of whether any biological evidence for causal relationships exists by analysis of all relevant evidence from human, animal, and *in vitro* systems. Since critical experiments may involve nonhuman systems, and most animal and mechanistic studies are conducted under relatively severe conditions compared with actual conditions experienced by humans, the second component attempts to understand to what degree causal relationships might pertain in human populations under actual exposure situations. Our published analyses primarily focused on the first component, although we briefly alluded to implications of results of those analyses for public health. Below, several key issues relevant to public health for each of the four micronutrients are briefly discussed.

24.4.1 DOCOSAHEXAENOIC ACID

There is no officially recommended intake level for DHA. It is not technically considered to be an essential micronutrient because it can be synthesized by the body from the essential fatty acid α-linolenic acid. However, some have suggested it should be considered conditionally essential because of exceptional needs during the brain growth spurt. We have argued that, despite failure of RCTs to consistently observe differences in cognitive performance of children bottle-fed with or without DHA, it is only prudent to make sure formula-fed babies receive as much DHA as breast-fed babies [12,13]: (a) DHA-supplemented infant formula does not adversely affect growth; (b) infants who consume unsupplemented formula are very likely to have less DHA in their brains than infants who consume supplemented formula; (c) DHA plays an important role in brain function. Supplementing formula with DHA is not only common sense, it is also consistent with the available scientific evidence. Now, several years after the appearance of our review, DHA can be found in almost all infant formulas.

24.4.2 CHOLINE

It is not obvious how to extrapolate the very interesting findings in rodents that increasing the dietary supply of choline during gestation by only about 2.5 times significantly improves cognitive performance in offspring. In the absence of any human studies, it is not clear what these results imply about public health. Since current recommended intakes for choline are not based on functional considerations but solely on the mean intake of breast-fed infants, there is no evidence to indicate whether these current intakes are optimal for brain function. Human studies are clearly needed.

24.4.3 IRON

Iron is unique among the four micronutrients because there is direct evidence in humans that iron deficiency adversely impacts cognitive/behavioral performance. Furthermore, as discussed, this evidence is supported and clarified by animal and mechanistic studies. Recommended intakes for iron are based on levels required to maintain minimal iron stores in healthy individuals. Iron-deficiency prevalence estimates indicate widespread intakes below these recommended levels in the developing world and also to some extent in the West. Some 2 billion women and children are iron-deficient worldwide. The greatest prevalence of ID±A in the United States is among adolescent girls (9–16%) and children during the brain growth spurt (7%). Based on the available evidence, these levels of deficiency could be having a significant impact on brain health.

24.4.4 VITAMIN D

Although there is no convincing evidence that low vitamin D status in humans adversely affects cognitive/behavioral performance, this possibility cannot be excluded because of the limited number of studies, together with the biological evidence that vitamin D is involved in critical brain functions. Currently, recommended intakes for vitamin D assume no exposure to sunlight, attempt to ensure sufficient dietary intake to prevent rickets, and incorporate an additional safety factor. Until demonstrated otherwise, it is prudent to assume these intake levels are sufficient to ensure optimal brain function. Hence, it could be of concern that there is widespread vitamin D deficiency in the United States, particularly among nursing infants, the elderly, and African Americans who are two to eight times more likely to be insufficient compared with age-matched whites. Supplementation to ensure adequacy in these populations is prudent.

24.5 NEW DIRECTIONS

Much of the population is inadequate in one or more micronutrients. One of us (B.N.A.) recently proposed a mechanism (the "triage theory") to explain how moderate chronic inadequacy of a micronutrient could lead to insidious biochemical changes that, over time, result in increased risk of chronic diseases of aging (including cognitive dysfunction) [6]. The same principle may apply during development, as it is known that dietary effects during gestation that do not interfere with a normal birth can affect susceptibility to some chronic diseases later in life (e.g., diabetes [18]). What is clearly needed is more research to determine micronutrient levels required during development for optimal intelligence in humans.

24.6 SUMMARY

In this chapter, based on our four previously published critical reviews, scientific evidence linking deficiencies during brain development to later deficits in cognitive or behavioral function is briefly summarized for each of four micronutrients: DHA, choline, iron, and vitamin D. Emphasis is placed on the degree to which the scientific evidence bases available for each micronutrient satisfy five criteria for making causal inferences. Possible public health consequences of deficiency in any of these four micronutrients during development are also briefly addressed. Given that much of the population is inadequate in one or more of these micronutrients, we conclude that ensuring adequate intake for optimal brain development should be a major public health goal.

REFERENCES

1. McCann, J. C., and Ames, B. N., Is docosahexaenoic acid, an n-3 long chain polyunsaturated fatty acid, required for the development of normal brain function? An overview of evidence from cognitive and behavioral tests in humans and animals. *Am J Clin Nutr* 82, 281, 2005.
2. McCann, J. C., and Ames, B. N., An overview of evidence for a causal relation between iron deficiency during development and deficits in cognitive or behavioral function. *Am J Clin Nutr* 85, 931, 2007.
3. McCann, J. C., and Ames, B. N., Is there convincing biological or behavioral evidence linking vitamin D deficiency to brain dysfunction? *FASEB J* 22, 982, 2008.
4. McCann, J. C., Hudes, M., and Ames, B. N., An overview of evidence for a causal relationship between dietary availability of choline during development and cognitive function in offspring. *Neurosci Biobehav Rev* 30, 696, 2006.
5. Ames, B. N., Increasing longevity by tuning-up metabolism. *EMBO Reports* 6, S20, 2005.
6. Ames, B. N., Low micronutrient intake may accelerate the degenerative diseases of aging through allocation of scarce micronutrients by triage. *PNAS* 103, 17589, 2006.
7. Ames, B. N., Atamna, H., and Killilea, D. W., Mineral and vitamin deficiencies can accelerate the mitochondrial decay of aging. *Mol Aspects Med* 26, 363, 2005.
8. Ames, B. N., et al., Evidence-based decision making on micronutrients and chronic disease: Long-term randomized controlled trials are not enough. *Am J Clin Nutr* 86, 522, 2007.
9. Wilkins, C. H., et al., Vitamin D deficiency is associated with low mood and worse cognitive performance in older adults. *Am J Geriatr Psychiatry* 14, 1032, 2006.
10. Wang, T. T., et al., Large-scale in silico and microarray-based identification of direct 1,25-dihydroxyvitamin D3 target genes. *Mol Endocrinol* 19, 2685, 2005.
11. Eyles, D., et al., Developmental vitamin D deficiency alters the expression of genes encoding mitochondrial, cytoskeletal and synaptic proteins in the adult rat brain. *J Steroid Biochem Mol Biol* 2007.
12. McCann, J., and Ames, B. N., DHA and cognitive development: An update on the science. *Pediatric Basics* 117, 17, 2007.
13. McCann, J. C., and Ames, B. N., Reply to P. Wainwright (Scientific precision and deciding whether to add LCPUFAS to infant formula). *Am J Clin Nutr* 83, 920, 2006a.
14. Greiner, R. S., et al., Docosapentaenoic acid does not completely replace DHA in n-3 FA-deficient rats during early development. *Lipids* 38, 431, 2003.
15. Sarkadi-Nagy, E., et al., The influence of prematurity and long chain polyunsaturate supplementation in 4-week adjusted age baboon neonate brain and related tissues. *Pediatr Res* 54, 244, 2003.
16. Sanders, L. M., and Zeisel, S. H., Choline: Dietary requirements and role in brain development. *Nutr Today* 42, 181, 2007.
17. Lozoff, B., et al., Dose-response relationships between iron deficiency with or without anemia and infant social-emotional behavior. *J Pediatr* 152, 696, 2008.
18. Patel, M. S., Srinivasan, M., and Laychock, S. G., Nutrient-induced maternal hyperinsulinemia and metabolic programming in the progeny. *Nestle Nutr Workshop Ser Pediatr Program* 55, 137, 2005.

25 Omega-3 Fatty Acids and Brain Function in Older People

Ricardo Uauy[1,2] and Alan D. Dangour[1]

[1]Nutrition and Public Health Intervention Research Unit, London School of Hygiene and Tropical Medicine, London, UK

[2]Institute of Nutrition and Food Technology (INTA), University of Chile, Santiago, Chile

CONTENTS

25.1 NUTRITION AND HEALTH OF OLDER PEOPLE

The progressive gain in mean human life expectancy is possibly one of the great biomedical achievements of modern times. The United Nations estimates that by the year 2050, individuals aged 60 years or older will represent 22% of the world's population, or about 2 billion people (up from 10% or 600 million people in 2000). The growth in the population of the proportion of the "oldest old" is of particular significance: the number of individuals aged 80 years or older is projected to more than triple in the period 2000–2050, from 73 million people (1.2%) to 400 million people (4.3%).[1] Declining fertility together with increased survival of infants and young children translates, for virtually all countries, to a rapid growth in the absolute number and in most cases also in the proportion of people reaching or surpassing 65 years of age. However, the promise of extending the upper limits of human life span remains largely unfulfilled; the fountain of eternal youth, a human quest since ancient times, is more an inspiration for fables than an evidence-based proposition.

While increased longevity is a triumph, the global burden of disease and disability associated with old age is also likely to remain one of the great challenges for human health and social well-being in the foreseeable future.[2] Most countries are undergoing a rapid epidemiological and nutrition transition characterized by the relative preponderance of disability and death caused by noncommunicable chronic diseases; these often result from changes in diet and physical activity patterns induced by economic affluence and other factors related to modern life. Increased life expectancy is creating entirely new challenges for society in all spheres. In the biomedical arena, an obvious outcome of this trend is the growing numbers of people presenting with age-related frailty, mental and physical disability, and multiple chronic diseases that compromise the quality of life. Consequently, there is a profound need to understand the mechanisms of ageing in order to define effective intervention strategies to revert or at least prevent specific age-related functional declines. The potential contributions to this process of complementary factors such as nutrition, lifestyle, socioeconomic factors, genetics, environment, human behaviors, bioengineering, and information technology may be of major significance.

In the case of nutrition, multiple essential nutrients and some nonessential nutrients are known to affect both functional development and ageing of the brain. The mature adult brain has traditionally been considered less vulnerable to the effects of diet, although extreme famine conditions have been demonstrated to induce functional alterations that are mostly reversible upon nutritional restoration. This is not the case during early development where malnutrition can result in abnormal brain structures and function, which in some cases have lasting or permanent effects. Folate and several other vitamins (especially vitamins A, B6, B12, C, and E) are known to affect brain development; other essential micronutrients such as iodine, iron, and zinc also play important roles. More recently, choline and omega-3 long-chain essential fatty acids derived from linolenic acid have been shown to be conditionally essential during early development and may also play a role in the prevention of functional decline during aging. Essential nutrients can interact with nonessential nutrients and toxicants that will contribute to defining normal and abnormal rates of functional decline with advancing age. Accumulating evidence suggest that specific genetic polymorphism affecting nutrient metabolism, transport, or tissue concentration is important in modulating these effects and may also modify an individual's responses to nutritional supplementation.

Figure 25.1 provides a schematic representation of the possible role of nutrient sufficiency in early brain development and in modulating the subsequent rate of functional decline. The timing of any nutritional deprivation in early development defines its impact, so that nutritional deficiencies during the so-called sensitive periods may have more permanent effects in terms of what is achieved during development. Once maturity and peak function are reached between the ages of 20 and 40 years, individuals or populations may present different rates of functional decline based on environmental conditions. In general, individuals who reach higher levels of function can decline for longer before their functions become compromised. Nutrition, toxic agents, or specific environmental conditions interact to prevent or enhance the loss of function and contribute in defining the age-related

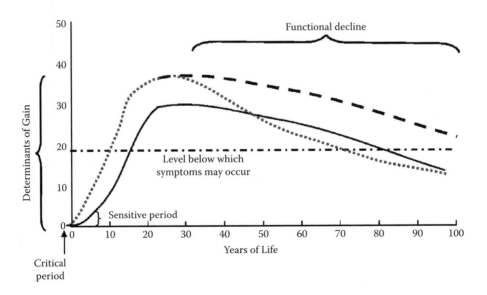

FIGURE 25.1 Functional gain and loss during brain development and aging are illustrated by solid line, potential positive impact of nutrition is illustrated by dotted line. Critical period starts from conception (before birth), sensitive period corresponds to time where nutrition has significant potential impact (first 3–6 years of life depending on nutrient). Once peak function has been attained decline may proceed with advancing age at a faster or slower rate than normal as illustrated by the dotted and dashed lines, respectively.

decline that becomes evident in later life. In this paper, we will review the current evidence linking nutrition and specifically omega-3 fatty acids (now more commonly known as n-3 fatty acids) with measures of brain function in older people. Given space limitations, our review is nonsystematic, but reference is made, where possible, to current relevant systematic reviews.

25.2 GLOBAL TRENDS IN COGNITIVE DECLINE AND DEMENTIA: CAN WE PREVENT OR DELAY THE ONSET OF DEMENTIA?

To serve the needs of older people, we must understand their changing functional status and disease risks, which in turn define their demand for health and care services affecting their inclusion within society. Several hundred studies over the past decade have reported on the prevalence of cognitive decline as well as Alzheimer's disease and other forms of dementia. As more people survive to older ages, there will be a continued rise in degenerative diseases related to ageing.

A landmark article published recently in the *Lancet* provides coherent data from a cross section of 15,000 people 65 years or older, studied at 11 sites in 7 low- and middle-income countries.[3] The collaborators at all sites used a prevalidated, well-standardized algorithm for the diagnosis of dementia (10/66 dementia) and also applied a computerized version of the dementia criterion from the *Diagnostic and Statistical Manual of Mental Disorders, Fourth Edition* (*DSM-IV*). The prevalence of dementia based on *DSM-IV* varied widely, from 0.3% in rural India to 6.3% in Cuba after adjusting for age and sex. *DSM-IV* prevalence in urban Latin American sites was 20% lower than that in Europe, while in China, the prevalence was only half, and in India and rural Latin America, a quarter or less of the European prevalence. The 10/66 dementia prevalence estimate was higher than that of *DSM-IV* criteria but much more consistent across sites, varying between 5.6% in rural China and 11.7% in the Dominican Republic.

Lower age-specific incidence rates of Alzheimer's disease and other dementias had previously been reported in developing compared with industrialized countries. Alzheimer's Disease International (ADI) recently commissioned an international group of experts to review all available data and reach a consensus on dementia prevalence in 14 WHO geographic regions.[4] The results published in 2005 suggested that 24.2 million people live with dementia worldwide, with 4.6 million new cases every year. A trend toward a lower prevalence in less-developed regions than in developed settings was noted, at least for sub-Saharan Africa and south Asia. Nonetheless, given global population distributions, most people with dementia live in low- and middle-income countries: 60% of the global total in 2001, predicted to rise to 71% by 2040. The total number of individuals affected is predicted to double every 20 years with a greater increase forecast in developing than in developed regions, reaching a global figure of more than 80 million individuals by 2040. Recent estimates of the burden of disability from neurodegenerative diseases are presented in Table 25.1.[5]

The risk factors for cognitive decline and dementia have been evaluated in several large epidemiological studies over the past few decades, and many have highlighted the important role of

TABLE 25.1
Global Annual Disability Adjusted Life Years Lost to Neurodegenerative Diseases

Condition	All	Males	Females
Alzheimer's disease and other dementias	17,108	6092	11,016
Epilepsy	6223	3301	2922
Parkinson's disease	2325	1124	1202
Cerebrovascular disease	72,024	35,482	36,542

Source: Mathers et al.[5]

nutrition. For example, a recent Finnish cohort study has shown that maintenance of healthy body weight, low cholesterol, low blood pressure, and regular physical activity were associated with significantly lower risk of incident dementia.[6] There are many other classical risk factors for dementia and Alzheimer's disease, which include increasing age, positive family history of dementia, female sex, lower level of education, medical conditions (vascular diseases), apolipoprotein E4 genetic polymorphism (ApoE4), and exposure to toxicants (organic solvents and aluminum). There are varying degrees of evidence for these risk factors; some may be only associations while others may be causally related.

Dementia most likely results from the interaction of an underlying genetic predisposition with environmental factors. For example, the ApoE4 polymorphism has been consistently shown to confer increased risk of developing dementia. ApoE4 interacts with lifestyle and health factors such as smoking, diet, atherosclerosis, and diabetes to increase dementia risk. There are also numerous other candidate genes under investigation. In the future, screening for genetic susceptibility may help identify those individuals that should be targeted for preventive strategies such as avoidance of exposure to specific environmental factors, including specific food components. This approach may lead to targeted interventions in the form of specific drugs, the need for additional specific essential nutrients, and the avoidance of toxicants or other dietary factors. Epidemiological research may help identify additional environmental factors, nutrients, or toxicants that contribute to dementia.

Much of the available evidence supports a major contribution of vascular risk factors and lifestyle-related risks in the development of dementias. The Rotterdam cognitive decline study prospectively examined the association between silent brain infarcts and the risk of dementia and cognitive decline in 1015 participants, who were 60 to 90 years of age and free of dementia and stroke at baseline.[7] Participants underwent neuropsychological testing and cerebral MRI in 1995–1996 and again in 1999–2000 and were monitored for dementia throughout the study period. The study demonstrated that older people who experienced silent brain infarcts during the follow-up period had an increased risk of dementia and more rapid decline in cognitive function, while those with no detectable microinfarcts over the study period showed no loss in cognitive function or in speed of processing. Vascular disease appears to have an additive effect with Alzheimer's disease pathology.[8] People with Alzheimer's disease pathology are more likely to show cognitive symptoms and decline faster if they also have vascular-related brain damage. Modifiable risk factors for vascular disease include high blood pressure, cholesterol and homocysteine plasma concentrations, diabetes, smoking, and obesity.

Population-based strategies to reduce heart disease (general screening, dietary interventions, education, pharmacotherapy) have resulted in reduced mortality and cardiovascular risk and have been demonstrated to be cost-effective. However, most of these studies have been conducted in adults younger than 65 years and few have examined effects on cognition. Nonetheless, such strategies, if adopted, may well contribute in delaying the onset of cognitive decline and dementia.

Epidemiological studies assessing the role of specific dietary factors in the etiology of dementia have been previously reviewed by Luchsinger and Mayeux[9] and more recently by the International Academy of Nutrition and Ageing [10] among others. Overall, results of many of the studies have been conflicting, but some factors appear relatively more frequently.

A large amount of epidemiological data suggests that there may be a protective role of the B vitamins on cognitive decline and dementia. Results from small nonrandomized trials in people with low concentrations of vitamin B12 suggest that supplementation improves cognitive performance. One large trial of folate supplementation also reported strong positive results.[11] However, care must be taken in the generalizability of findings since many trials have been conducted in individuals with low nutrient status, and supplements may not be similarly efficacious in individuals with plasma concentrations of these vitamins in the normal range; defining normality based only on biochemical measurements is also problematic since the relationship with functional indicators weakens as plasma values approach the normal range. We are presently conducting a randomized controlled trial—the OPEN study—with a 12-month intervention to assess whether older subjects who have no specific signs or symptoms of deficiency but have low concentrations of vitamin B12 will benefit from supplementation in terms of cognitive capacity, transcranial stimulated motor-evoked responses, sensory processing peripheral neuropathy, and other electrophysiological outcomes beyond the prevention of pernicious anemia. For now, however, there are no data from randomized trials to support the use of B vitamins for the primary prevention of cognitive decline.[12,13]

A possible role for antioxidant nutrients in protection from dementia also emerges, albeit weakly, from the literature, although some concerns exist about the potential adverse effects of intakes of large doses of some of these nutrients.[14] Data from several cohort studies suggest that consumption of fruits and vegetables, which are frequently rich in antioxidant nutrients, may be associated with slower rate of cognitive decline. There is even some suggestion that this may have particular relevance in ApoE4 carriers.[15] However, again the evidence from randomized controlled trials is less convincing.[16]

A final theme emerging from the studies is that there does appear to be evidence that protection against loss of cognitive function and dementia is provided by consumption of long-chain n-3 polyunsaturated fatty acids, most commonly found in fish. The evidence in support of this theme will be discussed in the next section of this chapter.

In conclusion, while there is no lack of hypotheses and even some supportive data linking diet to cognitive function and prevention of dementia in older age, there is still much work to be done. Specifically, the IANA taskforce stated a need for large prospective studies of adequate duration, including subjects whose diet is monitored at a sufficiently early stage or at least before the onset of disease or cognitive decline, and larger randomized controlled trials among older people.[10] Given the wide range of methods for the assessment of cognitive function and availability of various end points for clinical trials in dementia research, greater standardization of trial methodologies, and of measurement of relevant outcomes, is also urgently required.[17]

25.3 BIOLOGICAL BASIS FOR THE POTENTIAL USE OF N-3 FATTY ACIDS IN THE PREVENTION OF COGNITIVE DECLINE

For nearly 80 years, researchers have known that specific components of fat are essential for the proper growth and development of animals and humans,[18] and the concept of the essentiality of specific polyunsaturated fatty acids is now well defined. Dietary n-3 polyunsaturated fatty acids were found to revert subtle clinical signs and symptoms such as dry and scaly skin, abnormal retinal and visual function, and peripheral neuropathy in humans. These findings underpinned the concept that n-3 fatty acids were essential for normal retinal and CNS development and function. The relatively high concentrations of n-3 long-chain polyunsaturated fatty acids (n-3 LCPs), specifically docosahexaenoic acid (DHA), in the retina and cerebral cortex further supported this view.[19] On a dry weight basis, the human brain is predominantly lipid, with 22% of the cerebral cortex and 24% of white matter consisting of phospholipids; these lipid moieties are selectively enriched with DHA. The fact that brain phospholipid fatty acid composition during development and possibly during aging is modifiable by dietary fatty acid content is of great importance and provides an important clue to the potential role of polyunsaturated fatty acids in brain development and function.

LCP concentrations in brain tissues appear to decrease with age,[20] and it has been proposed that these changes in composition are associated with changes in central nervous system function. While the underlying cause of these changes in n-3 LCP concentrations in the ageing brain is largely unknown, it is likely that dietary intake plays an important role throughout life in determining the lipid composition of brain tissues. For example, consumption of large amounts of n-6 fats can reduce synthesis of n-3 LCPs, while consumption of diets containing 20- or 22-carbon polyunsaturated fatty acids might increase n-3 LCP brain concentrations. There are data from animal studies to support this proposal.[21]

Essential fatty acids are presently known to play crucial structural roles in brain tissue, especially in the structure and function of cell membranes, and much research has been conducted into functional implications associated with diet-induced compositional changes. Furthermore, the oxidative products of polyunsaturated fatty acids, such as eicosanoids and docosanoids, act as key cellular mediators of inflammation, allergy and immunity, oxidative stress, bronchial constriction, vascular responses, and thrombosis.[22] There is also a growing body of evidence to show that LCPs can affect the expression of genes that regulate cell differentiation and growth and may thereby have a profound and long lasting impact on human health. Indeed, via this mechanism, early diet may influence the structural development of organs, as well as the establishment of neural pathways necessary for sensory functions and central processing. In contrast to their proposed actions in childhood, where n-3 LCPs are required for healthy development of brain tissue, in older people, n-3 LCPs are more likely to act in a protective and health-maintaining manner. For example, n-3 LCPs are known to inhibit hepatic triglyceride synthesis and modify eicosanoid function, promoting vascular relaxation, diminished inflammatory responses, and decreased platelet aggregation.

Recently, additional protective actions of DHA have been discovered that may well be directly related to its effects in maintaining cognitive health in older age. In a mouse model of ischemic stroke, a bioactive docosanoid derived from DHA was found to inhibit two of the major causes of poststroke neuronal injury, lipid peroxidation and inflammatory response tracked by leukocyte infiltration.[23] This novel oxidative product of DHA (a docosanoid: 10-17S-docosatriene) named neuroprotectin D1[24] appears to potently down-regulate proinflammatory gene expression in cultured neuronal cells. DHA is also known to protect *in vitro*-cultured hippocampal neurons grown in a serum free media from apoptosis (programmed cell death),[25] potentially by the effects of neuroprotectin D1 down-regulating proapoptotic and up-regulating antiapoptotic genes, as well as protecting DNA in cells from oxidative damage.

Animal studies of Alzheimer's disease are of limited relevance to human disease because the pathological changes are characterized by loss of neurons and cortical tissue that occur late in the human disease and leave limited room to assess the consequences. However, a recent study examining the interaction of a low n-3 polyunsaturated fatty acid/high n-6 polyunsaturated fatty acid diet in an Alzheimer's disease mouse model (Tg2576 AD) provided the opportunity to study the disease before there is significant neuronal loss.[26] This animal model not only has similar pathological findings as those found in human Alzheimer's disease but also demonstrates early compromise in recent memory and learning processes. Tg2576 AD animals, as observed in the human Alzheimer's disease brain, had an 80%–90% loss of the p85 subunit of PI 3-kinase, accumulation of fractin (fragmented actin), and a loss of cortical drebin, a postsynaptic processing-related protein. These ultrastructural changes were associated with a loss of learning memory before there was any evidence of neuronal loss in the hippocampus or the frontal cortex in the animals. A high n-6/n-3 ratio diet with low docosahexaenoic acid (DHA, an n-3 LCP) induced an important depletion of frontal cortex DHA and a rise in the n-6 DPA (docosapentaenoic acid, characteristic of n-3 deficiency), while dietary supplementation with a high-DHA diet was able to prevent brain DHA depletion in the frontal cortex in the Alzheimer's disease mouse model and also preserve the learning capacity of the animals. Treatment of n-3 fatty acid-restricted mice with DHA protected them from the adverse consequences of the Alzheimer's disease genotype. Table 25.2 shows the interaction between the dietary interventions and the genetic type in defining brain lipid composition in these mature animals; these findings have potentially major implications for human neurodegenerative diseases, especially for Alzheimer's disease. A dietary intervention, if provided early enough, might prevent ultrastructural damage and functional losses before there is neuronal death; conversely, if progression leads to neuronal death, cognitive decline is likely to be irreversible.

Epidemiological evidence for an association between n-3 LCP consumption and cognitive function is limited but promising in terms of potential prevention of age-related cognitive decline. One of the first cohort studies on this subject was reported by Barbered-Gateau and colleagues.[27] The study traced 1416 older people who had previously provided information on habitual dietary intake

TABLE 25.2
Fatty Acid Composition of Brain (Frontal Cortex) Phospholipids of Rats According to Diets Fed and Genotype

Diet Groups (n-6/n-3 ratio)	Wild Type			Tg2576 AD Model		
	DHA[1]	DPA[2]	ARA[3]	DHA[1]	DPA[2]	ARA[3]
Control (7:1)	19.2 ± 0.3	0.52 ± 0.02	9.6 ± 0.1	19.5 ± 0.5	0.59 ± 0.03	9.5 ± 0.1
Low DHA (85:1)	18.0 ± 0.5	1.43 ± 0.12[a]	9.8 ± 0.1	16.4 ± 0.8[a,c]	1.81 ± 0.20[a]	9.9 ± 0.2
High DHA (5:1)	20.5 ± 0.5[b]	0.29 ± 0.05[b]	8.4 ± 0.2[b]	21.3 ± 0.6[b]	0.22 ± 0.02[b]	7.9 ± 0.2[b]

Values are mean + SD % of total fatty acids.

DHA, C22:6 n-3 docosahexaenoic acid; DPA, C22:5 n-6 docosapentaenoic acid; ARA, C20:4 n-6 araquidonic acid.

[a] Values significantly different from animals fed control diet ($p < .01$).

[b] Values significantly different from low DHA diet ($p < .01$).

[c] Values significantly different from wild type (normal genotype) animals ($p < .05$).

Source: Calon et al.[26]

of fish and found that those individuals who reported eating fish or seafood at least once a week were at significantly lower risk of developing dementia (Table 25.3). A recent systematic review identified only four cohort studies and one randomized control trial that investigated the effect of n-3 LCPs on cognitive health.[28] The four cohort studies all suggested a positive impact of increased n-3 LCP (fish and total n-3) consumption on risk of impaired cognitive function, dementia, or Alzheimer's disease. The one trial included in the review was small, conducted among demented older people and was of poor quality.

Since the publication of this review, the results of several more cohort studies have been reported on this topic. A 5-year follow-up of 210 surviving males in the Zutphen Elderly Study cohort who were aged 70 to 89 years at baseline demonstrated that fish consumers had significantly less 5-year cognitive decline than non-fish consumers and that this could be directly related to n-3 LCP intake. Older men who consumed approximately 400 mg n-3 LCP a day had less cognitive decline (1.1 point less on the 30-point Mini Mental State Examination [MMSE] scale) than men who consumed only 20 mg n-3 LCP a day.[29] While this analysis from the long-running Zutphen Elderly Study presented very detailed information on dietary intake, it did not present any information on biochemical markers of fatty acid status.

A recent report from the Framingham Heart study presented data on a 9-year follow-up of 899 men and women with a median age of 76 years at baseline.[30] Plasma fatty acid status was assessed at

TABLE 25.3
Incidence of Dementia in 1416 French Older People According to Reported Frequency of Fish Consumption

Fish Consumption	n	Cases	Incidence/100 Person Years Mean (95% Confidence Interval)
Once a day	19	1	1.00 (0.00–2.97)
At least once a week	1122	124	2.05 (1.69–2.41)
Sometimes	240	35	2.90 (1.94–3.87)
Never	35	10	6.61 (2.51–10.70)

Source: Barberger-Gateau et al.[27]

baseline, and the primary outcome for the analysis was the development of all-cause dementia and Alzheimer's disease. Over the follow-up period, 99 new cases of dementia (including 77 cases of Alzheimer's disease) were diagnosed, and study participants in the highest quartile of plasma phosphatidylcholine DHA concentration were 47% less likely to develop all-cause dementia and 39% less likely to develop Alzheimer's disease than participants in the lowest three quartiles. Highest quartile intakes were in the region of 200 mg DHA per day.[30]

A more recent study demonstrated a 26% lower prevalence of subclinical infarcts detected by MRI in subjects consuming tuna or other oily fish three times per week, compared with once a month.[31] Additionally, higher intakes of tuna and other oily fish were associated with MRI evidence of better status of some, but not all, markers of brain atrophy after adjustment for multiple risk factors. The authors conclude based on these observations that among older adults, modest consumption of tuna or other oily fish, but not fried fish, is associated with lower prevalence of subclinical infarcts and white matter abnormalities. Further analysis of data from individuals recruited to take part in the FACIT trial suggested that higher baseline plasma n-3 LCP content was associated with lower rate of decline in speed-related cognitive domains over 3 years.[32] However, plasma n-3 LCP content did not predict changes in memory, information-processing speed, or word fluency over the same period. These results from cohort studies suggesting a link between n-3 LCP status and cognitive health need to be confirmed by randomized controlled trials before firm recommendations can be established. A Cochrane review published in 2006 found no randomized controlled trials investigating the effect of n-3 LCP supplementation on cognitive function among healthy older adults.[33] However, since the publication of the Cochrane review, two trials have been published that are worthy of comment.

The first of these trials recruited 174 adults with mild to moderate Alzheimer's disease and a mean age of 74 years to receive either 2.3 g n-3 LCP or placebo for 6 months.[34] There was no evidence of a difference in the rate of decline in cognitive function between the intervention and control arms in the trial at the end of the 6-month intervention. However, in a subgroup of 32 participants with very mild cognitive function loss at baseline (MMSE of greater than 27 out of 30), the rate of cognitive decline in the intervention arm was significantly slower than in the placebo arm.

Very recently, the results of the first trial investigating n-3 LCP supplementation and cognitive function among healthy older people were published.[35] This was a 6-month randomized controlled trial investigating the effect of high- and low-dose n-3 LCP supplementation versus placebo on cognitive performance in a group of 302 healthy adults. At baseline, participants were aged 65 or older and had an MMSE score of greater than 21. Cognitive performance was assessed using an extensive neuropsychological test battery that included the cognitive domains of attention, sensorimotor speed, memory, and executive function. The results revealed that although plasma concentrations of n-3 LCPs increased significantly in the intervention arms over the course of the trial, there were overall no significant differential changes in any of the cognitive domains for either low- or high-dose fish oil supplementation compared with placebo.

While there is some evidence from observational studies that raised n-3 LCP intake can delay cognitive function loss, it must be confirmed by larger and more long-term randomized controlled trials. We are currently conducting a large randomized controlled trial—the OPAL study—which is investigating the effect of 0.7 g n-3 LCP supplementation for 24 months on a cohort of more than 860 healthy adults aged 70 to 79 years at baseline.[36] The results of the trial are due in late 2008.

In summary, there is a growing body of evidence that supports the hypothesis that n-3 LCPs are crucial for brain development and for the maintenance of good cognitive function in later life. This brief review, however, points to the many unanswered questions, specifically, the lack of high-quality population-based effectiveness trials of nutrition in the prevention of cognitive decline. With global population ageing continuing apace and a concomitant increase in the number of individuals experiencing poor cognitive health, we must speed up the pace of our search for cost-effective solutions.

25.4 SIGNIFICANCE OF INTERVENTIONS TO PREVENT COGNITIVE DECLINE

We are convinced that interventions to enable people to age healthily, specifically preserving cognitive function, are relevant throughout the life course and have the potential to deliver in a variety of important areas. First, successful interventions will enhance the quality of life of older people with an associated improved potential for independent living, thereby significantly reducing the need for costly investments in elderly care. Second, interventions may enhance occupational fitness of the labor force in later life with the associated highly significant additional contribution to national economic development and gross domestic product. Finally, older people with preserved cognitive function provide essential sustainability to the knowledge economy, increasing productivity of the adult workforce, and building greater personal wealth to support their lengthened retirement.

Putting a value on improvements in health and longevity permits the estimation of economic gains from the past and present, as well as the prospective future changes in longevity and mortality rates. The extension of healthy life years plus gains in productivity defines the gains in economic and quality of life that can be anticipated from effective interventions to prevent cognitive decline. The detailed analysis of how to estimate the economic gains from increased healthy life years and continued labor productivity can be found in Chapters X (McFadden) and Y (Bharghava) in this volume.

In addition, research on determinants of the ageing process provides a significant opportunity for industry to develop scientifically based products and services to address the challenges of healthy ageing for the growing number of older people globally. However, the necessary science is relatively new and even within the established scientific research communities, there is as yet limited awareness of its potential impact for social and economic well-being.

REFERENCES

1. United Nations. World population prospects, the 2000 revision: Highlights. New York: United Nations, 2001.
2. Westendorp RG. What is healthy aging in the 21st century? *Am J Clin Nutr* 2006;**83**(2):404S–9S.
3. Llibre Rodriguez JJ, Ferri CP, Acosta D, Guerra M, Huang Y, Jacob KS, Krishnamoorthy ES, Salas A, Sosa AL, Acosta I, Dewey ME, Gaona C, Jotheeswaran AT, Li S, Rodriguez D, Rodriguez G, Kumar PS, Valhuerdi A, Prince M. Prevalence of dementia in Latin America, India, and China: A population-based cross-sectional survey. *Lancet* 2008;**372**(9637):464–74.
4. Ferri CP, Prince M, Brayne C, Brodaty H, Fratiglioni L, Ganguli M, Hall K, Hasegawa K, Hendrie H, Huang Y, Jorm A, Mathers C, Menezes PR, Rimmer E, Scazufca M. Global prevalence of dementia: A Delphi consensus study. *Lancet* 2005;**366**(9503):2112–7.
5. Mathers CD, Lopez AD, Murray CJL. The burden of disease and mortality by condition: Data, methods and results for 2001. In: Lopez AD, Mathers CD, Ezzati M, Murray CJL, Jamison DT, eds. Global burden of disease and risk factors. New York: Oxford University Press, 2006: 45–240.
6. Kivipelto M, Ngandu T, Laatikainen T, Winblad B, Soininen H, Tuomilehto J. Risk score for the prediction of dementia risk in 20 years among middle aged people: A longitudinal, population-based study. *Lancet Neurol* 2006;**5**(9):735–41.
7. Vermeer SE, Prins ND, den Heijer T, Hofman A, Koudstaal PJ, Breteler MM. Silent brain infarcts and the risk of dementia and cognitive decline. *N Engl J Med* 2003;**348**(13):1215–22.
8. Kalaria RN, Maestre GE, Arizaga R, Friedland RP, Galasko D, Hall K, Luchsinger JA, Ogunniyi A, Perry EK, Potocnik F, Prince M, Stewart R, Wimo A, Zhang ZX, Antuono P. Alzheimer's disease and vascular dementia in developing countries: Prevalence, management, and risk factors. *Lancet Neurol* 2008;**7**(9):812 26.
9. Luchsinger JA, Mayeux R. Dietary factors and Alzheimer's disease. *Lancet Neurol* 2004;**3**(10):579–87.
10. Gillette Guyonnet S, Abellan Van Kan G, Andrieu S, Barberger Gateau P, Berr C, Bonnefoy M, Dartigues JF, de Groot L, Ferry M, Galan P, Hercberg S, Jeandel C, Morris MC, Nourhashemi F, Payette H, Poulain JP, Portet F, Roussel AM, Ritz P, Rolland Y, Vellas B. IANA task force on nutrition and cognitive decline with aging. *J Nutr Health Aging* 2007;**11**(2):132–52.

11. Durga J, van Boxtel MP, Schouten EG, Kok FJ, Jolles J, Katan MB, Verhoef P. Effect of 3-year folic acid supplementation on cognitive function in older adults in the FACIT trial: A randomised, double blind, controlled trial. *Lancet* 2007;**369**(9557):208–16.

12. Balk EM, Raman G, Tatsioni A, Chung M, Lau J, Rosenberg IH. Vitamin B6, B12, and folic acid supplementation and cognitive function: A systematic review of randomized trials. *Arch Intern Med* 2007;**167**(1):21–30.

13. Rodriguez-Martin JL, Qizilbash N, Lopez-Arrieta JM. Thiamine for Alzheimer's disease. *Cochrane Database Syst Rev* 2001(2):CD001498.

14. Miller ER, 3rd, Pastor-Barriuso R, Dalal D, Riemersma RA, Appel LJ, Guallar E. Meta-analysis: High-dosage vitamin E supplementation may increase all-cause mortality. *Ann Intern Med* 2005;**142**(1):37–46.

15. Dai Q, Borenstein AR, Wu Y, Jackson JC, Larson EB. Fruit and vegetable juices and Alzheimer's disease: The Kame Project. *Am J Med* 2006;**119**(9):751–9.

16. Tabet N, Birks J, Evans JG, Orrel M, Spector A. Vitamin E for Alzheimer's disease (Cochrane Review) Issue 1. Oxford: Update Software, 2003.

17. Vellas B, Andrieu S, Sampaio C, Coley N, Wilcock G. Endpoints for trials in Alzheimer's disease: A European task force consensus. *Lancet Neurol* 2008;**7**(5):436–50.

18. Burr GO, Burr MM. Nutrition classics from *The Journal of Biological Chemistry* 82:345–67, 1929. A new deficiency disease produced by the rigid exclusion of fat from the diet. *Nutr Rev* 1973;**31**(8):248–9.

19. Ballabriga A. Essential fatty acids and human tissue composition. An overview. *Acta Paediatr Suppl* 1994;**402**:63–8.

20. Soderberg M, Edlund C, Kristensson K, Dallner G. Lipid compositions of different regions of the human brain during aging. *J Neurochem* 1990;**54**(2):415–23.

21. Farquharson J, Cockburn F, Patrick WA, Jamieson EC, Logan RW. Infant cerebral cortex phospholipid fatty-acid composition and diet. *Lancet* 1992;**340**(8823):810–3.

22. Uauy R, Dangour AD. Nutrition in brain development and aging: Role of essential fatty acids. *Nutr Rev* 2006;**64**(5 Pt 2):S24–33; discussion S72–91.

23. Marcheselli VL, Hong S, Lukiw WJ, Tian XH, Gronert K, Musto A, Hardy M, Gimenez JM, Chiang N, Serhan CN, Bazan NG. Novel docosanoids inhibit brain ischemia-reperfusion-mediated leukocyte infiltration and pro-inflammatory gene expression. *J Biol Chem* 2003;**278**(44):43807–17.

24. Mukherjee PK, Marcheselli VL, Serhan CN, Bazan NG. Neuroprotectin D1: A docosahexaenoic acid–derived docosatriene protects human retinal pigment epithelial cells from oxidative stress. *Proc Natl Acad Sci U S A* 2004;**101**(22):8491–6.

25. Kim HY, Akbar M, Lau A, Edsall L. Inhibition of neuronal apoptosis by docosahexaenoic acid (22:6n-3). Role of phosphatidylserine in antiapoptotic effect. *J Biol Chem* 2000;**275**(45):35215–23.

26. Calon F, Lim GP, Yang F, Morihara T, Teter B, Ubeda O, Rostaing P, Triller A, Salem N, Jr, Ashe KH, Frautschy SA, Cole GM. Docosahexaenoic acid protects from dendritic pathology in an Alzheimer's disease mouse model. *Neuron* 2004;**43**(5):633–45.

27. Barberger-Gateau P, Letenneur L, Deschamps V, Peres K, Dartigues JF, Renaud S. Fish, meat, and risk of dementia: Cohort study. *BMJ* 2002;**325**(7370):932–3.

28. Issa AM, Mojica WA, Morton SC, Traina S, Newberry SJ, Hilton LG, Garland RH, Maclean CH. The efficacy of omega-3 fatty acids on cognitive function in aging and dementia: A systematic review. *Dement Geriatr Cogn Disord* 2006;**21**(2):88–96.

29. van Gelder BM, Tijhuis M, Kalmijn S, Kromhout D. Fish consumption, n-3 fatty acids, and subsequent 5-y cognitive decline in elderly men: The Zutphen Elderly Study. *Am J Clin Nutr* 2007;**85**(4):1142–7.

30. Schaefer EJ, Bongard V, Beiser AS, Lamon-Fava S, Robins SJ, Au R, Tucker KL, Kyle DJ, Wilson PW, Wolf PA. Plasma phosphatidylcholine docosahexaenoic acid content and risk of dementia and Alzheimer disease: The Framingham Heart Study. *Arch Neurol* 2006;**63**(11):1545–50.

31. Virtanen JK, Siscovick DS, Longstreth WT, Jr, Kuller LH, Mozaffarian D. Fish consumption and risk of subclinical brain abnormalities on MRI in older adults. *Neurology* 2008;**71**(6):439–46.

32. Dullemeijer C, Durga J, Brouwer IA, van de Rest O, Kok FJ, Brummer RJ, van Boxtel MP, Verhoef P. n 3 fatty acid proportions in plasma and cognitive performance in older adults. *Am J Clin Nutr* 2007;**86**(5):1479–85.

33. Lim WS, Gammack JK, Van Niekerk J, Dangour AD. Omega 3 fatty acid for the prevention of dementia. *Cochrane Database Syst Rev* 2006(1):CD005379.

34. Freund-Levi Y, Eriksdotter-Jonhagen M, Cederholm T, Basun H, Faxen-Irving G, Garlind A, Vedin I, Vessby B, Wahlund LO, Palmblad J. Omega-3 fatty acid treatment in 174 patients with mild to moderate Alzheimer disease: OmegAD study: A randomized double-blind trial. *Arch Neurol* 2006;**63**(10):1402–8.

35. van de Rest O, Geleijnse JM, Kok FJ, van Staveren WA, Dullemeijer C, Olderikkert MG, Beekman AT, de Groot CP. Effect of fish oil on cognitive performance in older subjects: A randomized, controlled trial. *Neurology* 2008;**71**(6):430–8.

36. Dangour AD, Clemens F, Elbourne D, Fasey N, Fletcher AE, Hardy P, Holder GE, Huppert FA, Knight R, Letley L, Richards M, Truesdale A, Vickers M, Uauy R. A randomised controlled trial investigating the effect of n-3 long-chain polyunsaturated fatty acid supplementation on cognitive and retinal function in cognitively healthy older people: The Older People And n–3 Long-chain polyunsaturated fatty acids (OPAL) study protocol [ISRCTN72331636]. *Nutr J* 2006;**5**:20.

26 Iron and Monoamine Oxidase in Brain Function and Dysfunction

Development of Neuroprotective-Neurorescue Drugs

Orly Weinreb, Tamar Amit, Silvia Mandel, and Moussa B. H. Youdim

Eve Topf and USA NPF Centers of Excellence, Department of Pharmacology, Technion-Faculty of Medicine, Haifa, Israel

CONTENTS

26.1 INTRODUCTION

The etiology of neurodegenerative diseases, such as Parkinson's disease (PD), Alzheimer's disease (AD), amyotrophic lateral sclerosis (ALS), and other neurodegenerative diseases, is not yet well understood. However, over the last two decades, significant accumulating evidence has shown that iron-dependent oxidative stress (OS), increased levels of iron [1, 2] and monoamine oxidase B(MAO-B) activity, and depletion of antioxidants in the brain may be major pathogenic factors in PD, AD, and other neurodegenerative diseases [1, 3]. Iron is an essential cofactor for many key proteins involved in the normal function of neuronal tissue. It is normally involved in oxygen transport, storage, and activation, electron transport, and many important metabolic processes. Yet, there is increasing evidence that iron accumulation and deposition in the brain can cause a vast range of disorders of the central nervous system (CNS) [1, 4] and progressively accumulates in the brain with age [5]. Free iron induces OS because of its interaction with hydrogen peroxide (Fenton reaction), resulting in an increased formation of hydroxyl free radicals. Free radical-related OS causes molecular damage that can then lead to a critical failure of biological functions and ultimately cell death [6–8]. In neurodegenerative diseases such as, PD, AD, Huntington's disease, ALS, and multiple sclerosis, iron accumulates at the site of the lesion and is thought to participate in the

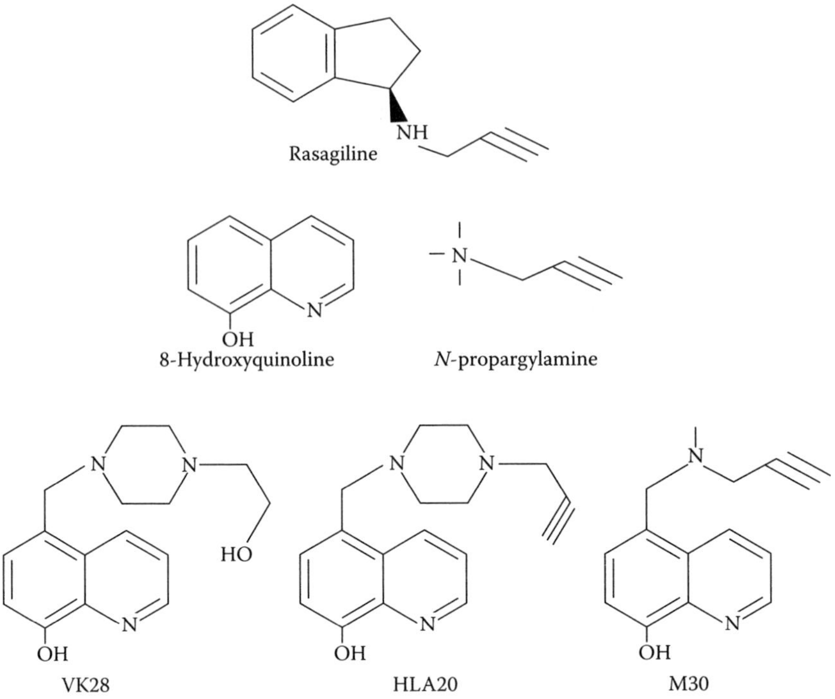

FIGURE 26.1 Structure of the novel multifunctional iron chelator–MAO inhibitor drugs and the molecules from which they are derived.

neurodegenerative process [9]. Evidence has also shown that significant accumulation of iron in white matter tracts and nuclei throughout the brain precedes the onset of neurodegeneration and movement disorder symptoms [10].

In view of the above, we recently proposed a new neuroprotective strategy, suggesting that neurodegenerative diseases may require a drug that combines iron chelating with antioxidant capacity and MAO-B inhibitory properties [2, 11, 12]. We have developed a number of iron chelators from our prototype brain-permeable iron chelator, VK-28 (5-[4-(2-hydroxyethyl)piperazine-1-ylmethyl]-quinoline-8-ol) (Figure 26.1) [13], which has been shown to have neuroprotective activity in an animal model of PD [13] but does not have any appreciable MAO inhibitory activity [14]. Therefore, in searching for new antioxidant–iron chelators with MAO-B inhibitory activity, we have designed a new series of chelators, the M-30 series [15], which also possess the propargyl moiety of the novel anti-PD drug, rasagiline (Azilect, Teva Pharmaceutical) [16]. Our previous studies have shown that the propargyl moiety is crucial for the potent MAO-B inhibitory and neuroprotective activities of rasagiline [17, 18].

The newly designed multifunctional chelators [19], which contain the neuroprotective propargylamine moiety of rasagiline, M-30 (5-[N-methyl-N-propargylaminomethyl]-8-hydroxyquinoline) and HLA-20 (5-[4-propargylpiperazin-1-ylmethyl]-8-hydroxyquinoline) (Figure 26.1), possess the following characteristics: first, they chelate iron and thus may prevent iron accumulation in the brain; second, they prevent iron-induced OS and thereby may protect neurons from oxidative injuries; third, these chelators might act as free radical scavengers to inhibit lipid peroxidation; lastly, they possess MAO-B and MAO-A inhibitory activity of the propargyl moiety and thus may be neuroprotective. This review will discuss recent novel studies of the mechanism of the multimodal action of these novel neuroprotective iron chelators, which are recommended as future therapeutic compounds for clinical trial of neurodegenerative diseases.

26.2 IRON, MONOAMINE OXIDASE INHIBITION, AND PD

The iron-related neurodegenerative disorders may be divided into two main categories: (1) those that result from iron accumulation in specific brain regions and (2) those that result from defective iron metabolism and homeostasis. These disorders frequently involve protein modification, misfolding, and aggregation, leading to the formation of the intracellular inclusion bodies that are the post-mortem hallmark of many neurodegenerative diseases [1, 4, 11]. In addition, in PD, the brain's defensive mechanisms against the formation of oxygen free radicals are impaired [8, 20]. In the substantia nigra of parkinsonian brains, there is a drastic depletion of endogenous antioxidants, such as reduced glutathione (GSH) [20–22]. Studies have shown that iron concentrations are significantly elevated in parkinsonian substantia nigra pars compacta (SNpc) and within the melanized dopamine (DA) neurons [23, 24]. Similar results have also been shown in 6-hydroxydopamine (6-OHDA) and N-methyl-4-phenyl-1,2,3,6-tetrahydropyridine (MPTP) models [25, 26].

Additional support for the role of iron in neurodegeneration comes from a previous work that has developed a strain of mouse with a targeted disruption of the gene encoding Irp2 (Ireb2) [10]. These mice misregulate iron metabolism in the central nervous system and with aging develop a movement disorder characterized by ataxia, bradykinesia, and tremor. Iron accumulates in the brains of these animals, preceding the onset of neurodegeneration and symptoms of movement disorder and mimicking to some degree symptoms reported for PD [27]. It is noteworthy that iron and ferritin accumulation occurs within the neurons and oligodendrocytes in distinctive regions of the brain with ubiqutin-positive inclusion bodies. Thus, misregulation of iron metabolism apparently leads to neurodegenerative disease in Ireb2(–/–) mice, suggesting a similar contribution to the pathogenesis of comparable human neurodegenerative diseases. In a recent study from the same group [28], it was reported that iron misregulation associated with the loss of IRP-2 protein in the Ireb2(–/–) mice affects DA metabolism in the striatum resulting from the loss of DA and DA-regulating proteins, further supporting the view that the IRP-2 –/– genotype may enable neurobiological events associated with aging and neurodegeneration.

Iron homeostasis is maintained by the interaction of several proteins. These include the product of the hemochromatosis gene (HFE), transferrin (Tf), the transferrin receptor (TfR), and iron regulatory proteins (IRPs) in the crypts of Lieberkühn, which determine the amount of iron that is allowed to cross the enterocyte and enter the bloodstream [29]. The brain is protected against plasma iron by the blood-brain barrier (BBB). Little is known about the mechanism of iron release into the brain or the regulation of the iron transport mechanism, and therefore, the reason for iron accumulation in the brain in an age- and region-dependent process remains to be clarified. However, three proteins are known to be capable of iron binding and transport in the brain: Tf, lactoferrin, and ferritin [1].

It is well established that iron participates in the Fenton chemistry, reacting with hydrogen peroxide to produce the most reactive of all reactive oxygen species (ROS), the hydroxyl radical. The formation of the latter, combined with depletion of endogenous antioxidants, particularly tissue GSH, the most common pathway of iron disposition in the brain, leads to OS [22, 30–33]. Iron also facilitates the decomposition of lipid peroxides to produce highly cytotoxic oxygen-related free radicals. Free radical-related OS causes damage to DNA, lipids, proteins, and ultimately cell death associated with neurodegenerative diseases [34]. Moreover, recent evidence suggests that iron, together with hydrogen peroxide, in the presence of DA, can induce the formation of the neurotoxic 6-OHDA [35–37]. 6-OHDA may be also formed from l-dihydroxyphenylalanine (l-DOPA) in the presence of iron and hydrogen peroxide [38]. Iron-dependent OS tends to be tissue-specific, owing to differential cell susceptibility. This vulnerability is affected by three main parameters: (1) axonal length and thickness, (2) axonal sprouting, and (3) thickness of the myelin sheath [39].

Additional contribution to OS stems from elevated MAO-B levels, leading to increased DA production [34]. This results in increased DA oxidation both via MAO and auto-oxidation, leading to elevated levels of hydrogen peroxide, which, in turn, participates in Fenton chemistry, thus creating a vicious circle. Indeed, post-mortem studies report that MAO-B in the human brain increases with age and in neurodegenerative diseases, in particular in PD [40–42]. The increased level of MAO-B in the aged brain is thought to be associated with OS, which may play a role in the vulnerability of the DA system and age-related degeneration [41]. These findings suggest that iron chelators, antioxidants, and MAO-B inhibitors may have great therapeutic potential against neurodegenerative diseases [13, 14, 43–45]. In fact, a number of iron chelators, antioxidants, and MAO-B inhibitors have been shown to possess neuroprotective activity in animal models of PD. For example, the MAO-B inhibitors rasagiline and selegiline have been shown to protect against the toxic damage induced by MPTP and 6-OHDA animal models of PD [46–48].

Chelation therapy as a concept is not new. Deferrioxamine (DFO), a prototype iron chelator drug with no oral activity, when injected intracerebroventricularly, protected against the dopaminergic neurodegeneration induced by 6-OHDA [49, 50]. Furthermore, the neuroprotective activity of DFO in preventing iron- and MPTP-induced neurotoxicity in mice has been reported [51, 52]. More recently, the antibiotic iron chelator 5-chloro-7-iodo-8-hydroxyquinoline (clioquinol) has been shown to be able to prevent MPTP neurotoxicity in mice [45]. In addition, our novel brain-permeable iron chelator VK-28 has been successfully used in reversal of 6-OHDA-induced detrimental effects in the rat [13, 49] and neural cell [53] models. Accumulated iron in the SNpc of animals treated with 6-OHDA or MPTP (or in the SNpc of PD patients) is thought to be in a labile pool, not bound to ferritin. It is this unbound iron that likely plays a pivotal role in the induction of OS-dependent neurodegeneration of DA neurons via Fenton chemistry. This hypothesis provides a likely explanation for the effectiveness of iron chelators in the 6-OHDA and MPTP animal models, because it is highly unlikely that chelators would affect ferritin-bound iron. The hypothesis that an iron chelator that can penetrate the BBB may be useful for treatment of neurodegenerative diseases is also supported by several clinical results. For example, in recent years, a successful treatment of Wilson's disease has emerged, namely, d-penicillamine, a copper chelator. Furthermore, chronic DFO treatment for aceruloplasminemia was shown by functional magnetic resonance imaging (MRI) to remove iron and improve the neurological aspects of the disorder [54]. However, clioquinol is highly toxic, but the hydrophilic nature of DFO and its large molecular size limit absorption across the gastrointestinal tract and prevent it from penetrating the BBB [55].

26.3   NEUROPROTECTIVE IRON CHELATOR/MAO-B INHIBITORS FOR THE TREATMENT OF PD

Several therapeutic strategies to prevent and/or treat PD have been suggested including use of iron chelation, antioxidants, and selective MAO-B inhibition. Indeed, several such compounds show neuroprotection in animal models, but most of them fail in the clinic [45, 56]. Recently, we have proposed a novel combination strategy of neuroprotection in neurodegenerative diseases that requires a drug combining iron chelation with antioxidant capacity and MAO-B inhibitory properties [2]. This has been achieved by introducing the neuroprotective and MAO inhibitory moiety, propargylamine of the anti-Parkinson MAO-B inhibitor drugs rasagiline and selegiline [57], into the pharmacophore of our prototype brain-permeable neuroprotective iron chelator VK-28 [49, 58]. The compound M-30 was developed for two purposes. On the one hand, it was designed to prevent the ability of iron to induce OS, as a consequence of reactive hydroxyl radical generation via its interaction with hydrogen peroxide (Fenton reaction). On the other hand, M-30 was designed to inhibit the formation of reactive hydroxyl radical from hydrogen peroxide generated by MAO and potentiate the pharmacological action of accumulated DA formed from l-DOPA.

In the search for superlative neuroprotective agents in a series of multifunctional iron chelators [14, 15, 58], selected novel chelators were screened *in vitro* for their ability to inhibit iron-induced lipid peroxidation. Three drugs, HLA-20, M-30, and VK-28 (Figure 26.1), were found to be the most effective inhibitors of lipid peroxidation with higher IC_{50} values, comparable with that of DFO, a potent inhibitor of lipid peroxidation and strong iron chelator. *In vitro* experiments demonstrated that one compound, M-30, holds promise for further *in vivo* study on MAO-A and MAO-B inhibitory activities and neuroprotective activity [59]. M-30, which possesses the propargyl MAO inhibitory moiety of the anti-PD drugs selegiline and rasagiline, shows highly potent inhibition of both MAO-A and MAO-B activities *in vitro* [14]. Nevertheless, *N*-desmethylation of M-30 results in a significant loss of MAO inhibitory activity. This effect is similar to what has previously been reported for desmethylated selegiline, rasagiline, and clorgyline [60]. Another compound that exhibits a moderate selective MAO-B inhibitory activity *in vitro* is HLA20 [59]. This may be related to its tertiary amine structure with the propargyl moiety being in the piperazine and not be available for oxidation, similar to MPTP, which also acts as a weak MAO-B inhibitor [61]. The high MAO inhibitory activity of M-30 may be attributed to the propargyl group in the molecule since propargyl-containing compounds such as rasagiline, pargyline, clorgyline, and deprenyl show high MAO inhibitory activity; removal of the propargyl moiety abolishes the inhibitory activity [62–65]. It is known that the propargyl moiety binds covalently with the N-5 of FAD cofactor at the active site of MAO enzyme to inhibit it [62].

One major finding with M-30 is the poor inhibition of the enzymes in the liver and small intestine. This is an important pharmacological advantage of the drug, as irreversible inhibition of MAO-A, which is prominent in these tissues, is associated with the "cheese reaction" (potentiation of tyramine-induced cardiovascular activity), as has been observed with tranylcypromine, iproniazid, and clorgyline [66, 67]. The exact mechanism underlying the preference of M-30 for the brain enzyme inhibition is not known. It is possible that in the brain, the inhibitor is metabolized to active metabolite(s) that accumulate and are retained by the brain. Another possibility is that this drug is metabolized by a brain-specific cytochrome P-450, which at present is being investigated.

The ability of chelators to inhibit lipid peroxidation may result from two processes: iron chelation and free radical scavenging. It is well established that strong iron chelators could form inert complexes with iron and interfere with the Fenton reaction, leading to a decrease in hydroxyl free radical production and thus blocking lipid peroxidation. The novel chelator M-30, which has been shown to possess high iron binding capacity [14, 15, 58], may also act by a similar mechanism to inhibit free radical formation. In addition, the chelators may act as radical scavengers to directly block formation of the free radical, as confirmed by the spin trapping of hydroxyl radical by 5.5-dimethyl-*l*-pyrroline-*N*-oxide (DMPO) measured in the electron paramagnetic resonance spectra of the resulting DMPO-hydroxyl radical spin adduct. The results [68] showed that the novel chelators can significantly reduce the DMPO-hydroxyl radical signal generated by the photolysis of H_2O_2. This suggests that these chelators work as radical scavengers to directly scavenge the hydroxyl radical because the photolysis of H_2O_2 generates hydroxyl radical independently from metal ions. The mechanism by which the novel multifunctional chelators act as radical-scavenging antioxidants is not well understood. Previous studies have shown that some phenolic and polyphenolic compounds such as vitamin E, catechin gallates, and green tea polyphenol (–)-epigallocatechin-3-gallate (EGCG) possess free radical-scavenging properties [69]. Their scavenging activities are believed to be due to the presence of the phenolic moiety [70]. Although both HLA-20 and M-30 contain a phenolic moiety, they exhibit different capacities in scavenging hydroxyl radical. Their mechanisms as radical-scavenging antioxidants need to be further investigated.

Additionally, cell culture neuroprotection studies revealed that the chelators HLA-20, VK-28, and M-30 afford neuroprotection in neuronal cell culture, at least as potent as that of the anti-PD drug rasagiline [71, 72]. They exhibit a biphasic response in PC12 cells: maximal protective effects at 1 µM and less potent protective effects at 10 µM. This biphasic pattern is typical of antioxidative drugs, such as the iron chelator EGCG [73, 74], vitamin C [75], R-apomorphine [76], and DA [77], being neuroprotective at low (1–10 µM) concentrations while having less potent protective actions or even toxic at higher (10–50 µM) concentrations because they become prooxidant and proapoptotic [74]. In fact, previous studies have provided novel insights into the gene mechanisms involved in both the neuroprotective and proapoptotic activities of antioxidative drugs. It has been found [74] that the high concentrations of DA (500 µM), R-apomorphine (50 µM), melatonin (50 µM), and EGCG (50 µM) exhibited a similar profile of proapoptotic gene expression, increasing the level of bax, caspase 6, fas ligand, and the cell cycle inhibitor GADD45 genes while decreasing antiapoptotic bcl-2 and bcl-xL. Conversely, the low concentrations (1–10 µM) of DA, R-apomorphine, and melatonin induced an immediate expression of antiapoptotic bcl-xL and/or bcl-2 mRNAs, while bax mRNA was reduced. Similarly, the potent antioxidant iron chelator EGCG at low concentrations (1–10 µM) decreased proapoptotic bax and caspase 6. In addition, the toxicity of strong iron chelators at the high concentrations (>10 µM) may also be related to their interaction with essential iron-containing proteins, such as Tf, ribonucleotide reductase, and tyrosine hydroxylase, resulting in the mobilization of iron from these enzymes and proteins, as reported for iron chelators, such as DFO and hydroxypyridinones [78, 79].

This neuroprotective property does not appear to be related to the MAO inhibitory activity of these compounds. Its support has come from previous structural activity studies in which we determined neuroprotection. We demonstrated that while propargylamine derivatives with little or no MAO inhibitory activity retained the neuroprotective property, in absence of the propargyl moiety, no neuroprotective action was noted [80–82]. Previous studies have clearly shown that the neuroprotective action of these propargylamine-containing compounds is related to their activation of and interaction with Bcl-2 family proteins bcl-2 and bcl-xL, and protein kinases C α and ε [17, 18].

The mechanism by which the novel iron chelators protect PC12 against cell death induced by serum withdrawal or by 6-OHDA is not well established. Because HLA-20 and M-30 show similar iron-binding capacity and antioxidative property [19, 83], the similar neuroprotective effects among these chelators, despite their remarkable difference in MAO-A and MAO-B inhibitory activity, imply a close correlation between the neuroprotective potency and the iron-chelating and antioxidative properties. It also suggests that neuroprotection may be mediated by their iron-chelating and antioxidative properties and the propargyl moiety. In fact, there is considerable evidence that iron chelators, such as DFO and VK-28 [13, 43, 44, 84], show neuroprotection against 6-OHDA-induced neuronal death in rats. 6-OHDA is a highly reactive neurotoxin, which is readily autooxidized and oxidatively deaminated by MAO, giving rise to H_2O_2 and ROS. It also potently inhibits mitochondrial respiratory chain complex I and IV activity, resulting in intracellular ROS. The redox iron released from ferritin due to ROS-related OS, via the Fenton reaction, exacerbates formation of the highly toxic hydroxyl radical, which in turn causes cascade of events of neuronal death [85]. The cell death induced by serum deprivation is also related to OS because studies have shown that serum deprivation can significantly increase the level of ROS in PC12 cells and rat cortical neurons [86]. Strong iron chelators, such as DFO, can bind iron to form stable iron complexes that block the hydroxyl radical production-induced cell death. Therefore, the novel compounds M-30 and HLA-20, which have proved to be good iron chelators, may exert their neuroprotective action by the same mechanism. However, a radical-scavenging mechanism may also play a role in the neuroprotective activity of these chelators because free radical-related OS is involved in the PC12 cell death induced by serum deprivation or 6-OHDA. As previously reported, radical-scavenging antioxidants, such as N-acetylcysteine or dithiothreitol, protect PC12 cells against oxidative stress [87, 88]. Apomorphine, a highly potent free radical scavenger, prevents H_2O_2- and 6-OHDA-induced cell death [89]. Nitrone spin trap, α-phenyl-N-tert-butylnitrone, and its derivatives, which react covalently with ROS including hydroxyl radical, have shown efficacy in a variety of animal models of CNS injury [90]. Therefore, the iron chelators M-30 and HLA-20 may also exert their neuroprotective effects via their capacity to directly scavenge free radicals. Indeed, an in vivo study in MPTP mice model [19, 83] has shown that M-30 displays the ability to increase striatal DA, serotonin, and noradrenaline, thus preventing MPTP neurotoxicity via the prevention of MPTP conversion to its neurotoxic metabolite, MPP+ [19, 83]. M-30 prevents MPTP-induced nigrostriatal dopaminergic depletion in mice by protecting against the loss of DA and decrease DA turnover, similar to nonselective MAO inhibitors and selective MAO-B inhibitors such as selegiline, rasagiline, lazabemide, and milacemide, thus making M-30 as a highly potential multifunctional anti-PD drug.

26.4 IRON AND ALZHEIMER'S DISEASE PATHOLOGY

In AD pathology, iron is significantly concentrated in and around amyloid senile plaques and neurofibrillary tangles (NFTs), leading to alterations in the pattern of the interaction between iron regulatory proteins (IRP) and their iron-responsive element (IRE) and disruption in the sequestration and storage of iron [91, 92]. Also, high levels of iron have been reported in the amyloid plaques of the Tg2576 mouse model for AD, resembling those seen in the brains of AD patients [93]. In addition to the accumulation of iron in senile plaques, it was demonstrated that the amount of iron present in the AD neuropil is twice that found in the neuropil of nondemented brains [91]. Further studies have suggested that accumulated iron supports the AD pathology as a possible source of OS-dependent reactive oxygen radicals, demonstrating that neurons in AD brains experience high oxidative load [94–96]. Post-mortem analyses of Alzheimer patients' brains have revealed activation of two enzymatic indicators of cellular OS: heme oxygenase (HO-1) [97] and NADPH oxidase [98]. Also, HO-1 was greatly enhanced in neurons and astrocytes of the hippocampus and cerebral cortex of Alzheimer's subjects, colocalizing to senile plaques and NFTs. A recent study reported that ribosomal RNA provided a binding site for redox-active iron and serves as a redox center within the cytoplasm of vulnerable neurons in AD brain, preceding the appearance of morphological changes indicating neurodegeneration [99]. In addition, other evidence suggests that the metabolism of iron is disrupted in AD. For example, the location of the iron-transport protein, transferrin, in senile plaques, instead of its regular location in the cytosol of oligodendrocytes, indicated that transferrin becomes trapped within plaques while transporting iron between cells [100]. The mediator of iron uptake by cells, melanotransferrin, and the iron-storage protein ferritin are altered in AD and are expressed within reactive microglial cells that are present both in and around senile plaques [101].

Previous studies assessing the effect of certain genes coding proteins involved in iron metabolism, such as hemochromatosis (HFE) and Tf genes, on the onset of AD have been contradictory [101]. At the biochemical level, iron was demonstrated to facilitate the aggregation of β-amyloid peptide (Aβ) and increase its toxicity and induce aggregation of hyperphosphorylated τ, the major constituent of NFTs [102]. Indeed, DFO prevented the formation of β-pleated sheets of $Aβ_{1-42}$ and dissolved preformed β-pleated sheets of plaquelike amyloid [103]. Some success has also been reported with the copper chelator clioquinol in AD [104]. Oral administration of clioquinol was reported to inhibit Aβ accumulation in an AD transgenic mouse model via its actions as a metal chelator [104]. However, none of these antioxidants and iron chelators has been successfully introduced for clinical use [104].

A direct link between iron metabolism and AD pathogenesis was provided recently by Rogers et al. who described the presence of an IRE in the 5′ untranslated region (5′-UTR) of the amyloid precursor protein (APP) transcript [105]. Thus, APP 5′-UTR is selectively responsive to intracellular iron levels in a pattern that reflects iron-dependent regulation of intracellular APP synthesis. Indeed, iron levels were shown to regulate mRNA translation of APP holoprotein in astrocytes [105, 106] and neuroblastoma cells by a pathway similar to iron control in the translation of the ferritin L, and H mRNAs by IREs in their 5′-UTRs.

Two main aspects consider the link between iron and AD, in relation to the recently discovered IRE in the 5′-UTR of APP mRNA, and thus iron chelators, as high-potential drugs for the treatment of AD. (1) The physiological aspect considers the neuroprotective response of Aβ in reducing iron-induced neurotoxicity. Consequently, given that Aβ possesses iron-chelating sites, it may be hypothesized that OS-induced intracellular iron levels stimulates APP holoprotein translation (via the APP 5′-UTR) and subsequently the generation of its cleavage product, Aβ peptide, as a compensatory response that eventually reduces OS. (2) The pathological aspect considers iron chelator compounds targeting the APP 5′-UTR, thus possessing the capacity to reduce APP translation and, subsequently, Aβ generation levels.

26.5 IRON-CHELATING STRATEGY FOR THE TREATMENT OF ALZHEIMER'S DISEASE

Under extreme pathological conditions, i.e., at some threshold level of ROS generation, it appears that the main role of Aβ switches from neuroprotective to dyshomeostatic in terms of cerebral biometals and APP/Aβ/metal-redox complexes, leading to a vicious cycle of increased ROS production and Aβ generation. Chelation has the potential to prevent iron-induced ROS, OS, and Aβ aggregation, and therefore, chelation therapy may be considered a valuable therapeutic strategy for AD. In fact, intramuscular administration of DFO, a potent iron chelator, slowed the clinical progression of AD dementia [107], and some success has also been achieved with clioquinol, another metal complexing agent [104, 108].

The identification of an IRE in the 5′-UTR of the APP transcript led to a novel therapeutic approach aimed at reducing amyloidosis by FDA preapproved drugs targeted to the IRE in the APP mRNA 5′-UTR [105]. For example, the APP 5′-UTR-directed drugs DFO (Fe^{3+} chelator), tetrathiomolybdate (Cu^{2+} chelator), and dimercaptopropanol (Pb^{2+} and Hg^{2+} chelator) were found to suppress APP holoprotein expression and lower Aβ peptide secretion [105, 109]. In addition, the bifunctional molecule, XH-1, which contains both amyloid-binding and metal-chelating moieties, was shown to reduce APP expression in SH-SY5Y cells and attenuate cerebral Aβ in PS1/APP transgenic mice [110]. Additional drug classes were also reported to suppress the APP 5′-UTR and limit APP expression including antibiotics, selective serotonin reuptake inhibitors, and other selective receptor antagonists and agonists [109].

The concept of metal chelators for clinical use in neurological disorders that could remove excess iron in the brain recently led our group to develop nontoxic, lipophilic, and brain-permeable iron chelators for AD. The novel iron chelators, VK-28 [13], and the multifunctional drugs, HLA-20 and M-30 (Figure 26.1) [14], were recently shown to induce a significant down-regulation of membrane-associated holo-APP levels in the mouse hippocampus and in human SH-SY5Y neuroblastoma cells, presumably by chelating intracellular iron pools [111]. Indeed, the iron chelator drugs, VK-28, HLA-20, and M-30, were found to suppress translation of a luciferase reporter mRNA via the APP 5′-UTR sequence [111]. Furthermore, M-30 markedly reduced the levels of the amyloidogenic Aβ in the medium of CHO cells stably transfected with the APP "Swedish" mutation (CHO/ΔNL) [111]. In addition, naturally occurring polyphenols (e.g., EGCG, curcumin) might be used as another novel and promising therapeutic approach for treating AD. Both compounds have well characterized antioxidant and metal-chelating (iron and copper) activities [112, 113] and have been demonstrated to exert neuroprotective activity against a variety of neurotoxic insults as well as to regulate APP processing and Aβ burden in cell culture and *in vivo* [114]. EGCG treatment led to a reduction of Aβ levels in CHO/ΔNL [115], murine neuronlike cells (N2a) transfected with the APP "Swedish" mutation and primary neurons derived from "Swedish" mutant APP-overexpressing mice [116]. *In vivo*, EGCG significantly reduced cerebral Aβ levels concomitant with reduced Aβ-amyloid plaques in TgAPPsw transgenic mice overproducing Aβ [116].

Our recent studies have shown that prolonged administration of EGCG to mice induced a reduction in holo-APP levels in the hippocampus [117]. In SH-SY5Y cells, EGCG significantly reduced both the mature and full-length cellular holo-APP without altering APP mRNA levels, as shown by two-dimensional gel electrophoresis, suggesting a posttranscriptional action [115]. Indeed, we demonstrated that EGCG reduced the translation of a luciferase reporter gene fused to the APP 5′-mRNA UTR that includes the APP IRE [111, 115]. The observation that the alteration in APP following treatment with EGCG was blocked by exogenous iron provides further support to the implication of metal-chelating property of EGCG in the regulation of iron homeostasis-associated proteins.

Finally, a new, additional aspect of iron chelator compounds in the etiology of AD therapy is related to their ability to abort anomalous cell cycle reactivation in postmitotic degenerating neurons. Indeed, during the last few years, accumulating evidence for an activated cell cycle in the vulnerable neuronal population in AD has suggested a crucial role for cell cycle abnormalities in

AD pathogenesis. Therefore, therapeutic interventions targeted toward ameliorating mitotic changes would be predicted to have positive impact on AD progression. Previous studies have shown that the reactivation of the cell cycle is an obligatory component of the apoptotic pathway evoked by Aβ peptides [118, 119]. Recently, we have found that M-30 (0.1 μM) significantly reduced the percentage of neurons in S phase while reraised their relative cell number in G_0/G_1 phase and lowered apoptotic levels after exposure to $Aβ_{25-35}$ in primary cultures of rat cortical neurons. In support, the novel iron chelator drugs were previously shown to induce cell cycle arrest; M-30 and HLA-20 increased the number of PC12 cells in G_0/G_1, and decreased the cell number in S phase as well as the proportion of cells in the G_2 phase, further indicating that both compounds inhibited cell progress beyond the G_0/G_1 phase.

We recently presented a novel neuroprotective target for iron chelators regarding the aberrant cell cycle reentry of postmitotic neurons in AD. Accordingly, similar to cancer drug therapy, a newly therapeutic strategy for neurodegenerative diseases is currently directed at interfering with mitogenic signaling and cell cycle progression to ameliorate cell death because iron chelators have been shown to affect critical regulatory molecules involved in cell cycle arrest and proliferation [111, 120]. Indeed, our studies [111, 115] revealed that the multifunctional iron chelators M-30 and HLA-20 as well as EGCG have a profound impact that induced differentiation features in neuroblastoma and PC12 cells, including cell body elongation, stimulation of neurite outgrowth, and up-regulation of the growth-associated protein 43 (GAP-43). Taken together, the data suggest that iron chelators may be considered potential therapeutic agents in AD, targeting early cell cycle anomalies and reestablishing the synaptic connection lost in the injured neuronal cells.

26.6 SUMMARY

It has become apparent that iron progressively accumulates in the brain with age and neurodegeneration and thus has a role in the pathology of neurodegenerative disease. MAO activity is increased in brain of PD patients. Iron and MAO activity are associated with auto-oxidation and oxidative deamination of DA, resulting in the generation of ROS and the onset of OS, which induce neurodegeneration. Iron chelators (DFO, VK-28, and clioquinol) but not copper chelators have been shown to be neuroprotective in the 6-OHDA and MPTP models of PD, as well as the irreversible MAO-B inhibitor, rasagiline. These findings prompted the development of multifunctional anti-PD drugs possessing iron-chelating pharmacophore of VK-28 and the propargylamine-MAO inhibitory activity of rasagiline. M-30 is a potent, nontoxic iron chelator, radical scavenger, and brain-selective irreversible MAO-A and MAO-B inhibitor, with little inhibition of peripheral MAO. It has neuroprotective activity in *in vitro* and *in vivo* models of PD, and unlike selective MAO-B inhibitors, it increases brain DA, serotonin, and noradrenaline. These findings indicate that besides its anti-PD action, it may also possess antidepressant activity, similar to selective MAO-A and nonselective MAO inhibitors.

In addition, there is a direct link between iron metabolism and AD pathogenesis, demonstrated by the presence of an iron-responsive element in the 5′-UTR of the APP transcript. Indeed, we have recently demonstrated M-30's capacity to lower the expression of APP and the generation of Aβ. Considering a recent report, describing a putative IRE in the 5′-UTR of APP mRNA and PD-related α-synuclein mRNA [121], a parallel can be drawn between APP and α-synuclein both in the physiological and pathological aspects with respect to iron regulation. Indeed, it can be predicted that this RNA structure may have the potential to function as a posttranscriptional regulator of α-synuclein protein synthesis by an age-related iron and redox pathophysiology upstream of neurodegeneration. These latest findings implicate the therapeutic potential of our multifunctional iron chelator/MAO-B inhibitors as a neuroprotective-neurorescue drug (Figure 26.2 and Figure 26.3) candidate for treatment of neurodegenerative diseases.

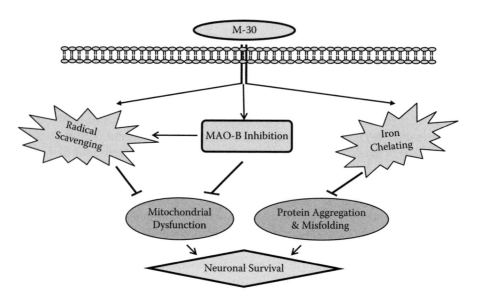

FIGURE 26.2 Schematic illustration of the multifactorial effects involved in the neuroprotective mechanism of action of the iron chelator, M-30.

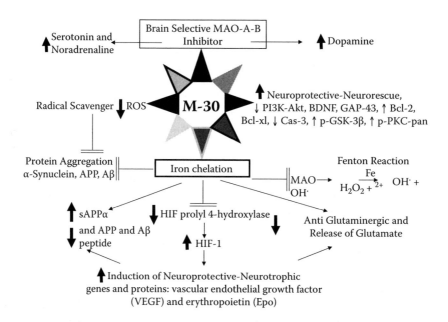

FIGURE 26.3 Multi-Targets for M-30, found to be a most potent iron chelator, possessing a highly effective inhibition of both MAO-A and MAO-B activities *in vitro* and *in vivo*. In addition, M-30 was demonstrated to possess a wide range of pharmacological activities, including neuritogenic effects associated with regulation of neurotrophic factors, cell signaling pathway and differentiation; activate a hypoxia signal transduction pathway and regulatory action on the amyloidogenic A_ peptide.

ACKNOWLEDGMENTS

We are grateful to the Alzheimer's Drug Discovery Foundation (ADDF), the Institute for the Study of Aging (New York), the Parker and Rochlin Funds, and Technion-Research and Development (Haifa, Israel) for the support of this work.

REFERENCES

1. Zecca, L., et al., Iron, brain ageing and neurodegenerative disorders. *Nat Rev Neurosci*, 5(11): 863. 2004.
2. Youdim, M. B., and J. J. Buccafusco, Multifunctional drugs for various CNS targets in the treatment of neurodegenerative disorders. *Trends Pharmacol Sci*, 26(1): 27. 2005.
3. Weinreb, O., et al., The application of proteomics and genomics to the study of age-related neurodegeneration and neuroprotection. *Antioxid Redox Signal*, 9(2): 169. 2007.
4. Zecca, L., et al., The role of iron and copper molecules in the neuronal vulnerability of locus coeruleus and substantia nigra during aging. *Proc Natl Acad Sci U S A*, 101(26): 9843. 2004.
5. Bartzokis, G., et al., Brain ferritin iron as a risk factor for age at onset in neurodegenerative diseases. *Ann N Y Acad Sci*, 1012: 224. 2004.
6. Halliwell, B., Role of free radicals in the neurodegenerative diseases: Therapeutic implications for antioxidant treatment. *Drugs Aging*, 18(9): 685. 2001.
7. Sayre, L. M., M. A. Smith, and G. Perry, Chemistry and biochemistry of oxidative stress in neurodegenerative disease. *Curr Med Chem*, 8(7): 721. 2001.
8. Przedborski, S., et al., The parkinsonian toxin MPTP: Action and mechanism. *Restor Neurol Neurosci*, 16(2): 135. 2000.
9. Mattson, M. P., Metal-catalyzed disruption of membrane protein and lipid signaling in the pathogenesis of neurodegenerative disorders. *Ann N Y Acad Sci*, 1012: 37. 2004.
10. LaVaute, T., et al., Targeted deletion of the gene encoding iron regulatory protein–2 causes misregulation of iron metabolism and neurodegenerative disease in mice. *Nature Genet*, 27(2): 209. 2001.
11. Youdim, M. B., M. Fridkin, and H. Zheng, Bifunctional drug derivatives of MAO-B inhibitor rasagiline and iron chelator VK-28 as a more effective approach to treatment of brain ageing and ageing neurodegenerative diseases. *Mech Ageing Dev*, 126(2): 317. 2005.
12. Youdim, M. B., My love with monoamine oxidase, iron and Parkinson's disease. *J Neural Transm Suppl*, (71): V. 2006.
13. Ben-Shachar, D., et al., Neuroprotection by a novel brain permeable iron chelator, VK-28, against 6-hydroxydopamine lesion in rats. *Neuropharmacology*, 46(2): 254. 2004.
14. Zheng, H., et al., Novel multifunctional neuroprotective iron chelator–monoamine oxidase drugs for neurodegenerative diseases: I. *In vitro* studies on antioxidant activity, prevention of lipid peroxide formation and monoamine oxidase inhibition. *J Neurochem*, 95(1): 68. 2005.
15. Shachar, D. B., et al., Neuroprotection by a novel brain permeable iron chelator, VK-28, against 6-hydroxydopamine lession in rats. *Neuropharmacology*, 46(2): 254. 2004.
16. Youdim, M. B., The path from anti Parkinson drug selegiline and rasagiline to multifunctional neuroprotective anti Alzheimer drugs ladostigil and m30. *Curr Alzheimer Res*, 3(5): 541. 2006.
17. Weinreb, O., et al., Neuroprotection via pro-survival protein kinase C isoforms associated with Bcl-2 family members. *FASEB J*, 18(12): 1471. 2004.
18. Bar-Am, O., et al., Regulation of Bcl-2 family proteins, neurotrophic factors, and APP processing in the neurorescue activity of propargylamine. *FASEB J* 19(13): 1899. 2005.
19. Gal, S., et al., Novel multifunctional neuroprotective iron chelator–monoamine oxidase inhibitor drugs for neurodegenerative diseases. In vivo selective brain monoamine oxidase inhibition and prevention of MPTP-induced striatal dopamine depletion. *J Neurochem*, 95(1): 79. 2005.
20. Riederer, P., et al., Neurochemical perspectives to the function of monoamine oxidase. *Acta Neurol Scand Suppl*, 126: 41. 1989.
21. Jenner, P., and C. W. Olanow, Oxidative stress and the pathogenesis of Parkinson's disease. *Neurology*, 47(6 Suppl 3): S161. 1996.
22. Bharath, S., et al., Glutathione, iron and Parkinson's disease. *Biochem Pharmacol*, 64(5–6): 1037. 2002.
23. Gerlach, M., et al., Potential sources of increased iron in the substantia nigra of parkinsonian patients. *J Neural Transm Suppl*, (70): 133. 2006.

24. Gotz, M. E., et al., The relevance of iron in the pathogenesis of Parkinson's disease. *Ann N Y Acad Sci,* 1012: 193. 2004.

25. Gerlach, M., et al., Strategies for the protection of dopaminergic neurons against neurotoxicity. *Neurotox Res,* 2(2–3): 99. 2000.

26. Youdim, M. B., G. Stephenson, and D. Ben Shachar, Ironing iron out in Parkinson's disease and other neurodegenerative diseases with iron chelators: A lesson from 6-hydroxydopamine and iron chelators, desferal and VK-28. *Ann N Y Acad Sci,* 1012: 306. 2004.

27. Jellinger, K., et al., Brain iron and ferritin in Parkinson's and Alzheimer's diseases. *J Neural Transm,* 2: 327. 1990.

28. Salvatore, M. F., et al., Neurochemical investigations of dopamine neuronal systems in iron-regulatory protein 2 (IRP-2) knockout mice. *Brain Res Mol Brain Res,* 139(2): 341. 2005.

29. Crichton, R. R., and J. L. Pierre, Old iron, young copper: From Mars to Venus. *Biometals,* 14(2): 99. 2001.

30. Ye, F. Q., P. S. Allen, and W. R. Martin, Basal ganglia iron content in Parkinson's disease measured with magnetic resonance. *Mov Disord,* 11(3): 243. 1996.

31. Lan, J., and D. H. Jiang, Desferrioxamine and vitamin E protect against iron and MPTP-induced neuro-degeneration in mice. *J Neural Transm,* 104(4–5): 469. 1997.

32. Han, J., et al., Inhibitors of mitochondrial respiration, iron (II), and hydroxyl radical evoke release and extracellular hydrolysis of glutathione in rat striatum and substantia nigra: Potential implications to Parkinson's disease. *J Neurochem,* 73(4): 1683. 1999.

33. Youdim, M. B., Y. S. Bakhle. Monoamine oxidase: isoforms and inhibitors in Parkinson's disease and depressive illness. *Br J Pharmacol* 147 Suppl 1: S287–296. 2006.

34. Youdim, M. B. H., and P. F. Riederer, A review of the mechanisms and role of monoamine oxidase inhibi-tors in Parkinson's disease. *Neurology,* 63(7 Suppl 1): S32. 2004.

35. Jellinger, K., et al., Chemical evidence for 6-hydroxydopamine to be an endogenous toxic factor in the pathogenesis of Parkinson's disease. *J Neural Transm Suppl,* 46: 297. 1995.

36. Linert, W., et al., Dopamine, 6-hydroxydopamine, iron, and dioxygen—Their mutual interactions and possible implication in the development of Parkinson's disease. *Biochim Biophys Acta,* 1316(3): 160. 1996.

37. Napolitano, M., et al., Experimental parkinsonism modulates multiple genes involved in the transduction of dopaminergic signals in the striatum. *Neurobiol Dis,* 10(3): 387. 2002.

38. Maharaj, H., et al., l-DOPA administration enhances 6-hydroxydopamine generation. *Brain Res,* 1063(2): 180. 2005.

39. Braak, H., and K. Del Tredici, Poor and protracted myelination as a contributory factor to neurodegenera-tive disorders. *Neurobiol Aging,* 25(1): 19. 2004.

40. Cohen, G., Oxidative stress, mitochondrial respiration, and Parkinson's disease. *Ann N Y Acad Sci,* 899: 112. 2000.

41. Fowler, J. S., et al., PET imaging of monoamine oxidase B in peripheral organs in humans. *J Nucl Med,* 43(10): 1331. 2002.

42. Fowler, C. J., et al., The effect of age on the activity and molecular properties of human brain monoamine oxidase. *J Neural Transm,* 49(1–2): 1. 1980.

43. Ben-Shachar, D., and M. B. Youdim, Intranigral iron injection induces behavioral and biochemical "parkinsonism" in rats. *J Neurochem,* 57(6): 2133. 1991.

44. Gassen, M., and M. B. Youdim, The potential role of iron chelators in the treatment of Parkinson's dis-ease and related neurological disorders. *Pharmacol Toxicol,* 80(4): 159. 1997.

45. Kaur, D., et al., Genetic or pharmacological iron chelation prevents MPTP-induced neurotoxicity in vivo: A novel therapy for Parkinson's disease. *Neuron,* 37(6): 899. 2003.

46. Mytilineou, C., and G. Cohen, Deprenyl protects dopamine neurons from the neurotoxic effect of 1-methyl-4-phenylpyridinium ion. *J Neurochem,* 45(6): 1951. 1985.

47. Sagi, Y., et al., Activation of tyrosine kinase receptor signaling pathway by rasagiline facilitates neurores-cue and restoration of nigrostriatal dopamine neurons in post-MPTP–induced parkinsonism. *Neurobiol Dis,* 25(1): 35. 2007.

48. Sagi, Y., M. Weinstock, and M. B. H. Youdim, Attenuation of MPTP-induced dopaminergic neurotoxicity by TV3326, a cholinesterase-monoamine oxidase inhibitor. *J Neurochem,* 2(86): 290. 2003.

49. Ben-Shachar, D., et al., The iron chelator desferrioxamine (Desferal) retards 6-hydroxydopamine-induced degeneration of nigrostriatal dopamine neurons. *J Neurochem,* 56(4): 1441. 1991.

50. Ben-Shachar, D., et al., Role of iron and iron chelation in dopaminergic-induced neurodegeneration: Implication for Parkinson's disease. *Ann Neurol,* 32 Suppl: S105. 1992.

51. Lan, J., and D. H. Jiang, Desferrioxamine and vitamin E protect against iron and MPTP-induced neuro-degeneration in mice. *J Neural Transm (Budapest),* 104(4–5): 469. 1997.

52. Lan, J., and D. H. Jiang, Excessive iron accumulation in the brain: A possible potential risk of neurode-generation in Parkinson's disease. *J Neural Transm,* 104(6–7): 649. 1997.

53. Sangchot, P., et al., Deferoxamine attenuates iron-induced oxidative stress and prevents mitochondrial aggre-gation and a-synuclein translocation in SK-N-SH Cells in Culture. *Dev Neurosci,* 24(2–3): 143. 2002.

54. Cario, H., et al., Recent developments in iron chelation therapy. *Klin Padiatr,* 219(3): 158. 2007.

55. Richardson, D. R., Novel chelators for central nervous system disorders that involve alterations in the metabolism of iron and other metal ions. *Ann N Y Acad Sci,* 1012: 326. 2004.

56. Gilgun-Sherki, Y., E. Melamed, and D. Offen, Oxidative stress induced-neurodegenerative diseases: The need for antioxidants that penetrate the blood brain barrier. *Neuropharmacology,* 40(8): 959. 2001.

57. Youdim, M. B., M. Fridkin, and H. Zheng, Novel bifunctional drugs targeting monoamine oxidase inhibi-tion and iron chelation as an approach to neuroprotection in Parkinson's disease and other neurodegen-erative diseases. *J Neural Transm,* 111(10–11): 1455. 2004.

58. Zheng, H., et al., Design, synthesis and evaluation of novel bifunctional iron-chelators as potential agents for neuroprotection in Alzheimer's, Parkinson's, and other neurodegenerative diseases. *Bioorg Med Chem.* 2005 13(3): 773–83.

59. Gal, S., et al., Novel multifunctional neuroprotective iron chelator-monoamine oxidase inhibitor drugs for neurodegenerative diseases. II. In vivo selective brain monoamine oxidase inhibition and prevention of MPTP induced striatal dopamine depletion. *J Neurochem,* 95(1): 79. 2005.

60. Kalir, A., A. Sabbagh, and M. B. H. Youdim, Selective acetylenic "suicide" and reversible inhibitors of monoamine oxidase types A and B. *Br J Pharmacol,* 73(1): 55. 1981.

61. Melamed, E., and M. B. H. Youdim, Prevention of dopaminergic toxicity of MPTP in mice by phenyleth-ylamine, a specific substrate of type B monoamine oxidase. *Br J Pharmacol,* 86(3): 529. 1985.

62. Youdim, M. B. H., The active centers of monoamine oxidase types "A" and "B": Binding with (14C)-clorgyline and (14C)-deprenyl. *J Neural Transm,* 43(3–4): 199. 1978.

63. Binda, C., et al., Crystal structures of monoamine oxidase B in complex with four inhibitors of the *N*-propargylaminoindan class. *J Med Chem,* 47(7): 1767. 2004.

64. Youdim, M. B. H., and M. Weinstock, Novel Neuroprotective anti-Alzheimer drugs with antidepressant activity derived from the anti-Parkinson drug, rasagiline. *Mech Ageing Dev,* 123(8): 1081. 2002.

65. Hubalek, F., et al., Inactivation of purified human recombinant monoamine oxidases A and B by rasagi-line and its analogues. *J Med Chem,* 25;47(7): 1760. 2004.

66. Weinstock, M., et al., TV3326, a novel neuroprotective drug with cholinesterase and monoamine oxidase inhibitory activities for the treatment of Alzheimer's disease. *J Neural Transm Suppl,* 60: S157. 2000.

67. Weinstock, M., et al., Effect of TV3326, a novel monoamine-oxidase cholinesterase inhibitor, in rat mod-els of anxiety and depression. *Psychopharmacology (Berlin),* 160(3): 318. 2002.

68. Borisenko, G. G., et al., Interaction between 6-hydroxydopamine and transferrin: "Let my iron go". *Biochemistry,* 39(12): 3392. 2000.

69. Tian, B., et al., Chemiluminescence analysis of the prooxidant and antioxidant effects of epigallocate-chin-3-gallate. *Asia Pac J Clin Nutr,* 16 Suppl 1: 153. 2007.

70. van Acker, S. A., et al., Structural aspects of antioxidant activity of flavonoids. *Free Radic Biol Med,* 20(3): 331. 1996.

71. Elmer, L., et al., Rasagiline-associated motor improvement in PD occurs without worsening of cognitive and behavioral symptoms. *J Neurol Sci,* 248(1–2): 78. 2006.

72. A randomized placebo-controlled trial of rasagiline in levodopa-treated patients with Parkinson disease and motor fluctuations: The PRESTO study. *Arch Neurol,* 62(2): 241. 2005.

73. Levites, Y., et al., Green tea polyphenol (−)-Epigallocatechin-3-gallate prevents *N*-methyl-4-phenyl-1,2,3,6-tetrahydropyridine-induced dopaminergic neurodegeneration. *J Neurochem,* 78: 1073. 2001.

74. Weinreb, O., S. Mandel, and M. B. H. Youdim, cDNA gene expression profile homology of antioxidants and their antiapoptotic and proapoptotic activities in human neuroblastoma cells. *FASEB J,* 17(8): 935. 2003.

75. Halliwell, B., Vitamin C: Antioxidant or pro-oxidant in vivo? *Free Radic Res,* 25(5): 439. 1996.

76. Gassen, M., B. Pinchasi, and M. B. Youdim, Apomorphine is a potent radical scavenger and protects cul-tured pheochromocytoma cells from 6-OHDA and H_2O_2-induced cell death. *Adv Pharmacol (New York),* 42: 320. 1998.

77. Gassen, M., A. Gross, and M. B. H. Youdim, Apomorphine, a dopamine receptor agonist with remarkable antioxidant and cytoprotective properties. *Adv Neurol,* 80: 297. 1999.

78. Porter, J. B., et al., Kinetics of removal and reappearance of non-transferrin-bound plasma iron with deferoxamine therapy. *Blood,* 88(2): 705. 1996.

79. Singh, S., et al., Therapeutic iron chelators and their potential side-effects. *Biochem Soc Symp,* 61: 127. 1995.

80. Maruyama, W., M. Naoi, and M. B. H. Youdim, Propargylamines protect dopamine cells from apoptosis induced by a neurotoxin, *N*-methyl (*R*) salsolinol, in neuromelanin, may mediate neurotoxicity via its interaction with redox active iron, S. Alexander and M. A. Gollins, eds., *Neuromelanin may mediate neurotoxicity via its interaction with redox active iron.* Kluwer Academic/Plenum Publishers: New York. p. 321. 2000.

81. Maruyama, W., et al., *N*-Propargyl-1 (*R*)-aminoindan, rasagiline, increases glial cell line-derived neurotrophic factor (GDNF) in neuroblastoma SH-SY5Y cells through activation of NF-kappaB transcription factor. *Neurochem Int,* 44(6): 393. 2004.

82. Maruyama, W., et al., Neuroprotection by propargylamines in Parkinson's disease. Suppression of apoptosis and induction of prosurvival genes. *Neurotoxicol Teratol,* 24(5): 675. 2002.

83. Gal, S., et al., M30, a novel multifunctional neuroprotective drug with potent iron chelating and brain selective monoamine oxidase–ab inhibitory activity for Parkinson's disease. *J Neural Transm Suppl,* (70): 447. 2006.

84. Cherny, R. A., et al., Treatment with a copper-zinc chelator markedly and rapidly inhibits beta-amyloid accumulation in Alzheimer's disease transgenic mice. *Neuron,* 30(3): 665. 2001.

85. Glinka, Y., M. Gassen, and M. B. Youdim, Mechanism of 6-hydroxydopamine neurotoxicity. *J Neural Transm Suppl,* 50: 55. 1997.

86. Atabay, C., et al., Removal of serum from primary cultures of cerebellar granule neurons induces oxidative stress and DNA fragmentation: Protection with antioxidants and glutamate receptor antagonists. *J Neurosci Res,* 43(4): 465. 1996.

87. Offen, D., et al., Prevention of dopamine-induced cell death by thiol antioxidants: Possible implications for treatment of Parkinson's disease. *Exp Neurol,* 141(1): 32. 1996.

88. Chiou, T. J., Y. T. Wang, and W. F. Tzeng, DT-diaphorase protects against menadione-induced oxidative stress. *Toxicology,* 139(1–2): 103. 1999.

89. Grunblatt, E., et al., Effects of *R*-apomorphine and *S*-apomorphine on MPTP-induced nigro-striatal doamine neuronal loss. *J Neurochem,* 77(1): 146. 2001.

90. Thomas, C. E., et al., Radical trapping and inhibition of iron-dependent CNS damage by cyclic nitrone spin traps. *J Neurochem,* 68(3): 1173. 1997.

91. Lovell, M. A., et al., Copper, iron and zinc in Alzheimer's disease senile plaques. *J Neurol Sci,* 158(1): 47. 1998.

92. Pinero, D. J., J. Hu, and J. R. Connor, Alterations in the interaction between iron regulatory proteins and their iron responsive element in normal and Alzheimer's diseased brains. *Cell Mol Biol (Noisy-Le-Grand),* 46(4): 761. 2000.

93. Smith, M. A., et al., Amyloid-beta deposition in Alzheimer transgenic mice is associated with oxidative stress. *J Neurochem,* 70(5): 2212. 1998.

94. Moreira, P. I., et al., Oxidative stress mechanisms and potential therapeutics in Alzheimer disease. *J Neural Transm,* 112(7): 921. 2005.

95. Castellani, R. J., et al., Contribution of redox-active iron and copper to oxidative damage in Alzheimer disease. *Ageing Res Rev,* 3(3): 319. 2004.

96. Honda, K., et al., Oxidative stress and redox-active iron in Alzheimer's disease. *Ann N Y Acad Sci,* 1012: 179. 2004.

97. Takeda, A., et al., Overexpression of heme oxygenase in neuronal cells, the possible interaction with Tau. *J Biol Chem,* 275(8): 5395. 2000.

98. Shimohama, S., et al., Activation of NADPH oxidase in Alzheimer's disease brains. *Biochem Biophys Res Commun,* 273(1): 5. 2000.

99. Honda, K., et al., Ribosomal RNA in Alzheimer disease is oxidized by bound redox-active iron. *J Biol Chem,* 280(22): 20978. 2005.

100. Connor, J. R., et al., Regional distribution of iron and iron-regulatory proteins in the brain in aging and Alzheimer's disease. *J Neurosci Res,* 31(2): 327. 1992.

101. Moalem, S., et al., Are hereditary hemochromatosis mutations involved in Alzheimer disease? *Am J Med Genet,* 93(1): 58. 2000.

102. Bush, A. I., The metallobiology of Alzheimer's disease. *Trends Neurosci,* 26(4): 207. 2003.

103. House, E., et al., Aluminium, iron, zinc and copper influence the *in vitro* formation of amyloid fibrils of Abeta42 in a manner which may have consequences for metal chelation therapy in Alzheimer's disease. *J Alzheimers Dis,* 6(3): 291. 2004.

104. Cherny, R. A., et al., Aqueous dissolution of Alzheimer's disease Abeta amyloid deposits by biometal depletion. *J Biol Chem,* 274(33): 23223. 1999.

105. Rogers, J. T., and D. K. Lahiri, Metal and inflammatory targets for Alzheimer's disease. *Curr Drug Targets*, 5(6): 535. 2004.
106. Rogers, J., et al., Inflammation and Alzheimer's disease pathogenesis. *Neurobiol Aging*, 17(5): 681. 1996.
107. Crapper McLachlan, D. R., et al., Intramuscular desferrioxamine in patients with Alzheimer's disease. *Lancet*, 337(8753): 1304. 1991.
108. Ritchie, C. W., et al., Metal-protein attenuation with iodochlorhydroxyquin (clioquinol) targeting Abeta amyloid deposition and toxicity in Alzheimer disease: A pilot phase 2 clinical trial. *Arch Neurol*, 60(12): 1685. 2003.
109. Payton, S., et al., Drug discovery targeted to the Alzheimer's APP mRNA 5′-untranslated region: The action of paroxetine and dimercaptopropanol. *J Mol Neurosci*, 20(3): 267. 2003.
110. Dedeoglu, A., et al., Preliminary studies of a novel bifunctional metal chelator targeting Alzheimer's amyloidogenesis. *Exp Gerontol*, 39(11–12): 1641. 2004.
111. Avramovich-Tirosh, Y., et al., Therapeutic targets and potential of the novel brain-permeable multifunctional iron chelator–monoamine oxidase inhibitor drug, M-30, for the treatment of Alzheimer's disease. *J Neurochem*, 100(2): 490. 2007.
112. Guo, Q., et al., Studies on protective mechanisms of four components of green tea polyphenols against lipid peroxidation in synaptosomes. *Biochim Biophys Acta*, 1304(3): 210. 1996.
113. Kumamoto, M., et al., Effects of pH and metal ions on antioxidative activities of catechins. *Biosci Biotechnol Biochem*, 65(1): 126. 2001.
114. Levites, Y., et al., Involvement of protein kinase C activation and cell survival/cell cycle genes in green tea polyphenol (−)-epigallocatechin-3-gallate neuroprotective action. *J Biol Chem*, 277(34): 30574. 2002.
115. Reznichenko, L., et al., Reduction of iron-regulated amyloid precursor protein and beta-amyloid peptide by (−)-epigallocatechin-3-gallate in cell cultures: Implications for iron chelation in Alzheimer's disease. *J Neurochem*, 97(2): 527. 2006.
116. Rezai-Zadeh, K., et al., Green tea epigallocatechin-3-gallate (EGCG) modulates amyloid precursor protein cleavage and reduces cerebral amyloidosis in Alzheimer transgenic mice. *J Neurosci*, 25(38): 8807. 2005.
117. Avramovich-Tirosh, Y., et al., Neurorescue activity, APP regulation and amyloid-beta peptide reduction by novel multi-functional brain permeable iron-chelating-antioxidants, M-30 and green tea polyphenol, EGCG. *Curr Alzheimer Res*, 4(4): 403. 2007.
118. Copani, A., et al., Mitotic signaling by beta-amyloid causes neuronal death. *FASEB J*, 13(15): 2225. 1999.
119. Wu, Q., et al., Beta-amyloid activated microglia induce cell cycling and cell death in cultured cortical neurons. *Neurobiol Aging*, 21(6): 797. 2000.
120. Amit, T., et al., Targeting multiple Alzheimer's disease etiologies with multimodal neuroprotective and neurorestorative iron chelators. *FASEB J*, 22(5): 1296. 2008.
121. Friedlich, A. L., R. E. Tanzi, and J. T. Rogers, The 5′-untranslated region of Parkinson's disease alpha-synuclein messenger RNA contains a predicted iron responsive element. *Mol Psychiatry*, 12(3): 222. 2007.

27 Antioxidative Defense of Brain Microglial Cells

Ralf Dringen[1,2] and Johannes Hirrlinger[3]

[1]Centre for Biomolecular Interactions Bremen,
University of Bremen, Bremen, Germany

[2]School of Psychology, Psychiatry, and Psychological Medicine,
Monash University, Clayton, Victoria, Australia

[3]Interdisciplinary Centre for Clinical Research,
University of Leipzig, Leipzig, Germany

CONTENTS

27.1 INTRODUCTION

Microglial cells are the neuroprotective immunocompetent cells in the central nervous system and have important functions in the normal as well as in the injured and inflamed brain.[1-4] In the healthy brain, microglial cells perform a continuous active tissue scanning for an altered microenvironment.[5] If this environment is disturbed, microglial cells respond with a program of protective activities that are part of the innate defense mechanisms of the brain and assist in specific immune functions and repair processes. After this activation, microglial cells produce and release a large number of neuroactive substances that include various cytokines, prostanoids, and proteases as well as radicals such as nitric oxide and superoxide.[1,2,6] Microglial activation primarily aims at the protection of the brain, since microglia-derived compounds contribute to the defense of the brain against pathogens and are important for the stimulation of repair processes after brain damage. Superoxide and NO support the cellular defense but have also the potential to harm the reactive oxygen species (ROS)-producing cell as well as neighboring cells. In addition, inappropriate activation can lead to cell degeneration. A variety of neurotoxins have been reported to cause neurotoxicity through the ROS that are generated by microglial nicotinamide adenine dinucleotide phosphate (reduced form) (NADPH) oxidase (Nox).[2] Therefore, microglial cells have also been considered as the enemy within the brain that is harmful for other brain cells due to the release of inflammatory cytokines and toxic ROS. The various aspects of microglial functions and dysfunctions for physiological and pathological situations in brain as well as the potential of microglial cells as pharmacological target for diseases have recently been reviewed in detail.[1–4,7–9] Here we will focus on the antioxidative potential of microglial cells that contributes to the self-defense of microglial cells against the ROS that these cells produce after activation.

27.2 GENERATION OF REACTIVE OXYGEN SPECIES BY MICROGLIAL CELLS

ROS are continuously produced during normal oxidative metabolism in all cells that use oxygen. Such ROS are normally detoxified by the endogenous cellular antioxidative machinery to avoid ROS-mediated cellular damage caused by lipid peroxidation, protein modification, and DNA strand breaks. However, cell types such as peripheral leucocytes or brain microglial cells make use of the toxic potential of ROS to defend the body against invading microorganisms. Contact to substances that signal cell damage or presence of pathogens activates microglial cells in brain to generate radicals such as superoxide and NO that are the products of the reactions catalyzed by the enzymes Nox and nitric oxide synthase (NOS), respectively.[6,10–13] The radicals superoxide and NO can directly react with cellular components. In addition, these radicals are substrates for the formation of other ROS. Superoxide can spontaneously disproportionate to H_2O_2 and oxygen, while NO reacts easily with other radicals. For example, NO and superoxide form the highly toxic peroxynitrite in a diffusion-limited reaction.[14]

Superoxide is produced by activated microglial cells by the enzymatic activity of Nox, which releases superoxide to the extracellular space. Active Nox is a multisubunit protein complex that reduces molecular oxygen to superoxide. The electron required for this reduction is derived from cellular NADPH. Activation of cells leads to the assembly of cytosolic protein subunits of Nox with the membrane-associated Nox proteins to the active complex that produces superoxide.[12,15,16] Expression of mRNAs of all Nox subunits and presence of most of the Nox proteins have been reported for microglial cells.[16–20] Essential for the superoxide production by Nox are especially the subunits gp91phox and p47phox. Microglial cultures that are derived from mice that are deficient in one of these two subunits are unable to produce superoxide following stimulation.[19,21] Within minutes after activation, cultured microglial cells produce substantial amounts of superoxide[18,22], demonstrating that the assembly of the subunits of Nox to the active enzyme is a very rapid process.

NO is produced in an enzymatic reaction that is catalyzed by NOS.[23] NOS use arginine and molecular oxygen as substrates to generate NO and citrulline in a five-electron transfer reaction in which NADPH serves as electron donor. Of the three isoforms of NOS, iNOS is induced upon activation in microglial cells *in vitro* and *in vivo*.[10,11] In contrast to the activation of superoxide production by Nox, the induction of iNOS requires protein synthesis that causes a delay of several hours before a sustained microglial NO production can be observed.[24]

27.3 ANTIOXIDATIVE DEFENSE IN MICROGLIAL CELLS

Oxidative stress is the consequence of an imbalance in the cellular generation and disposal of ROS that leads to oxidative damage by an elevated intracellular concentration of ROS. Since activated microglial cells are likely to encounter high concentrations of the generated superoxide and NO as well as of their reactive reaction products H_2O_2 and peroxynitrite, these cells contain various antioxidative mechanisms that remove ROS or prevent their generation to avoid oxidative damage of microglial cells themselves.[10] In addition to the low molecular weight antioxidants such as

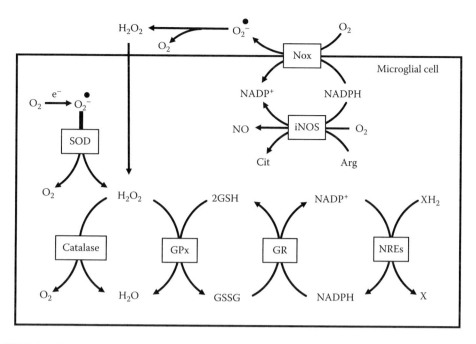

FIGURE 27.1 Generation and disposal of superoxide and hydrogen peroxide in microglial cells. Superoxide that is a product of NADPH oxidase (Nox) and by-product of the mitochondrial respiratory chain is converted to oxygen and H_2O_2 in spontaneous or SOD-catalyzed reaction. The peroxide is disposed of by catalase and/or glutathione peroxidases (GPx). Glutathione (GSH) serves as electron donor for the reactions catalyzed by GPx and is oxidized to glutathione disulfide (GSSG). The GSH consumed in the GPx reaction is regenerated by glutathione reductase (GR). NADPH-regenerating enzymes (NREs) oxidize their substrates (XH_2) to reduce $NADP^+$. Inducible NO synthase (iNOS) uses arginine (Arg) and oxygen as substrates to produce NO and citrulline (Cit) in an NADPH-dependent reaction.

glutathione (GSH), ascorbate, and α-tocopherol, microglial cells contain antioxidative enzymes such as superoxide dismutases (SODs), catalase, glutathione peroxidase(s) (GPx), and glutathione reductase (GR) that contribute substantially to the cellular defense against ROS.

Figure 27.1 gives an overview on the most important mechanisms that are involved in the formation and detoxification of ROS in microglial cells. Superoxide that is generated extracellularly by Nox[16] or in mitochondria as a side product of the respiratory chain[25] is converted to oxygen and H_2O_2 either spontaneously or by SOD-catalyzed disproportionation. Mitochondrial superoxide is a substrate of the manganese-containing SOD (MnSOD), while cytosolic and extracellular superoxide are substrates of SODs that contain copper and zinc (Cu/ZnSODs).[26,27] The H_2O_2 that is generated from the Nox-derived superoxide will be cleared after entering the cells by catalase and/or by the GSH-dependent reduction catalyzed by GPx (Figure 27.1). In the GPx reaction, GSH is oxidized to glutathione disulfide (GSSG), which is reduced again to GSH by the NADPH-consuming GR. Thus, cells require an efficient supply of NADPH for efficient peroxide reduction by GSH redox cycling via GPx and GR.

27.3.1 Low Molecular Weight Antioxidants in Microglial Cells

The first line of cellular defense against radicals are low molecular weight antioxidants like GSH, ascorbate, α-tocopherol, and urate, since these compounds react directly and quickly with radicals such as superoxide, NO, or the hydroxyl radical. GSH is an essential cellular antioxidant that is present in all types of brain cells.[28,29] Cell culture studies revealed that microglial cells have a high intracellular concentration of GSH. Enriched microglial cultures contain even a higher specific GSH content than cultured astrocytes, neurons, and oligodendrocytes.[29–34] The high intracellular GSH concentration in microglial cells is likely to contribute strongly to the defense against radical- and peroxide-mediated damage. The GSH content of cultured microglial cells increases strongly after exposure to tumor necrosis factor α (TNF-α) and lipopolysaccharide (LPS),[35–37] while induction of iNOS lowers the cellular GSH content in microglial cells.[38] In addition to GSH, microglial cells contain and release substantial amounts of the low molecular weight antioxidant uric acid.[39]

The antioxidative potential of α-tocopherol and ascorbate in microglial cells has not been studied in detail so far. However, these vitamins have interesting effects on microglial behavior and functions. For example, ascorbate inhibits cyclooxygenase 2 activity and prostaglandin E_2 synthesis in LPS-treated rat microglial cells.[40] Treatment of microglial cultures with α-tocopherol and ascorbate improves the survival of the cells, induces a ramified microglial morphology, and down-regulates adhesion molecules.[41] α-Tocopherol is also a highly potent stimulator of the proliferation of cultured microglial cells[42] and improves the degradation of proteins in these cells.[43] In addition, treatment with α-tocopherol lowers the cellular peroxide concentration, reduces LPS-induced lipid peroxidation in cultured microglial cells and brain,[44] and prevents microglial superoxide production by Nox.[45] Moreover, presence of tocopherol long chain fatty alcohols decreases the production of TNF-α and NO by activated microglial cells.[46] Thus, α-tocopherol and ascorbate are likely to play an important role in microglial redox signaling as well as in prevention of oxidative damage of these cells.

27.3.2 Antioxidative Enzymes in Microglial Cells

The expression of several antioxidative enzymes has been reported for microglial cells in brain sections as well as in brain cell cultures.

27.3.2.1 Superoxide Dismutase

Microglial cells in rat brain slices show a low level of immunocytochemically detectable cytosolic Cu/ZnSOD, while the mitochondrial MnSOD was not detectable in these cells.[47] However, after excitotoxically induced neurodegeneration by injection of quinolinic acid into rat brain, MnSOD-immunolabeled mitochondria were observed in activated microglial cells.[48] In addition, both Cu/ZnSOD and MnSOD were detected in microglial cells after transient cerebral ischemia in rat hippocampus.[49] Microglial cells in cell cultures also contain both isoforms of SOD.[32,47,50,51] The specific activity of MnSOD in microglial cultures is 20 and 4 times higher than that in cultured astrocytes and oligodendrocytes, respectively, while specific activities of Cu/ZnSOD are almost identical in these three types of neural cell cultures.[32]

The content of MnSOD in microglial cells is up-regulated by activation and oxidative stress. Treatment of mixed glial cell cultures with a ROS-generating system increased MnSOD immunoreactivity in microglial cells,[50] while stimulation of cultured microglial cells with LPS and interferon γ or with TNF-α caused induction of mitochondrial MnSOD.[35,51] An increase in the activity of MnSOD in microglial cells is likely to improve the potential of the cells to decompose mitochondrial superoxide. Since SODs compete for superoxide with the iron-catalyzed Haber-Weiss cycle and with NO, an elevated activity of MnSOD in activated microglial cells is likely to reduce the risk of mitochondrial damage by superoxide-derived hydroxyl radicals and peroxynitrite.

27.3.2.2 Glutathione Peroxidase

Immunohistochemical studies revealed strong immunoreactivity for GPx in microglial cells in rat[52] and human brain.[53] In rat brain, GPx was found up-regulated after administration of quinolinic acid, which was discussed as an important mechanism to withstand oxidative stress.[52] GPx was also localized in microglial cells by immunocytochemical staining of glial primary cell cultures.[51] The specific activity of GPx in microglial cultures is similar to that in cultured astrocytes, lower than that of cultured oligodendrocytes, but higher than the GPx activity of cultured neurons.[29,31,32]

27.3.2.3 Glutathione Reductase

In contrast to SODs and GPx, little is known on GR in microglial cells. Immunocytochemical staining demonstrated that among the different glial cells types in astroglia-rich cultures, microglial cells stained strongly for GR.[54] The prominent expression of GR in microglial cells was confirmed by activity measurements. Microglial cultures contain higher specific GR activities than cultured neurons and astrocytes but less than cultured oligodendrocytes.[29,31]

27.3.2.4 Catalase

The peroxisomal catalase is an important enzyme in cellular H_2O_2 clearance. In sections of human brain, catalase immunoreactivity was found for all cell types including microglial cells.[55] Also, microglial cells in mixed glial cultures from rat brain were immunopositive for catalase.[51] This immunocytochemical evidence for expression of catalase was confirmed by activity measurements. The specific catalase activity of microglia cultures was reported to be similar[32] to or significantly lower[29,31,34] than that of cultured neurons, astrocytes, and oligodendrocytes.

27.3.2.5 NADPH Regenerating Enzymes

NADPH is required for efficient regeneration by GR of GSH from the GSSG that accumulates during GSH-dependent detoxification of ROS. However, the microglial prooxidative enzymes Nox and iNOS also use NADPH to generate superoxide and NO, respectively. Several NADPH-regenerating enzymes maintain a high ratio of cellular NADPH to nicotinamide adenine dinucleotide phosphate (oxidized form) ($NADP^+$). In addition to the two pentose phosphate pathway enzymes glucose-6-phosphate dehydrogenase (G6PDH) and 6-phosphogluconate dehydrogenase (6PGDH), malic enzymes (MEs), $NADP^+$-dependent isocitrate dehydrogenases (ICDHs), and mitochondrial transhydrogenase also contribute to the NADPH-formation in brain cells.[56] Of these enzymes, G6PDH and 6PGDH are exclusively localized in the cytosol,[56,57] while transhydrogenase is exclusively localized in mitochondria.[58] In contrast, both cytosolic and mitochondrial isoforms of the enzymes ME and ICDH have been reported to be expressed in brain cells.[59,60]

In the cytosol of microglial cells, NADPH is likely to be regenerated by G6PDH and 6PGDH, since substantial activities of G6PDH and 6PGDH were measured in cultured microglial cells, while ME activity was not detectable in these cultures (L. Kussmaul and R. Dringen, unpublished results). In addition, cultured microglial cells contain cytosolic ICDH.[60] Thus, at least two cytosolic pathways can contribute to the NADPH regeneration in microglial cells.

The mitochondrial GSH system is considered to contribute strongly to the antioxidative potential of microglial mitochondria.[10] Since the supply of NADPH as electron donor is essential for GSH redox cycling, the availability of this reduced cosubstrate is crucial for efficient mitochondrial peroxide removal. Mitochondrial transhydrogenase as well as the mitochondrial isoforms of ME and ICDH contribute to the NADPH regeneration during GSSG reduction in isolated brain mitochondria.[61] Since microglial cells contain substantial activities of mitochondrial ICDH[60] and are known to express transhydrogenase,[58] at least mitochondrial ICDH and transhydrogenase are likely to supply the NADPH for mitochondrial GSSG reduction in microglial cells.

In activated microglial cells, NADPH supplies also the essential reduction equivalents for the reactions catalyzed by the prooxidative enzymes iNOS and Nox. The consequence of insufficient NADPH regeneration in activated microglial cells would be a competition of the prooxidative enzymes Nox and iNOS with the antioxidative enzyme GR for the substrate NADPH. Thus, efficient replenishment of the cellular pool of NADPH is crucial in activated microglial cells for both prooxidative and antioxidative enzymes.

27.3.2.6   Other Enzymes

Several other enzymes have been considered to contribute to the antioxidative defense of different types of cells; however, little is known about the expression and antioxidative functions of these proteins in microglial cells. For example, peroxiredoxin 1 is strongly expressed in cultured microglial cells, while little immunoreactivity for this enzyme was detected in microglial cells in brain sections.[62] However, peroxiredoxin 1 is strongly induced in microglial cells after hemorrhage.[63] The bilirubin-generating enzyme heme oxygenase 1 has been reported to be induced in brain microglial cells following transient focal brain ischemia,[64] intracerebral hemorrhage,[63] cortical lesions,[65] or thiamine-deficiency-induced neurodegeneration.[66] Interestingly, heme oxygenase 1 expression in microglial cells is induced by an astrocyte-derived factor.[67] These observations indicate that peroxiredoxin 1 as well as heme oxgenase 1 might contribute to microglial redox homeostasis at least in the activated state.

27.4   PEROXIDE METABOLISM

After activation, microglial cells are exposed to a substantial extracellular concentration of H_2O_2[68] that is generated spontaneously from the superoxide that is produced by Nox. In culture, microglial cells clear exogenous H_2O_2 in a reaction that follows first-order kinetics.[31] The specific H_2O_2 detoxification rate constant of cultured microglial cells is almost identical to those calculated for astrocyte and neuron cultures but lower than that of cultured oligodendrocytes.[29,34]

Exposure of microglial cells in culture to H_2O_2 or to a H_2O_2-generating system increases the microglial GSSG content to 30% of the total cellular glutathione,[31] demonstrating that GPx is involved in the peroxide clearance by microglial cells. The increase in microglial GSSG after H_2O_2 exposure is transient. Removal of extracellular H_2O_2 leads to a rapid GR-mediated restoration of the initial high GSH-to-GSSG ratio.[31] The importance of the GSH system in the protection of the cells against ROS is also demonstrated by the higher resistance of microglial cells that contain an elevated GSH content toward peroxide-induced toxicity.[36] In addition, since microglial cells contain catalase,[31,32,51] this enzyme is also likely to contribute to the clearance of H_2O_2.

In addition to their toxic potential,[10] ROS also contribute to cellular signaling,[69] which is most likely also the case for microglial cells. For example, the proliferation of microglia after stimulation by the proinflammatory cytokines IL-1β or TNF-α depends on extracellular H_2O_2 that is produced by the Nox of these cells.[70] In addition, release of H_2O_2 by microglial cells is stimulated by fibrillar β-amyloid peptide in a Nox-dependent process[71] as well as by exposure to $MnCl_2$ by a Nox-independent mechanism.[72] Thus, the potential of microglial cells to dispose of extracellular H_2O_2 has to be well balanced to avoid peroxide-mediated oxidative damage but allow signaling functions of the peroxide.

27.5   SUMMARY

Activated microglial cells release superoxide and NO that are products of Nox and iNOS. To prevent oxidative damage by these radicals and their reactive reaction products H_2O_2 and peroxynitrite, microglial cells contain a high cellular concentration of GSH and express substantial activities of various antioxidative enzymes. This prominent antioxidative potential is likely to protect activated microglial cells against oxidative damage and subsequently protects against the loss of important microglial functions in defense and repair of the brain.

ACKNOWLEDGMENTS

We very much like to thank Dr. Petra G. Hirrlinger for critically reading the manuscript and for valuable comments.

REFERENCES

1. Garden, G. A., and Moller, T., Microglia biology in health and disease, *J. Neuroimmun. Pharmacol.*, 1, 127, 2006.
2. Block, M. L., and Hong, J. S., Chronic microglial activation and progressive dopaminergic neurotoxicity, *Biochem. Soc. Trans.*, 35, 1127, 2007.
3. Hanisch, U. K., and Kettenmann, H., Microglia: Active sensor and versatile effector cells in the normal and pathologic brain, *Nat. Neurosci.*, 10, 1387, 2007.
4. Hailer, N. P., Immunosuppression after traumatic or ischemic CNS damage: It is neuroprotective and illuminates the role of microglial cells, *Prog. Neurobiol.*, 84, 211, 2008.
5. Nimmerjahn, A., Kirchhoff, F., and Helmchen, F., Resting microglial cells are highly dynamic surveillants of brain parenchyma *in vivo*, *Science*, 308, 1314, 2005.
6. Brown, G. C., Mechanisms of inflammatory neurodegeneration: iNOS and NADPH oxidase, *Biochem. Soc. Trans.*, 35, 1119, 2007.
7. Rock, R. B., and Peterson, P. K., Microglia as a pharmacological target in infectious and inflammatory diseases of the brain, *J. Neuroimmun. Pharmacol.*, 1, 117, 2006.
8. Bessis, A., et al., Microglial control of neuronal death and synaptic properties, *Glia*, 55, 233, 2007.
9. Napoli, I., and Neumann, H., Microglial clearance function in health and disease, *Neuroscience*, 158, 1030, 2009.
10. Dringen, R., Oxidative and antioxidative potential of brain microglial cells, *Antioxid. Redox. Signal.*, 7, 1223, 2005.
11. Saha, R. N., and Pahan, K., Regulation of inducible nitric oxide synthase gene in glial cells, *Antioxid. Redox. Signal.*, 8, 929, 2006.
12. Sumimoto, H., Structure, regulation and evolution of Nox-family NADPH oxidases that produce reactive oxygen species, *FEBS J.*, 275, 3249, 2008.
13. Nauseef, W. M., Biological roles for the NOX family NADPH oxidases, *J. Biol. Chem.*, 283, 16961, 2008.
14. Bal-Price, A., Matthias, A., and Brown, G. C., Stimulation of the NADPH oxidase in activated rat microglia removes nitric oxide but induces peroxynitrite production, *J. Neurochem.*, 80, 73, 2002.
15. Bokoch, G. M., and Knaus, U. G., NADPH oxidases: Not just for leukocytes anymore!, *Trends Biochem. Sci.*, 28, 502, 2003.
16. Infanger, D. W., Sharma, R. V., and Davisson, R. L., NADPH oxidases of the brain: Distribution, regulation, and function, *Antioxid. Redox. Signal.*, 8, 1583, 2006.
17. Green, S. P., et al., Induction of gp91-phox, a component of the phagocyte NADPH oxidase, in microglial cells during central nervous system inflammation, *J. Cereb. Blood Flow Metab.*, 21, 374, 2001.
18. Sankarapandi, S., et al., Measurement and characterization of superoxide generation in microglial cells: Evidence for an NADPH oxidase-dependent pathway, *Arch. Biochem. Biophys.*, 353, 312, 1998.
19. Lavigne, M. C., et al., Genetic requirement of p47phox for superoxide production by murine microglia, *FASEB J.*, 15, 285, 2001.
20. Zekry, D., Epperson, T. K., and Krause, K. H., A role for NOX NADPH oxidases in Alzheimer's disease and other types of dementia?, *IUBMB Life*, 55, 307, 2003.
21. Qin, L., et al., NADPH oxidase mediates lipopolysaccharide-induced neurotoxicity and proinflammatory gene expression in activated microglia, *J. Biol. Chem.*, 279, 1415, 2004.
22. Colton, C. A., and Gilbert, D. L., Production of superoxide anions by a CNS macrophage, the microglia, *FEBS Lett.*, 223, 284, 1987.
23. Murphy, S., and Gibson, C. L., Nitric oxide, ischaemia and brain inflammation, *Biochem. Soc. Trans.*, 35, 1133, 2007.
24. Ding, M., et al., Inducible nitric-oxide synthase and nitric oxide production in human fetal astrocytes and microglia. A kinetic analysis, *J. Biol. Chem.*, 272, 11327, 1997.
25. Adam-Vizi, V., Production of reactive oxygen species in brain mitochondria: Contribution by electron transport chain and non-electron transport chain sources, *Antioxid. Redox. Signal.*, 7, 1140, 2005.
26. Zelko, I. N., Mariani, T. J., and Folz, R. J., Superoxide dismutase multigene family: A comparison of the CuZn-SOD (SOD1), Mn-SOD (SOD2), and EC-SOD (SOD3) gene structures, evolution, and expression, *Free Radic. Biol. Med.*, 33, 337, 2002.
27. McCord, J. M., and Edeas, M. A., SOD, oxidative stress and human pathologies: A brief history and a future vision, *Biomed. Pharmacother.*, 59, 139, 2005.
28. Dringen, R., Metabolism and functions of glutathione in brain, *Prog. Neurobiol.*, 62, 649, 2000.

29. Dringen, R., Pawlowski, P. G., and Hirrlinger, J., Peroxide detoxification by brain cells, *J. Neurosci. Res.*, 79, 157, 2005.

30. Chatterjee, S., et al., Glutathione levels in primary glial cultures: Monochlorobimane provides evidence of cell type-specific distribution, *Glia*, 27, 152, 1999.

31. Hirrlinger, J., et al., Microglial cells in culture express a prominent glutathione system for the defense against reactive oxygen species, *Dev. Neurosci.*, 22, 384, 2000.

32. Hollensworth, S. B., et al., Glial cell type–specific responses to menadione-induced oxidative stress, *Free Radic. Biol. Med.*, 28, 1161, 2000.

33. Noack, H., et al., Nitrosative stress in primary glial cultures after induction of the inducible isoform of nitric oxide synthase (i-NOS), *Toxicology*, 148, 133, 2000.

34. Hirrlinger, J., et al., Oligodendroglial cells in culture effectively dispose of exogenous hydrogen peroxide: Comparison with cultured neurones, astroglial and microglial cells, *J. Neurochem.*, 82, 635, 2002.

35. Dopp, J. M., et al., Expression of the p75 TNF receptor is linked to TNF-induced NFkappaB translocation and oxyradical neutralization in glial cells, *Neurochem. Res.*, 27, 1535, 2002.

36. Persson, M., et al., Microglial glutamate uptake is coupled to glutathione synthesis and glutamate release, *Eur. J. Neurosci.*, 24, 1063, 2006.

37. Persson, M., et al., Microglial GLT-1 is upregulated in response to herpes simplex virus infection to provide an antiviral defence via glutathione, *Glia*, 55, 1449, 2007.

38. Chatterjee, S., et al., Induction of nitric oxide synthesis lowers intracellular glutathione in microglia of primary glial cultures, *Glia*, 29, 98, 2000.

39. Lekishvili, T., et al., BSE and vCJD cause disturbance to uric acid levels, *Exp. Neurol.*, 190, 233, 2004.

40. Fiebich, B. L., et al., Synergistic inhibitory effect of ascorbic acid and acetylsalicylic acid on prostaglandin E2 release in primary rat microglia, *J. Neurochem.*, 86, 173, 2003.

41. Heppner, F. L., et al., Vitamin E induces ramification and downregulation of adhesion molecules in cultured microglial cells, *Glia*, 22, 180, 1998.

42. Flanary, B. E., and Streit, W. J., Alpha-tocopherol (vitamin E) induces rapid, nonsustained proliferation in cultured rat microglia, *Glia*, 53, 669, 2006.

43. Stolzing, A., et al., Tocopherol-mediated modulation of age-related changes in microglial cells: Turnover of extracellular oxidized protein material, *Free Radic. Biol. Med.*, 40, 2126, 2006.

44. Godbout, J. P., et al., alpha-Tocopherol reduces lipopolysaccharide-induced peroxide radical formation and interleukin-6 secretion in primary murine microglia and in brain, *J. Neuroimmunol.*, 149, 101, 2004.

45. Egger, T., et al., Modulation of microglial superoxide production by alpha-tocopherol *in vitro*: Attenuation of p67(phox) translocation by a protein phosphatase-dependent pathway, *J. Neurochem.*, 79, 1169, 2001.

46. Muller, T., et al., Tocopherol long chain fatty alcohols decrease the production of TNF-alpha and NO radicals by activated microglial cells, *Bioorg. Med. Chem. Lett.*, 14, 6023, 2004.

47. Lindenau, J., et al., Cellular distribution of superoxide dismutases in the rat CNS, *Glia*, 29, 25, 2000.

48. Noack, H., et al., Differential expression of superoxide dismutase isoforms in neuronal and glial compartments in the course of excitotoxically mediated neurodegeneration: Relation to oxidative and nitrergic stress, *Glia*, 23, 285, 1998.

49. Liu, X. H., et al., An immunohistochemical study of copper/zinc superoxide dismutase and manganese superoxide dismutase in rat hippocampus after transient cerebral ischemia, *Brain Res.*, 625, 29, 1993.

50. Pinteaux, E., Perraut, M., and Tholey, G., Distribution of mitochondrial manganese superoxide dismutase among rat glial cells in culture, *Glia*, 22, 408, 1998.

51. Noack, H., et al., Peroxynitrite mediated damage and lowered superoxide tolerance in primary cortical glial cultures after induction of the inducible isoform of NOS, *Glia*, 28, 13, 1999.

52. Lindenau, J., et al., Enhanced cellular glutathione peroxidase immunoreactivity in activated astrocytes and in microglia during excitotoxin induced neurodegeneration, *Glia*, 24, 252, 1998.

53. Hirato, J., et al., Encephalopathy in megacystis-microcolon-intestinal hypoperistalsis syndrome patients on long-term total parenteral nutrition possibly due to selenium deficiency, *Acta Neuropathol.*, 106, 234, 2003.

54. Gutterer, J. M., et al., Purification of glutathione reductase from bovine brain, generation of an antiserum, and immunocytochemical localization of the enzyme in neural cells, *J. Neurochem.*, 73, 1422, 1999.

55. Fouquet, F., et al., Expression of the adrenoleukodystrophy protein in the human and mouse central nervous system, *Neurobiol. Dis.*, 3, 271, 1997.

56. Dringen, R., et al., Pentose phosphate pathway and NADPH metabolism, in *Handbook of Neurochemistry and Molecular Neurobiology. Vol. 5: Neural Energy Utilization*, Dienel, G., and Gibson, G., Eds., Springer Verlag, Heidelberg, 2007, 41.

57. Baquer, N. Z., Hothersall, J. S., and McLean, P., Function and regulation of the pentose phosphate pathway in brain, *Curr. Top. Cell Regul.*, 29, 265, 1988.

58. Arkblad, E. L., et al., Expression of proton-pumping nicotinamide nucleotide transhydrogenase in mouse, human brain and *C elegans*, *Comp Biochem. Physiol B Biochem. Mol. Biol.*, 133, 13, 2002.

59. Vogel, R., et al., Mitochondrial malic enzyme: Purification from bovine brain, generation of an antiserum, and immunocytochemical localization in neurons of rat brain, *J. Neurochem.*, 71, 844, 1998.

60. Minich, T., Yokota, S., and Dringen, R., Cytosolic and mitochondrial isoforms of NADP+-dependent isocitrate dehydrogenases are expressed in cultured rat neurons, astrocytes, oligodendrocytes and microglial cells, *J. Neurochem.*, 86, 605, 2003.

61. Vogel, R., et al., The regeneration of reduced glutathione in rat forebrain mitochondria identifies metabolic pathways providing the NADPH required, *Neurosci. Lett.*, 275, 97, 1999.

62. Kim, S. U., et al., Peroxiredoxin I is an indicator of microglia activation and protects against hydrogen peroxide-mediated microglial death, *Biol. Pharm. Bull.*, 31, 820, 2008.

63. Nakaso, K., et al., Co-induction of heme oxygenase-1 and peroxiredoxin I in astrocytes and microglia around hemorrhagic region in the rat brain, *Neurosci. Lett.*, 293, 49, 2000.

64. Koistinaho, J., et al., Long-term induction of haem oxygenase-1 (HSP-32) in astrocytes and microglia following transient focal brain ischaemia in the rat, *Eur. J. Neurosci.*, 8, 2265, 1996.

65. Bidmon, H. J., et al., Heme oxygenase–1 (HSP-32) and heme oxygenase-2 induction in neurons and glial cells of cerebral regions and its relation to iron accumulation after focal cortical photothrombosis, *Exp. Neurol.*, 168, 1, 2001.

66. Ke, Z. J., et al., Reversal of thiamine deficiency-induced neurodegeneration, *J. Neuropathol. Exp. Neurol.*, 62, 195, 2003.

67. Min, K. J., et al., Astrocytes induce hemeoxygenase-1 expression in microglia: A feasible mechanism for preventing excessive brain inflammation, *J. Neurosci.*, 26, 1880, 2006.

68. Twig, G., et al., Real-time detection of reactive oxygen intermediates from single microglial cells, *Biol. Bull.*, 201, 261, 2001.

69. D'Autreaux, B., and Toledano, M. B., ROS as signalling molecules: Mechanisms that generate specificity in ROS homeostasis, *Nat. Rev. Mol. Cell Biol.*, 8, 813, 2007.

70. Mander, P. K., Jekabsone, A., and Brown, G. C., Microglia proliferation is regulated by hydrogen peroxide from NADPH oxidase, *J. Immunol.*, 176, 1046, 2006.

71. Jekabsone, A., et al., Fibrillar beta-amyloid peptide Abeta1-40 activates microglial proliferation via stimulating TNF-alpha release and H2O2 derived from NADPH oxidase: A cell culture study, *J. Neuroinflam.*, 3, 24, 2006.

72. Zhang, P., Hatter, A., and Liu, B., Manganese chloride stimulates rat microglia to release hydrogen peroxide, *Toxicol. Lett.*, 173, 88, 2007.

28 Branched-Chain Amino Acids and Brain Metabolism

Radovan Murín and Bernd Hamprecht

Interfaculty Institute for Biochemistry, University
of Tuebingen, Tuebingen, Germany

CONTENTS

28.1 INTRODUCTION TO THE METABOLISM OF BRANCHED-CHAIN AMINO ACIDS

Valine, isoleucine, and leucine, amino acids with a characteristic methyl group branching of their carbon skeleton (Figures 28.1 through 28.3), constitute the group of branched-chain amino acids (BCAA). BCAAs are essential amino acids for mammals since there is no anabolism of their carbon skeleton in mammalian tissues. Therefore, the metabolism of BCAAs, widespread among mammalian cells, includes intermediary metabolism and catabolism. Depending on the metabolic context, the first, reversible step of BCAA degradation is a transamination reaction (step A in Figures 28.1 through 28.3). This can be considered as the intermediary metabolism of BCAAs and their corresponding branched-chain 2-oxo acids (branched-chain α-keto acids [BCKAs]). The next reaction is catalyzed by the BCKA dehydrogenase complex (BCKDH; step B in Figures 28.1 through 28.3). These two enzymes comprise the first stage of the catabolic pathways of BCAAs. In the second stage of the catabolic pathways, the carbon skeletons of BCAAs are eventually fragmented, giving rise to propionyl-CoA, acetyl-CoA, and acetoacetate. These three compounds are valuable substrates for brain energy metabolism. Propionyl-CoA is a catabolite of both valine (Figure 28.1) and isoleucine (Figure 28.2), while the degradation of leucine and isoleucine gives rise to acetyl-CoA

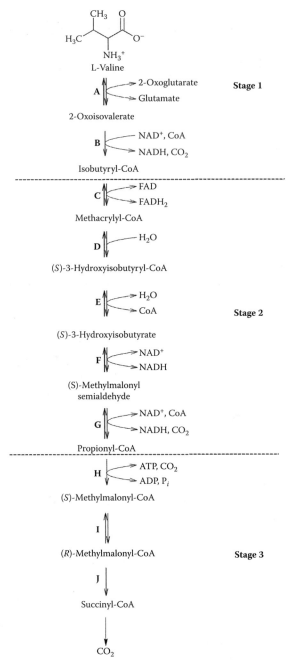

FIGURE 28.1 Scheme of the pathway of L-valine catabolism (modified version of the pathway presented by Murín et al. [21]). The first two enzymes of the pathway, (A) BCAA transaminase and (B) branched-chain 2-oxo acid dehydrogenase, are common for all three BCAAs and their cognate branched-chain 2-oxo acids and comprise the first stage of valine catabolism. In the second stage of valine catabolism, isobutyryl-CoA is converted to propionyl-CoA in a sequence of the enzymatic reactions catalyzed by (C) acyl-CoA dehydrogenase, (D) enoyl-CoA hydratase, (E) 3-hydroxyisobutyryl-CoA hydrolase, (F) 3-hydroxyisobutyrate dehydrogenase, and (G) methylmalonate semialdehyde dehydrogenase. The third stage of valine catabolism, the transformation of propionyl-CoA to succinyl-CoA, a member of the citric acid cycle, comprises the series of enzymatic reactions catalyzed by (H) propionyl-CoA carboxylase, (I) methylmalonyl-CoA racemase, and (J) methylmalonyl-CoA mutase. The third stages for both the valine and the isoleucine catabolic pathways are identical.

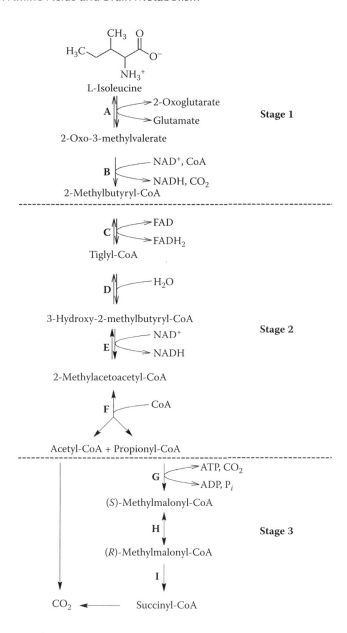

FIGURE 28.2 Scheme of the catabolic pathway of L-isoleucine catabolism (modified version of the pathway presented by Murín et al. [23]). Since (A) branched-chain amino acid transaminase and (B) branched-chain 2-oxo acid dehydrogenase are common in the catabolism of BCAAs and the latter is a common enzyme in the catabolism of the corresponding branched-chain 2-oxo acids, respectively, they are comprised as the first stage of isoleucine metabolism. 2-Methylbutyryl-CoA is subsequently converted to the products of isoleucine degradation, acetyl-CoA and propionyl-CoA, in a sequence of the four enzymatic reactions catalyzed by (C) 2-methylacyl-CoA dehydrogenase, (D) enoyl-CoA hydratase, (E) 3-hydroxyacyl-CoA dehydrogenase, and (F) acetyl-CoA C-acyltransferase. These enzymes are identical with the enzymes occurring in the 3-methyl branched-chain fatty acid β-oxidation pathway and comprise the second stage of the isoleucine degradative pathway. Acetyl-CoA enters the general metabolism directly. Propionyl-CoA is transformed to the member of the citric acid cycle, succinyl-CoA, by the series of enzymatic reactions constituting the third stage of isoleucine catabolism. These reactions are catalyzed by (G) propionyl-CoA carboxylase, (H) methylmalonyl-CoA racemase, and (I) methylmalonyl-CoA mutase.

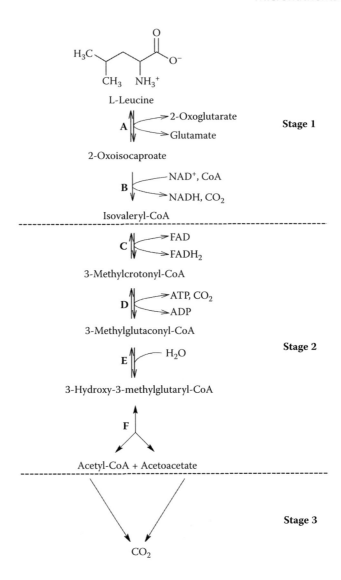

FIGURE 28.3 Scheme of the enzymatic degradation of L-leucine to acetoacetate and acetyl-CoA (scheme modified according to Murín and Hamprecht [28]). The enzymes of the first stage, (A) branched-chain amino acid transaminase and (B) branched chain 2-oxo acid dehydrogenase, are common for all BCAAs. The four enzymes, (C) isovaleryl-CoA dehydrogenase, (D) 3-methylcrotonyl-CoA carboxylase, (E) 3-methylglutaconyl-CoA hydratase, and (F) 3-hydroxy-3-methyl-glutaryl-CoA lyase, constitute the second stage of the leucine catabolic pathway and are involved in enzymatic conversion of isovaleryl-CoA to acetyl-CoA and acetoacetate.

(Figures 28.1 and 28.2). Acetoacetate is a direct catabolite of leucine only (Figure 28.3). Leucine is considered to be a solely ketogenic amino acid since a ketone body or a precursor of it—acetoacetate and acetyl-CoA, respectively—are the products of its degradation (Figure 28.3). Propionyl-CoA is converted to succinyl-CoA, a member of the citric acid cycle. Therefore, valine and isoleucine can serve an anaplerotic function. Leucine can be considered as an endogenous source of ketone bodies in the brain that can supplement glucose as a source of oxidatively generated energy [1] or as substrate for the biosynthesis of cholesterol [2] and fatty acids [3].

28.2 METABOLISM OF BCAAS IN THE BRAIN

28.2.1 UTILIZATION OF BCAAS BY THE BRAIN AND NEURAL CELLS

Only a restricted number of substances, which are transported through the blood-brain barrier (BBB), participate in brain energy metabolism. While glucose is firmly established as the major "fuel" molecule for brain energy metabolism, also some other blood-borne substances can be used by the brain for sustaining its energy demands. For the metabolism of such substances, two essential prerequisites must be fulfilled: transport through the BBB and an appropriate enzymic "equipment" of neural cells. Thus, in addition to fatty acids [4] and ketone bodies [1], amino acids are also considered as alternate, blood-borne substrates for brain energy metabolism that can meet the brain energy demands depending on the developmental and physiological status. Even the amino acid residues contained in proteins of brain cells appear to be useable as fuel substances [5].

BCAAs are considered as fuel molecules for brain energy metabolism since the early observation that rat brain slices readily utilize BCAAs in addition to glucose [6, 7]. Furthermore, brain can make use of the acetyl residues of the acetyl-CoA generated by the degradation of leucine for the synthesis of other amino acids [8, 9], cholesterol [10], and other lipids [11, 12].

The intricate cytoarchitecture of the brain formed by the diverse cell types organized into assemblies and patterns of specialized components affords complex metabolic cooperation among them. Therefore, the degradation capacity for BCAAs of some of the neural cells constituting the brain has been analyzed in cellular systems of varying complexity. Studies on the utilization of particular BCAAs have shown that astroglial and neuronal cultures are capable of metabolizing leucine [3, 14–19]. By metabolizing leucine, several compounds from its catabolic pathway can be found to be released from astroglial cells. Detailed studies with leucine showed that compounds such as 2-oxoisocaproate, acetoacetate, and β-hydroxybutyrate are released into the culture medium to a considerable extent by astroglial cells [14–16, 18]. Cultured neurons are also capable of actively metabolizing leucine [16, 19].

In comparison to the metabolism of leucine by brain cells, the metabolism of valine and isoleucine has not yet been studied that intensively. *In vitro* experiments with slices from cerebral cortex showed the capacity of brain cells to oxidize 2-oxoisovalerate, the cognate 2-oxo acid of valine [20]. In addition to their capacity of potently utilizing leucine, astroglial and neuronal cultures, they are also capable of metabolizing valine [13, 19, 21] and isoleucine [13, 22, 23]. Cultured astroglial cells release some intermediates of valine [21] and isoleucine [23] catabolism into their incubation media.

28.2.2 TRANSPORT OF BCAAS AND INTERMEDIATES OF BCAA METABOLISM

Although being zwitterions in the physiological pH range, BCAAs enter the brain [24, 25] by facilitated transport through the endothelial cells of the blood vessels [26]. In the brain, in addition to their catabolism, they serve also several other functions [27, 28]. The import of BCAAs through the BBB into the brain parenchyma is rapid, especially of leucine, the influx rate of which exceeds that of all other neutral L-amino acids [26, 29].

Structural similarities of BCAAs characterize their transport through the BBB. The main system transporting BCAAs, in competition with the two other BCAAs and aromatic amino acids, is the sodium-independent L-type transporter [30]. This permease is present in both luminal and abluminal membranes of capillary endothelial cells [31]. The abluminal membrane also contains two other, albeit sodium-dependent, systems capable of transporting leucine: a transporter preferring alanine, serine, and cysteine [32] and a transporter for large neutral amino acids [33]. Additionally to BCAA, their 2-oxo acid analogues can also be imported into the brain parenchyma. Monocarboxylate transporters located in capillary endothelial cells enable the brain to take up BCKAs from the blood stream [34].

In brain parenchyma, BCAAs are transported through the plasma membranes by both sodium-independent [35] and sodium-dependent transport systems [36, 37]. The proposed intercellular exchange of BCKAs between astrocytes and neurons [38] is facilitated by monocarboxylate transporters of different isoforms, which are expressed among astrocytes and neurons in a cell-specific manner [39, 40].

28.3 THE ENZYMES OF BCAA CATABOLISM

After transport into the cells, BCAAs can be converted to three components of general metabolism, i.e., propionyl-CoA, acetyl-CoA, and acetoacetate, by their respective catabolic pathways (Figures 28.1 through 28.3). The catabolic pathways of BCAAs can be subdivided into three stages. In the first stage of BCAA metabolism, they all share the two enzymatic steps catalyzed by BCAA transaminase (BCAT) and BCKDH. In the course of the second stage of these catabolic pathways, the branched-acyl-CoAs, the products of oxidative decarboxylation by BCKDH, are converted by a series of enzymatic reactions into compounds of the intermediary cellular metabolism such as acetyl-CoA, propionyl-CoA, and acetoacetate. The third stage of the catabolism of BCAAs consists of further metabolism of acetyl-CoA, propionyl-CoA, and acetoacetate including their ultimate oxidation to CO_2; in some cases, metabolites of these compounds are released from one type of brain cells to serve as fuel material in another type of brain cells as detailed below.

28.3.1 BRANCHED-CHAIN AMINO ACID AMINOTRANSFERASE

The enzyme BCAT catalyzes the reversible transamination of BCAAs to their cognate BCKAs. The amino group acceptor is 2-oxoglutarate, which is converted to glutamate. BCAT has a preference for leucine, as can be derived from the kinetic parameters [41]. The specific activity of BCAT in rat, human, and monkey brains amounts to approximately 2, 0.5, and 0.4 U/g of wet tissue [42]. In the brain, two BCAT isoforms are expressed, a cytosolic and a mitochondrial one. They share similar substrate specificity and possess only subtle differences in their steady state kinetic parameters [41, 43]. Both mammalian BCAT isoforms contain the consensus sequence motif CXXC, which in the case of the mitochondrial isoform is regulating the enzymatic activity by sensing the cellular redox status [44].

The expression of both BCAT isoforms was studied among cultured neural cells as well as in rat brain sections. In astroglia-rich primary cultures derived from newborn rat brains, immunocytochemical analysis revealed that astrocytes express both isoforms of BCAT [38, 45] with the mitochondrial isozyme being the predominant one [38]. Cultured neurons and oligodendrocytes contain solely the cytosolic isoform, whereas microglial cells express the mitochondrial isoform only [38, 45]. In the brain and spinal cord, expression of cytosolic BCAT is widespread among neurons with some variation depending on the tissue and cellular region [46, 47]. Astrocytes in the brain express the mitochondrial isoform of BCAT [48].

28.3.2 BRANCHED-CHAIN 2-OXO ACID DEHYDROGENASE

The product of BCAA transamination, BCKA, enters the irreversible part of catabolism in a step of oxidative decarboxylation catalyzed by BCKDH. This multienzyme complex is structurally similar to the 2-oxoglutarate and the pyruvate dehydrogenase complexes [49, 50]. BCKDH is composed of three protomers: branched-chain 2-oxo acid decarboxylase/dehydrogenase (EC 1.2.4.4), dihydrolipoamide acyltransferase (EC 1.8.1.4), and dihydrolipoamide dehydrogenase (EC 2.3.1.168). Since by passing through the BCKDH complex the carbon skeleton of BCAAs becomes committed to the irreversible parts of the degradative pathways, the enzymatic activity of the complex is tightly regulated via phosphorylation of the 2-oxo acid decarboxylase/dehydrogenase subunit, catalyzed by a BCKDH-specific kinase [51].

In addition, BCKDH can assemble with mBCAT, resulting in a BCAA metabolon capable of substrate channeling and thus achieving both regulation and high efficiency of pathway flux [52]. The formation and maintenance of this metabolic operation unit is dependent on the mitochondrial redox state [52]. In cultured neural cells, BCKDH is ubiquitously expressed [45]. Therefore, the expression of BCKDH is not correlated with the distribution of the mitochondrial BCAT isoform and is not determined by the capability to form such a complex. In the brain, the actual enzymatic activity of BCKDH was estimated to arise to somewhat more than 1% of BCAT activity [42].

Recently, mice modified in genes pertinent to leucine metabolism were generated by knockout of the gene for the dihydrolipoamide acyltransferase subunit of BCKDH [53] or for BCKDH kinase [54]. The former appears to be an excellent animal model for maple syrup urine disease. The latter demonstrates impressively the necessity for tight control of BCAA degradation.

28.3.3 THE SECOND AND THIRD STAGES OF THE VALINE CATABOLIC PATHWAY

In a series of enzymically catalyzed steps, isobutyryl-CoA, a valine-specific branched acyl-CoA, the intermediate emerging from the first stage of the valine catabolic pathway is converted in stage 2 to propionyl-CoA (Figure 28.1). This sequence of reactions is catalyzed by 2-methylacyl-CoA dehydrogenase (EC 1.3.99.12), enoyl-CoA hydratase (EC 4.2.1.17), 3-hydroxyisobutyryl-CoA hydrolase (EC 3.1.2.4), 3-hydroxyisobutyrate dehydrogenase (EC 1.1.1.31), and methylmalonate semialdehyde dehydrogenase (EC 1.2.1.27). Propionyl-CoA, the product of the latter enzymic reaction, is subsequently transformed to succinyl-CoA, a member of the citric acid cycle, by the series of enzymatic reactions constituting the third stage of the valine catabolic pathway. These reactions are catalyzed by propionyl-CoA carboxylase (EC 6.4.1.3), methylmalonyl-CoA racemase (EC 5.1.99.11), and methylmalonyl-CoA mutase (EC 5.4.99.2).

Expression among brain cells of the enzymes of the second and third stages of the valine catabolic pathway can be deduced from the ability of brain slices to oxidize valine to CO_2 [7]. The capability of valine metabolism is also preserved among cultured astroglial and neuronal cells [13, 15, 19]. Further studies on cultured neurons, astrocytes, oligodendrocytes, microglia, and ependymocytes showed that, in addition to enzymes comprising the first stage of the valine catabolic pathway, these cells invariably express an enzyme from the second stage, 3-hydroxyisobutyrate dehydrogenase [21]. The cell type-specific expression of other enzymes from the second stage of the valine catabolic pathway has not yet been investigated in detail.

The enzymes of the third part of valine catabolism are identical with the enzymes involved in propionate metabolism and are therefore not discussed here.

28.3.4 THE SECOND AND THIRD STAGES OF THE ISOLEUCINE CATABOLIC PATHWAY

The second stage of isoleucine catabolism ends by yielding the two compounds, acetyl-CoA and propionyl-CoA (Figure 28.2). The latter compound, as in valine catabolism, is carboxylated, racemized, and then isomerized to form succinyl-CoA by the enzymes of the third stage of the isoleucine catabolic pathway. The enzymatic conversion of 2-methylbutyryl-CoA in the second stage of catabolism is analogous to reactions in the β-oxidation pathway of 2-methyl branched-chain fatty acids and shares the appropriate enzymes with this pathway. In addition to the two enzymes, 2-methylacyl-CoA dehydrogenase (EC 1.3.99.12) and enoyl-CoA hydratase (EC 4.2.1.17), that operate also in the degradation of valine, 3-hydroxyacyl-CoA dehydrogenase (EC 1.1.1.178) and acetyl-CoA acetyltransferase are required to generate propionyl-CoA and acetyl-CoA from isoleucine. While the expression of all these enzymes among neural cells has not yet been studied in detail, their presence could be deduced from the results of experiments in which carbon atoms from [13]C- or ^{14}C-labeled isoleucine were incorporated into amino acids [22, 23, 55] or appeared in CO_2 [7].

28.3.5 THE SECOND AND THIRD STAGES OF THE LEUCINE CATABOLIC PATHWAY

In the second stage of leucine catabolism, isovaleryl-CoA is converted to acetoacetate and acetyl-CoA by passing through four enzymatically catalyzed reactions (Figure 28.3). The first three enzymes of this stage, isovaleryl-CoA dehydrogenase (EC 1.3.99.10), β-methylcrotonyl-CoA carboxylase (EC 6.4.1.4), and β-methylglutaconyl-CoA hydratase (EC 4.2.1.18) are specific for the conversion of the leucine metabolites they are named after. The fourth enzyme, 3-hydroxy-3-methylglutaryl-CoA lyase (EC 4.1.3.4), is also part of another pathway, the 3-hydroxy-3-methylglutaryl-CoA cycle [56], which plays a key role in the metabolism of ketone bodies.

The expression of these enzymes in the brain and cultured brain cells can be inferred from the fact that leucine can be completely oxidized by brain slices [7] and by astroglia-rich primary cultures [15, 18]. In the brain, the presence of the mRNAs for isovaleryl-CoA dehydrogenase [57], 3-methylcrotonyl-CoA carboxylase [58, 59], and the mitochondrial isoform of 3-hydroxy-3-methylglutaryl-CoA lyase [60] has been shown. The expression in the brain of 3-methylglutaconyl-CoA hydratase [61] and 3-methylcrotonyl-CoA carboxylase [58, 59] has been established. Data concerning enzymatic activities and cell-specific distribution of these enzymes in cultured neural cells or in the brain are apparently still missing, although with some exceptions. 3-Methylcrotonyl-CoA carboxylase was shown to be ubiquitously expressed in many of the neurons encountered in cultures [62] and in all types of cultured glial cells [58, 62]. In the brain, the somata of some of the neurons in pons, amygdala, inferior colliculus, and deep cerebellar nuclei are intensively stained with antibodies against 3-methylcrotonyl-CoA carboxylase. The nonneuronal cells with high expression levels of 3-methylcrotonyl-CoA carboxylase are astrocytes located near the brain surface, oligodendrocytes, ependymocytes, choroid plexus cells, and meningeal cells [59].

Of the mRNAs for the two mammalian isoforms of 3-hydroxy-3-methlyglutaryl-CoA lyase, mitochondrial and peroxisomal, the mitochondrial isozyme is present in the cerebellum, cortex, medulla, and midbrain regions of rat brain and in cultured cortical astrocytes [60]. Together with the other mitochondrially located enzymes acetoacetyl-CoA thiolase and 3-hydroxy-3-methylglutaryl-CoA synthase, mitochondrial 3-hydroxy-3-methylglutaryl-CoA lyase constitutes the ketogenic 3-hydroxy-3-methylglutaryl-CoA cycle [56]. The levels of gene transcripts for the first two of these three enzymes drop during brain development from 11-day-old suckling rats to 28-day-old weaned rats, whereas the level of 3-hydroxy-3-methylglutaryl-CoA lyase mRNA remains almost constant [60]. Apparently, this points to a constitutive expression of 3-hydroxy-3-methlyglutaryl-CoA lyase [60] that may be beneficial for neural cells, since it could provide them with ketone bodies originating from leucine.

28.3.6 BCAA Metabolites as Fuel Material in Metabolic Astroglia-Neuron Cooperation

BCAA degradation by cultured astrocytes generates metabolites that can be released from the cells into the extracellular space. If this also happens in the brain, these metabolites could be taken up by neighboring cells such as neurons, oligodendrocytes, and microglial cells. In these cells, they would serve as fuel materials for the oxidative generation of energy. All these compounds have in common that they are monocarboxylates, which can be easily transported by monocarboxylate transporters present in the plasma membranes of these cells [39, 40, 63]. The following major BCAA metabolites and their follow-up products could principally be serving in such a metabolic cooperation between neural cells: 2-oxoisovalerate, 2-oxo-3-methylvalerate, 2-oxoisocaproate, 3-hydroxyisobutyrate, propionate, acetoacetate, 3-hydroxybutyrate, and lactate [15, 18, 21, 23, 28]. It appears important to also consider the compounds that were expected to be released from the cultured cells on degradation of BCAAs but did not appear in the incubation medium. The catabolism of isoleucine generates acetyl-CoA (Figure 28.2). This may be expected to be converted to the ketone body acetoacetate, which in turn, by reduction could give rise to another ketone body, 3-hydroxybutyrate. Neither of these monocarboxylates could be detected in the incubation medium [23], although the methods used would allow detection as demonstrated in the case of leucine [15, 18]. The difference between the cases of isoleucine and leucine is that in the latter the ketone bodies are derived from

3-hydroxy-3-methylglutaryl-CoA that is a direct metabolite of leucine. As in the case of isoleucine, no ketone bodies are observed as a result of oxidative glucose metabolism that would also lead to acetyl-CoA. From this can be concluded that the formation of ketone bodies from acetyl-CoA via the 3-hydroxy-3-methylglutaryl-CoA cycle is not operative in these cultures. Since the cultures can oxidize octanoate [64], they must have acetyl-CoA C-acetyl transferase (β-ketothiolase), and since they can degrade 3-hydroxy-3-methylglutaryl-CoA generated from leucine, they must express the corresponding lyase. Thus the lack of appearance of ketone bodies during catabolism of isoleucine may be due to the lack of or too low activity of 3-hydroxy-3-methylglutaryl-CoA synthase. Alternatively, the steady-state level of acetyl-CoA could be too low for the generation of 3-hydroxy-3-methylglutaryl-CoA, possibly due to consumption of acetyl-CoA in the citric acid cycle.

28.4   OTHER FUNCTIONS OF BCAAS IN THE BRAIN AND NEURAL CELLS

Besides being the "minor" substrates for energy metabolism, BCAAs serve also in several other roles: (1) BCAAs are essential amino acids and are indispensable for protein synthesis; (2) BCAAs have a significant impact on nitrogen import into the brain parenchyma [65]; (3) together with their cognate 2-oxo acids, BCAAs (especially leucine) are involved in amino-group shuttling in a BCAA/BCKA cycle that accompanies the glutamate/glutamine cycle operating between neurons and astrocytes [38, 66]. In addition, leucine plays two other roles in the brain: (1) it participates in the regulation of mammalian target of rapamycin (mTOR) signaling, which, in turn, regulates food intake, and (2) it allosterically regulates the activity of certain enzymes. Furthermore, degradation of valine and isoleucine provides propionyl-CoA, which is subsequently converted to succinyl-CoA and thus increases the concentration of members of the citric acid cycle. Thus, these two amino acids can play an anaplerotic role.

The dual role of BCAAs in brain nitrogen metabolism, import of amino nitrogen into brain and participation of their cognate BCKAs in the glutamate/glutamine cycle, may be a consequence of (a) the high rates of their import through the BBB [24–26] into the brain parenchyma, (b) the high enzymatic activity of BCAT for reversible transamination of BCAA with 2-oxoglutarate, and (c) the cell-specific expression of monocarboxylate transporter isoforms among neurons and astrocytes [39, 40]. In the brain, BCAAs, especially leucine, may significantly contribute to the synthesis of glutamate and glutamine by providing their amino group in a transamination reaction [27, 65]. All BCAAs may contribute to glutamate synthesis to a certain extent. Under *in vitro* conditions, when neural cells are exposed to glutamate, leucine is also a major amino group donor for the formation of glutamate and glutamine [65]. If the conditions of glutamatergic neurotransmission are mimicked by the application of pulses of glutamate to astrocytes, valine specifically becomes the predominant amino group donor for the synthesis of glutamate and glutamine [19].

In addition, leucine exerts a regulatory function in glutamate metabolism. At elevated concentrations, leucine allosterically activates the two isoforms of glutamate dehydrogenase expressed in human brain, GLUD-1 and GLUD-2 [67], although the availability of leucine has a more profound effect on the brain-specific isoform GLUD-2 [68, 69]. Even at low concentrations, leucine acts synergistically with ADP in the activation of glutamate dehydrogenase. Therefore, this allosteric activation appears to bear physiological relevance [70]. As a signal for ample availability of fuel material, leucine would, by activation of glutamate dehydrogenase, raise the level of 2-oxoglutarate, a member of the citric acid cycle. This could accelerate the turn of the cycle, thereby raising the capacity of energy production. In addition, the accompanying increased availability of ammonia would facilitate the formation of glutamine and thus provide a basis for increasing the activity of the glutamate/glutamine cycle [69, 71].

In contrast to the other two BCAAs, leucine has been identified as an important messenger molecule reflecting the nutritional status of the organism, thus regulating food intake [72]. In fact, at elevated concentrations, leucine activates in hypothalamic neurons the mTOR signaling cascade that controls appetite [73]. In cells of the piriform cortex, decreased availability of leucine and of some other essential amino acids is associated with an increased level of uncharged tRNAs, resulting in dephosphorylation of translation initiation factor 2, eventually leading to the behavioral trait of searching for food of high nutritional value [74].

The catabolism of the other two BCAAs, valine and isoleucine, gives rise to the formation of succinyl-CoA, a member of the citric acid cycle. During pathophysiological conditions induced by liver diseases, e.g., cerebral hyperammonemia/hepatic encephalopathy, patients of low protein tolerance may be treated with a dietary supplement of BCAAs [75]. Although the mechanism by which BCAAs exert their beneficial effect remains to be fully elucidated, experimental data suggest that the catabolism of BCAAs may help to overcome the inhibition by ammonia of the pyruvate dehydrogenase complex [76, 77] and the 2-oxoglutarate dehydrogenase complex [78–80].

REFERENCES

1. Morris, A. A., Cerebral ketone body metabolism, *J. Inherit. Metab. Dis.*, 28, 109, 2005.
2. Lopes-Cardozo, M., et al., Acetoacetate is a cholesterogenic precursor for myelinating rat brain and spinal cord. Incorporation of label from [3-^{14}C]acetoacetate, [^{14}C]glucose and ^{3}H$_2$O, *Biochim. Biophys. Acta*, 794, 350, 1984.
3. Lopes-Cardozo, M., Larsson, O. M., and Schousboe, A., Acetoacetate and glucose as lipid precursors and energy substrates in primary cultures of astrocytes and neurons from mouse cerebral cortex. *J. Neurochem.*, 46, 773, 1986.
4. Gnaedinger, J. M., et al., Cerebral metabolism of plasma [14C]palmitate in awake, adult rat: Subcellular localization. *Neurochem. Res.*, 13, 21, 1988.
5. Abood, L.G., and Geiger, A., Breakdown of proteins and lipids during glucose-free perfusion of the cat's brain, *Am. J. Physiol.*, 182, 577, 1955.
6. Roberts, S., Seto, K., and Hanking, B. M., Regulation of cerebral metabolism of amino acids. I. Influence of phenylalanine deficiency on oxidative utilization *in vitro*, *J. Neurochem.*, 9, 493, 1962.
7. Swaiman, K. F., and Milstein, J. M., Oxidation of leucine, isoleucine and related ketoacids in developing rabbit brain, *J. Neurochem.*, 12, 981, 1965.
8. Berl, S., and Frigyesi, T. L., Metabolism of [^{14}C]leucine and [^{14}C]acetate in sensorimotor cortex, thalamus, caudate nucleus and cerebellum of the cat, *J. Neurochem.*, 15, 965, 1968.
9. Roberts, S., and Morelos, B. S., Regulation of cerebral metabolism of amino acids IV. Influence of amino acid levels on leucine uptake, utilization and incorporation into protein in vivo, *J. Neurochem.*, 12, 373, 1965.
10. Kabara, J. J., and Okita, G., Brain cholesterol: Biosynthesis with selected precursors in vivo, *J. Neurochem.*, 7, 298, 1961.
11. Patel, M. S., and Owen, O. E., The metabolism of leucine by developing rat brain: Effect of leucine and 2-oxo-4-methylvalerate on lipid synthesis from glucose and ketone bodies, *J. Neurochem.*, 30, 775, 1978.
12. Dhopeshwarkar, G. A., and Subramanian, C. Lipogenesis in the developing brain: Utilization of radioactive leucine, isoleucine, octanoic acid and beta-hydroxybutyric acid, *Lipids*, 14, 47, 1979.
13. Murthy, C. R., and Hertz, L., Comparison between acute and chronic effects of ammonia on branched-chain amino acid oxidation and incorporation into protein in primary cultures of astrocytes and of neurons. *J. Neurosci. Res.*, 17, 271, 1987.
14. Yudkoff, M., et al., Interrelationships of leucine and glutamate metabolism in cultured astrocytes, *J. Neurochem.*, 62, 1192, 1994.
15. Bixel, M. G., and Hamprecht, B., Generation of ketone bodies from leucine by cultured astroglial cells, *J. Neurochem.*, 65, 2450, 1995.
16. Hutson, S. M., et al., Role of branched-chain aminotransferase isoenzymes and gabapentin in neurotransmitter metabolism, *J. Neurochem.*, 71, 863, 1998.
17. Honegger, P., et al., Alteration of amino acid metabolism in neuronal aggregate cultures exposed to hypoglycaemic conditions, *J. Neurochem.*, 81, 1141, 2002.
18. Bixel, M. G., et al., Metabolism of [U-^{13}C]leucine in cultured astroglial cells, *Neurochem. Res.*, 29, 2057, 2004.
19. Bak, L. K., et al., Among the branched-chain amino acids, only valine metabolism is up-regulated in astrocytes during glutamate exposure, *J. Neurosci. Res.*, 85, 3465, 2007.

20. Brand, K., Metabolism of 2-oxoacid analogues of leucine, valine and phenylalanine by heart muscle, brain and kidney of the rat, *Biochim. Biophys. Acta*, 677, 126, 1981.

21. Murín, R., et al., Expression of 3-hydroxyisobutyrate dehydrogenase in cultured neural cells, *J. Neurochem.*, 105, 1176, 2008.

22. Johansen, M. L., et al., The metabolic role of isoleucine in detoxification of ammonia in cultured mouse neurons and astrocytes, *Neurochem. Int.*, 50, 1042, 2007.

23. Murín, R., et al., Glial metabolism of isoleucine, *Neurochem. Res.*, 34, 194, 2009.

24. Oldendorf, W. H., Brain uptake of radiolabeled amino acids, amines, and hexoses after arterial injection, *Am. J. Physiol.*, 221, 1629, 1971.

25. Oldendorf, W. H., Uptake of radiolabeled essential amino acids by brain following arterial injection, *Proc. Soc. Exp. Biol. Med.*, 136, 385, 1971.

26. Smith, Q. R., et al. Kinetics of neutral amino acid transport across the blood-brain barrier, *J. Neurochem.*, 49, 1651, 1987.

27. Yudkoff, M., Brain metabolism of branched-chain amino acids, *Glia*, 21, 92, 1997.

28. Murín, R., and Hamprecht, B., Metabolic and regulatory roles of leucine in neural cells, *Neurochem. Res.*, 33, 279, 2008.

29. Smith, Q. R., Transport of glutamate and other amino acids at the blood-brain barrier, *J. Nutr.*, 130, 4 Suppl., 1016S, 2000.

30. Christensen, H. N., Role of amino acid transport and counter transport in nutrition and metabolism, *Physiol. Rev.*, 70, 43, 1990.

31. Hawkins, R. A., et al., Structure of the blood-brain barrier and its role in the transport of amino acids, *J. Nutr.*, 136, 1 Suppl., 218S, 2006.

32. Hargreaves, K. M., and Pardridge, W. M., Neutral amino acid transport at the human blood-brain barrier, *J. Biol. Chem.*, 263, 19392, 1988.

33. O'Kane, R. L., and Hawkins, R. A., Na$^+$-dependent transport of large neutral amino acids occurs at the abluminal membrane of the blood-brain barrier, *Am. J. Physiol. Endocrinol. Metab.*, 285, E1167, 2003.

34. Steele, R. D. Blood-brain barrier transport of the alpha-keto acid analogs of amino acids, *Fed. Proc.*, 45, 2060, 1986.

35. Kim, D. K., et al., System L–amino acid transporters are differently expressed in rat astrocyte and C6 glioma cells, *Neurosci. Res.*, 50, 437, 2004.

36. Takanaga, H., et al., Characterization of a branched-chain amino-acid transporter SBAT1 (SLC6A15) that is expressed in human brain, *Biochem. Biophys. Res. Commun.*, 337, 892, 2005.

37. Bröer, A., et al., The orphan transporter v7-3 (slc6a15) is a Na$^+$-dependent neutral amino acid transporter (B0AT2), *Biochem. J.*, 393, 421, 2006.

38. Bixel, M. G., Hutson, S. M., and Hamprecht, B., Cellular distribution of branched-chain amino acid aminotransferase isoenzymes among rat brain glial cells in culture, *J. Histochem. Cytochem.*, 45, 685, 1997.

39. Bröer, S., et al., Characterization of the monocarboxylate transporter 1 expressed in *Xenopus laevis* oocytes by changes in cytosolic pH, *Biochem. J.*, 333, 167, 1998.

40. Pierre, K., and Pellerin, L., Monocarboxylate transporters in the central nervous system: Distribution, regulation and function, *J. Neurochem.*, 94, 1, 2005.

41. Hall, T. R., et al., Branched chain aminotransferase isoenzymes. Purification and characterization of the rat brain isoenzyme, *J. Biol. Chem.*, 268, 3092, 1993.

42. Suryawan, A., et al., A molecular model of human branched-chain amino acid metabolism, *Am. J. Clin. Nutr.*, 68, 72, 1998.

43. Wallin, R., Hall, T. R., and Hutson, S. M., Purification of branched chain aminotransferase from rat heart mitochondria, *J. Biol. Chem.*, 265, 6019, 1990.

44. Yennawar, N. H., et al., Human mitochondrial branched chain aminotransferase isozyme: Structural role of the CXXC center in catalysis, *J. Biol. Chem.*, 281, 39660, 2006.

45. Bixel, M., et al., Distribution of key enzymes of branched-chain amino acid metabolism in glial and neuronal cells in culture, *J. Histochem. Cytochem.*, 49, 407, 2001.

46. Sweatt, A. J., et al., Branched-chain amino acids and neurotransmitter metabolism: Expression of cytosolic branched-chain aminotransferase (BCATc) in the cerebellum and hippocampus, *J. Comp. Neurol.*, 477, 360, 2004.

47. Garcia-Espinosa, M. A., et al., Widespread neuronal expression of branched-chain aminotransferase in the CNS: Implication for leucine/glutamate metabolism and for signaling by amino acids, *J. Neurochem.*, 100, 1458, 2007.

48. Hutson, S. M., Lieth, E., and LaNoue, K. F., Function of leucine in excitatory neurotransmitter metabolism in the central nervous system, *J. Nutr.*, 131, 846S, 2001.

49. Pettit, F. H., Yeaman, S. J., and Reed, L., Purification and characterization of branched chain α-keto acid dehydrogenase complex of bovine kidney, *Proc. Natl. Acad. Sci. U.S.A.*, 75, 4881, 1978.

50. Danner, D. J., et al., Purification and characterization of branched-chain α-keto acid dehydrogenase from bovine liver mitochondria, *J. Biol. Chem.*, 254, 5522, 1979.

51. Popov, K. M., et al., Branched-chain alpha-ketoacid dehydrogenase kinase. Molecular cloning, expression, and sequence similarity with histidine protein kinases, *J. Biol. Chem.*, 267, 13127, 1992.

52. Islam, M. M., et al., A novel branched-chain amino acid metabolon: Protein-protein interactions in a supramolecular complex, *J. Biol. Chem.*, 282, 11893, 2007.

53. Homanics, G. E., et al., Production and characterization of murine models of classic and intermediate maple syrup urine disease, *BMC Med. Genet.*, 7, 33, 2006.

54. Joshi, M. A., et al. Impaired growth and neurological abnormalities in branched-chain alpha-keto acid dehydrogenase kinase-deficient mice, *Biochem. J.*, 400, 153, 2006.

55. Nguyen, N. H. T., et al., Propionate increases neuronal histone acetylation, but is metabolized oxidatively by glia. Relevance for propionic acidemia, *J. Neurochem.*, 101, 806, 2007.

56. Lynen, F., et al., Der chemische Mechanismus der Acetessigsäurebildung in der Leber, *Biochem. Z.*, 330, 269, 1958.

57. Matsubara, Y., et al., Molecular cloning and nucleotide sequence of cDNAs encoding the precursors of rat long chain acyl-coenzyme A, short chain acyl-coenzyme A, and isovaleryl-coenzyme A dehydrogenases. Sequence homology of four enzymes of the acyl-CoA dehydrogenase family, *J. Biol. Chem.*, 264, 16321, 1989.

58. Murín, R., et al., Immunocytochemical localization of 3-methylcrotonyl-CoA carboxylase in cultured ependymal, microglial and oligodendroglial cells, *J. Neurochem.*, 97, 1393, 2006.

59. Murín, R., and Hamprecht, B., Distribution of 3-methylcrotonyl-CoA carboxylase in rat brain, *J. Neurochem.*, 94, Suppl. 2, 242, 2005.

60. Cullingford, T. E., et al., Molecular cloning of rat mitochondrial 3-hydroxy-3-methylglutaryl-CoA lyase and detection of the corresponding mRNA and of those encoding the remaining enzymes comprising the ketogenic 3-hydroxy-3-methylglutaryl-CoA cycle in central nervous system of suckling rat, *Biochem. J.*, 329, 373, 1998.

61. Nakagawa, J., et al., AUH, a gene encoding an AU-specific RNA binding protein with intrinsic enoyl-CoA hydratase activity, *Proc. Natl. Acad. Sci. U.S.A.*, 92, 2051, 1995.

62. Bixel, M. G., and Hamprecht, B., Immunocytochemical localization of β-methylcrotonyl-CoA carboxylase in astroglial cells and neurons in culture, *J. Neurochem.*, 74, 1059, 2000.

63. Bröer, S., et al., Comparison of lactate transport in astroglial cells and monocarboxylate transporter 1 (MCT 1) expressing *Xenopus laevis* oocytes. Expression of two different monocarboxylate transporters in astroglial cells and neurons, *J. Biol. Chem.*, 272, 30096, 1997.

64. Auestad, N., et al. Fatty acid oxidation and ketogenesis by astrocytes in primary culture. *J. Neurochem.*, 56, 1376, 1991.

65. Yudkoff, M., et al., [^{15}N]leucine as a source of [^{15}N]glutamate in organotypic cerebellar explants, *Biochem. Biophys. Res. Commun.*, 115, 174, 1983.

66. Yudkoff, M., et al., Astrocyte leucine metabolism: Significance of branched-chain amino acid transamination, *J. Neurochem.*, 66, 378, 1996.

67. Shashidharan, P., et al., Novel human glutamate dehydrogenase expressed in neural and testicular tissues and encoded by an X-linked intronless gene, *J. Biol. Chem.*, 269, 16971, 1994.

68. Erecińska, M., and Nelson, D., Activation of glutamate dehydrogenase by leucine and its nonmetabolizable analogue in rat brain synaptosomes, *J. Neurochem.*, 54, 1335, 1990.

69. Plaitakis, A., and Zaganas, I., Regulation of human glutamate dehydrogenases: implications for glutamate, ammonia and energy metabolism in brain, *J. Neurosci. Res.*, 66, 899, 2001.

70. Plaitakis, A., Metaxari, M., and Shashidharan, P., Nerve tissue-specific (GLUD2) and housekeeping (GLUD1) human glutamate dehydrogenases are regulated by distinct allosteric mechanisms: Implications for biologic function, *J. Neurochem.*, 75, 1862, 2000.

71. Liao, C. L., Herman, M. M., and Bensch, K. G., Prolongation of G1 and S phase in C-6 glioma cells treated with maple syrup urine disease metabolites. Morphologic and cell cycle studies, *Lab. Invest.*, 38, 122, 1978.

72. Laviano, A., et al., Role of leucine in regulating food intake, *Science*, 313, 1236, 2006.

73. Cota, D., et al., Hypothalamic mTOR signaling regulates food intake, *Science*, 312, 927, 2006.

74. Hao, S., et al., Uncharged tRNA and sensing of amino acid deficiency in mammalian piriform cortex, *Science*, 307, 1776, 2005.
75. Charlton, M., Branched-chain amino acid enriched supplements as therapy for liver disease, *J. Nutr.*, 136, 295S, 2006.
76. Zwingmann, C., et al., Selective increase of brain lactate synthesis in experimental acute liver failure: Results of a [H-C] nuclear magnetic resonance study, *Hepatology*, 37, 420, 2003.
77. Zwingmann, C., The anaplerotic flux and ammonia detoxification in hepatic encephalopathy, *Metab. Brain Dis.*, 22, 235, 2007.
78. Lai, J. C., and Cooper, A. J., Brain alpha-ketoglutarate dehydrogenase complex: Kinetic properties, regional distribution, and effects of inhibitors, *J. Neurochem.*, 47, 1376, 1986.
79. Cooper, A. J., and Plum, F., Biochemistry and physiology of brain ammonia, *Physiol. Rev.*, 67, 440, 1987.
80. Hertz, L., et al., Some metabolic effects of ammonia on astrocytes and neurons in primary cultures, *Neurochem. Pathol.*, 6, 97, 1987.

Index